Lecture Notes in Computer Science 5157

Commenced Publication in 1973
Founding and Former Series Editors:
Gerhard Goos, Juris Hartmanis, and Jan van Leeuwen

Editorial Board

David Hutchison
Lancaster University, UK

Takeo Kanade
Carnegie Mellon University, Pittsburgh, PA, USA

Josef Kittler
University of Surrey, Guildford, UK

Jon M. Kleinberg
Cornell University, Ithaca, NY, USA

Alfred Kobsa
University of California, Irvine, CA, USA

Friedemann Mattern
ETH Zurich, Switzerland

John C. Mitchell
Stanford University, CA, USA

Moni Naor
Weizmann Institute of Science, Rehovot, Israel

Oscar Nierstrasz
University of Bern, Switzerland

C. Pandu Rangan
Indian Institute of Technology, Madras, India

Bernhard Steffen
University of Dortmund, Germany

Madhu Sudan
Massachusetts Institute of Technology, MA, USA

Demetri Terzopoulos
University of California, Los Angeles, CA, USA

Doug Tygar
University of California, Berkeley, CA, USA

Gerhard Weikum
Max-Planck Institute of Computer Science, Saarbruecken, Germany

Lecture Notes in Computer Science

David Wagner (Ed.)

Advances in Cryptology – CRYPTO 2008

28th Annual International Cryptology Conference
Santa Barbara, CA, USA, August 17-21, 2008
Proceedings

 Springer

Volume Editor

David Wagner
University of California, Berkeley
Berkeley CA 94720-1776, USA
E-mail: daw@cs.berkeley.edu

Library of Congress Control Number: 2008932556

CR Subject Classification (1998): E.3, G.2.1, F.2.1-2, D.4.6, K.6.5, C.2, J.1

LNCS Sublibrary: SL 4 – Security and Cryptology

ISSN   0302-9743
ISBN-10   3-540-85173-9 Springer Berlin Heidelberg New York
ISBN-13   978-3-540-85173-8 Springer Berlin Heidelberg New York

This work is subject to copyright. All rights are reserved, whether the whole or part of the material is
concerned, specifically the rights of translation, reprinting, re-use of illustrations, recitation, broadcasting,
reproduction on microfilms or in any other way, and storage in data banks. Duplication of this publication
or parts thereof is permitted only under the provisions of the German Copyright Law of September 9, 1965,
in its current version, and permission for use must always be obtained from Springer. Violations are liable
to prosecution under the German Copyright Law.

Springer is a part of Springer Science+Business Media

springer.com

© International Association for Cryptologic Research 2008
Printed in Germany

Typesetting: Camera-ready by author, data conversion by Scientific Publishing Services, Chennai, India
Printed on acid-free paper   SPIN: 12453214   06/3180   5 4 3 2 1 0

Preface

CRYPTO 2008, the 28th Annual International Cryptology Conference, was sponsored by the International Association for Cryptologic Research (IACR) in cooperation with the IEEE Computer Society Technical Committee on Security and Privacy and the Computer Science Department of the University of California at Santa Barbara. The conference was held in Santa Barbara, California, August 17–21, 2008. Susan Langford served as the General Chair of CRYPTO 2008, and I had the privilege of serving as the Program Chair.

The conference received 184 submissions, and all were reviewed by the Program Committee. Each paper was assigned at least three reviewers, while submissions co-authored by Program Committee members were reviewed by at least five people. All submissions were anonymous, and the identity of the authors were not revealed to committee members. During the first phase of the review process, the Program Committee, aided by reports from 142 external reviewers, produced a total of 611 reviews in all. Then, committee members discussed these papers in depth over a period of 8 weeks using an electronic messaging system, in the process writing 1,400 discussion messages. After careful deliberation, the Program Committee selected 32 papers for presentation. The authors of accepted papers were given 5 weeks to prepare final versions for these proceedings. These revised papers were not subject to editorial review and the authors bear full responsibility for their contents.

Gilles Brassard delivered the 2008 IACR Distinguished Lecture. The Best Paper Award was announced at the conference. Dan Bernstein served as the chair of the Rump Session, a forum for short and entertaining presentations on recent work of both a technical and non-technical nature.

I would like to thank everyone who contributed to the success of CRYPTO 2008. Shai Halevi provided software for facilitating the reviewing process that was of great help throughout the Program Committee's work, and I am especially grateful for his assistance. Alfred Menezes and Shai Halevi served as advisory members of the Program Committee, and I am grateful to them, and Arjen Lenstra and Bart Preneel, for their cogent advice. Susan Langford and others played a vital role in organizing the conference. Also, I am deeply grateful to the Program Committee for their hard work, enthusiasm, and conscientious efforts to ensure that each paper received a thorough and fair review. Thanks also to the external reviewers, listed on the following pages, for contributing their time and expertise. Finally, I would like to thank all the authors who submitted papers to CRYPTO 2008 for submitting their best research.

August 2008   David Wagner

CRYPTO 2008

August 17–21, 2008, Santa Barbara, California, USA

Sponsored by the
International Association for Cryptologic Research (IACR)

in cooperation with
IEEE Computer Society Technical Committee on Security and Privacy,
Computer Science Department, University of California, Santa Barbara

General Chair

Susan Langford, Hewlett-Packard Company

Program Chair

David Wagner, UC Berkeley

Program Committee

Boaz Barak	Princeton University
John Black	University of Colorado at Boulder
Xavier Boyen	Voltage Security
Melissa Chase	Brown University
Jean-Sebastien Coron	University of Luxembourg
Yevgeniy Dodis	New York University
Orr Dunkelman	KU Leuven
Matt Franklin	UC Davis
Craig Gentry	Stanford University
Henri Gilbert	Orange Labs
Kristian Gjosteen	Norwegian University of Science and Technology
Louis Granboulan	European Aeronautic Defence and Space Company
Danny Harnik	IBM Haifa Research Lab
Susan Hohenberger	Johns Hopkins University
Nick Hopper	University of Minnesota
Yuval Ishai	Technion Institute and UCLA
Thomas Johansson	Lund University
Ari Juels	RSA Laboratories

Lars Knudsen	DTU Mathematics
Kristin Lauter	Microsoft Research
Yehuda Lindell	Bar Ilan University
Tal Malkin	Columbia University
Manoj Prabhakaran	University of Illinois, Urbana-Champaign
Zulfikar Ramzan	Symantec
Renato Renner	ETH Zurich
Matt Robshaw	Orange Labs
Alon Rosen	Herzliya Interdisciplinary Center
Amit Sahai	UCLA
Hovav Shacham	UC San Diego
Tom Shrimpton	Portland State University and University of Lugano
Adam Smith	Pennsylvania State University
Serge Vaudenay	EPFL
Brent Waters	SRI International
Lisa Yin	Independent Consultant

Advisory Members

Alfred Menezes	
(CRYPTO 2007 Program Chair)	University of Waterloo
Shai Halevi	
(CRYPTO 2009 Program Chair)	IBM Research

External Reviewers

Michel Abdalla	Denis Charles	Marc Girault
Tolga Acar	Seung Geol Choi	Sharon Goldberg
Joel Alwen	Carlos Cid	Mark Gondree
Thomas Baignères	Martin Cochran	Vipul Goyal
Zuzana Beerliova	Roger Colbeck	Matthew Green
Amos Beimel	Scott Contini	Jens Groth
Mihir Bellare	Scott Coull	Venkatesan Guruswami
Josh Benaloh	Dána Dachman-Soled	Shai Halevi
Come Berbain	Oscar Dahlsten	Michael Hamburg
Olivier Billet	Jintai Ding	Carmit Hazay
Alexandra Boldyreva	Glenn Durfee	Martin Hirt
Dan Boneh	Ariel Elbaz	Thomas Holenstein
Colin Boyd	Jean-Charles Faugère	Mariusz Jakubowski
Emmanuel Bresson	Serge Fehr	Stas Jarecki
Reinier Broker	Marc Fischlin	Antoine Joux
Lennart Brynielsson	Pierre-Alain Fouque	Jonathan Katz
Ran Canetti	Martin Gagne	John Kelsey
Yaniv Carmeli	Juan Garay	Aggelos Kiayias
Rafik Chaabouni	Praveen Gauravaram	Yongdae Kim

Markulf Kohlweiss
Gillat Kol
Hugo Krawczyk
Alptekin Küpçü
Eyal Kushilevitz
Homin Lee
Matt Lepinski
Huijia Lin
Satya Lokam
Steve Lu
Vadim Lyubashevsky
Philip Mackenzie
Mohammad Mahmoody-
Ghidary
Krystian Matusiewicz
Daniele Micciancio
Ilya Mironov
Payman Mohassel
David Molnar
Tal Moran
Volker Muller
Sean Murphy
Yusuke Naito
Yassir Nawaz
Phong Nguyen
Jesper Buus Nielsen
Kobbi Nissim
Alina Oprea
Kazu Ota

Khaled Ouafi
Raphael Overbeck
Carles Padro
Pascal Paillier
Sylvain Pasini
Jacques Patarin
Chris Peikert
Christophe Petit
Thomas Peyrin
Duong Hieu Phan
David Pointcheval
Prashant Puniya
Tal Rabin
Mario Di Raimondo
Dominic Raub
Oded Regev
Omer Reingold
Renato Renner
Leonid Reyzin
Thomas Ristenpart
Matthieu Rivain
Phillip Rogaway
Guy Rothblum
Peter Ryan
Kazuo Sakiyama
Yu Sasaki
Michael Scott
Gil Segev
Yannick Seurin

abhi shelat
Igor Shparlinski
Nigel Smart
John Steinberger
Ron Steinfeld
Mike Szydlo
Stefano Tessaro
Soren Thomsen
Nikos Triandopoulos
Eran Tromer
Salil Vadhan
Vinod Vaikuntanathan
Martin Vuagnoux
Shabsi Walfish
Andrew Wan
Lei Wang
Hoeteck Wee
Enav Weinreb
Steve Weis
Daniel Wichs
Stefan Wolf
Duncan Wong
Juerg Wullschleger
Aaram Yun
Gideon Yuval
Erik Zenner
Yunlei Zhao
Vassilis Zikas

Table of Contents

Cryptanalysis I

Multiparty Computation I

Cryptanalysis II

Public-Key Crypto II

Hash Functions II

Privacy

Multiparty Computation II

Zero Knowledge

Oblivious Transfer

The Random Oracle Model and the Ideal Cipher Model Are Equivalent

Jean-Sébastien Coron[1], Jacques Patarin[2], and Yannick Seurin[2,3]

[1] University of Luxembourg
[2] University of Versailles
[3] Orange Labs

Abstract. The Random Oracle Model and the Ideal Cipher Model are two well known idealised models of computation for proving the security of cryptosystems. At Crypto 2005, Coron *et al.* showed that security in the random oracle model implies security in the ideal cipher model; namely they showed that a random oracle can be replaced by a block cipher-based construction, and the resulting scheme remains secure in the ideal cipher model. The other direction was left as an open problem, *i.e.* constructing an ideal cipher from a random oracle. In this paper we solve this open problem and show that the Feistel construction with 6 rounds is enough to obtain an ideal cipher; we also show that 5 rounds are insufficient by providing a simple attack. This contrasts with the classical Luby-Rackoff result that 4 rounds are necessary and sufficient to obtain a (strong) pseudo-random permutation from a pseudo-random function.

1 Introduction

Modern cryptography is about defining security notions and then constructing schemes that provably achieve these notions. In cryptography, security proofs are often relative: a scheme is proven secure, assuming that some computational problem is hard to solve. For a given functionality, the goal is therefore to obtain an efficient scheme that is secure under a well known computational assumption (for example, factoring is hard). However for certain functionalities, or to get a more efficient scheme, it is sometimes necessary to work in some idealised model of computation.

The well known *Random Oracle Model* (ROM), formalised by Bellare and Rogaway [1], is one such model. In the random oracle model, one assumes that some hash function is replaced by a publicly accessible random function (the random oracle). This means that the adversary cannot compute the result of the hash function by himself: he must query the random oracle. The random oracle model has been used to prove the security of numerous cryptosystems, and it has lead to simple and efficient designs that are widely used in practice (such as PSS [2] and OAEP [3]). Obviously, a proof in the random oracle model is not fully satisfactory, because such a proof does not imply that the scheme will remain secure when the random oracle is replaced by a concrete hash function

D. Wagner (Ed.): CRYPTO 2008, LNCS 5157, pp. 1–20, 2008.
© International Association for Cryptologic Research 2008

(such as SHA-1). Numerous papers have shown artificial schemes that are provably secure in the ROM, but completely insecure when the RO is instantiated with any function family (see [7]). Despite these separation results, the ROM still appears to be a useful tool for proving the security of cryptosystems. For some functionalities, the ROM construction is actually the only known construction (for example, for non-sequential aggregate signatures [6]).

The *Ideal Cipher Model* (ICM) is another idealised model of computation, similar to the ROM. Instead of having a publicly accessible random function, one has a publicly accessible random block cipher (or ideal cipher). This is a block cipher with a κ-bit key and a n-bit input/output, that is chosen uniformly at random among all block ciphers of this form; this is equivalent to having a family of 2^κ independent random permutations. All parties including the adversary can make both encryption and decryption queries to the ideal block cipher, for any given key. As for the random oracle model, many schemes have been proven secure in the ICM [5,11,14,16]. As for the ROM, it is possible to construct artificial schemes that are secure in the ICM but insecure for any concrete block cipher (see [4]). Still, a proof in the ideal cipher model seems useful because it shows that a scheme is secure against generic attacks, that do not exploit specific weaknesses of the underlying block cipher.

A natural question is whether the random oracle model and the ideal cipher model are equivalent models, or whether one model is strictly stronger than the other. Given a scheme secure with random oracles, is it possible to replace the random oracles with a block cipher-based construction, and obtain a scheme that is still secure in the ideal cipher model? Conversely, if a scheme is secure in the ideal cipher model, is it possible to replace the ideal cipher with a construction based on functions, and get a scheme that is still secure when these functions are seen as random oracles?

At Crypto 2005, Coron *et al.* [9] showed that it is indeed possible to replace a random oracle (taking arbitrary long inputs) by a block cipher-based construction. The proof is based on an extension of the classical notion of indistinguishability, called *indifferentiability*, introduced by Maurer *et al.* in [18]. Using this notion of indifferentiability, the authors of [9] gave the definition of an "indifferentiable construction" of one ideal primitive (F) (for example, a random oracle) from another ideal primitive (G) (for example an ideal block cipher). When a construction satisfies this notion, any scheme that is secure in the former ideal model (F) remains secure in the latter model (G), when instantiated using this construction. The authors of [9] proposed a slight variant of the Merkle-Damgård construction to instantiate a random oracle (see Fig. 1). Given any scheme provably secure in the random oracle model, this construction can replace the random oracle, and the resulting scheme remains secure in the ideal cipher model; other constructions have been analysed in [8].

The other direction (constructing an ideal cipher from a random oracle) was left as an open problem in [9]. In this paper we solve this open problem and show that the Luby-Rackoff construction with 6 rounds is sufficient to instantiate an ideal cipher (see Fig. 2 for an illustration). Actually, it is easy to see that it is

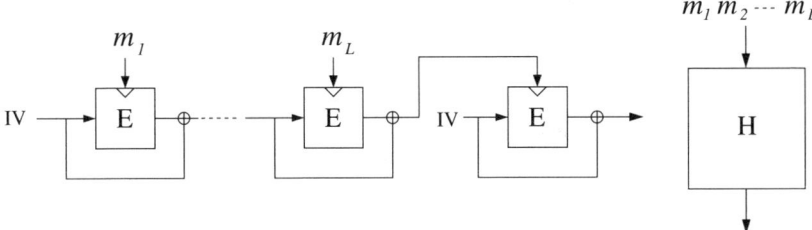

Fig. 1. A Merkle-Damgård like construction [9] based on ideal cipher E (left) to replace random oracle H (right). Messages blocks m_i's are used as successive keys for ideal-cipher E. IV is a pre-determined constant.

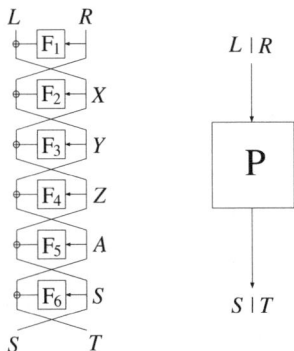

Fig. 2. The Luby-Rackoff construction with 6 rounds (left), to replace a random permutation P (right)

enough to construct a random *permutation* instead of an ideal cipher; namely, a family of 2^κ independent random permutations (*i.e.*, an ideal block cipher) can be constructed by simply prepending a k-bit key to the inner random oracle functions F_i's. Therefore in this paper, we concentrate on the construction of a random permutation. We also show that 5 rounds Luby-Rackoff is insecure by providing a simple attack; this shows that 6 rounds is actually optimal.

Our result shows that the random oracle model and the ideal cipher model are actually equivalent assumptions. It seems that up to now, many cryptographers have been reluctant to use the Ideal Cipher Model and have endeavoured to work in the Random Oracle Model, arguing that the ICM is richer and carries much more structure than the ROM. Our result shows that it is in fact not the case and that designers may use the ICM when they need it without making a stronger assumption than when working in the random oracle model. However, our security reduction is quite loose, which implies that in practice large security parameters should be used in order to replace an ideal cipher by a 6-round Luby-Rackoff.

We stress that the "indifferentiable construction" notion is very different from the classical indistinguishability notion. The well known Luby-Rackoff result that 4 rounds are enough to obtain a strong pseudo-random permutation from pseudo-random functions [17], is proven under the classical indistinguishability notion. Under this notion, the adversary has only access to the input/output of the Luby-Rackoff (LR) construction, and tries to distinguish it from a random permutation; in particular it does not have access to the input/output of the inner pseudo-random functions. On the contrary, in our setting, the distinguisher can make oracle calls to the inner round functions F_i's (see Fig. 2); the indifferentiability notion enables to accommodate these additional oracle calls in a coherent definition.

1.1   Related Work

One of the first paper to consider having access to the inner round functions of a Luby-Rackoff is [20]; the authors showed that Luby-Rackoff with 4 rounds remains secure if adversary has oracle access to the middle two round functions, but becomes insecure if adversary is allowed access to any other round functions.

In [15] a random permutation oracle was instantiated for a specific scheme using a 4-rounds Luby-Rackoff. More precisely, the authors showed that the random permutation oracle P in the Even-Mansour [14] block-cipher $E_{k_1,k_2}(m) = k_2 \oplus P(m \oplus k_1)$ can be replaced by a 4-rounds Luby-Rackoff, and the block-cipher E remains secure in the random oracle model; for this specific scheme, the authors obtained a (much) better security bound than our general bound in this paper.

In [12], Dodis and Puniya introduced a different model for indifferentiability, called indifferentiability in the *honest-but-curious* model. In this model, the distinguisher is not allowed to make direct calls to the inner hash functions; instead he can only query the global Luby-Rackoff construction and get all the intermediate results. The authors showed that in this model, a Luby-Rackoff construction with a super-logarithmic number of rounds can replace an ideal cipher. The authors also showed that indifferentiability in the honest-but-curious model implies indifferentiability in the general model, for LR constructions with up to a logarithmic number of rounds. But because of this gap between logarithmic and super-logarithmic, the authors could not conclude about general indifferentiability of Luby-Rackoff constructions. Subsequent work by Dodis and Puniya [13] studied other properties (such as unpredictability and verifiablity) of the Luby-Rackoff construction when the intermediate values are known to the attacker.

We have an observation about indifferentiability in the honest-but-curious model: general indifferentiability does not necessarily imply indifferentiability in the honest-but-curious model. More precisely, we show in Appendix B that LR constructions with up to logarithmic number of rounds are *not* indifferentiable from a random permutation in the honest-but-curious model, whereas our main result in this paper is that 6-rounds LR is indifferentiable from a random permutation in the general model.

2   Definitions

In this section, we recall the notion of indifferentiability of random systems, introduced by Maurer *et al.* in [18]. This is an extension of the classical notion of indistinguishability, where one or more oracles are publicly available, such as random oracles or ideal ciphers.

We first motivate why such an extension is actually required. The classical notion of indistinguishability enables to argue that if some system S_1 is indistinguishable from some other system S_2 (for any polynomially bounded attacker), then any application that uses S_1 can use S_2 instead, without any loss of security; namely, any non-negligible loss of security would precisely be a way of distinguishing between the two systems. Since we are interested in replacing a random permutation (or an ideal cipher) by a Luby-Rackoff construction, we would like to say that the Luby-Rackoff construction is "indistinguishable" from a random permutation. However, when the distinguisher can make oracle calls to the inner round functions, one cannot say that the two systems are "indistinguishable" because they don't even have the same interface (see Fig. 2); namely for the LR construction the distinguisher can make oracle calls to the inner functions F_i's, whereas for the random permutation he can only query the input and receive the output and vice versa. This contrasts with the setting of the classical Luby-Rackoff result, where the adversary has only access to the input/output of the LR construction, and tries to distinguish it from a random permutation. Therefore, an extension of the classical notion of indistinguishability is required, in order to show that some ideal primitive (like a random permutation) can be constructed from another ideal primitive (like a random oracle).

Following [18], we define an *ideal primitive* as an algorithmic entity which receives inputs from one of the parties and delivers its output immediately to the querying party. The ideal primitives that we consider in this paper are random oracles and random permutations (or ideal ciphers). A *random oracle* [1] is an ideal primitive which provides a random output for each new query. Identical input queries are given the same answer. A *random permutation* is an ideal primitive that contains a random permutation $P : \{0,1\}^n \rightarrow \{0,1\}^n$. The ideal primitive provides oracle access to P and P^{-1}. An *ideal cipher* is an ideal primitive that models a random block cipher $E : \{0,1\}^\kappa \times \{0,1\}^n \rightarrow \{0,1\}^n$. Each key $k \in \{0,1\}^\kappa$ defines a random permutation $E_k = E(k, \cdot)$ on $\{0,1\}^n$. The ideal primitive provides oracle access to E and E^{-1}; that is, on query $(0, k, m)$, the primitive answers $c = E_k(m)$, and on query $(1, k, c)$, the primitive answers m such that $c = E_k(m)$. These oracles are available for any n and any κ.

The notion of indifferentiability [18] is used to show that an ideal primitive $\mathcal{P}$ (for example, a random permutation) can be replaced by a construction C that is based on some other ideal primitive $\mathcal{F}$ (for example, C is the LR construction based on a random oracle F):

Definition 1 ([18]). *A Turing machine C with oracle access to an ideal primitive $\mathcal{F}$ is said to be $(t_D, t_S, q, \varepsilon)$-indifferentiable from an ideal primitive $\mathcal{P}$ if there exists a simulator S with oracle access to $\mathcal{P}$ and running in time at most t_S, such that for any distinguisher D running in time at most t_D and making at most q queries, it holds that:*

$$\left| \Pr\left[D^{C^{\mathcal{F}}, \mathcal{F}} = 1 \right] - \Pr\left[D^{\mathcal{P}, S^{\mathcal{P}}} = 1 \right] \right| < \varepsilon$$

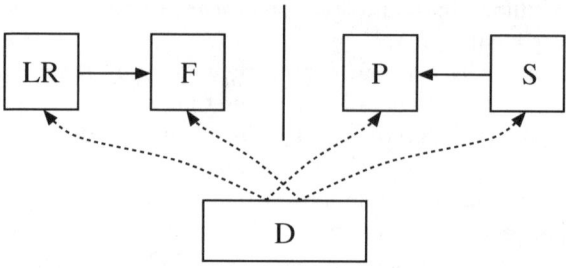

Fig. 3. The indifferentiability notion

$C^{\mathcal{F}}$ *is simply said to be indifferentiable from* $\mathcal{F}$ *if* ε *is a negligible function of the security parameter* n, *for polynomially bounded* q, t_D *and* t_S.

The previous definition is illustrated in Figure 3, where $\mathcal{P}$ is a random permutation, C is a Luby-Rackoff construction LR, and F is a random oracle. In this paper, for a 6-round Luby-Rackoff, we denote these random oracles $F_1, \ldots, F_6$ (see Fig. 2). Equivalently, one can consider a single random oracle F and encode in the first 3 input bits which round function $F_1, \ldots, F_6$ is actually called. The distinguisher has either access to the system formed by the construction LR and the random oracle F, or to the system formed by the random permutation P and a simulator $\mathcal{S}$. In the first system (left), the construction LR computes its output by making calls to F (this corresponds to the round functions F_i's of the Luby-Rackoff); the distinguisher can also make calls to F directly. In the second system (right), the distinguisher can either query the random permutation P, or the simulator that can make queries to P. We see that the role of the simulator is to simulate the random oracles F_i's so that no distinguisher can tell whether it is interacting with LR and F, or with P and S. In other words, 1) the output of S should be indistinguishable from that of random oracles F_i's and 2) the output of S should look "consistent" with what the distinguisher can obtain from P. We stress that the simulator does not see the distinguisher's queries to P; however, it can call P directly when needed for the simulation. Note that the two systems have the same interface, so now it makes sense to require that the two systems be indistinguishable.

To summarise, in the first system the random oracles F_i are chosen at random, and a permutation $C = LR$ is constructed from them with a 6 rounds Luby-Rackoff. In the second system the random permutation P is chosen at random and the inner round functions F_i's are simulated by a simulator with oracle access to P. Those two systems should be indistinguishable, that is the distinguisher should not be able to tell whether the inner round functions were chosen at random and then the Luby-Rackoff permutation constructed from it, or the random permutation was chosen at random and the inner round functions then "tailored" to match the permutation.

It is shown in [18] that the indifferentiability notion is the "right" notion for substituting one ideal primitive with a construction based on another ideal primitive. That is, if $C^{\mathcal{F}}$ is indifferentiable from an ideal primitive $\mathcal{P}$, then $C^{\mathcal{F}}$ can replace $\mathcal{P}$ in any cryptosystem, and the resulting cryptosystem is at least as secure in the $\mathcal{F}$ model as in the $\mathcal{P}$ model; see [18] or [9] for a proof. Our main result in this paper is that the 6 rounds Luby-Rackoff construction is indifferentiable from a random permutation; this implies that such a construction can replace a random permutation (or an ideal cipher) in any cryptosystem, and the resulting scheme remains secure in the random oracle model if the original scheme was secure in the random permutation (or ideal cipher) model.

3 Attack of Luby-Rackoff with 5 Rounds

In this section we show that 5 rounds are not enough to obtain the indifferentiability property. We do this by exhibiting for the 5 rounds Luby-Rackoff (see Fig. 4) a property that cannot be obtained with a random permutation.

Let Y and Y' be arbitrary values, corresponding to inputs of F_3 (see Fig. 4); let Z be another arbitrary value, corresponding to input of F_4. Let $Z' = F_3(Y) \oplus F_3(Y') \oplus Z$, and let:

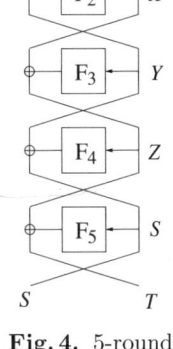

$$X = F_3(Y) \oplus Z = F_3(Y') \oplus Z' \quad (1)$$
$$X' = F_3(Y') \oplus Z = F_3(Y) \oplus Z' \quad (2)$$

From X, X', Y and Y' we now define four couples (X_i, Y_i) as follows:

$$(X_0, Y_0) = (X, Y), \quad (X_1, Y_1) = (X', Y)$$
$$(X_2, Y_2) = (X', Y'), \quad (X_3, Y_3) = (X, Y')$$

and we let $L_i \| R_i$ be the four corresponding plaintexts; we have:

$$R_0 = Y_0 \oplus F_2(X_0) = Y \oplus F_2(X)$$
$$R_1 = Y_1 \oplus F_2(X_1) = Y \oplus F_2(X')$$
$$R_2 = Y_2 \oplus F_2(X_2) = Y' \oplus F_2(X')$$
$$R_3 = Y_3 \oplus F_2(X_3) = Y' \oplus F_2(X)$$

Fig. 4. 5-rounds Luby-Rackoff

Let Z_0, Z_1, Z_2, Z_3 be the corresponding values as input of F_4; we have from (1) and (2):

$$Z_0 = X_0 \oplus F_3(Y_0) = X \oplus F_3(Y) = Z, \quad Z_1 = X_1 \oplus F_3(Y_1) = X' \oplus F_3(Y) = Z'$$
$$Z_2 = X_2 \oplus F_3(Y_2) = X' \oplus F_3(Y') = Z, \quad Z_3 = X_3 \oplus F_3(Y_3) = X \oplus F_3(Y') = Z'$$

Finally, let $S_i \| T_i$ be the four corresponding ciphertexts; we have:

$$S_0 = Y_0 \oplus F_4(Z_0) = Y \oplus F_4(Z), \quad S_1 = Y_1 \oplus F_4(Z_1) = Y \oplus F_4(Z')$$
$$S_2 = Y_2 \oplus F_4(Z_2) = Y' \oplus F_4(Z), \quad S_3 = Y_3 \oplus F_4(Z_3) = Y' \oplus F_4(Z')$$

We obtain the relations:

$$R_0 \oplus R_1 \oplus R_2 \oplus R_3 = 0, \quad S_0 \oplus S_1 \oplus S_2 \oplus S_3 = 0$$

Thus, we have obtained four pairs (plaintext, ciphertext) such that the xor of the right part of the four plaintexts equals 0 and the xor of the left part of the four ciphertexts also equals 0. For a random permutation, it is easy to see that such a property can only be obtained with negligible probability, when the number of queries is polynomially bounded. Thus we have shown:

Theorem 1. *The Luby-Rackoff construction with* 5 *rounds is not indifferentiable from a random permutation.*

This contrasts with the classical Luby-Rackoff result, where 4 rounds are enough to obtain a strong pseudo-random permutation from pseudo-random functions.

4   Indifferentiability of Luby-Rackoff with 6 Rounds

We now prove our main result: the Luby-Rackoff construction with 6 rounds is indifferentiable from a random permutation.

Theorem 2. *The LR construction with* 6 *rounds is* $(t_D, t_S, q, \varepsilon)$*-indifferentiable from a random permutation, with* $t_S = \mathcal{O}(q^4)$ *and* $\varepsilon = 2^{18} \cdot q^8 / 2^n$*, where* n *is the output size of the round functions.*

Note that here the distinguisher has unbounded running time; it is only bounded to ask q queries. As illustrated in Figure 3, we must construct a simulator S such that the two systems formed by (LR, F) and (P, S) are indistinguishable. The simulator is constructed in Section 4.1, while the indistinguishability property is proved in Section 4.2.

4.1   The Simulator

We construct a simulator S that simulates the random oracles $F_1, \ldots, F_6$. For each function F_i the simulator maintains an history of already answered queries. We write $x \in F_i$ when x belongs to the history of F_i, and we denote by $F_i(x)$ the corresponding output. When we need to obtain $F_i(x)$ and x does not belong to the history of F_i, we write $F_i(x) \leftarrow y$ to determine that the answer to F_i query x will be y; we then add $(x, F_i(x))$ to the history of F_i. We denote by n the output size of the functions F_i's. We denote by LR and LR^{-1} the 6-round Luby-Rackoff construction as obtained from the functions F_i's.

We first provide an intuition of the simulator's algorithm. The simulator must make sure that his answers to the distinguisher's F_i queries are coherent with the answers to P queries that can be obtained independently by the distinguisher. In other words, when the distinguisher makes F_i queries to the simulator (possibly in some arbitrary order), the output generated by the corresponding Luby-Rackoff must be the same as the output from P obtained independently by the distinguisher. We stress that those P queries made by the distinguisher cannot be seen by the simulator; the simulator is only allowed to make his own P queries (as illustrated in Fig. 3). In addition, the simulator's answer to F_i queries must be statistically close to the output of random functions.

The simulator's strategy is the following: when a "chain of 3 queries" has been made by the distinguisher, the simulator is going to define the values of all the other F_i's corresponding to this chain, by making a P or a P^{-1} query, so that the output of LR and the output of P are the same for the corresponding message. Roughly speaking, we say that we have a chain of 3 queries (x, y, z) when x, y, z are in the history of F_k, F_{k+1} and F_{k+2} respectively and $x = F_{k+1}(y) \oplus z$.

For example, if a query X to F_2 is received, and we have $X = F_3(Y) \oplus Z$ where Y, Z belong to the history of F_3 and F_4 respectively, then the triple (X, Y, Z) forms a 3-chain of queries. In this case, the simulator defines $F_2(X) \xleftarrow{\$} \{0,1\}^n$ and computes the corresponding $R = Y \oplus F_2(X)$. It also lets $F_1(R) \xleftarrow{\$} \{0,1\}^n$ and computes $L = X \oplus F_1(R)$. Then it makes a P-query to get $S\|T = P(L\|R)$. It also computes $A = Y \oplus F_4(Z)$. The values of $F_5(A)$ and $F_6(S)$ are then "adapted" so that the 6-round LR and the random permutation provide the same output, *i.e.* the simulator defines $F_5(A) \leftarrow Z \oplus S$ and $F_6(S) \leftarrow A \oplus T$, so that $\mathsf{LR}(L\|R) = P(L\|R) = S\|T$. In summary, given a F_2 query, the simulator looked at the history of (F_3, F_4) and adapted the answers of (F_5, F_6).

More generally, given a query to F_k, the simulator proceeds according to Table 1 below; we denote by $+$ for looking downward in the LR construction and by $-$ for looking upward. The simulator must first simulate an additional call to F_i (column "Call"). Then the simulator can compute either $L\|R$ or $S\|T$ (as determined in column "Compute"). Given $L\|R$ (resp. $S\|T$) the simulator makes a P-query (resp. a P^{-1}-query) to obtain $S\|T = P(L\|R)$ (resp. $L\|R = P^{-1}(S\|T)$). Finally Table 1 indicates the index j for which the output of (F_j, F_{j+1}) is adapted (column "Adapt").

Given a query x to F_k, with $2 \le k \le 3$, the simulator (when looking downward) must actually consider all 3-chains formed by (x, y, z) where $y \in F_{k+1}$ and $z \in F_{k+2}$. Therefore, for $k \le 2 \le 3$, one defines the following set:

$$\mathsf{Chain}(+1, x, k) = \{(y, z) \in (F_{k+1}, F_{k+2}) \mid x = F_{k+1}(y) \oplus z\}$$

where $+1$ corresponds to looking downward in the Luby-Rackoff construction. This corresponds to Lines $(F_2, +)$ and $(F_3, +)$ in Table 1.

Similarly, given a query t to F_k, with $4 \le k \le 5$, when looking upward the simulator must consider all 3-chains formed by (y, z, t) where $y \in F_{k-2}$ and $z \in F_{k-1}$; one defines the following set for $4 \le k \le 5$:

$$\mathsf{Chain}(-1, t, k) = \{(y, z) \in (F_{k-2}, F_{k-1}) \mid t = F_{k-1}(z) \oplus y\}$$

This corresponds to Lines $(F_4, -)$ and $(F_5, -)$ in Table 1.

Table 1. Simulator's behaviour

Query	Dir	History	Call	Compute	Adapt
F_1	$-$	(F_5, F_6)	F_4	$S\|T$	(F_2, F_3)
F_2	$+$	(F_3, F_4)	F_1	$L\|R$	(F_5, F_6)
F_2	$-$	$(\tilde{F}_6, F_1)$	F_5	$L\|R$	(F_3, F_4)
F_3	$+$	(F_4, F_5)	F_6	$S\|T$	(F_1, F_2)
F_4	$-$	(F_2, F_3)	F_1	$L\|R$	(F_5, F_6)
F_5	$+$	$(F_6, \tilde{F}_1)$	F_2	$S\|T$	(F_3, F_4)
F_5	$-$	(F_3, F_4)	F_6	$S\|T$	(F_1, F_2)
F_6	$+$	(F_1, F_2)	F_3	$L\|R$	(F_4, F_5)

Additionally one must consider the 3-chains obtained from a F_6 query S and looking in (F_1, F_2) history, with Line $(F_6, +)$:

$$\mathsf{Chain}(+1, S, 6) = \left\{ (R, X) \in (F_1, F_2) \mid \exists T, \ P\left(F_1(R) \oplus X \| R\right) = S\|T \right\} \qquad (3)$$

and symmetrically the 3-chains obtained from a F_1 query R and looking in (F_5, F_6) history, with Line $(F_1, -)$:

$$\mathsf{Chain}(-1, R, 1) = \left\{ (A, S) \in (F_5, F_6) \mid \exists L, \ P^{-1}(S\|F_6(S) \oplus A) = L\|R \right\} \qquad (4)$$

One must also consider the 3-chains associated with (F_1, F_6) history, obtained either from a F_2 query X or a F_5 query A, with Lines $(F_2, -)$ and $(F_5, +)$. Given a F_2 query X, we consider all $R \in F_1$, and for each corresponding $L = X \oplus F_1(R)$, we compute $S\|T = P(L\|R)$ and determine whether $S \in F_6$. Additionally, we also consider "virtual" 3-chains, where $S \notin F_6$, but S is such that $P(L'\|R') = S\|T'$ for some $(R', X') \in (F_1, F_2)$, with $L' = X' \oplus F_1(R')$ and $X' \neq X$. Formally, we denote :

$$\mathsf{Chain}(-1, X, 2) = \left\{ (R, S) \in (F_1, \tilde{F}_6) \mid \exists T, \ P(X \oplus F_1(R)\|R) = S\|T \right\} \qquad (5)$$

where $\tilde{F}_6$ in $\mathsf{Chain}(-1, X, 2)$ is defined as:

$$\tilde{F}_6 = F_6 \cup \left\{ S \mid \exists T', (R', X') \in (F_1, F_2 \setminus \{X\}), P(X' \oplus F_1(R')\|R') = S\|T' \right\}$$

and symmetrically:

$$\mathsf{Chain}(+1, A, 5) = \left\{ (R, S) \in (\tilde{F}_1, F_6) \mid \exists L, \ P^{-1}(S\|A \oplus F_6(S)) = L\|R \right\} \qquad (6)$$

$$\tilde{F}_1 = F_1 \cup \left\{ R \mid \exists L', (A', S') \in (F_5 \setminus \{A\}, F_6), \ P^{-1}(S'\|A' \oplus F_6(S')) = L'\|R \right\}$$

When the simulator receives a query x for F_k, it then proceeds as follows:

Query(x, k):

1. If x is in the history of F_k then go to step 4.
2. Let $F_k(x) \xleftarrow{\$} \{0,1\}^n$ and add $(x, F_k(x))$ to the history of F_k.
3. Call ChainQuery(x, k)
4. Return $F_k(x)$.

The ChainQuery algorithm is used to handle all possible 3-chains created by the operation $F_k(x) \xleftarrow{\$} \{0,1\}^n$ at step 2:

ChainQuery(x, k):

1. If $k \in \{2, 3, 5, 6\}$, then for all $(y, z) \in$ Chain$(+1, x, k)$:
   (a) Call CompleteChain$(+1, x, y, z, k)$.
2. If $k \in \{1, 2, 4, 5\}$, then for all $(y, z) \in$ Chain$(-1, x, k)$:
   (a) Call CompleteChain$(-1, x, y, z, k)$.

The CompleteChain(b, x, y, z, k) works as follows: it computes the message $L\|R$ or $S\|T$ that corresponds to the 3-chain (x, y, z) given as input, without querying (F_j, F_{j+1}), where j is the index given in Table 1 (column "Adapt"). If $L\|R$ is first computed, then the simulator makes a P query to obtain $S\|T = P(L\|R)$; similarly, if $S\|T$ is first computed, then the simulator makes a P^{-1} query to obtain $L\|R = P^{-1}(S\|T)$. Eventually the output of functions (F_j, F_{j+1}) is adapted so that $\mathsf{LR}(L\|R) = S\|T$.

CompleteChain(b, x, y, z, k):

1. If $(b, k) = (-1, 2)$ and $z \notin F_6$, then call Query$(z, 6)$, without considering in ChainQuery$(z, 6)$ the 3-chain that leads to the current 3-chain (x, y, z).
2. If $(b, k) = (+1, 5)$ and $y \notin F_1$, then call Query$(y, 1)$, without considering in ChainQuery$(y, 1)$ the 3-chain that leads to the current 3-chain (x, y, z).
3. Given (b, k) and from Table 1:
   (a) Determine the index i of the additional call to F_i (column "Call").
   (b) Determine whether $L\|R$ or $S\|T$ must be computed first.
   (c) Determine the index j for adaptation at (F_j, F_{j+1}) (column "Adapt").
4. Call Query(x_i, i), where x_i is the input of F_i that corresponds to the 3-chain (x, y, z), without considering in ChainQuery(x_i, i) the 3-chain that leads to the current 3-chain (x, y, z).
5. Compute the message $L\|R$ or $S\|T$ corresponding to the 3-chain (x, y, z).
6. If $L\|R$ has been computed, make a P query to get $S\|T = P(L\|R)$; otherwise, make a P^{-1} query to get $L\|R = P^{-1}(S\|T)$.
7. Now all input values $(x_1, \ldots, x_6)$ to $(F_1, \ldots, F_6)$ corresponding to the 3-chain (x, y, z) are known. Additionally let $x_0 \leftarrow L$ and $x_7 \leftarrow T$.
8. If x_j is in the history of F_j or x_{j+1} is in the history of F_{j+1}, abort.
9. Define $F_j(x_j) \leftarrow x_{j-1} \oplus x_{j+1}$
10. Define $F_{j+1}(x_{j+1}) \leftarrow x_j \oplus x_{j+2}$
11. Call ChainQuery(x_j, j) and ChainQuery$(x_{j+1}, j+1)$, without considering in ChainQuery(x_j, j) and ChainQuery(x_{j+1}, j) the 3-chain that leads to the current 3-chain (x, y, z).

Additionally the simulator maintains an upper bound B_{max} on the size of the history of each of the F_i's; if this bound is reached, then the simulator aborts; the value of B_{max} will be determined later. This terminates the description of the simulator.

We note that all lines in Table 1 are necessary to ensure that the simulation of the F_i's is coherent with what the distinguisher can obtain independently from P. For example, if we suppress the line $(F_2, +)$ in the table, the distinguisher can make a query for Z to F_4, then Y to F_3 and $X = F_3(Y) \oplus Z$ to F_2, then $A = F_4(Z) \oplus Y$ to F_5 and since it is not possible anymore to adapt the output of (F_1, F_2), the simulator fails to provide a coherent simulation.

Our simulator makes recursive calls to the Query and ChainQuery algorithms. The simulator aborts when the history size of one of the F_i's is greater than B_{max}. Therefore we must prove that despite these recursive calls, this bound B_{max} is never reached, except with negligible probability, for B_{max} polynomial in the security parameter. The main argument is that the number of 3-chains in the sets Chain(b, x, k) that involve the P permutation (equations (3), (4), (5) and (6)), must be upper bounded by the number of P/P^{-1}-queries made by the distinguisher, which is upper bounded by q. This gives an upper bound on the number of recursive queries to F_3, F_4, which in turn implies an upper bound on the history of the other F_i's. Additionally, one must show that the simulator never aborts at Step 8 in the CompleteChain algorithm, except with negligible probability. This is summarised in the following lemma:

Lemma 1. *Let q be the maximum number of queries made by the distinguisher and let $B_{max} = 5q^2$. The simulator $\mathcal{S}$ runs in time $\mathcal{O}(q^4)$, and aborts with probability at most $2^{14} \cdot q^8/2^n$, while making at most $105 \cdot q^4$ queries to P or P^{-1}.*

Proof. Due to space restriction, in Appendix A we only show that the simulator's running time is $\mathcal{O}(q^4)$ and makes at most $105 \cdot q^4$ queries to P/P^{-1}. The full proof of Lemma 1 is given in the full version of this paper [10].

4.2   Indifferentiability

We now proceed to prove the indifferentiability result. As illustrated in Figure 3, we must show that given the previous simulator $\mathcal{S}$, the two systems formed by (LR, F) and $(P, \mathcal{S})$ are indistinguishable.

We consider a distinguisher $\mathcal{D}$ making at most q queries to the system (LR, F) or $(P, \mathcal{S})$ and outputting a bit γ. We define a sequence Game$_0$, Game$_1$, ... of modified distinguisher games. In the first game Game$_0$, the distinguisher interacts with the system formed by the random permutation P and the previously defined simulator $\mathcal{S}$. In the subsequent games the system is modified so that in the last game the distinguisher interacts with (LR, F). We denote by S_i the event in game i that the distinguisher outputs $\gamma = 1$.

Game$_0$: the distinguisher interacts with the simulator $\mathcal{S}$ and the random permutation P.

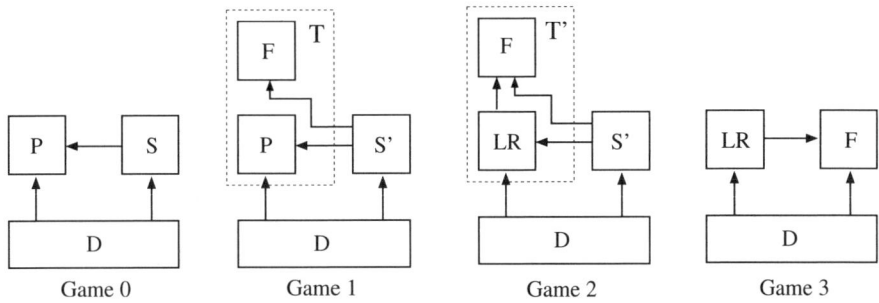

Fig. 5. Sequence of games for proving indifferentiability

Game_1: we make a minor change in the way F_i queries are answered by the simulator, to prepare a more important step in the next game. In Game_0 we have that a F_i query for x can be answered in two different ways: either $F_i(x) \xleftarrow{\$} \{0,1\}$, or the value $F_i(x)$ is "adapted" by the simulator. In Game_1, instead of letting $F_i(x) \xleftarrow{\$} \{0,1\}$, the new simulator $\mathcal{S}'$ makes a query to a random oracle F_i which returns $F_i(x)$; see Fig. 5 for an illustration. Since we have simply replaced one set of random variables by a different, but identically distributed, set of random variables, we have:

$$\Pr[S_0] = \Pr[S_1]$$

Game_2: we modify the way P and P^{-1} queries are answered. Instead of returning $P(L\|R)$ with random permutation P, the system returns $\mathsf{LR}(L\|R)$ by calling the random oracles F_i's (and similarly for P^{-1} queries).

We must show that the distinguisher's view has statistically close distribution in Game_1 and Game_2. For this, we consider the subsystem $\mathcal{T}$ with the random permutation P/P^{-1} and the random oracles F_i's in Game_1, and the subsystem $\mathcal{T}'$ with Luby-Rackoff LR and random oracle F_i's in Game_2 (see Fig. 5). We show that the output of systems $\mathcal{T}$ and $\mathcal{T}'$ is statistically close; this in turn shows that the distinguisher's view has statistically close distribution in Game_1 and Game_2.[1]

In the following, we assume that the distinguisher eventually makes a sequence of F_i-queries corresponding to all previous P/P^{-1} queries made by the distinguisher; this is without loss of generality, because from any distinguisher $\mathcal{D}$ we can build a distinguisher $\mathcal{D}'$ with the same output that satisfies this property.

The outputs to F_i queries provided by subsystem $\mathcal{T}$ in Game_1 and by subsystem $\mathcal{T}'$ in Game_2 are the same, since in both cases these queries are answered by random oracles F_i. Therefore, we must show that the output to P/P^{-1} queries provided by $\mathcal{T}$ and $\mathcal{T}'$ have statistically close distribution, when the outputs to F_i queries provided by $\mathcal{T}$ or $\mathcal{T}'$ are fixed.

---

[1] We do not claim that subsystems $\mathcal{T}$ and $\mathcal{T}'$ are indistinguishable for any possible sequence of queries (this is clearly false); we only show that $\mathcal{T}$ and $\mathcal{T}'$ have statistically close outputs for the particular sequence of queries made by the simulator and the distinguisher.

We can distinguish two types of P/P^{-1} queries to $\mathcal{T}$ or $\mathcal{T}'$:

- Type I: P/P^{-1} queries made by the distinguisher, or by the simulator during execution of the CompleteChain algorithm. From Lemma 1 there are at most $B_{max} + q \leq 6q^2$ such queries.
- Type II: P/P^{-1} queries made by the simulator when computing the sets Chain$(+1, S, 6)$, Chain$(-1, R, 1)$, Chain$(+1, A, 5)$ and Chain$(-1, X, 2)$, which are not of Type I. From Lemma 1 there are at most $Q_P = 105 \cdot q^4$ such queries.

We first consider Type I queries. Recall that the distinguisher is assumed to eventually make all the F_i queries corresponding to his P/P^{-1} queries; consequently at the end of the distinguisher's queries, the CompleteChain algorithm has been executed for all 3-chains corresponding to P/P^{-1} queries of Type I. We consider one such P query $L\|R$ (the argument for P^{-1} query is similar) of Type I. In Game$_2$ the answer $S\|T$ can be written as follows:

$$(S, T) = (L \oplus r_1 \oplus r_3 \oplus r_5, R \oplus r_2 \oplus r_4 \oplus r_6) \qquad (7)$$

where $r_1 = F_1(R)$, $r_2 = F_2(X)$, $r_3 = F_3(Y)$, $r_4 = F_4(Z)$, $r_5 = F_5(A)$ and $r_6 = F_6(S)$, and (X, Y, Z, A) are defined in the usual way.

Let j be the index used at steps 9 and 10 of the corresponding CompleteChain execution, and let x_j, x_{j+1} be the corresponding inputs. If the simulator does not abort during CompleteChain, this implies that the values $r_j = F_j(x_j)$ and $r_{j+1} = F_{j+1}(x_{j+1})$ have not appeared before in the simulator's execution. This implies that $r_j = F_j(x_j)$ and $r_{j+1} = F_{j+1}(x_{j+1})$ have not appeared in a previous P/P^{-1}-query (since otherwise it would have been defined in the corresponding CompleteChain execution), and moreover $F_j(x_j)$ and $F_{j+1}(x_{j+1})$ have not been queried before to subsystem $\mathcal{T}'$. Since the values $r_j = F_j(x_j)$ and $r_{j+1} = F_{j+1}(x_{j+1})$ are defined by the simulator at steps 9 and 10 of CompleteChain, these values will not be queried later to $\mathcal{T}'$. Therefore we have that $r_j = F_j(x_j)$ and $r_{j+1} = F_{j+1}(x_{j+1})$ are not included in the subsystem $\mathcal{T}'$ output; $\mathcal{T}'$ output can only include randoms in $(r_1, \ldots, r_{j-1}, r_{j+2}, \ldots, r_6)$. Therefore, we obtain from equation (7) that for fixed randoms $(r_1, \ldots, r_{j-1}, r_{j+2}, \ldots, r_6)$ the distribution of $S\|T = \mathsf{LR}(L\|R)$ in Game$_2$ is uniform in $\{0, 1\}^{2n}$ and independent from the output of previous P/P^{-1} queries.

In Game$_1$, the output to query $L\|R$ is $S\|T = P(L\|R)$; since there are at most $q + B_{max} \leq 6 \cdot q^2$ Type I queries to P/P^{-1}, the statistical distance between $P(L\|R)$ and $\mathsf{LR}(L\|R)$ is at most $6 \cdot q^2 / 2^{2n}$. This holds for a single P/P^{-1} query of Type I. Since there are at most $6 \cdot q^2$ such queries, we obtain the following statistical distance δ between outputs of systems $\mathcal{T}$ and $\mathcal{T}'$ to Type I queries, conditioned on the event that the simulator does not abort:

$$\delta \leq 6 \cdot q^2 \cdot \frac{6 \cdot q^2}{2^{2n}} \leq \frac{36 \cdot q^4}{2^{2n}} \qquad (8)$$

We now consider P/P^{-1} queries of Type II; from Lemma 1 there are at most $Q_P = 105 \cdot q^4$ such queries. We first consider the sets Chain$(+1, S, 6)$

and $\mathsf{Chain}(-1, X, 2)$, and we consider a corresponding query $L \| R$ to P, where $L = F_1(R) \oplus X$. By definition this query is not of Type I, so no $\mathsf{CompleteChain}$ execution has occurred corresponding to this query. Given $(R, X) \in \mathsf{Chain}(+1, S, 6)$ or for $(R, S) \in \mathsf{Chain}(-1, X, 2)$, we let $Y = F_2(X) \oplus R$. If Y is not in the history of F_3, we let $F_3(Y) \xleftarrow{\$} \{0, 1\}^n$; in this case, $Z = X \oplus F_3(Y)$ has the uniform distribution in $\{0, 1\}^n$; this implies that Z belongs to the history of F_4 with probability at most $|F_4|/2^n \leq 2q/2^n$. If Y belongs to the history of F_3, then we have that Z cannot be in the history of F_4, otherwise 3-chain (X, Y, Z) would already have appeared in $\mathsf{CompleteChain}$ algorithm, from Line $(F_2, +)$ and $(F_4, -)$ in Table 1. Therefore, we have that for all P queries $L \| R$ of Type II, no corresponding value of Z belongs to the history of F_4, except with probability at most $Q_P \cdot 2q/2^n$.

We now consider the sequence (L_i, R_i) of distinct P-queries of Type II corresponding to the previous sets $\mathsf{Chain}(+1, S, 6)$ and $\mathsf{Chain}(-1, X, 2)$. We must show that in Game_2 the output (S_i, T_i) provided by $\mathcal{T}'$ has a distribution that is statistically close to uniform, when the outputs to F_i queries provided by $\mathcal{T}'$ are fixed. We consider the corresponding sequence of (Y_i, Z_i); as explained previously, no Z_i belongs to the simulator's history of F_4, except with probability at most $Q_P \cdot 2q/2^n$. We claim that $F_4(Z_i) \oplus Y_i \neq F_4(Z_j) \oplus Y_j$ for all $1 \leq i < j \leq Q_P$, except with probability at most $(Q_P)^2/2^n$. Namely, if $Z_i = Z_j$ for some $i < j$, then $F_4(Z_i) \oplus Y_i = F_4(Z_j) \oplus Y_j$ implies $Y_i = Y_j$, which gives $(L_i, R_i) = (L_j, R_j)$, a contradiction since we have assumed the (L_i, R_i) queries to be distinct. Moreover, for all $i < j$ such that $Z_i \neq Z_j$, we have that $F_4(Z_i) \oplus Y_i = F_4(Z_j) \oplus Y_j$ happens with probability at most 2^{-n}; since there are at most $(Q_P)^2$ such i, j, we have that $F_4(Z_i) \oplus Y_i = F_4(Z_j) \oplus Z_j$ for some $i < j$ happens with probability at most $(Q_P)^2/2^n$.

This implies that the elements $A_i = Y_i \oplus F_4(Z_i)$ are all distinct, except with probability at most $(Q_P)^2/2^n$. Therefore elements $S_i = Z_i \oplus F_5(A_i)$ are uniformly and independently distributed in $\{0, 1\}^n$; this implies that elements S_i are all distinct, except with probability at most $(Q_P)^2/2^n$, which implies that elements $T_i = A_i \oplus F_6(S_i)$ are uniformly and independently distributed in $\{0, 1\}^n$. For each (S_i, T_i), the statistical distance with $P(L_i \| R_i)$ in Game_1 is therefore at most $Q_P/2^{2n}$. The previous arguments are conditioned on the event that no A_i or S_i belongs to the simulator's history for F_5 and F_6, which for each A_i or S_i happens with probability at most $B_{max}/2^n$. The reasoning for the sets $\mathsf{Chain}(-1, R, 1)$, $\mathsf{Chain}(+1, A, 5)$ is symmetric so we omit it. We obtain that the statistical distance δ_2 between the output of Type II P/P^{-1} queries in Game_1 and Game_2 is at most (conditioned on the event that the simulator does not abort):

$$\delta_2 \leq 2 \cdot \left(\frac{Q_P \cdot 2q}{2^n} + 2 \cdot \frac{(Q_P)^2}{2^n} + \frac{(Q_P)^2}{2^{2n}} + \frac{Q_P \cdot B_{max}}{2^n} \right) \leq \frac{2^{16} \cdot q^8}{2^n} \quad (9)$$

Let denote by Abort the event that the simulator aborts in Game_1; we obtain from Lemma 1 and inequalities (8) and (9) :

$$|\Pr[S_2] - \Pr[S_1]| \leq \Pr[\mathsf{Abort}] + \delta + \delta_2 \leq \frac{2^{14} \cdot q^8}{2^n} + \frac{36 \cdot q^4}{2^{2n}} + \frac{2^{16} \cdot q^8}{2^n} \leq \frac{2^{17} \cdot q^8}{2^n}$$

Game$_3$: the distinguisher interacts with random system (LR, F). We have that system (LR, F) provides the same outputs as the system in Game$_2$ except if the simulator fails in Game$_2$. Namely, when the output values of (F_j, F_{j+1}) are adapted (steps 9 and 10 of CompleteChain algorithm), the values $F_j(x_j)$ and $F_{j+1}(x_{j+1})$ are the same as the one obtained directly from random oracles F_j and F_{j+1}, because in Game$_2$ the P/P^{-1} queries are answered using LR/LR^{-1}. Let denote by Abort$_2$ the event that simulator aborts in Game$_2$; we have:

$$\Pr[\mathsf{Abort}_2] \leq \Pr[\mathsf{Abort}] + \delta + \delta_2 \leq \frac{2^{14} \cdot q^8}{2^n} + \frac{36 \cdot q^4}{2^{2n}} + \frac{2^{16} \cdot q^8}{2^n} \leq \frac{2^{17} \cdot q^8}{2^n}$$

which gives:

$$|\Pr[S_3] - \Pr[S_2]| \leq \Pr[\mathsf{Abort}_2] \leq \frac{2^{17} \cdot q^8}{2^n}$$

From the previous inequalities, we obtain the following upper bound on the distinguisher's advantage:

$$|\Pr[S_3] - \Pr[S_0]| \leq \frac{2^{18} \cdot q^8}{2^n}$$

which terminates the proof of Theorem 2.

5   Conclusion and Further Research

We have shown that the 6 rounds Feistel construction is indifferentiable from a random permutation, a problem that was left open in [9]. This shows that the random oracle model and the ideal cipher model are equivalent models. A natural question is whether our security bound in $q^8/2^n$ is optimal or not. We are currently investigating:

- a better bound for 6 rounds (or more),
- best exponential attacks against 6 rounds (or more),
- other models of indifferentiability with possibly simpler proofs.

Acknowledgements. we would like to thank the anonymous referees of Crypto 2008 for their useful comments.

References

1. Bellare, M., Rogaway, P.: Random oracles are practical: A paradigm for designing efficient protocols. In: Proceedings of the 1st ACM Conference on Computer and Communications Security, pp. 62–73 (1993)
2. Bellare, M., Rogaway, P.: The exact security of digital signatures - How to sign with RSA and Rabin. In: Maurer, U.M. (ed.) EUROCRYPT 1996. LNCS, vol. 1070, pp. 399–416. Springer, Heidelberg (1996)
3. Bellare, M., Rogaway, P.: Optimal Asymmetric Encryption. In: De Santis, A. (ed.) EUROCRYPT 1994. LNCS, vol. 950, pp. 92–111. Springer, Heidelberg (1995)

4. Black, J.: The Ideal-Cipher Model, Revisited: An Uninstantiable Blockcipher-Based Hash Function. In: Robshaw, M. (ed.) FSE 2006. LNCS, vol. 4047, pp. 328–340. Springer, Heidelberg (2006)
5. Black, J., Rogaway, P., Shrimpton, T.: Black-Box Analysis of the Block Cipher-Based Hash-Function Constructions from PGV. In: Yung, M. (ed.) CRYPTO 2002. LNCS, vol. 2442. Springer, Heidelberg (2002)
6. Boneh, D., Gentry, C., Shacham, H., Lynn, B.: Aggregate and Verifiably Encrypted Signatures from Bilinear Maps. In: Biham, E. (ed.) EUROCRYPT 2003. LNCS, vol. 2656, pp. 416–432. Springer, Heidelberg (2003)
7. Canetti, R., Goldreich, O., Halevi, S.: The random oracle methodology, revisited. In: Proceedings of the 30th ACM Symposium on the Theory of Computing, pp. 209–218. ACM Press, New York (1998)
8. Chang, D., Lee, S., Nandi, M., Yung, M.: Indifferentiable Security Analysis of Popular Hash Functions with Prefix-Free Padding. In: Lai, X., Chen, K. (eds.) ASIACRYPT 2006. LNCS, vol. 4284, pp. 283–298. Springer, Heidelberg (2006)
9. Coron, J.S., Dodis, Y., Malinaud, C., Puniya, P.: Merkle-Damgård Revisited: How to Construct a Hash Function. In: Shoup, V. (ed.) CRYPTO 2005. LNCS, vol. 3621, pp. 430–448. Springer, Heidelberg (2005)
10. Coron, J.S., Patarin, J., Seurin, Y.: The Random Oracle Model and the Ideal Cipher Model are Equivalent. Full version of this paper. Cryptology ePrint Archive, Report 2008/246, http://eprint.iacr.org/
11. Desai, A.: The security of all-or-nothing encryption: Protecting against exhaustive key search. In: Bellare, M. (ed.) CRYPTO 2000. LNCS, vol. 1880. Springer, Heidelberg (2000)
12. Dodis, Y., Puniya, P.: On the Relation Between the Ideal Cipher and the Random Oracle Models. In: Halevi, S., Rabin, T. (eds.) TCC 2006. LNCS, vol. 3876, pp. 184–206. Springer, Heidelberg (2006)
13. Dodis, Y., Puniya, P.: Feistel Networks Made Public, and Applications. In: Naor, M. (ed.) EUROCRYPT 2007. LNCS, vol. 4515, pp. 534–554. Springer, Heidelberg (2007)
14. Even, S., Mansour, Y.: A construction of a cipher from a single pseudorandom permutation. In: Matsumoto, T., Imai, H., Rivest, R.L. (eds.) ASIACRYPT 1991. LNCS, vol. 739, pp. 210–224. Springer, Heidelberg (1993)
15. Gentry, C., Ramzan, Z.: Eliminating Random Permutation Oracles in the Even-Mansour Cipher. In: Lee, P.J. (ed.) ASIACRYPT 2004. LNCS, vol. 3329. Springer, Heidelberg (2004)
16. Kilian, J., Rogaway, P.: How to protect DES against exhaustive key search (An analysis of DESX). Journal of Cryptology 14(1), 17–35 (2001)
17. Luby, M., Rackoff, C.: How to construct pseudorandom permutations from pseudorandom functions. SIAM Journal of Computing 17(2), 373–386 (1988)
18. Maurer, U., Renner, R., Holenstein, C.: Indifferentiability, Impossibility Results on Reductions, and Applications to the Random Oracle Methodology. In: Naor, M. (ed.) TCC 2004. LNCS, vol. 2951, pp. 21–39. Springer, Heidelberg (2004)
19. Patarin, J.: Pseudorandom Permutations Based on the DES Scheme. In: Charpin, P., Cohen, G. (eds.) EUROCODE 1990. LNCS, vol. 514, pp. 193–204. Springer, Heidelberg (1991)
20. Ramzan, Z., Reyzin, L.: On the Round Security of Symmetric-Key Cryptographic Primitives. In: Bellare, M. (ed.) CRYPTO 2000. LNCS, vol. 1880. Springer, Heidelberg (2000)

A   Proof of Lemma 1

We first give an upper bound on the total number of executions of algorithm CompleteChain(b, x, y, z, k) for $(b, k) \in \{(+1, 6), (-1, 2), (-1, 1), (+1, 5)\}$. We first consider the set:

$$\mathsf{Chain}(+1, S, 6) = \left\{ (R, X) \in (F_1, F_2) \mid \exists T, \ P(X \oplus F_1(R) \| R) = S \| T \right\}$$

that generates executions of CompleteChain$(+1, S, R, X, 6)$. We denote by bad_6 the event that CompleteChain$(+1, S, R, X, 6)$ was called while $X \oplus F_1(R) \| R$ has not appeared in a P/P^{-1} query made by the distinguisher.

Similarly, considering the set $\mathsf{Chain}(-1, X, 2)$, we denote by bad_2 the event that CompleteChain$(-1, X, R, S, 2)$ was called while $X \oplus F_1(R) \| R$ has not appeared in a P/P^{-1} query made by the distinguisher. Symmetrically, we denote by bad_1 and bad_5 the corresponding events for CompleteChain$(-1, R, A, S, 1)$ and CompleteChain$(+1, A, R, S, 5)$. We denote $\mathsf{bad} = \mathsf{bad}_1 \vee \mathsf{bad}_2 \vee \mathsf{bad}_5 \vee \mathsf{bad}_6$.

Lemma 2. *The total number of executions of* CompleteChain(b, x, y, z, k) *for* $(b, k) \in \{(+1, 6), (-1, 2), (-1, 1), (+1, 5)\}$ *is upper bounded by* q, *unless event* bad *occurs, which happens with probability at most:*

$$\Pr[\mathsf{bad}] \leq \frac{5 \cdot (B_{max})^4}{2^n} \tag{10}$$

Proof. If event bad has not occurred, then the distinguisher has made a P/P^{-1} query corresponding to all pairs (x, y) in CompleteChain(b, x, y, z, k) for $(b, k) \in \{(+1, 6), (-1, 2), (-1, 1), (+1, 5)\}$; since the distinguisher makes at most q queries, the total number of executions is then upper bounded by q.

We first consider event bad_6 corresponding to $\mathsf{Chain}(+1, S, 6)$. If $L \| R$ with $L = X \oplus F_1(R)$ has never appeared in a P/P^{-1} query made by the distinguisher, then the probability that $P(L \| R) = S \| T$ for some T is at most 2^{-n}. For a single S query to F_6, the probability that bad_6 occurs is then at most $|F_1| \cdot |F_2|/2^n \leq (B_{max})^2/2^n$. Since there has been at most B_{max} such queries to F_6, this gives:

$$\Pr[\mathsf{bad}_6] \leq \frac{(B_{max})^3}{2^n}$$

Symmetrically, the same bound holds for event bad_1.

Similarly, for event bad_2, if the distinguisher has not made a query for $P(L \| R)$ where $L = X \oplus F_1(R)$, then the probability that $P(L \| R) = S \| T$ with $S \in \tilde{F}_6$ is at most $|\tilde{F}_6|/2^n$, where $|\tilde{F}_6| \leq |F_6| + |F_1| \cdot |F_2| \leq 2 \cdot (B_{max})^2$. For a single X query, this implies that event bad_2 occurs with probability at most $|F_1| \cdot |\tilde{F}_6|/2^n$; since there are at most B_{max} such queries, this gives:

$$\Pr[\mathsf{bad}_2] \leq B_{max} \cdot \frac{|F_1| \cdot |\tilde{F}_6|}{2^n} \leq \frac{2 \cdot (B_{max})^4}{2^n}$$

Symmetrically, the same bound holds for event bad_5. From the previous inequalities we obtain the required bound for $\Pr[\mathsf{bad}]$. $\square$

Lemma 3. *Taking $B_{max} = 5 \cdot q^2$, the history size of the simulator F_i's does not reach the bound B_{max}, unless event* **bad** *occurs, which happens with probability at most:*

$$\Pr[\text{bad}] \leq \frac{2^{12} \cdot q^8}{2^n} \tag{11}$$

Proof. The 3-chains from Lines $(F_6, +)$, $(F_2, -)$, $(F_1, -)$ and $(F_5, +)$ in Table 1 are the only ones which can generate recursive calls to F_3 and F_4, since the other 3-chains from Lines $(F_2, +)$, $(F_3, +)$, $(F_4, -)$ and $(F_5, -)$ always include elements in F_3 and F_4 histories. Moreover from Lemma 2 the total number of corresponding executions of CompleteChain(b, x, y, z, k) where $(b, k) \in \{(+1, 6), (-1, 2), (-1, 1), (+1, 5)\}$ is upper bounded by q, unless event **bad** occurs. This implies that at most q recursive queries to F_3 and F_4 can occur, unless event **bad** occurs. Since the distinguisher himself makes at most q queries to F_3 and F_4, the total size of F_3 and F_4 histories in the simulator is upper bounded by $q + q = 2 \cdot q$.

The 3-chains from Lines $(F_2, +)$, $(F_3, +)$, $(F_4, -)$ and $(F_5, -)$ always include elements from both F_3 and F_4 histories. Therefore, the number of such 3-chains is upper bounded by $(2q)^2 = 4 \cdot q^2$. This implies that the simulator makes at most $4q^2$ recursive queries to F_1, F_2, F_5 and F_6. Therefore, taking:

$$B_{max} = 5 \cdot q^2 \tag{12}$$

we obtain that the simulator does not reach the bound B_{max}, except if event **bad** occurs; from inequality (10) and equation (12) we obtain (11).   □

Lemma 4. *With $B_{max} = 5 \cdot q^2$, the simulator makes at most $105 \cdot q^4$ queries to P/P^{-1} and runs in time $\mathcal{O}(q^4)$.*

Proof. The simulator makes queries to P/P^{-1} when computing the four sets Chain$(+1, S, 6)$, Chain$(-1, R, 1)$, Chain$(+1, A, 5)$ and Chain$(-1, X, 2)$ and also when completing a 3-chain with CompleteChain algorithm. Since the history size of the F_i's is upper bounded by $B_{max} = 5 \cdot q^2$, we obtain that the number Q_P of P/P^{-1}-queries made by the simulator is at most:

$$Q_P \leq 4 \cdot (B_{max})^2 + B_{max} \leq 105 \cdot q^4 \tag{13}$$

From this we have that the simulator runs in time $\mathcal{O}(q^4)$.   □

Lemma 2, 3 and 4 complete the first part of the proof. The remaining part consists in showing that the simulator never aborts at Step 8 in algorithm CompleteChain, except with negligible probability, and appears in the full version of this paper [10].

B   A Note on Indifferentiability in the Honest-but-Curious Model

In this section, we show that LR with up to logarithmic number of rounds is *not* indifferentiable from a random permutation in the honest-but-curious model

[12]; combined with our main result in the general model, this provides a separation between the two models. Note that this observation does not contradict any result formally proven in [12]; it only shows that honest-but-curious indifferentiability is not necessarily weaker than general indifferentiability.

Roughly speaking, in the honest-but-curious indifferentiability model, the distinguisher cannot query the F_i's directly. It can only make two types of queries: direct queries to the LR/LR^{-1} construction, and queries to the LR/LR^{-1} construction where in addition the intermediate results of the F_i's is provided. When interacting with the random permutation P and a simulator S, the first type of query is sent directly to P, while the second type is sent to S who makes the corresponding query to P, and in addition provides a simulated transcript of intermediate F_i results. Note that the simulator S is not allowed to make additional queries to P apart from forwarding the queries from the distinguisher; see [12] for a precise definition.

The authors of [12] define the notion of a *transparent construction*. Roughly speaking, this is a construction C^F such that the value of random oracle $F(x)$ can be computed efficiently for any x, by making a polynomial number of queries to C^F and getting the F outputs used by C^F to answer each query. The authors show that Luby-Rackoff with up to logarithmic number of rounds is a transparent construction. Namely the authors construct an extracting algorithm E such that when given oracle access to LR and the intermediate values F_i used to compute LR, the value $F_i(x)$ can be computed for any x at any round i. We note that algorithm E does not make queries to LR^{-1}, only to LR.

Algorithm E implies that for a LR construction with up to logarithmic number of rounds, it is possible to find an input message $L\|R$ such that the value S in $S\|T = \mathsf{LR}(L\|R)$ has a predetermined value, by only making forward queries to LR; namely this is how algorithm E can obtain $F_\ell(S)$, where ℓ is the last round. But this task is clearly impossible with a random permutation P: it is infeasible to find $L\|R$ such that S in $S\|T = P(L\|R)$ has a pre-determined value while only making forward queries to P. This implies that a simulator in the honest-but-curious model will necessarily fail (recall that such a simulator only forwards queries from the distinguisher to P and cannot make his own queries to P/P^{-1}). Therefore, LR with up to logarithmic number of rounds is *not* indifferentiable from a random permutation in the honest-but-curious model. Since our main result is that LR with 6 rounds is indifferentiable from a random permutation in the general model, this provides a separation between the two models.

Programmable Hash Functions
and Their Applications

Dennis Hofheinz[*] and Eike Kiltz[**]

Cryptology and Information Security Research Theme
CWI Amsterdam, The Netherlands
{hofheinz,kiltz}@cwi.nl

Abstract. We introduce a new information-theoretic primitive called *programmable hash functions* (PHFs). PHFs can be used to *program* the output of a hash function such that it contains solved or unsolved discrete logarithm instances with a certain probability. This is a technique originally used for security proofs in the random oracle model. We give a variety of *standard model* realizations of PHFs (with different parameters).

The programmability of PHFs make them a suitable tool to obtain black-box proofs of cryptographic protocols when considering adaptive attacks. We propose generic digital signature schemes from the strong RSA problem and from some hardness assumption on bilinear maps that can be instantiated with any PHF. Our schemes offer various improvements over known constructions. In particular, for a reasonable choice of parameters, we obtain short standard model digital signatures over bilinear maps.

1 Introduction

1.1 Programmable Hash Functions

A group hash function is an efficiently computable function that maps binary strings into a group $\mathbb{G}$. We propose the concept of a *programmable hash function* which is a keyed group hash function that can behave in two indistinguishable ways, depending on how the key is generated. If the standard key generation algorithm is used, then the hash function fulfills its normal functionality, i.e., it properly hashes its inputs into a group $\mathbb{G}$. The alternative (trapdoor) key generation algorithm outputs a key that is *indistinguishable* from the one output by the standard algorithm. It furthermore generates some additional secret trapdoor information that depends on two generators g and h from the group. This trapdoor information makes it possible to relate the output of the hash function to g and h: for any input X, one obtains integers a_X and b_X such that the

---

[*] Supported by the Dutch Organization for Scientific Research (NWO).

[**] Supported by the research program Sentinels (http://www.sentinels.nl). Sentinels is being financed by Technology Foundation STW, the Netherlands Organization for Scientific Research (NWO), and the Dutch Ministry of Economic Affairs.

D. Wagner (Ed.): CRYPTO 2008, LNCS 5157, pp. 21–38, 2008.
© International Association for Cryptologic Research 2008

relation $\mathrm{H}(X) = g^{a_X} h^{b_X} \in \mathbb{G}$ holds. For the PHF to be (m, n)-programmable we require that *for all* choices of $X_1, \ldots, X_m$ and $Z_1, \ldots, Z_n$ with $X_i \neq Z_j$, it holds that $a_{X_i} = 0$ but $a_{Z_j} \neq 0$, with some non-negligible probability. Hence parameter m controls the number of elements X for which we can hope to have $\mathrm{H}(X) = h^{b_X}$; parameter n controls the number of elements Z for which we can hope to have $\mathrm{H}(Z) = g^{a_Z} h^{b_Z}$ for some $a_Z \neq 0$.

The concept becomes useful in groups with hard discrete logarithms and when the trapdoor key generation algorithm does not know the discrete logarithm of h to the basis g. It is then possible to program the hash function such that the hash images of all possible choices $X_1, \ldots, X_m$ of m inputs are $\mathrm{H}(X_i) = h^{b_{X_i}}$, i.e., they do not depend on g (since $a_{X_i} = 0$). At the same time the hash images of all possible choices $Z_1, \ldots, Z_n$ of n (different) inputs are $\mathrm{H}(Z_i) = g^{b_{Z_i}} \cdot h^{b_{Z_i}}$, i.e., they do depend on g in a known way (since $a_{Z_i} \neq 0$). Intuitively, this resembles a scenario we are often confronted with in "provable security": for some of the hash outputs we know the discrete logarithm, and for some we do not. This situation appears naturally during a reduction that involves an adaptive adversary. Concretely, knowledge of the discrete logarithms of some hash queries can be used to simulate, e.g., a signing oracle for an adversary (which would normally require knowledge of a secret signing key). On the other hand, once the adversary produces, e.g., a signature on its own, our hope is that this signature corresponds to a hash query for which the we do *not* know the discrete logarithm. This way, the adversary has produced a piece of nontrivial secret information which can be used to break an underlying computational assumption.

This way of "programming" a hash function is very popular in the context of random oracles [3] (which, in a sense, are ideally programmable hash functions), and has been used to derive proofs of the adaptive security of simple signature schemes [4]. An (m, poly)-PHF is a (m, n)-PHF for all polynomials n. A (poly, m)-PHF is defined the same way. Using this notation, a random oracle implies a $(\mathsf{poly}, 1)$-PHF.[1]

INSTANTIATIONS. As our central instantiation of a PHF we use the following function which was originally introduced by Chaum et. al. [13] as a collision-resistant hash function. The "multi-generator" hash function $\mathsf{H}^{\mathsf{MG}} : \{0, 1\}^\ell \to \mathbb{G}$ is defined as $\mathsf{H}^{\mathsf{MG}}(X) := h_0 \prod_{i=1}^\ell h_i^{X_i}$, where the h_i are public generators of the group and $X = (X_1, \ldots, X_\ell)$. After its discovery in [13] it was also used in other constructions (e.g., [2, 11, 14, 25]), relying on other useful properties beyond collision resistance. Specifically, in the analysis of his identity-based encryption scheme, Waters [25] implicitly proved that, using our notation, H^{MG} is a $(1, \mathsf{poly})$-programmable hash function.

Our main result concerning instantiations of PHFs is a new analysis of H^{MG} showing that it is also a $(2, 1)$-PHF. Furthermore, we can use our new techniques to prove better bounds on the $(1, \mathsf{poly})$-programmability of H^{MG}. We stress that

---

[1] By "programming" the random oracle as $\mathrm{H}(X) = g^{a_X} h^{b_X}$ (for random a_X, b_X) with some sufficiently small but noticeable probability p and $\mathrm{H}(X) = h^{b_X}$ with probability $1 - p$ [16].

our analysis uses random walk techniques and is different from the one implicitly given in [25].

Unfortunately, the PHF H^{MG} has a relatively large public evaluation key. (Its key consists of $\ell+1$ group elements.) In our main application, signature schemes, this will lead to a tradeoff between public key size and signature size: using PHFs decreases the signature size, at the price of an increased public key size. See below for more details.

VARIATIONS. The concept of PHFs can be extended to randomized programmable hash functions (RPHFs). An RPHF is like a PHF whose input takes an additional parameter, the randomness. Our main construction of a randomized hash function is $RH^{Poly}{}_m$, which is $(m, 1)$-programmable. Note that unlike H^{MG}, the construction of the hash function depends on the parameter m. In particular, the complexity of $RH^{Poly}{}_m$ grows quadratically in m.

In some applications (e.g., for RSA signatures) we need a special type a PHF which we call bounded PHF. Essentially, for bounded PHFs we need to know a certain upper bound on the $|a_X|$, for all X. Whereas H^{MG} is bounded, $RH^{Poly}{}_m$ is only bounded for $m = 1$.

1.2 Applications

COLLISION RESISTANT HASHING. We aim to use PHFs as a tool to provide black-box proofs for various cryptographic protocols. As a toy example let us sketch why, in prime-order groups with hard discrete logarithms, any $(1, 1)$-PHF implies collision resistant hashing. Setting up H using the trapdoor generation algorithm will remain unnoticed for an adversary, but any collision $H(X) = H(Z)$ with $X \neq Z$ gives rise to an equation $g^{a_X} h^{b_X} = H(X) = H(Z) = g^{a_Z} h^{b_Z}$ with known exponents. Since the hash function is $(1, 1)$-programmable we have that, with non-negligible probability, $a_X = 0$ and $a_Z \neq 0$. This implies $g = h^{(b_X - b_Z)/a_Z}$, revealing the discrete logarithm of h to the base g.

GENERIC BILINEAR MAP SIGNATURES. We propose the following generic Bilinear Maps signature scheme with respect to a group hash function H. The signature of a message X is defined as the tuple

$$SIG_{BM}[H] : \quad sig = (H(X)^{\frac{1}{x+s}}, s) \in \mathbb{G} \times \{0, 1\}^{\eta}, \tag{1}$$

where s is a random η bit-string. Here $x \in \mathbb{Z}_{|\mathbb{G}|}$ is the secret key. The signature can be verified with the help of the public key g, g^x and a bilinear map. Our main theorem concerning the Bilinear Map signatures states that if, for some $m \geq 1$, H is an $(m, 1)$-programmable hash function and the q-Strong Diffie-Hellman (q-SDH) assumption [6] holds, then the above signature scheme is unforgeable against chosen message attacks [23]. Here, the parameter m controls the size $\eta = \eta(m)$ of the randomness s. For "80-bit security" and assuming the scheme establishes no more than $q = 2^{30}$ signatures [4], we can choose $\eta = 30+80/m$ such that $\eta = 70$ is sufficient when using our $(2, 1)$-PHF H^{MG}. The total signature size amounts to $160+70 = 230$ bits. (See below for details.) Furthermore, our generic

Bilinear Map scheme can also be instantiated with any randomized PHF. Then the signature of $\mathsf{SIG}_{\mathrm{BM}}[\mathrm{RH}]$ is defined as $sig := (\mathrm{RH}(X;r)^{1/(x+s)}, s, r)$, where r is chosen from the PRHF's randomness space.

GENERIC RSA SIGNATURES. We propose the following generic RSA signature scheme with respect to a group hash function H. The signature of a message X is defined as the tuple

$$\mathsf{SIG}_{\mathrm{RSA}}[\mathrm{H}]: \quad sig = (\mathrm{H}(X)^{1/e}, r) \in \mathbb{Z}_N \times \{0,1\}^\eta, \tag{2}$$

where e is a η bit prime. The eth root can be computed using the factorization of $N = pq$ which is contained in the secret key. Our main theorem concerning RSA signatures states that if, for some $m \geq 1$, H is a bounded $(m, 1)$-programmable hash function and the strong RSA assumption holds, then the above signature scheme is unforgeable against chosen message attacks. Again, the parameter m controls the size of the prime as $\eta \approx 30 + 80/m$. Our generic RSA scheme can also be instantiated with a bounded randomized PHF.

OTHER APPLICATIONS. BLS signatures [8, 9] are examples of "full-domain hash" (FDH) signature schemes [4]. Using the properties of a $(\mathsf{poly}, 1)$-programmable hash function, one can give a black-box reduction from unforgeability of $\mathsf{SIG}_{\mathrm{BLS}}$ to breaking the CDH assumption. The same reduction also holds for all full-domain hash signatures, for example also RSA-FDH [4]. Unfortunately, we do not know of any standard-model instantiation of $(\mathsf{poly}, 1)$-PHFs. This fact may be not too surprising given the impossibility results from [18].[2]

It is furthermore possible to reduce the security of Waters signatures [25] to breaking the CDH assumption, when instantiated with a $(1, \mathsf{poly})$-programmable hash function. This explains Waters' specific analysis in our PHF framework. Furthermore, our improved bound on the $(1, \mathsf{poly})$-programmability of H^{MG} gives a (slightly) tighter security reduction for Waters IBE and signature scheme.

1.3   Short Signatures

Our main application of PHFs are short signatures in the standard model. We now discuss our results in more detail. We refer to [6, 9] for applications of short signatures.

THE BIRTHDAY PARADOX AND RANDOMIZED SIGNATURES. A signature scheme $\mathsf{SIG}_{\mathrm{Fisch}}$ by Fischlin [19] (itself a variant of the RSA-based Cramer-Shoup signatures [17]) is defined as follows. The signature for a message m is given by $sig := (e, r, (h_0 h_1^r h_2^{m+r \bmod 2^\ell})^{1/e} \bmod N)$, where e is a random η-bit prime and r is a random ℓ bit mask. The birthday paradox (for uniformly sampled primes)

---

[2] We remark that the impossibility results from [18] do not imply that $(\mathsf{poly}, 1)$-programmable hash functions do not exist since they only rule out the possibility of proving the security of such constructions based on any assumption which is satisfied by random functions, thus it might still be possible to construct such objects using, say homomorphic properties.

tells us that after establishing q distinct Fischlin signatures, the probability that there exist two signatures, (e, r_1, y_1) on m_1 and (e, r_2, y_2) on m_2, with the *same prime* e is roughly $q^2 \eta / 2^\eta$. One can verify that in case of a collision, $(e, 2r_1 - r_2, 2y_1 - y_2)$ is a valid signature on the "message" $2m_1 - m_2$ (with constant probability). Usually, for "k bit security" one requires the adversary's success ratio (i.e., the forging probability of an adversary divided by its running time) to be upper bounded by 2^{-k}. For $k = 80$ and assuming the number of signature queries is upper bounded by $q = 2^{30}$, the length of the prime must therefore be at least $\eta > 80 + 30 = 110$ bits to immunize against this birthday attack. We remark that for a different, more technical reason, Fischlin's signatures even require $\eta \geq 160$ bits.

BEYOND THE BIRTHDAY PARADOX. In fact, Fischlin's signature scheme can be seen as our generic RSA signatures scheme from (2), instantiated with a concrete $(1, 1)$-RPHF ($\mathsf{RH}^{\mathsf{Poly}_1}$). In our notation, the programmability of the hash function is used at the point where an adversary uses a given signature (e, y_1) to create a forgery (e, y) with the *same prime* e. A simulator in the security reduction has to be able to compute $y_1 = \mathsf{H}(X)^{1/e}$ but must use $y = \mathsf{H}(Z)^{1/e}$ to break the strong RSA challenge, i.e., to compute $g^{1/e'}$ and $e' > 1$ from g. However, since the hash function is $(1, 1)$-programmable we can program H with g and $h = g^e$ such that, with some non-negligible probability, $\mathsf{H}(X)^{1/e} = h^{bx} = g^{bx_1}$ can be computed but $\mathsf{H}(Z)^{1/e} = (g^{az} h^{bz})^{1/e} = g^{az/e} g^{bz}$ can be used to break the strong RSA assumption since $a_Z \neq 0$.

Our central improvement consists of instantiating the generic RSA signature scheme with a $(m, 1)$-PHF to break the birthday bound. The observation is that such hash functions can guarantee that after establishing up to m signatures with respect to the same prime, forging is still impossible. In analogy to the above, with a $(m, 1)$-PHF the simulation is successful as long as there are at most m many signatures that use the same prime as in the forgery. By the generalized birthday paradox we know that after establishing q distinct generic RSA signatures the probability that there exists m signatures with the same prime is roughly $q^{m+1}(\frac{\eta}{2^\eta})^m$. Again, the success ration has to be bounded by 2^{-80} for $q = 2^{30}$ which means that $\mathsf{SIG}_{\mathrm{RSA}}[\mathsf{H}]$ instantiated with a $(2, 1)$-PHF can have primes as small as $\eta = 80$ bits to be provably secure.

The security proof for the bilinear map scheme $\mathsf{SIG}_{\mathrm{BM}}[\mathsf{H}]$ is similar. Due to the extended birthday paradox (for uniform random strings), $\mathsf{SIG}_{\mathrm{BM}}[\mathsf{H}]$ instantiated with a $(m, 1)$-PHF only needs $\eta = 30 + 80/m$ bits of randomness to be provably secure. For example, with our $(2, 1)$-PHF H^{MG} we need 70 bits of randomness.

COMPARISON. Table 1 compares the signature sizes of our and known signatures assuming $q = 2^{30}$. For RSA signatures our scheme $\mathsf{SIG}_{\mathrm{RSA}}[\mathsf{H}^{\mathsf{MG}}]$ offers a short alternative to Fischlin's signature scheme. More importantly, generating a random 80 bit prime will be considerably faster than a 160 bit one. We expect that, compared to the one by Fischlin, our new scheme roughly halves the signing time.

The main advantage of our bilinear maps scheme $\mathsf{SIG}_{\mathrm{BM}}[\mathsf{H}^{\mathsf{MG}}]$ is its very compact signatures of only 230 bits. This saves 90 bits compared to the short

Table 1. Recommended signature sizes of different schemes. The parameters are chosen to provide unforgeability with $k = 80$ bits security after revealing maximal $q = 2^{30}$ signatures. RSA signatures are instantiated with a modulus of $|N| = 1024$ bits, bilinear maps signatures in asymmetric pairings with $|\mathbb{G}| = \log p = 160$ bits. We assume without loss of generality that messages are of size ℓ bits (otherwise, we can apply a collision-resistant hash function first), where ℓ must be in the order of $2k = 160$ in order to provide k bits of security.

Scheme	Type	Signature Size	Key Size
Boneh-Boyen [6]	Bilinear	$|\mathbb{G}| + |\mathbb{Z}_p| = 320$	$2|\mathbb{G}| = 320$
Ours: $\mathsf{SIG}_{\mathrm{BM}}[\mathrm{H}^{\mathsf{MG}}]$	Bilinear	$|\mathbb{G}| + |s| = 230$	$(\ell + 2)|\mathbb{G}| = 26\mathrm{k}$
Cramer-Shoup [17]	RSA	$2 \cdot |\mathbb{Z}_N| + |e| = 2208$	$3 \cdot |\mathbb{Z}_N| + |e| = 3232$
Fischlin [19] (=$\mathsf{SIG}_{\mathrm{RSA}}[\mathrm{RH}^{\mathsf{Poly}_1}]$)	RSA	$|\mathbb{Z}_N| + |r| + |e| = 1344$	$4 \cdot |\mathbb{Z}_N| = 4096$
Ours: $\mathsf{SIG}_{\mathrm{RSA}}[\mathrm{H}^{\mathsf{MG}}]$	RSA	$|\mathbb{Z}_N| + |e| = 1104$	$(\ell + 1)|\mathbb{Z}_N| = 164\mathrm{k}$

signatures scheme from Boneh-Boyen [6, 7] and is only 70 bits larger than the random oracle BLS signatures. However, a drawback of our constructions is the size of the verification key since it includes the group hash key κ. For example, for $\mathrm{H}^{\mathsf{MG}} : \{0, 1\}^\ell \to \mathbb{G}$, κ contains $\ell + 1$ group elements, where $\ell = 160$. Concretely, that makes a verification key of $26k$ bits compared to 320 bits from [6].

We remark that our concrete security reductions for the two generic schemes are not tight, i.e., the reductions roughly lose $\log(q/\delta)$ bits of security (cf. Theorems 10 and 13). Strictly speaking, a non-tight reduction has to be penalized by having to choose a larger group order. Even though this is usually not done in the literature [17, 19], we also consider concrete signature size when additionally taking the non-tight security reduction into account. Since all known RSA schemes [17, 19] have the same non-tight reduction as we have, we only consider schemes based on bilinear maps. A rigorous comparison appears in the full version.

RELATED SIGNATURE SCHEMES. Our generic bilinear map signature scheme belongs to the class of "inversion-based" signature schemes originally proposed in [24] and first formally analyzed in [6]. Other related standard-model schemes can be found in [10, 22]. We stress that our signatures derive from the above since the message does not appear in the denominator of the exponent. This is an essential feature to get around the birthday bounds. Our generic RSA signature scheme builds on [19] which itself is based on the early work by Cramer and Shoup [17]. Other standard-model RSA schemes are [12, 15, 21, 26].

1.4 Open Problems

We show that PHFs provide a useful primitive to obtain black-box proofs for certain signature schemes. We leave it for future research to extend the application of PHFs to other types of protocols.

We leave it as an open problem to prove or disprove the standard-model existence of (poly, 1)-RPHFs. (Note that a positive result would imply a security proof for FDH signatures like [9]). Moreover, we are asking for a concrete construction of a deterministic $(3, 1)$-PHF that would make it possible to shrink the signature size of $\mathsf{SIG}_{\mathrm{BM}}[\mathrm{H}]$ to ≈ 215 bits. A *bounded* $(10, 1)$-RPHF would make it possible to shrink the size of the prime in $\mathsf{SIG}_{\mathrm{RSA}}[\mathrm{RH}]$ to roughly 40 bits. This is interesting since generating random 40 bit primes is very inexpensive. Finally, a $(2, 1)$ or $(1, \mathsf{poly})$-PHF with more compact parameters would have dramatic impact on the practicability of our signature schemes or Waters' IBE scheme [25].

2   Preliminaries

NOTATION. If x is a string, then $|x|$ denotes its length, while if S is a set then $|S|$ denotes its size. If $k \in \mathbb{N}$ then 1^k denotes the string of k ones. For $n \in \mathbb{N}$, we write $[n]$ shorthand for $\{1, \ldots, n\}$. If S is a set then $s \xleftarrow{\$} S$ denotes the operation of picking an element s of S uniformly at random. We write $\mathcal{A}(x, y, \ldots)$ to indicate that $\mathcal{A}$ is an algorithm with inputs $x, y, \ldots$ and by $z \xleftarrow{\$} \mathcal{A}(x, y, \ldots)$ we denote the operation of running $\mathcal{A}$ with inputs $(x, y, \ldots)$ and letting z be the output. With PPT we denote probabilistic polynomial time. For random variables X and Y, we write $X \overset{\gamma}{\equiv} Y$ if their statistical distance is at most γ.

DIGITAL SIGNATURES. A digital signature scheme SIG consists of the PPT algorithms. The key generation algorithm generates a secret signing and a public verification key. The signing algorithm inputs the signing key and a message and returns a signature. The deterministic verification algorithm inputs the verification key and returns accept or reject. We demand the usual correctness properties. We recall the definition for unforgeability against chosen-message attacks (UF-CMA), played between a challenger and a forger $\mathcal{F}$:

1. The challenger generates verification/signing key, and gives the verification key to $\mathcal{F}$;
2. $\mathcal{F}$ makes a number of *signing queries* to the challenger; each such query is a message m_i; the challenger signs m_i, and sends the result sig_i to $\mathcal{F}$;
3. $\mathcal{F}$ outputs a message m and a signature sig.

We say that forger $\mathcal{F}$ wins the game if sig is a valid signature on m and it has not queried a signature on m before. Forger $\mathcal{F}$ (t, q, ϵ)-breaks the UF-CMA security of SIG if its running time is bounded by t, it makes at most Q signing queries, and the probability that it wins the above game is bounded by ϵ. Finally, SIG is UF-CMA secure if no forger can (t, q, ϵ)-break the UF-CMA security of SIG for polynomial t and q and non-negligible ϵ.

PAIRING GROUPS AND THE q-SDH ASSUMPTION. Our pairing schemes will be defined on families of bilinear groups $(\mathbb{PG}_k)_{k \in \mathbb{N}}$. A pairing group $\mathbb{PG} = \mathbb{PG}_k = (\mathbb{G}, \mathbb{G}_T, p, \hat{e}, g)$ consist of a multiplicative cyclic group $\mathbb{G}$ of prime order p, where $2^k < p < 2^{k+1}$, a multiplicative cyclic group $\mathbb{G}_T$ of the same order, a generator $g \in \mathbb{G}$, and a non-degenerate bilinear pairing $\hat{e} \colon \mathbb{G} \times \mathbb{G} \to \mathbb{G}_T$. See [6] for a description of the properties of such pairings. We say an adversary $\mathcal{A}$ (t, ϵ)-breaks the q-strong Diffie-Hellman (q-SDH) assumption if its running time is bounded by t and

$$\Pr[(s, g^{\frac{1}{x+s}}) \xleftarrow{\$} \mathcal{A}(g, g^x, \ldots, g^{x^q})] \geq \epsilon,$$

where $g \xleftarrow{\$} \mathbb{G}_T$ and $x \xleftarrow{\$} \mathbb{Z}_p^*$. We require that in $\mathbb{PG}$ the q-SDH [5] assumption holds meaning that no adversary can (t, ϵ) break the q-SDH problem for a polynomial t and non-negligible ϵ.

RSA GROUPS AND THE STRONG RSA ASSUMPTION. Our RSA schemes will be defined on families of RSA groups $(\mathbb{RG}_k)_{k \in \mathbb{N}}$. A safe RSA group $\mathbb{RG} = \mathbb{RG}_k = (p, q)$ consists of two distinct safe prime p and q of $k/2$ bits. Let QR_N denote the cyclic group of quadratic residues modulo an RSA number $N = pq$. We say an adversary $\mathcal{A}$ (t, ϵ)-breaks the strong RSA assumption if its running time is bounded by t and

$$\Pr[(e > 1, z^{1/e}) \xleftarrow{\$} \mathcal{A}(N = pq, z)] \geq \epsilon,$$

where $z \xleftarrow{\$} \mathbb{Z}_N$. We require that in $\mathbb{RG}$ the strong RSA assumption [1, 20] holds meaning that no adversary can (t, ϵ)-break the strong RSA problem for a polynomial t and non-negligible ϵ.

3   Programmable Hash Functions

A *group family* $G = (\mathbb{G}_k)$ is a family of cyclic groups $\mathbb{G}_k$, indexed by the security parameter $k \in \mathbb{N}$. When the reference to the security parameter k is clear, we will simply write $\mathbb{G}$ instead of $\mathbb{G}_k$. A *group hash function* $\mathrm{H} = (\mathsf{PHF.Gen}, \mathsf{PHF.Eval})$ for a group family $G = (\mathbb{G}_k)$ and with input length $\ell = \ell(k)$ consists of two PPT algorithms. For security parameter $k \in \mathbb{N}$, a key $\kappa \xleftarrow{\$} \mathsf{PHF.Gen}(1^k)$ is generated by the key generation algorithm $\mathsf{PHF.Gen}$. This key κ can then be used for the deterministic evaluation algorithm $\mathsf{PHF.Eval}$ to evaluate H via $y \leftarrow \mathsf{PHF.Eval}(\kappa, X) \in \mathbb{G}$ for any $X \in \{0,1\}^\ell$. We write $\mathrm{H}_\kappa(X) = \mathsf{PHF.Eval}(\kappa, X)$.

Definition 1. *A group hash function* H *is an* (m, n, γ, δ)-*programmable hash function if there are PPT algorithms* $\mathsf{PHF.TrapGen}$ *(the trapdoor key generation algorithm) and* $\mathsf{PHF.TrapEval}$ *(the deterministic trapdoor evaluation algorithm) such that the following holds:*

Syntactics: *For group elements* $g, h \in \mathbb{G}$, *the trapdoor key generation* $(\kappa', t) \xleftarrow{\$}$ $\mathsf{PHF.TrapGen}(1^k, g, h)$ *produces a key* κ' *along with a trapdoor* t. *Moreover,* $(a_X, b_X) \leftarrow \mathsf{PHF.TrapEval}(t, X)$ *produces* $a_X, b_X \in \mathbb{Z}$ *for any* $X \in \{0,1\}^\ell$.
Correctness: *We demand* $\mathrm{H}_{\kappa'}(X) = \mathsf{PHF.Eval}(\kappa', X) = g^{a_X} h^{b_X}$ *for all generators* $g, h \in \mathbb{G}$ *and all possible* $(\kappa', t) \xleftarrow{\$} \mathsf{PHF.TrapGen}(1^k, g, h)$, *for all* $X \in \{0,1\}^\ell$ *and the corresponding* $(a_X, b_X) \leftarrow \mathsf{PHF.TrapEval}(t, X)$.
Statistically close trapdoor keys: *For all generators* $g, h \in \mathbb{G}$ *and for* $\kappa \xleftarrow{\$}$ $\mathsf{PHF.Eval}(1^k)$ *and* $(\kappa', t) \xleftarrow{\$} \mathsf{PHF.Eval}(1^k, g, h)$, *the keys* κ *and* κ' *are statistically* γ-*close:* $\kappa \overset{\gamma}{\equiv} \kappa'$.

Well-distributed logarithms: *For all generators* $g, h \in \mathbb{G}$ *and all possible* κ' *in the range of (the first output component of)* PHF.TrapGen($1^k, g, h$), *for all* $X_1, \ldots, X_m, Z_1, \ldots, Z_n \in \{0,1\}^\ell$ *such that* $X_i \neq Z_j$ *for any* i, j, *and for the corresponding* $(a_{X_i}, b_{X_i}) \leftarrow$ PHF.TrapEval(t, X_i) *and* $(a_{Z_i}, b_{Z_i}) \leftarrow$ PHF.TrapEval(t, Z_i), *we have*

$$\Pr\left[a_{X_1} = \ldots = a_{X_m} = 0 \quad \wedge \quad a_{Z_1}, \ldots, a_{Z_n} \neq 0\right] \geq \delta, \tag{3}$$

where the probability is over the trapdoor t *that was produced along with* κ'.

We simply say that H *is an* (m, n)-*programmable hash function if there is a negligible* γ *and a noticeable* δ *such that* H *is* (m, n, γ, δ)-*programmable. Furthermore, we call* H (poly, n)-*programmable if* H *is* (q, n)-*programmable for every polynomial* $q = q(k)$. *We say that* H *is* $(m, $poly$)$-*programmable (resp.* (poly, poly)-*programmable) if the obvious holds.*

Note that a group hash function can be a (m, n)-programmable hash function for different parameters m, n with different trapdoor key generation and trapdoor evaluation algorithms.

In our RSA application, the following additional definition will prove useful:

Definition 2. *In the situation of Definition 1, we say that* H *is* β-*bounded* (m, n, γ, δ)-*programmable if* $|a_x| \leq \beta(k)$ *always.*

As a first example, note that a (programmable) random oracle $\mathcal{O}$ (i.e., a random oracle which we can completely control during a proof) is trivially a $(1, $poly$)$ or (poly, 2)-programmable hash function: given generators g and h, we simply define the values $\mathcal{O}(X_i)$ and $\mathcal{O}(Z_j)$ in dependence of the X_i and Z_j as suitable expressions $g^a h^b$. (For example, by using Coron's method [16].)

As already mentioned in the introduction, we can show a positive and natural result with a similar reduction on the discrete logarithm problem: any (non-trivially) programmable hash function is collision-resistant (a proof appears in the full version).

Theorem 3. *Assume* $|\mathbb{G}|$ *is known and prime, and the discrete logarithm problem in* $\mathbb{G}$ *is hard. Let* H *be a* $(1, 1)$-*programmable hash function. Then* H *is collision-resistant.*

We will now give an example of a programmable hash function in the standard model.

Definition 4. *Let* $G = (\mathbb{G}_k)$ *be a group family, and let* $\ell = \ell(k)$ *be a polynomial. Then,* $\mathrm{H}^{\mathsf{MG}} = ($PHF.Gen, PHF.Eval$)$ *is the following group hash function:*

- PHF.Gen(1^k) *returns a uniformly sampled* $\kappa = (h_0, \ldots, h_\ell) \in \mathbb{G}^{\ell+1}$.
- PHF.Eval(κ, X) *parses* $\kappa = (h_0, \ldots, h_\ell) \in \mathbb{G}^{\ell+1}$ *and* $X = (x_1, \ldots, x_\ell) \in \{0,1\}^\ell$ *computes and returns*

$$\mathrm{H}^{\mathsf{MG}}_\kappa(X) = h_0 \prod_{i=1}^{\ell} h_i^{x_i}$$

Essentially this function was already used, with an objective similar to ours in mind, in a construction from [25]. Here we provide a new use case and a useful abstraction of this function; also, we shed light on the properties of this function from different angles (i.e., for different values of m and n). In [25], it was implicitly proved that H^{MG} is a $(1, \mathsf{poly})$-PHF:

Theorem 5. *For any fixed polynomial $q = q(k)$ and group $\mathbb{G}$ with known order, the function H^{MG} is a $(1, q)$-programmable hash function with $\gamma = 0$ and $\delta = 1/8(\ell + 1)q$.*

The proof builds upon the fact that $m = 1$ and does not scale in the m-component. With a completely different analysis, we can show that

Theorem 6. *For any group $\mathbb{G}$ with known order, the function H^{MG} is a $(2, 1)$-programmable hash function with $\gamma = 0$ and $\delta = O(1/\ell)$.*

Proof. We give only the intuition here. The full (and somewhat technical) proof appears in the full version. Consider the following algorithms:

- PHF.TrapGen$(1^k, g, h)$ chooses $a_0, \ldots, a_\ell \in \{-1, 0, 1\}$ uniformly and independently, as well as random group exponents[3] $b_0, \ldots, b_\ell$. It sets $h_0 = g^{a_0 - 1} h^{b_0}$ and $h_i = g^{a_i} h^{b_i}$ for $1 \leq i \leq \ell$ and returns $\kappa = (h_0, \ldots, h_\ell)$ and $t = (a_0, b_0, \ldots, a_\ell, b_\ell)$.
- PHF.TrapEval(t, X) parses $X = (x_1, \ldots, x_\ell) \in \{0, 1\}^\ell$ and returns $a = a_0 - 1 + \sum_{i=1}^{\ell} a_i x_i$ and $b = b_0 + \sum_{i=1}^{\ell} b_i x_i$.

It is clear that this fulfills the syntactic and correctness requirements of Definition 1. Also, since the b_i are chosen independently and uniformly, so are the h_i, and the trapdoor keys indistinguishability requirement follows. It is more annoying to prove (3), and we will only give an intuition here. First, note that the $X_1, X_2, Z_1 \in \{0, 1\}^\ell$ from (3) (for $m = 2$, $n = 1$) are independent of the a_i, since they are masked by the b_i in $h_i = g^{a_i} h^{b_i}$. Hence, if we view, e.g., X_1 as a subset of $[\ell]$ (where we define $i \in X_1$ iff the i-th component x_{1i} of X_1 is 1), then

$$a_{X_1} = a_0 - 1 + \sum_{i=1}^{\ell} a_i x_{1i} = -1 + a_0 + \sum_{i \in X_1} a_i$$

essentially[4] constitutes a random walk of length $|X_1| \leq \ell$. Theory says that it is likely that this random walk ends up with an a_{X_1} of small absolute value. That is, for any r with $|r| = O(\sqrt{\ell})$, the probability that $a_{X_1} = r$ is $\Theta(1/\sqrt{\ell})$. In particular, the probability for $a_{X_1} = 0$ is $\Theta(1/\sqrt{\ell})$. Now if X_1 and X_2 were disjoint and there was no a_0 in the sum, then a_{X_1} and a_{X_2} would be independent and we would get that $a_{X_1} = a_{X_2} = 0$ with probability $\Theta(1/\ell)$. But even if

---

[3] If $|\mathbb{G}|$ is not known, this may only be possible approximately.

[4] Usually, random walks are formalized as a sum of independent values $a_i \in \{-1, 1\}$; for us, it is more convenient to assume $a_i \in \{-1, 0, 1\}$. However, this does not change things significantly.

$X_1 \cap X_2 \neq \emptyset$, and taking into account a_0, we can conclude similarly by lower bounding the probability that $a_{X_1 \setminus X_2} = a_{X_2 \setminus X_1} = -a_{X_1 \cap X_2}$.

The additional requirement that $a_{Z_1} \neq 0$ with high probability is intuitively much more obvious, but also much harder to formally prove. First, without loss of generality, we can assume that $Z_1 \subseteq X_1 \cup X_2$, since otherwise, there is a "partial random walk" $a_{Z_1 \setminus (X_1 \cup X_2)}$ that contributes to a_{Z_1} but is independent of a_{X_1} and a_{X_2}. Hence, even when already assuming $a_{X_1} = a_{X_2} = 0$, a_{Z_1} still is sufficiently randomized to take a nonzero value with constant probability. Also, we can assume Z_1 not to "split" X_1 in the sense that $Z_1 \cap X_1 \in \{\emptyset, X_1\}$ (similarly for X_2). Otherwise, even assuming a fixed value of a_{X_1}, there is still some uncertainty about $a_{Z_1 \cap X_1}$ and hence about a_{Z_1} (in which case with some probability, a_{Z_1} does not equal any fixed value). The remaining cases can be handled with a similar "no-splitting" argument. However, note that the additional "-1" in the g-exponent of h_0 is essential: without it, picking X_1 and X_2 disjoint and setting $Z_1 = X_1 \cup X_2$ achieves $a_{Z_1} = a_{X_1} + a_{X_2} = 0$. A full proof is given in the full version.

Furthermore, using techniques from the proof of Theorem 6, we can asymptotically improve the bounds from Theorem 5 as follows (a proof can be found in the full version):

Theorem 7. *For any fixed polynomial $q = q(k)$ and group $\mathbb{G}$ with known order, the function H^{MG} is a $(1, q)$-programmable hash function with $\gamma = 0$ and $\delta = O(\frac{1}{q\sqrt{\ell}})$.*

One may wonder whether the scalability of H^{MG} with respect to m reaches further. Unfortunately, it does not (the proof for the following theorem appears in the full version):

Theorem 8. *Assume $\ell = \ell(k) \geq 2$. Say $|\mathbb{G}|$ is known and prime, and the discrete logarithm problem in $\mathbb{G}$ is hard. Then H^{MG} is not $(3, 1)$-programmable.*

If the group order $\mathbb{G}$ is not known (as will be the case in our upcoming RSA-based signature scheme), then it may not even be possible to sample group exponents uniformly. However, for the special case where $\mathbb{G} = \mathrm{QR}_N$ is the group of quadratic residues modulo $N = pq$ for safe distinct primes p and q, we can approximate a uniform exponent with a random element from $\mathbb{Z}_{N^2}$. In this case, the statistical distance between keys produced by $\mathsf{PHF.Gen}$ and those produced by $\mathsf{PHF.TrapGen}$ is smaller than $(\ell + 1)/N$. We get

Theorem 9. *For the group $\mathbb{G} = \mathrm{QR}_N$ of quadratic residues modulo $N = pq$ for safe distinct primes p and q, the function H^{MG} is $(\ell + 2)$-bounded $(1, q, (\ell + 1)/N, 1/8(\ell+1)q)$-programmable and also $(\ell+2)$-bounded $(2, 1, (\ell+1)/N, O(1/\ell))$-programmable.*

Similarly, one can show analogues of Theorem 8 and Theorem 3 for $\mathbb{G} = \mathrm{QR}_N$, only with the strong RSA assumption in place of the discrete log problem. We omit the details.

Randomized Programmable Hash Functions (RPHFs). In the full version we further generalize the notion of PHFs to randomized programmable hash functions (RPHFs). Briefly, RPHFs are PHFs whose evaluation is randomized, and where this randomness is added to the image (so that verification is possible). We show how to adapt the PHF definition to the randomized case, in a way suitable for the upcoming applications. We also give instantiations of RPHFs for parameters for which we do not know how to instantiate PHFs.

4 Applications of PHFs

4.1 Generic Signatures from Bilinear Maps

Let $\mathbb{PG} = (\mathbb{G}, \mathbb{G}_T, p = |\mathbb{G}|, g, \hat{e} : \mathbb{G} \times \mathbb{G} \to \mathbb{G}_T)$ be a pairing group. Let $n = n(k)$ and $\eta = \eta(k)$ be two arbitrary polynomials. Our signature scheme signs messages $m \in \{0,1\}^n$ using randomness $s \in \{0,1\}^\eta$.[5] Let a group hash function $H = (\mathsf{PHF.Gen}, \mathsf{PHF.Eval})$ with inputs from $\{0,1\}^n$ and outputs from $\mathbb{G}$ be given. We are ready to define our generic bilinear map signature scheme $\mathsf{SIG}_{\mathrm{BM}}[H]$.

Key-Generation: Generate $\mathbb{PG}$ such that H can be used for the group $\mathbb{G}$. Generate a key for H via $\kappa \xleftarrow{\$} \mathsf{PHF.Gen}(1^k)$. Pick a random index $x \in \mathbb{Z}_p$ and compute $X = g^x \in \mathbb{G}$. Return the public verification key $(\mathbb{PG}, X, \kappa)$ and the secret signing key x.

Signing: To sign $m \in \{0,1\}^n$, pick a random η-bit integer s and compute $y = H_\kappa(m)^{\frac{1}{x+s}} \in \mathbb{G}$. The signature is the tuple $(s, y) \in \{0,1\}^\eta \times \mathbb{G}$.

Verification: To verify that $(s, y) \in \{0,1\}^\eta \times \mathbb{G}$ is a correct signature on a given message m, check that s is of length η, and that

$$\hat{e}(y, X \cdot g^s) = \hat{e}(H(m), g).$$

Theorem 10. *Let H be a $(m, 1, \gamma, \delta)$-programmable hash function. Let $\mathcal{F}$ be a (t, q, ϵ)-forger in the existential forgery under an adaptive chosen message attack experiment with $\mathsf{SIG}_{\mathrm{BM}}$. Then there exists an adversary $\mathcal{A}$ that (t', ϵ')-breaks the q-SDH assumption with $t' \approx t$ and*

$$\epsilon \leq \frac{q}{\delta} \cdot \epsilon' + \frac{q^{m+1}}{2^{m\eta-1}} + \gamma.$$

We remark that the scheme can also be instantiated in asymmetric pairing groups where the pairing is given by $\hat{e} : \mathbb{G}_1 \times \mathbb{G}_2 \to \mathbb{G}_T$ and $\mathbb{G}_1 \neq \mathbb{G}_2$. In that case we let the element y from the signature be in $\mathbb{G}_1$ such that y can be represented in 160 bits [6]. Also, in asymmetric pairings, verification can equivalently check if $\hat{e}(y, X) = \hat{e}(H(m) \cdot y^{-1/s}, g)$. This way we avoid any expensive exponentiation in $\mathbb{G}_2$ and verification time becomes roughly the same as in the Boneh-Boyen short

---

[5] For signing arbitrary bitstrings, a collision resistant hash function $\mathsf{CR} : \{0,1\}^* \to \{0,1\}^n$ can be applied first. Due to the birthday paradox we choose $n = 2k$ when k bits of security are actually desired.

signatures [6]. It can be verified that the following proof also holds in asymmetric pairing groups (assuming there exists an efficiently computable isomorphism $\psi : \mathbb{G}_2 \to \mathbb{G}_1$).

An efficiency comparison of the scheme instantiated with the $(2, 1)$-PHF H^{MG} from Definition 4 appears in the full version.

Proof (Theorem 10). Let $\mathcal{F}$ be the adversary against the signature scheme. Throughout this proof, we assume that H is a (m, n, γ, δ)-programmable hash function. Furthermore, we fix some notation. Let m_i be the i-th query to the signing oracle and (s_i, y_i) denote the answer. Let m and (s, y) be the forgery output by the adversary. We introduce two types of forgers:

Type I: It always holds that $s = s_i$ for some i.
Type II: It always holds that $s \neq s_i$ for all i.

By $\mathcal{F}_1$ (resp., $\mathcal{F}_2$) we denote the forger who runs $\mathcal{F}$ but then only outputs the forgery if it is of type I (resp., type II). We now show that both types of forgers can be reduced to the $q+1$-SDH problem. Theorem 10 then follows by a standard hybrid argument.

Type I forgers

Lemma 11. *Let $\mathcal{F}_1$ be a forger of type I that (t_1, q, ϵ_1)-breaks the existential unforgeability of $\mathsf{SIG}_{\mathrm{BM}}[\mathsf{H}]$. Then there exists an adversary $\mathcal{A}$ that (t', ϵ')-breaks the q-SDH assumption with $t' \approx t$ and*

$$\epsilon' \geq \frac{\delta}{q}\left(\epsilon_1 - \frac{q^{m+1}}{2^{mn}} - \gamma\right).$$

To prove the lemma we proceed in games. In the following, X_i denotes the probability for the adversary to successfully forge a signature in Game i.

Game 0. Let $\mathcal{F}_1$ be a type I forger that (t_1, q, ϵ_1)-breaks the existential unforgeability of $\mathsf{SIG}_{\mathrm{BM}}[\mathsf{H}]$. By definition, we have

$$\Pr[X_0] = \epsilon_1. \tag{4}$$

Game 1. We now generate trapdoor keys $(\kappa', t) \overset{\$}{\leftarrow} \mathsf{PHF.TrapGen}(1^k, g, h)$ for uniformly selected generators $g, h \in \mathbb{G}$ to generate a H-key for public verification key of $\mathsf{SIG}_{\mathrm{BM}}[\mathsf{H}]$. By the programmability of H,

$$\Pr[X_1] \geq \Pr[X_0] - \gamma. \tag{5}$$

Game 2. Now we select the random values s_i used for answering signing queries not upon each signing query, but at the beginning of the experiment. Since the s_i were selected independently anyway, this change is only conceptual. Let $E = \bigcup_{i=1}^{q}\{s_i\}$ be the set of all s_i, and let $E^i = E \setminus \{i\}$. We also change the selection of the elements g, h used during $(\kappa', t) \overset{\$}{\leftarrow} \mathsf{PHF.TrapGen}(1^k, g, h)$ as follows. First, we uniformly choose $i^* \in [q]$ and a generator $\tilde{g} \in \mathbb{G}$. Define

$p^*(\eta) = \prod_{t \in E^*}(\eta + t)$ and $p(\eta) = \prod_{t \in E}(\eta + t)$ and note that $\deg(p^*) \leq q - 1$ and $\deg(p) \leq q$. Hence the values $g = \tilde{g}^{p^*(x)}$, $h = \tilde{g}^{p(x)}$, and $X = g^x = \tilde{g}^{xp^*(x)}$ can be computed from $\tilde{g}, \tilde{g}^x, \ldots, \tilde{g}^{x^q}$. Here the index $x \in \mathbb{Z}_{|G|}$ is the secret key of the scheme. We then set $E^* = E \setminus \{s_{i^*}\}$, $E^{*,i} = E^* \setminus \{i\}$, and

$$g = \tilde{g}^{p^*(x)} = \tilde{g}^{\prod_{t \in E^*}(x-t)}, \quad h = \tilde{g}^{p(x)} = \tilde{g}^{\prod_{t \in E}(x-t)}.$$

Note that we can compute $(x + s_i)$-th roots for $i \neq i^*$ from g and for all i from h. This change is purely conceptual:

$$\Pr[X_2] = \Pr[X_1]. \tag{6}$$

Observe also that i^* is independent of the adversary's view.

Game 3. In this game, we change the way signature requests from the adversary are answered. First, observe that the way we modified the generation of g and h in Game 2 implies that for any i with $s_i \neq s_{i^*}$, we have

$$
\begin{aligned}
y_i = H_{\kappa'}(m_i)^{\frac{1}{x+s_i}} &= \left(g^{a_{m_i}} h^{b_{m_i}}\right)^{\frac{1}{x+s_i}} \\
&= \left(\tilde{g}^{a_{m_i} \prod_{t \in E^*}(x-t)} \tilde{g}^{b_{m_i} \prod_{t \in E}(x-t)}\right)^{\frac{1}{x+s_i}} = \tilde{g}^{a_{m_i} \prod_{t \in E^{*,i}}(x-t)} \tilde{g}^{b_{m_i} \prod_{t \in E^i}(x-t)}
\end{aligned} \tag{7}
$$

for $(a_{m_i}, b_{m_i}) \leftarrow \mathsf{PHF.TrapEval}(t, m_i)$. Hence for $i \neq i^*$, we can generate the signature (s_i, y_i) without explicitly knowing the secret key x, but instead using the right-hand side of (7) for computing y_i. Obviously, this change in computing signatures is only conceptual, and so

$$\Pr[X_3] = \Pr[X_2]. \tag{8}$$

Observe that i^* is still independent of the adversary's view.

Game 4. We now abort and raise event $\mathsf{abort}_{\mathsf{coll}}$ if an s_i occurs more than m times, i.e., if there are pairwise distinct indices $i_1, \ldots, i_{m+1}$ with $s_{i_1} = \ldots = s_{i_{m+1}}$. There are $\binom{q}{m+1}$ such tuples $(i_1, \ldots, i_m)$. For each tuple, the probability for $s_{i_1} = \ldots = s_{i_{m+1}}$ is $1/2^{mn}$ A union bound shows that a $(m+1)$-wise collision occurs with probability at most

$$\Pr[\mathsf{abort}_{\mathsf{coll}}] \leq \binom{q}{m+1} \frac{1}{2^{m\eta}} \leq \frac{q^{m+1}}{2^{m\eta}}.$$

Hence,

$$\Pr[X_4] \geq \Pr[X_3] - \Pr[\mathsf{abort}_{\mathsf{coll}}] > \Pr[X_3] - \frac{q^{m+1}}{2^{m\eta}}. \tag{9}$$

Game 5. We now abort and raise event $\mathsf{abort}_{\mathsf{bad.s}}$ if the adversary returns an $s \in E^*$, i.e., the adversary returns a forgery attempt (s, y) with $s = s_i$ for some i, but $s \neq s_{i^*}$. Since i^* is independent from the adversary's view we have $\Pr[\mathsf{abort}_{\mathsf{bad.s}}] \leq 1 - 1/q$ for any choice of the s_i, so we get

$$\Pr[X_5] = \Pr[X_4 \wedge \neg\mathsf{abort}_{\mathsf{bad.s}}] \geq \frac{1}{q}\Pr[X_4]. \tag{10}$$

Game 6. We now abort and raise event $\mathsf{abort_{bad.a}}$ if there is an index i with $s_i = s_{i^*}$ but $a_{m_i} \neq 0$, or if $a_m = 0$ for the adversary's forgery message. In other words, we raise $\mathsf{abort_{bad.a}}$ iff we do not have $a_{m_i} = 0$ for all i with $s_i = s_{i^*}$ and $a_{m_i} \neq 0$. Since we have limited the number of such i to m in Game 4, we can use the programmability of H. We hence have $\Pr[\mathsf{abort_{bad.a}}] \leq 1 - \delta$ for any choice of the m_i and s_i, so we get

$$\Pr[X_6] \geq \Pr[X_5 \wedge \neg\mathsf{abort_{bad.a}}] \geq \delta \cdot \Pr[X_5]. \tag{11}$$

Note that in Game 6, the experiment never really uses secret key x to generate signatures: to generate the y_i for $s_i \neq s_{i^*}$, we already use (7), which requires no x. But if $\mathsf{abort_{bad.a}}$ does not occur, then $a_{m_i} = 0$ whenever $s_i = s_{i^*}$, so we can also use (7) to sign without knowing x. On the other hand, if $\mathsf{abort_{bad.a}}$ does occur, we must abort anyway, so actually no signature is required.

This means that Game 6 does not use knowledge about the secret key x. On the other hand, the adversary in Game 6 produces (whenever X_6 happens, which implies $\neg\mathsf{abort_{bad.a}}$ and $\neg\mathsf{abort_{bad.s}}$) during a forgery

$$y = \mathsf{H}_{\kappa'}(m)^{1/(x+s)} = \left(\tilde{g}^{a_m \prod_{t \in E^*}(x+t)} \tilde{g}^{b_m \prod_{t \in E}(x+t)}\right)^{\frac{1}{x+s}} = \tilde{g}^{\frac{a_m p^*(x)}{x+s}} \tilde{g}^{b_m p^*(x)}.$$

From y and its knowledge about h and the s_i, the experiment can derive

$$y' = \left(\frac{y}{g^{b_m}}\right)^{1/a_m} = \tilde{g}^{\frac{p^*(x)}{x+s}}.$$

Since $\gcd(\eta + s, p^*(\eta)) = 1$ (where we interpret $\eta + s$ and $p^*(\eta)$ as polynomials in η), we can write $p^*(\eta)/(\eta + s) = p'(\eta) + q_0/(\eta + s)$ for some polynomial $p'(\eta)$ of degree at most $q - 2$ and some $q_0 \neq 0$. Again, we can compute $g' = \tilde{g}^{p'(x)}$. We finally obtain

$$y'' = (y'/g')^{1/q_0} = \left(\tilde{g}^{\frac{p(x)}{(x+s)} - p'(x)}\right)^{1/q_0} = \tilde{g}^{\frac{1}{x+s}}.$$

This means that the from the experiment performed in Game 6, we can construct an adversary $\mathcal{A}$ that (t', ϵ')-breaks the q-SDH assumption. $\mathcal{A}$'s running time t' is approximately t plus a small number of exponentiations, and $\mathcal{A}$ is successful whenever X_6 happens:

$$\epsilon' \geq \Pr[X_6]. \tag{12}$$

Putting (4-12) together yields Lemma 11.

Type II forgers

Lemma 12. *Let $\mathcal{F}_2$ be a forger of type II that (t_1, q, ϵ_1)-breaks the existential unforgeability of $\mathsf{SIG_{BM}}[\mathsf{H}]$. Then there exists an adversary $\mathcal{A}$ that (t', ϵ')-breaks the q-SDH assumption such that $t' \approx t$ and*

$$\epsilon' \geq \frac{\delta}{2}(\epsilon_2 - \gamma).$$

The difference is that a type II forger returns a valid signature (s, y) with $s \notin \{s_1, \ldots, s_q\}$. The idea of the reduction is that the simulation can be setup such that from this forgery an element $\tilde{g}^{\frac{1}{x+s}}$ can be computed which breaks the q-SDH assumption. The simulation of the signature queries is simular the one for type I forgers, where now we only have to use the $(1, 1)$-programmability of H. A detailed proof is given in the full version.

4.2   Generic Signatures from RSA

Let $\mathbb{G} = \mathrm{QR}_N$ be the group of quadratic residues modulo an RSA number $N = pq$, where p and q are safe primes. Let $n = n(k)$ and $\eta = \eta(k)$ be two polynomials. Let a group hash function $\mathrm{H} = (\mathsf{PHF.Gen}, \mathsf{PHF.Eval})$ with inputs from $\{0, 1\}^n$ and outputs from $\mathbb{G}$ be given. We are ready to define our generic RSA-based signature scheme $\mathsf{SIG}_{\mathrm{RSA}}[\mathrm{H}]$:

Key-Generation: Generate $N = pq$ for safe distinct primes $p, q \geq 2^{\eta+2}$, such that H can be used for the group $\mathbb{G} = \mathrm{QR}_N$. $\kappa \overset{\$}{\leftarrow} \mathsf{PHF.Gen}(1^k)$. Return the public verification key (N, κ) and the secret signing key (p, q).

Signing: To sign $m \in \{0, 1\}^n$, pick a random η-bit prime e and compute $y = \mathrm{H}_\kappa(m)^{1/e} \bmod N$. The e-th root can be computed using p and q. The signature is the tuple $(e, y) \in \{0, 1\}^\eta \times \mathbb{Z}_N$.

Verification: To verify that $(e, y) \in \{0, 1\}^\eta \times \mathbb{Z}_N$ is a correct signature on a given message m, check that e is odd and of length η, and that $y^e = \mathrm{H}(m) \bmod N$.

Theorem 13. *Let H be a β-bounded $(m, 1, \gamma, \delta)$-programmable hash function for $\beta \leq 2^\eta$ and $m \geq 1$. Let $\mathcal{F}$ be a (t, q, ϵ)-forger in the existential forgery under an adaptive chosen message attack experiment with $\mathsf{SIG}_{\mathrm{RSA}}[\mathrm{H}]$. Then there exists an adversary $\mathcal{A}$ that (t', ϵ')-breaks the strong RSA assumption with $t' \approx t$ and*

$$\epsilon = \Theta\left(\frac{q}{\delta} \cdot \epsilon'\right) + \frac{q^{m+1}(\eta + 1)^m}{2^{m\eta - 1}} + \gamma .$$

The proof is similar to the case of bilinear maps (Theorem 10). However, due to the fact that the group order is not known some technical difficulties arise which is the reason why we need the PHF to be β-bounded for some $\beta \leq 2^\eta$. The full formal proof appears in the full version.

Let us again consider the instantiation $\mathsf{SIG}_{\mathrm{RSA}}[\mathrm{H}^{\mathsf{MG}}]$ for the $(2, 1)$-PHF H^{MG}. Plugging in the values from Theorem 9 the reduction from Theorem 13 leads to $\epsilon = \Theta(q\ell\epsilon') + \frac{q^3(\eta+1)^2}{2^{2\eta-1}}$. As explained in the introduction, for $q = 2^{30}$ and $k = 80$ bits we are now able to choose $\eta \approx 80$ bit primes.

References

1. Bari, N., Pfitzmann, B.: Collision-free accumulators and fail-stop signature schemes without trees. In: Fumy, W. (ed.) EUROCRYPT 1997. LNCS, vol. 1233. Springer, Heidelberg (1997)

2. Bellare, M., Goldreich, O., Goldwasser, S.: Incremental cryptography: The case of hashing and signing. In: Desmedt, Y.G. (ed.) CRYPTO 1994. LNCS, vol. 839, pp. 216–233. Springer, Heidelberg (1994)
3. Bellare, M., Rogaway, P.: Random oracles are practical: A paradigm for designing efficient protocols. In: Ashby, V. (ed.) ACM CCS 1993, Fairfax, Virginia, USA, November 3–5, 1993, pp. 62–73. ACM Press, New York (1993)
4. Bellare, M., Rogaway, P.: The exact security of digital signatures: How to sign with RSA and Rabin. In: Maurer, U.M. (ed.) EUROCRYPT 1996. LNCS, vol. 1070, pp. 399–416. Springer, Heidelberg (1996)
5. Boneh, D., Boyen, X.: Efficient selective-ID secure identity based encryption without random oracles. In: Cachin, C., Camenisch, J.L. (eds.) EUROCRYPT 2004. LNCS, vol. 3027, pp. 223–238. Springer, Heidelberg (2004)
6. Boneh, D., Boyen, X.: Short signatures without random oracles. In: Cachin, C., Camenisch, J.L. (eds.) EUROCRYPT 2004. LNCS, vol. 3027, pp. 56–73. Springer, Heidelberg (2004)
7. Boneh, D., Boyen, X.: Short signatures without random oracles and the SDH assumption in bilinear groups. Journal of Cryptology 21(2), 149–177 (2008)
8. Boneh, D., Lynn, B., Shacham, H.: Short signatures from the Weil pairing. In: Boyd, C. (ed.) ASIACRYPT 2001. LNCS, vol. 2248, pp. 514–532. Springer, Heidelberg (2001)
9. Boneh, D., Lynn, B., Shacham, H.: Short signatures from the Weil pairing. Journal of Cryptology 17(4), 297–319 (2004)
10. Boyen, X.: General ad hoc encryption from exponent inversion IBE. In: Naor, M. (ed.) EUROCRYPT 2007. LNCS, vol. 4515, pp. 394–411. Springer, Heidelberg (2007)
11. Brands, S.: An efficient off-line electronic cash system based on the representation problem. Report CS-R9323, Centrum voor Wiskunde en Informatica (March 1993)
12. Camenisch, J., Lysyanskaya, A.: A signature scheme with efficient protocols. In: Cimato, S., Galdi, C., Persiano, G. (eds.) SCN 2002. LNCS, vol. 2576, pp. 268–289. Springer, Heidelberg (2003)
13. Chaum, D., Evertse, J.-H., van de Graaf, J.: An improved protocol for demonstrating possession of discrete logarithms and some generalizations. In: Chaum, D., Price, W.L. (eds.) EUROCRYPT 1987. LNCS, vol. 304, pp. 127–141. Springer, Heidelberg (1988)
14. Chaum, D., van Heijst, E., Pfitzmann, B.: Cryptographically strong undeniable signatures, unconditionally secure for the signer. In: Feigenbaum, J. (ed.) CRYPTO 1991. LNCS, vol. 576, pp. 470–484. Springer, Heidelberg (1992)
15. Chevallier-Mames, B., Joye, M.: A practical and tightly secure signature scheme without hash function. In: Abe, M. (ed.) CT-RSA 2007. LNCS, vol. 4377, pp. 339–356. Springer, Heidelberg (2006)
16. Coron, J.-S.: On the exact security of full domain hash. In: Bellare, M. (ed.) CRYPTO 2000. LNCS, vol. 1880, pp. 229–235. Springer, Heidelberg (2000)
17. Cramer, R., Shoup, V.: Signature schemes based on the strong RSA assumption. In: ACM CCS 1999, Kent Ridge Digital Labs, Singapore, November 1–4, 1999, pp. 46–51. ACM Press, New York (1999)
18. Dodis, Y., Oliveira, R., Pietrzak, K.: On the generic insecurity of the full domain hash. In: Shoup, V. (ed.) CRYPTO 2005. LNCS, vol. 3621, pp. 449–466. Springer, Heidelberg (2005)
19. Fischlin, M.: The Cramer-Shoup strong-RSA signature scheme revisited. In: Desmedt, Y. (ed.) PKC 2003. LNCS, vol. 2567, pp. 116–129. Springer, Heidelberg (2002)

20. Fujisaki, E., Okamoto, T.: Statistical zero knowledge protocols to prove modular polynomial relations. In: Kaliski Jr., B.S. (ed.) CRYPTO 1997. LNCS, vol. 1294, pp. 16–30. Springer, Heidelberg (1997)
21. Gennaro, R., Halevi, S., Rabin, T.: Secure hash-and-sign signatures without the random oracle. In: Stern, J. (ed.) EUROCRYPT 1999. LNCS, vol. 1592, pp. 123–139. Springer, Heidelberg (1999)
22. Gentry, C.: Practical identity-based encryption without random oracles. In: Vaudenay, S. (ed.) EUROCRYPT 2006. LNCS, vol. 4004, pp. 445–464. Springer, Heidelberg (2006)
23. Goldwasser, S., Micali, S., Rivest, R.L.: A digital signature scheme secure against adaptive chosen-message attacks. SIAM Journal on Computing 17(2), 281–308 (1988)
24. Sakai, R., Ohgishi, K., Kasahara, M.: Cryptosystems based on pairing. In: SCIS 2000, Okinawa, Japan (January 2000)
25. Waters, B.R.: Efficient identity-based encryption without random oracles. In: Cramer, R. (ed.) EUROCRYPT 2005. LNCS, vol. 3494, pp. 114–127. Springer, Heidelberg (2005)
26. Zhu, H.: New digital signature scheme attaining immunity to adaptive chosen-message attack. Chinese Journal of Electronics 10(4), 484–486 (2001)

One-Time Programs

Shafi Goldwasser[1,3,*], Yael Tauman Kalai[2,**], and Guy N. Rothblum[3,***]

[1] Weizmann Institute of Science, Rehovot, Israel
shafi@theory.csail.mit.edu
[2] Georgia Tech, Atlanta, USA
yael@cc.gatech.edu
[3] MIT, Cambridge, USA
rothblum@csail.mit.edu

Abstract. In this work, we introduce *one-time programs*, a new computational paradigm geared towards security applications. A one-time program can be executed on a *single* input, whose value can be specified at run time. Other than the result of the computation on this input, nothing else about the program is leaked. Hence, a one-time program is like a black box function that may be evaluated once and then "self destructs." This also extends to k-time programs, which are like black box functions that can be evaluated k times and then self destruct.

One-time programs serve many of the same purposes of program obfuscation, the obvious one being software protection, but also including applications such as temporary transfer of cryptographic ability. Moreover, the applications of one-time programs go well beyond those of obfuscation, since one-time programs can only be executed once (or more generally, a limited number of times) while obfuscated programs have no such bounds. For example, one-time programs lead naturally to electronic cash or token schemes: coins are generated by a program that can only be run once, and thus cannot be double spent.

Most significantly, the new paradigm of one-time computing opens new avenues for conceptual research. In this work we explore one such avenue, presenting the new concept of "one-time proofs," proofs that can only be verified once and then become useless and unconvincing.

All these tasks are clearly impossible using software alone, as any piece of software can be copied and run again, enabling the user to execute the program on more than one input. All our solutions employ a secure memory device, inspired by the cryptographic notion of interactive oblivious transfer protocols, that stores two secret keys (k_0, k_1). The device takes as input a single bit $b \in \{0, 1\}$, outputs k_b, and then self destructs. Using such devices, we demonstrate that for every input length, any standard program (Turing machine) can be efficiently compiled into a functionally equivalent one-time program. We also show how this memory device can

---

[*] Supported by NSF Grants CCF-0514167, CCF-0635297, NSF-0729011, the RSA chair, and by the Weizmann Chais Fellows Program for New Scientists.
[**] Supported in part by NSF grant CCF-0635297.
[***] Supported by NSF Grants CCF-0635297, NSF-0729011, CNS-0430336 and by a Symantec Graduate Fellowship.

© International Association for Cryptologic Research 2008

be used to construct one-time proofs. Specifically, we show how to use this device to efficiently convert a classical witness for any NP statement, into "one-time proof" for that statement.

1   Introduction

In our standard computing world (and the standard theoretical computing models), computer programs can be copied, analyzed, modified, and executed in an arbitrary manner. However, when we think of security applications, such complete transfer of code is often undesirable, as it complicates and inhibits tasks such as revocation of cryptographic ability, temporary transfer of cryptographic ability, and preventing double- spending of electronic cash. Other tasks such as general program obfuscation [BGI+01, GK05] are downright impossible.

In this paper, we propose to study a new type of computer program, called a *one-time program*. Such programs can be executed only once, where the input can be chosen at any time. This notion extends naturally to k-time programs which can be executed at most k times on inputs that can be chosen by the user at any time. These programs have immediate applications to software protection, electronic tokens and electronic cash. Even more interestingly, they open new avenues for conceptual contributions. In particular, in this work they allow us to conceive of, define and realize the new cryptographic concept of *one-time zero-knowledge proofs*: zero-knowledge proofs that can be verified exactly once, by any verifier, without the prover being present. After the proof is verified once by any single verifier, it becomes useless and cannot be verified again.

Clearly a one-time program cannot be solely software based, as software can always be copied and run again, enabling a user to execute the program more times than specified. Instead, we suggest the use of a secure memory devices, *one-time memory* (OTM), as part of the one-time program. In general, when using a hardware device it is crucial that the device be as simple as possible, so that it can be scrutinized more easily. In particular, side-channel attacks have emerged as a devastating threat to the security of hardware devices.

The memory device used in this work is very simple and withstands even extremely powerful side-channel attacks. An OTM does not perform any computation, but its memory contents are assumed to be somewhat protected, i.e. they cannot be read and written arbitrarily. All we assume is that memory locations that are *never accessed* by the device are *never leaked* via a side channel (whereas memory that *is* accessed may be immediately leaked), and that the device has a *single* tamper-proof bit, see Section 3 for a fuller discussion. In particular, our device meets the desirable property laid out in the work of Gunnar *et. al.* [GLM+04], and can be decoupled into two components: the first component is tamper-proof but readable, and consists of a single bit. The second component is tamperable but read-proof. As mentioned above, in our case the read-proof requirement is only for memory locations that are never accessed by the device. These assumptions seem minimal if any non-trivial use is to be made of the secure device (see the illuminating discussions in Micah and Renin [MR04] and in

Gunnar *et. al.* [GLM⁺04]). Also, the device is very inexpensive, low energy and disposable, much like RFID tags used in clothing. Thus, a one-time program can be realized by a combination of standard software and such minimally secure memory devices.

We construct a universal one-time compiler that takes *any* standard polynomial-time program and memory devices as above, and transforms it (in polynomial time) into a one-time program which achieves the same functionality, under the assumption that one-way functions exist. This compiler uses techniques from secure function evaluation, specifically Yao's garbled circuit method [Yao86], as its starting point. These techniques, however, only give solutions for settings in which adversaries are honest-but-curious, whereas we want the security of one-time programs to also hold against malicious adversaries. Unlike the setting of secure function evaluation, we need to overcome this difficulty without the benefit of interaction. This is accomplished (non-interactively) using the secure memory devices (see Section 4 for details).

While we cannot show that this compiler is optimal in terms of the efficiency of the one-time programs it produces, we do argue that significant improvements would resolve a central open problem in cryptography. Specifically, significant complexity improvements would imply significant improvements in the communication complexity of secure-function-evaluation protocols. See Section 4 for further details.

Continuing in the spirit of one-time computing, we also define and construct *one-time proofs*. A one-time proof system for an NP language L allows the owner of a witness for the membership of some input x in L to transform this witness into a one-time proof token (a device with the above secure memory components). This proof token can be given to any efficient prover, who does not know a witness for x. The prover can use this token *exactly once* to prove to any verifier that $x \in L$ using an interactive proof. The prover does not learn anything from the proof token, and in particular cannot prove that $x \in L$ a second time. The witness owner does not need to be involved in the interaction between the prover and verifier. We show how to construct a one-time proof system with negligible soundness for any NP language. Achieving constant soundness is relatively straightforward, but amplifying the soundness is not. The technical difficulties are similar to those encountered in parallel composition of zero-knowledge proofs. We are able to resolve these difficulties (again, in a non-interactive manner) using the secure memory devices. See Section 5.2 for an overview.

We proceed to describe our contributions in detail.

2   One-Time Programs and One Time Compilers

Informally, a *one-time program for a function* f: (1) Can be used to compute f on a single input x of one's choice. (2) No efficient adversary, given the one-time program, can learn more about f than can be learned from a single pair $(x, f(x))$, where x is chosen by the adversary. Hence, it acts like a black-box function that can only be evaluated once.

Several formal definitions can be proposed for condition 2 above. We chose a definition inspired by Goldreich and Ostrovsky's [GO96] work on software protection, and the work of Barak *et al.* [BGI+01] on program obfuscation. Informally, for every probabilistic polynomial time algorithm A given access to a one-time program for f on inputs of size n, there exists another probabilistic polynomial time algorithm $S(1^n)$ which can request to see the value $f(x)$ for an x of its choice where $|x| = n$, such that (for any f) the output distributions of A and S are computationally indistinguishable, even to a machine that knows f!

The notion of one-time programs extends naturally to k-time programs which can be provably executed at most k times on input values that can be chosen by the user at any time. For simplicity of exposition we mostly deal with the one-time case throughout this paper.

As previously mentioned, a one-time program cannot be solely software based, and we propose to use secure hardware devices as building blocks in the constructions. In general, we model *secure hardware devices* as black-boxes with internal memory which can be accessed only via its I/O interface. A *one-time program* is a combination of *hardware*: (one or many) hardware devices $\mathcal{H}_1, \ldots, \mathcal{H}_m$; and *software*: a (possibly non-uniform) Turing machine $\mathcal{M}$, where the machine $\mathcal{M}$ accesses the hardware devices via their I/O interface. It is important to note that the Turing machine software component *is not secure*: it can be read and modified by the user whereas the access to the secure hardware is only via the I/O interface. Thus, we view one-time programs as software-hardware packages. An execution of one-time program $\mathcal{P} = \mathcal{M}^{\mathcal{H}_1, \ldots, \mathcal{H}_m}$ on input $x \in \{0,1\}^n$, is a run of $\mathcal{M}^{\mathcal{H}_1, \ldots, \mathcal{H}_m}(x)$, where the contents of $\mathcal{M}$'s output tape is the output. Throughout this work we use a new type of secure hardware device which we name a *one-time memory* (OTM), see the introduction and Section 3 for more details.

One-Time Compiler. To transform a standard computer program into a one-time program computing the same functionality, we propose the notion of a *one-time compiler*. The compiler takes a computer program, modeled as a (possibly non-uniform) Turing machine $\mathcal{T}$, an input length n, and a collection of OTM devices (the number of devices may depend on n). The compiler then creates a one-time program computing the same functionality as $\mathcal{T}$ on inputs of length n by initializing the internal memory of the OTMs and also outputting (in the clear) a software component for the program. It is also important that the compiler be efficient, in the sense that its running time is no worse than polynomial in $\mathcal{T}$'s worst-case running time on inputs of length n.

In this work we construct a one-time compiler as above. The compiler transforms any Turing machine $\mathcal{T}$ into a one-time program $\mathcal{P}$ that satisfies the two intuitive properties outlined above:

1. **Functionality.** For any $x \in \{0,1\}^n$, when the program is run *once* on input x it outputs $\mathcal{T}(x)$; namely, $\mathcal{P}(x) = \mathcal{T}(x)$.
2. **One-Time Secrecy.** For any PPT adversary $\mathcal{A}$, there exists a PPT simulator $\mathcal{S}$ with *one-time* oracle access to the machine $\mathcal{T}$. The simulator gets the machine's output, running time and space usage on a single input of its

choice, and its running time is polynomial in the machine T's worst-case running time (and in n). We require that for any machine T the following two distributions are indistinguishable:

(a) The output of the adversary $\mathcal{A}$ when it is run with arbitrary access to the one-time program $\mathcal{P}$ (i.e. full access to the software component and black-box access to the hardware component).

(b) The output of the simulator with one-time oracle access to T.

Moreover, the indistinguishability holds even for a distinguisher who takes T as input. Note that it is crucial that the simulator $\mathcal{S}$ *only accesses its oracle once*.[1] Also note that the simulator cannot access any part of the actual one-time program, including the hardware, not even in a black-box manner.

See the full version of this work for a more rigorous treatment.

Remark. Note that in the above definition we only allow the adversary black-box access to the hardware component. Thus, we implicitly assume that the hardware devices withstand all side channel attacks. This is a very strong (and perhaps unreasonable) assumption. However, as we shall see next, the actual security assumptions we impose on the memory devices we use are much weaker, and in some sense are minimal.

3   One-Time-Memory Device (OTM)

Informally, a OTM is a memory device initialized with two keys (k_0, k_1). It takes as input a single bit $b \in \{0, 1\}$, outputs k_b and "self destructs". There are several ways to formalize the concept of self destruction. The first would be to erase both keys after outputting k_b. However, to circumvent side-channel attacks, we prefer that the device never access the key not retrieved. Instead, we choose the following formalism.

An OTM is initialized with two keys (k_0, k_1), and one additional tamper-proof bit set to 0. The OTM input is a single bit $b \in \{0, 1\}$. Upon receiving an input, the OTM verifies that the tamper-proof bit is set to 0, sets it to 1 and outputs k_b. If the tamper-proof bit is not 0, the device outputs an error symbol $\perp$. Thus, an OTM outputs one of its two keys, and the other key is irretrievably lost (and never accessed).

Our security assumptions from the OTM device are quite minimal:

1. The memory locations that are *never accessed* by the device are *never leaked* via a side channel, whereas a memory cell that *is* accessed may be immediately leaked.

2. The *single* bit b is tamper-proof (but is *readable*).

Intuitively, the above two assumptions imply that the device is as secure as a black-box. See the full version of this work for details.

---

[1] This guarantees that the one-time program cannot be duplicated and run more than once. The simulator certainly cannot duplicate, and thus an adversary who can obtain two of the program's outputs cannot be simulated.

OTM's are inspired by the cryptographic notion of *one out of two oblivious transfer* [Rab05, EGL85], where a sender holds two keys and lets a receiver receive one of them. The key not chosen is irrevocably lost, and the sender does not learn (is oblivious to) which key the receiver received.

Whereas an oblivious transfer is an interactive protocol, an OTM is a physical device. The important requirement from an OTM is that the user using the device (analogous to the "receiver" in an oblivious transfer protocol) learns only the secret of his choice. The requirement that the key generator (analogous to the "sender") is oblivious and does not learn which secret the user received, makes little sense in our setting as the OTM is at the hands of the "receiver" and the key generator is no longer present at the time that the OTM is used.

3.1   Using OTMs vs. Other Hardware

There are many possible secure hardware devices one could conceive of using, and it is not a-priori clear whether one device is better or worse than another. This is a central question in the study of one-time programs, and secure hardware in general. In this section we compare the use of OTM devices in one-time programs with alternative solutions that use different hardware.

Task Specific Hardware. A trivial solution would be to build for each function f a special-purpose task-specific secure hardware device which computes f for one input x and then refuses to work any longer. We find this solution highly unsatisfactory, for several reasons:

- *Universality.* First, this approach calls for building a different hardware device for each different function f. This may be worthwhile for some tasks, but is too costly for most tasks and thus infeasible in practice. Instead, we advocate that the secure hardware device of choice should be *universal*. Namely, that the hardware device be task-independent, and "programmable" so it can be used to construct a one-time program for *any* functionality (one-time programs for different functions will differ in their software component). In the case of secure hardware (rather than ordinary hardware), universality is particularly important, as each type of hardware device needs to be intensely scrutinized to try to guarantee that it is not susceptible to side channel attacks. This seems impossible to do on a function by function basis.

- *Simplicity.* Second, perhaps the most central measure of reasonability for a secure hardware device is its *simplicity*, and the trivial solution suggested above is potentially *complex* as it requires producing complex hardware for complex functions. Our search for *simple* hardware devices, which are easy to build, analyze and understand, is motivated by several concerns; (i) The assumption that a hardware device is secure and/or tamper-proof is a very strong assumption, as one has to consider all possible *physical attacks*. The simpler a hardware device is, the easier it is to scrutinize and analyze its security, and the more reasonable the assumption that it is secure becomes. (ii) Continuing in this vein, *side channel attacks* have emerged as a significant threat to the integrity of cryptographic algorithms and devices (see e.g. Anderson [And01]). It seems intuitive that the less computation the hardware preforms, the less susceptible it will be to potentially devastating side-channel attacks. Indeed, this is the guiding principle behind the theoretical approach to defining physically secure devices taken by [MR04, GLM$^+$04]. In the case of task-specific hardware, ad absurdum the entire the computation can be done by the hardware. (iii) Finally, and perhaps most obviously, the simpler a hardware device is, the easier and cheaper to build it will be.

Secure General Purpose CPU. An alternate solution would be to use the physically shielded full-blown CPU, which was proposed by Best [Bes79] and Kent [Ken80] and used in the work of Goldreich and Ostrovsky on software protection [GO96]. This CPU contains a protected ROM (read only memory) unit in which a secret decryption key is written. The I/O access to the shielded CPU is through fetch instructions. In each computation cycle, the CPU fetches an encrypted instruction, decrypts it, and executes it. The hardware security assumption here is that both the cryptographic operations (decryption, encryption etc.), as well as the general-purpose computation operations, are perfectly shielded. Each such shielded CPU was associated with a different decryption key, and the encrypted software executed on it was to be encrypted with the matching encryption key. Goldreich-Ostrovsky envisioned protecting a software program by encrypting it and packaging it with a physically shielded CPU with a matching decryption key. One-time programs can be easily built in the same manner (adding a counter to the CPU to limit the number of executions).

This solution is certainly universal, as the CPU can compute all tasks. Yet, we do not believe it is suitable for the one-time programs application:

- *Simplicity.* We consider a full blown protected CPU to be far from the goal of hardware simplicity, and so complex as to make the Goldreich-Ostrovsky approach unviable for the design of one-time programs. This is evidenced by the simple fact that although these devices were first proposed in the early 1980's, they still seem beyond the reach of current day technology in terms of cost. It seems that in particular for the application of one-time programs, using a full-blown shielded CPU for computing one task a limited number of times is an overkill.

- *Side Channel Attacks.* Secure CPUs perform complex computations (both cryptographic and otherwise), and are thus susceptible to side-channel attacks. If we assume, when modeling side channel attacks, that each computational step may leak information about bits accessed by the hardware, the [GO96] device becomes especially vulnerable: once the secret key (which is used in every step of the computation) leaks, the security of the entire construction falls apart.

Now, we can re-examine OTMs in light of the above alternative suggestions. As we show in Section 4, OTMs are *universal* task-independent devices that can be used to make any program one-time. Moreover, an OTM is also a *simple* device. Most importantly, even if we assume that side-channel adversaries can capture every bit accessed by the hardware during a computation step, the OTM construction remains secure, as long as there is a single (readable) tamper-proof bit! The OTM key that is not chosen is *never accessed*, and thus OTM constructions are secure under the (seemingly minimal, see [MR04]) assumption that untouched memory bits are not leaked.

So far, the comparison between OTM and other secure hardware was qualitative. We now present some several quantitative complexity measures for analyzing secure hardware devices and their use by a one-time program. In the next section we shall see how our solution fares in comparison to the [GO96] hardware with respect to this complexity measures.

- *Hardware Runtime.* The total combined running time of all the hardware devices used by the one-time program. This measures the amount of computation done by the secure hardware devices (e.g. number of operations done by their CPU), and *not* the amount of computation done by the one-time program's software component. Clearly, it is desirable for the hardware to do as little work as possible, both because simple devices will be computationally much weaker than the CPU of a modern personal computer and because the more computation is done on a device the more susceptible it may become to side-channel attacks.

We also consider the *total runtime* of the one-time program, which is the combined runtime of the hardware and software components.

- *Size.* The combined sizes of all the hardware devices used by the one-time program. The size of a hardware device is the size of its (persistent) memory together with the size of its control program. The smaller the hardware device, the better, as protecting smaller memories is easier.

- *Latency.* Number of times the one-time program $\mathcal{P}$ accesses its secure hardware devices. We assume that each time hardware devices are accessed, many of them may be queried in parallel, but we want to minimize the number of (adaptive) accesses to the hardware devices, both to guarantee that the hardware is not involved in complex computations and to optimize performance (as accessing hardware is expensive).

4   A One-Time Compiler

In what follows, we present an efficient one-time compiler that uses OTMs, give an overview of the construction, compare it to other solutions from the literature, and conclude with a discussion on the implications of improvements to our results.

The Construction. Building on the ideas in Yao's Garbled-Circuit construction [Yao86], we demonstrate that a universal one-time compiler exists using OTMs. First, convert the input (Turing machine) program into a Boolean circuit on inputs of length n. Second, garble it using Yao's method. And, finally, use n OTM's to transform the garbled circuit into a one-time program. We encounter an obstacle in this last step, as the security of Yao's construction is only against honest-but-curious adversaries, whereas the one-time program needs to be secure against any malicious adversary. In the secure function evaluation setting this is resolved using interaction (e.g. via zero-knowledge proofs, see [GMW91]), or using some global setup and non-interactive zero-knowledge proofs. Our setting of one-time programs, however, is not an interactive setting, and we cannot use these solutions. Instead, we present a solution to this problem that uses the OTM devices; see below.

Informal Theorem 1: Assume that one-way functions exist. There exists an efficient one-time compiler C that for input length n uses n OTMs: $B_1, \ldots, B_n$. For any (non-uniform) Turing machine T, with worst-case running time $t_T(n)$ (on inputs of length n), the compiler C, on input $1^n, 1^{t_T(n)}$, description of T and security parameter $1^{\kappa(n)}$, outputs a one-time program $P \triangleq M^{B_1(v_1), \ldots, B_n(v_n)}$ such that $P(x) = T(x)$ for inputs of length n. Let $t_T(x)$ denote T's running time on input $x \in \{0, 1\}^n$. Then, $P(x)$ achieves: *latency* 1; *hardware runtime* $n \cdot \kappa(n)$; *total running time* $\tilde{O}(t_T(x) \cdot \mathrm{poly}(\kappa, n))$; and *size* $\tilde{O}(t_T(n) \cdot \mathrm{poly}(\kappa))$.

Proof (Construction Overview). We begin by briefly reviewing Yao's Garbled Circuit construction. We assume (for the sake of simplicity) that the underlying Turing machine has a boolean (1 bit) output, but the construction is easily generalized to multi-bit outputs while maintaining the performance claimed in the theorem statement. The construction proceeds by converting the Turing machine T into a boolean circuit of size $\tilde{O}(t_T(n))$ and then garbling it using Yao's garbled circuit method. This gives a garbled circuit $G(T)$ of size $\tilde{O}(t_T(n) \cdot \mathrm{poly}(\kappa))$, together with n key-pairs $(k_1^0, k_1^1) \ldots (k_n^0, k_n^1)$. The construction guarantees both (i) **Correctness**: namely there is an efficient algorithm that for any input $x \in \{0, 1\}^n$ takes as input $G(T)$ and only the n keys $k_1^{x_1}, \ldots, k_n^{x_n}$ (one from each pair of keys), and outputs $T(x)$. The algorithm's running time is $\tilde{O}(t_T(x) \cdot \mathrm{poly}(\kappa))$. The construction also guarantees (ii) **Privacy**: an adversary cannot learn more from the garbled circuit together with one key out of each key-pair, say the x_i-th key from the i-th key pair, than the output of the machine on the input $x = x_1 \circ \ldots \circ x_n$. Formally, there exists an efficient simulator S such that for any machine T, for any output value b and for any input x such that $C(x) = b$, the simulator's output on input $(b, x, 1^{t_T(n)}, 1^\kappa)$ is indistinguishable from $(x, G(T), k_1^{x_1}, \ldots, k_n^{x_n})$.

We want to use Yao's garbled circuit to build one-time programs using OTMs. A deceptively straightforward idea for a one-time compiler is to use n OTMs: garble the machine T, and put the i-th key pair in the i-th OTM. To compute T's output on an input x a user can retrieve the proper key from each OTM and use the correctness of the garbled circuit construction to get the machine's output. Privacy may seem to follow from the privacy of the garbled circuit construction. Surprisingly, however, the above construction does *not* seem to guarantee privacy. In fact, it hides a subtle but inherent difficulty.

The difficulty is that the privacy guarantee given by the garbled circuit construction is too weak. At a higher level this is because the standard Yao construction is in the *honest-but curious setting*, whereas we want to build a program that is one-time even against *malicious* adversaries. More concretely, the garbled circuit simulator generates a dummy garbled circuit *after the input x is specified*, i.e. only after it knows the circuit's output $T(x)$. This suffices for honest-but-curious two-party computation, but it is not sufficient for us. The (malicious) one-time program adversary may be adaptive in its choice of x: *the choice of x* could depend on the garbling itself, as well as the keys revealed as the adversary accesses the OTMs. This poses a problem, as the simulator, who wants to generate a dummy garbling and then run the adversary on it, does not know in advance on which input the adversary will choose to evaluate the garbled circuit. On closer examination, the main problem is that the simulator does not know in advance the circuit's output on the input the adversary will choose, and thus it does not know what the dummy garbling's output should be. Note that we are not in an interactive setting, and thus we cannot use standard solutions such as having the adversary commit to its input *x* before seeing the garbled circuit.

To overcome this difficulty, we need to change the naive construction that we got directly from the garbled circuit to give the simulator more power. Our objective is allowing the simulator to "hold off" choosing the output of the dummy garbling until the adversary has specified the input. We do this by "hiding" a random secret bit b_i in the i-th OTM, this bit is exposed no matter which secret the adversary requests. These n bits mask (via an XOR) the circuit's output, giving the simulator the flexibility to hold off "committing" to the *unmasked* garbled circuit's output until the adversary has completely specified its input *x* (by accessing all n of the OTMs). The simulator outputs a garbled dummy circuit that evaluates to some random value, runs the adversary, and once the adversary has completely specified *x* by accessing all n OTMs, the simulator can retrieve $\mathcal{T}(x)$ (via its one-time $\mathcal{T}$-oracle), and the last masking bit it exposes to the adversary (in the last OTM the adversary accesses) always unmasks the garbled dummy circuit's output to be equal to $\mathcal{T}(x)$.

Our Scheme vs. the [GO96] Scheme. Note that in the [GO96] scheme, both the hardware runtime and the latency is the same as the *entire* running time of the program, whereas in our scheme (using OTMs) the latency is 1 and the hardware runtime is $n \cdot \kappa(n)$ (independent of the program runtime). On the other hand, one advantage of the [GO96] scheme is that the size of the entire one-time program is proportional to the size of a single cryptographic key (independent of the program runtime), whereas in our scheme the size is quasi-linear in the (worst-case) runtime of the program.

4.1   Can We Get the Best of Both Worlds?

The primary disadvantage of our construction is the *size* of the one-time program. The "garbled circuit" part of the program (the software part) is as large as the (worst-case) running time of the original program. It is natural to ask whether this is inherent. In fact, it is not, as one-time programs based on the Goldreich-Ostrovsky construction (with a counter limiting the number of executions) have size only proportional to the *size* of the original program and a cryptographic key. However, as discussed above, the Goldreich-Ostrovsky solution requires complex secure hardware that runs the *entire computation* of the one-time program.

It remains to ask, then, whether it is possible to construct one-time programs that enjoy "the best of both worlds." I.e. to build *small* one-time programs with *simple hardware that does very little work.* This is a fundamental question in the study of one-time programs. Unfortunately, we show that building a one-time program that enjoys the best of both worlds (small size and hardware running time) is beyond the reach of current knowledge in the field of cryptography. This is done by showing that such a construction would resolve a central open problem in foundational cryptography: it would give a secure-function-evaluation protocols with sub-linear (in the computation size) communication complexity.

Informal Theorem 2: Assume that for every security parameter κ, there exists a secure oblivious transfer protocol for 1-bit message pairs with communication complexity $\text{poly}(\kappa)$. Fix any input length n and any (non-uniform) Turing machine $\mathcal{T}$. Suppose $\mathcal{P}$ is a one-time program corresponding to $\mathcal{T}$ (for inputs of length n). If $\mathcal{P}$ is of total size $s(n)$ and the worst-case (combined) running time of the secure hardware(s) on an n-bit input is $t(n)$, then there exists a secure function evaluation protocol where Alice has input $\mathcal{T}$, Bob has input $x \in \{0,1\}^n$, at the end of the protocol Bob learns $\mathcal{T}(x)$ but nothing else, Alice learns nothing, and the total communication complexity is $s(n) + O(t(n) \cdot \text{poly}(\kappa))$.

Let us examine the possibility of achieving the best of both worlds in light of this theorem. Suppose we had a one-time compiler that transforms any program $\mathcal{T}$ into a one-time program with simple secure hardware that does $O(n \cdot \kappa)$ work and has total size $O(|\mathcal{T}|+\kappa)$. By the above theorem, this would immediately give a secure function evaluation protocol where Alice has $\mathcal{T}$, Bob has x, Bob learns only $\mathcal{T}(x)$ and Alice learns nothing, *with linear in $(n, |\mathcal{T}|, \kappa)$ communication complexity!* All known protocols for this problem have communication complexity that is at least linear *in $\mathcal{T}$'s running time*, and constructing a protocol with better communication complexity is a central open problem in theoretical cryptography. For example, this is one of the main motivations for constructing a fully homomorphic encryption scheme. See the full version for more details.

5   One-Time Programs: Applications

One-time programs have immediate applications to software protection. They also enable new applications such as one-time proofs, outlined below. Finally, OTMs and one-time programs can be used to construct electronic cash and electronic token schemes [Cha82, Cha83]. The E-cash applications and the discussion of related work are omitted from this extended abstract for lack of space, see the full version for details.

5.1   Extreme Software Protection

By the very virtue of being one-time programs, they cannot be reverse engineered, copied, re-distributed or executed more than once.

Limiting the Number of Executions. A vendor can put an explicit restriction as to the number of times a program it sells can be used, by converting it into a one-time program which can be executed for at most k times. For example, this allows vendors to supply prospective clients with "preview" versions of software that can only be used a very limited number of times. Unlike techniques that are commonly employed in practice, here there is a *guarantee* that the software cannot be reverse-engineered, and the component that limits the number of executions cannot be removed. Moreover, our solution does not require trusting a system clock or communicating with the software over a network (as do many solutions employed in practice). This enables vendors to control (and perhaps charge for) the way in which users use their software, while completely maintaining the user's privacy (since the vendors never see users interacting with the programs). One-time programs naturally give solutions to such copy-protection and software protection problems, albeit at a price (in terms of complexity and distribution difficulty).

Temporary Transfer of Cryptographic Ability. As a natural application for one-time programs, consider the following setting, previously suggested by [GK05] in the context of program obfuscation. Alice wants to go on vacation for the month of September. While she is away, she would like to give her assistant Bob the power to decrypt and sign E-mails dated "September 2008" (and only those E-mails). Alice can now supply Bob with many one-time programs for signing and decrypting messages dated "September 2008". In October, when Alice returns, she is guaranteed that Bob will not be able to decrypt or sign any of her messages! As long as Alice knows a (reasonable) upper bound for the number of expected messages to be signed and decrypted, temporarily transferring her cryptographic ability to Bob becomes easy.

5.2   One-Time Proofs

The one-time paradigm leads to the new concept and constructions of *one-time proofs*: proof tokens that can be used to prove (or verify) an NP statement exactly once.

A one-time proof system for an NP language L consists of three entities: (i) a witness *owner* who has a witness to the membership of some element x in a language L, (ii) a prover, and (iii) a verifier, where the prover and verifier know the input x but do not know the witness to x's membership in L. A one-time proof system allows the witness owner to (efficiently) transform its NP witness into a hardware based *proof token*. The proof token can later be used by the efficient prover (who does not know a witness) to convince the verifier **exactly once** that the input x is in the language. The witness owner and the verifier are assumed to be "honest" and follow the prescribed protocols, whereas the prover may be malicious and deviate arbitrarily.[2]

In a one-time proof, the prover convinces the verifier by means of a standard interactive proof system. In particular, the verifier doesn't need physical access to the proof token (only the prover needs this access). After running the interactive proof and convincing the (honest) verifier once, the proof token becomes useless and cannot be used again. The point is that (i) the witness owner does not need

---

[2] The case where even verifiers do not behave honestly is also interesting, see the full version of this work for a discussion.

to be involved in the proof, beyond supplying the token (hence the proof system is off-line), and (*ii*) the prover, even though it convinces the verifier, learns nothing from interacting with the hardware, and in particular cannot convince the verifier a second time. Thus, for any NP statement, one-time proofs allow a witness owner to give other parties the capability to prove the statement in a controlled manner, without revealing to them the witness. A one-time proof system gives this "one-time proof" guarantee, as well as the more standard completeness and soundness guarantees, and a zero-knowledge guarantee (see [GMR89]), stating that anything that can be learned from the proof token, can be learned without it (by a simulator). Finally, a user who wants to use the one-time proof token to convince himself that $x \in L$ can do so without any interaction by running the interactive proof in his head (in this case the prover and verifier are the same entity).

Note that a prover who somehow does know a witness to x's membership can convince the verifier as many times as it wants. How then can one capture the one-time proof requirement? We do this by requiring that any prover who can use a single proof token to convince the verifier more than once, must in fact know a witness to x's membership in the language. Formally, the witness can be *extracted* from the prover in polynomial time. In particular, this means that if the prover can convince the verifier more than once using a single proof token, then the prover could also convince the verifier as many times as it wanted without ever seeing a proof token! In other words, the proof token does not help the prover prove the statement more than once.

Another natural setting where one-time proofs come up is in voting systems, where the goal is to ensure that voters can only vote once. In this setting, each voter will be given a one-time proof for possessing the right to vote (of course, one must also ensure that the proof tokens cannot be transferred from one voter to another). A similar application of one-time proofs is for electronic subway tokens. Here the subway operator wants to sell electronic subway tokens to passengers, where the tokens should be verifiable by the subway station turnstiles.[3] A passenger should only be able to use a token once to gain entrance to the subway, and after this the token becomes useless. This goal is easily realized in a natural way by one-time proofs. The subway operator generates a hard cryptographic instance, say a product $n = p \cdot q$ of two primes. Subway tokens are one-time proof tokens for proving that n is in fact a product of two large primes. The passengers play the role of the prover. The turnstiles are the verifier, and only let provers who can prove to them that n is a product of two primes into the subway station. Any passenger who can use a single token to gain entrance more than once, can also be used to find a witness, or the factorization of n, a task which we assume is impossible for efficient passengers.[4]

---

[3] This problem was originally suggested by Blum [Blu81]. A scheme using quantum devices (without a proof of security) was proposed by Bennett *et al.* [BBBW82].

[4] For simplicity we do not consider here issues of composability or maintaining passenger's anonymity.

More generally, one can view one-time proofs as a natural generalization of count-limited access control problems. In particular, we can convert any 3-round ID scheme (or any Σ-protocol) into a one-time proof of identity. See also the application to the E-token problem presented in the next section.

One-time proofs are different from non interactive zero knowledge (NIZK) proofs (introduced by [BFM88]). In both cases, the witness owner need not be present when the proof is being verified. However, in NIZK proof systems either the proof can be verified by arbitrary verifiers an unlimited number of times, and in particular is also not deniable [Pas03] (for example, NIZK proof systems in a CRS-like model, as in [BFM88]), or the proofs have to be tailored to a specific verifier and are useless to other verifiers (for example, NIZK proof systems in the pre-processing model [SMP88]). One-time zero knowledge proofs, on the other hand, can only be verified once, but by *any* user, and are later deniable. They also do not need a trusted setup, public-key infrastructure, or pre-processing, but on the other hand they do use secure hardware.

In the full version of this work we define one-time proofs and show (assuming one-way permutations) that any NP statement has an efficient one-time proof using OTMs. To attain small soundness we need to overcome problems that arise in the parallel composition of zero-knowledge proof (but in a non-interactive setting). This is accomplished by using the secure hardware to allow a delicate simulation argument. While we note that the general-purpose one-time compiler from the previous section can be used to construct a one-time proof,[5] this results in considerably less efficient (and less intuitively appealing) schemes.

Informal Theorem 3: Let κ be a security parameter and k a soundness parameter. Assume that there exists a one-way permutation on κ-bit inputs. Every NP language L has a one-time zero-knowledge proof with perfect completeness and soundness 2^{-k}. The proof token uses k OTMs (each of size $\mathrm{poly}(n, k, \kappa)$, where n is the input length).

Construction Overview. We construct a one-time proof for the NP complete language Graph Hamiltonicity, from which we can derive a one-time proof for any NP language. The construction uses ideas from Blum's [Blu87] protocol for Graph Hamiltonicity. The input is a graph $G = (V, E)$, the producer has a witness w describing a hamiltonian cycle in the graph. The one-time proof uses k OTMs to get a proof with soundness 2^{-k} (and perfect completeness).

The basic idea of Blum's zero-knowledge proof is for a prover to commit to a random permutation of the graph and send this commitment to the verifier. The verifier can then ask the prover wether to send it the permutation and all the de-commitments (openings of all the commitments), or to send de-commitments to a Hamiltonian cycle in the permuted graph. The proof is zero-knowledge, with soundness $1/2$.

A natural approach for our setting is for the witness owner to generate a proof token that has the committed permuted graph as a software component. The proof token also includes an OTM whose first secret is the permutation

---

[5] To do this, just generate a one-time program computing the prover's answers.

and all the de-commitments (the answer to one possible verifier query in Blum's protocol), and whose second secret is de-commitments to a Hamiltonian cycle in the permuted graph (the answer to the second verifier query). This indeed gives a (simple) one-time proof with soundness $1/2$ via standard arguments: the only thing a prover can learn from the token is one of the two possible de-commitment sequences, and we know (from Blum's zero-knowledge simulator), that the prover could generate this on its own. On the other hand, somewhat paradoxically, this proof token does allow a prover to convince a verifier that the graph has a Hamiltonian cycle in an interactive proof with perfect completeness and soundness $1/2$.

To amplify the soundness to 2^{-k}, a seemingly effective idea is to have the producer produce k such committed graphs and k such corresponding OTMs, each containing a pair of secrets corresponding to a new commitment to a random permutation of its graph. This idea is, however, problematic. The difficulty is that simulating the one-time proof becomes as hard as simulating parallel executions of Blum's zero-knowledge protocol. Namely, whoever gets the OTMs can choose which of the two secrets in each OTM it will retrieve as a function of *all* of the committed permuted graphs. In the standard interactive zero-knowledge proof setting this is resolved by adding interaction to the proof (see Goldreich and Kahan [GK96]). However, in our setting it is crucial to avoid adding interaction with the witness owner during the interactive proof phase, and thus known solutions do not apply. In general, reducing the soundness of Blum's protocol without adding interaction is a long-standing open problem (see e.g. Barak, Lindell and Vadhan [BLV06]). In our setting, however, we can use the secure hardware to obtain a simple solution.

To overcome the above problem, we use the secure hardware (OTMs) to "force" a user who wants to gain anything from the proof token, to access the boxes in sequential order, independently of upcoming boxes, as follows. The first committed graph C_1 is given to the user "in the clear", but subsequent committed graphs, C_i, $i \geq 2$, are not revealed until all of the prior $i-1$ OTMs (corresponding to committed graphs $C_1, ..., C_{i-1}$) have been accessed. This is achieved by, for each i: (1) splitting the description of the i-th committed graph C_i into $i-1$ random strings $m_i^1, ... m_i^{i-1}$ (or shares) such that their XOR is C_i; and (2) letting the j-th ROK output m_i^j for each $i \geq j+1$ as soon as the user accesses it (regardless of which input the user gives to the OTM). Thus by the time the user sees all shares of a committed graph, he has already accessed *all* the previous OTMs corresponding to the previous committed graphs. The user (information theoretically) does not know the i-th committed graph until he has accessed all the OTMs $1, ..., i-1$, and this forces "sequential" behavior that can be simulated. Of course, after accessing the boxes $1, ..., i-1$, the user can retrieve the committed graph and verify the proof's correctness as usual (i.e. completeness holds). See the full version for a formal theorem statement and proof.

We note that beyond being conceptually appealing, one-time proofs have obvious applications to identification and limited time (rather than revokable) credential proving applications. See the full version for formal definitions and further details.

6   Further Related Work

Using and Realizing Secure Hardware. Recently, Katz [Kat07], followed by Moran and Segev [MS08], studied the applications of secure hardware to constructing protocols that are secure in a concurrent setting. These works also model secure hardware as a "black box". Earlier work by Moran and Naor [MN05] showed how to construct cryptographic protocols based on "tamper-evident seals", a weaker secure hardware assumption that models physical objects such as sealed envelopes, see the full version of this work for a more detailed comparison.

Works such as Ishai, Sahai and Wagner [ISW03], Gennaro *et al.* [GLM+04] and Ishai, Prabhakaran Sahai and Wagner [IPSW06], aim at achieving the notion of "black-box" access to devices using only minimal assumptions about hardware (e.g. adversaries can read some, but not all, of the hardware's wires etc.). The work of Micali and Reyzin [MR04] was also concerned with realizing ideal "black-box" access to computing devices, but they focused on obtaining model-independent reductions between specific physically secure primitives.

Alternative Approaches to Software Protection. An alternative software-based approach to software protection and obfuscation was recently suggested by Dvir, Herlihy and Shavit [DHS06]. They suggest protecting software by providing a user with an incomplete program. The user can run the program only by communicating with a server, and the server provides the "missing pieces" of the program in a protected manner. The setting of one-time programs is different, as we want to restrict even the *number of times* a user can run the program.

Count-Limiting using a TPM. In recent (independent) work, Sarmenta, van Dijk, O'Donnel, Rhodes and Devadas [SvDO+06] explore cryptographic and system security-oriented applications of real-world secure hardware, the Trusted Platform Module (TPM) chip (see [TPM07]). They show how a single TPM chip can be used to implement a large number of trusted monotonic counters and also consider applications of such counters to goals such as e-cash, DRM, and count limited cryptographic applications. These have to do with count-limiting special tasks, specific functionalities or small groups of functionalities, whereas we focus on the question of count-limiting access to general-purpose programs (and its applications). Our OTM construction indicates that TPMs can be used to count-limit *any* efficiently computable functionality. Following our work in ongoing work Sarmenta *et al.* began to consider using TPMs to count-limit general-purpose programs.

Acknowledgements

We thank Ran Canetti for early collaboration on this work, for his interest, suggestions and support. We are grateful to Nir Shavit for his enthusiastic and pivotal support of the one-time program concept, for suggesting applications to time-independent off-line E-tokens, and commenting on drafts of this work. Illuminating discussions on secure hardware were very helpful, for which we thank Anantha Chandrakasan, as well as Srini Devadas and Luis Sarmenta, who also explained to us their work on trusted monotonic counters using TPMs.

We would also like to thank Oded Goldreich for helpful discussions regarding software protection, Susan Hohenberger for her continuous clear and helpful answers to our questions about electronic cash, Adam Kalai for his invaluable suggestions, Moni Naor for his advice and illuminating discussions, Adi Shamir for pointing out the importance of choosing the right order of operations on a ROM, and Salil Vadhan for his continuous help, insight and suggestions throughout various stages of this work.

Finally, we are especially indebted to Silvio Micali's crucial comments following the first presentation of this work, which helped us focus on the resistance of OTMs to side channel attacks. Thank you Silvio!

References

[And01]   Anderson., R.J.: Security Engineering: A Guide to Building Dependable Distributed Systems. Wiley, Chichester (2001)

[BBBW82]   Bennett, C.H., Brassard, G., Breidbard, S., Wiesner, S.: Quantum cryptography, or unforgeable subway tokens. In: CRYPTO 1982, pp. 267–275 (1982)

[Bes79]   Best, R.M.: Us patent 4,168,396: Microprocessor for executing enciphered programs (1979)

[BFM88]   Blum, M., Feldman, P., Micali, S.: Non-interactive zero-knowledge and its applications (extended abstract). In: STOC 1988, Chicago, Illinois, pp. 103–112 (1988)

[BGI+01]   Barak, B., Goldreich, O., Impagliazzo, R., Rudich, S., Sahai, A., Vadhan, S.P., Yang, K.: On the (im)possibility of obfuscating programs. In: Kilian, J. (ed.) CRYPTO 2001. LNCS, vol. 2139, pp. 1–18. Springer, Heidelberg (2001)

[Blu81]   Blum, M.: Personal communication (1981)

[Blu87]   Blum, M.: How to prove a theorem so no-one else can claim it. In: Proceedings of ICML, pp. 1444–1451 (1987)

[BLV06]   Barak, B., Lindell, Y., Vadhan, S.P.: Lower bounds for non-black-box zero knowledge. J. Comput. Syst. Sci. 72(2), 321–391 (2006)

[Cha82]   Chaum, D.: Blind signatures for untraceable payments. In: CRYPTO 1982, pp. 199–203 (1982)

[Cha83]   Chaum, D.: Blind signature systems. In: CRYPTO 1983, pp. 153–156 (1983)

[DHS06]   Dvir, O., Herlihy, M., Shavit, N.: Virtual leashing: Creating a computational foundation for software protection. Journal of Parallel and Distributed Computing (Special Issue) 66(9), 1233–1240 (2006)

[EGL85]   Even, S., Goldreich, O., Lempel, A.: A randomized protocol for signing contracts. Commun. ACM 28(6), 637–647 (1985)

[GK96]   Goldreich, O., Kahan, A.: How to construct constant-round zero-knowledge proof systems for np. J. Cryptology 9(3), 167–190 (1996)

[GK05]   Goldwasser, S., Kalai, Y.T.: On the impossibility of obfuscation with aux-
            iliary input. In: Tardos, É. (ed.) FOCS 2005, pp. 553–562. IEEE Computer
            Society, Los Alamitos (2005)
[GLM⁺04]   Gennaro, R., Lysyanskaya, A., Malkin, T., Micali, S., Rabin, T.: Algo-
            rithmic tamper-proof (atp) security: Theoretical foundations for secu-
            rity against hardware tampering. In: Naor, M. (ed.) TCC 2004. LNCS,
            vol. 2951, pp. 258–277. Springer, Heidelberg (2004)
[GMR89]   Goldwasser, S., Micali, S., Rackoff, C.: The knowledge complexity of inter-
            active proof-systems. SIAM Journal on Computing 18(1), 186–208 (1989)
[GMW91]   Goldreich, O., Micali, S., Wigderson, A.: Proofs that yield nothing but
            their validity, or all languages in np have zero-knowledge proof systems.
            Journal of the ACM 38(1), 691–729 (1991)
[GO96]   Goldreich, O., Ostrovsky, R.: Software protection and simulation on obliv-
            ious rams. Journal of the ACM 43(3), 431–473 (1996)
[IPSW06]   Ishai, Y., Prabhakaran, M., Sahai, A., Wagner, D.: Private circuits ii:
            Keeping secrets in tamperable circuits. In: Vaudenay, S. (ed.) EURO-
            CRYPT 2006. LNCS, vol. 4004, pp. 308–327. Springer, Heidelberg (2006)
[ISW03]   Ishai, Y., Sahai, A., Wagner, D.: Private circuits: Securing hardware
            against probing attacks. In: Boneh, D. (ed.) CRYPTO 2003. LNCS,
            vol. 2729, pp. 463–481. Springer, Heidelberg (2003)
[Kat07]   Katz, J.: Universally composable multi-party computation using tamper-
            proof hardware. In: Naor, M. (ed.) EUROCRYPT 2007. LNCS, vol. 4515,
            pp. 115–128. Springer, Heidelberg (2007)
[Ken80]   Kent, S.T.: Protecting Externally Supplied Software in Small Comput-
            ers. PhD thesis, Massachusetts Institute of Technology, Cambridge, Mas-
            sachusetts (1980)
[MN05]   Moran, T., Naor, M.: Basing cryptographic protocols on tamper-evident seals.
            In: Caires, L., Italiano, G.F., Monteiro, L., Palamidessi, C., Yung, M. (eds.)
            ICALP 2005. LNCS, vol. 3580, pp. 285–297. Springer, Heidelberg (2005)
[MR04]   Micali, S., Reyzin, L.: Physically observable cryptography (extended ab-
            stract). In: Naor, M. (ed.) TCC 2004. LNCS, vol. 2951, pp. 278–296.
            Springer, Heidelberg (2004)
[MS08]   Moran, T., Segev, G.: David and goliath commitments: Uc computation for
            asymmetric parties using tamper-proof hardware. In: Smart, N. (ed.) EURO-
            CRYPT 2008. LNCS, vol. 4965, pp. 527–544. Springer, Heidelberg (2008)
[Pas03]   Pass, R.: On deniability in the common reference string and random oracle
            model. In: Boneh, D. (ed.) CRYPTO 2003. LNCS, vol. 2729, pp. 316–337.
            Springer, Heidelberg (2003)
[Rab05]   Rabin, M.O.: How to exchange secrets with oblivious transfer. Cryptology
            ePrint Archive, Report 2005/187 (2005)
[SMP88]   De Santis, A., Micali, S., Persiano, G.: Non-interactive zero-knowledge
            with preprocessing. In: Goldwasser, S. (ed.) CRYPTO 1988. LNCS,
            vol. 403, pp. 269–282. Springer, Heidelberg (1990)
[SvDO⁺06]   Sarmenta, L.F.G., van Dijk, M., O'Donnell, C.W., Rhodes, J., Devadas,
            S.: Virtual monotonic counters and count-limited objects using a tpm
            without a trusted os (extended version). Technical Report 2006-064, MIT
            CSAIL Technical Report (2006)
[TPM07]   Trusted computing group trusted platform module (tpm) specifications
            (2007)
[Yao86]   Yao, A.C.: How to generate and exchange secrets. In: FOCS 1986, pp.
            162–167 (1986)

Adaptive One-Way Functions and Applications

Omkant Pandey[1], Rafael Pass[2,*], and Vinod Vaikuntanathan[3,**]

[1] UCLA
omkant@cs.ucla.edu
[2] Cornell University
rafael@cs.cornell.edu
[3] MIT
vinodv@mit.edu

Abstract. We introduce new and general complexity theoretic hardness assumptions. These assumptions abstract out concrete properties of a random oracle and are significantly stronger than traditional cryptographic hardness assumptions; however, assuming their validity we can resolve a number of long-standing open problems in cryptography.

Keywords: Cryptographic Assumptions, Non-malleable Commitment, Non-malleable Zero-knowledge.

1 Introduction

The state-of-the-art in complexity theory forces cryptographers to base their schemes on unproven hardness assumptions. Such assumptions can be general (e.g., the existence of one-way functions) or specific (e.g., the hardness of RSA or the Discrete logarithm problem). Specific hardness assumptions are usually stronger than their general counterparts; however, as such assumptions consider primitives with more structure, they lend themselves to constructions of more efficient protocols, and sometimes even to the constructions of objects that are not known to exist when this extra structure is not present. Indeed, in recent years, several new and more exotic specific hardness assumptions have been introduced (e.g., [4, 11, 12]) leading to, among other things, signatures schemes with improved efficiency, but also the first provably secure construction of identity-based encryption.

In this paper, we introduce a new class of strong but *general* hardness assumptions, and show how these assumptions can be used to resolve certain long-standing open problems in cryptography. Our assumptions are all abstractions of concrete properties of a random oracle. As such, our results show that for the problems we consider, random oracles are not necessary; rather, provably secure constructions can be based on concrete hardness assumptions.

* Supported by NSF CAREER Grant No. CCF-0746990, AFOSR Award No. FA9550-08-1-0197, BSF Grant No. 2006317.
** Supported in part by NSF Grant CNS-0430450.

D. Wagner (Ed.): CRYPTO 2008, LNCS 5157, pp. 57–74, 2008.
© International Association for Cryptologic Research 2008

1.1 Adaptive Hardness Assumptions

We consider *adaptive* strengthenings of standard general hardness assumptions, such as the existence of one-way functions and pseudorandom generators. More specifically, we introduce the notion of collections of adaptive 1-1 one-way functions and collections of adaptive pseudorandom generators. Intuitively,

- A *collection of adaptively 1-1 one-way functions* is a family of 1-1 functions $\mathcal{F}_n = \{f_{\mathsf{tag}} : \{0,1\}^n \mapsto \{0,1\}^n\}$ such that for every tag, it is hard to invert $f_{\mathsf{tag}}(r)$ for a random r, even for an adversary that is granted access to an "inversion oracle" for $f_{\mathsf{tag}'}$ for every $\mathsf{tag} \neq \mathsf{tag}'$. In other words, the function f_{tag} is one-way, even with access to an oracle that invert all the other functions in the family.
- A collection of *adaptive pseudo-random generators* is a family of functions $\mathcal{G}_n = G_{\mathsf{tag}} : \{0,1\}^n \mapsto \{0,1\}^m$ such that for every tag, G_{tag} is a pseudorandom even if given access to an oracle that decides whether given y is in the range of $G_{\mathsf{tag}'}$ for $\mathsf{tag}' \neq \mathsf{tag}$.

Both the above assumptions are strong, but arguably not "unrealistically" strong. Indeed, both these assumptions are satisfied by a (sufficiently) length-extending random oracle.[1] As such, they provide concrete mathematical assumptions that can be used to instantiate random oracles in certain applications. We also present some concrete candidate instantiations of these assumptions. For the case of adaptive 1-1 one-way functions, we provide construction based on the the "adaptive security" of Factoring, or the Discrete Log problem. For the case of adaptive PRGs, we provide a candidate construction based on a generalization of the advanced encryption standard (AES).

Related Assumptions in the Literature. Assumptions of a related flavor have appeared in a number of works. The class of "one-more" assumptions introduced by Bellare, Namprempre, Pointcheval and Semanko [4] are similar in flavor. Informally, the setting of the one-more RSA-inversion problem is the following: The adversary is given values $z_1, z_2, \ldots, z_k \in \mathbb{Z}_N^*$ (for a composite $N = pq$, a product of two primes) and is given access to an oracle that computes RSA inverses. The adversary wins if the number of values that it computes an RSA inverse of, exceeds the number of calls it makes to the oracle. They prove the security of Chaum's blind-signature scheme under this assumption. This flavor of assumptions has been used in numerous other subsequent works [5, 6].

Even more closely related, Prabhakaran and Sahai [31] consider an assumption of the form that there are collision-resistant hash functions that are secure even if the adversary has access to a "collision-sampler". In a related work, Malkin, Moriarty and Yakovenko [24] assume that the discrete logarithm problem in $\mathbb{Z}_p^*$ (where p is a k-bit prime) is hard even for an adversary that has access to an oracle that computes discrete logarithms in $\mathbb{Z}_q^*$ for any k-bit prime $q \neq p$. Both these works use the assumption to achieve secure computation in a relaxation

---

[1] Note that a random function over, say, $\{0,1\}^n \to \{0,1\}^{4n}$ is 1-1 except with exponentially small probability.

of the universal composability framework. (In a sense, their work couples the relaxed security notion to the hardness assumption. In contrast, we use adaptive hardness assumptions to obtain protocols that satisfy the traditional notion of security.)

1.2   Our Results

Non-Interactive Concurrently Non-Malleable Commitment Schemes. Non-malleable commitment schemes were first defined and constructed in the seminal paper of Dolev, Dwork and Naor [17]. Informally, a commitment scheme is non-malleable if no adversary can, upon seeing a commitment to a value v, produce a commitment to a related value (say $v - 1$). Indeed, non-malleability is crucial to applications which rely on the *independence* of the committed values. A stronger notion—called concurrent non-malleability–requires that no adversary, after receiving commitments of $v_1, \ldots, v_m$, can produce commitments to related values $\tilde{v}_1, \ldots, \tilde{v}_m$; see [23, 28] for a formal definition.

The first non-malleable commitment scheme of [17] was interactive, and required $O(\log n)$ rounds of interaction, where n is a security parameter. Barak [1] and subsequently, Pass and Rosen [28, 29] presented constant-round non-malleable commitment schemes; the protocols of [28, 29] are the most round-efficient (requiring 12 rounds) and the one of [28] is additionally concurrently non-malleable. We note that of the above commitment schemes, [17] is the only one with a black-box proof of security, whereas the schemes of [1, 28, 29] rely on the non-black-box proof technique introduced by Barak [1].[2]

Our first result is a construction of a *non-interactive, concurrently non-malleable* string commitment scheme, from a family of adaptive one-way permutations; additionally our construction only requires a black-box proof of security.

Theorem 1 (Informal). *Assume the existence of collections of adaptive 1-1 permutations. Then, there exists a non-interactive concurrently non-malleable string commitment scheme with a black-box proof of security.*

If instead assuming the existence of adaptive PRGs, we show the existence of 2-round concurrent non-malleable commitment with a black-box proof of security.

Theorem 2 (Informal). *Assume the existence of collections of adaptive PRGS. Then, there exists a 2-round concurrently non-malleable string commitment scheme with a black-box proof of security.*

Round-optimal Black-box Non-malleable Zero-knowledge. Intuitively, a zero-knowledge proof is *non-malleable* if a man-in-the-middle adversay, receiving a proof of a statement x, will not be able to provide a proof of a statement $x' \neq x$ unless he could have done so without hearing the proof of x. Dolev, Dwork and

---

[2] Subsequent to this work, Lin, Pass and Venkitasubramaniam [23] have presented constructions of concurrent non-malleable commitments using a black-box security proof, based on only one-way functions. Their construction, however, uses $O(n)$ communication rounds.

Naor [17] defined non-malleable zero-knowledge ($\mathcal{ZK}$) and presented an $O(\log n)$-round $\mathcal{ZK}$ proof system. Barak [1] and subsequently, Pass and Rosen [29] presented constant-round non-malleable $\mathcal{ZK}$ argument system. Again, the protocol of [17] is the only one with a black-box proof of security.

We construct a 4-*round non-malleable $\mathcal{ZK}$ argument* system with a black-box proof of security (that is, a black-box simulator). Four rounds is known to be optimal for black-box $\mathcal{ZK}$ [20] (even if the protocol is not required to be non-malleable) and for non-malleable protocols (even if they are not required to be $\mathcal{ZK}$) [22].

Theorem 3 (Informal). *Assume the existence of collections of adaptive 1-1 one-way function. Then, there exists a 4-round non-malleable zero-knowledge argument system with a black-box proof of security. Assume, instead, the existence of collections of adaptive one-way permutations. Then, there exists a 5-round non-malleable zero-knowledge argument system with a black-box proof of security.*

It is interesting to note that the (seemingly) related notion of concurrent zero-knowledge cannot be achieved in $o(\log n)$ rounds with a black-box proof of security. Thus, our result shows that (under our new assumptions), the notion of non-malleability and concurrency in the context of $\mathcal{ZK}$ are quantitatively different.

Efficient Chosen-Ciphertext Secure Encryption. Chosen ciphertext (CCA) security was introduced in the works of [26, 32] and has since been recognized as a *sine-qua-non* for secure encryption. Dolev, Dwork and Naor [17] gave the first construction of a CCA-secure encryption scheme based on general assumptions. Their construction, and the subsequent construction of Sahai [33], uses the machinery of non-interactive zero-knowledge proofs, which renders them less efficient than one would like. In contrast, the constructions of Cramer and Shoup [15, 16] are efficent, but are based on specific number-theoretic assumptions.

Bellare and Rogaway [7] proposed an encryption scheme that is CCA-secure in the random oracle model (see below for more details about the random oracle model). We show complexity-theoretic assumptions that are sufficient to replace the random oracle in this construction. We mention that, previously, Canetti [13] showed how to replace random oracles in a related construction to get a semantically secure encryption scheme, but without CCA security. In a more recent work, Boldyreva and Fischlin [10] also show how to obtain a weakened notion of non-malleability, but still without CCA security.

Interactive Arguments for which Parallel-repetition does not reduce the soundness error. A basic question regarding interactive proofs is whether parallel repetition of such protocols reduces the soundness error. Bellare, Impagliazzo and Naor [3] show that there are interactive *arguments* (i.e., computationally-sound) proofs in the Common Reference String (CRS) model, for which parallel-repetition does not reduce the soundness error. Their construction relies on

non-malleable encryption, and makes use of the CRS to select the public-key
for this encryption scheme. However, if instead relying on a non-interactive concurrent non-malleable commitment scheme in their construction, we can dispense
of the CRS altogether. Thus, by Theorem 1, assuming the existence of collections of adaptive 1-1 one-way functions, we show that there exists an interactive
argument for which parallel repetition does not reduce the soundness error. We
also mention that the same technique can be applied also to the strengthened
construction of [30].

Our Techniques. Our constructions are simple and efficient. In particular, for
the case of non-malleable commitment schemes, we show that appropriate instantiations of the Blum-Micali [9] or Naor [25] commitment schemes in fact are
non-malleable. The proof of these schemes are also "relatively straight-forward"
and follow nicely from the adaptive property of the underlying primitives.

Next, we show that by appropriately using our non-malleable commitment
protocols in the Feige-Shamir [18] $\mathcal{ZK}$ argument for $\mathcal{NP}$, we can also get a round-
optimal black-box non-malleable $\mathcal{ZK}$ proof for $\mathcal{NP}$. Although the construction
here is straight-forward, its proof of correctness is less so. In particular, to show
that our protocol is non-malleable, we rely on a techniques that are quite different from traditional proofs of non-malleability: in particular, the power of the
"adaptive" oracle will only be used inside hybrid experiments; the simulation,
on the other hand, will proceed by traditional rewinding. Interestingly, to get a
round-optimal solution, our proof inherently relies on the actual Feige-Shamir
protocol and high-lights some novel features of this protocol.

Interpreting Our Results. We offer two interpretations of our results:

- The *optimistic* interpretation: Although our assumptions are strong, they
  nonetheless do not (a priori) seem infeasible. Thus, if we believe that e.g.,
  AES behaves as an adaptively secure PRG, we show efficient solutions to
  important open questions.
- The *conservative* interpretation: As mentioned, our constructions are black-
  box; namely, both the construction of the cryptographic objects and the
  associated security proof utilize the underlying primitive—adaptive one-way
  permutations or adaptive PRGs—as a black-box, and in particular, do not
  refer to a specific implementation of these primitives. Thus, a conservative
  way to view our results is that to show even black-box lower-bounds and impossibility results for non-interactive concurrent non-malleable commitments
  and non-malleable zero-knowledge proofs, one first needs to to refute our assumptions. Analogously, it means that breaking our CCA-secure encryptions
  scheme, or proving a general parallel-repetition theorem for interactive arguments, first requires refuting our assumptions.

A cryptographer could choose to make "mild" assumptions such as $\mathcal{P} \neq \mathcal{NP}$,
"relatively mild" ones such as the existence of one-way functions, secure encryption schemes or trapdoor permutations, or "preposterous" ones such as "this
scheme is secure". Whereas preposterous assumptions clearly are undesirable,

mild assumptions are—given the state-of-the-art in complexity theory—too weak for cryptographic constructions of non-trivial tasks. Relatively mild assumptions, on the other hand, are sufficient for showing the feasibility of essentially all known cryptographic primitives.

Yet, to obtain efficient constructions, such assumptions are—given the current-state-of-art—not sufficient. In fact, it is a priori not even clear that although feasibility of a cryptographic task can be based on a relatively mild assumptions, that an "efficient" construction of the primitive is possible (at all!). One approach to overcome this gap is the random oracle paradigm, introduced in the current form by Bellare and Rogaway [7]: the proposed paradigm is to prove the security of a cryptographic scheme in the random-oracle model—where all parties have access to a truly random function—and next instantiate the random oracle with a concrete function "with appropriate properties". Nevertheless, as pointed out in [14] (see also [2, 21]) there are (pathological) schemes that can be proven secure in the random oracle model, but are rendered insecure when the random oracle is replaced by any concrete function (or family of functions).

In this work we, instead, investigate a different avenue for overcoming this gap between theory and practice, by introducing strong, but general, hardness assumption. When doing so, we, of course, need to be careful to make sure that our assumptions (although potentially "funky") are not preposterous. One criterion in determining the acceptability of a cryptographic assumption A is to consider (1) what the assumption is used for (for instance, to construct a primitive P, say) and (2) how much more "complex" the primitive P is, compared to A. For example, a construction of a pseudorandom generator assuming a one-way function is non-trivial, whereas the reverse direction is not nearly as interesting. Unfortunately, the notion of "complexity" of an assumption is hard to define. We here offer a simple interpretation: view complexity as "succinctness". General assumption are usually more succinct than specific assumptions, one-way functions are "easier" to define than, say, pseudorandom functions. Given this point of view, it seems that our assumptions are not significantly more complex than traditional hardness assumption; yet they allow us to construct considerably more complex objects (e.g., non-malleable zero-knowledge proofs).

On Falsifiability/Refutability of Our Assumptions. Note that the notions of non-malleable commitment and non-malleable zero-knowledge both are defined using simulation-based definitions. As such, simply assuming that a scheme is, say, non-malleable zero-knowledge, seems like a very strong assumption, which is hard to falsify[3]—in fact, to falsify it one needs to show (using a mathematical proof) that no Turning machine is a good simulator. In contrast, to falsify our assumptions it is sufficient to exhibit an attacker (just as with the traditional cryptographic hardness assumptions).

To make such "qualitative" differences more precise, Naor [27] introduced a framework for classifying assumptions, based on how "practically" an assumption can refuted. Whereas non-malleability, a priori, seems impossible to falsify

---

[3] Recall that falsifiability is Popper's classical criterion for distinguishing scientific and "pseudo-scientific" statements.

(as there a-priori is not a simple way to showing that no simulator exists). In contrast, traditional assumptions such as "factoring is hard" can be easily refuted simply by publishing challenges that a "falsifier" is required to solve. Our assumptions cannot be as easily refuted, as even if a falsifier exhibits an attack against a candidate adaptive OWF, it is unclear how to check that this attack works. However, the same can be said also for relatively mild (and commonly used) assumptions, such as "factoring is hard for subexponential-time".[4]

Additionally, we would like to argue that our assumptions enjoy a similar "win/win" situation as traditional cryptographic hardness assumptions. The adaptive security of the factoring or discrete logarithm problems seem like natural computational number theoretic questions. A refutation of our assumptions (and its implication to factoring and discrete logarithm problem) would thus be interesting in its own right. Taken to its extreme, this approach suggest that we might even consider assumptions that most probably are *false*, such as e.g., assuming that AES is an (adaptive one-way) *permutation*, as long as we believe that it might be hard to *prove* that the assumption is false.

2   New Assumptions and Definitions

The following sections introduce our definitions of adaptively secure objects—one-way functions, pseudorandom generators and commitment schemes—and posit candidate constructions for adaptively secure one-way functions and pseudorandom generators.

2.1   Adaptive One-Way Functions

In this paper, we define a *family* of adaptively secure injective one-way functions, where each function in the family is specified by an index $\mathsf{tag} \in \{0,1\}^n$. The adaptive security requirement says the following: consider an adversary that picks an index tag^* and is given $y^* = f_{\mathsf{tag}^*}(x^*)$ for a random x^* in the domain of f_{tag^*}, and the adversary is supposed to compute x^*. The adversary, in addition, has access to a "magic oracle" that on input (tag, y) where $\mathsf{tag} \neq \mathsf{tag}^*$, and get back $f_{\mathsf{tag}}^{-1}(y)$. In other words, the magic oracle helps invert all functions f_{tag} different from the "target function" f_{tag^*}. The security requirement is that the adversary have at most a negligible chance of computing x^*, even with this added ability. Note that the magic oracle is just a fictitious entity, which possibly does not have an efficient implementation (as opposed to the decryption oracle in the definition of CCA-security for encryption schemes which can be implemented efficiently given the secret-key). More formally,

---

[4] Note that the assumption that factoring is hard for subexponential-time can be falsified by considering a publishing a very "short" challenge (of length polylog$n$). However, in the same vein, our assumption can be falsified by considering challenges of length $\log n$; then it is easy to check if someone can exhibit an efficient attack on the adaptive security of an assumed one-way function, since the inverting oracle can also be efficiently implemented.

Definition 1 (Family of Adaptive One-to-one One-way Functions). *A family of injective one-way functions* $\mathcal{F} = \{f_{\mathsf{tag}} : D_{\mathsf{tag}} \mapsto \{0,1\}^*\}_{\mathsf{tag} \in \{0,1\}^n}$ *is called adaptively secure if,*

- (EASY TO SAMPLE AND COMPUTE.) *There is an efficient randomized* domain-sampler D, *which on input* $\mathsf{tag} \in \{0,1\}^n$, *outputs a random element in* D_{tag}. *There is a deterministic polynomial algorithm* M *such that for all* $\mathsf{tag} \in \{0,1\}^n$ *and for all* $x \in D_{\mathsf{tag}}$, $M(\mathsf{tag}, x) = f_{\mathsf{tag}}(x)$.
- (ADAPTIVE ONE-WAYNESS.) *Let* $\mathcal{O}(\mathsf{tag}, \cdot, \cdot)$ *denote an oracle that, on input* tag' *and* y *outputs* $f_{\mathsf{tag}'}^{-1}(y)$ *if* $\mathsf{tag}' \neq \mathsf{tag}$, $|\mathsf{tag}'| = |\mathsf{tag}|$ *and* $\bot$ *otherwise. The family* $\mathcal{F}$ *is adaptively secure if, for any probabilistic polynomial-time adversary* A, *there exists a negligible function* μ *such that for all* n, *and for all tags* $\mathsf{tag} \in \{0,1\}^n$,

$$\Pr[x \leftarrow D_{\mathsf{tag}} : A^{\mathcal{O}(\mathsf{tag}, \cdot, \cdot)}(\mathsf{tag}, f_{\mathsf{tag}}(x)) = x] \leq \mu(n)$$

where the probability is over the random choice of x *and the coin-tosses of* A.

A potentially incomparable assumption is that of an adaptively secure injective one-way function (as opposed to a family of functions); here the adversary gets access to an oracle that inverts the function on any y' that is different from the challenge y (that the adversay is supposed to invert). However, it is easy to see that an adaptively secure one-way function with subexponential security and a dense domain implies a family of adaptively secure one-way functions, as defined above. In fact, our construction of a family of adaptively secure one-way functions based on factoring goes through this construction.

Hardness Amplification. A strong adaptively secure one-way function is one where no adversary can invert the function with probability better than some negligible function in k (even with access to the inversion oracle). A weak one, on the other hand, only requires that the adversary not be able to invert the function with a probability better than $1 - 1/\mathrm{poly}(k)$ (even with access to the inversion oracle).

We remark that we can construct a collection of strong adaptively secure one-way function from a collection of weak adaptively secure one-way function. The construction is the same as Yao's hardness amplification lemma. We defer the details to the full version.

Candidates. We now present candidates for adaptively secure one-way functions, based on assumptions related to discrete-log and factoring.

Factoring. First, we show how to build an adaptively secure one-way function (not a family of functions) from the factoring assumption. Then, we show how to turn it into a family of functions, assuming, in addition, that factoring is subexponentially-hard.

The domain of the function f is $\{(p,q) \mid p, q \in \mathcal{P}_n, p < q\}$, where $\mathcal{P}_n$ is the set of all n-bit primes. Given this notation, $f(p,q)$ is defined to be pq. Assuming that it is hard to factor a number N that is a product of primes, even with access to an oracle that factors all other products of two primes, this function is adaptively secure.

We now show how to turn this into a family of adaptively secure one-way functions. The index is simply an $n' = n^{1/\epsilon}$-bit string (for some $\epsilon > 0$) $i = (i_1, i_2)$. The domain is the set of all strings (j_1, j_2) such that $p = i_1 \circ j_1$ and $q = i_2 \circ j_2$ are both n-bit primes. The function then outputs pq. Since we reveal the first $n' = n^{1/\epsilon}$ bits of the factors of $N = pq$, we need to assume that factoring is subexponentially hard (even with access to an oracle that factors other products of two primes). The function is clearly injective since factoring forms an injective function. In the full version, we additionally provide candidates for adaptive one-way functions based on the RSA and Rabin functions.

Discrete Logarithms. The family of adaptive OWFs $\mathcal{F}_{DL}$ is defined as follows: The domain of the function is a tuple (p, g, x) such that p is a $2n$-bit prime p whose first n bits equal the index i, g is a generator for $\mathbb{Z}_p^*$ and x is a $2n - 1$-bit number. The domain is easy to sample–the sampler picks a "long-enough" random string r and a $2n - 1$-bit number x. The function f_i uses r to sample a $2n$-bit prime p whose first n bits equal i (this can be done by repeated sampling, and runs in polynomial time assuming a uniformness conjecture on the density of primes in large intervals) and a generator $g \in \mathbb{Z}_p^*$. The output of the function on input (p, g, x) is $(p, g, g^x \bmod p)$. f_i is injective since the output determines p and g; given p and g, $g^x \bmod p$ next determines x uniquely since $x < 2^{2n-1}$ and p, being a $2n$-bit prime, is larger than 2^{2n-1}.

We also mention that the adaptive security of this family can be based on the subexponential adaptive security of the one-way function (as opposed to family) obtained by simply sampling random p, g, x (or even random p being a safe prime) and outputting p, g, g^x. (In the full version of the paper, we additionally show how to obtain our results under a different variant of *polynomial-time* adaptive hardness of the above one-way function; roughly speaking, the variant we require here is that the adversary gets access to an oracle that inverts the function on any input length.)

2.2 Adaptive Pseudorandom Generator

A family of adaptively secure pseudorandom generators $\mathcal{G} = \{G_{\mathsf{tag}}\}_{\mathsf{tag} \in \{0,1\}^*}$ is defined in a similar way to an adaptive one-way function. We require that the output of the generator G, on a random input x and an adversarially chosen tag be indistinguishable from uniform, even for an adversary that can query a magic oracle with a value (tag', y) (where $\mathsf{tag}' \neq \mathsf{tag}$) and get back 0 or 1 depending on whether y is in the range of $G_{\mathsf{tag}'}$ or not.

Definition 2 (Adaptive PRG). *A family of functions* $\mathcal{G} = \{G_{\mathsf{tag}} : \{0,1\}^n \mapsto \{0,1\}^{s(n)}\}_{\mathsf{tag} \in \{0,1\}^n}$ *is an adaptively secure pseudorandom generator (PRG) if* $|G_{\mathsf{tag}}(x)| = s(|x|)$ *for some function* s *such that* $s(n) \geq n$ *for all* n *and,*

- *(EFFICIENT COMPUTABILITY.) There is a deterministic polynomial-time algorithm* M_G *such that* $M_G(x, \mathsf{tag}) = G_{\mathsf{tag}}(x)$.

– (ADAPTIVE PSEUDORANDOMNESS.) *Let $\mathcal{O}(\mathsf{tag}, \cdot, \cdot)$ denote an oracle that, on input (tag', y) such that $\mathsf{tag}' \neq \mathsf{tag}$, $|\mathsf{tag}'| = |\mathsf{tag}|$, outputs 1 if y is in the range of $G_{\mathsf{tag}'}$ and 0 otherwise.*
The PRG G is adaptively secure if, for any probabilistic polynomial-time adversary A, there exists a negligible function μ such that for all n and for all tags $\mathsf{tag} \in \{0,1\}^n$,

$$\left| \Pr[y \leftarrow G_{\mathsf{tag}}(U_n) : A^{\mathcal{O}(\mathsf{tag}, \cdot, \cdot)}(y) = 1] - \Pr[y \leftarrow U_m : A^{\mathcal{O}(\mathsf{tag}, \cdot, \cdot)}(y) = 1] \right| \leq \mu(n)$$

where the probability is over the random choice of y and the coin-tosses of A.

Candidates. For the case of adaptive PRGs, we provide a candidate construction based on the advanced encryption standard (AES). AES is a permutation on 128 bits; that is, for a 128-bit seed s, AES_s is a permutation defined on $\{0,1\}^{128}$. However, due to the algebraic nature of the construction of AES, it can easily be generalized to longer input length. Let AES_n denote this generalized version of AES to n-bit inputs. Our candidate adaptive pseudorandom generator AESG_{tag} is simply $\mathsf{AESG}_{tag}(s) = \mathsf{AES}_s(tag \circ 0) \circ \mathsf{AES}_s(tag \circ 1)$.

2.3   Adaptively Secure Commitment Schemes

In this subsection, we define adaptively secure commitment schemes. Let $\{\mathrm{COM}_{\mathsf{tag}} = \langle S_{\mathsf{tag}}, R_{\mathsf{tag}} \rangle\}_{\mathsf{tag} \in \{0,1\}^*}$ denote a family of commitment protocols, indexed by a string tag. We require that the commitment scheme be secure, even against an adversary that can query a magic oracle on the transcript of a commitment interaction and get back a message that was committed to in the transcript. More precisely, the adversary picks an index tag and two equal-length strings x_0 and x_1 and gets a value $y_b = \mathrm{COM}_{\mathsf{tag}}(x_b; r)$, where b is a random bit and r is random. The adversary can, in addition, query a magic oracle on (y', tag') where $\mathsf{tag}' \neq \mathsf{tag}$ and get back the some x' such that $y' \in \mathrm{COM}_{\mathsf{tag}'}(x'; r')$ (if y' is a legal commitment) and $\perp$ otherwise. [5] The security requirement is that the adversary cannot distinguish whether y_b was a commitment to x_0 or x_1, even with this extra power.

Definition 3 (Adaptively-Secure Commitment). *A family of functions $\{\mathrm{COM}_{\mathsf{tag}}\}_{\mathsf{tag} \in \{0,1\}^*}$ is called an adaptively secure commitment scheme if S_{tag} and R_{tag} are polynomial-time and*

– STATISTICAL BINDING: *For any tag, over the coin-tosses of the receiver R, the probability that a transcript $\langle S^*, R_{\mathsf{tag}} \rangle$ has two valid openings is negligible.*
– ADAPTIVE SECURITY: *Let $\mathcal{O}(\mathsf{tag}, \cdot, \cdot)$ denote the oracle that, on input $\mathsf{tag}' \neq \mathsf{tag}$, $|\mathsf{tag}'| = |\mathsf{tag}|$ and c, returns an $x \in \{0,1\}^{\ell(n)}$ if there exists strings r_S and r_R, such that c is the transcript of the interaction between S with input x and random coins r_S and R with random coins r_R, and $\perp$ otherwise.*

---

[5] In case the transcript corresponds to the commitment of multiple messages, the oracle returns a canonical one of them. In fact, one of our commitment schemes is perfectly binding and thus, does not encounter this problem.

For any probabilistic polynomial-time oracle TM A, there exists a negligible function $\mu(\cdot)$ such that for all n, for all $\mathsf{tag} \in \{0,1\}^n$ and for all $x, y \in \{0,1\}^{\ell(n)}$,

$$\big| \Pr[c \leftarrow \langle S_{\mathsf{tag}}(x), R_{\mathsf{tag}} \rangle; A^{\mathcal{O}(\mathsf{tag}, \cdot, \cdot)}(c, \mathsf{tag}) = 1] -$$
$$\Pr[c \leftarrow \langle S_{\mathsf{tag}}(y), R_{\mathsf{tag}} \rangle; A^{\mathcal{O}(\mathsf{tag}, \cdot)}(c, \mathsf{tag}) = 1] \big| \leq \mu(n).$$

3   Non-malleable Commitment Schemes

In this section, we construct non-malleable string-commitment schemes. We first construct adaptively-secure bit-commitment schemes based on an adaptively secure injective OWF and an adaptively secure PRG—the first of these constructions is non-interactive and the second is a 2-round commitment scheme. We then show a simple "concatenation lemma", that constructs an adaptively secure string commitment scheme from an adaptively-secure bit-commitment scheme. Finally, we show that an adaptively secure commitment scheme are also concurrently non-malleable. The complete proofs are deferred to the full version.

Lemma 1. *Assume that there exists a family of adaptively secure injective one-way functions. Then, there exists an adaptively secure bit-commitment scheme. Furthermore, the commitment scheme is non-interactive.*

Further, assuming the existence of a family of adaptively secure pseudorandom generators, there exists a 2-round adaptively secure bit-commitment scheme.

The first of these constructions follows by replacing the injective one-way function in the Blum-Micali [9] commitment scheme, with an adaptively secure one, and the second follows from the Naor commitment scheme [25] in an analogous way.

Lemma 2 (Concatenation Lemma). *If there is an adaptively secure family of bit-commitment schemes, then there is an adaptively secure family of string-commitment schemes.*

The concatenation lemma follows by simply committing to each bit of the message independently using a single-bit commitment scheme $\mathrm{COM}_{\mathsf{tag}}$.

Finally, in the full version we show that any adaptively secure commitment scheme is concurrenly non-malleable according to the definition of [23]. The proof is essentially identical to the proof of [17] that any CCA-secure encryption scheme is also non-malleable.

Lemma 3. *If* $\{\mathrm{COM}_{\mathsf{tag}}\}_{\mathsf{tag} \in \{0,1\}^n}$ *is a tag-based adaptively secure commitment scheme, then it is also concurrently non-malleable.*

4   Four-Round Non-malleable Zero-Knowledge

In this section, we present a 4-round non-malleable zero-knowledge argument system. We start by reviewing the notion of non-malleable zero-knowledge [17] and refer the reader to [29] for a formal definition of the notion we consider in this work.

Non-malleable $\mathcal{ZK}$ proofs: An informal definition. Let Π_{tag} be a tag-based family of $\mathcal{ZK}$ proofs. Consider a man-in-the-middle adversary that participates in two interactions: in the left interaction the adversary A is verifying the validity of a statement x by interacting with an honest prover P using tag. In the right interaction A proves the validity of a statement x' to the honest verifier V using $\mathsf{tag}' \neq \mathsf{tag}$. The objective of the adversary is to convince the verifier in the right interaction. Π_{tag} is, roughly speaking, non-malleable, if for any man-in-the-middle adversary A, there exists a stand-alone prover S that manages to convince the verifier with essentially the same probability as A (without receiving a proof on the left).

Our protocol. The argument system is the Feige-Shamir protocol [18], compiled with an adaptively secure commitment scheme. In our analysis we rely on the following properties of the Feige-Shamir protocol:

- The first prover message is (perfectly) independent of the witness used by the prover (and even the statement). This property has previously been used to simplify analysis, but here we inherently rely on this property to *enable* our analysis.
- Given a random accepting transcript, and the *openings* of the commitments in the first message, it is possible to "extract a witness". In other words, any transcript implicitly defines a witness; additionally, given a random transcript, this witness will be valid with a high probability (if the transcript is accepting).

In what follows, we present a sketch of the protocol and the proof. The complete proof is deferred to the full version.

4.1   An Adaptively Secure WI Proof of Knowledge

The main component in the NMZK protocol is a three-round witness-indistinguishable (WI) proof of knowledge (POK); see [19] for a definition of witness indistinguishability and proof of knowledge. The protocol is simply a parallelization of the 3-round $\mathcal{ZK}$ proof $\tilde{\Pi}$ for the $\mathcal{NP}$-complete language of Hamiltonicity [8, 18], with the only change that the commitment scheme used in the proof is adaptively secure. Let Π_{tag} denote this family of protocols; it is a family which is parameterized by the tag of the adaptively secure commitment.
    We show that this family of protocols satisfy two properties:

- it has an "adaptive WI" property which, roughly stated, means that the transcripts of the protocol when the prover uses two different witnesses w_1 and w_2 are computationally indistinguishable, even if the distinguisher has access to a magic oracle that inverts all commitments $\mathrm{COM}_{\mathsf{tag}'}$, where $\mathsf{tag}' \neq \mathsf{tag}$.
- a random transcript of $\tilde{\Pi}_{\mathsf{tag}}$ uniquely defines a witness (even though not it is not computable in polynomial-time). We define this to be the *witness implicit in the transcript* in an instance of Π_{tag}. Furthermore, we show that the implicit witness in Π_{tag} is computable given access to $\mathcal{O}(\mathsf{tag}', \cdot, \cdot)$ for any $\mathsf{tag}' \neq \mathsf{tag}$.

4.2   The Non-malleable Zero-Knowledge Argument System

The non-malleable ZK protocol consists of two instances of the protocol Π_{tag} running in conjunction, one of them initiated by the verifier and the other initiated by the prover. We will denote the copy of Π_{tag} initiated by the verifier as Π_{tag}^V and the one initiated by the prover as Π_{tag}^P.

Recall that Π_{tag} is a parallelized version of a 3-round protocol $\tilde{\Pi}_{\mathsf{tag}}$; let A_i, C_i and Z_i denote the messages in the i'th repetion in these three rounds. In the description of the protocol, we let messages in the protocol Π_{tag}^V (resp. Π_{tag}^P) appear with a superscript of V (resp. P).

Theorem 4. *Assume that* COM *is a non-interactive adaptively secure commitment scheme. Then, the protocol in Figure 1 is a 4-round non-malleable zero-knowledge argument system.*

Proof (Sketch). Completeness, soundness and zero-knowledge properties of the protocol follow directly from the corresponding properties of the Feige-Shamir protocol. In Lemma 4, we show that the protocol non-malleable.

In other words, for every man-in-the-middle adversary A that interacts with the prover P_{tag} on a statement x and convinces the verifier $V_{\mathsf{tag}'}$ (for a $\mathsf{tag}' \neq \mathsf{tag}$) in a right-interaction on a statement x' (possibly the same as x), we construct a stand-alone prover that convinces the verifier on x' with the same probability as A, but *without access to the left-interaction*. The construction of the stand-alone prover in the proof of non-malleability (see Lemma 4) relies on the adaptive security of the commitment scheme $\mathrm{COM}_{\mathsf{tag}}$. It is important to note that the stand-alone prover itself runs in classical polynomial-time, and in particular does not use any oracles. Access to the commitment-inversion oracle is used only to show that the stand-alone prover works as expected (and in particular, that it convinces the verifier with the same probability as does the MIM adversary).

Lemma 4. *The protocol* $\mathrm{NM}_{\mathsf{tag}}$ *in Figure 1 is non-malleable.*

Proof (Sketch). For every man-in-the-middle adversary A, we construct a stand-alone prover S: the construction of the stand-alone prover S proceeds in three steps.

1. Run the adversary A with "honestly generated" verifier-messages on the right interaction, and extract the witness for the WIPOK Π_{tag}^V that the adversary initiates on the left interaction.
2. Use the witness thus obtained to simulate the left-interaction of the adversary A and rewind the WI proof of knowledge $\Pi_{\mathsf{tag}'}^P$ it initiates on the right interaction to extract the witness w' for the statement x'.
3. Finally provide an honest proof to the outside verifier of the statement x' using the tag tag' and witness w'.

Carrying out this agenda involves a number of difficulties. We first describe how to accomplish Step 1. This is done by invoking the simulator for the Feige-Shamir protocol, and is described below. Informally, S extracts the witness w' that the

---

Non-Malleable Zero-Knowledge Argument NM_{tag}

COMMON INPUT: An instance $x \in \{0,1\}^n$, presumably in the language L.

PROVER INPUT: A witness w such that $(x, w) \in R_L$.

ROUND 1: **(Verifier)** Pick w_1 and w_2 at random and compute $x_i = f(w_i)$ for $i \in \{1, 2\}$.

Let the $\mathcal{NP}$-relation $R_V = \{((x_1, x_2), w') \mid$ either $f(w') = x_1$ or $f(w') = x_2\}$.
Initiate the WI protocol Π_{tag}^V with the statement $(x_1, x_2) \in L_V$. In particular,
$\mathbf{V} \rightarrow \mathbf{P}$: Send (x_1, x_2) to P. Send $A_1^V, A_2^V, \ldots, A_n^V$ to P.

ROUND 2: **(Prover)** Let the $\mathcal{NP}$-relation R_P be

$$\{((x, x_1, x_2), w) \mid \text{ either } (x, w) \in R_L \text{ or } f(w) = x_1 \text{ or } f(w) = x_2\}$$

Initiate a WI protocol Π_{tag}^P with common input (x, x_1, x_2). Also, send the second-round messages of the protocol Π_{tag}^V. In particular,
(2a) $\mathbf{P} \rightarrow \mathbf{V}$: Send $A_1^P, A_2^P, \ldots, A_n^P$ to V.
(2b) $\mathbf{P} \rightarrow \mathbf{V}$: Send $C_1^V, C_2^V, \ldots, C_n^V$ to V.

ROUND 3: **(Verifier)** Send round-2 challenges of the protocol Π_{tag}^P and round-3 responses of Π_{tag}^V.
(3a) $\mathbf{V} \rightarrow \mathbf{P}$: Send $C_1^P, \ldots, C_n^P$ to P.
(3b) $\mathbf{V} \rightarrow \mathbf{P}$: Send $Z_1^V, \ldots, Z_n^V$ to P.

ROUND 4: **(Prover)** P verifies that the transcript $\{(A_i^V, C_i^V, Z_i^V)\}_{i \in [n]}$ is accepting for the subprotocol Π_{tag}^V. If not, abort and send nothing to V. Else,
$\mathbf{P} \rightarrow \mathbf{V}$: Send $Z_1^P, \ldots, Z_n^P$ to V.
V accepts iff the transcript $\{(A_i^P, C_i^P, Z_i^P)\}_{i \in [n]}$ is accepting for the subprotocol Π_{tag}^P.

Fig. 1. NON-MALLEABLE ZERO-KNOWLEDGE PROTOCOL NM_{tag} FOR A LANGUAGE L

MIM A uses in the subprotocol Π_{tag}^V in the left-interaction. Then, S acts as the honest prover using the witness w' in the protocol Π_{tag}^P.

We now describe how to carry out Step 2 of the agenda, and show that at the end of Step 2, S extracts a witness for the statement x' that the MIM adversary A uses in the right-interaction with essentially the same probability that A convinces the verifier on the right-interaction. S starts by running the protocol in the left-interaction using the witness w' it extracted using the strategy in Step 1. Consider the moment when A outputs the first message on the left (that is, the first message in the subprotocol Π_{tag}^V). Consider two cases.

Case One: In the first case, A has not yet received the round-3 messages in the right interaction (that is, the challenges in the subprotocol $\Pi_{tag'}^P$) (See Figure 2(i)). In this case, the Round-1 message that A sends on the left interaction is independent of the Round-3 message in the right interaction. Now, S proceeds as follows: S runs the left-interaction as a normal prover P_{tag} would with the fake-witness w', and rewinds the protocol $\Pi_{tag'}^P$ on the right-interaction to extract a witness for the statement x'. Since the rewinding process does not change the messages in the right-interaction before round 3, S can use w' to

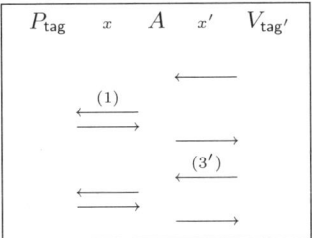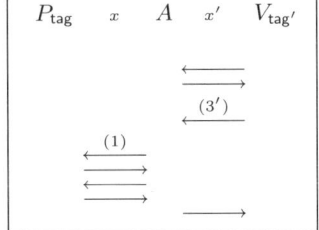

Fig. 2. Two scheduling strategies (i) on the left and (ii) on the right

produce the left-interaction just as an honest prover with witness w' would; note that we here rely on the property of the Feige-Shamir protocol that the first message sent by the prover (i.e., round 2) is independent of the statement and witness used.

Case Two: In the second case, A has already received the challenges in the subprotocol $\varPi^P_{\mathsf{tag}'}$ in the right interaction (See Figure 2(ii)). In this case, trying to rewind in the WIPOK $\varPi^P_{\mathsf{tag}'}$ on the right is problematic, since A could change the first message on the left, every time it is fed with a different challenge in round-3 on the right-interaction. In this case, S proceeds as follows: Every time the extractor for the WIPOK $\varPi^P_{\mathsf{tag}'}$ in the right-interaction rewinds, S repeats the entire procedure in Step 1 of the agenda to extract a witness w' corresponding to the (potentially new) Round-1 message in the left interaction. S then simulates the left-interaction with the witness thus extracted. Note that due to the particular scheduling, the extraction procedure on the right-interaction is unaffected by the rewinding on the left.

To analyze the correctness of the above simulator, we first show that the view generated by S following Step 1 of the agenda is indistinguishable from the view of A in a real interaction, even to a distinguisher that has access to the oracle $\mathcal{O}(\mathsf{tag}, \cdot, \cdot)$ that inverts $\mathrm{COM}_{\mathsf{tag}'}$ for any $\mathsf{tag}' \neq \mathsf{tag}$. Then, we use this to show that the *implicit witness* in the transcript of the subprotocol $\varPi^P_{\mathsf{tag}'}$ in the right-interaction is indistinguishable between the simulated and the real execution. This means that the witness that S extracts from the right interaction of A is computationally indistinguishable from the witness that A uses in the real interaction. We defer an analysis of the running-time to the full version; intuitively it follows that the running-time in expectation is polynomial since when performing rewinding on the right, we can *perfectly* emulate the messages on the left with the same distribution as when generating the initial view in Stage 1.

5   CCA2-Secure Encryption Scheme

Bellare and Rogaway [7] showed how to construct an efficient encryption scheme that is CCA2-secure in the random oracle model, starting from any trapdoor permutation. We show that the same scheme is CCA2-secure in the standard

model (that is, without assuming random oracles) by instantiation their scheme
with adaptively secure primitives.

To prove security of the construction, we assume an adaptively secure variant
of perfectly one-way hash functions (defined by Canetti [13]), and a family of
trapdoor permutations that is hard to invert even with access to an oracle that
inverts the perfectly one-way hash function. We note that Canetti [13] (define
and) use perfectly one-way hashing with auxiliary input to prove *IND-CPA*
security (semantic security) of the [7] construction.

We sketch the notion of adaptively secure perfectly one-way hashing w.r.t
auxiliary information and give a high-level intuition of the security proof; the
complete definition and proof is deferred to the full version. Consider a family
of functions $\mathcal{H}$ such that for a random function $H \leftarrow \mathcal{H}$, it is computationally
infeasible to distinguish between $h \leftarrow H(r; s)$ (for random r, s) and a random
value, even if the adversary is given (1) $g(r)$, where g is an uninvertible function
evaluated on the input r, and (2) access to an oracle that inverts every $h' \neq h$
(namely, the oracle, given any $h' \neq h$, computes (r', s') such that $h' = H(r'; s')$).

Theorem 5. *Let* TDPGen *be a family of trapdoor permutations that are uninvertible with access to the $\mathcal{H}$-inverting oracle, and let $\mathcal{H}$ be an adaptively secure
perfectly one-way hash family with auxiliary information. Then, the scheme in
Figure 3 is an IND-CCA2-secure encryption scheme.*

Proof (Idea). The proof is analogous to that for the two-key paradigm of Naor
and Yung [26]. The main idea of the proof is that there are two ways to decrypt a
ciphertext – the first is using the trapdoor f^{-1} (as the legal decryption algorithm
Dec does), and the second is using an oracle that inverts H. Given a ciphertext
$c = (c_0, c_1, c_2, s_1, s_2)$ and access to such an oracle, we first compute r' such that
$H((r', s_1, c_1); s_2) = c_2$, and check that $c_0 = f(r')$. If the check passes, output
$m = c_1 \oplus H(r'; s_1)$, otherwise output $\perp$. This allows the simulator to answer
the decryption queries of the adversary, given access to an oracle that inverts
H. Even with access to such an oracle, the adversary can neither (1) invert $f(r)$

Gen(1^n) : Run TDPGen(1^n) and get a pair (f, f^{-1}). Run PHGen(1^k) to get a
  perfectly one-way hash function H. Let PK $= (f, H)$ and SK $= f^{-1}$.
Enc(PK, m) :
  1. Pick random $r \leftarrow \{0, 1\}^n$. Compute $c_0 = f(r)$ and $c_1 = m \oplus H(r; s_1)$ for
     random s_1.
  2. Let $c' = (r, s_1, c_1)$. Compute $c_2 = H(c'; s_2)$ for random s_2.
  Output the ciphertext $\mathbf{c} = (c_0, c_1, c_2, s_1, s_2)$.
Dec(SK, c) : Parse $\mathbf{c}$ as $(c_0, c_1, c_2, s_1, s_2)$.
  1. Compute $r' = f^{-1}(c_0)$, and $m' = c_1 \oplus H(r'; s_1)$.
  2. Let $c' = (r', s_1, c_1)$. Output m' if $H(c'; s_2) = c_2$. Otherwise output $\perp$.

Fig. 3. AN IND-CCA2-SECURE ENCRYPTION SCHEME

on a new value r (since f is uninvertible even with access to the H-inverting oracle), nor (2) distinguish $H(r; \cdot)$ from random (since $\mathcal{H}$ is an adaptively secure perfectly one-way hash family). Thus, even with access to the decryption oracle, the scheme is semantically secure; that is to say that the scheme itself is IND-CCA2-secure.

Acknowledgements

We are very grateful to Yuval Ishai for illuminating discussions. First author thanks Vipul Goyal, Ivan Visconti, and Darrel Carbajal for their ideas/comments.

References

1. Barak, B.: Constant-round coin-tossing with a man in the middle or realizing the shared random string model. In: FOCS, pp. 345–355 (2002)
2. Bellare, M., Boldyreva, A., Palacio, A.: An uninstantiable random-oracle-model scheme for a hybrid-encryption problem. In: Cachin, C., Camenisch, J.L. (eds.) EUROCRYPT 2004. LNCS, vol. 3027, pp. 171–188. Springer, Heidelberg (2004)
3. Bellare, M., Impagliazzo, R., Naor, M.: Does parallel repetition lower the error in computationally sound protocols? In: FOCS, pp. 374–383 (1997)
4. Bellare, M., Namprempre, C., Pointcheval, D., Semanko, M.: The power of rsa inversion oracles and the security of chaum's rsa-based blind signature scheme. In: Syverson, P.F. (ed.) FC 2001. LNCS, vol. 2339, pp. 319–338. Springer, Heidelberg (2002)
5. Bellare, M., Neven, G.: Transitive signatures based on factoring and rsa. In: Zheng, Y. (ed.) ASIACRYPT 2002. LNCS, vol. 2501, pp. 397–414. Springer, Heidelberg (2002)
6. Bellare, M., Palacio, A.: Gq and schnorr identification schemes: Proofs of security against impersonation under active and concurrent attacks. In: Yung, M. (ed.) CRYPTO 2002. LNCS, vol. 2442, pp. 162–177. Springer, Heidelberg (2002)
7. Bellare, M., Rogaway, P.: Random oracles are practical: A paradigm for designing efficient protocols. In: First ACM Conference on Computer and Communications Security, Fairfax, pp. 62–73. ACM, New York (1993)
8. Blum, M.: How to prove a theorem so no one can claim it. In: Proc. of The International Congress of Mathematicians, pp. 1444–1451 (1986)
9. Blum, M., Micali, S.: How to generate cryptographically strong sequences of pseudo random bits. In: FOCS, pp. 112–117 (1982)
10. Boldyreva, A., Fischlin, M.: On the security of oaep. In: Lai, X., Chen, K. (eds.) ASIACRYPT 2006. LNCS, vol. 4284, pp. 210–225. Springer, Heidelberg (2006)
11. Boneh, D.: The decision diffie-hellman problem. In: Buhler, J.P. (ed.) ANTS 1998. LNCS, vol. 1423, pp. 48–63. Springer, Heidelberg (1998)
12. Boneh, D., Franklin, M.K.: Identity-based encryption from the weil pairing. In: Kilian, J. (ed.) CRYPTO 2001. LNCS, vol. 2139, pp. 213–229. Springer, Heidelberg (2001)
13. Canetti, R.: Towards realizing random oracles: Hash functions that hide all partial information. In: Kaliski Jr., B.S. (ed.) CRYPTO 1997. LNCS, vol. 1294, pp. 455–469. Springer, Heidelberg (1997)

14. Canetti, R., Goldreich, O., Halevi, S.: The random oracle methodology, revisited. J. ACM 51(4), 557–594 (2004)
15. Cramer, R., Shoup, V.: A practical public key cryptosystem provably secure against adaptive chosen ciphertext attack. In: Krawczyk, H. (ed.) CRYPTO 1998. LNCS, vol. 1462, pp. 13–25. Springer, Heidelberg (1998)
16. Cramer, R., Shoup, V.: Universal hash proofs and a paradigm for adaptive chosen ciphertext secure public-key encryption. In: Knudsen, L.R. (ed.) EUROCRYPT 2002. LNCS, vol. 2332, pp. 45–64. Springer, Heidelberg (2002)
17. Dolev, D., Dwork, C., Naor, M.: Nonmalleable cryptography. SIAM J. Comput. 30(2), 391–437 (2000)
18. Feige, U., Shamir, A.: Witness indistinguishable and witness hiding protocols. In: STOC, pp. 416–426 (1990)
19. Goldreich, O.: Foundations of Cryptography: Basic Tools. Cambridge University Press, Cambridge (2001), http://www.wisdom.weizmann.ac.il/~oded/frag.html
20. Goldreich, O., Krawczyk, H.: On the composition of zero-knowledge proof systems. SIAM J. Comput. 25(1), 169–192 (1996)
21. Goldwasser, S., Kalai, Y.T.: On the (in)security of the fiat-shamir paradigm. In: FOCS, p. 102–152 (2003)
22. Katz, J., Wee, H.: Black-box lower bounds for non-malleable protocols (2007)
23. Lin, H., Pass, R., Venkitasubramaniam, M.: Concurrent non-malleable commitments from any one-way function. In: Canetti, R. (ed.) TCC 2008. LNCS, vol. 4948, pp. 571–588. Springer, Heidelberg (2008)
24. Malkin, T., Moriarty, R., Yakovenko, N.: Generalized environmental security from number theoretic assumptions. In: Halevi, S., Rabin, T. (eds.) TCC 2006. LNCS, vol. 3876, pp. 343–359. Springer, Heidelberg (2006)
25. Naor, M.: Bit commitment using pseudorandomness. J. of Cryptology 4 (1991)
26. Naor, M., Yung, M.: Public-key cryptosystems provably secure against chosen ciphertext attacks. In: STOC 1990: Proceedings of the twenty-second annual ACM symposium on Theory of computing, pp. 427–437. ACM Press, New York (1990)
27. Naor, M.: On cryptographic assumptions and challenges. In: Boneh, D. (ed.) CRYPTO 2003. LNCS, vol. 2729, pp. 96–109. Springer, Heidelberg (2003)
28. Pass, R., Rosen, A.: Concurrent non-malleable commitments. In: FOCS, pp. 563–572 (2005)
29. Pass, R., Rosen, A.: New and improved constructions of non-malleable cryptographic protocols. In: STOC, pp. 533–542 (2005)
30. Pietrzak, K., Wikström, D.: Parallel repetition of computationally sound protocols revisited. In: Vadhan, S.P. (ed.) TCC 2007. LNCS, vol. 4392, pp. 86–102. Springer, Heidelberg (2007)
31. Prabhakaran, M., Sahai, A.: New notions of security: achieving universal composability without trusted setup. In: STOC, pp. 242–251 (2004)
32. Rackoff, C., Simon, D.R.: Cryptographic defense against traffic analysis. In: STOC 1993: Proceedings of the twenty-fifth annual ACM symposium on Theory of computing, pp. 672–681. ACM Press, New York (1993)
33. Sahai, A.: Non-malleable non-interactive zero knowledge and adaptive chosen-ciphertext security. In: FOCS, pp. 543–553 (1999)

Bits Security of the Elliptic Curve
Diffie–Hellman Secret Keys

Dimitar Jetchev[1] and Ramarathnam Venkatesan[2,3]

[1] Dept. of Mathematics, University of California at Berkeley, Berkeley, CA 94720
jetchev@math.berkeley.edu
[2] Microsoft Research, One Microsoft Way, Redmond WA 98052
[3] Microsoft Research India Private Limited, "Scientia", No:196/36,
2nd Main Road, Sadashivnagar, Bangalore – 560080, India
venkie@microsoft.com

Abstract. We show that the least significant bits (LSB) of the elliptic curve Diffie–Hellman secret keys are hardcore. More precisely, we prove that if one can efficiently predict the LSB with non-negligible advantage on a polynomial fraction of all the curves defined over a given finite field $\mathbb{F}_p$, then with polynomial factor overhead, one can compute the entire Diffie–Hellman secret on a polynomial fraction of all the curves over the same finite field. Our approach is based on random self-reducibility (assuming GRH) of the Diffie–Hellman problem among elliptic curves of the same order. As a part of the argument, we prove a refinement of H. W. Lenstra's lower bounds on the sizes of the isogeny classes of elliptic curves, which may be of independent interest.

1 Introduction

The Diffie–Hellman protocol for key exchange [16] is based on the hardness of computing the function $\mathrm{DH}_g(g^u, g^v) = g^{uv}$, where g is a fixed generator of the multiplicative group of a finite field $\mathbb{F}_p$, and $1 \leq u, v \leq p-1$ are integers. A natural question is whether one can compute some of the bits of g^{uv} given g, g^u, g^v. It is unknown if predicting partial information with significant advantage over a random guess will lead to a compromise of the Diffie–Hellman function. Boneh and Venkatesan [2], [25] have shown that if one is able to compute (in time polynomial in $\log p$) the $5\sqrt{\log p}$ most significant bits of g^{uv} for every input (g^u, g^v) then one can compute (in polynomial time) the entire shared secret key g^{uv}. For motivation, note that g^{uv} may be 1024 bits long, but one may want to use the least significant 128 bits of g^{uv} as a block cipher key. Thus, it is important to know that partial information is not computable or predictable with any significant advantage over a random guess. Another motivation stems from the fact that the methods used in [2] suggest attacks on cryptographic systems that reveal some information about g^{uv} to the attacker [8], [10], [18], [19], [20], [24], [26].

D. Wagner (Ed.): CRYPTO 2008, LNCS 5157, pp. 75–92, 2008.
© International Association for Cryptologic Research 2008

The analogous problem for elliptic curves studies the bit security of the following function:

Diffie–Hellman function: Let E be an elliptic curve over $\mathbb{F}_p$ and let $P \in E$ be a point of prime order q. We define the Diffie–Hellman function as

$$\mathrm{DH}_{E,P}(uP, vP) = uvP,$$

where $1 \leq u, v \leq q$ are integers. Moreover, we refer to the triple (P, uP, vP) as a *Diffie–Hellman triple* for E.

For simplicity, we restrict ourselves to *short Weierstrass equations (models)* of E, i.e., models of the form $y^2 = x^3 + ax + b$ with $a, b \in \mathbb{F}_p$ and $4a^3 + 27b^2 \neq 0$. By abuse of common terminology, an *elliptic curve* for us will be an $\mathbb{F}_p$-isomorphism class of short Weierstrass equations. It is not hard to see that the complexity of the Diffie–Hellman function is independent of the choice of the short Weierstrass equation for the elliptic curve E over $\mathbb{F}_p$. Indeed, given two different models W and W' for E over $\mathbb{F}_p$ and an explicit isomorphism $\varphi : W \to W'$ and its inverse $\varphi^{-1} : W' \to W$, a Diffie–Hellman triple (P, uP, vP) on W is mapped to a Diffie–Hellman triple $(\varphi(P), u\varphi(P), v\varphi(P))$ on W' and therefore, if one can compute $uv\varphi(P)$, one would know uvP. Yet, if one wants to formalize the notion of security of single bits of the Diffie–Hellman function, one needs to choose a short Weierstrass model (it is not necessarily true any more that if one knows one bit of the Diffie–Hellman secret $uv\varphi(P)$ on W' then one can compute the corresponding bit of uvP on W).

Boneh and Shparlinski [1] have reduced (in time polynomial in $\log p$) the Diffie–Hellman problem on an elliptic curve E to the problem of predicting the LSB of the secret key uvP with non-negligible advantage over a random guess on a polynomial fraction of all short Weierstrass models for E. Alternatively, if one looks for a polynomial time reduction of the Diffie–Hellman problem to the problem of predicting partial information on the *same* short Weierstrass model W, some results have been established using Gröbner bases [12].

A more general and natural situation would be to consider an oracle $\mathcal{A}$ that predicts the least significant bit of the Diffie–Hellman secret key for short Weierstrass models W chosen from a non-negligible subset G (i.e., from a $(\log p)^{O(1)}$-fraction) of all the short Weierstrass equations over $\mathbb{F}_p$ and arbitrary Diffie–Hellman triples on these models. Here, one encounters extra challenges. First, the set G may be distributed arbitrarily over all (exponentially many in $\log p$) isogeny classes of short Weierstrass models, where each isogeny class contains exponentially many isomorphism classes of short Weierstrass models, with each isomorphism class containing exponentially many short Weierstrass models. Second, relating the difficulty of computing the Diffie–Hellman function within each isogeny class is itself a nontrivial task: having an explicit (computable in time polynomial in $\log p$) isogeny from an elliptic curve E to another curve E' in the same class would achieve this task. By Tate's isogeny theorem [28], such a map exists if and only if E and E' have the same number of points (E and E' are said to be isogenous). Yet, such an isogeny can have large degree and it can take superpolynomial number of steps to compute it. Typically, isogeny computations are used in attacks such as the Weil descent attack [3], [7].

We show that such an oracle $\mathcal{A}$ is unlikely to exist by proving that its existence would imply the existence of a set S of polynomial (in $\log p$) fraction of all elliptic curves over $\mathbb{F}_p$ so that one can solve the Diffie–Hellman problem for every $E \in S$ and every Diffie–Hellman triple (P, uP, vP) for E. This is based on random self-reducibility among elliptic curves, which was first studied in[13]; by Tate's theorem achieving this via algebraic maps (isogenies) is possible only among those curves that have the same order (or trace). Thus our focus here is to identify the values of the trace for which the self-reducibility is applicable. This allows us to use Boneh-Shparlinski hard core bit result on isomorphism classes and enlarge the set of curves where it is applicable. For example, if on a specific isomorphism class their oracle algorithm does not apply, our random walk can (with a good probability) link it to another class where it applies. To show the hard core bit theorem for all the curves, one may consider the analysis based only on isomorphism classes, but the associated hardness assumption is clear and natural when restricted isogeny classes (in view of Tate's theorem). It will be interesting to see if one can develop new attacks, similar to the ones mentioned earlier for the finite field case. We remark that hard core bit theorems for finite field Diffie-Hellman function remain open and the best in this case is computing one bit (without error) is hard, if the generator is small [2].

2   Notation and Preliminaries

Throughout, $p \geq 5$ will be a prime number and $\tilde{\varepsilon} > 0$ will be a fixed real number. We will be considering the Diffie–Hellman problem for elliptic curves E over $\mathbb{F}_p$ and triples (P, uP, vP), where P is a point of prime order $q > (\log p)^{2+\tilde{\varepsilon}}$ and $1 \leq u, v \leq q$ are integers. We make this assumption because an isogeny $\phi : E \to E'$ of prime degree $\ell \leq (\log p)^{2+\tilde{\varepsilon}}$ will preserve the order of P and this assumption will be necessary for what follows.

We say that an oracle $\mathcal{B}$ computes the Diffie–Hellman function for E if for any point P of prime order $q > (\log p)^{2+\tilde{\varepsilon}}$,

$$\mathcal{B}(P, uP, vP) = uvP$$

holds with probability at least $1 - 1/p$ (here, the probability is taken over all possible choices of u and v).

Moreover, if z is a non-negative integer then LSB(z) will denote the least significant bit of z. To define the least significant bit of an element $x \in \mathbb{F}_p$, we first look at the identity map $\iota : \mathbb{F}_p \to \mathbb{Z}/p\mathbb{Z}$. If $0 \leq z \leq p - 1$ is the unique integer whose image is $\iota(x)$, we define LSB(x) = LSB(z). Also, if $Q \in E(\mathbb{F}_p)$ then $x(Q)$ and $y(Q)$ denote the x- and y-coordinates of Q, respectively.

Finally, let $H = \{t \in \mathbb{Z} : |t| \leq 2\sqrt{p}\}$ be the Hasse interval. For $t \in H$ one can write $t^2 - 4p$ uniquely as $d_t c_t^2$, where $d_t < 0$ is square-free and $c_t > 0$. We call c_t the conductor of t.

Advantage: Let $\mathcal{A}$ be an algorithm that, given a short Weierstrass equation W over $\mathbb{F}_p$, a point $P \in W(\mathbb{F}_p)$ of prime order $q > (\log p)^{2+\tilde{\varepsilon}}$ and two multiples uP and vP with $1 \leq u, v \leq q - 1$, outputs a single bit. We define the advantage $\mathrm{Adv}_{W,P}(\mathcal{A})$ of $\mathcal{A}$ as

$$\mathrm{Adv}_{W,P}(\mathcal{A}) := |\Pr_{u,v}[\mathcal{A}(P, uP, vP) = \mathrm{LSB}(x(uvP))] - \frac{1}{2}|.$$

We say that $\mathcal{A}$ has an advantage ε on W if $\mathrm{Adv}_{W,P}(\mathcal{A}) > \varepsilon$ holds for any point $P \in W(\mathbb{F}_p)$ of prime order $q > (\log p)^{2+\tilde{\varepsilon}}$.

3   The Main Result

For each prime p, let

$$\Gamma_p = \{W_{a,b} : (a, b) \in \mathbb{F}_p \times \mathbb{F}_p, \ 4a^3 + 27b^2 \neq 0\}$$

be the set of all short Weierstrass equations and let Ω_p be the set of all elliptic curves over $\mathbb{F}_p$ (i.e., $\Omega_p = \Gamma_p/\cong_{\mathbb{F}_p}$). Let $\Omega_p^{(t)}$ and $\Gamma_p^{(t)}$ denote the restriction to those curves with trace t.

Theorem 3.1. *Assume the Generalized Riemann Hypothesis (GRH) and let $c > 0$ be a fixed real. (a) For almost every t in the Hasse interval, the Diffie-Hellman problem is random self reducible among the set of elliptic curves with trace t. (b) Given a subset $G \subset \Gamma_p$, such that $|G| = \delta|\Gamma_p|$ for some $0 < \delta \leq 1$ with $1/\delta = O((\log p)^c)$, assume that there exists $\varepsilon > 0$ and an algorithm $\mathcal{A}$ running in time t that takes as input a short Weierstrass model W and a Diffie–Hellman triple (P, uP, vP) and outputs a single bit. Assume that $\mathcal{A}$ satisfies the following property: for any $W \in G$ and any point P of prime order $q > (\log p)^{2+\tilde{\varepsilon}}$ on W, $\mathrm{Adv}_{W,P}(\mathcal{A}) > \varepsilon$. Then there exists a subset $S \subseteq \Omega_p$ satisfying*

$$\frac{|\Omega_p|}{|S|} = O_c\left((\log p)^{\frac{3(c+1)}{2}}(\log\log p)^4\right),$$

and an algorithm $\mathcal{B}$ running in time $(\varepsilon^{-1}\log p)^{O(1)}$, such that $\mathcal{B}$ computes the entire Diffie–Hellman secret $\mathrm{DH}_{E,P}(uP, vP)$ for any $E \in S$ and any Diffie–Hellman triple (P, uP, vP) for E (Note that in the above displayed formula, the implied constant depends only on c). Moreover, these statements hold true with Ω_p and Γ_p replaced by $\Omega_p^{(t)}$ and $\Gamma_p^{(t)}$ for almost every value of the trace t.

Intuitively, (a) implies that an efficient algorithm for computing the Diffie–Hellman function in the average case would imply an efficient algorithm for the same function in the worst case (see Section 6 for the precise technical definition).

4   Counting Elliptic Curves

Let $p \geq 5$ be a prime and let $\Gamma_p = \{W_{a,b} : (a, b) \in \mathbb{F}_p \times \mathbb{F}_p, \ 4a^3 + 27b^2 \neq 0\}$ be the set of all short Weierstrass equations over $\mathbb{F}_p$. Explicitly, $W_{a,b}$ is the short

Weierstrass equation $y^2 = x^3 + ax + b$. Then $|\Gamma_p| = p(p-1)$ since the number of all pairs (a, b), such that $4a^3 + 27b^2 = 0$ is equal to p. Indeed, any such pair is parameterized by $a = -3c^2$ and $b = 2c^3$ for some $c \in \mathbb{F}_p$ and each such c is uniquely determined from (a, b).

4.1   Isomorphism Classes

Two short Weierstrass equations $W_{a,b}$ and $W_{a',b'}$ are isomorphic over $\mathbb{F}_p$ if there exists an element $u \in \mathbb{F}_p^\times$, such that $a' = u^4 a$ and $b' = u^6 b$. To count the elliptic curves E over $\mathbb{F}_p$, we observe that the number of short Weierstrass equations $W \in \Gamma_p$ for E is exactly $\dfrac{p-1}{\# \operatorname{Aut}(E)}$. In particular, this gives us the formula

$$\sum_E \frac{1}{\# \operatorname{Aut}(E)} = p,$$

where the sum is taken over all elliptic curves E over $\mathbb{F}_p$.

4.2   Isogeny Classes

Tate's isogeny theorem states that two elliptic curves E_1, E_2 over $\mathbb{F}_p$ are isogenous if and only if $\#E_1(\mathbb{F}_p) = \#E_2(\mathbb{F}_p)$. For any elliptic curve $E_{/\mathbb{F}_p}$ we have the Hasse bound $|p + 1 - \#E(\mathbb{F}_p)| \leq 2\sqrt{p}$. For an integer $t \in H$ consider the isogeny class of short Weierstrass equations

$$C_t = \{W_{a,b} \in \Gamma : \#W_{a,b}(\mathbb{F}_p) = p + 1 - t\}.$$

Our goal is to provide upper and lower bounds on the size of C_t for any $t \in H$. We show how to do this in the next two sections. A useful definition for what follows is the *weighted cardinality*:

Definition 4.1 (Weighted cardinality). *Let U be any set of elliptic curves over $\mathbb{F}_p$. We define the* weighted cardinality *to be the sum*

$$\#'U = \sum_{E \in U} \frac{1}{\# \operatorname{Aut}(E)}.$$

4.3   Lenstra's Upper Bound

Lemma 4.1. *Let Σ be a set of integers t satisfying $|t| \leq 2\sqrt{p}$. There exists an effectively computable constant c_u (independent of p), such that*

$$\sum_{t \in \Sigma} |C_t| \leq c_u |\Sigma| p^{3/2} (\log p)(\log \log p)^2.$$

Proof. By [15, Prop.1.9(a)], there exists an effective constant c, such that

$$\#'\{W \in \Gamma_p : 1 + p - \#W(\mathbb{F}_p) \in \Sigma\}_{/\cong_{\mathbb{F}_p}} \leq c|\Sigma| p^{1/2} (\log p)(\log \log p)^2.$$

Now, the lemma is a consequence of the fact that the weight of an elliptic curve E is $(\# \operatorname{Aut}(E))^{-1}$ (which is either $1/2$, $1/3$ or $1/6$) and that the isomorphism class corresponding to E contains $\dfrac{p-1}{\# \operatorname{Aut}(E)}$ short Weierstrass equations.

4.4   Refining Lenstra's Lower Bound

We need a simple refinement of the lower bound established by Lenstra in [15, Prop.1.9(b)] on the size of a collection of isogeny classes.

If $|t| \leq 2\sqrt{p}$, the weighted number of elliptic curves over $\mathbb{F}_p$ whose trace of Frobenius is t is equal to the *Kronecker class number* $H(t^2 - 4p)$ (see [4], [15, pp.654-655]). For a fixed integer $\Delta < 0$, $\Delta \equiv 0, 1 \mod 4$, the Kronecker class number $H(\Delta)$ is the weighted number of equivalence classes of binary quadratic forms of discriminant Δ (the weight of a quadratic form is defined to be inverse of the number of automorphisms of the form). Let Δ_0 be the fundamental discriminant associated with Δ and let χ_0 be the quadratic character associated to Δ_0. Using an analytic class number formula, one expresses $H(\Delta)$ in terms of the special value $L(1, \chi_0)$ of the L-function of the character χ_0 and the discriminant Δ. Thus, a lower bound for $H(\Delta)$ would follow from a lower bound on the special value of the above L-function. The following result is proved in [15, Prop.1.8]:

Lemma 4.2. *(i) There exists an effectively computable positive constant c_0, such that for each $z \in \mathbb{Z}_{>0}$, there exists $\Delta^* = \Delta^*(z)$, such that*

$$H(\Delta) \geq c_0 \frac{|\Delta|^{1/2}}{\log z},$$

for each Δ which satisfies $|\Delta| \leq z$, $\Delta < 0$, $\Delta \equiv 0, 1 \mod 4$ and $\Delta_0 \neq \Delta^*$.
(ii) Assume the Generalized Riemann Hypothesis. There exists an effectively computable constant $c_0' > 0$, such that for each $z \in \mathbb{Z}_{>0}$

$$H(\Delta) \geq c_0' \frac{|\Delta|^{1/2}}{\log \log z},$$

for each Δ which satisfies $|\Delta| \leq z$, $\Delta < 0$ and $\Delta \equiv 0, 1 \mod 4$.

The following refinement of Lenstra's Proposition 1.9(b) is necessary for our argument:

Proposition 4.1. *Let $0 < \mu < 1$ and let Σ be a set of integers t satisfying $|t| \leq 2\sqrt{p}(1-\mu)$. Let*

$$w_\Sigma = \#'\{E : E \text{ elliptic curve over } \mathbb{F}_p, 1 + p - \#E(\mathbb{F}_p) \in \Sigma\}/_{\cong_{\mathbb{F}_p}},$$

be the weighted cardinality of the short Weierstrass equations whose traces of Frobenius are in Σ.
(i) There exists an effectively computable constant $c_1 > 0$, such that

$$w_\Sigma \geq c_1(|\Sigma| - 2)\frac{\mu^{1/2}p^{1/2}}{\log p}.$$

(ii) Assume the Generalized Riemann Hypothesis. Then there exists an effectively computable constant $c_1' > 0$, such that

$$w_\Sigma \geq c_1'|\Sigma|\frac{\mu^{1/2}p^{1/2}}{\log \log p}.$$

Proof. One can express

$$w_\Sigma = \sum_{t \in \Sigma} H(t^2 - 4p).$$

i) We apply Lemma 4.2 with $z = 4p$ to get that there exists a constant $c_0 > 0$, such that $H(\Delta) \geq c_0 \dfrac{|\Delta|^{1/2}}{\log p}$ unless $\Delta_0 = \Delta^*$. As in the proof of Lenstra's Proposition 1.9(b), there are at most two values of t for which the fundamental discriminant of $t^2 - 4p$ is equal to Δ^*. Hence, it remains to estimate $|t^2 - 4p|$ from below to obtain a lower estimate on w_Σ. But if $t \in \Sigma$ then

$$|t^2 - 4p| \geq 4p - 4p(1 - \mu)^2 = 8\mu p - 4\mu^2 p > 8\mu p - 4\mu p = 4\mu p.$$

Thus, $|t^2 - 4p|^{1/2} \geq 2\mu^{1/2} p^{1/2}$. Hence, if $c_1 = c_0$ then

$$w_\Sigma \geq c_1 (|\Sigma| - 2) \frac{\mu^{1/2} p^{1/2}}{\log p}.$$

ii) The second part follows similarly except that we use the lower bound under the Generalized Riemann Hypothesis for the Kronecker class number $H(\Delta)$ from Lemma 4.2(ii).

5   Isogeny Graphs

We recall a construction for isogeny graphs for ordinary elliptic curves [13]. For an integer $t \in H$ consider the isogeny class $C_t \subset \Gamma_p$ of short Weierstrass equations over $\mathbb{F}_p$. Let $S_t = C_t / \sim$ be the corresponding isogeny class of elliptic curves (i.e., we identify two short Weierstrass equations $W_{a,b}, W_{a',b'} \in C_t$ if they are isomorphic over $\mathbb{F}_p$).

Throughout the whole paper, an *isogeny* between two elliptic curves will always mean an isogeny defined over $\mathbb{F}_p$.

5.1   Ordinary Isogeny Classes and Isogeny Volcanoes

1. Ordinary isogeny classes. Suppose that S_t is an isogeny class of ordinary elliptic curves. To understand the structure of S_t one looks at the endomorphism rings of the elliptic curves inside S_t. For any curve $E \in S_t$, the endomorphism ring $\text{End}(E)$ is an order in a quadratic imaginary field [27, §III.9]. Let $\pi : E \to E$ be the Frobenius endomorphism. The characteristic polynomial of π is $X^2 - tX + p = 0$, so we can regard π as an algebraic integer. It only depends on the class S_t. The following theorem is proved in [14, §4.2] (see also [13, Thm.2.1])

Theorem 5.1 (Kohel). *Let E and E' be two elliptic curves over $\mathbb{F}_p$ that are isogenous over $\mathbb{F}_p$, let K be the quadratic imaginary field $\text{End}(E) \otimes \mathbb{Q}$ and $\mathcal{O}_K$ be the maximal order of K.*

1. *We have $\mathbb{Z}[\pi] \subseteq \mathrm{End}(E) \subseteq \mathcal{O}_K$ and $\mathbb{Z}[\pi] \subseteq \mathrm{End}(E') \subseteq \mathcal{O}_K$.*
2. *The following are equivalent:*
    (a) *$\mathrm{End}(E) = \mathrm{End}(E')$*
    (b) *There exist isogenies $\phi : E \to E'$ and $\psi : E' \to E$ of relatively prime degree.*
    (c) *$(\mathcal{O}_K : \mathrm{End}(E)) = (\mathcal{O}_K : \mathrm{End}(E'))$.*
3. *Let $\phi : E \to E'$ be an isogeny from E to E' of prime degree ℓ defined over $\mathbb{F}_p$. Then one of the three cases occurs: i) $\mathrm{End}(E)$ contains $\mathrm{End}(E')$ with index ℓ; ii) $\mathrm{End}(E')$ contains $\mathrm{End}(E)$ with index ℓ; iii) $\mathrm{End}(E') = \mathrm{End}(E)$.*
4. *Let ℓ be a prime that divides exactly one of $(\mathcal{O}_K : \mathrm{End}(E))$ and $(\mathcal{O}_K : \mathrm{End}(E'))$. Then every isogeny $\phi : E \to E'$ has degree a multiple of ℓ.*

2. Isogeny volcanoes.

A convenient visualization of the elliptic curves in an isogeny class in the ordinary case together with the isogenies between them is given by *isogeny volcanoes* [5], [14]. The curves are represented in levels according to their endomorphism rings. Two curves E_1 and E_2 are in the same level if and only if $\mathrm{End}(E_1) \cong \mathrm{End}(E_2)$. Thus, every level corresponds to an order $\mathcal{O}$ in a fixed quadratic imaginary field K. The level corresponding to an order $\mathcal{O}$ is above the level corresponding to an order $\mathcal{O}'$ if $\mathcal{O} \supsetneq \mathcal{O}'$.

Following [5], [6] and [14], we distinguish among three types of isogenies $\phi : E \to E'$ of prime degree ℓ over $\mathbb{F}_p$:

1. ϕ is *horizontal* if $\mathrm{End}(E) = \mathrm{End}(E')$;
2. ϕ is *up* if $(\mathrm{End}(E') : \mathrm{End}(E)) = \ell$;
3. ϕ is *down* if $(\mathrm{End}(E) : \mathrm{End}(E')) = \ell$.

One can compute the number of horizontal, up and down isogenies of a given prime degree coming out of a particular ordinary elliptic curve E in terms of the degree and the Legendre symbol. The result (see [5, §2.1], [6, Thm.4] and [14, Ch.4, Prop.23]) is summarized in the following

Proposition 5.1. *Let E be an ordinary elliptic curve over $\mathbb{F}_p$, with endomorphism ring $\mathrm{End}(E)$ contained in the quadratic imaginary field K with fundamental discriminant $-D < 0$. Let ℓ be a prime different from p and let $c_\pi = (\mathcal{O}_K : \mathbb{Z}[\pi])$ and $c_E = (\mathcal{O}_K : \mathrm{End}(E))$. Then*

1. *Assume $\ell \nmid c_E$. Then there are exactly $1 + \left(\dfrac{-D}{\ell} \right)$ horizontal isogenies $\phi : E \to E'$ of degree ℓ over $\mathbb{F}_p$.*
    (a) *If $\ell \nmid c_\pi$, there are no other isogenies $E \to E'$ of degree ℓ over $\mathbb{F}_p$.*
    (b) *If $\ell \mid c_\pi$, there are $\ell - \left(\dfrac{-D}{\ell} \right)$ down isogenies of degree ℓ over $\mathbb{F}_p$.*
2. *Assume $\ell \mid c_E$. Then there is one up isogeny $E \to E'$ of degree ℓ over $\mathbb{F}_p$.*
    (a) *If $\ell \nmid \dfrac{c_\pi}{c_E}$ then there are no horizontal isogenies of degree ℓ over $\mathbb{F}_p$.*
    (b) *If $\ell \mid \dfrac{c_\pi}{c_E}$ then there are ℓ down isogenies of degree ℓ over $\mathbb{F}_p$.*

Finally, we say that two isomorphism classes of elliptic curves E_1 and E_2 in the same isogeny class belong to the same level of the isogeny volcano if and only if $c_{E_1} = c_{E_2}$.

5.2 Expander Graphs and a Rapid Mixing Lemma

Let k be a positive integer and let $\mathcal{I}$ be an infinite set of positive integers. Consider a sequence of graphs $\{G_h\}_{h \in \mathcal{I}}$, each of which is k-regular and connected, such that G_h has h vertices. Let A_h be the adjacency matrix of G_h. Since G_h is k-regular, the vector v_h consisting of 1's in each coordinate is an eigenvectors for A_h with eigenvalue $\lambda_{\text{triv}} = k$ and any other eigenvalue λ of A_h satisfies $|\lambda| \le k$. We refer to the eigenvalue λ_{triv} as the trivial eigenvalue. Furthermore, since G_h is connected, the eigenvalue k has multiplicity one.

Definition 5.1. *The sequence* $\{G_h\}_{h \in \mathcal{I}}$ *is called* a sequence of expander graphs *if there exists a constant $0 < \nu < 1$, such that for any h and any eigenvalue $\lambda \ne \lambda_{\text{triv}}$ of A_h, $|\lambda| \le \nu \lambda_{\text{triv}}$.*

The main application of expander graphs is to prove the rapid mixing of random walks provided we have a good upper bound on the spectral gap ν. The property is summarized in the following proposition which will be used in our particular application (see [13, Prop.3.1] for a proof):

Proposition 5.2. *Let G be a k-regular graph with h vertices. Assume that every eigenvalue $\lambda \ne \lambda_{\text{triv}}$ of G satisfies the bound $|\lambda| \le \nu \lambda_{\text{triv}}$ for some $0 < \nu < 1$. Let S be a set of vertices of G and let x be any vertex of G. Then a random walk of length at least $\dfrac{\log\left(\dfrac{2h}{|S|^{1/2}}\right)}{\log(\nu^{-1})}$ starting at x will land in S with probability at least $\dfrac{|S|}{2h}$.*

5.3 Isogeny Graphs in the Ordinary Case

Fix an isogeny class C_t of short Weierstrass equations for some $t \in H$ and the corresponding set S_t of elliptic curves. Following [13, §2.1] we define an isogeny graph to be a graph $\mathcal{G}$ whose vertices are all the elements of S_t that belong to a fixed level of the isogeny volcano for S_t.

Let $E_1, E_2 \in S_t$. Two isogenies $\phi : E_1 \to E_2$ and $\phi' : E_1 \to E_2$ are said to be equivalent if there exists an automorphism $\alpha \in \text{Aut}(E_2)$, such that $\phi' = \alpha\phi$ (see also [9, Prop.2.3]). The edges of the graph are equivalence classes of *horizontal* isogenies that have prime degrees at most $(\log p)^{2+\bar\varepsilon}$. The degree bound is chosen in such a way that it is small enough to allow the isogenies to be computed and large enough to allow the graph to be connected and to have rapid mixing properties.

The graph $\mathcal{G}$ is known to be isomorphic to a graph $\mathcal{H}$ whose vertices are elliptic curves $\mathbb{C}/\mathfrak{a}$ with complex multiplication by the order $\mathcal{O}$ corresponding to the level for the graph $\mathcal{G}$ in the isogeny volcano (here, $\mathfrak{a} \subset \mathcal{O}$ is an ideal) and whose edges are isogenies of the form $\mathbb{C}/\mathfrak{a} \to \mathbb{C}/\mathfrak{a}\mathfrak{l}^{-1}$, where $\mathfrak{l} \subset \mathcal{O}$ is an invertible prime ideal satisfying $N(\mathfrak{l}) \le (\log p)^{2+\bar\varepsilon}$ [6, §3], [7], [13, §2.1]. Equivalently, $\mathcal{H}$ is the Cayley graph of the Picard group $\text{Pic}(\mathcal{O})$ of the order $\mathcal{O}$ with respect to the generators $[\mathfrak{l}] \in \text{Pic}(\mathcal{O})$, where $\mathfrak{l}$ ranges over the invertible prime ideals of $\mathcal{O}$ whose norm is at most $(\log p)^{2+\bar\varepsilon}$.

5.4   The Spectral Gap of an Isogeny Graph

For a particular isogeny graph $\mathcal{G}$ of ordinary elliptic curves, one can bound the nontrivial eigenvalues via character sum estimates under the Generalized Riemann Hypothesis. This is done via spectral analysis of the corresponding Cayley graph $\mathcal{H}$. For what follows, it will be convenient to view the eigenvectors of the adjacency matrix of $\mathcal{H}$ as functions on the corresponding ideal classes of the Picard group. The following proposition is proven in [13, §4]:

Proposition 5.3. *Let* $m = (\log p)^{2+\bar{\varepsilon}}$ *and let* $e = \#\mathcal{O}^{\times}$.
(i) The graph $\mathcal{H}$ *has eigenfunctions equal to the characters* χ *of* $\mathrm{Pic}(\mathcal{O})$ *with corresponding eigenvalues the character sums*

$$\lambda_{\chi} = \sum_{p \leq m} \sum_{\substack{\mathfrak{a} \subset \mathcal{O}, \\ N\mathfrak{a} = p}} \chi(\mathfrak{a}).$$

(ii) Let $D < 0$ *and let* $\mathcal{O}$ *be an order of discriminant* D. *The trivial eigenvalue* λ_{triv} *is equal to the number of ideal classes of the form* $[\mathfrak{l}]$ *where* $\mathfrak{l}$ *invertible prime ideal of* $\mathcal{O}$ *of norm at most* m *(note that* λ_{triv} *is asymptotically equal to* $\dfrac{m}{e \log m}$ *where* $e = \#\mathcal{O}^{\times}$). *If* χ *is a nontrivial character of the Picard group* $\mathrm{Pic}(\mathcal{O})$, *then under the Generalized Riemann Hypothesis,*

$$\lambda_{\chi} = O(m^{1/2} \log |mD|).$$

Remark 5.1. Propositions 5.2 and 5.3 show the following: suppose that S is a set of elliptic curves belonging to the same level of the isogeny volcano, such that $|\mathcal{G}|/|S| = (\log p)^{O(1)}$ and such that one can efficiently compute $\mathrm{DH}_{E,P}(uP, vP)$ for every $E \in S$ and every Diffie–Hellman triple (P, uP, vP) for E. Then there is a random polynomial time reduction of the computation of the Diffie–Hellman function on an arbitrary curve $E \in V(\mathcal{G})$ to the Diffie–Hellman function on a curve in S. Hence, one can compute the Diffie–Hellman secret on any curve E in $V(\mathcal{G})$ with high probability in time polynomial in $\log p$.

Indeed, a random walk of length polynomial in $\log p$ will connect E to a curve in S with high probability (high probability means $1 - O(p^{-r})$ for some $r > 0$). Since any step in this random walk is an isogeny that is computable in time polynomial in $\log p$, the resulting composition of isogenies and their duals are computable in time polynomial in $\log p$ (even if the degree of the composition is large). Finally, if (P, uP, vP) is a Diffie–Hellman triple for E and $\phi : E \to E'$ is an isogeny to an elliptic curve $E' \in S$, one can consider the Diffie–Hellman triple $(\phi(P), u\phi(P), v\phi(P))$ on E' and compute the Diffie–Hellman function for that triple to obtain $uv\phi(P)$. After applying the dual isogeny, we obtain the point $duvP$, where d is the degree of the composition (note that the degree is polynomial in $\log p$). Finally, since we are in a prime-order subgroup, we compute e, such that de is congruent to 1 modulo the group order. The point $ed(uvP) = uvP$ is then the desired point.

Remark 5.2. There exist isogeny graphs for supersingular elliptic curves as well. These supersingular graphs were first considered in [11] and [17]. Their expansion properties were shown much later by Pizer [21], [22]. Given a prime p, the supersingular elliptic curves are always defined over $\mathbb{F}_{p^2}$. According to [17], all isomorphism classes of supersingular elliptic curves belong to the same isogeny class. In practice, we ignore supersingular curves in our argument for the main theorem. Yet, the corresponding isogeny graph is still an expander graphs.

6   Random Self-reducibility

We define random self-reducibility. Intuitively, we would like to prove that an efficient algorithm for the Diffie–Hellman function in the average case would imply an efficient algorithm in the worst case.

6.1   Smooth Isogeny Classes and Random Self-reducibility

Let R be a fixed polynomial. Consider the following properties of a set S of elliptic curves over $\mathbb{F}_p$:

1. There exists a subset $S' \subseteq S$ with $|S'|/|S| \geq R(\log p)^{-1}$.
2. There exists an algorithm $\mathcal{A}$, such that: i) $\mathcal{A}$ computes the Diffie–Hellman function on any elliptic curve $E \in S'$; ii) $\mathcal{A}$ produces random output whenever one feeds in a Diffie–Hellman triple for an elliptic curve $E \notin S'$.

Definition 6.1. *Let S be a set of elliptic curves that satisfies conditions 1. and 2. We call S random self-reducible with respect to R if given an elliptic curve $E \in S$, one can compute the Diffie–Hellman function for any triple (Q, uQ, vQ) on E with expected $(\log p)^{O(1)}$ queries to $\mathcal{A}$ on elliptic curves $E' \in S$ that are randomly distributed among all classes in S.*

6.2   Random Self-reducibility for Single Levels in the Isogeny Volcanoes

We first show that horizontal levels in the isogeny volcanoes with sufficiently many curves on which the Diffie–Hellman problem is solvable are random self-reducible:

Lemma 6.1. *Let $\mathcal{G}$ be the graph corresponding to a particular level of the isogeny volcano for some isogeny class of elliptic curves. Assume that the set of vertices $V(\mathcal{G})$ of $\mathcal{G}$ satisfies 1. and 2. for some polynomial R. Then $V(\mathcal{G})$ is random self-reducible with respect to R.*

Proof. Let E be any elliptic curve in $V(\mathcal{G})$ and (P, uP, vP) be any Diffie–Hellman triple for E. We will show how to connect this input to the Diffie–Hellman function to an input on a *random* elliptic curve E' from $V(\mathcal{G})$ via a sequence of isogenies that are computable in polynomial time. Let $S' \subset V(\mathcal{G})$ be the distinguished set from item 1 above. and let $\mu = |S'|/|V(\mathcal{G})|$. Let $E_0 = E$. We will use the fact that $\mathcal{G}$ is an expander graph. Let $\tau = \left\lfloor \dfrac{\log\left(\frac{2|V(\mathcal{G})|}{|S|^{1/2}}\right)}{\log(\nu^{-1})} \right\rfloor + 1$, where ν is the spectral gap for $\mathcal{G}$. Using the upper bound for ν from Proposition 5.2, we obtain that τ is polynomial in $\log p$, i.e., $\tau = (\log p)^{O(1)}$.

We repeat the following procedure $m \geq \dfrac{2}{\mu} \log p$ times:

1. Consider a random walk $E_0, E_1, \ldots, E_\tau$ on $\mathcal{G}$ of length τ. Let ϕ be the composition of the isogenies along the walk, $\hat{\phi}$ be the dual isogeny of ϕ and d be their degree. Compute $e = d^{-1}$ modulo q (recall that q is the prime order of the original point P).
2. If $E' = E_\tau$, query the oracle on the elliptic curve E' and the Diffie–Hellman triple $(\phi(P), u\phi(P), v\phi(P))$ under ϕ.
3. If the oracle returns the point Q on E', compute and return $e\hat{\phi}(Q) \in E(\mathbb{F}_p)$.

Since the computation of a single isogeny of degree ℓ takes $O(\ell^4)$ time (see [14]), each of the above steps runs in time $O((\log p)^{8+4\bar{\varepsilon}}\tau)$ which is polynomial in $\log p$ (as do all other steps below).

By Proposition 5.2, the probability that $E_\tau \notin S'$ is at most $1 - \dfrac{\mu}{2}$. Thus, if we repeat the above steps m times, the probability that none of the end points of the random walk is in S' is at most

$$\left(1 - \frac{\mu}{2}\right)^m \leq e^{-\frac{\mu m}{2}} \leq e^{-\frac{\mu \cdot 2/\mu \log p}{2}} = O(p^{-1}).$$

Therefore, the above procedure will produce a list $A = L(P, uP, vP)$ of points that contains the desired point uvP with high probability. To obtain the desired solution, we compute the list $B = L(P, (u+r)P, vP)$ for a random $r \in [1, q-1]$ as in the method of Shoup [23]. We check if A and $-rvP + B$ have a unique common element, and if so, we output it. Otherwise, we report a failure. The analysis of this last step is the same as in [23].

6.3   Random Self-reducibility for Multiple Levels in the Isogeny Volcanoes

Owing to space limitations we will only outline how one can apply the methods of the single level case to solve the case of multiple levels in the isogeny volcano. Outside of this section, we restrict our discussion to the case of a single level.

Definition 6.2. *Let B be a positive real number. An isogeny class S_t of elliptic curves is called B-smooth if its conductor c_t is B-smooth, i.e., if any prime divisor of c_t is at most B.*

The next lemma proves reducibility of the Diffie–Hellman problem for a whole isogeny class (not just a single level).

Lemma 6.2. *Let $r > 0$ be any real constant and assume that S_t satisfies i) and ii) for some polynomial R, and that S_t is $(\log p)^r$-smooth. Assuming the Generalized Riemann Hypothesis, any instance of the Diffie–Hellman problem on any elliptic curve $E \in S_t$ can be computed in time polynomial in $\log p$.*

The next lemma guarantees that the conductor c_t will have $O(\log \log p)$ distinct prime factors for almost all traces t in the Hasse interval. Let m be a positive integer such that $\log \log m > 1$ and let N_m be the number of traces $t \in H$, such that c_t has less than m distinct prime factors.

Lemma 6.3. *There exists a constant C (independent of m and p) such that*

$$N_m \geq (1 - e^{-Cm \log m})|H|.$$

Proof omitted.

Remark 6.1. Suppose that $c > 0$ is fixed. By choosing k large enough (independent of p) and applying the above lemma for $m = k \log \log p$, we can guarantee that $N_m = (1 - O((\log p)^{-c}))|H|$. This means that for most of the traces $t \in H$, c_t will have $O(\log \log p)$ distinct prime divisors.

For the classes S_t for which the volcano has multiple levels, we may not be able to exploit random self-reducibility in some of them. We can bound c_t to be small enough and having $O(\log \log p)$ prime divisors, so that starting from an arbitrary elliptic curve, we can reach the appropriate random self-reducible level in time polynomial in $\log p$ by searching through the levels via vertical isogenies and testing each level for random self-reducibility.

7 Proof of Theorem 3.1

7.1 Notation

Let $\mathcal{A}$ be the oracle from Theorem 3.1 and ε be the corresponding advantage. A short Weierstrass equation W is called LSB-*predictable*, if for any point $P \in W(\mathbb{F}_p)$ of prime order $q > (\log p)^{2+\bar{\varepsilon}}$, $\mathrm{Adv}_{W,P}(uP, vP) > \varepsilon$ (in other words, $\mathcal{A}$ predicts the least significant bit of the Diffie–Hellman function for W and the generator P with advantage ε).

More generally, if T is any set of short Weierstrass equations over $\mathbb{F}_p$ and $0 < \delta' < 1$ is a real number, we refer to T as δ'-*predictable* if at least $\delta'|T|$ elliptic curves in T are LSB-predictable.

7.2 Most of the Isogeny Classes Are Smooth

Let B be an arbitrary integer. The following lemma shows that almost all of the isogeny classes S_t of elliptic curves over $\mathbb{F}_p$ are B-smooth. The latter will be useful in applying the tunneling argument and Lemma 6.2.

Lemma 7.1. *The number of traces $t \in H$, such that the isogeny class S_t corresponding to t is B-smooth is at least $\left(1 - \dfrac{2}{B}\right)|H|$.*

Proof. Fix a prime ℓ, such that $B < \ell < \sqrt{p}$ and consider the solutions of the congruence

$$t^2 \equiv 4p \mod \ell^2$$

for $t \in H$. First, the congruence $t^2 \equiv 4p \mod \ell$ has exactly $1 + \left(\dfrac{4p}{\ell}\right)$ solutions.

Each such solution t lifts uniquely to a solutions $\tilde{t}$ modulo ℓ^2 by Hensel's lemma since the derivative of $f(x) = x^2 - 4p$ does not vanish modulo $\ell > 2$ at any such t. Thus,

$$\Pr_{t \in H}\left[c_t \text{ is not } B - \text{smooth}\right] \leq \sum_{B < \ell < \sqrt{p}} \frac{1}{\ell^2}\left[1 + \left(\frac{4p}{\ell}\right)\right] <$$

$$< \sum_{B < \ell < \sqrt{p}} \frac{2}{\ell^2} < \int_B^{\infty} \frac{2}{u^2}\,du = \frac{2}{B}.$$

7.3 Lower Bound on Smooth, Predictable Isogeny Classes

Here, we show that there is a polynomial fraction of traces $t \in H$ such that S_t is smooth and C_t contains sufficiently many LSB-predictable short Weierstrass equations.

Lemma 7.2. *Let δ and c be as in the statement of Theorem 3.1. There exists a constant c_1 (independent of p), such that the number of traces $t \in H$ for which S_t is $(\log p)^{c+2}$-smooth and C_t is $\delta/2$-predictable is at least $c_1 \dfrac{|H|}{(\log p)^{c+1}(\log\log p)^2}$*

Proof. Let

$$S_{\delta/2} = \{t \in H \ : \ C_t \text{ is } \delta/2\text{-predictable}\}$$

and

$$U = \{t \in H \ : \ S_t \text{ is } (\log p)^{c+2}\text{-smooth}\}.$$

By Lemma 7.1, $|U| \geq \left(1 - \dfrac{2}{(\log p)^{c+2}}\right)|H|$. We would like to estimate $|U \cap S_{\delta/2}|$. First, we need an estimate on $|S_{\delta/2}|$. For each $t \in S_{\delta/2}$, C_t contains at most $|C_t|$ LSB-predictable curves. For each $t \notin S_{\delta/2}$, C_t contains at most $(\delta/2)|C_t|$ LSB-predictable curves. Thus, we get the inequality

$$\sum_{t \in S_{\delta/2}} |C_t| + \sum_{t \notin S_{\delta/2}} \frac{\delta}{2}|C_t| \geq |G| = \delta|\Gamma_p|$$

We combine this with Lemma 4.1 to obtain

$$\delta|\Gamma_p| \leq \sum_{t \in S_{\delta/2}} |C_t| + \sum_{t \notin S_{\delta/2}} \frac{\delta}{2}|C_t| = \sum_{t} \frac{\delta}{2}|C_t| + \sum_{t \in S_{\delta/2}} \left(1 - \frac{\delta}{2}\right)|C_t| \leq$$

$$\leq \frac{\delta}{2}|\Gamma_p| + \left(1 - \frac{\delta}{2}\right) c_u |S_{\delta/2}| p^{3/2}(\log p)(\log \log p)^2.$$

Thus,

$$|S_{\delta/2}| \geq \left(\frac{\delta/2}{1 - \delta/2}\right) \frac{|\Gamma_p|}{c_u p^{3/2}(\log p)(\log \log p)^2} \geq c_1' \frac{|H|}{(\log p)^{c+1}(\log \log p)^2},$$

for some constant $c_1' > 0$ (since $\delta = O((\log p)^c)$). Hence,

$$|U \cap S_{\delta/2}| = |U| + |S_{\delta/2}| - |U \cup S_{\delta/2}| \geq \left(1 - \frac{2}{(\log p)^{c+2}}\right)|H| +$$

$$+ c_1' \frac{|H|}{(\log p)^{c+1}(\log \log p)^2} - |H| \geq c_1 \frac{|H|}{(\log p)^{c+1}(\log \log p)^2},$$

for some constant c_1 independent of p. This proves the lemma.

7.4 Predicting LSB within an Isomorphism Class

It was shown in [1] that within an isomorphism class of short Weierstrass equations, predicting the least significant bit on a non-negligible fraction of the short Weierstrass equations is at least as hard as computing the entire Diffie–Hellman secret key for the elliptic curve corresponding to this class.

For any short Weierstrass equation $W : y^2 = x^3 + ax + b$ and any $\lambda \in \mathbb{F}_p^\times$ we denote by W_λ the isomorphic curve $y^2 = x^3 + a\lambda^4 x + b\lambda^6$ and by $\phi_\lambda : W \to W_\lambda$ the isomorphism $(x, y) \mapsto (\lambda^2 x, \lambda^3 x)$. The result is summarized as follows:

Theorem 7.1 (Boneh-Shparlinski). *Let $0 < \varepsilon, \delta < 1$. Let p be a prime and W be a short Weierstrass equation over $\mathbb{F}_p$. Let $P \in W(\mathbb{F}_p)$ be a point of prime order. Suppose that there is a τ-time algorithm $\mathcal{A}$, such that $\mathrm{Adv}_{W_\lambda, \phi_\lambda(P)}(\mathcal{A}) > \varepsilon$ for at least δ-fraction of all $\lambda \in \mathbb{F}_p^\times$. Then the Diffie–Hellman function for W with respect to the generator P can be computed in expected time $\tau \cdot (\varepsilon^{-1}\delta^{-1}\log p)^{O(1)}$.*

7.5 Predictable Isomorphism Classes within a Predictable Isogeny Class

Lemma 7.3. *Let $0 < \beta < 1$, such that $1/\beta = O((\log p)^c)$, let $t \in H$ be a trace, such that C_t be a β-predictable isogeny class of short Weierstrass equations. There exists a constant $0 < c_2 < 1$, such that the number of $\beta/2$-predictable isomorphism classes of elliptic curve inside C_t is at least $c_2 \dfrac{|S_t|}{(\log p)^c}$.*

Proof. Let $T_{\beta/2}$ be the set of $\beta/2$-predictable isomorphism classes of short Weierstrass models contained C_t. Each isomorphism class $I \subset C_t$, $I \in T_{\beta/2}$ contains at most $|I|$ LSB-predictable elliptic curves and each isomorphism class $I \notin T_{\beta/2}$ contains at most $\frac{\beta}{2}|I|$ LSB-predictable elliptic curves. Thus,

$$\beta|C_t| \leq \sum_{\substack{I\subset C_t, \\ I\in T_{\beta/2}}} |I| + \sum_{\substack{I\subset C_t, \\ I\notin T_{\beta/2}}} \frac{\beta}{2}|I| = \sum_{I\subset C_t} \frac{\beta}{2}|I| + \sum_{\substack{I\subset C_t, \\ I\in T_{\beta/2}}} \left(1 - \frac{\beta}{2}\right)|I| \leq$$

$$\leq \frac{\beta}{2}|C_t| + \frac{(2-\beta)p}{4}|T_{\beta/2}|.$$

Therefore,

$$|T_{\beta/2}| \geq 2\left(\frac{\beta}{1-\beta/2}\right)\frac{|C_t|}{p} > c_2\frac{|S|}{(\log p)^c},$$

for some constant $c_2 > 0$ independent of p (since $1/\beta = O((\log p)^c)$).

7.6   Proof of Theorem 3.1

Proof (Proof of Theorem 3.1). According to Lemma 7.2, there exists a constant c_1 (independent of p), such that for at least $c_1\dfrac{|H|}{(\log p)^{c+1}(\log\log p)^2}$ traces $t \in H$, S_t is $(\log p)^{c+2}$-smooth and C_t is $\delta/2$-predictable. Let $0 < \mu < 1$ be the real number defined by $2\sqrt{p}\mu = \dfrac{c_1}{4}\cdot\dfrac{|H|}{(\log p)^{c+1}(\log\log p)^2}$. We will apply our refinement of Lenstra's lemma with this particular μ. Indeed, let Σ be the set of all traces $t \in H$ which satisfy $|t| \leq 2\sqrt{p}(1-\mu)$ and such that S_t is $(\log p)^{c+2}$-smooth and C_t is $\delta/2$-predictable. Then

$$|\Sigma| \geq \left\lceil \frac{c_1}{2}\cdot\frac{|H|}{(\log p)^{c+1}(\log\log p)^2}\right\rceil.$$

Since we have assumed the Generalized Riemann Hypothesis, Proposition 4.1(ii) implies that

$$\#'\{W \in C_t \ : \ t \in \Sigma\}_{/\cong_{\mathbb{F}_p}} \geq |\Sigma|\frac{\mu^{1/2}p^{1/2}}{\log\log p} \geq \tilde{c}\frac{p}{(\log p)^{\frac{3}{2}(c+1)}(\log\log p)^4},$$

for some constant $\tilde{c}$ independent of p. Let

$$S := \{W \in C_t \ : \ t \in \Sigma\}_{/\cong_{\mathbb{F}_p}}$$

Since the weighted cardinality of each isogeny class is $p/2$, $p/4$ or $p/6$, we obtain that there exists a constant $\tilde{c}'$ (independent of p), such that

$$|S| \geq \tilde{c}'\frac{|\Omega_p|}{(\log p)^{\frac{3}{2}(c+1)}(\log\log p)^4}.$$

We claim that S satisfies the properties of the theorem. Indeed, by Lemma 7.3 applied to $\beta = \delta/2$ we obtain that for each $t \in \Sigma$, C_t contains a polynomial fraction of $\delta/4$-predictable isomorphism classes. The result of Boneh and Shparlinski then implies that one can compute the Diffie–Hellman function on each of these isomorphism classes in time $\tau(\log p)^{O(1)}$ (since $1/\delta$ is polynomial in $\log p$). Finally, applying Lemma 6.2 we obtain that one can solve the Diffie–Hellman problem on any $E \in S$ in time $\tau(\log p)^{O(1)}$. That completes the proof.

Acknowledgements. We thank Dan Boneh, David Jao, Steve Miller, Bjorn Poonen and Ken Ribet for discussions.

References

1. Boneh, D., Shparlinski, I.: On the unpredictability of bits of elliptic curve Diffie-Hellman scheme. In: Kilian, J. (ed.) CRYPTO 2001. LNCS, vol. 2139, pp. 201–212. Springer, Heidelberg (2001)
2. Boneh, D., Venkatesan, R.: Hardness of computing the most significant bits of secret keys in Diffie-Hellman and related schemes. In: Koblitz, N. (ed.) CRYPTO 1996. LNCS, vol. 1109, pp. 129–142. Springer, Heidelberg (1996)
3. Cohen, H., Frey, G. (eds.): Handbook of elliptic and hyperelliptic curve cryptography, Theory and Practice (2005)
4. Deuring, M.: Die Typen der Multiplikatorenringe elliptischer Funktionenkörpen, vol. 14, pp. 197–272. Abh. Math. Sem. Hansischen Univ (1941)
5. Fouquet, M., Morain, F.: Isogeny volcanoes and the SEA algorithm. In: Fieker, C., Kohel, D.R. (eds.) ANTS 2002. LNCS, vol. 2369, pp. 276–291. Springer, Heidelberg (2002)
6. Galbraith, S.D.: Constructing isogenies between elliptic curves over finite fields. LMS J. Comput. Math. 2, 118–138 (1999) (electronic)
7. Galbraith, S.D., Hess, F., Smart, N.P.: Extending the GHS Weil descent attack. In: Knudsen, L.R. (ed.) EUROCRYPT 2002. LNCS, vol. 2332, pp. 29–44. Springer, Heidelberg (2002)
8. Gonzalez Vasco, M.I., Shparlinski, I.: Security of the most significant bits of the Shamir message passing scheme. Math. Comput. 71(237), 333–342 (2002)
9. Gross, B.H.: Heights and the special values of L-series, Number theory (Montreal, Que., 1985). In: CMS Conf. Proc., vol. 7, pp. 115–187. Amer. Math. Soc., Providence (1987)
10. Howgrave-Graham, N., Nguyen, P.Q., Shparlinski, I.: Hidden number problem with hidden multipliers, timed-release crypto, and noisy exponentiation. Math. Comput. 72(243), 1473–1485 (2003)
11. Ihara, Y.: Discrete subgroups of PL(2, $k_\wp$), Algebraic Groups and Discontinuous Subgroups. In: Proc. Sympos. Pure Math., Boulder, Colo., 1965, vol. IX, pp. 272–278. Amer. Math. Soc., Providence (1966)
12. Jao, D., Jetchev, D., Venkatesan, R.: On the security of certain partial Diffie–Hellman secrets. In: Srinathan, K., Pandu Rangan, C., Yung, M. (eds.) INDOCRYPT 2007. LNCS, vol. 4859. Springer, Heidelberg (2007)
13. Jao, D., Miller, S.D., Venkatesan, R.: Do all elliptic curves of the same order have the same difficulty of discrete log? In: Roy, B. (ed.) ASIACRYPT 2005. LNCS, vol. 3788, pp. 21–40. Springer, Heidelberg (2005)

14. Kohel, D.: Endomorphism rings of elliptic curves over finite fields. University of California, Berkeley, Ph.D. thesis (1996)
15. Lenstra, H.W.: Factoring integers with elliptic curves. Ann. of Math 126(2), 649–673 (1987)
16. Menezes, A.J., van Oorschot, P.C., Vanstone, S.A.: Handbook of applied cryptography. CRC Press, Inc., Boca Raton (1996)
17. Mestre, J.-F.: La méthode des graphes. Exemples et applications. In: Proceedings of the international conference on class numbers and fundamental units of algebraic number fields (Katata), pp. 217–242 (1986)
18. Nguyen, P.Q.: The dark side of the hidden number problem: Lattice attacks on DSA. In: Proc. Workshop on Cryptography and Computational Number Theory, pp. 321–330 (2001)
19. Nguyen, P.Q., Shparlinski, I.: The insecurity of the digital signature algorithm with partially known nonces. J. Cryptology 15(3), 151–176 (2002)
20. Nguyen, P.Q., Shparlinski, I.: The insecurity of the elliptic curve digital signature algorithm with partially known nonces. Des. Codes Cryptography 30(2), 201–217 (2003)
21. Pizer, A.K.: Ramanujan graphs and Hecke operators. Bull. Amer. Math. Soc (N.S.) 23(1), 127–137 (1990)
22. Pizer, A.K.: Ramanujan graphs, Computational perspectives on number theory (Chicago, IL, 1995). In: AMS/IP Stud. Adv. Math., vol. 7, pp. 159–178. Amer. Math. Soc., Providence (1998)
23. Shoup, V.: Lower bounds for discrete logarithms and related problems. In: Fumy, W. (ed.) EUROCRYPT 1997. LNCS, vol. 1233, pp. 256–266. Springer, Heidelberg (1997)
24. Shparlinski, I.: On the generalized hidden number problem and bit security of XTR. In: Bozta, S., Sphparlinski, I. (eds.) AAECC 2001. LNCS, vol. 2227, pp. 268–277. Springer, Heidelberg (2001)
25. Shparlinski, I.: Cryptographic applications of analytic number theory: Complexity lower bounds and pseudorandomness. PCS, vol. 22. Birkhäuser, Basel (2003)
26. Shparlinski, I., Winterhof, A.: A hidden number problem in small subgroups. Math. Comp. 74, 2073–2080 (2005)
27. Silverman, J.H.: The arithmetic of elliptic curves. Springer, New York (1992)
28. Tate, J.: Endomorphisms of abelian varieties over finite fields. Invent. Math. 2, 134–144 (1966)

Improved Bounds on Security Reductions for Discrete Log Based Signatures

Sanjam Garg[1,*], Raghav Bhaskar[2], and Satyanarayana V. Lokam[2]

[1] IIT, Delhi, India
sanjamg@yahoo.com
[2] Microsoft Research India, Bangalore, India
{rbhaskar,satya}@microsoft.com

Abstract. Despite considerable research efforts, no efficient reduction from the discrete log problem to forging a discrete log based signature (e.g. Schnorr) is currently known. In fact, negative results are known. Paillier and Vergnaud [PV05] show that the forgeability of several discrete log based signatures *cannot* be equivalent to solving the discrete log problem in the standard model, *assuming* the so-called one-more discrete log assumption and algebraic reductions. They also show, under the same assumptions, that, any security reduction in the Random Oracle Model (ROM) from discrete log to forging a Schnorr signature must lose a factor of at least $\sqrt{q_h}$ in the success probability. Here q_h is the number of queries the forger makes to the random oracle. The best known positive result, due to Pointcheval and Stern [PS00], also in the ROM, gives a reduction that loses a factor of q_h. In this paper, we improve the negative result from [PV05]. In particular, we show that any algebraic reduction in the ROM from discrete log to forging a Schnorr signature must lose a factor of at least $q_h^{2/3}$, assuming the one-more discrete log assumption. We also hint at certain circumstances (by way of restrictions on the forger) under which this lower bound may be tight. These negative results indicate that huge loss factors may be inevitable in reductions from discrete log to discrete log based signatures.

Keywords: Provable Security, Random Oracle Model, Schnorr Signature Scheme.

1 Introduction

Discrete Log (DL) based signatures, such as those proposed by Schnorr [Sch90], are among the simplest and the most efficient signature schemes. The small size of the produced signatures and the scope for pre-computation to efficiently generate signatures on-line make them particularly attractive for many applications. Though they have steadily withstood cryptanalytic attacks over the years, only recently has something been known about their *provable* security. Unfortunately, as we discuss below, this knowledge seems to derive largely from *negative* results.

---

* Part of the work done while visiting Microsoft Research India.

D. Wagner (Ed.): CRYPTO 2008, LNCS 5157, pp. 93–107, 2008.
© International Association for Cryptologic Research 2008

The best known positive result, due to Pointcheval and Stern [PS96, PS00], is a security reduction from the Discrete Log (DL) problem to forging a Schnorr signature *in the Random Oracle Model* (ROM) [BR93]. Their reduction rewinds a forger algorithm and uses a certain *Forking Lemma* to obtain two distinct forgeries on the same message which permits it to solve the discrete log problem. However, the reduction incurs a loss factor in efficiency in the sense that the obtained DL solver will lose a factor of q_h either in time complexity or success probability as compared to the forger. Here q_h is the number of queries the forger makes to the random oracle. Despite several efforts, no better reduction is known in the ROM. Nor is any reduction known at all in the Standard model. This situation remained until a major result was obtained in 2005 by Paillier and Vergnaud. In [PV05], they showed that no efficient reduction can exist in the standard model from DL to forging a Schnorr signature, *assuming*[1] (i) the so-called n-DL problem is hard (also called the one-more discrete log assumption) and (ii) the reduction is algebraic. (We explain both these notions a bit more in Section 2.) This indicates that the discrete log problem and forgeability of Schnorr signatures are unlikely to be equivalent in the standard model. A similar situation is known to exist in the case of RSA, by a result due to Boneh and Venkatesan [BV98] (where they also consider reductions similar to algebraic reductions). In the ROM, [PV05] also proved that any algebraic reduction must lose a factor of at least $\sqrt{q_h}$ in its success probability if it were to convert an efficient forger of the Schnorr signature scheme into an efficient DL-solver, again assuming that n-DL is hard. Thus, there remained a gap between the lower bound of $\sqrt{q_h}$ and the upper bound of q_h on the loss factor of algebraic reductions in the ROM from DL to forging Schnorr signatures. This paper is an attempt to close this gap.

Our Contributions: We improve the lower bound from $\sqrt{q_h}$ to $q_h^{2/3}$. More precisely, we show that any efficient algebraic reduction from DL to forging (a universal forger under key-only attack) a Schnorr signature scheme must lose a factor of $q_h^{2/3}$ in its success probability, assuming that the n-DL problem is hard. Our proof, as in [PV05], constructs a meta-reduction that uses the supposed algebraic reduction converting a forger into a DL-solver to solve the n-DL problem. In this process, the meta-reduction needs to simulate the forger (or adversary) that is used by the reduction. Our improvement hinges on a more careful construction of this simulation and a more refined analysis of the success probability of the meta-reduction in this simulation. In this analysis, we make use of known estimates [Pil90] on the expected length of a longest increasing subsequence of a random sequence. The adversary (simulated by the meta-reduction) in our lower bound proof has a certain structure. We observe that a reduction in the ROM that exploits an adversary *adhering to such structure* can indeed solve DL with a loss factor of *at most* $q_h^{2/3}$, i.e., under these restrictions on the forger, our lower bound is tight. These insights and our concrete lower bound indicate that

---

[1] Obviously, given our state of knowledge on lower bounds, some assumption is needed for such impossibility results.

huge loss factors may be inevitable, even in the ROM, in security reductions that convert a Schnorr-forger into a DL-solver. In other words, while forging Schnorr signatures and extracting discrete logs are known to be *equivalent in the sense of polynomial time reductions* in the ROM, the loss factors incurred in such reductions might be impractically large (under certain assumptions, as always). We note that proving negative results on security reductions (such as lower bounds on loss factors) in the Random Oracle Model is a harder task than in the standard model.

While we state and prove our results for the Schnorr signature scheme, they are valid for many schemes based on the discrete log problem. This follows from the same arguments as in [PV05] since we use the same meta-reduction.

The rest of the paper is organized as follows: In Section 2, we review some definitions and present technical preliminaries. In Section 3, we provide the main result which shows that any reduction from the Discrete Log problem to forgeability of Schnorr signatures must lose a factor of $q_h^{2/3}$ in its success probability and comment on the tightness of the result. In Section 4 we state conclusions and mention some open problems.

2 Preliminaries

Let $\mathbb{G} = \langle g \rangle$ be a group of prime order q generated by g.

Definition 1 (DL Problem). *Given $r \in \mathbb{G}$, computing $k \in \mathbb{Z}_q$ such that $r = g^k$ is known as the Discrete Log (DL) problem over the group $\mathbb{G}$.*

A probabilistic algorithm $\mathcal{A}$ is said to be an (ε, τ)-solver for DL if

$$\Pr_{k \xleftarrow{\$} \mathbb{Z}_q} \left[\mathcal{A}(g^k) = k \right] \geq \varepsilon,$$

where the probability is taken over the random tape of $\mathcal{A}$ and random choices of k and $\mathcal{A}$ stops after time at most τ.

The (ε, τ)-discrete log assumption (for group $\mathbb{G}$) says that no (ε, τ)-solver can exist for DL over $\mathbb{G}$. The (asymptotic) *DL-assumption* says that the (ε, τ)-discrete log assumption holds whenever $\tau = \text{poly}(\log q)$ and ε is a non-negligible function of $\log q$.

Definition 2 (n-DL Problem). *Given an $(n+1)$-tuple $(r_0, r_1, \ldots, r_n)$ of distinct elements in $\mathbb{G}$ and up to n queries to a discrete log oracle, computing the $(n+1)$-tuple of elements $(k_0, k_1, \ldots, k_n)$ $(k_i \in \mathbb{Z}_q)$ such that $r_i = g^{k_i}$ for $0 \leq i \leq n$ is known as the n-DL problem.*

A probabilistic algorithm $\mathcal{A}$ is said to be an (ε, τ)-solver for n-DL if

$$\Pr_{k_0, k_1, \ldots, k_n \xleftarrow{\$} \mathbb{Z}_q} \left[\mathcal{A}^{\text{DL-oracle}}(g^{k_0}, g^{k_1}, \ldots, g^{k_n}) = k_0, k_1, \ldots, k_n \right] \geq \varepsilon,$$

where the probability is taken over the random tape of $\mathcal{A}$ and random choices of the k_i and $\mathcal{A}$ stops after time at most τ.

The (ε, τ)-n-DL assumption says that no (ε, τ)-solver can exist for n-DL over $\mathbb{G}$ in time τ and with probability greater than ε. The (asymptotic) n-DL assumption says that the (ε, τ)-n-DL assumption holds whenever $\tau = \text{poly}(\log q)$ and ε is a non-negligible function of $\log q$.

Note that the DL problem is at least as hard as the n-DL problem. Hence the n-DL assumption is a stronger assumption than the DL assumption. It is not known if it is strictly stronger.

Definition 3 (Schnorr Signature Scheme). *Let p and q be primes such that $q \mid (p-1)$ Let g be a generator of the cyclic subgroup $\mathbb{G}$ of order q in $\mathbb{Z}_p^*$. Let H be a secure hash function with range $\{1, \ldots, q-1\}$. The Schnorr signature scheme consists of the following three algorithms:*

1. Key Generation: *Choose a random x with $0 < x < q$. x is the private key and $y := g^x$ is the public key.*
2. Signing: *Given the input message m, choose a random k mod q. Let $c := H(m, r)$, and $s := k + cx$. Return (c, s) as the signature.*
3. Verification: *Given the message m and the signature pair (c, s), calculate $r = g^s y^{-c}$. Let $c' = H(m, r)$. If $c = c'$ then return true else return false.*

Attack and Forgery types: An adversary can broadly mount two kinds of attacks against signature schemes: *Key-only attack (KOA*, also called no-message attack) and *Chosen message attack (CMA)*. In the first attack, the attacker knows only the public key of the signer while in the latter, the attacker can also obtain signatures on messages of his choice adaptively. The result of the attacks by the adversary are classified as follows:

1. Total Break - The adversary learns the secret key of the signer.
2. Universal Forgery (UF) - The adversary can produce a valid signature for any message.
3. Existential Forgery(EF) - The adversary can produce a new message signature pair.

Thus, by combining the attack type and the attack result, one can talk about various levels of security for digital signatures. For instance, a (ε, τ)-universal forger under key-only attack is an adversary who, knowing only the public key, can produce a signature on any given message with probability ε in time at most τ. An (ε, τ, q_h)-universal forger under key-only attack is the same adversary *in the Random Oracle Model* who, makes at most q_h hash queries to the random oracle. For details refer to [MvOV96, PV05].

Algebraic Reductions: We assume our reductions to be *algebraic algorithms*. An algorithm $\mathcal{R}$ is said to be algebraic *with respect to a group* $\mathbb{G}$ if the only operations $\mathcal{R}$ performs on group elements are group operations; on objects that are not group elements, $\mathcal{R}$ is allowed to perform any (standard basic) operations. In particular, given $g_1, g_2 \in \mathbb{G}$, $\mathcal{R}$ can only (i) check if g_1 and g_2 are the same, (ii) compute $g_1 \cdot g_2$ (we represent the group multiplicatively), and (iii) raise g_1 to a power (including to -1, thus computing inverse). Other natural variations on this definition are possible. Algebraic algorithms encompass many natural algorithms/reductions used in cryptography and impose weaker conditions than other known models such as the generic group model (GGM). They were originally and more generally defined by [BV98] in the context of the *ring* $\mathbb{Z}_{pq}$ in their study of RSA versus factoring.

A formal property characterizing an algebraic algorithm $\mathcal{R}$ may be described as follows. Suppose $\mathcal{R}$ takes group elements $g_1, \ldots g_k$ (and possibly other objects) and produces a group element h after τ steps. Then, there is an associated function $\texttt{Extract}$ that takes all of inputs to $\mathcal{R}$, the code/program of $\mathcal{R}$ (so $\texttt{Extract}$ could have non-black-box access to $\mathcal{R}$), and produces integers $\alpha_1, \ldots, \alpha_k$ such that $h = g_1^{\alpha_1} \cdots g_k^{\alpha_k}$ in time polynomial in τ and $|\mathcal{R}|$, where $|\mathcal{R}|$ denotes the length of the program of $\mathcal{R}$.

In particular, suppose $\mathcal{R}$ is algebraic with respect to group $\mathbb{G} = \langle g \rangle$. Suppose $\mathcal{R}$ produces elements $y_1, \ldots y_n$ during its computation. Then, given $\mathcal{R}$'s code and all its inputs, $\texttt{Extract}$ would be able to produce $x_1, \ldots, x_n$ such that $y_i = g^{x_i}$ in time polynomial in $\mathcal{R}$'s running time to produce y_i and its code length $|\mathcal{R}|$.

3   Improved Lower Bound

Theorem 1. *Suppose there exists an algebraic reduction $\mathcal{R}$ that converts an (ε, τ, q_h)-universal forger $\mathcal{A}$ under a key-only attack in the random oracle model on the Schnorr signature scheme into an (ε', τ')-solver for the discrete logarithm problem. Further, assume that $\mathcal{R}$ invokes $\mathcal{A}$ at most n times.*

Then there exists a probabilistic algorithm $\mathcal{M}$ that (ε'', τ'')-solves n-DL, where

$$\varepsilon'' \geq \varepsilon' \left(1 - \frac{2n^{3/2}}{q_h} - \frac{1}{q-1} \right) \quad and \tag{1}$$

$$\tau'' \leq \text{poly}(\tau', |\mathcal{R}|, n, q_h, \log q). \tag{2}$$

Recall from Section 2 that $|\mathcal{R}|$ denotes the length of the code of $\mathcal{R}$.

Corollary 1. *Under the n-DL assumption, any efficient algebraic reduction that converts a UF-KO attack on the Schnorr signature scheme to an algorithm for the discrete log problem must incur a loss factor of $q_h^{2/3}$ in its success probability or its running time.*

Proof. If there is such a reduction from a feasible attack, the time complexity τ'' of the meta-reduction $\mathcal{M}$ is polynomial (in $\log q$). Also, by the n-DL assumption, ε'' must be negligibly small. Suppose now that $\varepsilon' \leq \varepsilon/q_h^{2/3}$ (any non-negligible ε' will do). By the assumed feasibility of the attack, ε is non-negligible and hence so is ε'. Thus, we must have $\varepsilon''/\varepsilon'$ negligibly close to zero. From (1), we obtain $\frac{2n^{3/2}}{q_h} + \frac{1}{q-1}$ must be negligibly close to 1. Since q (the size of the group) is exponentially large, this means that $\frac{2n^{3/2}}{q_h}$ is negligibly close to 1. Hence $n = \Omega(q_h^{2/3})$. Since the reduction makes n calls to the attacker, we must then have $\tau' \geq n\tau = \Omega(q_h^{2/3}\tau)$.   □

Structure of the proof of Theorem 1: Our proof proceeds, as in [PV05], by constructing a meta reduction $\mathcal{M}$ that solves n-DL using the supposed algebraic reduction $\mathcal{R}$. Note that $\mathcal{R}$ in turn makes n calls to *any* given adversary (universal forger) $\mathcal{A}$ to solve the discrete log problem. Thus our meta reduction $\mathcal{M}$ will simulate a particular kind of adversary $\mathcal{A}$ that responds to $\mathcal{R}$ by giving signatures on $\mathcal{R}$'s chosen (public key, message) pairs. To generate these forgeries by the simulated adversary, $\mathcal{M}$ takes advantage of the fact that it gets to make n calls to a discrete log oracle. Using the algebraic nature of $\mathcal{R}$ and its ability to solve the discrete log problem (given n calls to the universal forger), $\mathcal{M}$ will be able to extract the discrete logs of all the $n + 1$ elements given as its own input. Note that, since the reduction $\mathcal{R}$ is in the Random Oracle Model, the hash function H to which the adversary $\mathcal{A}$ may make up to q_h calls is under the control (in the sense of simulation) of $\mathcal{R}$.

In the next two subsections, we will describe the meta-reduction $\mathcal{M}$ and analyze its performance. The bound on $\mathcal{M}$'s running time τ'' will follow easily from the construction of $\mathcal{M}$. The bound on its success probability ε'', however, needs an intricate analysis and differs significantly from that in [PV05].

3.1   The Meta-reduction $\mathcal{M}$

The meta-reduction $\mathcal{M}$ gets an $n + 1$-tuple $(r_0, \ldots, r_n)$ of group elements. It feeds r_0 to the reduction $\mathcal{R}$ on an arbitrary random tape. It uses the remaining n elements $r_1, \ldots, r_n$ to simulate $\mathcal{A}$. At the end of the simulation it gets the discrete log $k_0 = \log_g r_0$ from $\mathcal{R}$. It extracts $k_i = \log_g r_i$ for $1 \leq i \leq n$ from the transcript of $\mathcal{R}$ as described in Section 2.

Simulation of $\mathcal{A}$: The reduction $\mathcal{R}$ is allowed to invoke the universal forger $\mathcal{A}$ up to n times with freely chosen public keys $y_i = g^{x_i}$, messages m_i, and random tapes ϖ_i where $i = 1, \ldots, n$. We note that the input of ϖ_i is necessary to carry out an oracle replay attack. In fact, we can prove that an algebraic reduction cannot solve the discrete log problem using independent runs of the adversary (adversaries which derive their randomness from external sources) using techniques very similar to those in [PV05] for the standard model. Without loss of generality, we may assume that these n invocations of $\mathcal{A}$ are pairwise distinct, i.e., that two distinct executions of $\mathcal{A}$ differ in the value of the public key and/or the random tape, and/or at least one value returned by the random oracle H of $\mathcal{R}$. The adversary $\mathcal{A}$ is allowed to make up to q_h hash queries to the random oracle H which is under the control (in the sense of simulation) of the reduction $\mathcal{R}$. In its simulation of $\mathcal{A}$, the meta-reduction $\mathcal{M}$ makes $\mathcal{A}$ return a forgery on *a randomly chosen hash query from these q_h queries*.

We denote the i-th execution of the adversary by $\mathcal{A}_i$. The state of $\mathcal{A}_i$ at any given time is determined by its input (y_i, m_i, ϖ_i) and the return values to its hash queries till that time. Let us define $c_k(i)$ to be the response of the k-th hash query made by the the i-th execution of the adversary. Let us also define

$$\text{History}_h(i) := \langle (y_i, m_i, \varpi_i), c_1(i), c_2(i), \ldots, c_{h-1}(i) \rangle,$$

to be the history of computation of $\mathcal{A}_i$ up to the h-th query. If the j-th execution of the adversary and its i-th execution have the same history till before the h-th query, then the h-th hash query of both executions must be the same (however, the responses to these queries could differ in these two executions). This puts constraints on the way meta-reduction $\mathcal{M}$ should model the adversary $\mathcal{A}$. *In particular, all random selections made by $\mathcal{A}_i$ are in fact pseudo-random in History(i) when the selection takes place.*

During its simulation $\mathcal{M}$ can make up to n calls to the discrete log oracle DL_{OM}. Naturally, its queries to DL_{OM} will depend on the vector of group elements $\boldsymbol{r} = (r_1, \ldots, r_n)$. To economize on the number of its queries, $\mathcal{M}$ will call DL_{OM} through a discrete-log "stub-routine" DL_{stub}. The stub maintains a list of already asked queries and makes sure that a query asked multiple times is not actually queried to the oracle but answered by the stub itself.

The meta-reduction $\mathcal{M}$ will simulate the adversary with perfect forgeries (i.e. that succeed with probability 1). Since the reduction $\mathcal{R}$ solves the DL problem with probability at least ε' while interacting with an adversary that produces a forgery with probability at least ε, $\mathcal{R}$ will succeed with at least the same probability with a perfect forger.

Notation: For vectors $\boldsymbol{g} = (g_1, \ldots, g_w) \in \mathbb{G}^w$ and $\boldsymbol{b} = (b_1, \ldots, b_w) \in \mathbb{Z}$, we define $\boldsymbol{g^b}$ as $\boldsymbol{g^b} := \prod_{k=1}^{w} g_k^{b_k}$.

We now describe the simulation by $\mathcal{M}$ of $\mathcal{A}_i$ for $1 \leq i \leq n$:

1. Receive $(y_i, m_i, \varpi_i) \in \mathbb{G} \times \{0,1\}^* \times \{0,1\}^*$ from $\mathcal{R}$
2. For $h \in [1, q_h]$
    (a) Randomly[2] select $\boldsymbol{\alpha}_h \leftarrow (\mathbb{F}_q)^n$
    (b) Query H to get $c_h(i) = H(m_i, \boldsymbol{r^{\alpha_h}})$
3. (a) Randomly[3] select $l_i \leftarrow [1, q_h]$
    i. Set $c_i \leftarrow c_{l_i}(i)$ and $\boldsymbol{\beta}_i \leftarrow \boldsymbol{\alpha}_{l_i}$
    ii. Request $s_i \leftarrow DL_{stub}(\boldsymbol{r^{\beta_i}} \cdot y_i^{c_i})$
    iii. Append $(y_i, m_i, \varpi_i) \mapsto (s_i, c_i)$ and $(l_i, \boldsymbol{\beta}_i)$ to `Transcript` of $\mathcal{R}$
    (b) Return $\sigma_i = (s_i, c_i)$

Extraction of Discrete Logs: The reduction $\mathcal{R}$ uses the n simulations of its calls to $\mathcal{A}$ as described above and returns the discrete log k_0 of r_0 (with probability ε'). As described in Preliminaries, using the transcript of the algebraic reduction $\mathcal{R}$'s computation that produced k_0, we can extract the discrete logs $x_1, \ldots, x_n$ of $y_1, \ldots, y_n$, respectively, i.e., $Extract(k_0, \texttt{Trasncript}) = (x_1, \ldots, x_n)$. Now, $\mathcal{M}$ has the following system of linear equations[4] over $\mathbb{F}_q$:

$$\begin{cases} \boldsymbol{\beta_1} \cdot \boldsymbol{k} = s_1 - c_1 \cdot x_1 \\ \vdots \\ \boldsymbol{\beta_n} \cdot \boldsymbol{k} = s_n - c_n \cdot x_n \end{cases} \tag{3}$$

in the unknowns $\boldsymbol{k} = (k_1, \ldots, k_n)$, where $k_i = \log_g r_i$, $1 \leq i \leq n$.

---

[2] In fact, pseudorandomly in History$_h(i)$.
[3] In fact, pseudorandomly in History$_{q_h}(i)$.
[4] $\boldsymbol{a} \cdot \boldsymbol{b}$ denotes the dot product of the vectors $\boldsymbol{a}$ and $\boldsymbol{b}$.

Let $B = \begin{bmatrix} \beta_1 \\ \vdots \\ \beta_n \end{bmatrix}$ denote the $n \times n$ matrix over $\mathbb{F}_q$ with rows $\beta_i \in \mathbb{F}_q^n$ for $1 \le i \le n$.

If B is nonsingular $\mathcal{M}$ can directly solve for $\boldsymbol{k}$. If not, $\mathcal{M}$ may not be able to solve for $\boldsymbol{k}$, for instance, if the above system is inconsistent. We consider the following three[5] mutually exclusive events:

Event A: All the l_i are distinct, i.e., for $i \ne j$, $l_i \ne l_j$. In this case, $\mathcal{M}$ checks if the matrix B is nonsingular and if so, solves for $\mathbf{k}$ from (3). Otherwise $\mathcal{M}$ outputs FAIL.

Event B: For some $i < j$, $l_i = l_j$ and for every such pair i, j, $\beta_i = \beta_j$ implies $s_i - c_i x_i = s_j - c_j x_j$. In this case, we have $m \le n$ distinct equations and matrix B has m distinct rows.

If $\text{rank}(B) = m$, then we claim that $\mathcal{M}$ can obtain all discrete logs $k_1, \ldots, k_n$. If $m = n$, this is obvious. If $m < n$, then note that $\mathcal{M}$ has not used up all its n calls to the discrete log oracle DL_{OM}, thanks to the stub DL_{stub}. Indeed, since $s_i = \text{DLOG}(r^{\beta_i} \cdot g^{c_i x_i})$, it is easy to see that the i-th and j-th equations are identical if and only if $\beta_i = \beta_j$ and $c_i x_i = c_j x_j$. Hence, this can be detected by DL_{stub} by keeping track of its input arguments. It follows that $\mathcal{M}$ can ask $n - m$ more queries to DL_{stub} and get $n - m$ k_i's. Using these and the fact that $\text{rank}(B) = m$, $\mathcal{M}$ can compute all the k_i, $1 \le i \le n$.

If $\text{rank}(B) < m$, then $\mathcal{M}$ outputs FAIL.

Event C: For some $i < j$, $l_i = l_j$, $\beta_i = \beta_j$, but $s_i - c_i x_i \ne s_j - c_j x_j$. In this case, the system (3) is clearly inconsistent and so $\mathcal{M}$ cannot solve for $\boldsymbol{k}$. It outputs FAIL.

It is clear that $\mathcal{M}$ will correctly solve the n-DL problem on $(r_0, r_1, \ldots, r_n)$ except when it outputs FAIL. In the next subsection, we analyze the probability of failure of $\mathcal{M}$.

3.2   Analysis of $\mathcal{M}$

The bound on $\mathcal{M}$'s running time is easy. The only extra time it needs, compared to $\mathcal{R}$, is to compute the elements r^{β_i}, for using the pseudorandom generator in steps (2) and (3), for the extraction procedure on the program of $\mathcal{R}$, and to solve the linear system in Event A or Event B. We let $\mathcal{M}$ use an efficient and secure pseudorandom generator G in making its random choices from its input (y, m, ϖ). It is clear that all of these can be done in time $\text{poly}(\tau', |\mathcal{R}|, n, \log q, q_h)$.

From now on, we identify all pseudorandom choices of $\mathcal{M}$ with truly random choices for simplicity of analysis. The difference in the estimates is clearly negligible. Indeed, by standard arguments, if the difference were not negligible, then $\mathcal{M}$ can be used to efficiently distinguish truly random bits and the output of G contradicting the security of the pseudorandom generator G.

---

[5] Events A and B can actually be combined into one, albeit with a complicated definition, for the purposes of analysis. We separate them for clarity of presentation.

It is easy to estimate the probability of $\mathcal{M}$'s failure given Event B. The estimate in case of Event A is the same. In these cases, $\mathcal{M}$ fails if the rank of the $m \times n$ matrix B whose rows are the distinct β's is less than m. Since the β's are chosen randomly and independently, this probability is bounded by the probability that a random $m \times n$ matrix over $\mathbb{F}_q$ has rank $< m$. We recall this well-known estimate and prove it for completeness.

Lemma 1. Let $M \in \mathbb{F}_q{}^{m \times n}$, $m \leq n$, be a random matrix (its entries are uniformly and independently chosen from $\mathbb{F}_q$). Then

$$\Pr[\mathrm{rank}(M) < m] \leq \frac{q^{-(n-m)}(1 - q^{-m})}{q - 1}.$$

Proof. It is easy to see that the probability that the m rows of M are linearly independent is given by

$$
\begin{aligned}
\Pr[\mathrm{rank}(M) = m] &= \frac{(q^n - 1)(q^n - q) \cdots (q^n - q^{m-1})}{q^{mn}} \\
&= \left(1 - \frac{1}{q^n}\right)\left(1 - \frac{1}{q^{n-1}}\right) \cdots \left(1 - \frac{1}{q^{n-m+1}}\right) \\
&\geq 1 - \sum_{i=0}^{m-1} q^{-n+i} \\
&= 1 - q^{-n+m}\frac{1 - q^{-m}}{q - 1}.
\end{aligned}
$$
$\square$

We thus conclude that

$$\Pr[\mathcal{M} \text{ fails } |B] \leq \frac{q^{-(n-m)}(1 - q^{-m})}{q - 1} \leq \frac{1}{q - 1}. \tag{4}$$

Estimating the probability of $\mathcal{M}$'s failure given Event C takes more work. We state the bound as the following lemma and complete the analysis. We prove the lemma in Section 3.3.

Lemma 2. $\Pr[C] \leq \frac{2n^{3/2}}{q_h}$.

Clearly,

$$\Pr[\mathcal{M} \text{ fails }] = \Pr[\mathcal{M} \text{ fails } |A]\Pr[A] + \Pr[\mathcal{M} \text{ fails } |B]\Pr[B] + \Pr[C].$$

Note that Event A happens when all the randomly chosen $l_i \in_R [1..q_h]$, $1 \leq i \leq n$, are distinct and that Event B and Event C are subsumed by the event that there are collisions among the l_i, i.e., $\exists i \neq j$ such that $l_i = l_j$. Hence their probabilities can be fairly tightly bounded by using well-known estimates on the *Birthday Problem*. However, we do not need them here.

Combining the estimates (4) and Lemma 2, we obtain

$$\Pr[\mathcal{M} \text{ fails }] \leq \frac{1}{q - 1}(\Pr[A] + \Pr[B]) + \frac{2n^{3/2}}{q_h} \leq \frac{1}{q - 1} + \frac{2n^{3/2}}{q_h}.$$

Completing the proof of Theorem 1: Meta-reduction $\mathcal{M}$ will solve the n-DL instance $(r_0, \ldots, r_n)$ if $\mathcal{M}$ succeeds (i.e. does not fail in the above sense) given that $\mathcal{R}$ succeeds in solving the discrete log instance on r_0. We assumed that $\mathcal{R}$ solves the discrete log problem with probability at least ε' given the appropriate adversary (that we simulated using $\mathcal{M}$). Hence the probability ε'' with which $\mathcal{M}$ solves n-DL is bounded by

$$\varepsilon'' \geq \varepsilon' \left(1 - \frac{2n^{3/2}}{q_h} - \frac{1}{q-1} \right). \qquad \square$$

3.3   Proof of Lemma 2

Recall that Event C occurs when, for some $i < j$, $l_i = l_j$ and $\boldsymbol{\beta}_i = \boldsymbol{\beta}_j$, but $s_i - c_i x_i \neq s_j - c_j x_j$ resulting in an inconsistent system of equations in (3). Since $\boldsymbol{\beta}_i$ and $\boldsymbol{\beta}_j$ were chosen pseudorandomly in $\text{History}_{l_i}(i)$ and $\text{History}_{l_j}(j)$, apart from negligible differences in the probability estimates, we can assume that $\boldsymbol{\beta}_i = \boldsymbol{\beta}_j$ only if $\text{History}_{l_i}(i) = \text{History}_{l_j}(j)$. In particular, this implies we can assume that $y_i = y_j$ and hence $x_i = x_j$. Note further that since $s_i = \text{DLOG}(r^{\boldsymbol{\beta}_i} \cdot y_i^{c_i})$, $c_i x_i = c_j x_j$ if and only if $s_i = s_j$. Hence, given $\boldsymbol{\beta}_i = \boldsymbol{\beta}_j$, we can conclude (for the purposes of probability estimates) that $s_i - c_i x_i \neq s_j - c_j x_j$ if and only if $c_i \neq c_j$. Thus, Event C occurs only if two executions $\mathcal{A}_i$ and $\mathcal{A}_j$ of the adversary not only output forgeries at the same hash query ($l_i = l_j$) but also at that query instance $\boldsymbol{\beta}_i = \boldsymbol{\beta}_j$ and $c_i \neq c_j$. In particular, the two histories $\text{History}_{l_i}(i)$ and $\text{History}_{l_j}(j)$ are identical till the point l_i and then they diverge after the hash responses from the respective random oracles (controlled by $\mathcal{R}$) to the query at l_i. We call this point of divergence, the *forking point* between executions $\mathcal{A}_i$ and $\mathcal{A}_j$.

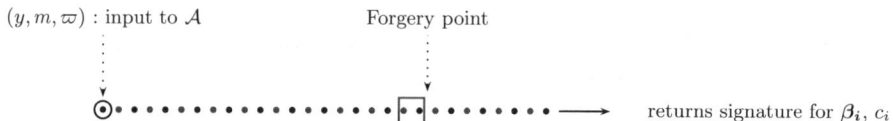

Fig. 1. A single execution of the adversary

Pictorially, we can represent an execution of the adversary by a line as in Fig. 1. The blue points symbolize inputs from $\mathcal{R}$. The first blue point is the input (y, m, ϖ) to $\mathcal{A}$ and the rest of the blue points are the responses by the random oracle (in control of $\mathcal{R}$). The red points represent the hash queries made by the adversary $\mathcal{A}$ (simulated by $\mathcal{M}$). On completing its execution, $\mathcal{A}$ returns a forgery for $(\boldsymbol{\beta}_i, c_i)$ pair for a random selection of $l_i \in [1..q_h]$. This is denoted by the red square and we will call this the *forgery point*. We also abbreviate this execution of the adversary as in Fig. 2. We will use two lines to represent two different executions of $\mathcal{A}$. Two lines with overlapping initial segments represent identical histories over the shared region. At the forking point, the responses from the respective random oracles are different. This is shown in Fig. 3.

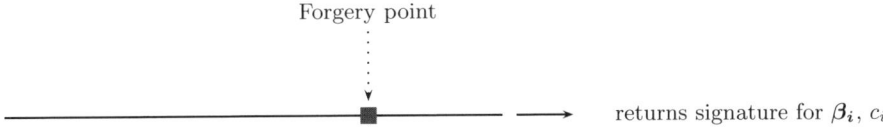

Fig. 2. A single execution of the adversary: a simplified view

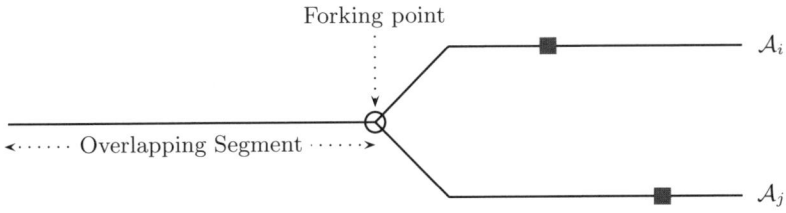

Fig. 3. Two executions sharing a common history

Thus, the simulation of an execution $\mathcal{A}_j$ fails if its forgery point l_j (the red square) is also a point of forking (a black circle), i.e., $c_{l_j} \neq c_{l_i}$, from a previous execution $\mathcal{A}_i$ that returned a forgery at that point, i.e., $l_i = l_j$.

Let F_i be the set of points on which if a forgery is returned by $\mathcal{A}_i$, then the meta-reduction $\mathcal{M}$ fails. We call F_i the set of *failure causing points* of $\mathcal{A}_i$. In other words, it is the set of points at which $\mathcal{A}_i$ forks from some previous executions of $\mathcal{A}$ and at each of these points, the corresponding previous executions of $\mathcal{A}$ have returned a forgery. More formally,

$$F_i := \{l_j : j < i \text{ such that History}_{l_j}(i) = \text{History}_{l_j}(j) \text{ and } c_{l_j}(i) \neq c_{l_j}(j)\}.$$

To illustrate, in Fig. 4, we see the execution $\mathcal{A}_i$ of the adversary and it can be seen that there are previous executions $\mathcal{A}_{j_1}$, $\mathcal{A}_{j_2}$ and $\mathcal{A}_{j_3}$ from which $\mathcal{A}_i$ forks off at the points l_{j_1}, y, and l_{j_3}, respectively. The executions $\mathcal{A}_{j_1}$, $\mathcal{A}_{j_2}$ and $\mathcal{A}_{j_3}$ had previously returned forgeries at points l_{j_1}, l_{j_2}, and l_{j_3}, respectively. Now, if $\mathcal{A}_i$ returns a forgery at l_{j_1} or l_{j_3}, then $\mathcal{M}$ will fail. Thus the set F_i consists of points l_{j_1} and l_{j_3}. Note that we can comment on the set F_i only when the forking of $\mathcal{A}_i$ from the previous executions of $\mathcal{A}$ is known.

If, for the ith execution of the adversary, F_i is the set of failure causing points, then we define $\mathfrak{F}_i$ as

$$\mathfrak{F}_i := \{x \mid x \in F_i \wedge x < l_i\} \cup \{l_i\}.$$

Let $\mathfrak{M}_i$ denote the collection of all $\mathfrak{F}_i$'s till the ith execution of the adversary, i.e., $\mathfrak{M}_i = \{\phi, \mathfrak{F}_1, \mathfrak{F}_2 \ldots \mathfrak{F}_i\}$.

The following claim explains why we include $\mathfrak{F}_i$ instead of $F_i \cup \{l_i\}$ in $\mathfrak{M}_i$. The main idea is to enable a clean description of F_{i+1} based on F_j for $j < i$.

Claim. For the $(i + 1)^{th}$ execution of the adversary, F_{i+1} will be one of the sets in $\mathfrak{M}_i$, i.e., $F_{i+1} \in \mathfrak{M}_i$.

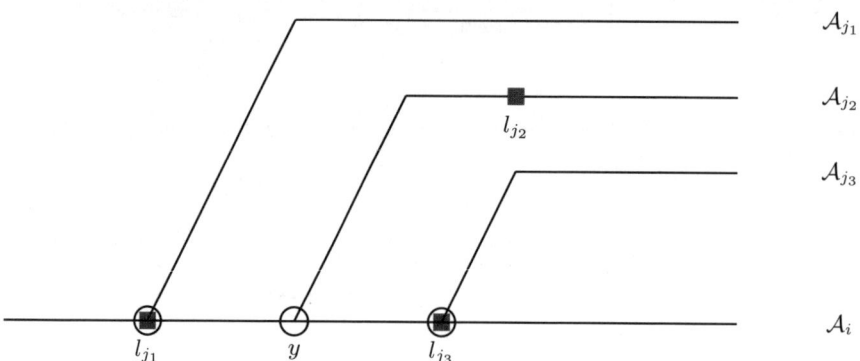

Fig. 4. Forging points, forking points, and the set of failure points

Proof of Claim: Let $F_{i+1} = \{l_{j_1} < l_{j_2} < \cdots < l_{j_t}\}$, where $j_1, j_2, \cdots, j_t < i+1$ are previous executions that forge and fork (from $\mathcal{A}_{i+1}$) at the locations $l_{j_1}, \ldots, l_{j_t}$ respectively. For simplicity of notation, let $k := j_t$. We claim that $F_{i+1} = \mathfrak{F}_k$.

First, we prove that $F_{i+1} \subseteq \mathfrak{F}_k$. Clearly, $l_k \in \mathfrak{F}_k$. Hence, let us assume $l_j \in F_{i+1}$, where $l_j < l_k$. It suffices to show that $l_j \in F_k$. Since $l_j \in F_{i+1}$, $j < i+1$, $\text{History}_{l_j}(j) = \text{History}_{l_j}(i+1)$, and $c_{l_j}(j) \neq c_{l_j}(i+1)$. On the other hand, since l_k is also in F_{i+1}, $k < i+1$ and $\text{History}_{l_k}(k) = \text{History}_{l_k}(i+1)$. But $l_j < l_k$. It follows that $\text{History}_{l_j}(k) = \text{History}_{l_j}(i+1) = \text{History}_{l_j}(j)$ and $c_{l_j}(k) = c_{l_j}(i+1) \neq c_{l_j}(j)$. Hence $l_j \in F_k$.

Next, we prove that $\mathfrak{F}_k \subseteq F_{i+1}$. By definition, $l_k \in \mathfrak{F}_k$. So, consider an $l_k \neq l_j \in \mathfrak{F}_k$. By construction of $\mathfrak{F}_k$, $l_j < l_k$ for some $j < k$ and $l_j \in F_k$. Hence $\text{History}_{l_j}(j) = \text{History}_{l_j}(k)$ and $c_{l_j}(j) \neq c_{l_j}(k)$. Since $l_k \in F_{i+1}$, $\text{History}_{l_k}(k) = \text{History}_{l_k}(i+1)$ and $c_{l_k}(k) \neq c_{l_k}(i+1)$. Since $l_j < l_k$, $\text{History}_{l_j}(k) = \text{History}_{l_j}(i+1)$ and $c_{l_j}(k) = c_{l_j}(i+1)$. Combining this with the previous relation, we obtain $\text{History}_{l_j}(j) = \text{History}_{l_j}(i+1)$ and $c_{l_j}(j) \neq c_{l_j}(i+1)$. It follows that $l_j \in F_{i+1}$.

This concludes the proof of the claim.

Clearly, the size of $\mathfrak{F}_i$ can be at most one greater than the maximum of the sizes of all $\mathfrak{F}_j$ ($1 \leq j < i$) since $|\mathfrak{F}_i| \leq |F_i|+1$ and $|F_i| \leq \max\{|\mathfrak{F}_j| : 1 \leq j \leq i-1\}$.

Thus, by construction, $\mathfrak{F}_i$ are sets comprising of integers in increasing order. The size of a $\mathfrak{F}_i$ increases only if it is the set of failure points for the current execution and the forgery point for this execution is greater than all elements in the set $\mathfrak{F}_i$. As the forgery point is randomly picked by $\mathcal{M}$ from the set $[1, q_h]$, the maximum of $|\mathfrak{F}_i|$ is at most the length of a longest increasing sub-sequence in a random sequence of n distinct integers from $[1, q_h]$ (we may assume $n \ll q_h$ as otherwise, we are already done). It is easy to see that any permutation of $[n]$ is equally represented by the ordering on such a random sequence. Let λ_n denote the random variable denoting the length of a longest increasing subsequence in a random permutation of $[n]$. The following result due to Kerov and Versik (see Pilpel's paper [Pil90]) will be useful.

Theorem 2 (Kerov-Versik 1977). *For sufficiently large n, $\mathbb{E}[\lambda_n] \leq 2\sqrt{n}$.*

We can now estimate $\Pr[C]$. Clearly, C occurs if for at least one execution $\mathcal{A}_i$, the forgery point l_i falls into the failure causing set of points F_i. Since l_i is chosen uniformly at random from $[1..q_h]$, we have $\Pr[C] \leq \sum_{i=1}^{n} |F_i|/q_h$. Now, $|F_i|$ is at most $\max_{j=1}^{i-1} |\mathfrak{F}_j| \leq \max_{j=1}^{n} |\mathfrak{F}_j|$. For every n-sequence (l_i) from $[1..q_h]^n$, $\max_{j=1}^{n} |\mathfrak{F}_j|$ is upper bounded by the length of the longest subsequence in the corresponding permutation of $[n]$. Hence $\mathbb{E}[\max_{j=1}^{n} |\mathfrak{F}_j|] \leq \mathbb{E}[\lambda_n]$. Thus,

$$
\begin{aligned}
\Pr[C] &= \sum_{t=1}^{n} \Pr[C \mid \max_{j=1}^{n} |\mathfrak{F}_j| = t] \cdot \Pr[\max_{j=1}^{n} |\mathfrak{F}_j| = t] \\
&\leq \sum_{t=1}^{n} \sum_{i=1}^{n} \frac{|F_i| \text{ given } \max_{j=1}^{n} |\mathfrak{F}_j| = t}{q_h} \cdot \Pr[\max_{j=1}^{n} |\mathfrak{F}_j| = t] \\
&\leq \frac{n}{q_h} \sum_{t=1}^{n} t \Pr[\max_{j=1}^{n} |\mathfrak{F}_j| = t] \\
&\leq \frac{n}{q_h} \mathbb{E}[\max_{j=1}^{n} |\mathfrak{F}_j|] \leq \frac{n}{q_h} \mathbb{E}[\lambda_n] \\
&\leq \frac{2n^{3/2}}{q_h} \quad \text{using Theorem 2.}
\end{aligned}
$$

This completes the proof of Lemma 2.   □

3.4   Remarks on Tightness of the Lower Bound

The adversary we have simulated in Section 3 behaves randomly in the sense that the probability of a forgery being returned for any of the hash queries in a given execution is the same. It also has the property that the probability of success is uniformly distributed across all executions. For this restricted adversary we claim that the lower bound of $q_h^{2/3}$ is indeed tight.

To justify our claim we construct a reduction $\mathcal{R}$ that, using the adversary $\mathcal{A}$ (an (ε, τ) UF-KOA 'uniform' adversary) breaks the Discrete Log problem. The reduction $\mathcal{R}$ tries to obtain two forgeries on the same message m (under the same public key y) but for different values of the hash responses. The adversary $\mathcal{A}$ accepts (y, m, ϖ) as input and then makes queries to the hash oracle. We represent the sequence of responses given by the oracle by $\boldsymbol{c} = \langle c_1(j), c_2(j), \ldots, c_{q_h}(j) \rangle$ and the subsequence of the first i responses by $\boldsymbol{c}_i = \langle c_1(j), c_2(j), \ldots, c_i(j) \rangle$. The reduction $\mathcal{R}$ calls the adversary $\mathcal{A}$ with the same (y, m, ϖ) but different $\boldsymbol{c}$. We define $Ind(\boldsymbol{c})$ as the index of the hash query on which the forgery is returned by an execution of the adversary. We let $Ind(\boldsymbol{c}) = \infty$ in case the forgery is returned for a hash value never obtained by calling the hash oracle (or if $\mathcal{A}$ fails).

Let $\mathcal{S} = \{\boldsymbol{c} \mid Ind(\boldsymbol{c}) < \infty\}$ and $\mathcal{S}_i = \{\boldsymbol{c} \mid Ind(\boldsymbol{c}) = i\}$. Then $Pr[\mathcal{S}] \geq \varepsilon - \frac{1}{q-1} = \nu$, as the probability that the adversary correctly guesses the hash value (without making the hash query) is $\frac{1}{q-1}$. Assuming that the adversary outputs its forgeries uniformly across the hash-query indices and its several executions i.e. $Ind(\boldsymbol{c})$ is pseudo-random in $\boldsymbol{c}$, we get $Pr[\mathcal{S}_i] \geq \nu/q_h, \forall i \in [1, q_h]$.

The reduction $\mathcal{R}$ divides the q_h hash query points into λ intervals of equal width q_h/λ as shown in Fig. 5. The circles represent the hash query points $[1, q_h]$ at which an instance of execution of the adversary could return a forgery and $\mathcal{K}_1, \mathcal{K}_2, \ldots, \mathcal{K}_\lambda$ are equal-sized partitions of these points. Then, $\Pr[Ind(\mathbf{c}) \in \mathcal{K}_i] \geq \nu/\lambda, \forall i \in [1, \lambda]$. $\mathcal{R}$ runs the adversary $\mathcal{A}$ in $\lambda + 1$ phases. In phase 1, $\mathcal{R}$ invokes the adversary at most λ/ν times (changing $\mathbf{c}$ each time) with the hope of getting a forgery in partition $\mathcal{K}_1$ and moves to the next phase as soon as it gets the desired forgery. The probability of obtaining a forgery in the first interval in λ/ν runs of $\mathcal{A}$ is $1 - (1 - \nu/\lambda)^{\lambda/\nu} \geq 1 - e^{-1}$. Let $\mathbf{c}^{(1)}$ denote the set of hash responses for the successful execution in phase 1 and l_1 be the index of the forgery for this execution. In case of success in phase 1, $\mathcal{R}$ calls the adversary in phase 2 at most λ/ν times with $\mathbf{c}$'s such that $c_{l_1} = c_{l_1}^{(1)}$ is satisfied for all. Otherwise in case of failure in phase 1 ($l_1 = \infty$), $\mathcal{R}$ executes the phase 2 in a similar fashion as phase 1 but hoping this time to get the forgery in the second partition $\mathcal{K}_2$. In general, in phase i ($i \in [2, \lambda]$), $\mathcal{R}$ calls the adversary at most λ/ν times with $\mathbf{c}$'s such that $c_{l_j} = c_{l_j}^{(j)}$ (l_j - index of hash query for the successful execution in the most recent successful phase (j) and $\mathbf{c}^{(j)}$ is the sequence of hash responses for that execution) with the hope of getting a forgery on some hash query $l_i \in \mathcal{K}_i$. The probability of getting a forgery in the interval $\mathcal{K}_i$ is also $1 - (1 - \nu/\lambda)^{\lambda/\nu} \geq 1 - e^{-1}$. Hence, at the end of phase λ, $\mathcal{R}$ expects to have $(1 - e^{-1}) \cdot \lambda$ forgeries in $\frac{\lambda}{\nu} \cdot \lambda$ executions of the adversary $\mathcal{A}$. Let the set of forgery points be $\mathcal{I}$. In phase $\lambda + 1$, $\mathcal{R}$ runs the adversary $\mathcal{A}$ with $\mathbf{c}$'s such that $c_{l_j} = c_{l_j}^{(j)}$, where j, l_j are defined as before. Then, $\Pr[Ind(\mathbf{c}) \in \mathcal{I}] \geq \nu\lambda(1 - e^{-1})/q_h$. The number of executions required to get a forgery on one of the points in $\mathcal{I}$ is $q_h/(\nu\lambda(1 - e^{-1}))$, and this happens with a probability $1 - e^{-1}$. The total number of required executions of $\mathcal{A}$ are $\frac{\lambda^2}{\nu} + \frac{q_h}{\nu\lambda(1 - e^{-1})}$, which takes the optimal value for $\lambda = \Theta(q_h^{1/3})$ for which the number of executions is $\Omega(\frac{q_h^{2/3}}{\nu})$. Thus at the end of phase $\lambda + 1$, $\mathcal{R}$ obtains two forgeries (c_1, s_1) and (c_2, s_2) on the same message m under the same public key y and the same randomness k (see Definition 3) but different hash responses c_1 and c_2. If the reduction R uses the discrete-log challenge g^x as the public key in the above interactions, it can obtain x as $\frac{s_1 - s_2}{c_1 - c_2}$.

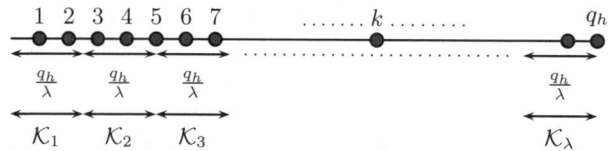

Fig. 5. Dividing the hash query points into equal intervals of width q_h/λ

4   Conclusions and Open Problems

In this paper we improved the lower bound from $q_h^{1/2}$ to $q_h^{2/3}$ on the loss factor in a security reduction that converts a forgery attack on Schnorr signature scheme to an algorithm to the discrete log problems. This means that to achieve the same level of security as before, one needs to employ larger parameters than before. We also presented a new attack strategy for solving the discrete log problem using a restricted class of Schnorr signature forgers more efficiently. This attack strategy indicates that the lower bound $q_h^{2/3}$ is tight for the restricted adversary we simulate. Since the lower bound proof relies on the n-DL assumption and restricts itself to algebraic reductions, the gap between the lower bound of $q_h^{2/3}$ and the upper bound of q_h may in some sense be inevitable.

One of the most interesting open problems is to prove that the lower bound of $q_h^{2/3}$ is tight for a general adversary. Another major open question is to understand relationship between the DL problem and the n-DL problem.

References

[BR93]   Bellare, M., Rogaway, P.: Random oracles are practical: A paradigm for designing efficient protocols. In: CCS 1993: Proceedings of the 1st ACM Conference on Computer and Communications Security, pp. 62–73. ACM Press, New York (1993)

[BV98]   Boneh, D., Venkatesan, R.: Breaking RSA not be equivalent to factoring. In: Nyberg, K. (ed.) EUROCRYPT 1998. LNCS, vol. 1403, pp. 59–71. Springer, Heidelberg (1998)

[MvOV96]   Menezes, A.J., van Oorschot, P.C., Vanstone, S.: Handbook of Applied Cryptography. CRC Press, Boca Raton (1996)

[Pil90]   Pilpel, S.: Descending subsequences of random permutations. J. Comb. Theory Ser. A 53(1), 96–116 (1990)

[PS96]   Pointcheval, D., Stern, J.: Security proofs for signature schemes. In: Maurer, U.M. (ed.) EUROCRYPT 1996. LNCS, vol. 1070, pp. 387–398. Springer, Heidelberg (1996)

[PS00]   Pointcheval, D., Stern, J.: Security arguments for digital signatures and blind signatures. Journal of Cryptology 13(3), 361–396 (2000)

[PV05]   Paillier, P., Vergnaud, D.: Discrete-log-based signatures not be equivalent to discrete log. In: Roy, B. (ed.) ASIACRYPT 2005. LNCS, vol. 3788, pp. 1–20. Springer, Heidelberg (2005)

[Sch90]   Schnorr, C.P.: Efficient identification and signatures for smart cards. In: Quisquater, J.-J., Vandewalle, J. (eds.) EUROCRYPT 1989. LNCS, vol. 434, pp. 688–689. Springer, Heidelberg (1990)

Circular-Secure Encryption from Decision Diffie-Hellman

Dan Boneh[1,*], Shai Halevi[2], Mike Hamburg[1], and Rafail Ostrovsky[3,**] ,

[1] Computer Science Dept., Stanford University
{dabo,mhamburg}@cs.stanford.edu
[2] IBM Research
shaih@alum.mit.edu
[3] Computer Science Dept. and Dept. of Math., UCLA
rafail@cs.ucla.edu

Abstract. We describe a public-key encryption system that remains secure even encrypting messages that depend on the secret keys in use. In particular, it remains secure under a "key cycle" usage, where we have a cycle of public/secret key-pairs (pk_i, sk_i) for $i = 1, \ldots, n$, and we encrypt each sk_i under $pk_{(i \bmod n)+1}$. Such usage scenarios sometimes arise in key-management systems and in the context of anonymous credential systems. Also, security against key cycles plays a role when relating "axiomatic security" of protocols that use encryption to the "computational security" of concrete instantiations of these protocols.

The existence of encryption systems that are secure in the presence of key cycles was wide open until now: on the one hand we had no constructions that provably meet this notion of security (except by relying on the random-oracle heuristic); on the other hand we had no examples of secure encryption systems that become demonstrably insecure in the presence of key-cycles of length greater than one.

Here we construct an encryption system that is circular-secure against chosen-plaintext attacks under the Decision Diffie-Hellman assumption (without relying on random oracles). Our proof of security holds even if the adversary obtains an encryption clique, that is, encryptions of sk_i under pk_j for all $1 \leq i, j \leq n$. We also construct a circular counterexample: a one-way secure encryption scheme that breaks completely if an encryption cycle (of any size) is published.

1 Introduction

Secure encryption is arguably the most basic task in cryptography, and significant work has gone into defining and attaining it. All commonly accepted definitions for secure encryption [2, 3, 8, 10, 16, 17, 19] assume that the plaintext messages to be encrypted cannot depend on the secret decryption keys themselves. The danger of encrypting messages that the adversary cannot find on its own was already noted more than two decades ago by Goldwasser and Micali [10, §5.1].

* Supported by NSF and the Packard Foundation.
** Partially supported by IBM Faculty Award, Xerox Innovation Group Award, NSF grants 0430254, 0716835, 0716389 and U.C. MICRO grant.

D. Wagner (Ed.): CRYPTO 2008, LNCS 5157, pp. 108–125, 2008.
© International Association for Cryptologic Research 2008

Over the last few years, however, it was observed that in some situations the plaintext messages do depend on the secret keys. An important example is when we have a cycle of public/secret key-pairs $(\text{pk}_i, \text{sk}_i)$ for $i = 1, \ldots, n$, and we encrypt each sk_i under $\text{pk}_{(i \bmod n)+1}$. Security in this more demanding setting was termed *key-dependent message security* (KDM-security) by Black et al. [4] and *circular security* by Camenisch and Lysyanskaya [6].

Such situations may arise due to careless key management, for example a backup system may store the backup encryption key on disk and then encrypt the entire disk, including the key, and backup the result. Another example is the BitLocker disk encryption utility (used in Windows Vista) where the disk encryption key can end up on disk and be encrypted along with the disk contents. There are also situations where circular security is needed "by design", e.g., Camenisch and Lysyanskaya used it in their anonymous credential system [6] to discourage users from delegating their secret keys. Finally, in the formal-methods community the notion of key-dependent security from [4] was used to prove equivalence between "computational security" and "axiomatic security" [1, 18].

Definitions of security for this setting were given by Black et al. [4], who defined models of KDM security in both the symmetric and public-key settings. In their public-key model the adversary is given public keys $\text{pk}_1, \ldots, \text{pk}_n$ and can access an oracle $\mathcal{O}$ that returns the encryption of $g(\text{sk}_1, \ldots, \text{sk}_n)$ under pk_i for any polynomial-time function g and any index $1 \leq i \leq n$ of the adversary's choosing. (A key-cycle can be obtained in this model when the adversary requests the encryption of sk_i under $\text{pk}_{(i \bmod n)+1}$ for all i.) The system is KDM-secure if the adversary cannot distinguish the oracle $\mathcal{O}$ from an oracle that always returns an encryption of (say) the all-zero string.

A simple example of KDM is when an encryption system is used to encrypt its own secret key (i.e., a cycle of size one). It is straightforward to construct a secure encryption scheme that becomes completely insecure once the adversary sees such self-referential ciphertext, and similarly it is straightforward to construct an encryption scheme that remains secure under such self-referential encryption [4].[1] The question becomes much harder when dealing with more complicated key-dependent messages, for example key-cycles of size more than one. For these cases, the problem has been wide open. On one hand, we had no examples of encryption systems that are secure without key-cycles but demonstrably insecure in the presence of a key cycle of size more than one. On the other hand, we had no constructions that can be proved to meet such notions of security (except by relying on the random-oracle heuristic). Some initial steps toward constructions in the standard model are given in [12] (who focused on other primitives such as PRFs) and [15] (who achieved weaker variants of these security notions).

---

[1] For the former, start from a secure encryption system (where secret keys are not valid ciphertexts), and modify the encryption algorithm so that when encrypting the secret key it outputs it in the clear. For the latter, modify the encryption algorithm so that when encrypting the secret key it outputs the encryption of a distinguished symbol $\perp$.

1.1 Our Results

Our main result is a public-key system that is circular-secure (or even "clique-secure") in the standard model under the Decision Diffie-Hellman assumption. That is, even an adversary who sees an encryption of sk_i under pk_j for all $1 \leq i, j \leq n$ cannot distinguish the ciphertexts from n^2 encryptions of (say) the all-zero string. In fact, we prove a slightly stronger result by showing that our system is KDM-secure against chosen-plaintext attacks in the model of Black et al. [4], when the adversary is restricted to affine functions of the keys. Hence, our system tolerates the adversary seeing encryption cliques (or even encryptions of more complicated functions of the secret keys) without compromising security.

The difficulty in constructing such a system is the simulation of an encryption clique without knowledge of any of the secret keys. We overcome this difficulty by having a system which is sufficiently homomorphic that such a clique can be constructed directly. We point out that one may be tempted to use a Cramer-Shoup-like construction and simulation [7] to prove n-circular security. After all, a Cramer-Shoup simulator is in possession of all secret keys (needed for responding to decryption queries) and can use them to create an encryption clique to give to the adversary. Unfortunately, we could not get this intuition to work. The problem is that the simulator has to embed the DDH challenge into the circular clique, but it is difficult to do so while creating a valid clique.

In Section 5 we also take a first step toward showing that standard security notions do not imply circular security. Specifically, we construct a very simple one-way encryption system that breaks completely as soon as a key-cycle of any size is published.

2 KDM Security: Definitions and Properties

We begin by reviewing the definitions of Key-Dependent Message security (KDM) in the public-key setting from Black et al. [4]. We use a small extension of the definition, used also in [12], that restricts the adversary to a particular set of functions.

A public-key encryption system $\mathcal{E}$ consists of three algorithms $(\mathsf{G}, \mathsf{E}, \mathsf{D})$ where G is a key-generation algorithm that takes as input a security parameter λ and outputs a public/secret key pair (pk, sk); $\mathsf{E}(pk, m)$ encrypts message m with public key pk; and $\mathsf{D}(sk, c)$ decrypts ciphertext c with secret key sk. We have the usual correctness condition, asserting that decryption correctly recovers the plaintext message from the ciphertext (with probability one).

We use S to denote the space of secret keys output by $\mathsf{G}()$ and use M to denote the message (plaintext) space. Throughout the paper we assume that $S \subseteq M$ so that any secret key sk can be encrypted using any public key pk'. All of these notations assume an implied security parameter λ.

2.1 KDM Security with Respect to a Set of Functions $\mathcal{C}$

Informally, KDM security implies that the adversary cannot distinguish the encryption of a key-dependent message from an encryption of 0. We define key-dependence relative to a fixed set of functions $\mathcal{C}$.[2] Let $n > 0$ be an integer and let $\mathcal{C}$ be a finite set of functions $\mathcal{C} := \{f : S^n \to M\}$. For each function $f \in \mathcal{C}$ we require that $|f(z)|$ is the same for all inputs $z \in S^n$ (i.e. the output length is independent of the input).

We define KDM security with respect to $\mathcal{C}$ using the following game that takes place between a challenger and an adversary $\mathcal{A}$. For an integer $n > 0$ and a security parameter λ the game proceeds as follows:

init. The challenger chooses a random bit $b \xleftarrow{R} \{0,1\}$. It generates $(\mathrm{pk}_1, \mathrm{sk}_1), \ldots, (\mathrm{pk}_n, \mathrm{sk}_n)$ by running $\mathsf{G}(\lambda)$ n times, and sends the vector $(\mathrm{pk}_1, \ldots, \mathrm{pk}_n)$ to $\mathcal{A}$.

queries. The adversary repeatedly issues queries where each query is of the form (i, f) with $1 \le i \le n$ and $f \in \mathcal{C}$. The challenger responds by setting

$$y \leftarrow f(\mathrm{sk}_1, \ldots, \mathrm{sk}_n) \in M \quad \text{and} \quad c \xleftarrow{R} \begin{cases} \mathsf{E}(\mathrm{pk}_i,\ y) & \text{if } b = 0 \\ \mathsf{E}(\mathrm{pk}_i,\ 0^{|y|}) & \text{if } b = 1 \end{cases}$$

and sends c to $\mathcal{A}$.

finish. Finally, the adversary outputs a bit $b' \in \{0,1\}$.

We say that $\mathcal{A}$ is a $\mathcal{C}$-KDM adversary and that $\mathcal{A}$ wins the game if $b = b'$. Let W be the event that $\mathcal{A}$ wins the game and define $\mathcal{A}$'s advantage as

$$\mathrm{KDM}^{(\mathrm{n})}\mathrm{Adv}[\mathcal{A}, \mathcal{E}](\lambda) := \left| \Pr[W] - \frac{1}{2} \right|$$

Definition 1. *We say that a public-key encryption scheme $\mathcal{E}$ is n-**way KDM-secure with respect to** $\mathcal{C}$ if $\mathrm{KDM}^{(\mathrm{n})}\mathrm{Adv}[\mathcal{A}, \mathcal{E}](\lambda)$ is a negligible function of λ for any adversary $\mathcal{A}$ that runs in expected polynomial time in λ.*

We are primarily interested in function classes $\mathcal{C}$ that imply that the public-key system $\mathcal{E}$ is circular secure. Specifically, we look for function classes $\mathcal{C} := \{f : S^n \to M\}$ that are non-trivial, in the sense that they contain:

- all $|M|$ constant functions $f : S^n \to M$ (recall that a constant function maps all inputs in S^n to some constant $m \in M$), and
- all n selector functions $f_i(x_1, \ldots, x_n) = x_i$ for $1 \le i \le n$.

It is easy to see that KDM-security with respect to such non-trivial function class implies standard semantic security (even for symmetric encryption), since the constant functions let the adversary obtain the encryption of any message of its choice. The selector functions imply *circular security* since they let the adversary obtain $\mathsf{E}(\mathrm{pk}_i,\ \mathrm{sk}_j)$ for all $1 \le i, j \le n$.

The main result in this paper is a public-key system that is KDM-secure relative to a non-trivial function class (and hence also circular-secure). Specifically, we prove security relative to the class of *affine functions* (over the group that is used in the system).

---

[2] Technically $\mathcal{C}$ is a family of sets, parameterized by the security parameter.

2.2   Decision Diffie-Hellman

Let $\mathbb{G}$ be a group of prime order q. We let $\mathcal{P}_{\text{DDH}}$ be the distribution (g, g^x, g^y, g^{xy}) in $\mathbb{G}^4$ where g is a random generator of $\mathbb{G}$ and x, y are uniform in $\mathbb{Z}_q$. We let $\mathcal{R}_{\text{DDH}}$ be the distribution (g, g^x, g^y, g^z), where g is a random generator of $\mathbb{G}$ and x, y, z are uniform in $\mathbb{Z}_q$ subject to $z \neq xy$. A DDH adversary $\mathcal{A}$ takes as input a tuple (g, h, u, v) in $\mathbb{G}^4$ and outputs 0 or 1. Define

$$\text{DDH Adv}[\mathcal{A}, \mathbb{G}] := \left| \Pr[x \xleftarrow{R} \mathcal{P}_{\text{DDH}} \; : \; \mathcal{A}(x) = 1] \; - \; \Pr[x \xleftarrow{R} \mathcal{R}_{\text{DDH}} \; : \; \mathcal{A}(x) = 1] \right|$$

Informally, we say that DDH holds in $\mathbb{G}$ if $\text{DDH Adv}[\mathcal{A}, \mathbb{G}]$ is negligible for all efficient $\mathcal{A}$.

3   A Circular-Secure Encryption Scheme

We build a circular-secure encryption system (for any n) based on the Decision Diffie-Hellman (DDH) assumption. The system is a generalization of the ElGamal system where the secret key is a bit vector rather than an element in $\mathbb{Z}_q$. Let $\mathbb{G}$ be a group of prime order q and g a fixed generator of $\mathbb{G}$. The size of $\mathbb{G}$ is determined by a security parameter λ, in particular, $1/q$ is negligible in λ.

The public-key encryption system $\mathcal{E}$:

- **Key Generation.** Let $\ell := \lceil 3 \log_2 q \rceil$. Choose random $g_1, \ldots, g_\ell$ in $\mathbb{G}$ and a random vector $\boldsymbol{s} = (s_1, \ldots, s_\ell)$ in $\{0, 1\}^\ell$. Let $h \leftarrow (g_1^{s_1} \cdots g_\ell^{s_\ell})^{-1}$ and define the public and secret keys to be

$$\text{pk} := (g_1, \ldots, g_\ell, h) \quad \text{and} \quad \text{sk} := (g^{s_1}, \ldots, g^{s_\ell})$$

  Note that the secret key sk is a random vector $\boldsymbol{s}$ in $\{0, 1\}^\ell$ encoded as a vector of ℓ group elements.
- **Encryption.** To encrypt a group element $m \in \mathbb{G}$, choose a random $r \xleftarrow{R} \mathbb{Z}_q$ and output the ciphertext

$$\left(g_1^r, \; \ldots, \; g_\ell^r, \; h^r \cdot m \right)$$

- **Decryption.** Let $(c_1, \ldots, c_\ell, \; d)$ be a ciphertext and $\text{sk} = (v_1, \ldots, v_\ell)$ a secret key. Do:
    - decode the secret key: for $i = 1, \ldots, \ell$ set $s_i \leftarrow 0$ if $v_i = 1$ and $s_i \leftarrow 1$ otherwise;
    - output $m \leftarrow d \cdot (c_1^{s_1} \cdots c_\ell^{s_\ell})$.

It is easy to verify that the system is correct, that is, the decryption algorithm decrypts properly constructed ciphertexts.

3.1   Discussion and Outline of the Security Proof

Proving that the system is circular secure is somewhat involved. Before proving security, we give some intuition for the construction and its proof. First, consider the basic ElGamal system. The public key is a pair $(g, g^x) \in \mathbb{G}^2$ and the secret key is $x \in \mathbb{Z}_q$. A 1-cycle for this system, namely $\mathsf{E}(\mathrm{pk}, \mathrm{sk})$, is a ciphertext $(g^r, \ e(x) \cdot g^{rx})$ where $e(\cdot)$ is some invertible encoding function mapping $\mathbb{Z}_q$ to $\mathbb{G}$. To prove 1-circular security we would need to show that the 4-tuple

$$\big(g, \ g^x, \ g^r, \ e(x) \cdot g^{rx}\big) \quad \in \mathbb{G}^4 \tag{1}$$

is indistinguishable from a random 4-tuple in $\mathbb{G}^4$, but this is unlikely to follow from DDH.

It is tempting to define the secret key as $\mathrm{sk} := v^x$ (for some generator v), in which case the 1-cycle becomes $(g^r, \ v^x \cdot g^{rx})$. The resulting system can be shown to be 1-circular secure under DDH. Unfortunately, the system does not work since one cannot decrypt ElGamal encryptions using the key $\mathrm{sk} = v^x$.

As a compromise between these two systems, we pick the secret key as an ℓ-bit vector $s \xleftarrow{R} \{0,1\}^\ell$ and store the key as $\mathrm{sk} := (g^{s_1}, \ldots, g^{s_\ell})$. We decrypt using sk by going back to the bit-vector representation. The challenge is to prove n-circular security from the DDH assumption.

Proof Outline. It is instructive to attempt a direct proof that the system $\mathcal{E}$ is 1-circular secure. Observe that in the system $\mathcal{E}$, it is easy to construct "ciphertext vectors" whose decryption are elements of the secret key: For every $1 \leq i \leq \ell$, the $(\ell + 1)$-vector $(1 \ldots 1g1 \ldots 1)$, with g in position i and 1's everywhere else, decrypts to the secret-key element g^{s_i}. Hence the simulator can generate "an encryption of the secret key" without knowing the secret key itself. This almost suffices for the proof, except that these vectors are not really valid ciphertext vectors, since the encryption algorithm would never output them.

We therefore begin by moving to an "expanded variant" of our system (which we call $\mathcal{E}_1$) that has the same decryption procedure, but where every $(\ell + 1)$-vector is a valid ciphertext (see description later in Section 3.2). Moreover, $\mathcal{E}_1$ has the same "blinding" properties as ElGamal, so the simulator can produce not just one encryption of the secret key but *a random encryption* of it. This is enough to prove that the expanded system $\mathcal{E}_1$ is 1-circular secure. Moving from 1-circular to n-circular security is done using homomorphic properties: $\mathcal{E}_1$ is homomorphic with respect to both the plaintext and secret key, so it is possible to translate an encryption of m with respect to secret key s to an encryption of $m \cdot d$ with respect to secret key $s \oplus \delta$, just by knowing d and δ. This allows the simulator to first produce a 1-cycle and then expand it into an n-clique. The circular security (or even clique-security) of $\mathcal{E}_1$ follows.

Finally we deduce the security of $\mathcal{E}$ from that of $\mathcal{E}_1$, roughly because the adversary gets "strictly less information" when attacking $\mathcal{E}$ than when attacking the expanded variant $\mathcal{E}_1$.

3.2 Proof of Circular-Security

We now prove that the system $\mathcal{E}$ provides circular security. As mentioned above, we actually prove a slightly stronger statement, namely that $\mathcal{E}$ is KDM-secure with respect to the set of affine functions.

Affine functions. The set of affine functions acting on S^n is defined as follows. Let $\text{sk}_1, \ldots, \text{sk}_n$ be n secret keys generated by G (each an ℓ-vector over $\mathbb{G}$). Let s be the vector in $\mathbb{G}^{n\ell}$ obtained by concatenating these n secret keys. For every $n\ell$-vector $\boldsymbol{u} = (u_i)$ over $\mathbb{Z}_q$ and every scalar $h \in \mathbb{G}$, there is a natural map from $\mathbb{G}^{n\ell}$ to $\mathbb{G}$, that can be informally described as $f_{\boldsymbol{u},h} : \quad \boldsymbol{s} \longrightarrow (\boldsymbol{u} \cdot \boldsymbol{s} + h)$. More precisely, we have

$$f_{\boldsymbol{u},h}(\boldsymbol{s}) \stackrel{\text{def}}{=} \prod_{i=1}^{n\ell} s_i^{u_i} \cdot h \quad \in \mathbb{G}.$$

We call $f_{\boldsymbol{u},h}$ an affine function from $\mathbb{G}^{n\ell}$ to $\mathbb{G}$.

Definition 2. *The set of affine functions $\mathcal{C}_{n\ell}$ is the set of all functions $f_{\boldsymbol{u},h}$, where $\boldsymbol{u} \in \mathbb{Z}_q^{n\ell}$ and $h \in \mathbb{G}$.*

The set $\mathcal{C}_{n\ell}$ acts on n-tuples of secret keys by viewing the n-tuple as a vector in $\mathbb{G}^{n\ell}$, and it maps every such vector to an element of $\mathbb{G}$.

KDM-security theorem with respect to $\mathcal{C}_{n\ell}$. The following theorem shows that if the DDH assumption holds in the group $\mathbb{G}$ then $\mathcal{E}$ is n-way KDM-secure with respect to the set $\mathcal{C}_{n\ell}$ of affine functions, for any $n = n(\lambda)$ that is polynomial in the security parameter. Circular security follows since $\mathcal{C}_{n\ell}$ contains the constant and selector functions.[3]

Theorem 1. *For any $n > 0$ and for any $\mathcal{C}_{n\ell}$-KDM adversary $\mathcal{A}$, there exists a DDH adversary $\mathcal{B}$ (whose running time is about the same as that of $\mathcal{A}$) such that*

$$\text{KDM}^{(\text{n})}\text{Adv}[\mathcal{A}, \mathcal{E}] \leq (3\ell - 2) \cdot \text{DDH Adv}[\mathcal{B}, \mathbb{G}] + 1/q$$

Note that this bound is independent of n.

Switching to Additive Notation. Our proof requires a fair amount of linear algebra. To make it clearer, we will be using additive notation for the group $\mathbb{G}$; note that $\mathbb{G}$ and $\mathbb{G}^k$ are vector spaces over $\mathbb{Z}_q$. (Recall that we already used additive notation when we informally described the class of affine functions above.) To avoid ambiguity, we use Latin letters for elements of $\mathbb{Z}_q$ and Greek letters for elements of $\mathbb{G}$. In particular, let $\mathbb{G}$ be generated by γ. We use lower-case letters

---

[3] The set $\mathcal{C}_{n\ell}$ includes "selector functions" for each element from each secret key (rather than "selector functions" that select entire keys). This makes no difference in our case, since our scheme encrypts element by element (so a secret key is encrypted as ℓ separate ciphertext vectors, one for each element of the secret key).

for scalars and column vectors, and upper-case letters for matrices. We write $\boldsymbol{\mu} \cdot \boldsymbol{v}$ for the inner product and $\boldsymbol{\mu} \times \boldsymbol{v}$ for the outer product:

$$\boldsymbol{\mu} \cdot \boldsymbol{v} := \boldsymbol{\mu}^{\top} \boldsymbol{v} = \sum_i \mu_i v_i \qquad \text{and} \qquad \boldsymbol{\mu} \times \boldsymbol{v} := \boldsymbol{\mu}\, \boldsymbol{v}^{\top} = (\mu_i v_j)_{i,j} \in \mathbb{G}^{\dim(\mu) \times \dim(v)}$$

When we write the product of a matrix or vector of elements of $\mathbb{G}$ with a matrix or vector of elements of $\mathbb{Z}_q$, we mean to use the standard formula. For example, by $\boldsymbol{\mu} \cdot \boldsymbol{v}$ we mean $\sum_i \mu_i v_i$, which would be written in multiplicative notation as $\prod_i \mu_i^{v_i}$. It is easily seen that the usual properties of vectors and matrices still hold in this notation (for any term involving at most one literal from $\mathbb{G}$ and all other literals from $\mathbb{Z}_q$).

We use 0 to denote both the number zero and the identity element in $\mathbb{G}$, the meaning will be clear from the context. We write 0^ℓ for a column vector of ℓ zeros, and $0^{k \times \ell}$ for a $k \times \ell$ matrix of zeros. We write Id_i for the identity matrix in $\mathbb{Z}_q^{i \times i}$. When A, B are two matrices with the same number of rows (and both over $\mathbb{G}$ or both over $\mathbb{Z}_q$), then we write $(A|B)$ for the augmented matrix consisting of all the columns of A followed by all the columns of B.

The definitions of linear independence of vectors, vector spaces and subspaces, rank of matrices, etc., are all standard, and we use the same notions for both $\mathbb{G}$ and $\mathbb{Z}_q$. We write $\mathrm{Rk}_i\left(\mathbb{Z}_q^{a \times b}\right)$ (resp $\mathrm{Rk}_i\left(\mathbb{G}^{a \times b}\right)$) for the set of matrices in $\mathbb{Z}_q^{a \times b}$ (resp $\mathbb{G}^{a \times b}$) with rank i. As a special case, we write $\mathrm{GL}_i(\mathbb{Z}_q)$ for the invertible $i \times i$ matrices over $\mathbb{Z}_q$.

The System $\mathcal{E}$ in Additive Notation

- **Key Generation.** Let $\ell := \lceil 3\log_2 q \rceil$. Choose a random nonzero vector $\boldsymbol{\psi} \xleftarrow{R} \mathbb{G}^\ell$ and a random vector $\boldsymbol{s}$ in $\{0,1\}^\ell \subset \mathbb{Z}_q^\ell$. Let $\delta \leftarrow -\boldsymbol{\psi} \cdot \boldsymbol{s} \in \mathbb{G}$ and define the public and secret keys to be

$$\mathrm{pk} := \left(\boldsymbol{\psi}^{\top} | \delta\right) \in \mathbb{G}^{1 \times (\ell+1)} \qquad \text{and} \qquad \mathrm{sk} := \boldsymbol{s}\gamma \in \mathbb{G}^\ell$$

Though the secret key is encoded as $\mathrm{sk} = \boldsymbol{s}\gamma \in \mathbb{G}^\ell$, below it will be convenient to consider also the decoded form. Specifically, we refer to the $\ell+1$ binary vector $\boldsymbol{s}' = \left(\boldsymbol{s}^{\top} | 1\right)^{\top} \in \mathbb{Z}_q^{\ell+1}$ as the *decoded secret key*.

- **Encryption.** To encrypt a message $\mu \in \mathbb{G}$, choose a random $r \xleftarrow{R} \mathbb{Z}_q^n$ and output the ciphertext row-vector

$$\boldsymbol{\xi}^{\top} \leftarrow \left(r\boldsymbol{\psi}^{\top} | r\delta + \mu\right) = r\,\mathrm{pk} + \left(0^{1 \times \ell} | \mu\right) \quad \in \mathbb{G}^{1 \times (\ell+1)} \tag{2}$$

- **Decryption.** Let $\boldsymbol{\xi}^{\top} \in \mathbb{G}^{1 \times (\ell+1)}$ be the ciphertext. Decryption is just an inner product between the ciphertext and the decoded secret key:

$$\mu \leftarrow \boldsymbol{\xi} \cdot \left(\boldsymbol{s}^{\top} | 1\right)^{\top}$$

Decryption works since the decoded secret key $\left(\boldsymbol{s}^{\top} | 1\right)^{\top}$ is orthogonal to the public key pk.

We observe that an $\ell + 1$ vector over $\mathbb{G}$ is decrypted to zero if and only if it belongs to the subspace orthogonal to the decoded secret key $\boldsymbol{s'}$, and every coset of this subspace is decrypted to a different element of $\mathbb{G}$. On the other hand, "valid encryptions of zero" (i.e., the ones obtained from the encryption algorithm) are taken from a small subspace of the vectors orthogonal to $\boldsymbol{s'}$, namely the one-dimensional subspace spanned by the public key pk. Similarly, "valid encryptions" of other elements are obtained as shifting this one-dimensional subspace by multiples of $\left(0^{1\times\ell}|1\right)$.

A Few Lemmata. We present some simple lemmata and facts about $\mathcal{E}$. First, we show that DDH implies that it is difficult to determine the rank of a matrix of group elements. In particular, it is difficult to distinguish a random matrix of rank r_1 from a random matrix of rank $r_2 > r_1$.

Lemma 1 (Matrix DDH). *Let* $1 \leq r_1 < r_2 \leq a, b$ *be positive integers, and let* $\mathcal{A} : \mathbb{G}^{a\times b} \to \{0,1\}$ *be a polynomial-time algorithm. Write*

$$P(\mathcal{A}, i) := \Pr\left[\Phi \xleftarrow{R} \mathrm{Rk}_i\left(\mathbb{G}^{a\times b}\right) \; : \; \mathcal{A}(\Phi) = 1\right]$$

Then there is a DDH adversary $\mathcal{B}$, *running in about the same time as* $\mathcal{A}$, *such that*

$$\left|P(\mathcal{A}, r_2) - P(\mathcal{A}, r_1)\right| \leq (r_2 - r_1)\, \mathrm{DDH\,Adv}[\mathcal{B}, \mathbb{G}]$$

Proof. We use a hybrid argument between the $r_2 - r_1 + 1$ distributions

$$\mathrm{Rk}_i\left(\mathbb{G}^{a\times b}\right) \quad \text{where} \quad i \in [r_1, r_2]$$

The algorithm $\mathcal{B}$ is given a DDH challenge $(\alpha_1, \alpha_2, \alpha_3, \alpha_4)$. It picks a random $i \xleftarrow{R} [r_1 + 1, r_2]$ and sets

$$\Phi_1 := \begin{pmatrix} \begin{array}{|c|c|c|c} \hline \alpha_1 & \alpha_2 & & \\ \hline \alpha_3 & \alpha_4 & & \\ \hline & & \gamma\,\mathrm{Id}_{i-2} & \\ \hline & & & 0^{(a-i)\times(b-i)} \end{array} \end{pmatrix} \in \mathbb{G}^{a\times b}$$

with all the other blocks zero. $\mathcal{B}$ then chooses

$$L \xleftarrow{R} \mathrm{GL}_a(\mathbb{Z}_q) \quad \text{and} \quad R \xleftarrow{R} \mathrm{GL}_b(\mathbb{Z}_q) \quad \text{and sets} \quad \Phi_2 := L\,\Phi_1\,R$$

$\mathcal{B}$ now calls $\mathcal{A}(\Phi_2)$ and outputs whatever $\mathcal{A}$ outputs.

Now if $(\alpha_1, \alpha_2, \alpha_3, \alpha_4)$ was drawn from $\mathcal{P}_{\mathrm{DDH}}$, then Φ_1 has rank $i - 1$, and Φ_2 is uniform in $\mathrm{Rk}_{i-1}\left(\mathbb{G}^{a\times b}\right)$. But if $(\alpha_1, \alpha_2, \alpha_3, \alpha_4)$ was drawn from $\mathcal{R}_{\mathrm{DDH}}$, then Φ_1 has rank i, and Φ_2 is uniform in $\mathrm{Rk}_i\left(\mathbb{G}^{a\times b}\right)$. The lemma then follows by the standard hybrid argument. $\qquad\square$

We will also need the following lemma on universal hashing. Recall that a distribution $\mathcal{D}$ on a set $\mathcal{X}$ is ϵ-uniform if $\sum_{x\in X}\left|\mathcal{D}(x) - \frac{1}{|\mathcal{X}|}\right| \leq \epsilon$.

Lemma 2 (Simplified left-over hash lemma). *Let $\mathcal{H}$ be a 2-universal hash family from a set $\mathcal{X}$ to a set $\mathcal{Y}$. Then the distribution*

$$(H, H(x)) \quad \text{where} \quad H \xleftarrow{R} \mathcal{H} \quad \text{and} \quad x \xleftarrow{R} \mathcal{X}$$

is $\sqrt{\frac{|\mathcal{Y}|}{4|\mathcal{X}|}}$-uniform on $\mathcal{H} \times \mathcal{Y}$.

Proof. This is an immediate corollary from [13] (see also [21, Theorem 6.21]). □

Recall that the secret key in our system is a vector in $\{0,1\}^\ell$ where $\ell = \lceil 3 \log_2 q \rceil$. We therefore obtain the following corollary of Lemma 2.

Corollary 1. *Let $r \xleftarrow{R} \mathbb{Z}_q^\ell$, and $s \xleftarrow{R} \{0,1\}^\ell$. Then $\left(r^\top \middle| -r \cdot s\right)^\top$ is $\frac{1}{q}$-uniform in $\mathbb{Z}_q^{n+1}$.*

Proof. Let $H_r(s) := -r \cdot s$. Then $\{H_r : r \in \mathbb{Z}_q^\ell\}$ is 2-universal, so that $\left(r^\top \middle| -r \cdot s\right)^\top$ is $\sqrt{\frac{q}{4 \cdot 2^\ell}}$-uniform in $\mathbb{Z}_q^{n+1}$. Since $\ell = \lceil 3 \log_2 q \rceil$ we have $\sqrt{\frac{q}{4 \cdot 2^\ell}} \leq \frac{1}{2q} < \frac{1}{q}$. □

We note that Erdös and Hall [9] proved a slightly stronger version of Corollary 1 — they obtain a similar result with a smaller ℓ (i.e. $\ell \approx \lceil 2 \log_2 q \rceil$). This enables us to slightly shorten our public and secret keys. However, the proof of Corollary 1 using the left over hash lemma is more general and enables us to prove security of an extension discussed in Section 4.

The Expanded System $\mathcal{E}_1$. As discussed in Section 3.1, a technical difficulty in the proof is that not every $(\ell + 1)$-vector over $\mathbb{G}$ is a valid ciphertext in $\mathcal{E}$. We therefore introduce an "expanded version" of our scheme (denoted $\mathcal{E}_1$) that has the same secret key and decryption procedure, but a larger public key. In this system every vector in $\mathbb{G}^{1 \times (\ell+1)}$ is a valid ciphertext. We later prove that $\mathcal{E}_1$ is n-way KDM-secure with respect to $\mathcal{C}_{n\ell}$, and then use it to deduce also the KDM-security of the original system $\mathcal{E}$.

- **Key Generation.** Let $\ell = \lceil 3 \log_2 q \rceil$. Choose a random secret key $s \xleftarrow{R} \{0,1\}^\ell \subset \mathbb{Z}_q^\ell$. Choose a random matrix $\Psi \xleftarrow{R} \mathrm{Rk}_\ell\left(\mathbb{G}^{(\ell+1) \times \ell}\right)$, and set $\Phi := \left(\Psi \middle| -\Psi s\right) \in \mathbb{G}^{(\ell+1) \times (\ell+1)}$. Define the public and secret keys to be

$$\mathrm{pk} := \Phi \quad \text{and} \quad \mathrm{sk} := s\gamma$$

  That is, the secret key is as in the system $\mathcal{E}$, but we use an expanded public key Φ, which is a matrix of $\ell + 1$ public keys from $\mathcal{E}$ (all with respect to the same secret key s).
- **Encryption.** To encrypt an element $\mu \in \mathbb{G}$, choose a random row vector $r \xleftarrow{R} \mathbb{Z}_q^{1 \times (\ell+1)}$ and output the ciphertext

$$\xi \leftarrow r\Phi + \left(0^{1 \times \ell} \middle| \mu\right) \in \mathbb{G}^{1 \times (\ell+1)}$$

  This is similar to the original system $\mathcal{E}$, except that instead of a random multiple of the public-key vector as in Eq. (2), here we use a random linear combination of all the rows of the expanded public key Φ.

- **Decryption.** Decryption is the same as in $\mathcal{E}$. Decryption works since the decoded secret key $\left(s^\top | 1 \right)^\top$ is orthogonal to all the rows of the expanded public key Φ.

We stress that the main difference between $\mathcal{E}$ and $\mathcal{E}_1$ is that in $\mathcal{E}$ the public key is just one vector orthogonal to the decoded secret key. In $\mathcal{E}_1$, on the other hand, the expanded public key spans the entire (ℓ-dimensional) subspace orthogonal to the decoded secret key. Jumping ahead, we will later show that under DDH, the adversary cannot distinguish between ciphertext vectors in $\mathcal{E}$ (taken from a 1-dimensional subspace) and ciphertext vectors in $\mathcal{E}_1$ (taken from an ℓ-dimensional subspace). Thus essentially the only difference from the adversary's perspective is that in $\mathcal{E}_1$ it sees more vectors in the public key.

In the proof below we use the following simple facts about $\mathcal{E}_1$:

Totality and uniformity. For any secret key sk with public key Φ and any element μ, if a ciphertext $\boldsymbol{\xi}$ decrypts to μ using sk, then $\boldsymbol{\xi}$ is a possible output of $\mathsf{E}(\Phi, \mu)$, i.e. a valid encryption of μ. Furthermore, all possible outputs of $\mathsf{E}(\Phi, \mu)$ are equally likely.

Public-key blinding. Let $\Phi \in \mathbb{G}^{(\ell+1)\times(\ell+1)}$ be a public key for some secret key sk and let R be a random invertible matrix, $R \xleftarrow{R} \mathrm{GL}_{\ell+1}(\mathbb{Z}_q)$. Then blind-pk$(\Phi) := R\Phi$ is a uniformly random public key for sk. Furthermore, encryption with Φ and with $R\Phi$ produce the same distribution of ciphertexts.

Ciphertext blinding. Let $\Phi \in \mathbb{G}^{(\ell+1)\times(\ell+1)}$ be a public key, and let $\boldsymbol{\xi}$ be any encryption of $\mu \in \mathbb{G}$ with respect to Φ. Let $\boldsymbol{r} \xleftarrow{R} \mathbb{Z}_q^{1\times(\ell+1)}$ be a random row vector, then blind-ct$(\Phi, \boldsymbol{\xi}) := \boldsymbol{r}\Phi + \boldsymbol{\xi}$ draws uniformly at random from $\mathsf{E}(\Phi, \mu)$.

Total blinding. If instead of being a valid public key, Φ is a matrix of full rank $\ell+1$, then the output of blind-ct$(\Phi, \boldsymbol{\xi})$ is uniformly random in $\mathbb{G}^{1\times(\ell+1)}$.

Self-referential encryption. Let sk $= (\gamma_1, \dots, \gamma_\ell)^\top \in \mathbb{G}^\ell$ be a secret key with public key Φ. Denoting by $e_i \in \{0,1\}^\ell$ the unit vector with 1 in position i and 0 elsewhere, we have that $\left(\gamma e_i | 0 \right)$ is an encryption of the secret-key element γ_i with respect to Φ.

Plaintext homomorphism. Let $f(\boldsymbol{\mu}) = \boldsymbol{a} \cdot \boldsymbol{\mu} + \beta$ be an affine function from $\mathbb{G}^n$ to $\mathbb{G}$. Fix some vector $\boldsymbol{\mu} \in \mathbb{G}^n$, let Φ be a public key, and let $\varXi \in \mathbb{G}^{n\times(\ell+1)}$ be a matrix whose i'th row is an encryption of μ_i with respect to Φ. Then $\boldsymbol{a}\varXi + \left(0^{1\times\ell} | \beta \right)$ is an encryption of $f(\boldsymbol{\mu})$ with respect to Φ.

Secret-key homomorphism. Let $\boldsymbol{s} \in \{0,1\}^\ell$ be used for a secret key with public key Φ, and let $\boldsymbol{\xi} \xleftarrow{R} \mathsf{E}(\Phi, \mu)$ be an encryption of an element $\mu \in \mathbb{G}$. Let $f(\boldsymbol{x}) = A\boldsymbol{x} + \boldsymbol{b}$ be an invertible affine function from $\mathbb{Z}_q^\ell$ to $\mathbb{Z}_q^\ell$, and set

$$M_f := \left(\begin{array}{c|c} A & \boldsymbol{b} \\ \hline 0^{1\times\ell} & 1 \end{array} \right) \quad \text{so that} \quad M_f \left(\boldsymbol{x}^\top | 1 \right)^\top = \left(f(\boldsymbol{x})^\top | 1 \right)^\top$$

Suppose that $f(\boldsymbol{s}) \in \{0,1\}^\ell$ (so $f(\boldsymbol{s})$ can be used for a secret key). Then ΦM_f^{-1} is a public key for $f(\boldsymbol{s})$, and $\boldsymbol{\xi} M_f^{-1}$ is an encryption of μ with public key ΦM_f^{-1}. In particular, extend the xor function $\oplus$ to $\mathbb{Z}_q \times \{0,1\} \to \mathbb{Z}_q$ by

$$x \oplus 0 := x \quad \text{and} \quad x \oplus 1 := 1 - x$$

and extend it to vectors by applying it element-wise. Then for a fixed $\boldsymbol{a} \in \{0,1\}^\ell$, the function $f(\boldsymbol{s}) := \boldsymbol{s} \oplus \boldsymbol{a}$ is an affine function, so we can compute a public key and ciphertext vectors for $\boldsymbol{s} \oplus \boldsymbol{a}$ from a public key and ciphertext vectors for $\boldsymbol{s}$.

$\mathcal{E}_1$ Is KDM-Secure with Respect to $\mathcal{C}_{n\ell}$

Theorem 2. *For any $\mathcal{C}_{n\ell}$-KDM-adversary $\mathcal{A}$ against $\mathcal{E}_1$ there exists a DDH-adversary $\mathcal{B}$ (whose running time is about the same as that of $\mathcal{A}$) such that*

$$\mathrm{KDM}^{(\mathrm{n})}\mathrm{Adv}[\mathcal{A}, \mathcal{E}_1] \leq (2\ell - 1)\,\mathrm{DDH}\,\mathrm{Adv}[\mathcal{B}, \mathbb{G}] + 1/q$$

Proof. We present this proof as a series of games, and we let w_i denote the probability that the adversary wins Game i.

Game 0. This game is identical to the $\mathcal{C}_{n\ell}$-KDM-security game defined in Section 2.1. By definition,

$$\left| w_0 - \frac{1}{2} \right| = \mathrm{KDM}^{(\mathrm{n})}\mathrm{Adv}[\mathcal{A}, \mathcal{E}_1] \tag{3}$$

Game 1. Game 1 looks the same as Game 0 to the adversary, but the challenger does not use the secret keys internally. For setup:

- The challenger generates a secret key $\boldsymbol{s} \xleftarrow{R} \{0,1\}^\ell$ with public key Φ, and then "forgets" $\boldsymbol{s}$. That is, the challenger does not use $\boldsymbol{s}$ for the rest of Game 1.
- The challenger chooses n random vectors $\boldsymbol{a}_1, \ldots, \boldsymbol{a}_n \xleftarrow{R} \{0,1\}^\ell$. It then produces a view to the adversary that is consistent with the n secret keys $\mathrm{sk}_i := (\boldsymbol{s} \oplus \boldsymbol{a}_i)\gamma$, but without ever using the "forgotten" $\boldsymbol{s}$.
- For each $i \in [1, n]$, the challenger uses the **secret-key homomorphism** and **public-key blinding** properties of $\mathcal{E}_1$ to generate a uniformly random public key pk_i for sk_i from $(\Phi, \boldsymbol{a}_i)$.

For brevity, let $\boldsymbol{\sigma} := (\; \mathrm{sk}_1^\top \mid \mathrm{sk}_2^\top \mid \ldots \mid \mathrm{sk}_n^\top \;)^\top$ denote the concatenation of the encoded secret keys (but the challenger does not use the value of $\boldsymbol{\sigma}$). To compute $\mathsf{E}(\mathrm{pk}_i, f(\boldsymbol{\sigma}))$ for an affine function f in $\mathcal{C}_{n\ell}$:

- For each $j \in [1, n]$, the challenger uses the **self-referential encryption** property to generate an encryption $\mathsf{E}(\mathrm{pk}_j, \mu)$ for every element $\mu \in \mathrm{sk}_j$, and uses **secret-key homomorphism** to transform it into an encryption under pk_i, $\mathsf{E}(\mathrm{pk}_i, \mu)$.
- The challenger concatenates these to obtain a matrix $\varXi$ of encryptions under pk_i of all the elements in $\boldsymbol{\sigma}$.

- The challenger uses the **plaintext homomorphism** property to generate an encryption $\xi \leftarrow \mathsf{E}(\mathrm{pk}_i, f(\boldsymbol{\sigma}))$.
- The challenger sends blind-ct$(\mathrm{pk}_i, \boldsymbol{\xi})$ to the adversary.

The distribution of secret keys, public keys and ciphertexts is identical to Game 0, so

$$w_1 = w_0 \tag{4}$$

Informally, the challenger has used a single public key Φ to generate an entire clique of ciphertexts, without knowing any of their secret keys. It remains to show formally that this gives the adversary no useful information.

The remaining games will be identical to Game 1, except that the initial public key Φ will be computed differently.

Game 2. In Game 2, the challenger does:

$$\Psi \xleftarrow{R} \mathrm{Rk}_1 \left(\mathbb{G}^{(\ell+1)\times\ell} \right) \quad \text{and} \quad \Phi \leftarrow \left(\Psi \big| -\Psi s \right) \in \mathbb{G}^{(\ell+1)\times(\ell+1)}$$

This is the same procedure used in Game 1, except that now Ψ has rank 1 instead of rank ℓ. Lemma 1 tells us that there is a DDH-adversary $\mathcal{B}$, running in about the same time as $\mathcal{A}$, such that

$$|w_2 - w_1| \leq (\ell - 1) \, \mathrm{DDH} \, \mathrm{Adv}[\mathcal{A}, \mathbb{G}] \tag{5}$$

Note that Ψ here may be computed by choosing random nonzero vectors $\psi \xleftarrow{R} \mathbb{G}^{\ell+1}$ and $r \xleftarrow{R} \mathbb{Z}_q^\ell$, and setting $\Psi \leftarrow \psi \times r$. Thus we see that $\Phi = \psi \times \left(r^\top \big| -r \cdot s \right)^\top$ is $(1/q)$-uniform in $\mathrm{Rk}_1 \left(\mathbb{G}^{(\ell+1)\times(\ell+1)} \right)$ by Corollary 1.

Game 3. Since Φ is $(1/q)$-uniform in $\mathrm{Rk}_1 \left(\mathbb{G}^{(\ell+1)\times(\ell+1)} \right)$, we can replace it by a random matrix in $\mathrm{Rk}_1 \left(\mathbb{G}^{(\ell+1)\times(\ell+1)} \right)$. Thus, Game 3 is the same as Game 2, except that $\Phi \xleftarrow{R} \mathrm{Rk}_1 \left(\mathbb{G}^{(\ell+1)\times(\ell+1)} \right)$. Then

$$|w_3 - w_2| \leq 1/q \tag{6}$$

Note that in Game 3 the secret s is not used anywhere.

Game 4. Game 4 is the same as Game 3, except that $\Phi \xleftarrow{R} \mathrm{Rk}_{\ell+1} \left(\mathbb{G}^{(\ell+1)\times(\ell+1)} \right)$. By the **total blinding** property of $\mathcal{E}_1$, the ciphertexts returned to the adversary are all uniformly random, regardless of the challenger's bit b. Therefore,

$$w_4 = \frac{1}{2} \tag{7}$$

On the other hand, by lemma 1, there exists a DDH-adversary $\mathcal{B}$, running in about the same time as $\mathcal{A}$, such that

$$|w_4 - w_3| \leq \ell \, \mathrm{DDH} \, \mathrm{Adv}[\mathcal{B}, \mathbb{G}] \tag{8}$$

Combining equations (3) through (8), we find that

$$\mathrm{KDM}^{(\mathrm{n})} \mathrm{Adv}[\mathcal{A}, \mathcal{E}_1] \leq (2\ell - 1) \, \mathrm{DDH} \, \mathrm{Adv}[\mathcal{B}, \mathbb{G}] + 1/q$$

This completes the proof of Theorem 2. $\qquad\qquad\square$

$\mathcal{E}$ **Is KDM-Secure with Respect to** $\mathcal{C}_{n\ell}$**.** We now deduce the KDM-security of $\mathcal{E}$ from that of $\mathcal{E}_1$.

Lemma 3. *For any* $\mathcal{C}_{n\ell}$*-KDM adversary* $\mathcal{A}$ *against* $\mathcal{E}$*, there is a DDH-adversary* $\mathcal{B}_1$ *and a* $\mathcal{C}_{n\ell}$*-KDM adversary* $\mathcal{B}_2$ *against* $\mathcal{E}_1$*, both running in about the same time as* $\mathcal{A}$*, such that*

$$\mathrm{KDM}^{(\mathrm{n})}\mathrm{Adv}[\mathcal{A}, \mathcal{E}] \leq (\ell - 1) \cdot \mathrm{DDH}\,\mathrm{Adv}[\mathcal{B}_1, \mathbb{G}] + \mathrm{KDM}^{(\mathrm{n})}\mathrm{Adv}[\mathcal{B}_2, \mathcal{E}_1]$$

Proof. We present the proof as a series of games. Let w_i denote the probability that the adversary $\mathcal{A}$ wins Game i.

Game 0. Game 0 is identical to the $\mathcal{C}_{n\ell}$-KDM-security game with respect to $\mathcal{E}$ defined in Section 2.1. By definition,

$$\left| w_0 - \frac{1}{2} \right| = \mathrm{KDM}^{(\mathrm{n})}\mathrm{Adv}[\mathcal{A}, \mathcal{E}]$$

Game 1. Game 1 is the same as Game 0, except that the challenger generates public keys and encryptions in a different but equivalent way. Specifically,

- The challenger chooses a random rank-1 matrix $\Psi_0 \xleftarrow{R} \mathrm{Rk}_1\left(G^{(\ell+1)\times\ell}\right)$.
- The challenger chooses n secret keys $s_i \xleftarrow{R} \{0, 1\}^\ell$, for $i = 1, \ldots, n$. It creates the corresponding n public keys as follows. For $i \in [1, n]$ generate the public key pk_i by choosing two random invertible matrices $L_i \xleftarrow{R} \mathrm{GL}_{\ell+1}(\mathbb{Z}_q)$ and $R_i \xleftarrow{R} \mathrm{GL}_\ell(\mathbb{Z}_q)$ and setting

$$\Psi_i \leftarrow L_i \Psi_0 R_i \qquad \text{and} \qquad \mathrm{pk}_i := \Phi_i \leftarrow \left(\Psi_i \big| -\Psi_i s_i\right).$$

  Note that the matrix Ψ_i is a uniformly random rank-1 matrix and is independent of Ψ_0.
- For each $i \in [1, n]$, the challenger chooses a random nonzero row of the public key Φ_i, and sends it to the adversary as the $\mathcal{E}$-public-key φ_i. This row is nonzero, random and orthogonal to s_i by construction, so it is a valid public key for s_i under $\mathcal{E}$.
- When answering queries, instead of encrypting a message μ with φ_i under the system $\mathcal{E}$, the challenger encrypts it under $\mathcal{E}_1$ using Φ_i as the public key. In other words, it responds with $R\Phi_i + \left(0|\mu\right)$ where $R \xleftarrow{R} \mathbb{Z}_q^{n\times(\ell+1)}$. Note that Φ_i is not a valid public key for $\mathcal{E}_1$, but only because it has rank 1 instead of rank ℓ.

Because Φ_i has rank 1, all rows of Φ_i are multiples of φ_i. Therefore, the distributions of ciphertexts $r \times \varphi_i + \left(0|\mu\right)$ under $\mathcal{E}$ and $R\Phi_i + \left(0|\mu\right)$ in Game 1 are identical. The distributions of public and secret keys are also identical, so the attacker sees the same distribution of messages as in Game 0. As a result, $w_1 = w_0$.

Game 2. Game 2 is the same as Game 1, except that the challenger chooses $\Psi_0 \xleftarrow{R} \mathrm{Rk}_\ell\left(\mathbb{G}^{(\ell+1)\times\ell}\right)$ so that Φ is a random, valid public key under $\mathcal{E}_1$. This is the only difference between Games 1 and 2. By Lemma 1, there is a DDH adversary $\mathcal{B}_1$, running in about the same time as $\mathcal{A}$, such that

$$|w_2 - w_1| \leq (\ell - 1) \, \text{DDH Adv}[\mathcal{B}_1, \mathbb{G}]$$

At this point the attacker is attacking $\mathcal{E}_1$, with all but one row of the public keys hidden. Call this process $\mathcal{B}_2$; then

$$\left| w_2 - \frac{1}{2} \right| = \text{KDM}^{(n)} \text{Adv}[\mathcal{B}_2, \mathcal{E}_1]$$

so that

$$\text{KDM}^{(n)} \text{Adv}[\mathcal{A}, \mathcal{E}] \leq (\ell - 1) \, \text{DDH Adv}[\mathcal{B}_1, \mathbb{G}] + \text{KDM}^{(n)} \text{Adv}[\mathcal{B}_2, \mathcal{E}_1]$$

as claimed.   □

Theorem 1 now follows by combining Theorem 2 with Lemma 3.   □

4   Extensions

Security under the linear assumption. The linear assumption, introduced in [5], is a weaker assumption than DDH. Weaker versions of the linear assumption were studied in [14, 20]. The proof of Theorem 2 generalizes easily to use these weaker versions of the linear assumption. In particular, to use the r-linear assumption one need only change the value of ℓ to $\ell := \lceil (r+2) \log_2 q \rceil$. This hurts efficiency, but bases security on a weaker assumption. Note that the DDH assumption is identical to the 1-linear assumption.

Shrinking the ciphertext and secret keys. Ciphertexts and secret keys in our system contain $\ell := \lceil 3 \log_2 q \rceil$ elements in $\mathbb{G}$ where $q = |\mathbb{G}|$. This size of ℓ is chosen so that secret keys have sufficient entropy to make the distribution in Corollary 1 be $(1/q)$-uniform.

Recall that the secret key sk in our system is an encoding of a vector $s \in \{0,1\}^\ell$, namely $\text{sk}_i := g^{s_i}$ for $i = 1, \ldots, \ell$. The vector s had to be binary for two reasons. First, during decryption we need to recover s from its encoding sk. Second, the proof of Theorem 2 relied on the fact that a vector $s \in \{0,1\}^\ell$ can be mapped to a random vector in $\{0,1\}^\ell$ using an appropriate random affine map (i.e. by xoring with a known random vector in $\{0,1\}^\ell$, which is an affine map).

Let T be the set of ℓ-tuples that contains all $\ell!$ permutations of $(1, 2, \ldots, \ell)$. It is not hard to see that T satisfies the two properties mentioned above: (1) if we encode an ℓ-tuple in T by exponentiation as before then decoding can be done efficiently during decryption, and (2) an element $s \in T$ can be mapped to a random element in T by the appropriate random affine transformation, namely a random permutation matrix. Hence, the proof of the main theorem (Theorem 1) will go through unchanged if algorithm G chooses s at random in the set T.

Since the set T is larger than the set $\{0,1\}^\ell$ — the former is of size $\ell!$ while the latter is of size 2^ℓ — we can use a smaller value of ℓ and still satisfy the entropy bounds of Corollary 1. In particular, it suffices to choose

$$\ell = \left\lceil \frac{4.5 \log_2 q}{\log_2 \log_2 q} \right\rceil \quad \text{so that} \quad \ell! > q^3$$

This shrinks ciphertexts and secret keys by a factor of $O(\log \log q)$ over the original system.

5   One-Way Encryption That Is Not 2-Circular Secure

Beyond constructing encryption systems for which we can prove circular security, one may ask the more fundamental question of "what does it really take" to get circular security. For example, can we obtain circular-secure encryption from CPA-secure encryption? Recent work casts doubt on our ability to prove such implications using standard tools [11], but does not shed light on the deeper question of the truth of it. In fact, today we cannot even rule out the possibility that every CPA-secure system is also n-circular secure for all $n \geq 2$.

In this section we try to make some progress toward ruling out this possibility. Ideally, one would like to exhibit a CPA-secure system that is not (say) 2-circular secure. Unfortunately, we did not find a candidate system. Instead, we show a weaker example of a one-way encryption system that breaks completely once an n-cycle of encryptions is published (for any n).

One-way encryption is a very weak notion of security, requiring only that an attacker cannot recover the *entire* plaintext after seeing the ciphertext. A little more precisely, an encryption scheme $(\mathsf{G}, \mathsf{E}, \mathsf{D})$ is one-way secure if for a random public key pk and an encryption of a random message, $c \leftarrow \mathsf{E}(\mathrm{pk}, m)$ for $m \xleftarrow{R} M$, no feasible adversary can recover m from (pk, c), except with insignificant probability.

Let $\mathcal{E} = (\mathsf{G}, \mathsf{E}, \mathsf{D})$ be a one-way secure system for message space M, and we assume that the secret keys are contained in M. Consider an encryption scheme $\bar{\mathcal{E}} = (\bar{\mathsf{G}}, \bar{\mathsf{E}}, \bar{\mathsf{D}})$ that operates on pairs of messages (i.e., has message space $M \times M$).

Key generation. Run G twice to generate two public/secret keys pairs $(\mathrm{pk}_1, \mathrm{sk}_1)$ and $(\mathrm{pk}_2, \mathrm{sk}_2)$. Output $\bar{\mathrm{pk}} := \mathrm{pk}_1$ as the public key and $\bar{\mathrm{sk}} := (\mathrm{sk}_1, \mathrm{sk}_2)$ as the secret key.

Encryption. An encryption of a message (m_1, m_2) under $\bar{\mathrm{pk}} = \mathrm{pk}_1$ is the pair $(m_1, \mathsf{E}_{\mathrm{pk}_1}(m_2))$.

Decryption. Given a ciphertext (a, b) and secret key $\bar{\mathrm{sk}} = (\mathrm{sk}_1, \mathrm{sk}_2)$, output the pair $(a, \mathsf{D}_{\mathrm{sk}_1}(b))$.

Claim 3. *The system $\bar{\mathcal{E}}$ above is a one-way encryption system if $\mathcal{E}$ is. However, an attacker seeing an encryption cycle (of any size) can find all the secret keys involved.*

The proof is straightforward, and is omitted here. The "However" part follows since an adversary seeing an encryption of a secret key $(\mathrm{sk}_1, \mathrm{sk}_2)$ under any public key gets sk_1 in the clear, and therefore can decrypt any message encrypted under the public key corresponding to $(\mathrm{sk}_1, \mathrm{sk}_2)$.

Remark: In $\bar{\mathcal{E}}$ the first half of the plaintext is transmitted in the clear. One can partially hide this part too, as follows:

Assume that we have a one-way permutation f on secret-keys of $\mathcal{E}$, and moreover that f is defined and is one-way on the entire message space of $\mathcal{E}$. Further assume that from any secret key we can efficiently compute a corresponding public key (which we denote by writing $\mathrm{pk} = P(\mathrm{sk})$). Then define a system $\tilde{\mathcal{E}} = (\tilde{\mathsf{G}}, \tilde{\mathsf{E}}, \tilde{\mathsf{D}})$ as follows:

Key generation. Run G twice to generate two public/secret keys pairs $(\mathrm{pk}_1, \mathrm{sk}_1)$ and $(\mathrm{pk}_2, \mathrm{sk}_2)$. Output $\tilde{\mathrm{pk}} := P(f(\mathrm{sk}_1))$ as the public key and $\tilde{\mathrm{sk}} := (\mathrm{sk}_1, \mathrm{sk}_2)$ as the secret key.

Encryption. An encryption of (m_1, m_2) under $\tilde{\mathrm{pk}}$ is $\left(f(m_1), \mathsf{E}_{\tilde{\mathrm{pk}}}(m_1), \mathsf{E}_{\mathrm{pk}}(m_2) \right)$.

Decryption. Given a ciphertext (a, b, c) and secret key $\tilde{\mathrm{sk}} = (\mathrm{sk}_1, \mathrm{sk}_2)$, compute $\mathrm{sk} = f(\mathrm{sk}_1)$ and output the pair $(\mathsf{D}_{\mathrm{sk}}(b), \mathsf{D}_{\mathrm{sk}}(c))$.

Again, proving a claim analogous to Claim 3 is straightforward.

6   Conclusions

We presented the first encryption system that can be proved to be n-circular secure under chosen-plaintext attack in the standard model. Security is based on the Decision Diffie-Hellman assumption and holds even if the adversary is given affine functions of the secret keys. In addition, we constructed in Section 5 a simple system that is weakly secure, but breaks completely once a key-cycle is published.

An important remaining problem is to obtain circular security against chosen ciphertext attacks. Other interesting problems are to improve the performance of our system, and to construct a semantically secure system that becomes insecure once an n-encryption cycle is published.

References

1. Adao, P., Bana, G., Herzog, J., Scedrov, A.: Soundness of formal encryption in the presence of key-cycles. In: di Vimercati, S.d.C., Syverson, P.F., Gollmann, D. (eds.) ESORICS 2005. LNCS, vol. 3679, pp. 374–396. Springer, Heidelberg (2005)
2. Bellare, M., Namprempre, C.: Authenticated encryption: relations among notions and analysis of the generic composition paradigm. In: Okamoto, T. (ed.) ASI-ACRYPT 2000. LNCS, vol. 1976, pp. 531–545. Springer, Heidelberg (2000)
3. Bellare, M., Rogaway, P.: Encode-then-encipher encryption: How to exploit nonces or redundancy in plaintexts for efficient encryption. In: Okamoto, T. (ed.) ASI-ACRYPT 2000. LNCS, vol. 1976, pp. 317–330. Springer, Heidelberg (2000)

4. Black, J., Rogaway, P., Shrimpton, T.: Encryption-scheme security in the presence of key-dependent messages. In: Nyberg, K., Heys, H.M. (eds.) SAC 2002. LNCS, vol. 2595, pp. 62–75. Springer, Heidelberg (2003)
5. Boneh, D., Boyen, X., Shacham, H.: Short group signatures. In: Franklin, M. (ed.) CRYPTO 2004. LNCS, vol. 3152, pp. 41–55. Springer, Heidelberg (2004)
6. Camenisch, J., Lysyanskaya, A.: An efficient system, for non-transferable anonymous credentials with optional anonymity revocation. In: Pfitzmann, B. (ed.) EUROCRYPT 2001. LNCS, vol. 2045, pp. 93–118. Springer, Heidelberg (2001)
7. Cramer, R., Shoup, V.: A practical cryptosystem provably secure under chosen ciphertext attack. In: Krawczyk, H. (ed.) CRYPTO 1998. LNCS, vol. 1462, pp. 13–25. Springer, Heidelberg (1998)
8. Dolev, D., Dwork, C., Naor, M.: Non-malleable cryptography. SIAM J. of Computing 30(2), 391–437 (2000)
9. Erdös, P., Hall, R.: Probabilistic methods in group theory II. Houston Math Journal 2, 173–180 (1976)
10. Goldwasser, S., Micali, S.: Probabilistic encryption. Jour. of Computer and System Science 28(2), 270–299 (1984)
11. Haitner, I., Holenstein, T.: On the (Im)Possibility of Key Dependent Encryption. Cryptology ePrint Archive (2008), http://eprint.iacr.org/2008/164
12. Halevi, S., Krawczyk, H.: Security under key-dependent inputs. In: proceedings of the 14th ACM conference on computer and communications security (CCS) (2007), http://eprint.iacr.org/2007/315
13. Hastad, J., Impagliazzo, R., Levin, L., Luby, M.: A pseudorandom generator from any one-way function. SIAM Journal on Computing 28(4), 1364–1396 (1999)
14. Hofheinz, D., Kiltz, E.: Secure hybrid encryption from weakened key encapsulation. In: Menezes, A. (ed.) CRYPTO 2007. LNCS, vol. 4622, pp. 553–571. Springer, Heidelberg (2007)
15. Hofheinz, D., Unruh, D.: Towards key-dependent message security in the standard mode. In: Smart, N. (ed.) EUROCRYPT 2008. LNCS, vol. 4965, pp. 108–126. Springer, Heidelberg (2008)
16. Katz, J., Yung, M.: Unforgeable encryption and adaptively secure modes of operation. In: Schneier, B. (ed.) FSE 2000. LNCS, vol. 1978, pp. 284–299. Springer, Heidelberg (2001)
17. Krawczyk, H.: The order of encryption and authentication for protecting communications (or: How secure is SSL?). In: Kilian, J. (ed.) CRYPTO 2001. LNCS, vol. 2139. Springer, Heidelberg (2001)
18. Laud, P., Corin, R.: Sound computational interpretation of formal encryption with composed keys. In: Lim, J.-I., Lee, D.-H. (eds.) ICISC 2003. LNCS, vol. 2971, pp. 55–66. Springer, Heidelberg (2004)
19. Rackoff, C., Simon, D.: Non-interactive zero-knowledge proof of knowledge and chosen ciphertext attack. In: Feigenbaum, J. (ed.) CRYPTO 1991. LNCS, vol. 576, pp. 433–444. Springer, Heidelberg (1992)
20. Shacham, H.: A Cramer-Shoup Encryption Scheme from the Linear Assumption and from Progressively Weaker Linear Variants. Cryptology ePrint Archive (2007) http://eprint.iacr.org/2007/074
21. Shoup, V.: A Computational Introduction to Number Theory and Algebra. Cambridge University Press, Cambridge (2005)

Public-Key Locally-Decodable Codes

Brett Hemenway[1],[*] and Rafail Ostrovsky[2],[**]

[1] Department of Mathematics, University of California, Los Angeles
bretth@math.ucla.edu
[2] Department of Computer Science and Department of Mathematics,
University of California, Los Angeles
rafail@cs.ucla.edu

Abstract. In this paper we introduce the notion of a Public-Key Encryption Scheme that is also a Locally-Decodable Error-Correcting Code (PKLDC). In particular, we allow any polynomial-time adversary to read the entire ciphertext, and corrupt a constant fraction of the bits of the *entire* ciphertext. Nevertheless, the decoding algorithm can recover any bit of the plaintext with all but negligible probability by reading only a sublinear number of bits of the (corrupted) ciphertext.

We give a general construction of a PKLDC from any Semantically-Secure Public Key Encryption (SS-PKE) and any Private Information Retrieval (PIR) protocol. Since Homomorphic encryption implies PIR, we also show a reduction from any Homomorphic encryption protocol to PKLDC.

Applying our construction to the best known PIR protocol (that of Gentry and Ramzan), we obtain a PKLDC, which for messages of size n and security parameter k achieves ciphertexts of size $\mathcal{O}(n)$, public key of size $\mathcal{O}(n + k)$, and locality of size $\mathcal{O}(k^2)$. This means that for messages of length $n = \omega(k^{2+\epsilon})$, we can decode a bit of the plaintext from a corrupted ciphertext while doing computation sublinear in n.

Keywords: Public Key Cryptography, Locally Decodable Codes, Error Correcting Codes, Bounded Channel Model, Chinese Remainder Theorem, Private Information Retrieval.

1 Introduction

Error correction has been an important field of research since Shannon laid the groundwork for a mathematical theory of communication in the nineteen forties, and active research continues until this day. An error correcting code is a pair of algorithms C and D such that given a message x, $C(x)$ is a codeword such that, given a string y, if the Hamming Distance between $d(C(x), y)$ is

[*] Part of this research was done while the author was visiting IPAM. This research was supported in part by VIGRE and NSF grants 0716835 and 0716389.
[**] Part of this research was done while visiting IPAM. This research was supported in part by IBM Faculty Award, Xerox Innovation Group Award, NSF grants 0430254, 0716835, 0716389 and U.C. MICRO grant.

D. Wagner (Ed.): CRYPTO 2008, LNCS 5157, pp. 126–143, 2008.
© International Association for Cryptologic Research 2008

"small", then $D(C(x)) = x$. When speaking of an error correcting code, two of its most important characteristics are the *information rate*, which is the ratio of the message size to the codeword size $\frac{|x|}{|C(x)|}$, and the *error rate* which is the smallest ϵ such that if $d(C(x), y) > \epsilon|C(x)|$ then $D(C(x))$ fails to recover x uniquely. Since the field's inception, many codes have been found that exhibit both constant information rate, and constant error rate, which, in a sense, is optimal. These codes all share the property that to recover even a small portion of the message x from the codeword y, the receiver must decrypt the entire codeword. In [1], Katz and Trevisan posed the question: can codes be found in which a single bit of the message can be recovered by decoding only a small number of bits from the codeword? Codes of this type are called *locally-decodable*, and would be immensely useful in encoding large amounts of data which only need to be recovered in small portions, for example any kind of database or archive. Currently the best known locally-decodable codes are due to Yekhanin [2]; they can tolerate a constant error rate, but achieve only slightly better than exponentially small information rates[1].

In 1994, Lipton examined the notion of error-correction in the computationally bounded channel model [3]. In this model, errors are not introduced in codewords at random, but in a worst case fashion *by a computationally bounded adversary* who can corrupt up to a constant fraction of the entire codeword. This realistic restriction on the power of the channel allowed for the introduction of cryptographic tools into the problem of error correction. In Lipton [3] and Gopalan, Lipton, Ding [4] it was shown how, assuming a shared private key, one can use hidden permutations to achieve improved error correcting codes in the private key setting. Recently, Micali, Peikert, Sudan and Wilson used the computationally bounded channel model to show how existing error correcting codes could be improved in the public-key setting [5]. After seeing the dramatic improvement of error-correcting codes in the computationally bounded channel model, a natural question then becomes whether locally-decodable codes can also be improved in this model.

The first progress in this setting was by Ostrovsky, Pandey and Sahai [6], where they construct a constant information-rate, constant error-rate locally-decodable code in the case where the sender and receiver share a private key. This left open the question whether the same can be accomplished in the Public-Key setting, which does not follow from their results. Indeed, a naïve proposal (that does not work) would be to encrypt the key needed by [6] separately and then switch to the private-key model already solved by [6]. This however leaves unresolved the following question: how do you encrypt the private key from [6] in a locally-decodable fashion? Clearly, if we allow the adversary to corrupt a constant fraction of all the bits (including encryption of the key and the message), and we encrypt the key separately, then the encryption of the key must consume a constant fraction of the message, otherwise it can be totally corrupted by an Adversary. But if this is the case all hope for local decodability is lost. Another

---

[1] Yekhanin achieves codewords of size $2^{n^{1/\log\log n}}$ for messages of length n, assuming there exist infinitely many Mersenne primes.

suggestion is to somehow hide the encryption of the key inside the encryption of the actual message, but it is not clear how this can be done.

A more sophisticated, but also flawed, idea is to use Lipton's code-scrambling approach [3]. In his paper, Lipton uses a private shared permutation to "scramble" the code and essentially reduce worst-case error to random error. A first observation is that we can use PIR to implement a random permutation in the public-key setting. We would then proceed as follows: the receiver would generate a random permutation $\sigma \in S_r$, and the receiver's public key would be a set of PIR queries $Q_1, \ldots, Q_r$, where Q_i is a PIR query for the $\sigma(i)$th block of an r block database, using some known PIR protocol. The sender would then break their message x into blocks, $x_1, \ldots, x_r$, apply standard error correction to each block, calculate the $Q_1, \ldots, Q_r$ on their message, apply standard error correction to each PIR response $R_i = Q_i(\mathsf{ECC}(x))$, and send the message $\mathsf{ECC}(R_1), \ldots, \mathsf{ECC}(R_r)$. If ECC and PIR have constant expansion rates, as is the case with many ECCs and the Gentry-Ramzan PIR [7], the resulting code has only constant expansion rate. But an adversary can still destroy a single block, by focusing damage on a single PIR response. If we add redundancy by copying the message c times, and publishing cr PIR queries, the adversary can still destroy a block with non-negligible probability by destroying constant number of blocks at random, and with non-negligible probability the adversary will destroy all c responses corresponding to the same block, and the information in that block will be lost. Recall that we demand that no bit of information should be destroyed except with negligible probability. Hence this method does not work. Of course, this can be fixed by increasing the redundancy beyond a constant amount, but then the codeword expansion becomes more than constant as does the public key size. Thus, this solution does not work, and new ideas are needed. Indeed, in this paper, we use PIR to implement a hidden permutation, but we achieve a PKLDC which can recover from constant error-rate with only *constant* ciphertext expansion.

1.1   Previous Work

The first work on error correction in the computationally bounded channel model was done by Lipton in [3]. In Lipton [3] and Gopalan, Lipton, Ding [4] it was shown how to use hidden permutations to achieve improved error correcting codes in the private key setting. In [5], Micali, Peikert, Sudan and Wilson demonstrate a class of binary error correcting codes with positive information rate, that can uniquely decode from $\frac{1}{2} - \epsilon$ error rate, under the assumption that one-way functions exist. These codes decode from an error rate *above* the proven upper bound of $\frac{1}{4} - \epsilon$ in the (unbounded) adversarial channel model. The first application of the computationally bounded channel to Locally Decodable Codes was given by Ostrovsky, Pandey and Sahai [6], although their work was in the private-key setting, and does not extend to the public-key setting.

In addition to extending the work in the computationally bounded channel model, our work draws heavily from the field of Computational Private Information Retrieval (PIR). The first computational PIR protocol was given by

Ostrovsky and Kushilevitz [8], and since then there has been much progress. For a survey of work relating to computational PIR see [9].

1.2   Our Results

In this paper, we present a general reduction from semantically-secure encryption and a PIR protocol to a Public Key Encryption system with local decodability (PKLDC). We also present a general reduction from any homomorphic encryption to a PKLDC. In §5 we present the first Locally Decodable Code with constant information-rate which does not require the sender and receiver to share a secret key. To achieve this, we work in the Computationally Bounded Channel Model, which allows us to use cryptographic tools that are not available in the Adversarial Channel Model. Our system presents an improvement in communication costs over the best codes in the information-theoretic setting. We create codes with constant information-rate, as compared with the best known locally decodable codes [2] in the information-theoretic setting which have an almost exponentially small information rate.

Informally, our results can be summarized as follows,

Main Theorem (informal). *Given a computational PIR protocol with query size $|Q|$, and response size $|R|$ which retrieves dk bits per query, and a semantically-secure encryption scheme, there exists a Public Key Locally Decodable Code which can recover from a constant error-rate in the bits of the message, which has public key size $\mathcal{O}(n|Q|/(dk^2) + k)$ and ciphertexts of size $\mathcal{O}(n|R|/(dk^2))$, where n is the size of the plaintext and k is the security parameter. The resulting code has locality $\mathcal{O}(|R|k/d)$, i.e. to recover a single bit from the message we must read $\mathcal{O}(|R|k/d)$ bits of the codeword.*

Combining the main theorem with the general reduction from homomorphic encryption to PIR, we obtain

Corollary 1. *Under any homomorphic encryption scheme which takes plaintexts of length m to ciphertexts of length αm, there is a Public-Key Locally Decodable Code which can recover from a constant error-rate in the bits of the message, with public key size $\mathcal{O}(nk\beta \sqrt[\beta]{n})$ and ciphertexts of size $\mathcal{O}(n\alpha^{\beta-1}k)$, for any $\beta \in \mathbb{N}$, where n is the size of the plaintext and k is the security parameter. The resulting code has locality $\mathcal{O}(\alpha^{\beta-1}k^2)$, i.e. to recover a single bit from the message we must read $\mathcal{O}(\alpha^{\beta-1}k^2)$ bits of the codeword.*

We can further improve efficiency if we have a Length-Flexible Additively Homomorphic Encryption like Dåmgard-Jurik [10], using this cryptosystem we obtain

Corollary 2. *Under the Decisional Composite Residuousity Assumption [11] there is a Public-Key Locally Decodable Code which can recover from a constant error-rate in the bits of the message, with public key size $\mathcal{O}(n\log^2(n) + k)$ and ciphertexts of size $\mathcal{O}(n\log(n))$, where n is the size of the plaintext and k is the security parameter. The resulting code has locality $\mathcal{O}(k^2\log(n))$, i.e. to recover a single bit from the message we must read $\mathcal{O}(k^2\log(n))$ bits of the codeword.*

We also give a specific construction of a system based on the Φ-hiding assumption first introduced by Cachin, Micali and Stadler in [12], and later used by Gentry and Ramzan in [7]. Under this assumption we obtain

Corollary 3. *Under the Small Primes Φ-Hiding Assumption there is a Public-Key Locally Decodable Code which can recover from a constant error-rate in the bits of the message, with public key size $\mathcal{O}(n)$ and ciphertexts of size $\mathcal{O}(n)$, where n is the size of the plaintext and k is the security parameter. The resulting code has locality $\mathcal{O}(k^2)$, i.e. to recover a single bit from the message we must read $\mathcal{O}(k^2)$ bits of the codeword.*

Note that in full generality, our main result requires two assumptions, the existence of a PIR protocol and a semantically-secure encryption protocol. In practice, however, two separate assumptions are usually not needed, and all the corollaries apply under a single hardness assumption.

Our construction does have a few disadvantages over the information-theoretic codes. First, our channel is computationally limited. This assumption is fairly reasonable, but it is also necessary one for any type of public key encryption. In [5], Micali et al. show that if a true adversarial channel exists, which can always introduce errors in a worst-case fashion, then one-way functions cannot exist. Second, our code has a larger "locality" than most information-theoretic codes. For example, in Yekhanin's Codes, the receiver is only required to read three letters of the codeword to recover one letter of the message. In our code in §5 the receiver must read $\mathcal{O}(k^2)$ bits to recover 1 bit of the plaintext, where k is the security-parameter. It should be noted, however, that to maintain the semantic security of the cryptosystem, the receiver must read $\omega(\log k)$ bits to recover any single bit of the message. It is an interesting question whether the locality of our code can be reduced from $\mathcal{O}(k^2)$ to $\mathcal{O}(k)$. For long messages (i.e. $n = \omega(k^{2+\epsilon})$) our code still presents a very significant improvement in locality over standard error correcting codes.

2   Computationally Locally Decodable Codes

2.1   Modelling Noisy Channels

When discussing error correcting, or locally-decodable codes, it is important to consider how the errors are introduced by the channel. While it may be natural to assume the errors are introduced "at random", small changes in the exact nature of these errors can result in substantial changes in the bounds on the best possible codes.

The first definition of a noisy channel is due to Claude Shannon [13]. Shannon defined the *symmetric channel* where each message symbol is independently changed to a random different symbol with some fixed probability, called the error rate. An alternative definition of a noisy channel is Hamming's *adversarial channel*, where one imagines an adversary corrupting bits of the message in a worst-case fashion, subject only to the total number of bits that can be corrupted per block.

In 1994, Lipton [3] observed that the adversarial channel model assumes that the adversarial channel itself is computationally unbounded. In that paper, Lipton proposed a new model of *computationally bounded noise*, which is similar to Hamming's adversarial channel, except the adversary is restricted to computation which is polynomial in the block length of the code. This restriction on the channel's ability to introduce error is a natural one, and it is implied by the existence of any one-way function [5]. Throughout this paper, we use Lipton's model.

2.2   Definitions

We use the standard definition of computational indistinguishability for public key encryption, where we also view the size of the plaintext as a function of the security parameter. That is, we set the plaintext x to be of length k^α, where k is the security parameter and $\alpha > 1$.

The primary difference between our definition and the standard definition of semantic security is the local decodability property of the cryptosystem. Roughly, this says that given an encryption c of a message x, and a corrupted encryption c' such that the hamming distance of c and c' is less than $\delta|c|$, the time it takes the decoder to decode any bit x_i of the plaintext x from c' is much shorter than the length of the message, and does not increase as the message length increases.

Definition 1. *We call a Public Key Cryptosystem semantically-secure (in the sense of indistinguishability) and δ-computationally locally-decodable if there is a triple of probabilistic polynomial-time algorithms (G, E, D), such that for all k and for all α sufficiently large*

- *$(PK, SK) \leftarrow G(1^k, \alpha)$,*
- *$c \leftarrow E(PK, x, r)$ (where $|x| = k^\alpha$ is a plaintext message of length polynomial in k, and r is the randomness of the encryption algorithm);*
- *$b' \leftarrow D(SK, c', i)$*

so that for all probabilistic polynomial-time adversaries A, A':

$$\Pr[(PK, SK) \leftarrow G(1^k, \alpha); \{x^0, x^1, \gamma\} \leftarrow A(PK); A'(E(PK, x^b, r), \gamma) = b] < \frac{1}{2} + \nu(k),$$

where x^0 and x^1 must both be of length k^α, and the probability is taken over the key generation algorithm's randomness, b, randomness r used in the encryption algorithm E and the internal randomness of A and A'.[2] Furthermore, it is computationally, locally-decodable. That is, for all probabilistic polynomial-time adversaries A'' and A''',

$$\Pr[(PK, SK) \leftarrow G(1^k, \alpha); (x, \gamma) \leftarrow A''(PK);$$
$$c \leftarrow E(PK, x, r); \{c', i\} \leftarrow A'''(c, \gamma):$$
$$D(SK, c', i) = x_i] > 1 - \nu(k),$$

---

[2] As is standard practice, we allow the adversary A to pass state information γ, which could include information about the plaintexts x^0, x^1, which might be of use in determining which plaintext is encrypted by $E(PK, x^b, r)$.

where x_i denotes the ith bit of x, x must be of the length k^α, c' and c must be of the same length and the hamming distance between c' and c is at most $\delta|c|$, and where the probability is taken over the key generation algorithm's randomness, the randomness r used in the encryption algorithm E and the internal randomness of both A'' and A'''. The information-rate is $\frac{|m|}{|c|}$ and we call the decryption algorithm locally-decodable *if its running time is* $o(k^\alpha)$, and the efficiency of the local decodability is measured as a function of k and α.

3   Building Blocks

Our construction relies on a number of standard cryptographic tools and for completeness we briefly review them here.

3.1   Private Information Retrieval

A computational Private Information Retrieval protocol (PIR) is a protocol in which a user or client wants to query a position from a database, while keeping the position queried hidden from the server who controls the database. In particular the user generates a decryption key D_{PIR}, picks a position j and generates a query Q_j. Then, given Q_j, the server who has a database (or message) x, can execute query Q_j on x and obtain a response R_j. The privacy requirement is that the server cannot guess the position j with probability noticeably greater than random. The correctness requirement is that given D_{PIR}, and R_j the user can correctly recover the jth position of the message x. The efficiency of a PIR protocol is measured in the communication complexity, i.e. the sizes of Q and R. Currently, the most efficient PIR protocol is that of Gentry and Ramzan [7], which has $|Q| = |R| = \mathcal{O}(k)$ where k is a security parameter, and each query successfully retrieves approximately $k/4$ bits of the message x.

Formal definitions and concrete constructions of computational Private Information Retrieval protocols can be found in [7], [8], [12], [14] or [15] .

3.2   Semantically-Secure Public Key Encryption

Our construction requires a semantically-secure encryption protocol, SSE. The only requirement we make on the protocol SSE, is that for a message x, $|\mathsf{SSE}(x)| = \mathcal{O}(|x|)$. For concreteness, we assume $|\mathsf{SSE}(x)| = c_1|x|$ for some constant c_1. This is achieved by many cryptosystems for example [11], [10], [16], [17], or the Φ-hiding based scheme in described §5.1.

To avoid making additional intractability assumptions, it is natural to choose a hardness assumption that yields both a semantically-secure encryption protocol as well as a PIR protocol. In practice this is almost always the case, for example Paillier's Cryptosystem [11] and Chang's PIR [15], or Gentry-Ramzan [7] (or Cachin-Micali-Stadler PIR [12]) and the encryption protocol outlined in Section 5.1. It is also worth noting that since [14] shows that any homomorphic encryption protocol immediately yields a PIR protocol, if we have a homomorphic encryption, we need not make an additional assumption to obtain a PIR protocol.

3.3   Reed-Solomon Codes

The Reed-Solomon Error Correcting Code (RS-ECC) works as follows: first we fix a prime p of length k, and all computations are done in the field $\mathbb{Z}/p\mathbb{Z}$. Then, given a plaintext x of length n, we represent x as a polynomial f_x of degree $n/k - 1$ over $\mathbb{Z}/p\mathbb{Z}$. This can be done in many ways, perhaps the simplest is to break x into blocks of size k and view these as the coefficients of f_x. Then, the encoding of x is simply the evaluation of f_x at a number of points in $\mathbb{Z}/p\mathbb{Z}$. We need at least n/k evaluations to uniquely determine a polynomial of degree $n/k - 1$, the RSECC adds redundancy by evaluating f_x at more points, $\mathsf{RSECC}(x) = (f_x(1), \ldots, f_x(\rho n/k))$ for some $\rho > 1$. For distinct plaintexts x, y, we have $f_x - f_y \neq 0$. Since a nonzero polynomial of degree $n/k - 1$ has at most $n/k - 1$ zeros, and $\mathsf{RSECC}(x)$ and $\mathsf{RSECC}(y)$ must have hamming distance at least $(\rho - 1)n/k + 1$, this code can recover from $(\rho - 1)n/(2k)$ errors in the evaluation points, i.e. it can recover from an error rate of $\frac{1}{2} - \frac{1}{2\rho}$ in the digits of the code.

From now on we will view $\mathsf{RSECC}(x)$ as a $\rho n/k$-tuple which can be successfully decoded from an error rate of $\frac{1}{2} - \frac{1}{2\rho}$ in its digits.

3.4   Binary Error Correction

A desirable property of any error-correcting code is the ability to recover from a constant fraction of errors among the *bits* of the codeword. A drawback of many error-correcting codes, and locally-decodable codes, is that they are defined over large alphabets, and can only recover from a constant fraction of errors in the alphabet of the code. The natural alphabet of the RSECC described above is the field $\mathbb{Z}/p\mathbb{Z}$. In practice, all these codes are implemented on computers, where the natural alphabet is $\{0, 1\}$. Thus when we say that a code like the Reed-Solomon code can tolerate a constant fraction of errors, we mean a constant fraction of errors in their natural alphabet. In the Reed Solomon code, if one bit of each evaluation point is corrupted, there are no guarantees that the message will not be corrupted. Binary error correcting codes do exist, but they are generally not as efficient as codes over larger alphabets.

To allow our code to tolerate a constant fraction of errors in the *bits* of the ciphertext, we will make use of a binary error correcting code ECC, with two properties. First, $|\mathsf{ECC}(x)| = c_2|x|$ for some constant c_2, and second ECC can recover from an error-rate of $\frac{1}{2} - \delta$ in the *bits* of $\mathsf{ECC}(x)$. Such codes exist, for $\delta > \frac{1}{4}$ in the unbounded adversarial channel model, and $\delta > 0$ in the computationally bounded channel model. See the full version of this paper for a more in-depth discussion.

4   Construction

4.1   High Level Outline of Our Construction

A public key will be a list of t PIR queries $Q_1, \ldots, Q_t$, along with the public key to the semantically-secure encryption SSE. The private key will be the private key for the semantically-secure encryption, the private key for the PIR protocol and a permutation $\sigma \in S_t$ such that Q_j is a query for the $\sigma(j)$th position of the message. To encrypt an n-bit message X, we first divide X into r blocks $X_1, \ldots, X_r$, then we encrypt each block using our semantically-secure encryption (this can be done by further subdividing the block if necessary). Then we encode each block using the Reed-Solomon code, thus obtaining a list of evaluation points that constitute the Reed-Solomon encoding of this block. Next, we concatenate the evaluation points for all the blocks, and, treating this list as a single database, we evaluate all t PIR queries on it. Finally, we encode each PIR response with a standard binary error correcting code ECC.

In more detail, we assume that when we evaluate a PIR query Q on a message X, the PIR response R encodes dk bits of X where k is our security parameter and d depends on the specific PIR protocol used. For example the Gentry-Ramzan protocol has $d \approx \frac{1}{4}$, while a PIR protocol like [12] which only retrieves a single bit at a time has $d = 1/k$. Next, we fix a prime p of length k which will determine the base-field of the RSECC. Then, we set $r = n/(\ell k)$, thus each block X_i has $|X_i| = \ell k$, where ℓ is the parameter that will determine the "spread" of our code. Next we encrypt each block X_i using SSE, obtaining $\mathsf{SSE}(X_1), \ldots, \mathsf{SSE}(X_r)$ where $|\mathsf{SSE}(X_i)| = c_1 \ell k$. Then we encode each encrypted block as $c_1 \rho \ell$ field elements in $\mathbb{Z}/p\mathbb{Z}$ using RSECC. Thus we can recover any block X_i as long as no more than $\frac{1}{2} - \frac{1}{2\rho}$ of the field elements that encode it are corrupted. Finally, we concatenate all $c_1 r \rho \ell$ field elements, thus at this point our "database" is $c_1 r \rho \ell k = c_1 n \rho$ bits. Next we evaluate all t queries $Q_1, \ldots, Q_t$ on this database. Since we wish to retrieve all the information in X, we need $t = c_1 n \rho / (dk)$. Thus we obtain t PIR responses $R_1, \ldots, R_t$. Finally, we send the t-tuple $(\mathsf{ECC}(R_1), \ldots, \mathsf{ECC}(R_t))$.

Thus our final encryption is of size $c_1 c_2 n \rho |R_j| / (dk)$. If $|R_j| \approx k$ as is case in [12], [15], [7], then our encryption will be of length $c_1 c_2 \rho n / d$. If we use the PIR protocol in [7] then, d will be constant, thus our code will have constant information rate. Notice that the spread parameter ℓ has no effect on the length of the encryption. This encryption is error correcting because as long as no more than $\frac{1}{2} - \frac{1}{2\rho}$ of the responses that encode a given block are corrupted, the block can be recovered correctly by first decoding each point using ECC, and then reconstructing the block using the RSECC. This cryptosystem is also locally-decodable since to decrypt a given block, it suffices to read the $\frac{c_1 \rho \ell}{dk}$ PIR responses that encode it.

4.2   Error Correcting Public Key Encryption

We now define a triple of algorithms G, E, D for our encryption scheme.

Key Generation: $G(1^k, \alpha)$.

- Fix a prime p of length k.
- Generate public-key private-key pair for SSE, PK_E, SK_E.
- Generate a PIR decryption key D_{PIR}.
- Generate a random permutation $\sigma \in S_t$.

- Generate t PIR queries $Q_1, \ldots, Q_t$, where Q_j queries the block of dk bits at position $(\sigma(j) - 1)c_1 dk + 1$ of a $c_1 n\rho$ bit database.

The public key will then be

$$PK = (PK_E, Q_1, \ldots, Q_t)$$

and the secret key will be

$$SK = (\sigma, SK_E, D_{PIR})$$

Thus the public key will be of length $t|Q| + |SK_E| = c_1 n\rho|Q|/(dk)$. If we use [7], then $|Q| = k$ and d is constant, so assuming $|SK_E| = \mathcal{O}(k)$, we obtain $|PK| = \mathcal{O}(n + k)$.

Encryption: given an n-bit message X,

- Break X into $r = \frac{n}{\ell k}$ blocks X_i of size ℓk.
- Encrypt each block using SSE. If SSE can only encrypt strings of length k, we simply divide X_i into shorter strings, encrypt the shorter strings and then concatenate the encryptions.
- For each encrypted block, $\mathsf{SSE}(X_i)$ we encode it as a list of $c_1 \rho \ell$ field elements $Z_{i,1}, \ldots, Z_{i,c_1\rho\ell}$ in $\mathbb{Z}/p\mathbb{Z}$ using the RSECC.
- Concatenate all the evaluations, creating $\tilde{X} = Z_{1,1}, \ldots, Z_{1,c_1\rho\ell}, \ldots, Z_{r,1}, Z_{r,c_1\rho\ell}$. Thus $|\tilde{X}| = rc_1\rho\ell k = c_1 n\rho$ bits, and we run each PIR query $\{Q_1, \ldots, Q_t\}$ on $\tilde{X}$ receiving responses $R_1, \ldots, R_t$. Since each PIR query recovers dk bits, we will need c_1/d queries to recover each field element Z.
- Encode each R_j individually using the binary error correcting code ECC.
- The encryption is then the t-tuple $(\mathsf{ECC}(R_1), \ldots, \mathsf{ECC}(R_t))$.

Decryption: to recover the ith block, of a message X from the t-tuple $(\mathsf{ECC}(R_1), \ldots, \mathsf{ECC}(R_t))$

- We wish to retrieve the encoding $Z_{i,1}, \ldots, Z_{i,c_1\rho\ell}$, which are the bits of $\tilde{X}$ in positions $(i-1)c_1\rho\ell/d + 1, \ldots, ic_1\rho\ell/d$, Thus we select the $c_1\rho\ell/d$ responses that encode X_i, $\{\mathsf{ECC}(R_{\sigma^{-1}((i-1)c_1\rho\ell/d+1)}), \ldots, \mathsf{ECC}(R_{\sigma^{-1}(ic_1\rho\ell/d)})\}$.
- Decode each $\mathsf{ECC}(R_j)$ to obtain $\{R_{\sigma^{-1}((i-1)c_1\rho\ell/d+1)}, \ldots, R_{\sigma^{-1}(ic_1\rho\ell/d)}\}$.
- Decode each of the $c_1\rho\ell/d$ PIR responses R_j to obtain $Z_{i,1}, \ldots, Z_{i,c_1\rho\ell}$.
- Using the RSECC reconstruct $\mathsf{SSE}(X_i)$ from $Z_{i,1}, \ldots, Z_{i,c_1\rho\ell}$.
- Decrypt $\mathsf{SSE}(X_i)$.

Notice that to recover block X_i we only need to read $c_1 c_2 |R|\rho\ell/d$ bits of the encryption. In the Gentry-Ramzan PIR $|R| = k$ and $d = 1/4$, so we are reading only $\mathcal{O}(\ell k)$ bits of the message. For correctness we will choose $\ell = k$, thus in this case our scheme will achieve locality $\mathcal{O}(k^2)$.

4.3 Local-Decodability

One of the most interesting features of our construction is the local-decodability. To recover a small portion of the message X, only a small portion of the cipher-text $(\mathsf{ECC}(R_1), \ldots, \mathsf{ECC}(R_t))$ needs to be decoded. During encryption the message X is broken into blocks of length ℓk bits, and this is the smallest number of bits that can be recovered at a time. To recover a single bit of X, or equivalently the entire block X_i that contains it, we must read $c_1 \rho \ell / d$ blocks of the ciphertext $\{\mathsf{ECC}(R_{\sigma^{-1}((i-1)c_1\rho\ell/d+1)}), \ldots, \mathsf{ECC}(R_{\sigma^{-1}(ic_1\rho\ell/d)})\}$. Since $|\mathsf{ECC}(R_j)| = c_2|R_j|$, we must read a total of $c_1 c_2 |R| \rho \ell / d$ bits. Since the probability of error will be negligible in ℓ, we will set $\ell = k$. Here c_2 and ρ are parameters that determine the error-rate of our code.

Using the Gentry-Ramzan PIR, we have $|R| = k$ and $d = 1/4$, so the locality is $\mathcal{O}(k^2)$. Using the Chang's PIR [15] based on Paillier's cryptosystem [11] we have $|R| = 2k$ and $d = 1/2$ so we achieve the same encryption size and locality, although in this situation the public key size is $\mathcal{O}(n^{3/2})$ instead of $\mathcal{O}(n)$ in the Gentry-Ramzan case.

4.4 Proof of Security

The semantic security of our scheme follows immediately from the semantic security of the underlying encryption SSE. The full proof of the correctness (i.e. local decodability) of our scheme requires some care. The formal proof can be found in the full version of this paper. Here, we outline only the high-level ideas of the proof. The structure of the proof is as follows. Given an encryption $(\mathsf{ECC}(R_1), \ldots, \mathsf{ECC}(R_t))$, the outer ECC forces an adversary to concentrate their errors among only a few R_j. Thus, we may assume that the adversary is only allowed to introduce errors into a constant fraction of the R_j. Then, we note that any polynomial-time adversary cannot tell which remainders R_j encode which block X_i by the privacy of the PIR protocol. Thus any errors introduced in the R_j will be essentially uniform among the Z's that make up the Reed-Solomon encryptions. Next, we show that our code has sufficient "spread" so that errors introduced uniformly among the R_j will cluster on the R_j encoding a given block X_i with only negligible probability. Finally, if the errors are not clustered among the R_j that encode a given block, we show that the RSECC will correctly recover that block.

Thus we arrive at the following result

Main Theorem. *Given a computational PIR protocol with query size $|Q|$, and response size $|R|$ which retrieves dk bits per query, and a semantically-secure encryption protocol SSE, there exists a Public Key Locally Decodable Code which can recover from a constant error-rate in the bits of the message, which has public key size $\mathcal{O}(n|Q|/(dk^2) + k)$ and ciphertexts of size $\mathcal{O}(n|R|/(dk^2))$, where n is the size of the plaintext and k is the security parameter. The resulting code has locality $\mathcal{O}(|R|k/d)$, i.e. to recover a single bit from the message we must read $\mathcal{O}(|R|k/d)$ bits of the codeword.*

4.5   Extensions

For convenience, in our proof of correctness, we set the parameter ρ equal to $1/2$. It should be clear that this value is somewhat arbitrary and that by increasing ρ we increase the error tolerance of the code along with the ciphertext expansion. Similarly, in our proof we set the parameter ℓ to be the security parameter k. We can change ℓ, and an increase in ℓ corresponds to a decrease in the probability that the channel succeeds in introducing an error, and a decrease in the locality of the code. In particular our code fails with probability that is negligible in ℓ, and the smallest number of bits that can be recovered from the message is $\mathcal{O}(\ell k)$.

Our protocol also benefits nicely from the idea of Batch Codes [18]. Since our protocol requires making multiple PIR queries to the same message, this is an ideal application of Batch Codes, which can be used to amortize the cost of making multiple PIR queries to a fixed database. By first "batching" the message $\tilde{X}$ in §4.2, we can significantly decrease server computation by slightly increasing ciphertext expansion, or we can decrease ciphertext expansion by paying a slight increase in server computation. It should be noted that batch codes are perfect, in the sense that batching the message in this way does not change the probability of correctness.

We can also increase the efficiency of our construction by further taking advantage of the bounded channel model. If in addition to the sender knowing the receiver's public key, we assume that the receiver knows the verification key to the senders signature algorithm (a reasonable assumption since anyone receiving messages from the sender should be able to verify them), our scheme benefits nicely from the sign and list-decode methods described in [5]. The use of digital signatures before applying the RSECC or the binary ECC has the effect of increasing the maximum tolerable error-rate, and decreasing the codeword expansion. Unlike the application of Batch Codes above, this sign and list-decode technique will slightly increase the probability that a message fails to decrypt, although it still remains negligible.

4.6   Constructions Based on Homomorphic Encryption

It was shown in [14] that any homomorphic encryption protocol yields a PIR protocol, thus our construction can be achieved based on any homomorphic encryption protocol. In this situation, it is unnecessary to first encrypt each block X_i before applying the RSECC since the PIR protocol described in [14] is already semantically-secure. Thus the idea of coupling encryption and error-correction is even more natural in this situation. Using the construction in [9] to construct a PIR protocol from a homomorphic encryption protocol and then applying our construction yields

Corollary 1. *Under any homomorphic encryption protocol which takes plaintexts of length m to ciphertexts of length αm, there is a Public-Key Locally Decodable Code which can recover from a constant error-rate in the bits of the message, with public key size $\mathcal{O}(nk\beta \sqrt[\beta]{n})$ and ciphertexts of size $\mathcal{O}(n\alpha^{\beta-1}k)$, for any $\beta \in \mathbb{N}$, where n is the size of the plaintext and k is the security parameter.*

The resulting code has locality $\mathcal{O}(\alpha^{\beta-1}k^2)$, i.e. to recover a single bit from the message we must read $\mathcal{O}(\alpha^{\beta-1}k^2)$ bits of the codeword.

Using a Length-Flexible Additively Homomorphic Encryption protocol such as the one described in [10] yields an even more efficient PIR protocol. Using the methods outlined in [9] and applying our construction we arrive at the following result

Corollary 2. *Under the Decisional Composite Residuousity Assumption [11] there is a Public-Key Locally Decodable Code which can recover from a constant error-rate in the bits of the message, with public key size $\mathcal{O}(n\log^2(n) + k)$ and ciphertexts of size $\mathcal{O}(n\log(n))$, where n is the size of the plaintext and k is the security parameter. The resulting code has locality $\mathcal{O}(k^2\log(n))$, i.e. to recover a single bit from the message we must read $\mathcal{O}(k^2\log(n))$ bits of the codeword.*

5   A Concrete Protocol Based on Φ-Hiding

We now present a concrete example of our reduction based on the Gentry-Ramzan [7] PIR protocol. A straightforward application of our main construction in §4.2 already yields a PKLDC with public key size $\mathcal{O}(n)$ and constant ciphertext expansion, but the Gentry-Ramzan PIR protocol has many nice properties which can be exploited to simplify the construction and further increase the efficiency of the protocol. The construction we present here differs from the straightforward application of our general reduction to the Gentry-Ramzan protocol in two ways. First, we are able to integrate the basic semantically-secure encryption protocol into our construction, thus reducing the ciphertext expansion by a constant factor, and eliminating the need for another hardness assumption. Second, we use the Chinese Remainder Theorem Error Correcting Code (CRT-ECC) instead of the Reed-Solomon code used in the general construction. This is because the Φ-hiding assumption allows us to do hidden chinese-remaindering, and so it is a more natural code to use in this context. This does not change the arguments in any substantial way, since from the ring-theoretic perspective, the CRT-ECC and the Reed-Solomon ECC are exactly the same.[3]

5.1   A Φ-Hiding Based Semantically-Secure Encryption Protocol

Here, we describe a simple semantically-secure public key encryption scheme, BasicEncrypt that will be an essential building block of our construction. The encryption protocol consists of three algorithms, G, E, D described below.

To generate the keys, $G(1^k)$ first selects a small prime-power π, then generates $m \in \mathcal{H}_k^\pi$, i.e. $m = pq$, where $p, q \in_R \mathcal{P}_k{}^4$, subject to $\pi \mid p - 1$. The public key will be $PK = (g, m, \pi)$ where g is a generator for the cyclic group G_m, and $SK = \frac{\varphi(m)}{\pi}$.

---

[3] See the full version for a more detailed discussion of this point.

[4] We use the notation $\in_R$ to denote selecting uniformly at random from a set.

To encrypt a message $x \in \mathbb{Z}/\pi\mathbb{Z}$, we have

$$E(x) - g^{x+\pi r} \mod m,$$

for a random $r \in \mathbb{Z}/m\mathbb{Z}$. To decrypt, we do

$$D(y) = y^{\varphi(m)/\pi} = g^{x\varphi(m)/\pi \mod \varphi(m)} \mod m = \left(g^{\varphi(m)/\pi}\right)^x \mod m,$$

then, using the Pohlig-Hellman algorithm to compute the discrete logarithm in the group $\langle g^{\varphi(m)/\pi} \rangle$, we can recover $x \mod \pi = x$. If a is a small prime, and $\pi = a^c$, the Pohlig-Hellman algorithm runs in time $c\sqrt{a}$. Thus the decryption requires $\mathcal{O}(\log(m/\pi)+c\sqrt{a})$ group operations in G_m which is acceptable for small primes a. In our locally decodable code, we will require multiple different prime powers $\pi_1, \ldots, \pi_t$, and we will choose the small primes a, as the first primes, i.e. $\pi_1 = 5^{e_1}, \pi_2 = 7^{e_2}, \pi_3 = 11^{e_3}$. If we require t prime powers π_i, the Prime Number Theorem implies that the largest a will be approximately $t \log t$. Since t will be less than the message length, n, $\sqrt{a}$ will be polynomial in the message length, and hence polynomial in the security parameter k.

It is worth noticing that this scheme is additively homomorphic over the group $\mathbb{Z}/\pi\mathbb{Z}$, although we do not have an explicit use for this property. When $\pi = 2$, this is just Goldwasser-Micali Encryption [19], for larger π it was described in [20] and [21]. An extension of this scheme is described in [16].

While this protocol is not new, none of the previous descriptions of this protocol make use of the Φ-hiding assumption, and instead their security is based on some form of composite residuousity assumption, i.e. it is impossible to tell whether a random group element h belongs to the subgroup of order π in G_m. We are able to prove security under the Φ-hiding assumption because the Φ-hiding assumption is strictly stronger than these other assumptions. The proof that this protocol is semantically-secure under the Φ-hiding assumption is in the full version [22].

5.2   Outline of Our Φ-Hiding Based Construction

We begin by fixing a list of t prime powers $\{\pi_1, \ldots, \pi_t\}$ as part of the public parameters. For concreteness we choose $\pi_1 = 5^{e_1}$, $\pi_2 = 7^{e_2}$, ... as in §5.1. A public key will be a list of t RSA moduli $\{m_1, \ldots, m_t\}$, such that each m_j Φ-hides some prime power $\pi_{j'}$. The Private key will be the factorizations of the m_j, more specifically $\varphi(m_1), \ldots, \varphi(m_t)$, along with a random permutation $\sigma \in S_t$ such that m_j Φ-hides $\pi_{\sigma(j)}$. To encrypt a message $X \in \{0,1\}^n$, we first divide X into blocks X_i of size ℓk. Where k is the security parameter, and ℓ is a parameter determining the "spread" of the code. As in the Gentry-Ramzan PIR scheme, we view each block as a number in the range $\{0 \ldots 2^{\ell k}\}$. Our public key will be $t = \frac{\rho n}{dk}$ RSA moduli $\{m_1, \ldots, m_{\frac{\rho n}{dk}}\}$ such that each modulus Φ-hides a prime power π_j. We will use $s = \lceil \rho \ell / d \rceil$ of the π_j to encode each block X_i. Since there are $\lceil n/\ell k \rceil$ blocks, and for each block we use $\lceil \rho \ell / d \rceil$ prime powers, we use a total of $\frac{n}{\ell k} \cdot \frac{\rho \ell}{d} = \frac{\rho n}{dk} = t$ prime powers. The parameter ρ determines the redundancy of the CRT-ECC, hence increasing ρ increases the error tolerance and also the ciphertext expansion. Recall that d is the information rate of the Gentry-Ramzan PIR, so d is some fixed constant less than $1/4$, for concreteness we may assume $d = 1/5$. Exactly which prime is hidden by which modulus will be chosen at random at the time of key generation, and is part of the receiver's secret key. For each block X_i, the sender encrypts X_i modulo the s prime powers $\{\pi_{(i-1)s+1}, \dots, \pi_{is}\}$, where each π_j is roughly of size dk. Notice here that we have used ρ times as many moduli π_j as necessary to encode each block, thus for each block X_i we have effectively calculated an encoding of X_i under the CRT-ECC which can tolerate $\left(\frac{1}{2} - \frac{1}{2\rho}\right)\frac{\ell}{d}$ corrupted moduli.[5] We do this for each block, and thus the resulting encryption is $\frac{\rho\ell}{d} \cdot \frac{n}{\ell k}$ residues. Since each residue is of size k, the encryption of the whole message is now $\frac{n}{\ell k}\frac{\rho\ell}{d} = \frac{\rho n}{dk}$ encryptions of size k. Finally, we encode each of the $\rho n/(kd)$ encryptions independently using the error correcting code in §3.4. So our final encryption is of size $\rho c_2 n/d$ bits, which is a constant multiple of n. This encryption is error correcting because as long as no more than $\frac{1}{2} - \frac{1}{2\rho}$ of the residues that encode a given block are corrupted, the block can be recovered correctly by first decrypting each residue, and then reconstructing the CRT-ECC. This cryptosystem is also locally-decodable since to decrypt a given block, it suffices to decrypt the $\frac{\rho\ell}{d}$ encryptions that encode it.

5.3 Error Correcting Public Key Encryption Based on Φ-Hiding

We now define a triple of algorithms G, E, D for our encryption scheme.

Key Generation: $G(1^k, \alpha)$.

- Let $p_1, \dots, p_t$ be primes with $5 \leq p_1 < p_2 < \cdots < p_t$, and choose $e_j = \left\lfloor \frac{k}{4 \log p_j} \right\rfloor$, thus e_j is the largest integer such that $\log\left(p_j^{e_j}\right) < dk$, for some $d < \frac{1}{4}$. Set $\pi_j = p_j^{e_j}$. To encrypt n-bit messages, we will need to choose $t = \frac{\rho n}{dk}$. Since we assume $n = k^\alpha$, this becomes $t = \frac{\rho k^{\alpha-1}}{d}$.
- Generate a random permutation $\sigma \in_R S_t$, the symmetric group on t elements.
- Generate moduli $m_1, \dots, m_t$ such that $m_j \in \mathcal{H}_k^{\pi_{\sigma(j)}}$, i.e. m_j Φ-hides $\pi_{\sigma(j)}$.
- Find generators $\{g_j\}$ of the cyclic groups $\{G_{m_j}\}$.

The public key will then be

$$PK = ((g_1, m_1, \pi_1), \dots, (g_t, m_t, \pi_t)),$$

and the secret key will be

$$SK = \left(\sigma, \frac{\varphi(m_1)}{\pi_{\sigma(1)}}, \dots, \frac{\varphi(m_t)}{\pi_{\sigma(t)}}\right).$$

---

[5] See the full version for a more in-depth discussion of the error tolerance of the CRT-ECC.

Encryption: given an n-bit message X,

- Break X into $\frac{n}{\ell k}$ blocks X_i of size ℓk, and treat each X_i as an integer in the range $\{0 \ldots 2^{\ell k}\}$.
- For block X_i, we will use the s prime powers $\pi_{(i-1)s+1}, \ldots, \pi_{is}$ to encode X_i. Since the moduli $m_{\sigma^{-1}((i-1)s+1)}, \ldots, m_{\sigma^{-1}(is)}$ that correspond to these π's is unknown to the sender, he must apply the Chinese Remainder Theorem using all the π_j's. Thus for each block X_i, using the CRT, the sender generates $\tilde{X}_i \in [1, \ldots, (\pi_1 \cdots \pi_t)]$, such that

$$\tilde{X}_i = \begin{cases} X_i \mod \pi_j & \text{for } j \in [(i-1)s+1, \ldots, is], \\ 0 \mod \pi_j & \text{for } j \in [1, \ldots, (i-1)s] \cup [is+1, \ldots, t]. \end{cases}$$

To recover from error-rate $\frac{1}{2} - \frac{1}{2\rho}$, we set $s = \frac{\rho \ell}{d}$.

- The sender then sets $\tilde{X} = \sum_{i=1}^{\frac{n}{\ell k}} \tilde{X}_i$. Thus for each j, $\tilde{X} = X_i \mod \pi_{\sigma(j)}$ for the unique i such that $(i-1)s+1 \leq \sigma(j) \leq is$.
- For $j \in [1, \ldots, t]$, generate a random $r_j \in \{0, \ldots, \pi_1 \cdots \pi_t\}$.
- Then calculate $h_j = g_j^{\tilde{X} + r_j \pi_1 \cdots \pi_t} \mod m_j$ for each $j \in \{1, \ldots, t\}$. Thus

$$h_j = E\left(\tilde{X} \mod \pi_{\sigma(j)}\right) = E(X_i \mod \pi_{\sigma(j)}),$$

where $(i-1)s+1 \leq \sigma(j) \leq is$, and E is the encryption protocol described in §5.1. At this point, partial information about the block X_i is spread over s of the h_j's.

- Apply the binary Error Correcting Code ECC to each h_j individually.
- The encryption is then the t-tuple $(\mathsf{ECC}(h_1), \mathsf{ECC}(h_2), \ldots, \mathsf{ECC}(h_t))$.

Decryption: to recover the ith block, of a message X from the t-tuple $(h_1, \ldots, h_t)$

- Select the s encryptions that encode X_i, $\{\mathsf{ECC}(h_{\sigma^{-1}((i-1)s+1)}), \ldots, \mathsf{ECC}(h_{\sigma^{-1}(is)})\}$.
- Decode each $\mathsf{ECC}(h_j)$ to find obtain $\{h_{\sigma^{-1}((i-1)s+1)}, \ldots, h_{\sigma^{-1}(is)}\}$.
- Decrypt each of the s encryptions using the decryption algorithm from §5.1. This gives $a_1, \ldots, a_s$ where $a_j = X_i \mod (\pi_{(i-1)s+j})$.
- Using the Chinese Remainder Code Decoding Algorithm, reconstruct X_i from the s remainders $a_1, \ldots, a_s$. Note that if there are no errors introduced, this step can be replaced by simple Chinese Remaindering.

5.4   Analysis

The proof of security remains essentially the same as in the general setting.

For the locality, we note that to recover a single bit of X, or equivalently the entire block X_i that contains it, we must read s blocks of the ciphertext $\{\mathsf{ECC}(h_{\sigma^{-1}((i-1)s+1)}), \ldots, \mathsf{ECC}(h_{\sigma^{-1}(is)})\}$. Since $|h_j| = k$ and $|\mathsf{ECC}(h_j)| = c_2 k$, we must read a total of $sc_2 k = \frac{\rho c_2 \ell k}{d}$ bits. Since the probability of error will be negligible in ℓ, we set $\ell \approx k$, and since $d < \frac{1}{4}$, we find that we need to read

$5c_2\rho k^2$ bits of the ciphertext to recover one bit of the plaintext, where c and ρ are parameters that determine the error-rate of our code. Thus our system only achieves local-decodability for $n = \Omega(k^{2+\epsilon})$. For $n \approx k^3$, our system already offers a significant improvement over standard error-correcting codes.

Thus we arrive at the following result

Corollary 3. *Under the Small Primes Φ-Hiding Assumption there is a Public-Key Locally Decodable Code which can recover from a constant error-rate in the bits of the message, with public key size $\mathcal{O}(n)$ and ciphertexts of size $\mathcal{O}(n)$, where n is the size of the plaintext and k is the security parameter. The resulting code has locality $\mathcal{O}(k^2)$, i.e. to recover a single bit from the message we must read $\mathcal{O}(k^2)$ bits of the codeword.*

References

1. Katz, J., Trevisan, L.: On the efficiency of local decoding procedures for error-correcting codes. In: STOC 2000: Proceedings of the 32nd Annual Symposium on the Theory of Computing, pp. 80–86 (2000)
2. Yekhanin, S.: Towards 3-Query Locally Decodable Codes of Subexponential Length. In: Proceedings of the 39th ACM Symposiom on the Theory of Computinng (STOC) (2007)
3. Lipton, R.J.: A new approach to information theory. In: Enjalbert, P., Mayr, E.W., Wagner, K.W. (eds.) STACS 1994. LNCS, vol. 775, pp. 699–708. Springer, Heidelberg (1994)
4. Gopalan, P., Lipton, R.J., Ding, Y.Z.: Error correction against computationally bounded adversaries (manuscript, 2004)
5. Micali, S., Peikert, C., Sudan, M., Wilson, D.A.: Optimal error correction against computationally bounded noise. In: Kilian, J. (ed.) TCC 2005. LNCS, vol. 3378, pp. 1–16. Springer, Heidelberg (2005)
6. Ostrovsky, R., Pandey, O., Sahai, A.: Private locally decodable codes. In: Arge, L., Cachin, C., Jurdziński, T., Tarlecki, A. (eds.) ICALP 2007. LNCS, vol. 4596, pp. 298–387. Springer, Heidelberg (2007)
7. Gentry, C., Ramzan, Z.: Single-database private information retrieval with constant communication rate. In: Caires, L., Italiano, G.F., Monteiro, L., Palamidessi, C., Yung, M. (eds.) ICALP 2005. LNCS, vol. 3580, pp. 803–815. Springer, Heidelberg (2005)
8. Kushilevitz, E., Ostrovsky, R.: Replication is not needed: Single database, computationally-private information retrieval. In: IEEE Symposium on Foundations of Computer Science, pp. 364–373 (1997)
9. Ostrovsky, R., Skeith III, W.E.: A survey of single-database private information retrieval: Techniques and applications. In: Skeith III, W.E. (ed.) PKC 2007. LNCS, vol. 4450, pp. 393–411. Springer, Heidelberg (2007)
10. Damgård, I., Jurik, M.: A generalisation, a simplification and some applications of paillier's probabilistic public-key system. In: Kim, K.-c. (ed.) PKC 2001. LNCS, vol. 1992, pp. 119–136. Springer, Heidelberg (2001)
11. Paillier, P.: Public-key cryptosystems based on composite degree residuosity classes. In: Stern, J. (ed.) EUROCRYPT 1999. LNCS, vol. 1592, pp. 223–238. Springer, Heidelberg (1999)

12. Cachin, C., Micali, S., Stadler, M.: Computationally private information retrieval with polylogarithmic communication. In: Stern, J. (ed.) EUROCRYPT 1999. LNCS, vol. 1592, pp. 402–414. Springer, Heidelberg (1999)

13. Shannon, C.E.: A Mathematical Theory of Communication. Bell System Technical Journal 27, 343–379, 623–656 (1948)

14. Ishai, Y., Kushilevitz, E., Ostrovsky, R.: Sufficient conditions for collision resistant hashing. In: Kilian, J. (ed.) TCC 2005. LNCS, vol. 3378, pp. 445–456. Springer, Heidelberg (2005)

15. Chang, Y.C.: Single database private information retrieval with logarithmic communication. In: Wang, H., Pieprzyk, J., Varadharajan, V. (eds.) ACISP 2004. LNCS, vol. 3108. Springer, Heidelberg (2004)

16. Naccache, D., Stern, J.: A new public key cryptosystem based on higher residues. In: CCS 1998: Proceedings of the 5th ACM conference on Computer and communications security, pp. 59–66. ACM Press, New York (1998)

17. Gamal, T.E.: A public key cryptosystem and a signature scheme based on discrete logarithms. In: CRYPTO 1984, pp. 10–18. Springer, New York (1985)

18. Ishai, Y., Kushilevitz, E., Ostrovsky, R., Sahai, A.: Batch codes and their applications. In: STOC 2004: Proceedings of the thirty-sixth annual ACM symposion the theory of computing, pp. 373–382. ACM Press, New York (2004)

19. Goldwasser, S., Micali, S.: Probabilistic Encryption. Journal of Computer and System Sciences 28 (2), 270–299 (1984)

20. Benaloh, J.D.C.: Verifiable secret-ballot elections. PhD thesis. Yale University (1987)

21. Benaloh, J.C.: Dense probabilistic encryption. In: Proceedings of the Workshop on Selected Areas in Cryptography, pp. 120–128 (1994)

22. Hemenway, B., Ostrovsky, R.: Public key locally decodable codes (2007), http://eprint.iacr.org/2007/083/

Key-Recovery Attacks on Universal Hash Function Based MAC Algorithms[*]

Helena Handschuh[1] and Bart Preneel[2,3]

[1] Spansion, 105 rue Anatole France 92684 Levallois-Perret Cedex, France
helena.handschuh@spansion.com
[2] Katholieke Universiteit Leuven, Dept. Electrical Engineering-ESAT/COSIC,
Kasteelpark Arenberg 10, bus 2446, B-3001 Leuven, Belgium
bart.preneel@esat.kuleuven.be
[3] IBBT, Van Crommenlaan, B-9000 Gent

Abstract. This paper discusses key recovery and universal forgery attacks on several MAC algorithms based on universal hash functions. The attacks use a substantial number of verification queries but eventually allow for universal forgeries instead of existential or multiple forgeries. This means that the security of the algorithms completely collapses once a few forgeries are found. Some of these attacks start off by exploiting a weak key property, but turn out to become full-fledged divide and conquer attacks because of the specific structure of the universal hash functions considered. Partial information on a secret key can be exploited too, in the sense that it renders some key recovery attacks practical as soon as a few key bits are known. These results show that while universal hash functions offer provable security, high speeds and parallelism, their simple combinatorial properties make them less robust than conventional message authentication primitives.

1 Introduction

Message Authentication Code (MAC) algorithms are symmetric cryptographic primitives that allow senders and receivers who share a common secret key to make sure the contents of a transmitted message has not been tampered with. Three main types of constructions for MAC algorithms can be found in the literature: constructions based on block ciphers, based on hash functions and based on universal hash functions. In this paper we focus on the third class. The interesting feature of these MAC algorithms is that they are secure against an opponent with unlimited computing power. Simmons performed seminal work in this area [35]. The most widely used schemes are constructed following the Wegman-Carter paradigm [37]: first, the input message is hashed to a short digest using a universal hash function indexed by a secret key. Such a hash function has the property that for any two distinct inputs the probability over

---

[*] This work was partially funded by the European Commission through the IST Programme under Contract IST-2002-507932 ECRYPT and by the Belgian Government through the IUAP Programme under contract P6/26 BCRYPT.

D. Wagner (Ed.): CRYPTO 2008, LNCS 5157, pp. 144–161, 2008.
© International Association for Cryptologic Research 2008

all keys that two inputs have a specific difference is small. Then, the resulting hash value is encrypted by adding a one-time key. This approach is provably secure in the information theoretic setting. Brassard proposed a construction of a computationally secure MAC algorithm from universal hash functions by replacing the one-time pad by the output of a pseudo-random function applied to a nonce [11]. Later work suggests to apply a pseudo-random function directly to the hash result. A second optimization is to also derive the key for the universal hash function from a short key; if the universal hash function key is large (it can be several Kbytes long), this would still be too inefficient; instead one reuses this key for multiple messages. This paper will show that this opens the way for new attacks. A large number of computationally secure MAC algorithms derived from universal hash functions have been proposed following this model. Such MAC algorithms and universal hash functions include UMAC [9], MMH [17], NMH [17], Square Hash [15], Poly1305-AES [6], CWC [25], GCM/GMAC [30, 31], and a recent polynomial variant by Bernstein [7]. They are seen to be attractive because of their high speed and the simple constructions and security analysis.

Our contribution. This paper explores the implications of the reuse of key material in constructions of MAC algorithms based on universal hash functions as suggested by Wegman and Carter [37]; our work leads to improved forgery attacks and shortcut attacks that recover secret keys or parts thereof with the same complexity as a forgery attack. From a provable security viewpoint, it may seem desirable that forgery is equivalent to key recovery (since it implies that a weaker forgery attack can be reduced to the most difficult attack, namely a key recovery attack). However, key recovery allows for an arbitrary number of selective forgeries, hence its practical implications are much more serious. While for some applications it may be acceptable that a MAC forgery occurs from time to time, being able to recover the secret key as soon as the forgery bound is reached, is clearly not. It is important to stress that our results do not violate the proven security bounds on forgery attacks. Key recovery attacks are only a problem because MAC keys are being reused, which was not the case in the initial approach for authentication based on universal hash functions.

As their name suggests, universal hash functions guarantee good behavior of the hash function for *any* input pair; however, this refers to an average behavior over all keys and does not guarantee that each key yields a hash function with a uniform output distribution. For some schemes we identify rather large classes of weak keys that allow to easily forge authentication tags by swapping two blocks or by assigning specific values to some message blocks. The use of a weak key can typically be detected with a single text/MAC pair: it is sufficient to modify the text and submit a verification query. In principle the parties could check for the presence of weak keys, but in some cases this will substantially increase the complexity of the key generation procedure since a large number of combinations need to be avoided.

While the forgery probability of these schemes can be bounded, we show that in several MAC algorithms a small number of forgeries leads to if not total at least partial key recovery, which in turn allows us to create an arbitrary number of forgeries – unlike conventional MAC algorithms such as CBC-MAC [18, 32] the security of MAC algorithms based on universal hash functions collapses once a few forgeries are found; sometimes even a single forgery suffices.

For some constructions, we present enhanced key recovery attacks. We guess part of the key and try to confirm this guess with a MAC verification query. In schemes with large keys consisting of many words (say of 32 bits each), this allows for an efficient word by word key recovery. Some of our key recovery attacks can take into account partial information we may have obtained on a key, for example if sender and verifier read out a few bytes of the key over the phone to make sure they are using the right key or when side-channel attacks reveal some key bits or relations among key bits or their Hamming weight.

A final class of attacks, described for the purpose of completeness, exploits birthday attacks on special types of messages. If the MAC computation is stateful, that is, if the MAC generation algorithm depends on a nonce (a random number or a counter), these attacks require reuse of nonces by the sender (counter misuse), which is typically not allowed by the security model. However, one should expect that in some cases attackers can 'reset' senders (think of a bug in the code or a fault injection that make the software used by the sender crash). It is an unfortunate property of these MAC algorithms that a limited number of sender resets results in serious security weaknesses.

Most of our attacks however do *not* require nonce reuse at all, since we mostly consider a very simple person-in-the-middle scenario: the sender chooses a unique nonce and creates a text/MAC pair. The attacker modifies the text and submits a verification query *without changing the tag or the nonce*. The idea behind our attacks is that messages for which the resulting hash values collide by construction can be substituted without knowing (or even re-using) the one-time value the hash result was encrypted with. This scenario clearly fits within the security model of MAC algorithms (see e.g., Bellare et al. [4]).

Some of the results may seem rather straightforward and a few of these observations have been made earlier on some schemes. However, designers of universal hash functions typically do not emphasize these weaknesses. To the best of our knowledge, this paper is the first work that systematically investigates how robust or brittle a broad series of universal hash functions is when some of the above observations are applied. It also shows how a simple weakness can easily be developed to a partial key recovery attack with devastating consequences. In particular, we show new findings on the following universal hash functions and MAC algorithms based thereupon: polynomial hash [6, 8, 13, 22, 25], MMH [17], Square Hash [15], NH/UMAC [9] and its variants VMAC [28] and WH/WMAC [23]. Due to space limitations, the results on Johansson's bucket hashing with a short key size [20] only appear in the long version of this paper.

Related work. McGrew and Fluhrer have observed in [29] that once a single forgery has been performed, additional forgeries become easier; more specifically, the forgery probability for MAC algorithms such as CBC-MAC and HMAC increases cubically with the number of known text-MAC pairs, while for universal hash functions the forgery probability increases only quadratically.

Black and Cochran have analyzed in [10] what happens after a first collision is observed in the output of the MAC algorithm; they show that subsequent forgeries or reforgeability becomes easier than expected; for two cases they also present a key recovery attack. They propose a solution to these reforgeability attacks in the form of the WMAC construction [10]; however, we will show in this paper that this construction does not offer any protection against most of our key recovery attacks.

There is a broad literature on key recovery attacks on MAC algorithms (e.g., [34]). The techniques used in this paper are based on classical divide-and-conquer techniques. Our attacks on GCM/GMAC [30, 31] extend earlier work by Ferguson [16] and Joux [21] by considering the case where the one-time pad (more precisely the addition of the output of a pseudo-random function) is replaced by the application of a pseudo-random function.

Organization of the paper. In Sect. 2 we provide some background on MAC algorithms, information theoretic authentication and universal hash functions. Section 3 describes our attacks and discusses a number of attacks on different schemes in detail. In Sect. 4 we summarize our results and present some open problems and directions for future work.

2 Background

2.1 MAC Algorithm

A MAC algorithm consists of three algorithms: 1) a key generation algorithm, that is, a randomized algorithm that generates a κ-bit key k; 2) a MAC generation algorithm, that on input a text x and a key k generates an m-bit tag (this algorithm can be randomized and/or can be stateful); 3) a MAC verification algorithm, that on input a text and a tag generates an answer true or false (1/0) (this algorithm is deterministic and typically stateless). The security of a MAC algorithm can be formalized in concrete complexity following Bellare et al. [5] (see also [33]).

Definition 1. *A MAC algorithm is $(\epsilon, t, q, q', q'', L)$ secure if, an adversary who does not know k, and*

- *can spend time t (operations);*
- *can obtain the MAC for q chosen texts;*
- *can observe the MAC for q' known texts; and*
- *can obtain the result of q'' verification queries on text-MAC pairs of his choice.*

(each text of length L), cannot produce an existential forgery with probability of success larger than ϵ.

Here known text-MAC pairs are generated according to a distribution chosen by the sender, while chosen text-MAC pairs are generated according to a distribution specified by an adversary.

There are two generic attacks on MAC algorithms: finding the key by exhaustive search, which requires κ/m known text-MAC pairs and on average $2^{\kappa-1}$ (offline) MAC evaluations and guessing the MAC value, which requires on average $\min(2^{m-1}, 2^{\kappa-1})$ MAC verifications. The second attack can be detected easily by the large number of wrong MAC verifications. Iterated MAC algorithms with an n-bit internal memory succumb to a birthday forgery attack that requires $2^{n/2}$ known text-MAC pairs and $\min(2^{n/2}, 2^{n-m})$ chosen text-MAC pairs [24, 34].

Typical MAC algorithms used in practice are CBC-MAC (the variants EMAC [18, 32] and CMAC [19] work for arbitrary length messages) and HMAC [2, 3].

2.2 Information Theoretic Authentication

Information theoretic authentication was developed in the 1970s by Simmons [35] and Carter and Wegman [12, 37]. It has several very attractive features. First and foremost, its security is not based on any complexity theoretic assumption and holds against opponents with unlimited computing power. In the last decade researchers have attempted to fine tune the speed of these constructions; the current state of the art is that they are up to 15 times faster than CBC-MAC based on AES or HMAC based on SHA-1. Moreover, if properly designed, these functions can be parallelizable and incremental; the latter property means that if a local change is made to a large message, the MAC value can be updated in time that depends on the size of the local update rather than on the overall message length. Information theoretic authentication also has serious disadvantages: as for the one-time pad, it is required that the key is used only once. Moreover, in order to get extreme performance (below 2 cycles/byte in software), very large keys are required which increases key storage costs and limits key agility. Finally, note that Simmons' theory shows that the security level in bits against forgery is at most half the key size in bits. This paper will demonstrate that these constructions have another disadvantage; however, this disadvantage will only become apparent if the information theoretic scheme is replaced by a computationally secure scheme with comparable efficiency.

One of the most elegant examples of an information theoretic authentication scheme is the polynomial construction (see e.g., [8, 13]). This construction uses two n-bit keys k and k', and operates on messages of bitlength $\ell = t \cdot n$. It starts by splitting the input message x into t blocks of bitlength n denoted x_1 through x_t. The x_i, k and k' are represented as elements of $GF(2^n)$. The authentication function $g_{k,k'}(x)$ can then be described as follows:

$$g_{k,k'}(x) = k' + \sum_{i=1}^{t} x_i \cdot k^i.$$

The probability of sending an acceptable tag without having observed a text/tag pair (the impersonation probability) equals 2^{-n}, while the probability of forging a text/tag pair after having observed one pair (the substitution probability) is equal to $(\ell/n)/2^n = t/2^n$. It has been pointed out that the value of k can be reused for multiple messages, as long as the value of k' is used only once (e.g., [13]).

2.3 MAC Algorithms Based on Universal Hashing

Following Carter and Wegman [12], a universal hash function is a family of functions indexed by a parameter called the key with the following property: for all distinct inputs, the probability over all keys that they collide is small.

Definition 2. *Let* $g_k : A \longrightarrow B$ *be a family of functions with* $a = |A|$ *and* $b = |B|$. *Let* ϵ *be any positive real number. Then* g_k *is an* ϵ-**almost universal** *class (or* ϵ-AU *class)* $\mathcal{G}$ *of hash functions if*

$$\forall\, x, x' \neq x \in A : \Pr_k \{g_k(x) = g_k(x')\} \leq \epsilon.$$

These functions can only be used for message authentication if the output is processed using another function. If one wants to use a universal hash function directly for message authentication, a stronger property is required, namely the hash function needs to be strongly universal [12, 37]. This concept has later on been generalized by Krawczyk [26].

Definition 3. *Let* $B, \star$ *be an Abelian group. Then* g_k *is an* ϵ-**almost** $\star$ **universal** *class (or* ϵ-A$\star$U *class) of hash functions if*

$$\forall\, x, x' \neq x \in A \text{ and } \forall \Delta \in B : \Pr_k \{g_k(x) = g_k(x') \star \Delta\} \leq \epsilon.$$

Many simple mathematical operations, such as polynomial evaluation, matrix-vector product and inner products, yield ϵ-AU or ϵ-A$\star$U hash functions.

A MAC algorithm based on universal hash functions consists of two building blocks: an efficient keyed compression function that reduces long inputs to a fixed length and a method to process the short hash result and an output transformation. In practical constructions, the encryption with the one-time pad (addition of k') is typically replaced by applying a pseudo-random function with secret key k'. In this case one obtains computational rather than unconditional security. Informally, a pseudo-random function family is a function that a computationally limited adversary cannot distinguish with probability substantially better than $1/2$ from a function chosen uniformly at random from all functions with the same range and domain.

The following compression functions have been proposed in practice. We consider an input that consists of t blocks of fixed length; if necessary, t will be assumed to be even. A simple ϵ-AU hash function is the polynomial evaluation function over a finite field:

- $g_k(x) = \sum_{i=0}^{t} x_i \cdot k^i$, $x_i, k \in \mathrm{GF}(2^n)$ or $\mathrm{GF}(p)$.

The following functions are ϵ-A$\star$U for $\star$ equal to addition modulo 2, 2^n or p (this is clear from the context):

- polynomial [6, 30, 31]: $g_k(x) = \sum_{i=1}^{t} x_i \cdot k^i$, $x_i, k \in \mathrm{GF}(2^n)$ or $\mathrm{GF}(p)$;
- MMH [17]: $g_k(x) = \left(\left(\left(\sum_{i=1}^{t} x_i \cdot k_i \right) \bmod 2^{64} \right) \bmod p \right) \bmod 2^{32}$, $x_i, k_i \in \mathbf{Z}_{2^{32}}$ and $p = 2^{32} + 15$;

- Square Hash [15]: $g_k(x) = \sum_{i=1}^{t} (x_i + k_i)^2 \bmod p$, $x_i, k_i \in \mathbf{Z}_{2^w}$;
- NMH [17, 37]: $g_k(x) = \left(\sum_{i=1}^{t/2} (x_{2i-1} + k_{2i-1}) \cdot (x_{2i} + k_{2i}) \right) \bmod p$, x_i, k_i $\in \mathbf{Z}_{2^{32}}$ and $p = 2^{32} + 15$;
- NH [9]: $g_k(x) = \left(\sum_{i=1}^{t/2} ((x_{2i-1} + k_{2i-1}) \bmod 2^w) \cdot ((x_{2i} + k_{2i}) \bmod 2^w) \right) \bmod$ 2^{2w}, $x_i, k_i \in \mathbf{Z}_{2^w}$;
- WH [23] $g_k(x) = \left(\sum_{i=1}^{t/2} (x_{2i-1} + k_{2i-1}) \cdot (x_{2i} + k_{2i}) x^{(t/2-i)w} \right) \bmod p(x)$, x_i, $k_i \in \mathrm{GF}(2^w)$ (polynomial arithmetic).

Note that one can also cascade multiple ϵ-AU hash functions followed by an ϵ-A$\star$U hash function to obtain more efficient constructions (see Stinson [36]).

For the output transformation, the options listed below can be considered; this list has been inspired by the slides of [6]. Here $f_{k'}()$ denotes a pseudo-random function family indexed by the key k'. The first option results in a stateless MAC algorithm, while the other two assume that the sender keeps the state with the counter n (which is called the nonce hereafter). The last two options offer a better security level but one needs to guarantee that n is not reused during the MAC generation. Nonce reuse during the MAC verification is typically allowed.

Option 1: $\mathrm{MAC}_{k||k'}(x) = f_{k'}(g_k(x))$ with g ϵ-AU.
Option 2: $\mathrm{MAC}_{k||k'}(x) = f_{k'}(n) \star g_k(x)$ with g ϵ-A$\star$U;
Option 3: $\mathrm{MAC}_{k||k'}(x) = f_{k'}(n||g_k(x))$ with g ϵ-AU; this variant needs a larger input of f.

A security analysis of Option 3 (under the name WMAC[1]) is provided by Black and Cochran in [10]. Due to space restrictions we omit an analysis of the randomized message preprocessing of Dodis and Pietrzak [14].

The MAC algorithms based on universal hash functions considered in standards are UMAC [27] and GCM [31]; both follow Option 2. ISO/IEC JTC1/SC27 is currently developing an international standard on MAC algorithms based on universal hash functions (part 3 of IS 9797).

3   Security Results

In this central section of the paper we present new results on the security of universal hash function based MAC algorithms. We present a detailed analysis of the vulnerability of several MAC algorithms with respect to three different scenarios. First we briefly describe each of the algorithms and immediately identify classes of weak keys for each of them. Subsequently we describe more powerful universal forgery and key recovery attacks based on partial key information leakage and the divide-and-conquer principle. Next we describe a key recovery attack on polynomial hash functions using the birthday paradox.

---

[1] Unfortunately this name has already been given by Kaps et al. to a completely different MAC algorithm one year earlier [23], cf. Sect. 3.1.

3.1 Weak Keys

In symmetric cryptology, a class of keys is called a weak key class if for the members of that class the algorithm behaves in an unexpected way *and* if it is easy to detect whether a particular unknown key belongs to this class. For a MAC algorithm, the unexpected behavior can be that the forgery probability for this key is substantially larger than average. Moreover, if a weak key class is of size C, one requires that identifying that a key belongs to this class requires testing fewer than C keys by exhaustive search and fewer than C verification queries.

The security proofs for universal hash functions bound the average forgery probability over all keys; this does not cause any problem if the key is used only once since for most applications a bound on the average rather than the worst case forgery probability is sufficient. However, if the key of the universal hash function is reused (which is the case in many current constructions), the fact that the weak keys are easy to recover can have dramatic implications, as key recovery allows for arbitrary forgeries.

Polynomial hash functions. We first study the ϵ-A$\oplus$U polynomial hash function $g_k(x) = \sum_{i=1}^{t} x_i \cdot k^i$ with k, $x_i \in \mathrm{GF}(2^n)$. The forgery probability of this scheme is equal to $t/2^n$ (cf. Sect. 2.2). This function forms the core of the GMAC (Galois MAC) and GCM (Galois/Counter Mode of operation) constructions for a MAC algorithm and for authenticated encryption by McGrew and Viega [30]; both have been standardized by NIST [31]. The GCM construction uses Option 2 of Sect. 2.3 with the pseudo-random value generated by a block cipher in counter (CTR) mode. It allows to truncate the output by selecting the leftmost τ bits.

One can also consider polynomials evaluated over $\mathrm{GF}(p)$ to take advantage of fast multiplication hardware available on current processors. The fastest scheme today is Poly1305-AES of Bernstein [6] that uses $p = 2^{130} - 5$.

Clearly $k = 0$ is a weak key for these polynomial hash functions. In this case (which appears with probability 2^{-n} respectively $1/p$) all messages map to 0 under g_k and all messages have the same valid tag, allowing for trivial forgery attacks: an attacker can take any valid text/MAC pair and substitute the text by any text of her choice (that is, the forgery probability is 1 independent of the output transformation). As 2^n and p are typically very large, this is not a realistic threat.

MMH. This inner product construction was introduced by Halevi and Krawczyk in 1997 [17]. It is an ϵ-A+U with $\epsilon = 1.5/2^{30}$.

Next we describe two weak key classes for MMH. A first class consists of the keys for which any k_i value is equal to 0. We present a more general description by considering a universal hash function family $g_k(x)$ with k consisting of t elements k_i of the ring R and x consisting of t elements x_i of the ring R.

Proposition 1 (type I). *If $g_k(x)$ is of the form $z_0\left(\sum_{i=1}^{t} z_1(k_i, x_i)\right)$ where $z_1(k_i, x_i)$ is a multiple of k_i and $z_0()$ is an arbitrary function, then any key k for which at least one $k_i = 0$ belongs to a weak key class. The fraction of weak* *keys is equal to* $1 - (1 - 1/|R|)^t \approx t/|R|$. *Membership in this class can be tested with* t *MAC queries and* t *verification queries.*

The proposition assumes that one asks for a text/MAC pair (if applicable with a given nonce), one modifies x_i, which implies that the MAC value does not change if $k_i = 0$, and submits a verification query with the modified x_i but the original MAC value. Note that one could reduce the number of text/MAC pairs to a single one by modifying always the same text/MAC pair. If a nonce is used, this requires the assumption that the verifier does not check repeating nonces (which is a standard assumption in MAC algorithms), but our attacks also apply (with a higher complexity) if the verifier does check for repeating nonces. Hence this attack is equally effective whether Option 1, 2, or 3 is used. This comment also applies to the other weak key classes identified below.

A second weak key class is based on the summation and exploits two equal subkeys:

Proposition 2 (type II). *If* $g_k(x)$ *is of the form* $z_0 \left(\sum_{i=1}^{t} z_2(k_i, x_i) \right)$ *for any functions* $z_2(k_i, x_i)$ *and* $z_0()$, *then any key* k *for which there exist* i *and* j *such that* $k_i = k_j$ *belongs to a weak key class. The fraction of weak keys is equal to* $R^t - R!/(R - t)! \approx t(t - 1)/(2|R|)$ *(for* t *small). Membership in this class can be tested with* $t(t - 1)/2$ *MAC queries and* $t(t - 1)/2$ *verification queries.*

This class can be further generalized: if $k_i = \alpha \cdot k_j$ (over the integers), this can be detected as follows: one asks for a message, substitutes the blocks x_i and x_j by $x_i' = x_j/\alpha$ and $x_j' = \alpha \cdot x_i$ and makes a verification query (this requires that α divides x_j and x_i is smaller than $2^{32}/\alpha$). Type II weak keys correspond to $\alpha = 1$; they can be identified by swapping two message blocks. The probability that two key words satisfy such a relation for a fixed value of α is equal to $1/(\alpha \cdot 2^{32})$.

Square Hash. This construction was introduced by Etzel, Patel and Ramzan [15] in 1999. Proposition 2 also results in weak keys for Square Hash.

NMH, NH and WH. The NMH construction by Wegman and Carter is a variant of MMH described in the same article [17]. It consists of an inner product of sums of message and key words. NMH requires a large key, but yields the fastest throughput for long messages on general processors as it can make use of multimedia instructions. NMH* is a variant of NMH in which the inner addition is done modulo 2^w for efficiency reasons, and the result of the modulo p operation is further reduced modulo 2^w. Typical choices are $w = 32$ and $p = 2^{32} + 15$.

NH is a yet another variant of NMH that uses additional modular reductions to improve the performance (this reduction corresponds to ignoring carry bits). It forms the crucial building block of the UMAC and VMAC algorithms. UMAC [9, 27] was introduced by Black, Halevi, Krawczyk, Krovetz and Rogaway in 1999. UMAC first applies a universal hash function called UHASH to the message in order to derive a shorter fixed-length hash value. In order to obtain very high efficiency, UHASH is constructed of three layers, the first of which is NH, the second one is a polynomial hash function over $\mathrm{GF}(p)$ and the third one an MMH construction. In order to obtain a MAC algorithm, Option 2 is selected, similar to the GCM and Poly1305-AES constructions. VMAC is a variant of UMAC that is optimized for 64-bit arithmetic [28]. WH is a variant of NH that replaces elements of $\mathbf{Z}_{2^w}$ by polynomials (elements of $GF(2^w)$) to reduce the cost of hardware implementations; the resulting MAC is called WMAC [23].

This class of hash functions has weak keys that are based on symmetry:

Proposition 3 (type III). *If $g_k(x)$ is of the form $z_0 \left(\sum_{i=1}^{t/2} z_3(k_{2i-1}, x_{2i-1}) \cdot z_3(k_{2i}, x_{2i}) \right)$ for any functions $z_3(k_i, x_i)$ and $z_0()$, then any key k for which there exists an i such that $k_{2i-1} = k_{2i}$ belongs to a weak key class. The fraction of weak keys is equal to $1 - (1 - 1/|R|)^{t/2} \approx t/2|R|$ (for t small). Membership in this class can be tested with $t/2$ MAC queries and $t/2$ verification queries.*

One can also apply Proposition 2 to the $t/2$ terms in the sum.

We conclude this section with a simple numerical example to illustrate the impact of these weak keys. If $t = 256$ and $|R| = 2^{32}$, the keys are 8192 bits long and the fraction of weak keys for Proposition 1, 2, and 3 is equal to 2^{-24}, 2^{-17}, and 2^{-25} respectively; membership in this class can be verified with 256, 32 640, and 128 queries.

3.2   Divide and Conquer Attacks and Partial Key Information Leakage

In this section we show how the existence of these weak key classes lead us to discover efficient key recovery attacks and how partial information on the key can be exploited to further speed up these attacks. The techniques are based on the construction of inner collisions in the universal hash function (following [34]) which can be confirmed with a single message substitution and a single verification query.

Polynomial hash. We can verify a guess for part of the key (namely k), even if we do not know the key k' to the pseudo-random function. Assume for example that $t \geq 2$, $k \neq 0$ and $x_i = 0$, for $3 \leq i \leq t$. Then for a given x_1 and x_2 and a guess for k, we can choose any x_2' and compute $x_1' = x_1 + (x_2 - x_2') \cdot k$. If the guess for k is correct, the message $x_1'||x_2'||0||\ldots$ will be accepted as a valid message for the same tag as message $x_1||x_2||0||\ldots$. Hence if this message is accepted, with high probability the guess for k was correct. It may also be that the guess for k was wrong but that due to a collision in the pseudo-random function $f_{k'}(.)$ the verification worked. This case can be ruled out by repeating the attack with a different message. Note that the attacker constructs here messages based on a text/MAC pair with a nonce selected by the sender, and the receiver uses each nonce only once. The GCM mode precludes this divide-and-conquer attack by deriving k from k'. In SNOW 3G [1] the key k is used only once, hence if the receiver checks for nonce reuse a correct guess for k can no longer be used.

Extensions of the Joux attacks on GCM variants. Joux presents an elegant attack on GCM which uses Option 2 with nonce reuse by the sender (the attack is described in Annex A). Here we show that we can extend his attack to Option 1 (even if classical encryption in ECB mode is used as suggested by Joux) or Option 3 (but without nonce reuse); this result shows that some of these weaknesses are inherent to polynomial hashing over $GF(2^n)$ and not to the use of the one-time pad or CTR mode as in GCM. In this case we have no access to the individual bits of $g_k(x) \oplus g_k(x')$ for two distinct messages x and x', but we can only test equality between the complete n-bit strings $g_k(x)$ and $g_k(x')$. We show that in this case a forgery also leads to key recovery, that is, recovering a key k requires an expected number of $\log_2(t) + 2^n/t$ verification queries. The attack goes as follows:

1. obtain a text x and the corresponding MAC value;
2. choose x' such that the polynomial with coefficients from $x - x'$ has t distinct roots (they should also be distinct from all previous roots used in this algorithm);
3. perform a MAC verification query for x'; if the result is incorrect, go to step 1;
4. after $2^n/t$ trials on average, you expect a correct MAC value and then you know that k is one of t values, that is, one of the at most t roots of a polynomial of step 2;
5. perform another $\log_2(t)$ MAC verification queries (binary search) to identify which one of the t roots is equal to k.

This attack is very similar to the key recovery attack by Black and Cochran [10]; the main difference is that [10] first assumes a forgery and then identifies the key, while here the forgery algorithm is optimized in function of the key recovery. We make the following comments:

- Exceptionally, this attack would also work with high probability if k is changed every time as for SNOW 3G [1] (in this case Step 5 would require nonce reuse by the receiver). The attack would then only be useful if the receiver could be convinced to reuse a key.
- If $n = 128$ as for GCM, this attack is clearly impractical. Even if $t = 2^{64}$ this would still require the verification of a large number of messages of length 2^{64}.
- The expected number of multiplications performed by the verification oracle is about $t \cdot 2^n/t = 2^n$, hence one may conclude that this attack is not better than brute force key search. First, it should be pointed out that this is a divide-and-conquer attack that applies independently of the key k' of the pseudo-random function. Second, this method may help to overcome limitations set on the number of incorrect MAC verifications, as it tests as many candidate key values as possible with a single verification query.
- It is straightforward to take into account in this attack any information one may have on the key k, such as the value of the first three bytes, the sum of any bits etc. Consider $n = 80$ and $t = 2^{18}$ and assume that we have an oracle that gives us $s = 24$ key bits. Then the attack requires 2^{38} text-MAC verifications to find the remaining 56 key bits. The same observation also

applies to the remainder of this section, since the attacker has to solve only very simple equations in the words of the key.

MMH. For the case of MMH, assume that one knows a multiplicative relation between two key words, say $k_i = (\alpha \cdot k_j) \bmod p$ for an $\alpha \in \mathbf{Z}_p$. Consider w.l.o.g. messages for which only x_i and x_j are non-zero. Then if one replaces (x_i, x_j) by (x_i', x_j') the hash value will be unchanged provided that $x_j' = (x_j + (x_i - x_i') \cdot \alpha) \bmod p$. We require that the $\bmod 2^{64}$ operation can be ignored, which holds if $x_i \cdot k_i + x_j \cdot k_j < 2^{64}$ and $x_i' \cdot k_i + x_j' \cdot k_j < 2^{64}$. One way to ensure that the first inequality is met is by choosing $x_i, x_j < 2^{31}$; for the second inequality one chooses $x_i' < 2^{31}$. Then $x_j' < 2^{31}$ with probability approximately $1/2$, hence it is easy to find tuples (x_i, x_j) and (x_i', x_j') for which these equations hold. Note that for every value of α one can precompute solutions without knowing the key. This implies that a single verification query after a message substitution allows to verify a guess for α. Overall this means that 32 bits of information on the key can be found with approximately 2^{32} verification queries; subsequently arbitrary forgeries are possible. Recovery of the t key words requires $t \cdot 2^{32}$ verification queries.

Square Hash. Square hash has the weakness that one can verify a guess for any key word k_i of w bits with a single verification query: it suffices to modify x_i into $x_i' = (-2k_i - x_i) \bmod p$. This follows immediately from the fact that $(-x)^2 \equiv x^2 \bmod p$. This attack requires approximately p MAC verifications to find a single key word and $t \cdot p$ MAC verifications to identify the correct key (a divide-and-conquer approach). Note that this attack could be precluded by ensuring that the hash function is restricted to input values less than $p/2$. This attack is simpler than the collision-based key recovery attack of Black and Cochran [10].

NMH, NH and WH. The following attack applies to all these constructions: it is sufficient to find a message for which one factor becomes 0 (e.g., for NMH $(x_{2i-1} + k_{2i-1}) = 0 \bmod p$ and for NH $(x_{2i-1} + k_{2i-1}) = 0 \bmod 2^w$); in this case $g_k(x)$ is independent of x_{2i}. This attack requires p or 2^w text-MAC queries and a similar number of MAC verification queries to find a single key word k_i. With a divide-and-conquer approach recovering the complete key requires t times more work.

A slightly more efficient attack is described next. Assume that one knows an additive relation between two key words, say $k_{2i} = (k_{2i-1} + \Delta) \bmod 2^w$ for a $\Delta \in \mathbf{Z}_{2^w}$. Then if one replaces (x_{2i-1}, x_{2i}) by $((x_{2i} + \Delta) \bmod 2^w, (x_{2i-1} - \Delta) \bmod 2^w)$ the hash value will be unchanged. This implies that a single verification query allows to verify a guess for Δ. By trying all 2^w values of Δ, it follows that with 2^w verification queries one finds a relation between k_{2i} and k_{2i-1}. This relation is sufficient to perform subsequent forgeries. In order to find k_{2i} (and k_{2i-1}), one can apply the method of the previous paragraph; recovering the t key words requires only $t \cdot 2^w$ verification queries and $t \cdot 2^{w-2}$ MAC queries. Alternatively one can choose $x_{2i-1} = x_{2i} + \Delta$ (which makes the two factors equal) and apply the techniques of Square Hash with chosen texts.

If these functions are used with large word lengths (64 bits or more such as for VMAC), these observations do *not* pose a serious problem. However, NH has variants with $w = 32$ (for UMAC) and WH is being proposed with values of w equal to 16, 32 and 64. If one wants to obtain higher security levels, one can use two parallel instantiations and in this case the function will behave as if the word length was twice as large. However, if one wants to use single instantiations of the small word sizes for very high performance, a security level against forgery of $c \cdot 2^{-w}$ for a small constant c may be acceptable, but a partial key recovery attack with complexity on the order of 2^w clearly is not.

The above attacks shed a new light on the security of UMAC; even if the UMAC RFC [27] warns to restrict the number of MAC verifications, the document explicitly states that no key recovery attack is known that can exploit a forgery.

In order to conclude this section, we point out that the above attacks (except for the attack by Joux on GCM) apply to the three options and do not require nonce reuse.

3.3   Birthday Collision Attacks on Polynomial Hash Functions

In this section we consider attacks on polynomial hash functions; these attacks require nonce reuse by the sender for Option 2 and 3 of Sect. 2.3; this implies that these attacks violate the standard security assumptions for these MAC algorithms. However, it is important to note that Option 1 has no nonce, hence it is obvious that for this case the attacks stay within the model.

If nonces are reused in polynomial hash schemes over $GF(p)$, it is easy to recover the key k (Bernstein [6] is also very clear about the fact that in this case no security guarantees can be provided).

For polynomial hash over $GF(2^n)$ and GCM, if we assume that somehow the sender can be convinced to reuse a nonce, or if we are using Option 1, we can enhance our previous attack by using the special messages of the Ferguson attack (we refer the reader to Annex A for a description of the original attack). As we no longer have access to the individual bits of $g_k(x) \oplus g_k(x')$ for two distinct messages x and x', we can only test equality between the complete n-bit strings $g_k(x)$ and $g_k(x')$. The birthday paradox allows to extend the attack.

1. consider messages with $x_i = 0$ except if $i = 2^j$ for some $j \geq 0$. As squaring in $GF(2^n)$ is a linear operation in $GF(2^n)$, the hash function becomes linear over this subset of inputs, hence we can write the bits of the hash result as follows:
$$g_k(x)[.] = \sum_{i*,j*,u*} x_{i*}[j*] \cdot k[u*].$$

2. by guessing the linear combination of s ($1 \leq s < n$) well-chosen bits of the key k, we can generate a set of λ messages for which the hash result is restricted to a subspace of size 2^{n-s} (e.g., the first s bits are equal to 0) (for details, see [16]);

3. now collect $\lambda = 2^{(n-s)/2+1}$ messages/MAC pairs (with the same nonce!) resulting in 2 collisions for the MAC value; each collision yields $n - s$ linear equations for the remaining $n - s$ key bits.

If we choose $n = 80$, $s = 24$, then $n - s = 56$: 2^{29} messages yield 2 collisions, resulting in 112 linear equations in the remaining 56 key bits. Note that even in the limit case of $s = 0$, this attack can recover the key k in $2^{n/2+1}$ chosen text/MAC pairs independently of k'.

4   Conclusion and Directions for Further Work

Table 1 presents an overview of the attacks. The security analysis of the many schemes and variants under multiple attacks is rather complex; we attempt here to summarize the main points. All universal hash functions except for the poly-nomial hash functions have large classes of weak keys, which have more serious implications than what was believed so far. All the universal hash functions that are based on key words that are substantially smaller than the output size (e.g, MMH, Square Hash and NMH with w-bit subkeys) allow – in spite of their large internal keys – for efficient divide-and-conquer key recovery attacks with ap-proximately $t \cdot 2^w$ MAC verifications. While it is well understood that 2^w MAC verifications allow for a single forgery and – in the case of key reuse – for multiple forgeries [10], the implications of a key recovery attack are much more serious as they allow for arbitrary forgery later on. Most of the hash functions (except for polynomial hashing over $GF(p)$) allow for improved attacks if an oracle provides partial information on the secret key. It is important to stress that most of our attacks work for the three options (with and without nonces) and do not require nonce reuse: they are thus completely within the standard model for MAC algo-rithms. It is surprising that the more expensive Option 3 (called WMAC in [10]) does not offer an increased protection against key recovery attacks.

Overall, the polynomial hash functions over $GF(p)$ seem to present fewer problems, but they are extremely vulnerable to nonce reuse, in particular if Option 2 (addition of a pseudo-random string) is used.

While universal hash functions have very attractive performance and provable security, our attacks demonstrate that most published MAC algorithms based on universal hash functions can be very brittle because of their simple combinatorial properties. Even within the security model, key recovery attacks are very efficient in spite of the large internal keys. Moreover, for those schemes for which no key recovery attack exists, a small violation of secure usage principles results in a collapse of security. This can be noted from very strong warnings in the specifications about nonce reuse and the detection of a large number of incorrect MAC verifications. A similar comment holds for partial key leakage (which can occur e.g., under side channel attacks). As it is very difficult to predict how a cryptographic algorithm will be used and abused, this is a highly undesirable property.

Table 1. Summary of our findings on MAC algorithms based on universal hash functions; if the key consists of small words, a word length of w is assumed. The third column indicates which part of k one needs to guess to allow for a verification with a single query; the fourth column indicates whether our attacks can exploit partial information on the key; the last column indicates the applicability of the new birthday attacks presented in this paper.

	number of weak keys	divide and conquer attacks	partial key information	new birthday attack
Polynomial hash $GF(2^n)$	1	k only	yes	yes
Polynomial hash $GF(p)$	1	k only	?	
Bucket hashing with small key	type I/II	subkey k_j	yes	yes
MMH	type I/II	w-bit subkey k_i	yes	
Square Hash	type II	w-bit subkey k_i	yes	
NMH/NH/WH	type II/III	w-bit subkey k_i	yes	

We present the following recommendations for MAC algorithms based on universal hash functions:

- Avoid reusing keys; while this may not be an option for schemes such as UMAC, for the polynomial hash functions, the overhead to generate a new value k for the universal hash function is quite small and brings a substantial increase of security and robustness as it will render most key recovery attacks useless; this approach is taken by SNOW 3G [1].
- In environments where side channel attacks are a concern, additional measures need to be taken which may negatively influence the performance.
- Sender and receiver need to guarantee/check uniqueness of nonces. The requirement for the sender seems to be more stringent in particular if a crash of the device needs to be considered. Appropriate implementation measures are necessary, such as storing counters in non-volatile memory. At the receiver side, restricting the number of MAC verifications with a single key seems more stringent than checking uniqueness. In any case, if random numbers are used to instantiate nonces, checking uniqueness for the receiver is problematic.

As an alternative for universal hash function based MACs, we recommend an AES-based scheme such as EMAC [18, 32]. It is somewhat slower but more "robust" in the following sense:

- Internal collisions (2^{64} texts) lead to forgeries, but not to key recovery.
- There is no known way to use an oracle that gives access to 32 or 64 key bits to speed-up key recovery by more than 2^{32} or 2^{64} operations.
- The algorithm benefits of a faster key setup.

We are currently investigating other universal hash functions, such as the new function proposed by Bernstein [7]. We also believe that it would be valuable to formalize the "regularity" of a universal hash function which could alleviate concerns related to weak keys.

Acknowledgements. The authors would like to thank the anonymous referees for helpful comments.

References

1. 3GPP TS 35.216, Specification of the 3GPP Confidentiality and Integrity Algorithms UEA2 & UIA2; Document 2: SNOW 3G specification (March 2006)
2. Bellare, M.: New Proofs for NMAC and HMAC: Security without Collision-Resistance. In: Dwork, C. (ed.) CRYPTO 2006. LNCS, vol. 4117, pp. 602–619. Springer, Heidelberg (2006)
3. Bellare, M., Canetti, R., Krawczyk, H.: Keying Hash Functions for Message Authentication. In: Koblitz, N. (ed.) CRYPTO 1996. LNCS, vol. 1109, pp. 1–15. Springer, Heidelberg (1996)
4. Bellare, M., Goldreich, O., Mityagin, A.: The Power of Verification Queries in Message Authentication and Authenticated Encryption. November 18 (2004), http://eprint.iacr.org/2004/309
5. Bellare, M., Kilian, J., Rogaway, P.: The Security of Cipher Block Chaining. In: Desmedt, Y.G. (ed.) CRYPTO 1994, vol. 839, pp. 341–358. Springer, Heidelberg (1994)
6. Bernstein, D.J.: The Poly1305-AES message-authentication code. In: Gilbert, H., Handschuh, H. (eds.) FSE 2005. LNCS, vol. 3557, pp. 32–49. Springer, Heidelberg (2005), http://cr.yp.to/talks/2005.02.15/slides.pdf
7. Bernstein, D.J.: Polynomial Evaluation and Message Authentication, October 22 (2007) http://cr.yp.to/papers.html#pema
8. Bierbrauer, J., Johansson, T., Kabatianskii, G., Smeets, B.: On Families of Hash Functions via Geometric Codes and Concatenation. In: Stinson, D.R. (ed.) CRYPTO 1993. LNCS, vol. 773, pp. 331–342. Springer, Heidelberg (1994)
9. Black, J., Halevi, S., Krawczyk, H., Krovetz, T., Rogaway, P.: UMAC: Fast and Secure Message Authentication. In: Wiener, M. (ed.) CRYPTO 1999. LNCS, vol. 1666, pp. 216–233. Springer, Heidelberg (1999)
10. Black, J., Cochran, M.: MAC Reforgeability, November 27 (2007), http://eprint.iacr.org/2006/095
11. Brassard, G.: On Computationally Secure Authentication Tags Requiring Short Secret Shared Keys. In: Chaum, D., Rivest, R.L., Sherman, A.T. (eds.) Crypto 1982, pp. 79–86. Plenum Press, New York (1983)
12. Carter, J.L., Wegman, M.N.: Universal classes of hash functions. Journal of Computer and System Sciences 18, 143–154 (1979)
13. den Boer, B.: A Simple and Key-Economical Unconditional Authentication Scheme. Journal of Computer Security 2, 65–71 (1993)
14. Dodis, Y., Pietrzak, K.: Improving the Security of MACs via Randomized Message Preprocessing. In: Biryukov, A. (ed.) FSE 2007. LNCS, vol. 4593, pp. 414–433. Springer, Heidelberg (2007)
15. Etzel, M., Patel, S., Ramzan, Z.: Square Hash: Fast Message Authentication via Optimized Universal Hash Functions. In: Wiener, M. (ed.) CRYPTO 1999. LNCS, vol. 1666, pp. 234–251. Springer, Heidelberg (1999)
16. Ferguson, N.: Authentication Weaknesses in GCM (May 20, 2005), http://csrc.nist.gov/groups/ST/toolkit/BCM/documents/comments/CWC-GCM/Ferguson2.pdf

17. Halevi, S., Krawczyk, H.: MMH: Software Message Authentication in the Gbit/second Rates. In: Biham, E. (ed.) FSE 1997. LNCS, vol. 1267, pp. 172–189. Springer, Heidelberg (1997)
18. ISO/IEC 9797, Information Technology – Security Techniques – Message Authentication Codes (MACs) – Part 1: Mechanisms using a Block Cipher, ISO/IEC (1999)
19. Iwata, T., Kurosawa, K.: OMAC: One-Key CBC MAC. In: Johansson, T. (ed.) FSE 2003. LNCS, vol. 2887, pp. 129–153. Springer, Heidelberg (2003)
20. Johansson, T.: Bucket Hashing with a Small Key Size. In: Fumy, W. (ed.) EUROCRYPT 1997. LNCS, vol. 1233, pp. 149–162. Springer, Heidelberg (1997)
21. Joux, A.: Authentication Failures in NIST Version of GCM (2006), http://csrc.nist.gov/CryptoToolkit/modes/
22. Kabatianskii, G.A., Johansson, T., Smeets, B.: On the Cardinality of Systematic A-codes via Error Correcting Codes. IEEE Trans. on Information Theory IT42(2), 566–578 (1996)
23. Kaps, J.-P., Yüksel, K., Sunar, B.: Energy Scalable Universal Hashing. IEEE Trans. on Computers 54(12), 1484–1495 (2005)
24. Knudsen, L.: Chosen-text Attack on CBC-MAC. Electronics Letters 33(1), 48–49 (1997)
25. Kohno, T., Viega, J., Whiting, D.: CWC: A High-Performance Conventional Authenticated Encryption Mode. In: Roy, B., Meier, W. (eds.) FSE 2004. LNCS, vol. 3017, pp. 408–426. Springer, Heidelberg (2004)
26. Krawczyk, H.: LFSR-based Hashing and Authentication. In: Desmedt, Y.G. (ed.) CRYPTO 1994. LNCS, vol. 839, pp. 129–139. Springer, Heidelberg (1994)
27. Krovetz, T.: UMAC: Message Authentication Code using Universal Hashing. IETF, RFC 4418 (informational) (March 2006)
28. Krovetz, T.: Message Authentication on 64-bit Architectures. In: Biham, E., Youssef, A.M. (eds.) SAC 2006. LNCS, vol. 4356, pp. 327–341. Springer, Heidelberg (2007)
29. McGrew, D.A., Fluhrer, S.: Multiple Forgery Attacks against Message Authentication Codes, http://eprint.iacr.org/2005/161
30. McGrew, D.A., Viega, J.: The Security and Performance of the Galois/Counter Mode (GCM) of Operation. In: Canteaut, A., Viswanathan, K. (eds.) INDOCRYPT 2004. LNCS, vol. 3348, pp. 343–355. Springer, Heidelberg (2004)
31. National Institute of Standards and Technology (NIST), SP 800-38D, Recommendation for Block Cipher Modes of Operation: Galois/Counter Mode (GCM) and GMAC, November 2007 (earlier drafts published in May 2005, April 2006, June 2007)
32. Petrank, E., Rackoff, C.: CBC MAC for Real-time Data Sources. Journal of Cryptology 13(3), 315–338 (2000)
33. Preneel, B., Bosselaers, A., Govaerts, R., Vandewalle, J.: A Chosen Text Attack on The Modified Cryptographic Checksum Algorithm of Cohen and Huang. In: Brassard, G. (ed.) CRYPTO 1989. LNCS, vol. 435, pp. 154–163. Springer, Heidelberg (1990)
34. Preneel, B., van Oorschot, P.C.: On the Security of Iterated Message Authentication Codes. IEEE Trans. on Information Theory IT-45(1), 188–199 (1999)
35. Simmons, G.J.: A Survey of Information Authentication. In: Simmons, G.J. (ed.) Contemporary Cryptology: The Science of Information Integrity, pp. 381–419. IEEE Press, Los Alamitos (1991)
36. Stinson, D.R.: Universal Hashing and Authentication Codes. Designs, Codes, and Cryptography 4(4), 369–380 (1994)
37. Wegman, M.N., Carter, J.L.: New Hash Functions and their Use in Authentication and Set Equality. Journal of Computer and System Sciences 22(3), 265–279 (1981)

A   The Joux and Ferguson Attacks on GCM

The Joux attacks on GCM. In his comment on the April 2006 draft of the NIST GCM mode [21], Joux presents a very elegant key recovery attack if the polynomial hash function is used with Option 2 under the condition of nonce reuse by the sender. The attack requires two text/MAC pairs with the same nonce and computes the exor of the tags, which is equal to $g_k(x) \oplus g_k(x')$ (the encryption with a one-time pad or pseudo-random function cancels out). This is a polynomial of degree at most t for which k is one of the roots. As there are at most t roots, it is easy to recover k, either by asking for a second set of text/MAC pairs (as proposed by Joux) or by using the technique of the first paragraph of Section 3.2; the latter approach requires at most t text/MAC pairs and t MAC verification queries (but without nonce reuse). Joux also shows that a small discrepancy in the specification between the May 2005 draft of [31] and the original GCM proposal of [31] leads to an efficient recovery of k *without nonce reuse*; this issue is solved in the June 2007 draft of [31]. In his comments, Joux suggests that *"replacing the counter encryption for MACs by the classical encryption with the block cipher usually used with Wegman-Carter MACs seems a safe option."* He also proposes a further mitigation of the security risks by using a stronger key derivation and by using separate keys for the different components of GCM.

The Ferguson attack on GCM. Ferguson points out another clever attack on GCM that exploits the arithmetic properties $\text{GF}(2^n)$ with the truncation option. His attack consists of the following steps:

1. consider messages with $x_i = 0$ except if $i = 2^j$ for some $j \geq 0$. As squaring in $\text{GF}(2^n)$ is a linear operation in $\text{GF}(2^n)$, the hash function becomes linear over this subset of inputs (that is, the hash computation can be written as a matrix-vector product over $\text{GF}(2)$);
2. for a single chosen text x of this form, modify the message into x' (again of the same form), where x' is chosen such that the first $s \leq \tau$ bits of $g_k(x)$ and $g_k(x')$ are equal, independent of the value of k. This can be achieved with some simple linear algebra with the constraint that $s < \log_2(t)$;
3. submit x' for verification: the probability of a successful forgery is $1/2^{\tau-s}$.

Moreover, a successful forgery results in $\tau - s$ additional linear equations in the bits of k, which makes it easier to obtain an additional forgery and thus even more information on the key.

Note that the GCM final standard [31] still uses Option 2 (in order to save a single encryption and avoid pipeline stalls) and still allows for truncation. On the other hand, it adds very explicit warnings about the risks of nonce reuse and truncation (in Appendices A and C respectively).

Cryptanalysis of the GOST Hash Function

Florian Mendel[1], Norbert Pramstaller[1], Christian Rechberger[1],
Marcin Kontak[2], and Janusz Szmidt[2]

[1] Institute for Applied Information Processing and Communications (IAIK),
Graz University of Technology, Inffeldgasse 16a, 8010 Graz, Austria
`Florian.Mendel@iaik.tugraz.at`
[2] Institute of Mathematics and Cryptology, Faculty of Cybernetics,
Military University of Technology, ul. Kaliskiego 2, 00-908 Warsaw, Poland

Abstract. In this article, we analyze the security of the GOST hash function. The GOST hash function, defined in the Russian standard GOST 34.11-94, is an iterated hash function producing a 256-bit hash value. As opposed to most commonly used hash functions such as MD5 and SHA-1, the GOST hash function defines, in addition to the common iterative structure, a checksum computed over all input message blocks. This checksum is then part of the final hash value computation.

As a result of our security analysis of the GOST hash function, we present the first collision attack with a complexity of about 2^{105} evaluations of the compression function. Furthermore, we are able to significantly improve upon the results of Mendel et al. with respect to preimage and second preimage attacks. Our improved attacks have a complexity of about 2^{192} evaluations of the compression function.

Keywords: cryptanalysis, hash function, collision attack, second preimage attack, preimage attack.

1 Introduction

A cryptographic hash function H maps a message M of arbitrary length to a fixed-length hash value h. Informally, a cryptographic hash function has to fulfill the following security requirements:

- *Collision resistance:* it is practically infeasible to find two messages M and M^*, with $M^* \neq M$, such that $H(M) = H(M^*)$.
- *Second preimage resistance:* for a given message M, it is practically infeasible to find a second message $M^* \neq M$ such that $H(M) = H(M^*)$.
- *Preimage resistance:* for a given hash value h, it is practically infeasible to find a message M such that $H(M) = h$.

The resistance of a hash function to collision and (second) preimage attacks depends in the first place on the length n of the hash value. Regardless of how a hash function is designed, an adversary will always be able to find preimages or second preimages after trying out about 2^n different messages. Finding collisions

D. Wagner (Ed.): CRYPTO 2008, LNCS 5157, pp. 162–178, 2008.
© International Association for Cryptologic Research 2008

requires a much smaller number of trials: about $2^{n/2}$ due to the birthday paradox. If the internal structure of a particular hash function allows collisions or (second) preimages to be found more efficiently than what could be expected based on its hash length, then the function is considered to be broken. For a formal treatment of the security properties of cryptographic hash functions we refer to [18,22].

Recent cryptanalytic results on hash functions mainly focus on collision attacks. Collisions have been shown for many commonly used hash functions (see for instance [5,6,16,24,25,26]), but we are not aware of any published collision attack on the GOST hash function. In this article, we will present a security analysis of the GOST hash function with respect to both collision and (second) preimage resistance. The GOST hash function is widely used in Russia and is specified in the Russian national standard GOST 34.11-94 [3]. This standard has been developed by *GUBS of Federal Agency Government Communication and Information* and *All-Russian Scientific and Research Institute of Standardization*. The GOST hash function is the only hash function that can be used in the Russian digital signature algorithm GOST 34.10-94 [2]. Therefore, it is also used in several RFCs and implemented in various cryptographic applications (as for instance openSSL).

The GOST hash function is an iterated hash function producing a 256-bit hash value. As opposed to most commonly used hash functions such as MD5 and SHA-1, the GOST hash function defines, in addition to the common iterative structure, a checksum computed over all input message blocks. This checksum is then part of the final hash value computation. The GOST standard also specifies the GOST block cipher [1], which is the main building block of the hash function. Therefore, it can be considered as a block-cipher-based hash function. While there have been published several cryptanalytic results regarding the block cipher (see for instance [4,11,14,19,20]), only a few results regarding the hash function have been published to date. Note that for the remainder of this article we refer to the GOST hash function simply as GOST.

Related Work. In [7], Gauravaram and Kelsey show that the generic attacks on hash functions based on the Merkle-Damgård design principle can be extended to hash functions with linear/modular checksums independent of the underlying compression function. Hence, second preimages can be found for long messages (consisting of 2^t message blocks) for GOST with a complexity of 2^{n-t} evaluations of the compression function.

At FSE 2008, Mendel *et al.* have presented the first attack on GOST exploiting the internal structure of the compression function. The authors exploit weaknesses in the internal structure of GOST to construct pseudo-preimages for the compression function of GOST with a complexity of about 2^{192} compression function evaluations. Furthermore, they show how the attack on the compression function of GOST can be extended to a (second) preimage attack on the hash function. The attack has a complexity of about 2^{225} evaluations of the compression function of GOST. Both attacks are structural attacks in the sense that they are independent of the underlying block cipher.

Our Contribution. We improve upon the state of the art as follows. First, we show that for plaintexts of a specific structure, we can construct fixed-points in the GOST block cipher efficiently. Second, based on this property in the GOST block cipher we then show how to construct collisions in the compression function of GOST with a complexity of 2^{96} compression function evaluations. This collision attack on the compression function is then extended to a collision attack on the GOST hash function. The extension is possible by combining a multicollision attack and a generalized birthday attack on the checksum. The attack has a complexity of about 2^{105} evaluations of the compression function of GOST. Furthermore, we show that due to the generic nature of our attack we can construct meaningful collisions, *i.e.* collisions in the chosen-prefix setting with the same complexity. Note that in most cases constructing meaningful collisions is more complicated than constructing (random) collisions (see for instance MD5 [21]). Third, we show how the (second) preimage attack of Mendel *et al.* can be improved by additionally exploiting weaknesses in the GOST block cipher. The new improved (second) preimage attack has a complexity of 2^{192} evaluations of the compression function of GOST.

Table 1. Comparison of results for the GOST hash function

source	attack complexity	attack
Gauravaram and Kelsey CT-RSA 2008 [7]	2^{256-t}	second preimages for long messages (2^t blocks)
Mendel *et al.* FSE 2008 [15]	2^{225}	preimages and second preimages
	2^{105}	collisions
this work	2^{105}	meaningful collisions (chosen-prefix)
	2^{192}	preimages and second preimages

The remainder of this article is structured as follows. In Section 2, we give a short description of the GOST hash function. In Section 3, we describe the GOST block cipher and show how to construct fixed-points efficiently. We use this in the collision attack on the hash function in Section 4. In Section 5, we show a new improved (second) preimage attack for the hash function. Finally, we present conclusions in Section 6.

2   The Hash Function GOST

GOST is an iterated hash function that processes message blocks of 256 bits and produces a 256-bit hash value. If the message length is not a multiple of 256, an unambiguous padding method is applied. For the description of the padding

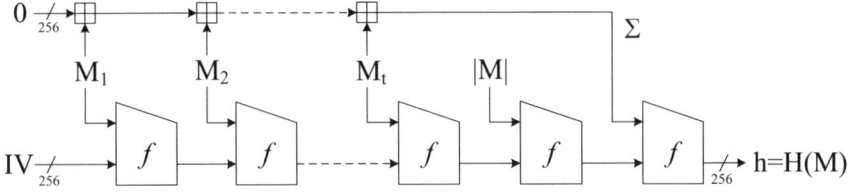

Fig. 1. Structure of the GOST hash function

method we refer to [3]. Let $M = M_1 \| M_2 \| \cdots \| M_t$ be a t-block message (after padding). The hash value $h = H(M)$ is computed as follows (see Fig. 1):

$$H_0 = IV \tag{1}$$

$$H_i = f(H_{i-1}, M_i) \quad \text{for } 0 < i \le t \tag{2}$$

$$H_{t+1} = f(H_t, |M|) \tag{3}$$

$$H_{t+2} = f(h_{t+1}, \Sigma) = h , \tag{4}$$

where $\Sigma = M_1 \boxplus M_2 \boxplus \cdots \boxplus M_t$, and $\boxplus$ denotes addition modulo 2^{256}. IV is a predefined initial value and $|M|$ represents the bit-length of the entire message prior to padding. As can be seen in (4), GOST specifies a checksum (Σ) consisting of the modular addition of all message blocks, which is then input to the final application of the compression function. Computing this checksum is not part of most commonly used hash functions such as MD5 and SHA-1.

The compression function f of GOST basically consist of three parts (see also Fig. 2): the state update transformation, the key generation, and the output transformation. In the following, we will describe these parts in more detail.

2.1   State Update Transformation

The state update transformation of GOST consists of 4 parallel instances of the GOST block cipher, denoted by E. The intermediate hash value H_{i-1} is split into four 64-bit words $h_3 \| h_2 \| h_1 \| h_0$. Each 64-bit word is used in one stream of the state update transformation to construct the 256-bit value $S = s_3 \| s_2 \| s_1 \| s_0$ in the following way:

$$s_0 = E(k_0, h_0) \tag{5}$$

$$s_1 = E(k_1, h_1) \tag{6}$$

$$s_2 = E(k_2, h_2) \tag{7}$$

$$s_3 = E(k_3, h_3) \tag{8}$$

where $E(K, P)$ denotes the encryption of the 64-bit plaintext P under the 256-bit key K. We refer to Section 3, for a detailed description of the GOST block cipher.

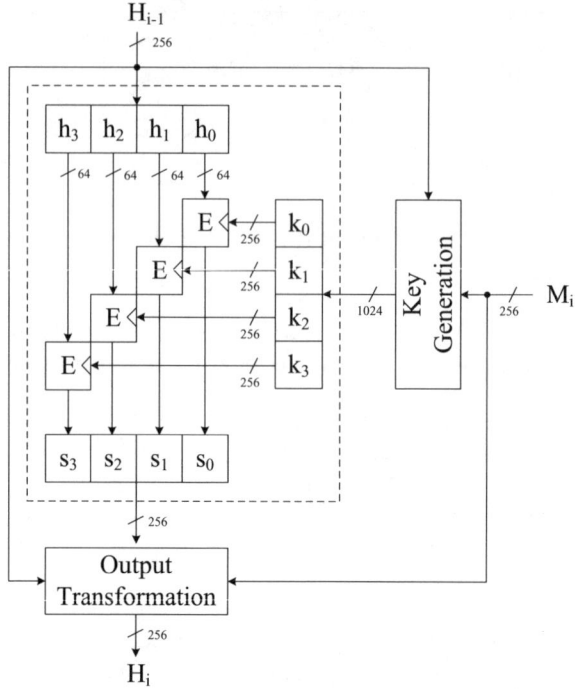

Fig. 2. The compression function of GOST

2.2 Key Generation

The key generation of GOST takes as input the intermediate hash value H_{i-1} and the message block M_i to compute a 1024-bit key K. This key is split into four 256-bit keys k_i, *i.e.* $K = k_3 \| \cdots \| k_0$, where each key k_i is used in one stream as the key for the GOST block cipher E in the state update transformation. The four keys k_0, k_1, k_2, and k_3 are computed in the following way:

$$k_0 = P(H_{i-1} \oplus M_i) \tag{9}$$
$$k_1 = P(A(H_{i-1}) \oplus A^2(M_i)) \tag{10}$$
$$k_2 = P(A^2(H_{i-1}) \oplus \texttt{Const} \oplus A^4(M_i)) \tag{11}$$
$$k_3 = P(A(A^2(H_{i-1}) \oplus \texttt{Const}) \oplus A^6(M_i)) \tag{12}$$

where A and P are linear transformations and $\texttt{Const}$ is a constant. Note that $A^2(x) = A(A(x))$. For the definition of the linear transformation A and P as well as the value of $\texttt{Const}$, we refer to [3], since we do not need it for our analysis.

2.3 Output Transformation

The output transformation of GOST combines the intermediate hash value H_{i-1}, the message block M_i, and the output of the state update transformation S to compute the output value H_i of the compression function. It is defined as follows.

$$H_i = \psi^{61}(H_{i-1} \oplus \psi(M_i \oplus \psi^{12}(S))) \qquad (13)$$

The linear transformation $\psi : \{0,1\}^{256} \to \{0,1\}^{256}$ is given by:

$$\psi(\Gamma) = (\gamma_0 \oplus \gamma_1 \oplus \gamma_2 \oplus \gamma_3 \oplus \gamma_{12} \oplus \gamma_{15}) \| \gamma_{15} \| \gamma_{14} \| \cdots \| \gamma_1 \qquad (14)$$

where Γ is split into sixteen 16-bit words, *i.e.* $\Gamma = \gamma_{15} \| \gamma_{14} \| \cdots \| \gamma_0$.

3 The GOST Block Cipher

The GOST block cipher is specified by the Russian government standard GOST 28147-89 [1]. Several cryptanalytic results have been published for the block cipher (see for instance [4,11,14,19,20]). However, if the block cipher is used in a hash function then we are facing a different attack scenario: the attacker has full control over the key. First results considering this fact for the security analysis of hash functions have been presented for instance in [13]. We will exploit having full control over the key for constructing fixed-points for the GOST block cipher.

3.1 Description of the Block Cipher

The GOST block cipher is a 32 round Feistel network with a block size of 64 bits and a key length of 256 bits. The round function of the GOST block cipher consists of a key addition, eight different 4×4 S-boxes S_j ($0 \le j < 8$) and a cyclic rotation (see also Figure 3). For the definition of the S-boxes we refer to [3], since we do not need them for our analysis.

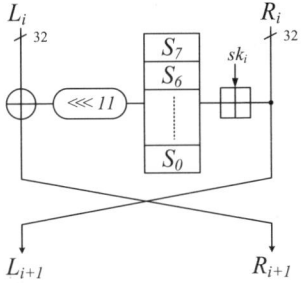

Fig. 3. One round of the GOST block cipher

The key schedule of the GOST block cipher defines the subkeys sk_i derived from the 256-bit $K = k_7 \| k_6 \| \cdots \| k_0$ as follows

$$sk_i = \begin{cases} k_{i \bmod 8}, & i = 0, \ldots, 23, \\ k_{7-(i \bmod 8)}, & i = 24, \ldots, 31. \end{cases} \qquad (15)$$

3.2   Constructing a Fixed-Point

In the following, we will show how to efficiently construct fixed-points in the GOST block cipher. It is based on the following observation. Note that a similar observation was used by Kara in [10] for a chosen plaintext attack on the GOST block cipher.

Observation 1. *Assume we are given a plaintext $P = L_0 \| R_0$ with $L_0 = R_0$. Then we can construct a fixed-point for the block cipher by constructing a fixed-point in the first 8 rounds.*

In the following, we refer to a plaintext $P = L_0 \| R_0$ with $L_0 = R_0$ as a *symmetric* plaintext (or for short as symmetric). Note that by using the block cipher for constructing a hash function, an attacker has full control over the key. Furthermore, each word of the key is only used once in the first 8 rounds of the block cipher. Hence, constructing a fixed-point in the first 8 rounds can be done efficiently. First, we choose random values for the first 6 words of the key (subkeys $sk_0, \ldots, sk_5$) and compute L_6 and R_6. Next, we choose the last 2 words of the key (subkeys sk_6 and sk_7) such that $L_8 = L_0$ and $R_8 = R_0$. With this method we can construct a fixed-point in the first 8 rounds of the block cipher with a computational cost of 8 round computations.

It is easy to see that if we have a fixed-point in the first 8 rounds, then this is also a fixed-point for rounds 9-16 and 17-24 since the same subkeys are used in these rounds. In the last 8 rounds the subkey is put in the opposite order, see (15). However, since the GOST block cipher is a Feistel network, we have here (rounds 25-32) a decryption if $L_{24} = R_{24}$. This implies that we have a fixed-point for the GOST block cipher (for all 32 rounds) if the plaintext is symmetric. Hence, for symmetric plaintexts we can efficiently construct fixed-points for the GOST block cipher.

4   Collision Attack on GOST

In this section, we present a collision attack on the GOST hash function with a complexity of about 2^{105} evaluations of the compression function. First, we will show how to construct collisions for the compression function of GOST, and based on this attack we then describe the collision attack for the hash function. For the remainder of this article we follow the notation of [15].

4.1   Constructing a Collision in the Compression Function

In the following, we show how to construct a collision in the compression function of GOST. The attack is based on structural weaknesses of the compression function. These weaknesses have been used in [15] to construct pseudo-collisions and pseudo-preimages for the compression function of GOST with a complexity of 2^{96} and 2^{192}, respectively.

Now we show a collisions attack on the compression function by additionally exploiting weaknesses in the underlying GOST block cipher. Since the transformation ψ is linear, (13) can be written as:

$$H_i = \psi^{61}(H_{i-1}) \oplus \psi^{62}(M_i) \oplus \psi^{74}(S) \tag{16}$$

Furthermore, ψ is invertible and hence (16) can be written as:

$$\underbrace{\psi^{-74}(H_i)}_{X} = \underbrace{\psi^{-13}(H_{i-1})}_{Y} \oplus \underbrace{\psi^{-12}(M_i)}_{Z} \oplus S \tag{17}$$

Note that Y depends linearly on H_{i-1} and Z depends linearly on M_i. As opposed to Y and Z, S depends on both H_{i-1} and M_i processed by the block cipher E. For the following discussion, we split the 256-bit words X, Y, Z defined in (17) into 64-bit words:

$$X = x_3 \| x_2 \| x_1 \| x_0 \quad Y = y_3 \| y_2 \| y_1 \| y_0 \quad Z = z_3 \| z_2 \| z_1 \| z_0$$

Now, (17) can be written as:

$$x_0 = y_0 \oplus z_0 \oplus s_0 \tag{18}$$
$$x_1 = y_1 \oplus z_1 \oplus s_1 \tag{19}$$
$$x_2 = y_2 \oplus z_2 \oplus s_2 \tag{20}$$
$$x_3 = y_3 \oplus z_3 \oplus s_3 \tag{21}$$

Now assume, that we can find 2^{96} message blocks M_i^j, where $M_i^k \neq M_i^t$ with $k \neq t$, such that all message blocks produce the same value x_0. Then we know that due to the birthday paradox two of these message blocks also lead to the same values x_1, x_2, and x_3. In other words, we have constructed a collision for the compression function of GOST. The attack has a complexity of about 2^{96} evaluations of the compression function of GOST.

Based on this short description, we will show now how to construct message blocks M_i^j, which all produce the same value x_0. Assume, we want to keep the value s_0 in (18) constant. Since $s_0 = E(k_0, h_0)$ and k_0 depends linearly on the message block M_i, we have to find keys k_0^j and hence, message blocks M_i^j, which all produce the same value s_0. This can be done by exploiting the fact that in the GOST block cipher fixed-points can be constructed efficiently for symmetric plaintexts (see Section 3.2). In other words, if h_0 is symmetric then we can construct 2^{96} message blocks M_i^j where $s_0 = h_0$, and (18) becomes

$$x_0 = y_0 \oplus z_0 \oplus h_0 . \tag{22}$$

However, to find message blocks M_i^j for which x_0 has the same value, we still have to ensure that also the term $y_0 \oplus z_0$ in (22) has the same value for all message blocks. Therefore, we get the following equation (64 equations over $GF(2)$)

$$y_0 \oplus z_0 = c \tag{23}$$

where c is an arbitrary 64-bit value. We know that y_0 depends linearly on H_{i-1} and z_0 depends linearly on M_i, see (17). Therefore, the choice of the message block M_i and accordingly, the choice of the key k_0, is restricted by 64 equations over $GF(2)$. Hence, for constructing a fixed-point in the GOST block cipher we have to consider these restrictions. For the following discussion let

$$A \cdot k_0 = d \tag{24}$$

denote the set of 64 equations over $GF(2)$ which restricts the choice of the key k_0, where A is a 64×256 matrix over $GF(2)$ and d is a 64-bit vector. It follows from Observation 1 that for constructing a fixed-point in the GOST block cipher (for symmetric plaintexts), it is sufficient to construct a fixed-point in the first 8 rounds. Hence, one method to construct an appropriate fixed-point would be to construct many arbitrary fixed-points and then check if (24) holds. With this method we find an appropriate fixed-point with a complexity of about 2^{64}. Since we need 2^{96} such fixed-points for the collision attack, this would lead to a complexity of 2^{160} evaluations of the compression function of GOST. However, we can improve this complexity by using a meet-in-the-middle approach (see also Fig. 4).

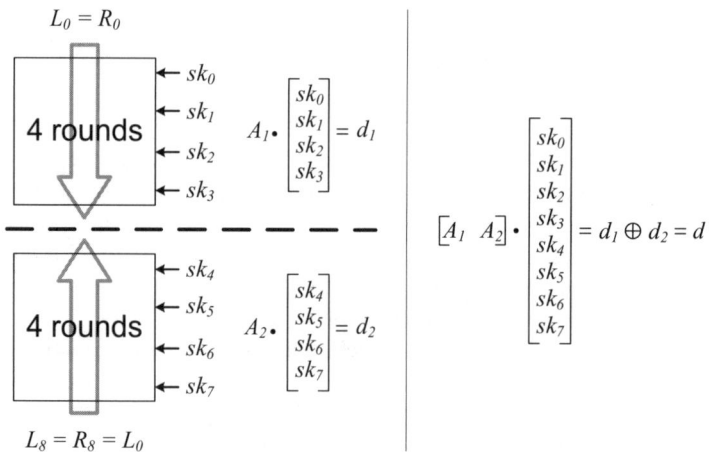

Fig. 4. Constructing a fixed-point in the GOST block cipher

We split the first 8 rounds of the GOST block cipher into 2 parts P_1 (rounds 1-4) and P_2 (rounds 5-8). Since the subkey used in the first 8 rounds is restricted by $A \cdot k_0$, we also split this system of 64 equations over $GF(2)$ into two parts:

$$A_1 \cdot \begin{bmatrix} sk_0 \\ sk_1 \\ sk_2 \\ sk_3 \end{bmatrix} = d_1 \qquad A_2 \cdot \begin{bmatrix} sk_4 \\ sk_5 \\ sk_6 \\ sk_7 \end{bmatrix} = d_2 \tag{25}$$

where $A = [A_1 \, A_2]$ and $d = d_1 \oplus d_2$. Now we can apply a meet-in-the-middle attack to construct 2^{64} appropriate fixed-points for the GOST block cipher with a complexity of 2^{64}. It can be summarized as follows.

1. Choose a random value for d_1. This determines also $d_2 = d \oplus d_1$.
2. For all 2^{64} subkeys $sk_0, \ldots, sk_3$ which fulfill (25) compute L_4, R_4 and save the result in the list L.
3. For all 2^{64} subkeys $sk_4, \ldots, sk_7$ which fulfill (25) compute rounds 4-8 backward to get L_4, R_4 and check for a matching entry in the list L. Note that since there are 2^{64} entries in the list L we expect to always find a matching entry in the list L. Hence, we get 2^{64} appropriate fixed-points for the GOST block cipher with a complexity of about 2^{64} and memory requirements of $2^{64} \cdot 40 \approx 2^{70}$ bytes.

By repeating this attack about 2^{32} times for different choices of d_1, we get 2^{96} appropriate fixed-points. In other words, we found 2^{96} keys k_0^j which all produce the same value $s_0 = E(k_0^j, h_0)$ and additionally fulfill (24). Consequentially, we have 2^{96} message blocks M_i^j which all result in the same value x_0 with $X = \psi^{-74}(H_i)$. By applying a birthday attack we will find two message blocks M_i^k and M_i^t with $k \neq t$ where also x_1, x_2, and x_3 are equal. In other words, we can find a collision for the compression function of GOST with a complexity of about 2^{96} instead of 2^{128} evaluations of the compression function of GOST.

4.2   Constructing Collisions for the Hash Function

In this section, we show how the collision attack on the compression function can be extended to the hash function. The attack has a complexity of about 2^{105} evaluations of the compression function of GOST. Note that the hash function defines, in addition to the common iterative structure, a checksum computed over all input message blocks which is then part of the final hash computation. Therefore, to construct a collision in the hash function we have to construct a collision in the iterative structure (*i.e.* chaining variables) as well as in the checksum. To do this we use multicollisions.

A multicollision is a set of messages of equal length that all lead to the same hash value. As shown in [9], constructing a 2^t collision, *i.e.* 2^t messages consisting of t message blocks which all lead to the same hash value, can be done with a complexity of about $t \cdot 2^x$ for any iterated hash function, where 2^x is the cost of constructing a collision in the compression function. As shown in Section 4.1, collisions for the compression function of GOST can be constructed with a complexity of 2^{96} if h_0 is symmetric in $H_{i-1} = h_3 \| h_2 \| h_1 \| h_0$. Note that by using an additional message block M_{i-1} we find a chaining variable $H_{i-1} = f(H_{i-2}, M_{i-1})$, where h_0 is symmetric with a complexity of 2^{32} compression function evaluations. Hence, we can construct a 2^{128} collision with a complexity of about $128 \cdot (2^{96} + 2^{32}) \approx 2^{103}$ evaluations of the compression function of GOST. With this method we get 2^{128} messages M^* that all lead to the same value H_{256} as depicted in Figure 5.

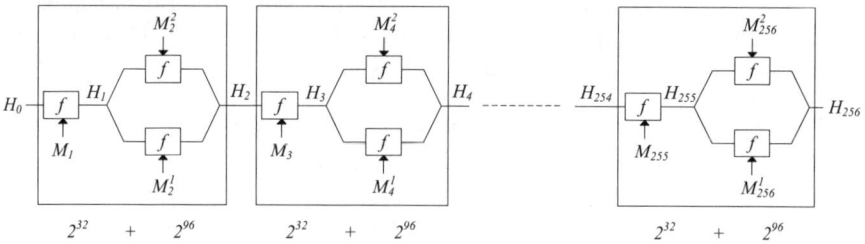

Fig. 5. Constructing a multicollision for GOST

To construct a collision in the checksum of GOST we have to find 2 distinct messages which produce the same value $\Sigma = M_1 \boxplus M_2^{j_2} \boxplus \cdots \boxplus M_{255} \boxplus M_{256}^{j_{256}}$ with $j_2, j_4, \ldots j_{256} \in \{1, 2\}$. By applying a birthday attack we can find these 2 messages with a complexity of about 2^{127} additions over $GF(256)$ and memory requirements of 2^{134} bytes. Due to the high complexity and memory requirements of the birthday attack, one could see this part as the bottleneck of the attack. However, the runtime and memory requirements can significantly be reduced by applying a generalized birthday attack introduced by Wagner in [23]. Wagner shows that if ℓ is a power of two then the memory requirements and the running time for the generalized birthday problem is given by $2^{n/(1+\lg \ell)}$ and $\ell \cdot 2^{n/(1+\lg \ell)}$, respectively. Note that in the standard birthday attack we have $\ell = 2^1$.

Let us now consider the case $\ell = 2^3$. Then the birthday attack in the second part of the attack has a complexity of $2^3 \cdot 2^{256/4} = 2^{67}$ and uses lists of size $2^{256/4} = 2^{64}$. In detail, we need to construct 8 lists of size 2^{64} in the first step of the attack. Hence, we need to construct a $2^{8 \cdot 64}$ collision in the first part of the attack to get 8 lists of the needed size. Constructing this multicollision has a complexity of about $8 \cdot 64 \cdot (2^{32} + 2^{96}) = 2^{105}$ compression function evaluations and memory requirements of $8 \cdot 64 \cdot (2 \cdot 64) = 2^{16}$ bytes. Hence, we can construct a collision for the GOST hash function with a complexity of about 2^{105} and memory requirements of $2^{64} \cdot 2^6 = 2^{70}$ bytes by using a generalized birthday attack with $\ell = 8$ lists. Furthermore, the colliding message pair consists of $8 \cdot (2 \cdot 64) = 1024$ message blocks. Note that $\ell = 8$ is the best choice for the attack. On one hand if we choose $\ell > 8$ then the memory requirements of the attack would decrease but the attack complexity would increase. Since we need about 2^{70} bytes of memory for constructing fixed-points in the GOST block cipher, this does not improve the attack. On the other hand if we choose $\ell < 8$ then the memory requirements of the attack would be significantly higher.

A Remark on the Length Extension Property. Once, we have found a collision, *i.e.* collision in the iterative part (chaining variables) and the checksum, we can construct many more collisions by appending an arbitrary message block. Note that this is not necessarily the case for a straight-forward birthday attack.

By applying a birthday attack we construct a collision in the final hash value (after the last iteration) and appending a message block is not possible. Hence, we need a collision in the iterative part as well as in the checksum for the extension property. Note that by combining the generic birthday attack and multicollisions, one can construct collisions in both parts with a complexity of about $128 \cdot 2^{128} = 2^{135}$ while our attack has a complexity of 2^{105}.

A Remark on Meaningful Collisions. In a chosen-prefix setting, an attacker searches for a message pair (M, M^*) such that for a given hash function H

$$H(M_{pre}\|M) = H(M^*_{pre}\|M^*) \tag{26}$$

for any pair (M_{pre}, M^*_{pre}). In [21], Stevens et al. show that such a more powerful attack exists for MD5. Furthermore, they describe an application of this attack for colliding X.509 certificates. Note that their attack (in a chosen-prefix setting) has a complexity of 2^{50}, while the currently best published collision attack for MD5 has a complexity of about 2^{30} evaluations of the compression function [12].

However, in the case of GOST the collision attack in the chosen-prefix setting has the same complexity as the collision attack. Due to the generic nature of the collision attack, differences in the chaining variables can be canceled efficiently. Assume that the chosen prefix (M_{pre}, M^*_{pre}) consists of t message blocks resulting in the chaining variables H_t and H_t^*. Then the attack can be summarized as follows.

1. We have to find two message blocks M_{t+1} and M^*_{t+1} such that $h_0 = h_0^* = 0$, where $H_{t+1} = h_3\|h_2\|h_1\|h_0$ and $H^*_{t+1} = h_3^*\|h_2^*\|h_1^*\|h_0^*$. This has a complexity of about $2 \cdot 2^{64}$ evaluations of the compression function of GOST.
2. Now we have to find two message blocks M_{t+2} and M^*_{t+2} such that $H_{t+2} = H^*_{t+2}$. This can be done similar as constructing a collision in the compression function of GOST (see Section 4.1). First, we choose a random value for c in (22) and construct 2^{96} message blocks M_{t+2}, where x_0 is equal. Second, we construct 2^{96} message blocks M^*_{t+2}, where $x_0^* = x_0$. To guarantee that $x_0 = x_0^*$ we have to adjust c^* in (22) such that the following equation holds.

$$x_0 = x_0^* = y_0^* \oplus z_0^* \oplus h_0^* = c^* \oplus h_0^*$$

   By applying a meet-in-the-middle attack we will find two message blocks M_{t+2} and M^*_{t+2} which produce the same chaining variables $(H_{t+2} = H^*_{t+2})$. This step of the attack has a complexity of $2 \cdot 2^{96}$ evaluations of the compression function of GOST.
3. Once we have constructed a collision in the iterative part (chaining variables), we have to construct a collision in the checksum as well. Therefore, we proceed as described in Section 4.2. By generating a 2^{512} collision and applying a generalized birthday attack with $\ell = 8$ we can construct a collision in the checksum of GOST with a complexity of 2^{105} compression function evaluations and memory requirements of 2^{70} bytes.

Hence, we can construct meaningful collisions, *i.e.* collisions in the chosen-prefix setting, for the GOST hash function with a complexity of about 2^{105} compression function evaluations.

5   Improved Preimage Attack for the Hash Function

In a preimage attack, we want to find, for a given hash value h, a message M such that $H(M) = h$. As we will show in the following, for GOST we can construct preimages of h with a complexity of about 2^{192} evaluations of the compression function of GOST. Before describing the attack, we will first show how to construct preimages for the compression function of GOST. Based on this attack we then present the preimage attack for the hash function.

5.1   A Preimage Attack for the Compression Function

In a similar way as we have constructed a collision in Section 4.1, we can construct a preimage for the compression function of GOST. In the attack, we have to find a message block M_i, such that $f(H_{i-1}, M_i) = H_i$ for the given values H_{i-1} and H_i. Note that the value of H_i determines $x_3, \ldots, x_0$, since $X = \psi^{-74}(H_i)$. Furthermore, assume that h_0 (in $H_{i-1} = h_3\|\cdots\|h_0$) is symmetric. Then the attack can be summarized as follows.

1. Since we will construct fixed-points for the GOST block cipher such that $s_0 = E(k_0, h_0) = h_0$, we have to adjust c in (22) such that

$$x_0 = y_0 \oplus z_0 \oplus h_0 = c \oplus h_0$$

   holds with $X = \psi^{-74}(H_i)$. Once c is fixed, this also determines d in (24).
2. Choose a random value for d_1 (this also determines $d_2 = d \oplus d_1$) and apply a meet-in-the-middle attack to obtain 2^{64} message blocks M_i^j for which x_0 is correct. Note that this step of the attack has memory requirements of 2^{70} bytes.
3. For each message block compute X and check if x_3, x_2, and x_1 are correct. This holds with a probability of 2^{-192}. Thus, after testing all 2^{64} message blocks, we will find a correct message block with a probability of $2^{-192} \cdot 2^{64} = 2^{-128}$. Note that we can repeat the attack about 2^{64} times for different choices of d_1.

Hence, we will find a preimage for the compression function of GOST with a probability of about 2^{-64} and a complexity of about 2^{128} evaluations of the compression function of GOST and memory requirements of 2^{70} bytes.

5.2   Extending the Attack to the Hash Function

Now, we show how the preimage attack on the compression function can be extended to the GOST hash function. The attack has a complexity of about 2^{192} evaluations of the compression function of GOST. Moreover, the preimage consists of 257 message blocks, *i.e.* $M = M_1\|\cdots\|M_{257}$. The preimage attack consists of four steps as also shown in Figure 6.

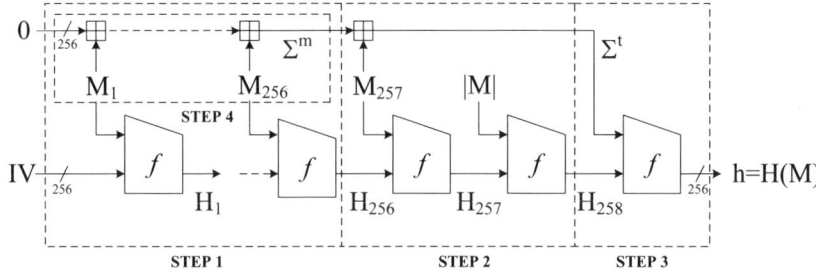

Fig. 6. Preimage Attack on GOST

STEP 1: Multicollisions for GOST. For the preimage attack on the hash function, we construct a 2^{256} collision. This means, we have 2^{256} messages $M^* = M_1^{j_1} \| M_2^{j_2} \| \cdots \| M_{256}^{j_{256}}$ for $j_1, j_2, \ldots, j_{256} \in \{1, 2\}$ consisting of 256 blocks that all lead to the same hash value H_{256}. This results in a complexity of about $256 \cdot 2^{128} = 2^{136}$ evaluations of the compression function of GOST. Furthermore, the memory requirement is about $2 \cdot 256$ message blocks, *i.e.* we need to store 2^{14} bytes. With these multicollisions, we are able to construct the needed value of Σ^m in STEP 4 of the attack (where the superscript m stands for 'multicollision').

STEP 2: Constructing H_{258} Including the Length Encoding. In this step, we have to find a message block M_{257} such that for the given H_{256} determined in STEP 1, and for $|M|$ determined by our assumption that we want to construct preimages consisting of 257 message blocks, we find a $H_{258} = h_3 \| \cdots \| h_0$ where h_0 is symmetric. Note that since we want to construct a message that is a multiple of 256 bits, we choose M_{257} to be a full message block and hence no padding is needed. We proceed as follows. Choose an arbitrary message block M_{257} and compute H_{258} as follows:

$$H_{257} = f(H_{256}, M_{257})$$
$$H_{258} = f(H_{257}, |M|)$$

where $|M| = (256 + 1) \cdot 256$. Then we check if h_0 in the resulting value H_{258} is symmetric. This has a probability of 2^{-32}. Hence, this step of the attack requires $2 \cdot 2^{32}$ evaluations of the compression function of GOST.

STEP 3: Preimages for the Last Iteration. To construct a preimage for the last iteration of GOST we proceed as described in Section 5.1. Since h_0 in H_{258} is symmetric, we will find a preimage for the last iteration of GOST with a probability of 2^{-64} (and a complexity of about 2^{128}). Therefore, we have to repeat this step of the attack about 2^{64} times for different values of H_{258} (where h_0 is symmetric) to find a preimage for the last iteration. Hence, finishing this step of the attack has a complexity of about $2^{64} \cdot (2 \cdot 2^{32} + 2^{128}) \approx 2^{192}$ evaluations of the compression function of GOST. Once we have found a preimage for the last iteration, also the value Σ^m is determined, since $\Sigma^m = \Sigma^t \boxminus M_{257}$.

STEP 4: Constructing Σ^m. In STEP 1, we constructed a 2^{256} collision in the first 256 iterations of the hash function. From this set of messages that all lead to the same H_{256}, we now have to find a message $M^* = M_1^{j_1} \| M_2^{j_2} \| \cdots \| M_{256}^{j_{256}}$ for $j_1, j_2, \ldots, j_{256} \in \{1, 2\}$ that leads to the value of $\Sigma^m = \Sigma^t \boxminus M_{257}$. This can be done by applying a meet-in-the-middle attack. First, we save all values for $\Sigma_1 = M_1^{j_1} \boxplus M_2^{j_2} \boxplus \cdots \boxplus M_{128}^{j_{128}}$ in the list L. Note that we have in total 2^{128} values in L. Second, we compute $\Sigma_2 = M_{129}^{j_{129}} \boxplus M_{130}^{j_{130}} \boxplus \cdots \boxplus M_{256}^{j_{256}}$ and check if $\Sigma^m \boxminus \Sigma_2$ is in the list L. After testing all 2^{128} values, we expect to find a matching entry in the list L and hence a message $M^* = M_1^{j_1} \| M_2^{j_2} \| \cdots \| M_{256}^{j_{256}}$ that leads to $\Sigma^m = \Sigma^t \boxminus M_{257}$. This step of the attack has a complexity of 2^{128} and a memory requirement of $2^{128} \cdot 2^5 = 2^{133}$ bytes. Once we have found M^*, we found a preimage for GOST consisting of 256+1 message blocks, namely $M^* \| M_{257}$.

The Attack Complexity. The complexity of the preimage attack is determined by the computational effort of STEP 2 and STEP 3, *i.e.* a preimage of h can be found in about 2^{192} evaluations of the compression function. The memory requirements for the preimage attack are determined by finding M^* in STEP 4, since we need to store 2^{133} bytes for the meet-in-the-middle attack. Due to the high memory requirements of STEP 4, one could see this part as the bottleneck of the attack. However, the memory requirements of STEP 4 can significantly be reduced by applying a memory-less variant of the meet-in-the-middle attack introduced by Quisquater and Delescaille in [17]. Hence, a preimage can be constructed for the GOST hash function with a complexity of 2^{192} evaluations of the compression function and memory requirements of about 2^{70} bytes.

5.3   A Remark on Second Preimages

Note that the presented preimage attack on GOST also implies a second preimage attack. In this case, we are not given only the hash value h but also a message M that results in this hash value. We can construct for any given message a second preimage in the same way as we construct preimages. The difference is, that the second preimage will always consist of at least 257 message blocks. Thus, we can construct a second preimage for any message M (of arbitrary length) with a complexity of about 2^{192} evaluations of the compression function of GOST.

6   Conclusion

In this article, we have presented a collision attack and a (second) preimage attack on the GOST hash function. Both the collision and the (second) preimage attack are based on weaknesses in the GOST block cipher, namely fixed-points can be constructed efficiently for plaintexts of a specific structure. The internal structure of the compression function allows to construct collisions with a complexity of about 2^{96} evaluations of the compression function. This alone would not render the hash function insecure. The fact that we can construct multicollisions for any iterated hash function including the GOST hash function and the possibility of applying a (generalized) birthday attack to construct also a collision in the checksum make the collision attack on the hash function possible. The attack has a complexity of about 2^{105} compression function evaluations. Furthermore, the generic nature of the attack allows us to construct meaningful collisions, *i.e.* collisions in the chosen-prefix setting, with the same complexity. In a similar way as we construct collisions for the hash function, we can construct (second) preimages for the hash function with a complexity of about 2^{192} evaluations of the compression function. This improves the previous (second) preimage attack of Mendel *et al.* by a factor of 2^{33}. Even though the complexities of our attacks are far beyond of being practical, they point out weaknesses in the design principles of the hash function GOST.

Acknowledgements

The authors wish to thank Mario Lamberger, Vincent Rijmen, and the anonymous referees for useful comments and discussions.

The work in this paper has been supported in part by the Secure Information Technology Center - Austria (A-SIT) and by the Austrian Science Fund (FWF), project P19863.

References

1. GOST 28147-89, Systems of the Information Treatment. Cryptographic Security. Algorithms of the Cryptographic Transformation (1989) (in Russian)
2. GOST 34.10-94, Information Technology Cryptographic Data Security Produce and Check Procedures of Electronic Digital Signature Based on Asymmetric Cryptographic Algorithm (1994) (in Russian)
3. GOST 34.11-94, Information Technology Cryptographic Data Security Hashing Function (1994) (in Russian)
4. Biryukov, A., Wagner, D.: Advanced Slide Attacks. In: Preneel, B. (ed.) EURO-CRYPT 2000. LNCS, vol. 1807, pp. 589–606. Springer, Heidelberg (2000)
5. De Cannière, C., Mendel, F., Rechberger, C.: Collisions for 70-Step SHA-1: On the Full Cost of Collision Search. In: Adams, C.M., Miri, A., Wiener, M. (eds.) SAC 2007. LNCS, vol. 4876, pp. 56–73. Springer, Heidelberg (2007)
6. De Cannière, C., Rechberger, C.: Finding SHA-1 Characteristics: General Results and Applications. In: Lai, X., Chen, K. (eds.) ASIACRYPT 2006. LNCS, vol. 4284, pp. 1–20. Springer, Heidelberg (2006)
7. Gauravaram, P., Kelsey, J.: Linear-XOR and Additive Checksums Don't Protect Damgård-Merkle Hashes from Generic Attacks. In: Malkin, T. (ed.) CT-RSA 2008. LNCS, vol. 4964, pp. 36–51. Springer, Heidelberg (2008)
8. Joscák, D., Tuma, J.: Multi-block Collisions in Hash Functions Based on 3C and 3C+ Enhancements of the Merkle-Damgård Construction. In: Rhee, M.S., Lee, B. (eds.) ICISC 2006. LNCS, vol. 4296, pp. 257–266. Springer, Heidelberg (2006)
9. Joux, A.: Multicollisions in Iterated Hash Functions. Application to Cascaded Constructions. In: Franklin, M.K. (ed.) CRYPTO 2004. LNCS, vol. 3152, pp. 306–316. Springer, Heidelberg (2004)

10. Kara, O.: Reflection Attacks on Product Ciphers. Cryptology ePrint Archive, Report 2007/043 (2007), http://eprint.iacr.org/
11. Kelsey, J., Schneier, B., Wagner, D.: Key-Schedule Cryptoanalysis of IDEA, G-DES, GOST, SAFER, and Triple-DES. In: Koblitz, N. (ed.) CRYPTO 1996. LNCS, vol. 1109, pp. 237–251. Springer, Heidelberg (1996)
12. Klima, V.: Tunnels in Hash Functions: MD5 Collisions Within a Minute. Cryptology ePrint Archive, Report 2006/105 (2006), http://eprint.iacr.org/
13. Knudsen, L.R., Rijmen, V.: Known-Key Distinguishers for Some Block Ciphers. In: Kurosawa, K. (ed.) ASIACRYPT 2007. LNCS, vol. 4833, pp. 315–324. Springer, Heidelberg (2007)
14. Ko, Y., Hong, S., Lee, W., Lee, S., Kang, J.-S.: Related Key Differential Attacks on 27 Rounds of XTEA and Full-Round GOST. In: Roy, B.K., Meier, W. (eds.) FSE 2004. LNCS, vol. 3017, pp. 299–316. Springer, Heidelberg (2004)
15. Mendel, F., Pramstaller, N., Rechberger, C.: A (Second) Preimage Attack on the GOST Hash Function. In: Nyberg, K. (ed.) FSE. LNCS, vol. 5086, pp. 224–234. Springer, Heidelberg (2008)
16. Mendel, F., Rijmen, V.: Cryptanalysis of the Tiger Hash Function. In: Kurosawa, K. (ed.) ASIACRYPT 2007. LNCS, vol. 4833, pp. 536–550. Springer, Heidelberg (2007)
17. Quisquater, J.-J., Delescaille, J.-P.: How Easy is Collision Search. New Results and Applications to DES. In: Brassard, G. (ed.) CRYPTO 1989. LNCS, vol. 435, pp. 408–413. Springer, Heidelberg (1990)
18. Rogaway, P., Shrimpton, T.: Cryptographic Hash-Function Basics: Definitions, Implications, and Separations for Preimage Resistance, Second-Preimage Resistance, and Collision Resistance. In: Roy, B.K., Meier, W. (eds.) FSE 2004. LNCS, vol. 3017, pp. 371–388. Springer, Heidelberg (2004)
19. Saarinen, M.-J.O.: A chosen key attack against the secret S-boxes of GOST (unpublished manuscript, 1998)
20. Seki, H., Kaneko, T.: Differential Cryptanalysis of Reduced Rounds of GOST. In: Stinson, D.R., Tavares, S.E. (eds.) SAC 2000. LNCS, vol. 2012, pp. 315–323. Springer, Heidelberg (2001)
21. Stevens, M., Lenstra, A.K., de Weger, B.: Chosen-Prefix Collisions for MD5 and Colliding X.509 Certificates for Different Identities. In: Naor, M. (ed.) EUROCRYPT 2007. LNCS, vol. 4515, pp. 1–22. Springer, Heidelberg (2007)
22. Stinson, D.R.: Some Observations on the Theory of Cryptographic Hash Functions. Des. Codes Cryptography 38(2), 259–277 (2006)
23. Wagner, D.: A Generalized Birthday Problem. In: Yung, M. (ed.) CRYPTO 2002. LNCS, vol. 2442, pp. 288–303. Springer, Heidelberg (2002)
24. Wang, X., Lai, X., Feng, D., Chen, H., Yu, X.: Cryptanalysis of the Hash Functions MD4 and RIPEMD. In: Cramer, R. (ed.) EUROCRYPT 2005. LNCS, vol. 3494, pp. 1–18. Springer, Heidelberg (2005)
25. Wang, X., Yin, Y.L., Yu, H.: Finding Collisions in the Full SHA-1. In: Shoup, V. (ed.) CRYPTO 2005. LNCS, vol. 3621, pp. 17–36. Springer, Heidelberg (2005)
26. Wang, X., Yu, H.: How to Break MD5 and Other Hash Functions.. In: Cramer, R. (ed.) EUROCRYPT 2005. LNCS, vol. 3494, pp. 19–35. Springer, Heidelberg (2005)

Preimages for Reduced SHA-0 and SHA-1

Christophe De Cannière[1,2] and Christian Rechberger[3]

[1] Département d'Informatique École Normale Supérieure
christophe.decanniere@ens.fr
[2] Katholieke Universiteit Leuven, Dept. ESAT/SCD-COSIC, and IBBT
[3] Graz University of Technology
Institute for Applied Information Processing and Communications (IAIK)
christian.rechberger@iaik.tugraz.at

Abstract. In this paper, we examine the resistance of the popular hash function SHA-1 and its predecessor SHA-0 against dedicated preimage attacks. In order to assess the security margin of these hash functions against these attacks, two new cryptanalytic techniques are developed:

- **Reversing the inversion problem:** the idea is to start with an impossible expanded message that would lead to the required digest, and then to correct this message until it becomes valid without destroying the preimage property.
- **P^3graphs:** an algorithm based on the theory of random graphs that allows the conversion of preimage attacks on the compression function to attacks on the hash function with less effort than traditional meet-in-the-middle approaches.

Combining these techniques, we obtain preimage-style shortcuts attacks for up to 45 steps of SHA-1, and up to 50 steps of SHA-0 (out of 80).

Keywords: hash function, cryptanalysis, preimages, SHA-0, SHA-1, directed random graph.

1 Introduction

Until recently, most of the cryptanalytic research on popular dedicated hash functions has focused on collisions resistance, as can be seen from the successful attempts to violate the collision resistance property of MD4 [10], MD5 [32, 34], SHA-0 [6] and SHA-1 [13, 21, 33] using the basic ideas of differential cryptanalysis [2]. The community developed a wealth of fairly sophisticated tools that aid this type of analysis, including manual [33] and automated [7, 8, 20] methods to search and evaluate characteristics optimized for differential cryptanalysis of the used building blocks.

This wealth of results stands in stark contrast to what is known about the preimage and second preimage resistance of these hash functions. This is especially unsatisfying since most applications of hash functions actually rely more on preimage and second preimage resistance than on collision resistance.

Some of the Main Features of Our Results: All currently known generic preimage attacks require either impractically long first preimages [15], a first

D. Wagner (Ed.): CRYPTO 2008, LNCS 5157, pp. 179–202, 2008.
© International Association for Cryptologic Research 2008

preimage lying in a very small subset of the set of all possible preimages [35], or a target digest constructed in a very special way [14].

In this work, we study the resistance of SHA-0 and SHA-1 against dedicated cryptanalytic attacks in settings where only relatively short preimages are allowed and a first preimage might not be available. An example of a very common use case of hash functions that relies on the resistance against these kind of attacks: hashed passwords. Especially SHA-1 is ubiquitously used, and will continue to be recommended by NIST even after 2010 outside the application of digital signatures [24], e.g., as RNG or KDF.

We exploit weak diffusion properties in the step transformation and in the message expansion to divide the effort to find a preimage, and consider only one or a small number of bits at a given time. In particular we present two new cryptanalytic tools. Firstly a compression function attack by means of correcting invalid messages, described in Sect. 3. Secondly, an algorithm based on the theory of random graphs that allows an efficient conversion of preimage attacks on the compression function to attacks on the hash function is presented in Sect. 4.

Later, in Sect. 5 we will discuss the results of combining these methods. This results in cryptanalytic shortcuts attacks for up to 50 step of SHA-0 (out of 80) and 45 steps of SHA-1. As a proof-of-concept we give a preimage for the 33-step SHA-0 compression function and also a second preimage of an ASCII text under the SHA-0 hash function reduced to 31 steps in Appendix B.

2   The SHA Family

In this paper, we will focus on the hash function SHA-1 and its predecessor SHA-0. The SHA-1 algorithm, designed by the US National Security Agency (NSA) and adopted as a standard in 1995, is widely used, and is representative for a large class of hash functions which started with MD4 and includes most algorithms in use today. In this section, we only briefly review a few features of the SHA design which are important for the techniques presented in this paper. For a complete description we refer to the specifications [25].

SHA-0 and SHA-1 consist of the iterative application of a compression function (denoted by f in Fig. 1), which transforms a 160-bit chaining variable h_{j-1} into h_j, based on a 512-bit message block m_j. At the core of the compression function lies a block cipher g which is used in Davies-Meyer mode (see Fig. 2).

The block cipher itself consists of two parts: a message expansion and a state update transformation.

The purpose of the message expansion is to expand a single 512-bit input message block into eighty 32-bit words $W_0, \ldots, W_{79}$. This is done by splitting the message block into sixteen 32-bit words $M_0, \ldots, M_{15}$, which are then expanded linearly according to the following recursive rule:

$$W_i = \begin{cases} M_i & \text{for } 0 \le i < 16, \\ (W_{i-3} \oplus W_{i-8} \oplus W_{i-14} \oplus W_{i-16}) \lll s & \text{for } 16 \le i < 80. \end{cases}$$

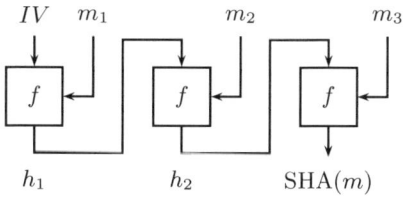

h_1 h_2 $\mathrm{SHA}(m)$

Fig. 1. An iterated hash function

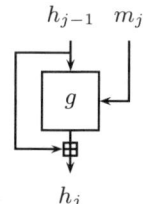

Fig. 2. The Davies-Meyer mode

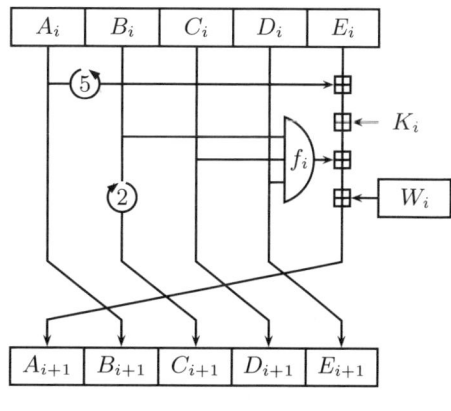

Fig. 3. A single state update step

The only difference between SHA-0 and SHA-1 lies in the rotation value s, which is 0 for SHA-0, and 1 for SHA-1.

The state update transformation takes as input a 160-bit chaining variable h_{j-1} which is used to initialize five 32-bit registers $A, B, \ldots, E$. These registers, referred to as state variables, are then iteratively updated in 80 steps, one of which is shown in Fig. 3. Note that the state update transformation can also be described recursively in terms of A_i only: after introducing $A_{-1} = B_0$, $A_{-2} = C_0 \lll 2$, $A_{-3} = D_0 \lll 2$, and $A_{-4} = E_0 \lll 2$, we can write:

$$A_{i+1} = (A_i \lll 5) + W_i + f(A_{i-1}, A_{i-2} \ggg 2, A_{i-3} \ggg 2) + (A_{i-4} \ggg 2) + K_i.$$

Because of this property, we will only consider the state variable A_i in the remainder of this paper.

3   Inverting the Compression Function

Before devising (second-) preimage attacks against the complete SHA function, we first focus on its compression function, and develop inverting methods which will be used as building blocks afterwards.

3.1   Possible Approaches

The recent successes in constructing collisions in SHA-0 and SHA-1 raise the natural question whether the differential techniques developed for collision attacks could also be used for constructing preimages. The question is especially pertinent in the case of second preimages, which are in fact just special types of collisions.

A first straightforward approach would consist in reusing the differential characteristics used in collision attacks by applying the corresponding message difference to the given message. If the characteristic is followed, then this will yield a second preimage. While this approach was applied to MD4 by Yu et al. [35], and to SHA-1 reduced to 53 steps by Rechberger and Rijmen [29, 30], it has some serious limitations when trying to find second preimages of reasonably short messages. The main problem is that, since the starting message is already fixed, the probability of the characteristic directly translates into the success probability of the attack (instead of determining the number of trials, as in collision attacks). This probability is further reduced by the fact that we lose the possibility to influence the difference propagation by fixing bits of the message to special values. In the case of MD4 and 53-step SHA-1, this results in attacks which only succeed with a probability of 2^{-56} and $2^{-151.5}$ respectively.

A second approach, which was recently proposed by Leurent in [17], relies on the existence of special messages which can simultaneously be combined with a large number of different characteristics, resulting in a large set of related messages. The idea is to compute the hash value of such a special message, and then apply the appropriate differences in order to steer this value towards the target value. Similar strategies have previously been used in practical second preimage attacks on SMASH by Lamberger et al. [16], and more recently in preimage attacks on GOST by Mendel et al. [18, 19]. In the case of MD4, this approach does not require a first preimage to start with, and results in a preimage attack against full MD4 with a complexity of 2^{100}.

It is not clear, however, how these ideas could efficiently be applied to hash functions such as SHA-0 or SHA-1, which, while still being vulnerable, show much more resistance to differential cryptanalysis than MD4. In the next sections, we will therefore study a completely different approach, which, as will be seen, has little in common with the techniques used in collision attacks.

3.2   Turning the Function Around

The problem we are trying to solve in this section is the following: given a 160-bit target value h_1, and a 160-bit chaining input h_0, find a 512-bit message input m_0 such that $f(h_0, m_0) = h_1$, or equivalently that $g(h_0, m_0) = h_1 - h_0$. Since the size of the message is much larger than the size of the output, we expect this equation to have a very large number of solutions. The difficulty in determining the 512 unknown input bits, however, lies in the fact that each of the 160 bit-conditions imposed at the output, depends in a complicated way on all 512 input bits.

The main observation on which the inversion method proposed in this paper is based, is that we can obtain a larger, but considerably less interconnected system of equations by expressing the problem in terms of internal state variables, rather than in terms of message words. That is, instead of trying to tweak a message in the hope to be able to control its effect on the output after being expanded and fed through several iterations of the state update transformation, we will start from state variables which already produce the correct output, and modify them

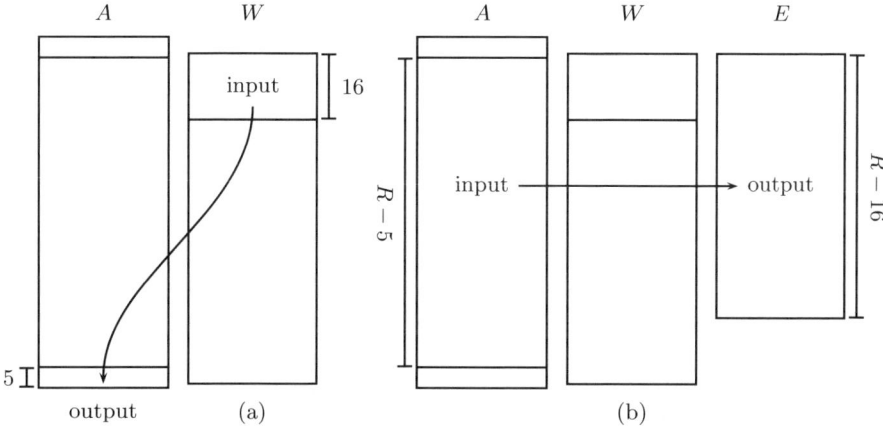

Fig. 4. Two equivalent descriptions of the inversion problem for a compression function reduced to R rounds

in such a way that the expanded message words, which can easily be derived from them, satisfy the linear recursion of the message expansion.

The idea is illustrated in Fig. 4. Instead of considering the function which maps $M_0 \dots M_{15}$ to $A_{76} \dots A_{80}$ as in Fig. 4(a), we will first fix $A_{76} \dots A_{80}$ to the target value determined by $h_1 - h_0$, and then analyze the function in Fig. 4(b) which maps $A_1 \dots A_{75}$ to error words $E_0 \dots E_{64}$, where

$$E_i = W_i \oplus W_{i+2} \oplus W_{i+8} \oplus W_{i+13} \oplus (W_{i+16} \ggg s), \quad \text{and}$$
$$W_i = A_{i+1} - (A_i \lll 5) - f(A_{i-1}, A_{i-2} \ggg 2, A_{i-3} \ggg 2) - (A_{i-4} \ggg 2) - K_i.$$

Clearly, finding an input which maps to $h_1 - h_0$ in Fig. 4(a) is equivalent to the problem of finding an input which maps to zero in Fig. 4(b).

The potential advantages of this alternative approach are clearly seen when analyzing how flipping a single bit in the input affects the output in both cases. In the first case, illustrated in Fig. 5(a), a single flip in the message quickly propagates through both the expanded message and the state, resulting in a completely uncontrollable pattern of changes at the output. In the second case, however, a bit-flip in the state propagates to the output in a very predictable way, as shown in Fig. 5(b). A change in the state affects at most 6 consecutive expanded messages words, and at most 22 words of the output. More importantly, depending on the position of the flipped bit in the state word, it will leave the least significant bits of all W_i and E_i untouched. The downside is that both the input and the output of the function to invert are considerably larger.

3.3 Fixing Problems Column by Column

Let us now analyze in a little bit more detail how state bits affect the output words in our new function. In order to simplify the analysis, we will for now assume that we deal with a variant of SHA-0 reduced to R rounds.

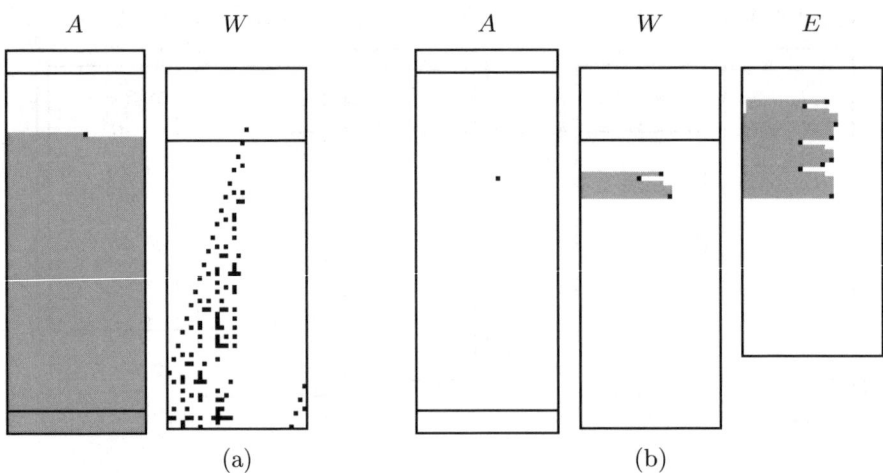

Fig. 5. Bits affected by a single bit flip at the input (SHA-1). Black bits are guaranteed to flip; gray bits may be flipped; white bits are unaffected.

Suppose that we restrict ourselves to the first $j + 1$ bits of each expanded message word W_i (denoted by $W_i^{j\cdots0}$), and that we keep all state bits constant except for those at bit position $j + 2$ (referred to as a_i^{j+2}). In this case, we can derive a simple relation (by collecting all constant parts into a $j + 1$-bit word $C_i^{j\cdots0}$ and a 1-bit variable c_i^j), which holds as long as $0 \le j < 25$:

$$W_i^{j\cdots0} = C_i^{j\cdots0} - (f(c_i^j, a_{i-2}^{j+2}, a_{i-3}^{j+2}) \ll j) - (a_{i-4}^{j+2} \ll j). \tag{1}$$

The interesting property of this relation is that the effect of the state bits a_i^{j+2} is confined to the most significant bit of $W_i^{j\cdots0}$. Furthermore, this effect is linear in all rounds where f_{XOR} or f_{IF} is used. Since the words E_i in SHA-0 are just a bitwise XOR of expanded message words W_i, this property holds for those words as well.

We can now use this observation to gradually fix the bits of E_i to zero, column by column. We start by determining $a_1^2 \ldots a_{R-5}^2$ such that the least significant bits of all $R - 16$ output words E_i are zero. Since we have $R - 5$ degrees of freedom and only need to satisfy $R-16$ conditions, we expect to find 2^{11} different solutions. Thanks to the special structure of the equations, these solutions can be found recursively with a computation effort which is linear in the number of rounds R. Next, we use $a_1^3 \ldots a_{R-5}^3$ (which, as indicated by (1), will not affect the least significant bits) to correct the second least significant bits. We proceed this way as long as (1) holds, and eventually we will only be left with non-zero bits in the 7 most significant bits of the $R - 16$ output words.

In order to eliminate the remaining non-zero bits, we could just repeat the previous procedure with different solutions for the state bits, until these non-zero bits disappear by themselves. This would require in the order of $2^{7\cdot(R-16)}$ trials. In the next section, we will show how this number can be reduced.

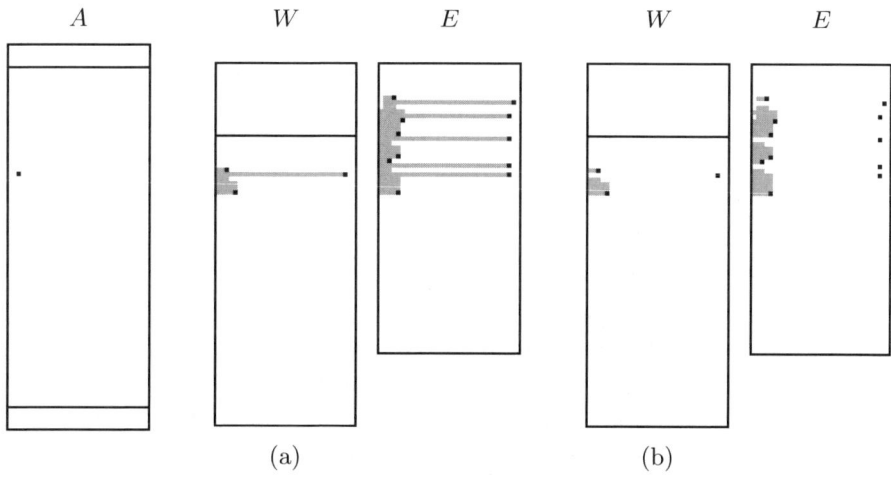

Fig. 6. Flipping state bit 29 with (a) and without (b) carries (SHA-1)

3.4 Preventing Carries

A natural way to improve the previous attack is to try to extent the property found in (1) to the case $j \geq 25$. The problem however is that the equation gets an extra term for $25 \leq j < 30$:

$$W_i^{j\cdots 0} = C_i^{j\cdots 0} - (a_i^{j+2} \lll j - 25) - (f(c_i^j, a_{i-2}^{j+2}, a_{i-3}^{j+2}) \lll j) - (a_{i-4}^{j+2} \lll j).$$

Hence, when trying to fix the output bits in column j, we have to make sure that this extra term at position $j - 25$ does not reintroduce errors in the previously fixed columns. In order to do so, we will first try to confine the potential trouble caused by this term to a single column by preventing the propagation of carries to other columns (the idea is shown in Fig. 6). This can easily be achieved by noting that the 5 most significant bits of A_i, which we are currently trying to determine, affect the least significant part of W_i through the equation $W_i = X_i - (A_i \lll 5)$, where

$$X_i = A_{i+1} - f(A_{i-1}, A_{i-2} \ggg 2, A_{i-3} \ggg 2) - (A_{i-4} \ggg 2) - K_i.$$

If we now choose the 7 least significant columns of the state in beforehand in such a way that there are no zeros in the 5 least significant bits of X_i, then no carries (borrows) will appear later on when the 5 most significant bits of A_i are modified. Once these 7 columns have been determined, we start correcting the output columns for $5 \leq j < 25$ in exactly the same way as explained in the previous section.

When we arrive at $j \geq 25$, we will try to use the state bits at position $j + 2$ to simultaneously correct columns j and $j - 25$ of the output. This time, we have $R - 5$ degrees of freedom to satisfy $2 \times (R - 16)$ conditions, and hence we will still have to rely on chance for $R - 27$ of these conditions. In total, we will leave $5 \times (R - 27)$ uncorrected output bits in columns 25–29 and $2 \times (R - 16)$ in columns 30–31. As a consequence, we will need to perform 2^c trials with $c = 2 \cdot (R - 16) + 5 \cdot (R - 27)$ in order for all non-zero bits to be eliminated.

3.5   Relaxing the Problem: Partial-Pseudo-Preimages

In the previous section, we had to leave a number of output bits uncorrected because of a lack of degrees of freedom in the state bits in columns 27–31. One way to create up to 10 additional degrees of freedom in each of these 5 columns is to allow the attacker to modify bits $a_{-4}^j \ldots a_0^j$ and/or $a_{R-4}^j \ldots a_R^j$ as well. In this case, the input and the output of the compression function will only partially match h_0 and h_1, and we call this a partial-pseudo-preimage. It is easy to see that each additional degree of freedom will reduce the cost by a factor two, i.e., if we allow $b_1 \leq 25$ input bits and $b_2 \leq 25$ output bits to deviate from their original target, then the computation effort of finding a partial-pseudo-preimage will be given by

$$2^c, \quad \text{where} \quad c = 2 \cdot (R - 16) + 5 \cdot (R - 27) - (b_1 + b_2) \,.$$

3.6   Application to SHA-1

The techniques explained for SHA-0 can be applied to SHA-1 in a relatively straightforward way. The only difference is that affected bits in W_i, with $i \geq 16$, will not only propagate to the corresponding columns in the error words, but also to the columns shifted by one position to the right. In order to compensate for this, it suffices to consider different state bits when correcting the columns, i.e., instead of using $a_1^{j+2} \ldots a_{R-5}^{j+2}$ to correct column j (and $j-25$ if $j \geq 25$), we will now use the state bits $a_1^{j+2} \ldots a_{11}^{j+2}$ and $a_{12}^{j+3} \ldots a_{R-5}^{j+3}$. This works fine as long as $j < 29$. The bits $a_{12}^{j+3} \ldots a_{R-5}^{j+3}$ cannot be used anymore when $j = 29$, though. Since we lose $R - 16$ degrees of freedom for fixing the last pair of columns (columns 29 and 4), the computational effort increases to:

$$2^c, \quad \text{where} \quad c = 3 \cdot (R - 16) + 5 \cdot (R - 27) - (b_1 + b_2) \,.$$

In addition to this, and for the same reason, we can now only fully exploit 20 additional degrees of freedom at the output, i.e., $b_2 \leq 20$. We still have $b_1 \leq 25$, though.

4   Preimages from Partial-Pseudo-Preimages – P³graphs

For the discussion in this section, let's assume we are given a method to produce partial-pseudo-preimages that is faster than a method to find preimages directly.

We first discuss a number of well understood methods in Sect. 4.1 that transform such attacks on the compression function into a preimage attack on the hash function by means of meet-in-the-middle and tree building techniques. Next, in Sect. 4.2 we discuss a new method using so-called P³graphs, that makes it possible to exploit the existence of such weaker attacks more directly.

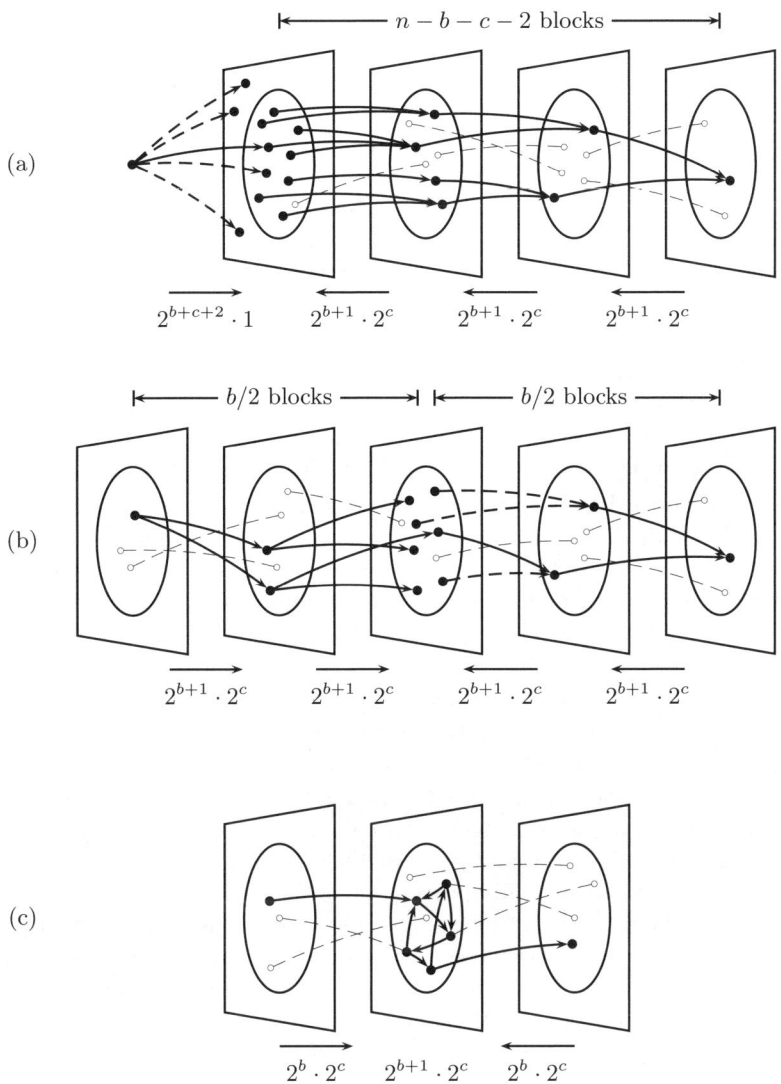

Fig. 7. Three different ways to build preimages from partial pseudo-preimages

4.1 Meet-in-the-Middle and Tree Based Methods

Inverting a Davis-Meyer compression function is the problem of finding a pair (h, m) such that $g(h, m) + h$ equals a given digest d. It was shown that no black-box attack can give a preimage faster than essentially 2^n [3, 27]. Inverting a Merkle-Damgård hash function is the problem of, given an initial chaining input h_0, finding an (almost arbitrarily large) number x of message blocks $m_0 \ldots m_x$ such that h_x equals a given digest d.

In the following, we assume that a part of the chaining input (say $n - b_1$ out of the n bits) can be chosen by the attacker, or in other words: the attacker can control all but b_1 bits of the chaining input (always the same $n - b_1$ bits). Let's further assume that a partial preimage attack on the compression function (of cost 2^c) has the property that a preimage can be found where all but b_2 out of n bits match the targeted digest d (again always the same $n - b_2$ bits). In addition to the parameters b_1 and b_2 introduced in Sect. 3.5, we will denote the number of bit positions of the chaining variable which can be controlled both from the input and from the output by $n - b$. All the following methods yield a preimage of the hash function for any given digest d

- **Meet-in-the-middle approach 1.** A basic unbalanced meet-in-the-middle approach that does not take advantage of the b bits that overlap has runtime $2^{(n+c)/2+1}$ and memory costs of $2^{(n-c)/2}$. The balanced case appeared already in [9], memoryless variants appear to have been first proposed in [23, 28].

- **Meet-in-the-middle approach 2.** By using the fact that both in forwards, and backwards direction, only b bits need to meet, the runtime requirement improves to $2^{b/2+c_1} + 2^{b/2+c_2}$, where c_1 denotes the cost of a partial-preimage attack (the forward part, if no compression function attack is available, a brute force attack with this property has cost 2^{n-b}), and c_2 denotes the cost of the pseudo-preimage attack (this is equivalent to calling the partial-pseudo-preimage attack 2^b times at the cost of 2^{b+c}). The total runtime is hence $2^{3b/2+c+1}$, the memory requirement is 2^b.

- **Layered Tree method due to Leurent, see Fig. 7(a).** In [17] the following tree method was proposed. Starting from the target hash d, produce two different pseudo preimages with cost 2^{b+c+1}. As a next step, produce four different pseudo preimages with the same cost that target both new target chaining values. This process is continued for $n - b - c - 2$ blocks and needs about $2^{n-b-c-1}$ of storage. For a fixed length preimage, only the last layer of the tree can be used for random trials in the forward direction, amounting to 2^{b+c+2} trials. Variants with a different branch number, or with less restrictions on the way the tree grows are thinkable [17].

- **Alternative Backward-Forward Tree method, see Fig. 7(b).** Similar to the approach above, one could let the tree grow in the backward direction for $b/2$ blocks, regardless of the time complexity of the compression function attack. In the forward direction we rely on using the partial-pseudo-preimage on the compression function of cost 2^c again, now having to call it 2^b times to have a partial-preimage. Using this, the tree grows in the forwards direction in exactly the same manner as in the backwards direction. Because of the birthday effect, both trees have at least one connection with high probability. The total runtime is $b \cdot 2^{b+c+1}$, the memory requirement is $2^{b/2}$.

- **Tree Method due to Mendel and Rijmen, see Fig. 7(c).** In [22] a tree-based method was proposed that has the same runtime and memory requirements as the new graph-based method we are about to introduce in the following section.

4.2   A Graph Based Approach

The meet-in-the-middle method discussed above requires the generation of many partial-preimages for the first part of the preimage and many pseudo-preimages for the second part of the preimage. The new method based on random directed graphs we are about to introduce allows to reduce the number of partial-preimages needed at the beginning and pseudo-preimages needed at the end to 1, at the cost of a number of partial-pseudo-preimages (each 2^c) in between. Hence the name P^3graph method, see also Fig. 7(c). We first outline the proposed method, and give time and memory complexities. Afterwards we discuss and compare it with other methods.

Edges of P^3graph: Using a partial pseudo preimage algorithm, generate 2^{b+1} tuples $(h_{(i)}, m_{(i)})$, at cost 2^{b+c+1}. All these tuples, which map $h_{(i)}$ to $f(h_{(i)}, m_{(i)})$, can be seen as the 2^{b+1} edges of a directed graph consisting of 2^b nodes. As explained in Appendix A, we expect the majority of those nodes to be part of a large densely interconnected component.

First Message Block, Forward Direction: Using the partial preimage generation method, generate a single tuple (h_0, m_0) that hits this component. The expected work is in the order of 2^{b+c}.

Last Message Block, Backward Direction: Also here, generate a single tuple (h_x, m_x) such that $f(h_x, m_x) = \mathsf{d}$ and that h_x falls into the interconnected part of the graph. The expected work is again in the order of: 2^{b+c}.

Connection: What remains to be found is a connection (a path) between these nodes (the entry node and the exit node) in the graph. Given the number of edges in the graph, such a path is very likely to exist, as we discuss in detail in Appendix A. Total expected work: $2^{b+c+1} + 2^{b+c} + 2^{b+c} = 2^{b+c+2}$

On Exploiting Precomputation. The computations for constructing the first message block and the P^3graph do not need to be repeated when attacking a different digest. The effort for every additional preimage attack is only 2^{b+c}.

4.3   Discussion

There are a number of useful and distinctive properties of the P^3graph method. Firstly, the graph approach does not impose any structure on the connections of partial-pseudo-preimages, which is an intuitive explanation of the efficiency again compared to the L-Tree and the BF-Tree methods. Secondly, the P^3graph is friendlier towards precomputation: Whereas the full P^3graph (potentially in such a way that the IV of the hash function is one of the nodes) can be precomputed, it is not possible to precompute the backwards tree for the L-Tree and the BF-Tree method. Another advantage of the P^3graph method over all other known methods is that paths (and hence preimages) of almost any length have high probability to exist. There is no upper limit, the lower bound is discussed in

Appendix A. This property will be useful when dealing with the padding in a preimage attack on the hash function (see Sect 5.1).

One drawback of the P^3graph method can be the higher memory requirements. Storage requirements for all the edges is exponential in the number of bits b that can not be controlled. Hence the runtime gain of the P^3graph method is useful in practice if the compression function attack allows to choose a reasonable small b. The P^3graph method allows time/memory tradeoffs that resemble e.g., the BF-Tree method. Space constraints do not allow us to discuss them here. In Table 1 we summarize and compare the meet in the middle approach with the P^3graph method.

5   Putting Everything Together

We have now set the state to talk about the security margin of the SHA-0 and SHA-1 hash function against the new cryptanalytic methods. We do this by combining the compression function attack from Sect. 3 and the P^3graph method from Sect. 4.

5.1   Padding

So far, we neglected the fact that in a preimage attack on SHA-0 and SHA-1, the padding fixes a part of the input message of the last message block. Hence, without being able to cope with such a restriction, our attack would only be a second preimage attack, but not a preimage attack. We discuss here several possibilities to produce a correctly padded last message block without a first preimage.

- **Restrict the Degrees of Freedom in the Compression Function Attack:** In order to fix a particular value for the message length, at least the last 65 bits of the last message block need to be fixed. Among them are 25 bits whose freedom is needed in the compression function attack (for both SHA-0 and SHA-1), hence fixing them results (without further optimizations) in a slowdown of the compression function attack by up to a factor of 2^{25}. In detail, these bits are M_{14}^0, $M_{15}^{0...4,24...31}$ and $M_{16}^{0...4,24...31}$.
- **Expandable Messages:** By making sure that every message length can be constructed after the compression function attacks have been performed, almost no additional degrees of freedom need to be spent for a correct padding.

Table 1. Comparison of the meet-in-the-middle approach, various tree approaches, and the P^3graph method. All numbers are exponents of base 2.

	MITM2	L(ayered)-Tree	BF-Tree	MR-Tree	P^3graph
total work	$3b/2+c+1$	$b+c+1+log_2(n-b-c)$	$b+log_2(b)+c+1$	$b+c+2$	$b+c+2$
total mem.	b or less	$n-b-c-1$	$b/2$	$b+1$	$b+1$
onl. work	$b+c$	-	-	$b+c$	$b+c$
offl. work	$2b+c$	-	-	$b+c+log_2(3)$	$b+c+log_2(3)$
memory	b	-	-	$b+1$	$b+1$
flexible len.	no	no	no	no	yes

Using any of the following methods will hence return preimages of uncontrollable length. The only two property that the compression function attack needs to have, are as follows. Firstly, to make sure that the end of the message (before the length encoding, i.e., the LSB of M_{13}) is a '1'. Secondly, make sure that the length is a exact multiple of the block length, i.e., fix the last nine bits of M_{15} to '110111111' (447). In total ten bits need to be fixed for this, which will result (without further optimizations) in a slowdown of the compression function attack by a factor 2^6. In detail, the six crucial bits are M_{14}^0 and $M_{16}^{0\cdots4}$. Possibilities to construct expandable messages are as follows.

- **Multicollisions:** As soon as the compression function attack has a complexity slightly above the birthday bound $(2^{n/2+log_2(n)})$, the multicollision idea [12] can be used to construct expandable messages [15] without being the bottleneck.
- **Flexibility of the P³graph Method (cycles):** In the random directed graph used in the P³graph method of Sect. 4.2, we expect to have many cycles, also on the path between entry- and exit node. As detailed in Appendix A, we hence expect to find paths of any length longer than some lower bound that connect any entry- and exit node with high probability.

5.2 Summary of Attacks

From Sect. 3 we learn that $b_1 = b_2 = 25$ is a straight-forward choice for the case of SHA-0. Since the method allows us to pick the same bit positions, we also have $b = 25$. Since $b_2 \leq 20$ for SHA-1, we will have to restrict ourselves to $b = 20$ in this case. Note that for seriously reduced SHA-0 and SHA-1, less degrees of freedom are of use in the compression function attack, and hence b can be smaller. A quick check in Table 1 will convince the reader that memory requirements will not be a problem in the practical implementation of such an attack, even with the most time efficient P³graph method.

In order to illustrate our results we consider SHA-0 and SHA-1 reduced to concrete numbers of steps, and give attack complexities in Figure 8. We combine the attacks on the compression function as given in Sect. 3 with the different generic ways of turning them into a preimage attack as outlined in Sect. 4.2. In our implementation of this attack the memory requirements are negligible. Additionally, we also give attack complexities in Table 2. For both SHA-0 and SHA-1, the number of steps for which we list results are chosen as follows. To compare (lack of) resistance against the new attack of similarly reduced primitives, we pick 32 steps in all cases. Additionally, we give results for the highest number of steps for which the attack would be better than the birthday bound and an actual brute force attack, respectively. Our approach takes less than 2^{160} hash evaluations for SHA-0 reduced to up to 50 steps and for SHA-1 reduced to up to 45 steps. Note that inverting the hash function also implies the ability to construct a fixed point.

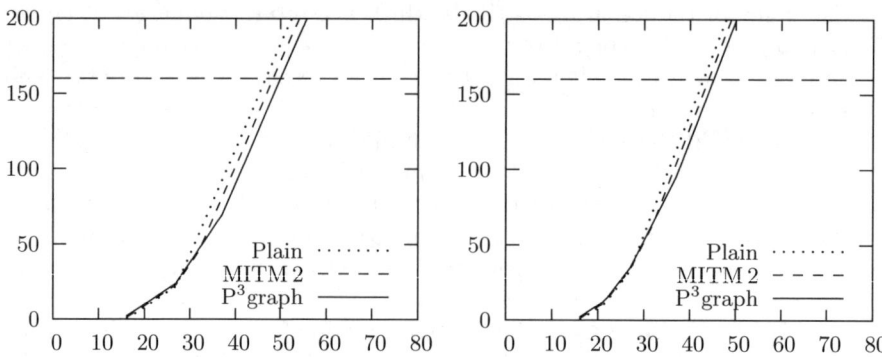

Fig. 8. Complexities of second preimage attacks against reduced SHA-0 (left) and SHA-1 (right). The line 'Plain' refers to a direct preimage attack using only a single block. The line 'MITM 2' refers to a meet-in-the-middle approach where partial-preimages in the forward direction are combined with pseudo-preimages in the backwards direction. The line 'P^{3}graph' refers to the new graph based method.

Table 2. Exemplification of new preimage attacks on reduced SHA-0 (left table) and SHA-1 (right table). Efforts are expressed in terms of time complexity; memory and communication costs can be considered negligible. For ideal building blocks, all these attacks would require a 2^{160} effort. For simplification, the small constant factor between the numbers given here and a naive brute force search is neglected. We give the total runtime for attacking the first target digest; attacks on subsequent targets will be faster.

type of attack	building block	steps	b	effort with new attack	building block	steps	b	effort with new attack
inv. compression f.	SHA-0	32	25	2^{32}	SHA-1	32	20	2^{53}
inv. compression f.	SHA-0	38	25	2^{74}	SHA-1	35	20	2^{77}
inv. compression f.	SHA-0	50	25	2^{158}	SHA-1	45	20	2^{157}
2nd preimage of hash	SHA-0	32	12	2^{47}	SHA-1	32	10	2^{65}
2nd preimage of hash	SHA-0	38	25	2^{76}	SHA-1	34	14	2^{77}
2nd preimage of hash	SHA-0	49	25	2^{153}	SHA-1	45	20	2^{159}
preimage of hash	SHA-0	37	25	2^{75}	SHA-1	34	17	2^{80}
preimage of hash	SHA-0	49	25	2^{159}	SHA-1	44	20	2^{157}

6   Conclusions and Outlook

The first method to construct preimages for SHA-0 and SHA-1 reduced to a nontrivial number of steps (up to 50 out of 80) is presented. The impossible message approach we proposed exploits weak diffusion properties in the step transformation and in the message expansion, which allows to divide the work and consider only one or a small number of column at a given time. Both, the impossible message approach, and the P^{3}graph we introduced to efficiently transform attacks on the compression function to attacks on the hash function, are rather generic and await to be applied to other settings and hash functions as well.

Our results shed some light on the security margin offered by SHA-0 and SHA-1 when only preimage attacks are of a concern. However, several aspects of this work suggest that the security margin might be smaller. Let's compare the result of this work on cryptanalytic preimage attacks to the situation of collision search attacks in 2004 and early 2005:

- **Step-Reduced Variants:** Work on SHA-1 resulted in theoretical collision attacks for up to 58 steps [1, 31]. Our preimage attacks cover slightly less steps but are on a comparable magnitude.
- **Degrees of Freedom:** Whereas in the most recent collision search attacks on SHA-1 the availability of degrees of freedom is the limiting factor for further improvements, this was of no concern in earlier work. The fact that not all degrees of freedom are used in our new preimage attacks suggests that further improvements are possible.
- **Sensitivity for Different Choices of Rotation Constants:** The state update transformation of SHA-0 and SHA-1 uses the fixed set of rotation constants $(5, -2)$. A study of the effect of different choices of rotation constants on earlier collision search strategies [26] concluded that already a slightly different choice would impact the performance significantly, although in a complex way. In our attack, we observe a similar situation: The attack complexity directly depends on the used rotation constants and would be lower or higher, depending on the actual choice. The most recent collision search attacks on SHA-1 do not show such a strong dependency on the choice of rotation constants. Again, this suggests that further improvements on the preimage attack presented in this paper is an interesting open problem.

Acknowledgements

The authors wish to thank Florian Mendel, Adi Shamir, Yiqun Lisa Yin and the anonymous reviewers for their useful comments. The work in this paper has been supported in part by the Fund for Scientific Research (FWO), the Chaire France Telecom pour la sécurité des réseaux, the Secure Information Technology Center-Austria (A-SIT), by the Austrian Science Fund (FWF), project P19863, and by the IAP Programme P6/26 BCRYPT of the Belgian State (Belgian Science Policy).

References

1. Biham, E., Chen, R., Joux, A., Carribault, P., Lemuet, C., Jalby, W.: Collisions of SHA-0 and Reduced SHA-1. In: Cramer, R. (ed.) EUROCRYPT 2005. LNCS, vol. 3494, pp. 36–57. Springer, Heidelberg (2005)
2. Biham, E., Shamir, A.: Differential Cryptanalysis of DES-like Cryptosystems. In: Menezes, A., Vanstone, S.A. (eds.) CRYPTO 1990. LNCS, vol. 537, pp. 2–21. Springer, Heidelberg (1991)

3. Black, J., Rogaway, P., Shrimpton, T.: Black-Box Analysis of the Block-Cipher-Based Hash-Function Constructions from PGV. In: Yung, M. (ed.) CRYPTO 2002. LNCS, vol. 2442, pp. 320–335. Springer, Heidelberg (2002)

4. Bollobás, B.: Random Graphs. Academic Press, London (1985)

5. Bollobás, B.: Modern Graph Theory. Springer, Heidelberg (1998)

6. Chabaud, F., Joux, A.: Differential Collisions in SHA-0. In: Krawczyk, H. (ed.) CRYPTO 1998. LNCS, vol. 1462, pp. 56–71. Springer, Heidelberg (1998)

7. De Cannière, C., Mendel, F., Rechberger, C.: Collisions for 70-Step SHA-1: On the Full Cost of Collision Search. In: Adams, C.M., Miri, A., Wiener, M.J. (eds.) SAC 2007. LNCS, vol. 4876, pp. 56–73. Springer, Heidelberg (2007)

8. De Cannière, C., Rechberger, C.: Finding SHA-1 Characteristics: General Results and Applications. In: Lai, X., Chen, K. (eds.) ASIACRYPT 2006. LNCS, vol. 4284, pp. 1–20. Springer, Heidelberg (2006)

9. Diffie, W., Hellman, M.: Exhaustive cryptanalysis of the NBS Data Encryption Standard. Computer 10(6), 74–84 (1977)

10. Dobbertin, H.: Cryptanalysis of MD4. J. Cryptology 11(4), 253–271 (1998)

11. Erdös, P., Rènyi, A.: On random graphs. Publicationes Mathematicae 6, 290–297 (1959)

12. Joux, A.: Multicollisions in Iterated Hash Functions. Application to Cascaded Constructions. In: Franklin, M.K. (ed.) CRYPTO 2004. LNCS, vol. 3152, pp. 306–316. Springer, Heidelberg (2004)

13. Joux, A., Peyrin, T.: Hash Functions and the (Amplified) Boomerang Attack. In: Menezes, A. (ed.) CRYPTO 2007. LNCS, vol. 4622, pp. 244–263. Springer, Heidelberg (2007)

14. Kelsey, J., Kohno, T.: Herding Hash Functions and the Nostradamus Attack. In: Vaudenay, S. (ed.) EUROCRYPT 2006. LNCS, vol. 4004, pp. 183–200. Springer, Heidelberg (2006)

15. Kelsey, J., Schneier, B.: Second Preimages on n-Bit Hash Functions for Much Less than 2^n Work. In: Cramer, R. (ed.) EUROCRYPT 2005. LNCS, vol. 3494, pp. 474–490. Springer, Heidelberg (2005)

16. Lamberger, M., Pramstaller, N., Rechberger, C., Rijmen, V.: Second Preimages for SMASH. In: Abe, M. (ed.) CT-RSA 2007. LNCS, vol. 4377, pp. 101–111. Springer, Heidelberg (2006)

17. Leurent, G.: MD4 is Not One-Way. In: Nyberg, K. (ed.) FSE 2008. LNCS. Springer, Heidelberg (to appear, 2008)

18. Mendel, F., Pramstaller, N., Rechberger, C.: A (Second) Preimage Attack on the GOST Hash Function. In: Nyberg, K. (ed.) FSE. LNCS, vol. 5086, pp. 224–234. Springer, Heidelberg (2008)

19. Mendel, F., Pramstaller, N., Rechberger, C., Kontac, M., Szmidt, J.: Cryptanalysis of the GOST Hash Function. In: Wagner, D. (ed.) Proceedings of CRYPTO 2008. LNCS. Springer, Heidelberg (to appear, 2008)

20. Mendel, F., Pramstaller, N., Rechberger, C., Rijmen, V.: The Impact of Carries on the Complexity of Collision Attacks on SHA-1. In: Robshaw, M.J.B. (ed.) FSE 2006. LNCS, vol. 4047, pp. 278–292. Springer, Heidelberg (2006)

21. Mendel, F., Rechberger, C., Rijmen, V.: Update on SHA-1. In: Rump Session of CRYPTO 2007 (2007)

22. Mendel, F., Rijmen, V.: Weaknesses in the HAS-V Compression Function. In: Nam, K.-H., Rhee, G. (eds.) ICISC 2007. LNCS, vol. 4817, pp. 335–345. Springer, Heidelberg (2007)

23. Morita, H., Ohta, K., Miyaguchi, S.: A Switching Closure Test to Analyze Cryptosystems. In: Feigenbaum, J. (ed.) CRYPTO 1991. LNCS, vol. 576, pp. 183–193. Springer, Heidelberg (1992)
24. National Institute of Standards and Technology. NIST's Policy on Hash Functions (2006), http://csrc.nist.gov/groups/ST/hash/policy.html
25. National Institute of Standards and Technology (NIST). FIPS-180-2: Secure Hash Standard (August 2002), http://www.itl.nist.gov/fipspubs/
26. Pramstaller, N., Rechberger, C., Rijmen, V.: Impact of Rotations in SHA-1 and Related Hash Functions. In: Preneel, B., Tavares, S.E. (eds.) SAC 2005. LNCS, vol. 3897, pp. 261–275. Springer, Heidelberg (2006)
27. Preneel, B., Govaerts, R., Vandewalle, J.: Hash Functions Based on Block Ciphers: A Synthetic Approach. In: Stinson, D.R. (ed.) CRYPTO 1993. LNCS, vol. 773, pp. 368–378. Springer, Heidelberg (1994)
28. Quisquater, J.-J., Delescaille, J.-P.: How Easy is Collision Search. New Results and Applications to DES. In: Brassard, G. (ed.) CRYPTO 1989. LNCS, vol. 435, pp. 408–413. Springer, Heidelberg (1990)
29. Rechberger, C., Rijmen, V.: On Authentication with HMAC and Non-random Properties. In: Dietrich, S., Dhamija, R. (eds.) FC 2007 and USEC 2007. LNCS, vol. 4886, pp. 119–133. Springer, Heidelberg (2007)
30. Rechberger, C., Rijmen, V.: New Results on NMAC/HMAC when Instantiated with Popular Hash Functions. Journal of Universal Computer Science (JUCS), Special Issue on Cryptography in Computer System Security 14(3), 347–376 (2008)
31. Rijmen, V., Oswald, E.: Update on SHA-1. In: Menezes, A. (ed.) CT-RSA 2005. LNCS, vol. 3376, pp. 58–71. Springer, Heidelberg (2005)
32. Stevens, M., Lenstra, A.K., de Weger, B.: Chosen-Prefix Collisions for MD5 and Colliding X.509 Certificates for Different Identities. In: Naor, M. (ed.) EUROCRYPT 2007. LNCS, vol. 4515, pp. 1–22. Springer, Heidelberg (2007)
33. Wang, X., Yin, Y.L., Yu, H.: Finding Collisions in the Full SHA-1. In: Shoup, V. (ed.) CRYPTO 2005. LNCS, vol. 3621, pp. 17–36. Springer, Heidelberg (2005)
34. Wang, X., Yu, H.: How to Break MD5 and Other Hash Functions. In: Cramer, R. (ed.) EUROCRYPT 2005. LNCS, vol. 3494, pp. 19–35. Springer, Heidelberg (2005)
35. Yu, H., Wang, G., Zhang, G., Wang, X.: The Second-Preimage Attack on MD4. In: Desmedt, Y.G., Wang, H., Mu, Y., Li, Y. (eds.) CANS 2005. LNCS, vol. 3810, pp. 1–12. Springer, Heidelberg (2005)

A   Some Useful Properties of Random Graphs

In this appendix, we briefly review some properties of random graphs which are relevant to the graph based approach proposed in Sect. 4.2. For a more rigorous and comprehensive treatment of random graph theory we refer to [4, 11] and [5, Chapt. VII.5].

A.1   Following Edges in a Random Directed Graph

Let G be a large directed graph consisting of n nodes and $m = c \cdot n$ randomly selected edges. On average, each node has c outgoing edges, and we denote the probability that a given ordered pair of nodes is connected by an edge by:

$$p_c = \frac{m}{n^2} = \frac{c}{n} .$$

Let us now study what happens when we start from an arbitrary node a and construct sets of nodes $S_0, S_1, S_2, \ldots$ where $S_0 = \{a\}$, and S_i contains all nodes that can be reached from a in exactly i hops. If we eventually end up with an empty set, the initial node a is called a "dying" node. In the opposite case, a is said to "explode". Clearly, if there exists an edge from a to b, and b is an exploding node, then a must be exploding as well. Conversely, a node a can only die if none of the n nodes in the graph are both connected to a and exploding. Hence, the probability p_e that a node explodes must satisfy:

$$1 - p_e = (1 - p_c \cdot p_e)^n \approx e^{-c \cdot p_e} .$$

From this expression we can deduce that p_e must necessarily be 0 as long as $c \leq 1$. However, when $c > 1$, the equation $1 - x = e^{-c \cdot x}$ does have a non-zero (and positive) solution, which we will refer to as $\gamma(c)$.[1]

Assuming that the sets S_i reach some moderately large size (i.e., a does not die), we can write a simple recursive relation between the expected sizes $E(|S_i|)$ of successive sets by computing the probability that an arbitrary node is connected to at least one node of S_i:

$$E(|S_{i+1}|) = n \cdot \left[1 - (1 - p_c)^{E(|S_i|)} \right] \approx n \cdot \left[1 - e^{-c \cdot E(|S_i|)/n} \right] . \qquad (2)$$

Note that we can apply the same reasoning to obtain an almost identical recursive relation between successive values of $E(|S_0 \cup S_1 \cdots S_i|)$. By filling in $i = \infty$, we find that the expected size of the sets converges to:

$$E(|S_\infty|) \approx E(|S_0 \cup S_1 \cdots S_\infty|) \approx n \cdot \gamma(c) .$$

A.2   Connecting Two Given Nodes

In the previous section, we argued that a node a explodes with probability $p_e = \gamma(c)$, and that a fraction $\gamma(c)$ of all nodes can be reached from it if it does. Similarly, if a dies, it can be shown that only a negligible fraction of nodes will be reached. The probability p_p that two given nodes a and b are connected by a path is hence:

$$p_p = \gamma(c)^2 .$$

In the context of the attack proposed in this paper, we are interested in the expected number of random edges $\overline{m}$ that need to be added to a graph in order to find a path between two given nodes a and b. Suppose our current graph has $m > n$ edges. In that case we know that with probability $1 - \gamma(m/n)^2$ there will be no path between a and b, in which case we will need at least one more edge. Repeating this reasoning, we find;

$$\overline{m} \approx n + \sum_{m=n}^{n^2} \left[1 - \gamma(m/n)^2 \right] .$$

---

[1] One can show that $\gamma(c) = 1 + W(-c \cdot e^{-c})/c$, where $W(x)$ is Lambert's W function.

We can approximate this sum by an integral, and after a few changes of variables, we eventually obtain:

$$\overline{m} \approx n + n \cdot \int_1^\infty \left[1 - \gamma(c)^2\right] dc$$

$$= n + n \cdot \int_0^1 \left(1 - \gamma^2\right) \cdot \frac{dc}{d\gamma} d\gamma$$

$$= 2 \cdot n .$$

This result, which states that, in order to connect two given nodes, we need on average twice as many edges as nodes (i.e., $c = 2$), is the main property used in Sect. 4.2.

A.3 Path Lengths

If we want to apply our graph based attack to a hash function which includes the message length in the padding block, then we not only need to make sure that there exists a path between two given nodes; we would also like to know in advance how long this path will be.

In order to estimate how many nodes can be reached for a fixed path length, we need to solve the recursive relation of (2). A closed form solution probably does not exist, but we can find a very good approximation:

$$E(|S_i|) \approx n \cdot \gamma \cdot \frac{\left[\alpha^{2 \cdot (i - \delta)} + 1\right]^\beta}{\alpha^{(i - \delta)} + 1} ,$$

where $\alpha = c \cdot (1 - \gamma)$, $\alpha^{2 \cdot \beta - 1} = c$, and $n \cdot \gamma \cdot c^{-\delta} = 1$. For $c = 2$, we find that $\gamma = 0.80$, $\alpha = 0.41$, $\beta = 0.12$, and

$$\delta = \frac{1}{\log_2 c} \cdot (\log_2 n + \log_2 \gamma) = \log_2 n - 0.33 .$$

We can now compute the minimal path length l for which we expect that S_l includes all reachable nodes (i.e., $S_l = S_\infty$). By solving the inequality $E(|S_\infty|) - E(|S_l|) < 1$, we obtain:

$$l > \left[\frac{1}{\log_2 \alpha} - \frac{1}{\log_2 c}\right] \cdot (\log_2 n + \log_2 \gamma) = 1.77 \cdot \log_2 n - 0.58 .$$

In other words, if we are given a random graph with n nodes and $2 \cdot n$ edges, and if this graph connects two arbitrary nodes a and b (this occurs with probability $\gamma^2 = 0.63$), then we expect to find paths from a to b of length l for any l exceeding $1.77 \cdot \log_2 n$.

B Proof-of-Concept Examples

As a proof-of-concept, we give examples of an implementation of the described methods. We chose two examples. The first is a preimage for the 33-step SHA-0 compression function. The second is also a second preimage of a (roughly) 1KB ASCII text for the 31-step SHA-0 hash function, using the P³graph method.

B.1   A Preimage for the 33-Step Compression Function of SHA-0

As a proof-of-concept, we consider the compression function of SHA-0 reduced to 33 steps. In Figure 9 we give a preimage for the all-1 output. $A_{-4} \ldots A_0$ and $W_0 \ldots W_{15}$ represent the input to the compression function. Computing $A_{-4} + A_{29} \ldots A_0 + A_{33}$ results in the all-1 output.

i	A_i	W_i
-4:	00110111111111111111111111111100	
-3:	11010111111111111111111111111100	
-2:	00100111111111111111111111111100	
-1:	00100111111111111111111111111111	
0:	10110111111111111111111111111111	10100111011111011000111010001001
1:	00100010000000000000100000010110	01100111100011001010011000011011
2:	11000010000100000010001001110110	01010000100000010111100010000111
3:	11100001000010000100000011110110	01000001100001011000100101100011
4:	00110101000000000000000101100100	10110010111111010101011101011001
5:	01000100000000000000000000001100	10100010011110010111101001010111
6:	10110110000000000000000000111010	11011111101101110110011001001001
7:	01100111000000000000000000001110	00001111111110110111010000110011
8:	00011100000000000000000000011000	10000111001111011000001011111100
9:	10100100000000000000000000000000	01000001111111011000011010001011
10:	11100111000000000000000001000001	10011100101111010111111010000011
11:	10100010100000000000000001101001	10101101000000111111101001001011
12:	00010010010001101000000100100001	01011101010110010110110100111101
13:	00110001001011000000101011111110	00011011111010010011001011011001
14:	00101110011011010000110001001000	00000011001110111111110010011010
15:	11101101100111111111111110010000	11100001000001101011110110000010
16:	10100101000000101100100101011010	01101011001010000100011000101011
	· · ·	· · ·
28:	01110111001110111011010101110100	11111100111010011110011000110001
29:	11001000000000000000000000000011	01110110101111001110011000100110
30:	00101000000000000000000000000011	11011100010011000000000000111010
31:	11011000000000000000000000000011	10000010111111000100100010100100
32:	11011000000000000000000000000000	11011011010101110010011011100100
33:	01001000000000000000000000000000	

Fig. 9. A preimage of the all-1 output for the 33-step SHA-0 compression function

```
0000000:  416c 6963 6520 7761 7320 6265 6769 6e6e    Alice was beginn
0000010:  696e 6720 746f 2067 6574 2076 6572 7920    ing to get very
0000020:  7469 7265 6420 6f66 2073 6974 7469 6e67    tired of sitting
0000030:  2062 7920 6865 7220 7369 7374 6572 206f     by her sister o
0000040:  6e20 7468 6520 6261 6e6b 2c20 616e 6420    n the bank, and
0000050:  6f66 2068 6176 696e 6720 6e6f 7468 696e    of having nothin
0000060:  6720 746f 2064 6f3a 206f 6e63 6520 6f72    g to do: once or
0000070:  2074 7769 6365 2073 6865 2068 6164 2070     twice she had p
0000080:  6565 7065 6420 696e 746f 2074 6865 2062    eeped into the b
0000090:  6f6f 6b20 6865 7220 7369 7374 6572 2077    ook her sister w
00000a0:  6173 2072 6561 6469 6e67 2c20 6275 7420    as reading, but
00000b0:  6974 2068 6164 206e 6f20 7069 6374 7572    it had no pictur
00000c0:  6573 206f 7220 636f 6e76 6572 7361 7469    es or conversati
00000d0:  6f6e 7320 696e 2069 742c 2060 616e 6420    ons in it, 'and
00000e0:  7768 6174 2069 7320 7468 6520 7573 6520    what is the use
00000f0:  6f66 2061 2062 6f6f 6b2c 2720 7468 6f75    of a book,' thou
0000100:  6768 7420 416c 6963 6520 6077 6974 686f    ght Alice 'witho
0000110:  7574 2070 6963 7475 7265 7320 6f72 2063    ut pictures or c
0000120:  6f6e 7665 7273 6174 696f 6e3f 2720 536f    onversation?' So
0000130:  2073 6865 2077 6173 2063 6f6e 7369 6465     she was conside
0000140:  7269 6e67 2069 6e20 6865 7220 6f77 6e20    ring in her own
0000150:  6d69 6e64 2028 6173 2077 656c 6c20 6173    mind (as well as
0000160:  2073 6865 2063 6f75 6c64 2c20 666f 7220    she could, for
0000170:  7468 6520 686f 7420 6461 7920 6d61 6465    the hot day made
0000180:  2068 6572 2066 6565 6c20 7665 7279 2073     her feel very s
0000190:  6c65 6570 7920 616e 6420 7374 7570 6964    leepy and stupid
00001a0:  292c 2077 6865 7468 6572 2074 6865 2070    ), whether the p
00001b0:  6c65 6173 7572 6520 6f66 206d 616b 696e    leasure of makin
00001c0:  6720 6120 6461 6973 792d 6368 6169 6e20    g a daisy-chain
00001d0:  776f 756c 6420 6265 2077 6f72 7468 2074    would be worth t
00001e0:  6865 2074 726f 7562 6c65 206f 6620 6765    he trouble of ge
00001f0:  7474 696e 6720 7570 2061 6e64 2070 6963    tting up and pic
0000200:  6b69 6e67 2074 6865 2064 6169 7369 6573    king the daisies
0000210:  2c20 7768 656e 2073 7564 6465 6e6c 7920    , when suddenly
0000220:  6120 5768 6974 6520 5261 6262 6974 2077    a White Rabbit w
0000230:  6974 6820 7069 6e6b 2065 7965 7320 7261    ith pink eyes ra
0000240:  6e20 636c 6f73 6520 6279 2068 6572 2e20    n close by her.
0000250:  5468 6572 6520 7761 7320 6e6f 7468 696e    There was nothin
0000260:  6720 736f 2056 4552 5920 7265 6d61 726b    g so VERY remark
0000270:  6162 6c65 2069 6e20 7468 6174 3b20 6e6f    able in that; no
0000280:  7220 6469 6420 416c 6963 6520 7468 696e    r did Alice thin
0000290:  6b20 6974 2073 6f20 5645 5259 206d 7563    k it so VERY muc
00002a0:  6820 6f75 7420 6f66 2074 6865 2077 6179    h out of the way
00002b0:  2074 6f20 6865 6172 2074 6865 2052 6162     to hear the Rab
00002c0:  6269 7420 7361 7920 746f 2069 7473 656c    bit say to itsel
00002d0:  662c 2060 4f68 2064 6561 7221 204f 6820    f, 'Oh dear! Oh
00002e0:  6465 6172 2120 4920 7368 616c 6c20 6265    dear! I shall be
00002f0:  206c 6174 6521 2720 2877 6865 6e20 7368    late!' (when sh
```

Fig. 10. 31-round SHA-0: original message (part 1)

```
0000300: 6520 7468 6f75 6768 7420 6974 206f 7665    e thought it ove
0000310: 7220 6166 7465 7277 6172 6473 2c20 6974    r afterwards, it
0000320: 206f 6363 7572 7265 6420 746f 2068 6572     occurred to her
0000330: 2074 6861 7420 7368 6520 6f75 6768 7420     that she ought
0000340: 746f 2068 6176 6520 776f 6e64 6572 6564    to have wondered
0000350: 2061 7420 7468 6973 2c20 6275 7420 6174     at this, but at
0000360: 2074 6865 2074 696d 6520 6974 2061 6c6c     the time it all
0000370: 2073 6565 6d65 6420 7175 6974 6520 6e61     seemed quite na
0000380: 7475 7261 6c29 3b20 6275 7420 7768 656e    tural); but when
0000390: 2074 6865 2052 6162 6269 7420 6163 7475     the Rabbit actu
00003a0: 616c 6c79 2054 4f4f 4b20 4120 5741 5443    ally TOOK A WATC
00003b0: 4820 4f55 5420 4f46 2049 5453 2057 4149    H OUT OF ITS WAI
00003c0: 5354 434f 4154 2d50 4f43 4b45 542c 2061    STCOAT-POCKET, a
00003d0: 6e64 206c 6f6f 6b65 6420 6174 2069 742c    nd looked at it,
00003e0: 2061 6e64 2074 6865 6e20 6875 7272 6965     and then hurrie
00003f0: 6420 6f6e 2c20 416c 6963 6520 7374 6172    d on, Alice star
0000400: 7465 6420 746f 2068 6572 2066 6565 742c    ted to her feet,
0000410: 2066 6f72 2069 7420 666c 6173 6865 6420     for it flashed
0000420: 6163 726f 7373 2068 6572 206d 696e 6420    across her mind
0000430: 7468 6174 2073 6865 2068 6164 206e 6576    that she had nev
0000440: 6572 2062 6566 6f72 6520 7365 656e 2061    er before seen a
0000450: 2072 6162 6269 7420 7769 7468 2065 6974     rabbit with eit
0000460: 6865 7220 6120 7761 6973 7463 6f61 742d    her a waistcoat-
0000470: 706f 636b 6574 2c20 6f72 2061 2077 6174    pocket, or a wat
0000480: 6368 2074 6f20 7461 6b65 206f 7574 206f    ch to take out o
0000490: 6620 6974 2c20 616e 6420 6275 726e 696e    f it, and burnin
00004a0: 6720 7769 7468 2063 7572 696f 7369 7479    g with curiosity
00004b0: 2c20 7368 6520 7261 6e20 6163 726f 7373    , she ran across
00004c0: 2074 6865 2066 6965 6c64 2061 6674 6572     the field after
00004d0: 2069 742c 2061 6e64 2066 6f72 7475 6e61    it, and fortuna
00004e0: 7465 6c79 2077 6173 206a 7573 7420 696e    tely was just in
00004f0: 2074 696d 6520 746f 2073 6565 2069 7420    time to see it
0000500: 706f 7020 646f 776e 2061 206c 6172 6765    pop down a large
0000510: 2072 6162 6269 742d 686f 6c65 2075 6e64    rabbit-hole und
0000520: 6572 2074 6865 2068 6564 6765 2e20 496e    er the hedge. In
0000530: 2061 6e6f 7468 6572 206d 6f6d 656e 7420    another moment
0000540: 646f 776e 2077 656e 7420 416c 6963 6520    down went Alice
0000550: 6166 7465 7220 6974 2c20 6e65 7665 7220    after it, never
0000560: 6f6e 6365 2063 6f6e 7369 6465 7269 6e67    once considering
0000570: 2068 6f77 2069 6e20 7468 6520 776f 726c     how in the worl
0000580: 6420 7368 6520 7761 7320 746f 2067 6574    d she was to get
0000590: 206f 7574 2061 6761 696e 2e0a              out again..
```

Fig. 11. 31-round SHA-0: original message (part 2)

```
0000000: 6093 e793 8844 423f cf3e 4140 3479 5078   '....DB?.>A@4yPx
0000010: f8ac 0a92 7e6a 1956 d8b7 b004 1bf9 027f   ....~j.V........
0000020: 13fd 7b20 5cbd 783c 9b3d 78d2 e0bd 8106   ..{ \.x<.=x.....
0000030: fee5 2a1d 8efe 23eb 6bd8 7621 354f 0c9c   ..*...#.k.v!5O..
0000040: 9b86 3bbf 6469 db87 b11d 9195 707d 3f5a   ..;.di......p}?Z
0000050: 277b 582e 44fa 9440 a57c be61 14bc 7c39   '{X.D..@.|.a..|9
0000060: aabc 785e 3c7d 85ef 35bd 855d 1b7d 84fd   ..x~<}..5..].}..
0000070: a7d6 c497 a55a d1ae 21ea 5210 19cc f5e1   .....Z..!.R.....
0000080: b6a5 86d7 e20e 085d e7ab ab81 dd74 ffad   .......].....t..
0000090: 6a33 7421 b5cf 5fa2 c709 48b3 836d 6f2a   j3t!.._...H..mo*
00000a0: 8d3d 7e50 eefd 793c 2cbd 84ea d83d 78bc   .=~P..y<,....=x.
00000b0: 7d7b 64a9 483c 18f3 f559 a0d5 bf69 d5f8   }{d.H<...Y...i..
00000c0: 5e7d 920f 9cbe 10a2 0d5d 5bb1 453d 7b31   ^}.......][.E={1
00000d0: d03d 7f7f fe6d 019b 5fa4 fed5 fbf5 79dd   .=...m.._....y.
00000e0: 37bd 7ced ddfd 79aa 18fd 7da7 063d 8622   7.|...y...}..=."
00000f0: ece1 65d6 0372 499e 9c7c 8472 5267 8c88   ..e..rI..|.rRg..
0000100: fa9e 8747 255d a7e9 cafd 73dd b87d 3785   ...G%]....s..}7.
0000110: b63d 3c42 2e35 3292 771b 690c a41b 77f1   .=<B.52.w.i...w.
0000120: abfd 84fa d93d 8646 9c3d 7774 b23d 7c79   .....=.F.=wt.=|y
0000130: aef9 1db8 c192 413e d8ef 6d8b b39e f536   ......A>..m....6
0000140: 0fa1 c66f 3ffd 955e 6f3b c780 3265 afa6   ...o?..^o;..2e..
0000150: 76ac 6b63 fa32 6784 510b 5c5d cd0d 5413   v.kc.2g.Q.\]..T.
0000160: babd 6b15 c5fd 7cab b17d 7c12 a97d 7d5a   ..k...|..}|..}}Z
0000170: d313 a994 f376 99d2 49b4 e6df 154a 5d84   .....v..I....J].
0000180: 38a0 0a47 d12e 07c9 9065 778b 1b7d 7f34   8..G.....ew..}.4
0000190: 54bc dbfd 2cb4 96c2 0ebb 3db1 8afb 8442   T...,.....=....B
00001a0: 74bd 7b59 25fd 7951 86fd 7ff1 717d 78be   t.{Y%.yQ....q}x.
00001b0: 5357 37b3 6524 7861 6ab2 ec05 8f4c 966e   SW7.e$xaj....L.n
00001c0: ec5d 8b9f 2d7d 6fb7 f36b fba1 eb6d 7b34   .]..-}o..k...m{4
00001d0: bdc5 8179 08c5 5b61 89fd 3b15 2b7d 59ab   ...y..[a..;.+}Y.
00001e0: f07d 7fcc 36fd 7c85 3cbd 7eac 45fd 85c4   .}..6.|.<.~.E...
00001f0: 752d aeef df79 9808 a886 8285 a5dd ff34   u-...y........4
0000200: 5c8d 9e8f b2ba 8079 167d 657a c33d 43bc   \......y.}ez.=C.
0000210: 1db9 76d0 e3e9 70df 986d 7c1e 657d 8363   ..v...p..m|.e}.c
0000220: 613d 7750 3e3d 7944 fa7d 77a5 373d 7765   a=wP>=yD.}w.7=we
0000230: c560 ac62 e5b2 47dd 01fe aebe e8ac e99a   .'.b..G........
0000240: 887d 930f 5f7c 0fc3 f789 7790 de7d 7f71   .}.._|....w..}.q
0000250: b4bd 7ba9 4d3d 6c8a 1579 75b8 c439 84d2   ..{.M=l..yu..9..
0000260: 513d 7b27 a3bd 7f43 357d 7fa9 e9bd 7704   Q={'...C5}....w.
0000270: ff1d 6a35 02bd 3859 2703 d027 4915 5452   ..j5..8Y'..'I.TR
0000280: dd05 9eb7 577a 8263 01a2 a46f d8bd 5daa   ....Wz.c...o..].
0000290: eebd 72a2 21db 732a 98b3 f657 d033 fb18   ..r.!.s*...W.3..
00002a0: 987d 82f5 f2bd 7c08 2dfd 85c8 38fd 82ca   .}....|.-...8...
00002b0: 5939 ee8e 140f 5b3d 0cc9 9c81 9c92 5965   Y9....[=......Ye
00002c0: 3b9d 96af 8b47 7d9f e2ff 8392 c6ac ff71   ;....G}........q
00002d0: b5f3 81bd d482 750b 5749 f1aa 4cfc e77a   ......u.WI..L..z
00002e0: b1fd 7ead e23d 7900 aabd 7f55 3cbd 83f5   ..~..=y....U<...
00002f0: 97bb e4dd 6941 50cd 567f 37d0 3e5c 9e26   ....iAP.V.7.>\.&
```

Fig. 12. 31-round SHA-0: second preimage (part 1)

```
0000300:  7a23 d3cf cdbc 6851 fc6b 6fdc 0a73 e75c  z#....hQ.ko..s.\
0000310:  5c53 e94b c211 c83c 9d3b 59c7 77fd 7a5a  \S.K...<.;Y.w.zZ
0000320:  9afd 7b0b 883d 835f c8fd 7f30 98bd 7f34  ..{..=._...0...4
0000330:  570a e920 9bc7 4e38 9d9f 7faa 7e51 9dbd  W.. .N8....~Q..
0000340:  0f0c c697 20e5 9f98 9c99 fff8 442d 7383  .... ......D-s.
0000350:  583a 2e86 7bc5 a5a9 48e1 57da 0675 61ce  X:..{...H.W..ua.
0000360:  1a3d 78d0 23bd 7ac5 24fd 804e 473d 7aa0  .=x.#.z.$..NG=z.
0000370:  b7c3 6cdc 9ce1 2251 87d2 dbef 4739 a47c  ..l..."Q....G9.|
0000380:  9d15 92a7 4a9c bcc5 74a9 579c 41dd 7e99  ....J...t.W.A.~.
0000390:  a8db 7a99 398f 4864 1fa4 54bd 9d6c 7c8e  ..z.9.Hd..T..1|.
00003a0:  57bd 7ac7 12fd 84b9 703d 7a02 9cbd 7c37  W.z.....p=z...|7
00003b0:  f88f b361 8ec1 1971 f419 9d71 beb2 f4ca  ...a...q...q....
00003c0:  1c42 eccf 31e1 3783 3e6d bf75 3765 83a6  .B..1.7.>m.u7e..
00003d0:  41cc 5f17 c588 0436 df79 4dd9 fafd 752f  A._....6.yM...u/
00003e0:  353d 7fcc fffd 79e5 057d 7cc1 c93d 84b5  5=....y..}|..=..
00003f0:  9080 9f98 75f5 c427 c6d3 ffbb 2d55 00d0  ....u..'....-U..
0000400:  3c01 d6c7 410b 7bcd 8d7c f79e c27d 7b5c  <...A.{..|...}{\
0000410:  f6dc 7047 4bd6 6e66 2ab7 84a2 2e7d 8676  ..pGK.nf*....}.v
0000420:  b1fd 795b dbbd 7e58 043d 82bf 9b3d 836b  ..y[..~X.=...=.k
0000430:  fbc6 0485 29f2 5213 6b02 b802 3b6a 30df  ....).R.k...;j0.
0000440:  fa7d 8887 177d 4027 298e 7ba9 145b 7aed  .}...}@').{..[z.
0000450:  303d 8219 9cbd 7c5f 1cf9 36b5 b439 3dee  0=....|_..6..9=.
0000460:  b63d 76d4 9bfd 7b6c bdbd 83b8 7e3d 8463  .=v...{1....~=.c
0000470:  93b0 32ab c928 2966 29aa ae16 6ec5 9ad0  ..2..()f)...n....
0000480:  067e 86bf 306d 7b87 f77d ffb8 446d 7bcf  .~..0m{..}..Dm{.
0000490:  143d 35b6 e879 39cf d7b9 5c05 79bd 571f  .=5..y9...\.y.W.
00004a0:  2cfd 8640 4f7d 80d7 bf3d 85b5 7d7d 7e35  ,..@0}...=..}}~5
00004b0:  ef2e 8255 95e1 8361 6086 946e e1ce 3da9  ...U...a'..n..=.
00004c0:  e88c eab7 23f1 0da3 261b 7baf ce35 6bae  ....#...&.{..5k.
00004d0:  2f39 e040 12a1 a732 463d 693f d915 7566  /9.@...2F=i?..uf
00004e0:  bfbd 7d9d 853d 7bee f6bd 7d1e 1e3d 7afe  ..}..={...}..=z.
00004f0:  8ecb 8c22 62eb 7e25 7d3d fbc1 0f75 350d  ..."b.~%}=...u5.
0000500:  d281 c797 9775 6000 77df 9f95 3737 7fbb  .....u'.w...77..
0000510:  485c 79e1 0b9c 7585 0344 efea 56e4 f0e6  H\y...u..D..V...
0000520:  4b7d 78a6 2efd 7fc3 f03d 80c3 3f3d 827a  K}x......=..?=.z
0000530:  30c8 3047 1144 d3a9 104a 7c41 3947 4120  0.0G.D...J|A9GA
0000540:  49a0 8a9f 5c1d 026b e885 6374 2775 8269  I...\..k..ct'u.i
0000550:  cb7d 017c fcb4 c107 50fb 6c2e 37bb 71a6  .}.|....P.1.7.q.
0000560:  eb7d 821c d3bd 8633 6ffd 7cbd 81fd 77e7  .}.....3o.|...w.
0000570:  b2c4 fef3 1c48 7d72 136a 2995 0afe 99d5  .....H}r.j).....
0000580:  6420 7368 6520 7761 7320 746f 2067 6574  d she was to get
0000590:  206f 7574 2061 6761 696e 2e0a             out again..
```

Fig. 13. 31-round SHA-0: second preimage (part 2)

On the Power of Power Analysis in the Real World: A Complete Break of the KEELOQ Code Hopping Scheme

Thomas Eisenbarth[1], Timo Kasper[1], Amir Moradi[2,⋆], Christof Paar[1],
Mahmoud Salmasizadeh[2], and Mohammad T. Manzuri Shalmani[2]

[1] Horst Görtz Institute for IT Security
Ruhr University Bochum, Germany
[2] Department of Computer Engineering and Electronic Research Center
Sharif University of Technology, Tehran, Iran
{eisenbarth,tkasper,moradi,cpaar}@crypto.rub.de,
{salmasi,manzuri}@sharif.edu

Abstract. KEELOQ remote keyless entry systems are widely used for access control purposes such as garage openers or car door systems. We present the first successful differential power analysis attacks on numerous commercially available products employing KEELOQ code hopping. Our new techniques combine side-channel cryptanalysis with specific properties of the KEELOQ algorithm. They allow for efficiently revealing both the secret key of a remote transmitter and the manufacturer key stored in a receiver. As a result, a remote control can be cloned from only ten power traces, allowing for a practical key recovery in few minutes. After extracting the manufacturer key once, with similar techniques, we demonstrate how to recover the secret key of a remote control and replicate it from a distance, just by eavesdropping on at most two messages. This key-cloning without physical access to the device has serious real-world security implications, as the technically challenging part can be outsourced to specialists. Finally, we mount a denial of service attack on a KEELOQ access control system. All proposed attacks have been verified on several commercial KEELOQ products.

1 Motivation

The KEELOQ block cipher is widely used for security relevant applications, e.g., remote keyless entry (RKE) systems for car or building access, and passive radio frequency identification (RFID) transponders for car immobilizers [13]. In the course of the last year, the KEELOQ algorithm has moved into the focus of the international cryptographic research community. Shortly after the first cryptanalysis of the cipher [1], more analytical attacks were proposed [4,5], revealing

---

⋆ Amir Moradi performed most of the work described in this contribution as a visiting researcher at Ruhr University Bochum.

D. Wagner (Ed.): CRYPTO 2008, LNCS 5157, pp. 203–220, 2008.
© International Association for Cryptologic Research 2008

mathematical weaknesses of the cipher. The best known analytical attacks target the identify friend or foe (IFF) mode of KEELOQ and require at least 2^{16} plaintext-ciphertext pairs from one transponder. This allows, after several days of computations, for a simple cloning of the transponder and, only in case of a weak key derivation method[1], for obtaining the manufacturer key that is required to generate keys for new valid transponders. Despite the impressive contribution to the cryptanalysis of the cipher, the real-world impacts of the previous attacks are somewhat limited, as described in Sect. 2.3.

Motivated by the ongoing research we investigate the vulnerability of actual KEELOQ implementations with respect to side-channel analysis (SCA), in order to evaluate the security of all KEELOQ modes (IFF and code hopping) and all key derivation schemes. As a result, we present three very practical key recovery attacks and a denial of service attack with severe implications for RKE systems that are currently used in the field. These new attacks — which combine differential power analysis (DPA) with the extend-and-prune strategy of [3] — can be applied to various implementations of KEELOQ. In particular, we have been able to successfully attack hardware realizations, i.e., the Microchip HCSXXX family of integrated circuits (ICs), as well as software implementations running on Microchip PIC microcontrollers. In contrast to the hitherto existing attacks, the techniques proposed by us are also applicable in case of more sophisticated key derivation schemes (cf. Sect. 2.2) and are suitable for breaking both the KEELOQ code hopping mode and the IFF mode.

Since the introduction of DPA in 1999 [6], it has become an established method for extracting cryptographic keys from security devices by exploiting power consumption traces. However, almost ten years later, there is a surprising discrepancy between the well established theory of power analysis (cf., e.g., the CHES workshop proceedings since 1999) and the very few, if any, confirmed DPA attacks on real-world security systems. The targets considered in the literature are often home-made or known implementations on platforms that are well-known to the attacker, and are typically examined in an ideal environment [16,9,14], for example with an artificially generated trigger signal. The practical relevance of such a white box cryptanalysis for real-world realizations of cryptography sometimes remains an open question. During our investigations, we were confronted with a known cipher, but with a black box implementation, i.e., no knowledge or information about the devices except for the characterization in the data sheet. This demanded for some extra efforts and reverse engineering of the unknown targets. Despite these obstructions, we were able to mount highly effective attacks with considerable implications on the security of commercial KEELOQ code hopping systems.

The remainder of this contribution is structured as follows. After an introduction to the KEELOQ cipher and its key derivation schemes in Sect. 2, we elaborate in Sect. 3 on how the secret key of a transmitter can be revealed using SCA with as few as ten power traces and only minutes of computation time.

---

[1] If the key of the transmitter is derived from XORing a simple function of the device serial number with the manufacturer key, the latter can easily be obtained.

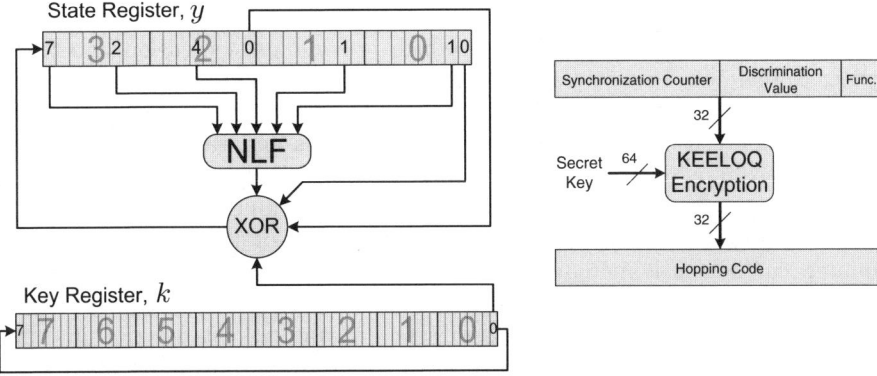

Fig. 1. Block diagram of the KEELOQ encryption

Fig. 2. Generation of KEELOQ hopping codes

Similarly, the manufacturer key used in a receiver is obtained in less than one day. In Sect. 4, we describe several real-world attacks which follow from the key extraction. First, remotes which are in the possession of an attacker can be cloned. The most devastating attack allows to recover the secret key of a transmitter from a distance, just by eavesdropping on at most two hopping code messages. It is perceivable that a technically experienced person (with malicious intent) will develop a machine that allows for automatically spoofing KEELOQ code hopping systems. With such a machine, a completely unskilled attacker could gain access to objects that are protected by these systems without leaving any traces. Finally, we detail on putting an RKE system out of service which would prevent the legitimate owner to open a car door or to access a garage. All our attacks have been extensively tested and verified. We present various experimental results and provide figures for power analysis both based on electric current and the electromagnetic (EM) emanation of different KEELOQ devices.

2   Background

KEELOQ is a block cipher with a 64 bit key and a block size of 32 bits. As illustrated in Fig. 1, it can be viewed as a non-linear feedback shift register (NLFSR) where the feedback depends linearly on two register bits, one key bit, and a non-linear function (NLF). The NLF maps five other register bits to a single bit [1,4,5]. Prior to an encryption, the secret key and plaintext are loaded in the key register and the state register, respectively. In each clock cycle, the key register is rotated to the right and the state register is shifted to the right so that the fresh bit prepared by the XOR function becomes part of the state. After 528 clock cycles, the state register contains the ciphertext. The decryption process is similar to the encryption, except for the direction of the shifts and the taps for the NLF and the XOR function.

2.1   Code Hopping Protocol

In addition to KEELOQ IFF systems which provide authentication of a transmitter to the main system using a simple challenge-response protocol, KEELOQ is used in code hopping (or rolling code) applications [10]. In this mechanism, which is widely used, e.g., in car anti-theft systems and garage door openers, the transmitter is equipped with an encoder and the receiver with a decoder. Both share a secret key and a fixed discrimination value, *disc*, with 10 or 12 bits. In addition, they are synchronized with a 16 bit or 18 bit synchronization counter, *cnt*, which is incremented in the encoder each time a hopping code is transmitted. According to Fig. 2, the transmitter constructs a hopping code by encrypting a 32 bit message formed of *disc*, *cnt* and a 4 bit function information. The latter determines the task desired by a remote control, for instance, it enables to open or close more than one door in a garage opener system.

One message sent via the radio frequency (RF) interface consists of a hopping code followed by the serial number of the transmitter. The receiver decrypts the hopping code using the shared secret key to obtain *disc* and the current *cnt*. The transmitter is authenticated if *disc* is identical to the shared one and *cnt* fits in a window of valid values. Three windows are defined for the counter. If the difference between a received *cnt* and the last stored value is within the first window, i.e., 16 codes, the intended function will be executed after a single button press. Otherwise, the second window containing up to 2^{15} codes[2] is examined. In this so-called resynchronization window, the desired function is carried out only if two consecutive counter values are within it, i.e., after pressing the button twice. The third window contains the rest of the counter space. Any transmission with a *cnt* value within this window will be ignored, to exclude the repetition of a previous code and thus prevent replay attacks.

2.2   Key Derivation Schemes

There are two types of keys involved in a typical KEELOQ application. The device key is unique for each remote control and is shared by the transmitter and the receiver. It is established during a learning phase. The manufacturer key is mainly used for deriving device keys. It is to our knowledge identical for all receivers of a given manufacturer and hence enables producing transmitters that cannot be cloned by competitors. Since the manufacturer's key is critical for the security of the product, it is stored in a read protected memory of the receiver. The known key derivation schemes are reviewed in the following:

(a) According to Fig. 3.a, the device key is obtained by two KEELOQ decryptions. The two functions F_1 and F_2 (which are usually simple paddings) are applied to the serial number of the transmitter to form the plaintexts for the decryptions.

---

[2] These window sizes are recommended by Microchip, but they can be altered to fit the needs of a particular system.

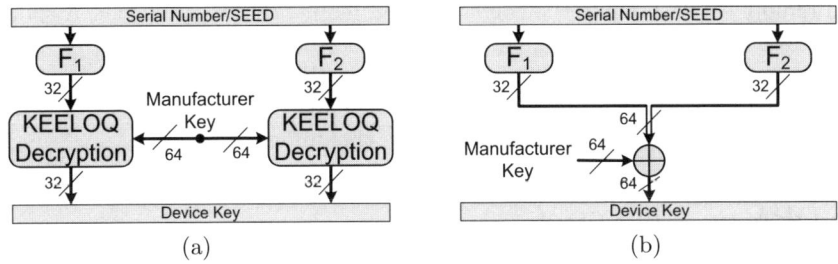

Fig. 3. Key derivation schemes

(b) The next key derivation scheme is similar to the previous one, except for a randomly generated seed value which is stored in the transmitter and is used instead of the serial number to generate the device key. During the learning phase, a transmitter can be forced to send its seed value.

(c) As presented in Fig. 3.b, sometimes the device key is generated from an XOR of a simple function of the serial number with the manufacturer key.

(d) The last scheme is similar to the third one. The device key is derived from an XOR of the manufacturer key and a simple function of the seed value of the transmitter.

Note that a manufacturer may develop a proprietary key derivation scheme not included in the above list.

2.3   Previous Work

The first two cryptanalytical attacks on the KEELOQ algorithm were published by Bogdanov [1]. One attack is based on slide and guess-and-determine techniques and needs about $2^{50.6}$ KEELOQ encryptions. The other one additionally uses a cycle structure analysis technique and requires 2^{37} encryptions. However, both attacks require the entire codebook, i.e., all 2^{32} plaintext-ciphertext pairs. Courtois *et al.* [4] proposed two attacks. One is a slide-algebraic attack demanding for $2^{51.4}$ KEELOQ encryptions and 2^{16} known plaintext-ciphertext pairs. The second slide attack can be carried out knowing almost the entire codebook. It reveals the secret key with a complexity of approximately 2^{27} KEELOQ encryptions. Recently, Indesteege *et al.* presented more practical attacks on the KEELOQ cipher [5]. All of them are based on slide and meet-in-the-middle attacks. The best one uses 2^{16} known plaintext-ciphertext pairs and has a complexity of $2^{44.5}$ KEELOQ encryptions. It can find the secret key in two days using 50 dual core computers.

The above attacks are applicable to KEELOQ IFF systems but they cannot be directly applied to the code hopping mode [10], which appears to be the dominant commercial application of KEELOQ. The required minimum of 2^{16} plaintext-ciphertext pairs cannot be obtained in case of a code hopping system, because an adversary has only access to the ciphertexts that are transmitted by a remote control, while the corresponding plaintexts are unknown. Although knowing a sequence of 2^{16} ciphertexts and the discrimination value of a code hopping encoder would be sufficient to perform the attack described in [5], the commercial products employing the KEELOQ code hopping protocol, i.e., HCS modules, do not allow an attacker to access this information.

3   DPA on KEELOQ

When we started to analyze the targets using KEELOQ, we were exposed to a "classical" situation for physical attacks: even though the algorithm was known, hardly anything was known about the implementation. We found that the transmitters usually employ HCSXXX modules of Microchip, featuring a hardware implementation of the cipher. The receivers we looked at are typically equipped with a read-protected PIC microcontroller on which a KEELOQ decryption routine is implemented in software. This section explains the details of DPA-attacking transmitters and receivers, starting with a general approach that is appropriate for both types of realizations.

Initial Cipher Analysis. Before being able to actually perform a DPA on a particular implementation of a cipher, one needs to make certain assumptions about the leakage produced by it. Then, a DPA scheme for exploiting that leakage must be developed, which depends on the cipher structure as well as on the particularities of the given implementation.

Measurement. The power traces are gathered by measuring the current via a shunt resistor connected to the ground pin of the target chip. In addition, we acquire the EM radiation of the device by means of near field probes. For convenience, we built a printed circuit board (PCB) that allows for emulating KEELOQ chips and for controlling a transmitter from a PC so that a measurement sequence can be executed automatically. The power traces were acquired using an Agilent Infiniium 54832D digital oscilloscope with a maximum sampling rate of 4 GS/s.

Data Pre-Processing and Alignment. One problem of aligning the power traces of an unknown implementation is the absence of a suitable trigger signal. The solution for this is target-specific and detailed in Sect. 3.2 and Sect. 3.3 for transmitters and receivers, respectively. Another problem is that all of the target devices are clocked by a noisy internal oscillator. Hence we had to find a way to remove the clock jitter. We know that most of the data-dependent leakage occurs in the instant when the registers are clocked, producing peaks with varying amplitudes in each clock period. The amplitudes of these peaks directly correspond to the dynamic power consumption of the target circuit and thus hold most of the relevant information. Accordingly, we extract the peaks from the power consumption in software, and base our DPA attack solely on the amplitudes of the peaks. This peak extraction step has two advantages for the subsequent analysis: (i) the amount of data is greatly reduced, which facilitates the post-processing and the data storage, and (ii) more importantly, the peak extraction allows for an accurate alignment of the traces. Other methods for removing the clock jitter, such as Fourier transform, filtering, etc., turned out to be less effective and more complicated.

Developing and Performing the DPA. After peak extraction and alignment steps, the traces can be processed by the DPA algorithm. For the transmitter modules we only knew the ciphertext and hence had to perform our attacks starting from the last round of the encryption. For the software implementation of the PICs we knew the plaintexts and started the attack of the first round of the decryption.

3.1   Building a Powerful DPA for KEELOQ

It is known that for successfully performing a DPA attack, some intermediate value of the cipher has to be identified that (i) depends on known data (like the plaintext or the ciphertext), (ii) depends on the key bits, and (iii) is easy to predict. Furthermore, it is advisable to choose a value that has a high degree of nonlinearity with respect to the key, to avoid so-called "ghost peaks" for "similar" keys [2]. For every DPA, a model for estimating the power consumption is needed. Compared to the two shift registers, the power consumption of the combinational part, i.e., a few XORs and the 5×1 non-linear function, is small and can be neglected. Note that the Hamming distance of the key register does not change, since the key is simply rotated. This leads to a theoretically constant power consumption of the key register in each clock cycle. Hence, we focus on the state register $\boldsymbol{y}$. We execute a correlation DPA attack (CPA) [2] based on the following hypothetical power model

$$\mathrm{P}_{Hyp}^{(i)} = \mathrm{HD}\left(\boldsymbol{y}^{(i)}, \boldsymbol{y}^{(i-1)}\right) = \mathrm{HW}\left(\boldsymbol{y}^{(i)} \oplus \boldsymbol{y}^{(i-1)}\right) \tag{1}$$

where $\mathrm{P}_{Hyp}^{(i)}$ denotes the hypothetical power consumption in the ith round, HD and HW are Hamming distance and Hamming weight, respectively, $\boldsymbol{y}^{(i)}$ indicates the content of the state register in the ith round, and $\oplus$ is a 32 bit XOR function. As mentioned before, the known ciphertext attack on the encryption is identical to the known plaintext attack on the decryption[3]. We describe the known ciphertext attack on the encryption. Starting from the 528th round, 32 bits of the final state $\boldsymbol{y}^{(528)} = \left(y_0^{(528)}, \ldots, y_{31}^{(528)}\right)$, are known. Furthermore, 31 bits of $\boldsymbol{y}^{(527)}$, i.e., $\left(y_1^{(527)}, \ldots, y_{31}^{(527)}\right)$, are known because they are identical to $\left(y_0^{(528)}, \ldots, y_{30}^{(528)}\right)$. Therefore, just $y_0^{(527)}$ is unknown. According to Fig. 1, we can write

$$y_{31}^{(i+1)} = k_0^{(i)} \oplus y_{16}^{(i)} \oplus y_0^{(i)} \oplus \mathrm{NLF}\left(y_{31}^{(i)}, y_{26}^{(i)}, y_{20}^{(i)}, y_9^{(i)}, y_1^{(i)}\right) \tag{2}$$

---

[3] Both attacks target state $\boldsymbol{y}^{(l)}$ of the decryption, which is the same as state $\boldsymbol{y}^{(528-l)}$ of the encryption.

where $k_0^{(i)}$ is the rightmost bit of the key register in the ith round. Knowing that $k_j^{(i)} = k_{(i+j) \bmod 64}$, we can rewrite Eq. (2) as

$$y_0^{(527)} = k_{15} \oplus y_{16}^{(527)} \oplus y_{31}^{(528)} \oplus \text{NLF}\left(y_{31}^{(527)}, y_{26}^{(527)}, y_{20}^{(527)}, y_9^{(527)}, y_1^{(527)}\right) \quad (3)$$

Thus, recovering $y_0^{(527)}$ directly reveals one bit of the key register. This process is the same for recovering the LSB of the state register of the previous rounds, i.e., $y_0^{(i)}$, $i = (526, 525, \ldots)$. However, Eq. (3), depends linearly on the key bit k_{15}. Above we stated that nonlinearity helps distinguishing correct key hypotheses from wrong ones. Hence, recovering the key bit-by-bit might not be the best choice[4]. Fortunately, according to Fig. 1, the LSB of the round state, $y_0^{(i)}$, enters the NLF leading to a nonlinear relation between the key bit k_{15} and the state $\boldsymbol{y}^{(526)}$. Accordingly, the nonlinearity for one key bit k_j increases in each round after it was clocked into the state.

Algorithm 1. A Scalable DPA for KEELOQ

Input: m : length of key guess, n: number of surviving key guesses, k: known previous
  key bits
Output: SurvivingKeys
  1: KeyHyp ← **all** $\{0,1\}^m$
  2: **for all** KeyHyp$_i$; $0 \leq i < 2^m$ **do**
  3:     Perform CPA on round $(528 - m)$ using P_{Hyp} and k
  4: **end for**
  5: SurvivingKeys ← n most probable partial keys of KeyHyp

Taking the increased nonlinearity in the successive rounds into account, we developed a scalable DPA, as described in Alg. 1, that allows for finding a subset n of surviving key candidates by guessing m bits of the key in an instant. Note that in step 3 of the algorithm the CPA is performed on round $(528 - m)$, hence taking advantage of a key bit passing the NLF m times. The significance of the known previous bits k will become clear below in the extended attack (Alg. 2), where Alg. 1 is executed repeatedly.

We performed simulations of the attack described in Alg. 1, assuming a Hamming distance leakage model. The simulated traces allow for testing our attacks and also to evaluate how well an attack would work under "perfect" conditions. We generated a set of encryption traces with random plaintext input and computed the Hamming distance of all registers for each round. We performed a correlation DPA where we predicted the Hamming distance of the state register of round 522, $\text{P}_{Hyp} = \text{HD}\left(\boldsymbol{y}^{(522)}\right)$. Fig. 4 shows the correlation for the $2^6 = 64$ key hypotheses over the first few rounds. Of course, the correlation is 1 for the

---

[4] Simulations show that an attack recovering the key bit by bit is much weaker than an attack that recovers several key bits at a time. Still, the key can also be recovered for single bit key guesses – in other words even a classical DPA on the LSB of the state register is feasible.

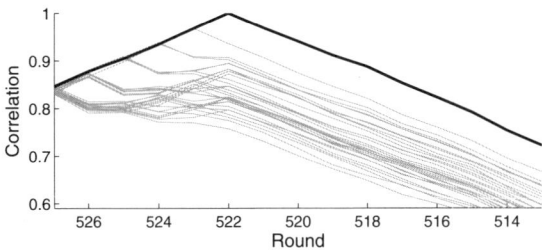

Fig. 4. Simulated correlation of key hypotheses as a function of KEELOQ rounds. Correct key guess (black solid line) vs. wrong key guesses (thin gray lines).

right key (thick solid line) in round 522. Unfortunately, some of the wrong key guesses (thin gray lines) also yield a high correlation. This is due to the high linearity between both the state and the key guesses, and between the different states. Furthermore we get a high correlation in the rounds before and after the predicted round. This is because most of the bits of the shift register remain unchanged in the nearby rounds. The most probable wrong key guess is always the one that differs only in the LSB. This underlines our expectation that the linearity increases the error probability of guessing the less significant key bits.

Algorithm 2. Pruning for the Best Key Hypothesis

Input: m : length of key guess, n: number of surviving key guesses
Output: K: recovered key
 1: $K \leftarrow$ Algorithm $1(m, n, \emptyset)$
 2: **for** $round = 1$ to $\lceil \frac{64}{m} \rceil$ **do**
 3:     $K' \leftarrow \emptyset$
 4:     **for all** $k_i \in K, 0 \leq i < n$ **do**
 5:         $K' \leftarrow K' \cup$ Algorithm $1(m, n, k_i)$
 6:     **end for**
 7:     $K \leftarrow n$ most probable keys of K'
 8: **end for**
 9: **return** K

To improve the strength of our attack and to take care of the misleading high correlations, we added another attack step. Alg. 1 can be repeated to guess all partial keys, one after the other. These iterations of the attack need to be done one after another, because we require the previous key bits and thus the state y as a known input for each execution of the algorithm. Since some of the bits of the previous key guess might be faulty, we keep a number n of the most probable partial key guesses as survivors. Wrong surviving candidates of the previous round will result in a misleading initial state y for the following attack round and hence strongly decrease the correlation of subsequent key guesses. This does not only allow for an assertion of the correct previous key guesses, but also for detecting faulty previous keys. Hence, the attack has an error-correcting property. If all key guesses of one round show a low correlation, we can go one step back and broaden the number of surviving key guesses n. Alg. 2 describes this procedure, which is similar to the "pruning process" described by Chari *et al.* in [3]. In the last round ($i = \lceil \frac{64}{m} \rceil$) the program verifies whether an error occurred and the key with the highest correlation coefficient is selected out of the n surviving keys. It will be shown in the following subsections that Alg. 2 results in a quite strong attack.

3.2   Details of the Hardware Attack

For attacking commercial KEELOQ code hopping encoders we first had to find the points in time in the power traces (Fig. 5.a) that correspond to the encryption function. We found that the encryption happens after writing to the EEPROM[5], i.e., in the time interval between 20.5 ms and 24 ms (Fig. 5.b). The power traces reveal that the frequency of the internal oscillators of the ICs is approximately 1.25 MHz.

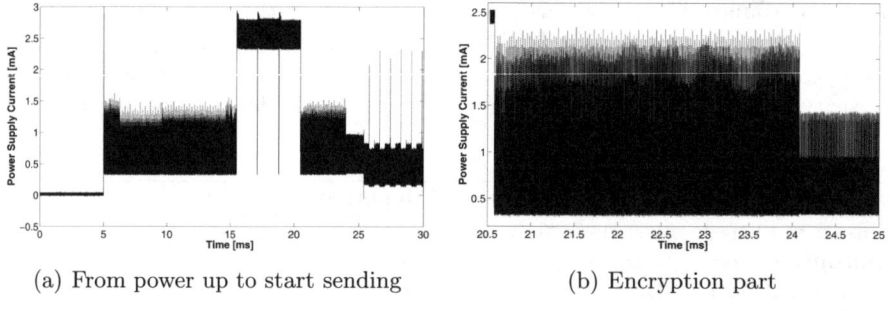

(a) From power up to start sending   (b) Encryption part

Fig. 5. Power consumption traces of a HCS module

We modified the attack described in Sect. 3.1 to correlate all known and predicted rounds to the corresponding power peaks. This is possible since we are able to locate the leakage of each round. The modified attack was performed on HCS200, HCS201, HCS300, HCS301, HCS361, HCS362, and HCS410 [11,12] in both DIP and SOIC packages. In the best case we were able to recover the secret key of DIP package ICs from only six power traces when sampling at a rate of 200 MS/s. At most 30 power traces are sufficient to reveal the secret key of an HCS module in an SOIC package, which has a lower power consumption, resulting in a worse signal-to-noise ratio (SNR) of the measurements. Fig. 6 compares the correlation coefficients of the correct key of HCS201 chips in DIP and SOIC packages as a function of the number of traces. The sudden increase of the correlation is due to the error-correcting property of our attack, and also due to the fact that we repeated the attack for all 528 rounds of the algorithm in order to verify the revealed key.

---

[5] The high amplitude periods of the power trace correspond to writing to the internal EEPROM.

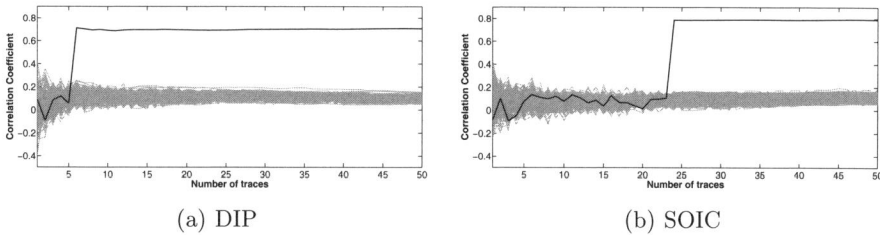

(a) DIP   (b) SOIC

Fig. 6. Correlation coefficients of key hypotheses of HCS201 ICs as a function of the number of measured traces

We repeated the above experiments with an EM near field probe RF-U 5-2 [8] to directly monitor the electromagnetic emanation, instead of measuring the electric current via a shunt resistor. The probe was positioned close to the ground pin of the HCS201 IC in a DIP package, in order to acquire the peaks of the EM field that are caused by the change of electric current. Compared to inserting a resistor in series to the device, this differential electromagnetic analysis (DEMA) can be regarded as non-invasive, as no modification of the PCB is necessary. Contrary to our assumption that the SNR would suffer from environmental noise and thus much more traces would be required to recover the key, the results obtained and the number of traces needed are very comparable to the case of power traces acquired by means of a resistor (Fig. 5). In the best case, we succeeded with recovering the key after only 10 DEMA measurements.

To estimate the minimum technical requirements for the SCA, we performed experiments with varying sampling rates and evaluated the number of power traces required for recovering the correct key. Fig. 7 shows the results for attacking a HCS201 chip in a DIP package in the case of current measurements via a resistor. We conclude that our attack can be carried out effectively even with low-cost equipment, e.g., an oscilloscope with a maximum sample rate as low as 50 MS/s enables finding the secret key from only 60 power traces.

3.3 Details of the Software Attack

The next target of our attack is the code hopping decoder implemented in the receiver. We recall that the receiver contains the manufacturer key, which is an attractive target for a complete break of the system. A PIC microcontroller handles the key management, controls for instance the motor of the garage door or the locking system of the car, and performs the KEELOQ decryption in software.

Receivers usually offer a so-called "learning mode". In this learning mode the user can register new transmitters to cooperate with the receiver. We were able to identify the key derivation scheme of the target receiver as scheme (a) of Sect. 2.2. Hence we can recover the manufacturer key k_M by performing a DPA key recovery on the KEELOQ decryption that is performed during learning mode.

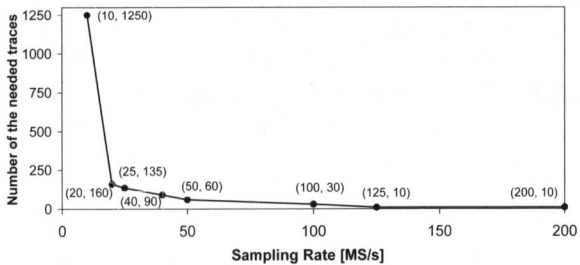

Fig. 7. Number of measurements required for revealing the secret key of a HCS201 IC in a DIP package as a function of the sampling rate. The numbers in parentheses give the exact coordinates of the points.

Before executing the DPA, we adapted the power model of the attack of Sect. 3.1 to a PIC software implementation. Typically, PIC microcontrollers leak the Hamming weight of the processed data [15]. Furthermore, one can assume that the state is stored in the 8 bit registers of the PIC microcontroller which are regularly accessed. Hence, instead of predicting the Hamming distance HD $\left(y^{(i)}, y^{(i-1)}\right)$ of the whole state – as was done for the hardware attack in Sect. 3.2 – we predict the Hamming weight of the least significant byte (LSB) of the KEELOQ state register:

$$P_{Hyp}^{(i)} = \text{HW}\left(y_{\text{LSB}}^{(i)}\right) = \sum_{k=0}^{7} y_k^{(i)}$$

We performed the attack by putting the receiver into learning mode and sending hopping code messages with random serial numbers to the receiver[6]. Lacking any information in the power consumption of the PIC that could have been used as trigger, we triggered the scope directly after transmitting the last bit via the RF interface. This results in our traces not being well-aligned, leading to a high number of power samples needed to perform a successful DPA attack.

While performing the attack we noticed that the correlation coefficient of the correct key become continuously worse with an increasing number of rounds. For the first few key bits, 1000 traces sampled at 125 MS/s are sufficient to find the key. Surprisingly, we need roughly ten times as many traces for recovering the full 64 bit key. This gradual decrease of the correlation is due to a misalignment that occurs during the execution of the KEELOQ algorithm. Hence, the problem is not a bad trigger condition, since the trigger affects all time instances in the same way. We assume that the program code is likely to have a data-dependent execution time for each round of KEELOQ, causing the increasing misalignment with an increasing number of rounds, and hence complicating the SCA.

---

[6] We emulated a remote control by connecting the RF interface of a transmitter to the parallel port of a PC.

4   Attacks and Implications

In the previous section we showed how the keys of hardware and software implementations of KEELOQ can be recovered by a skilled adversary using SCA. We will now evaluate the vulnerability of real-world systems to our attacks and illustrate the implications. We detail four different attack scenarios, which allow for breaking basically any system using KEELOQ with modest efforts. We focus on code hopping applications, since they are more commonly used and, due to the lack of known plaintexts, harder to cryptanalyze than IFF systems. Still, IFF systems are just as vulnerable to our DPA attacks as the code hopping devices. Some of the transmitters we analyzed even offer both operating modes. The success of some of our attacks depends on the knowledge about the particular key derivation scheme, as described in Sect. 2.2. However, they are appropriate for all key derivation schemes we are aware of.

Note that for all the attack scenarios described below it is very difficult, e.g., for an insurance or a prosecutor, to find evidence that a crime has been committed, as typically no traces are left when electronically circumventing an RKE system.

4.1   Cloning a Transmitter

For cloning a transmitter using power analysis, an adversary needs physical access to it to acquire at least 10 to 30 power traces. Hence, the button of the remote control has to be pressed several times, while measuring the power consumption and monitoring the transmitted hopping code messages. After recovering the device key k_{Dev} with the side-channel attack described in Sect. 3.2, the recorded messages can be decrypted, disclosing the discrimination and counter values of the original transmitter at the time of the attack. Now, the HCS module of a spare remote control can be programmed with the serial number, counter value and discrimination value of the master. Consequently, the freshly produced transmitter appears to be genuine to a receiver and allows for accessing the same target as the original.

Implications. This attack applies to scenarios in which a transmitter is handed over to an untrustworthy person for some minutes, e.g., car rental or cleaning personnel. While possessing the transmitter, an attacker could clone it for future reuse. Depending on the time interval between recovering the key and using the reproduced remote control, the attacker will have to press the button of the transmitter several times, for catching up with the counter value in the receiver which might have been increased meanwhile by the legitimate operator.

Given that a technically demanding SCA has to be carried out in order to clone just one remote control, and physical access to it is required, it can be speculated that the impact of this attack is limited. The cost-benefit ratio is typically too low, except for the case that very high monetary values are involved, e.g., rental of expensive cars. In most cases it is easier, e.g., to smash a window, unless an attacker intends to remain unnoticed or to gain access to an object repeatedly. It is important to stress that in most modern cars the door access mechanism and the immobilizer are separate systems. Thus, even if a car can be opened with our attack, this does not imply that a criminal can equally easily drive away with it.

4.2   Recovering a Manufacturer Key

The key recovery of the manufacturer key k_M depends on the particular key derivation scheme. If scheme (c) or (d) of Sect. 2.2 is used, i.e., an XOR of a known input and the manufacturer key k_M, disclosing the latter is trivial. After a successful key recovery attack on one transmitter (see above) of the same brand, k_M is found by reversing the XOR function. The known input is either part of each hopping code message, in case of the serial number, or can be obtained from the remote control, in case of a seed. The derived k_M can be verified with a second transmitter.

An adversary targeting the manufacturer key k_M for scheme (a) or (b) of Sect. 2.2 requires physical access to one receiver of that manufacturer. The key of the KEELOQ decryption performed inside the receiver during the learning phase is recovered with an SCA requiring several thousand power traces, as described in Sect. 3.3. The k_M obtained from the DPA can be verified with a single hopping code message.

Knowing the manufacturer key k_M, valid device keys for producing transmitters with arbitrary serial numbers can be generated, just by applying the key derivation. The counterfeited remote controls will be recognized as genuine by all receivers of that manufacturer.

Implications. The described key recovery requires access to a transmitter (key derivation by XOR) or a receiver (key derivation by KEELOQ) of the manufacturer to be attacked and, in case of a key derivation employing KEELOQ, a very skilled adversary for performing the SCA. The RKE devices can be simply purchased, e.g., from a hardware store, and even returned after extracting the key. The recovered k_M does not directly permit to unauthorizedly open a door secured by KEELOQ, because the newly produced transmitters need to be registered by the receiver first, implying physical access. Still, the described key recovery is highly relevant in the context of product piracy: the economic function of k_M is customer retention, e.g., a business model could comprise making the main profit on selling spare transmitters that operate with the receivers of that manufacturer. Without knowledge of the manufacturer key, valid transmitters cannot be produced to work with the receiver. However, a competitor possessing k_M could produce transmitters compatible to those receivers, or even produce whole RKE systems bearing the brand and being compatible to those of the original manufacturer, and hence severely affect the business of the latter. Due to the depicted high economic impact, it is very conceivable that this attack will be carried out by criminals sooner or later. In the worst case, this might result in publicly available lists of manufacturer keys on the web.

4.3   Cloning a Transmitter without Physical Access

Knowing the manufacturer key k_M, e.g., by performing the previous attack, and the key derivation method of a target device family, a remote control can be cloned by eavesdropping. The adversary has to intercept at most two hopping code messages, c_1 and c_2, sent by the target transmitter of the same brand. The process of finding the secret key of the eavesdropped transmitter depends on the key derivation schemes detailed in Sect. 2.2.

If the key is derived from the serial number of the transmitter (schemes (a) and (c)), finding its device key is straightforward, since the intercepted messages contain the serial number. The adversary can simply perform the key derivation using the known manufacturer key to obtain the device key. After decrypting one message c_i and thereby disclosing the current counter value, the adversary is able to generate valid hopping code messages for spoofing the receiver and gain access to a protected site. The computational complexity of this attack is a single KEELOQ decryption.

However, if a seed value is used during the key derivation (schemes (b) and (d)), recovering the secret key of the eavesdropped transmitter is more difficult. An exhaustive search needs to be performed to find the seed value. For recovering k_{Dev}, the adversary calculates $k_{Dev}^{seed} = \text{KeyDerivation}(k_M, seed)$ for each possible value of *seed* and decrypts the two intercepted messages c_1 and c_2 using k_{Dev}^{seed}:

$$(cnt_1, disc_1) = \text{KEELOQ}^{-1}\left(c_1, k_{Dev}^{seed}\right) \quad (cnt_2, disc_2) = \text{KEELOQ}^{-1}\left(c_2, k_{Dev}^{seed}\right)$$

Once both messages have the same discrimination value, i.e., $disc_1 = disc_2$, and similar counter values[7] cnt_1 and cnt_2, the correct device key is found[8].

There are three different seed sizes possible for KEELOQ systems, depending on the chip family. If a 32 bit seed value is used (e.g., HCS200), the adversary has to run on average 2^{32} KEELOQ decryptions to find the correct seed. According to our software implementations, this takes less than two hours on a 2.4 GHz Intel Core2 Quad PC. On a special-purpose computing machine such as COPACOBANA [7], the correct 32 bit seed value and hence the key can be recovered in just one second. In case of a 48 bit seed value (e.g., HCS360) it is not promising to recover the correct seed value using standard PCs. Still, it is possible to perform the 2^{48} required KEELOQ decryptions on average in about nine hours using COPACOBANA. However, chips like the HCS410, using a 60 bit seed, are not vulnerable to this attack. Running 2^{60} KEELOQ decryptions is not feasible in a reasonable time with currently existing equipment. We would like to mention that none of the real-world KEELOQ systems we analyzed used *any* seed. Moreover, if physical access to the transmitter is given, even 60 bit seed values are obtained by pressing one button.

---

[7] "Similar" counters means that the difference $cnt_2 - cnt_1$ is less than a small threshold, e.g., 16, depending on the period between the two eavesdrops.

[8] With a small probability we get false positives. These can easily be identified by sending one message to the target receiver.

Implications. This attack has the most devastating impact and it scales very well. It affects all KEELOQ RKE systems of a manufacturer, as soon as the respective k_M is known. Extracting k_M, as described in Sect. 4.2, can be out-sourced to criminal cryptographers who may construct (and sell) an easy-to-use machine that eavesdrops on two messages, automatically recovers the device key, and opens the target. Thus it enables a completely unskilled person to gain illicit access to objects secured with KEELOQ, at little cost and without leaving any traces. Building such an eavesdropping device is simple once the manufacturer key is available. It is sufficient to connect a legitimate receiver to a (laptop) PC and to monitor hopping codes from a range of up to several hundred meters, depending on the targeted RKE system.

4.4   Denial of Service

As detailed in Sect. 2.1 and in [10], the counters of a receiver and a transmitter are synchronized with every valid hopping code message received. This behavior can be exploited for putting an RKE system out of operation. We assume that the adversary has recovered the device key k_{Dev} of a target transmitter, e.g., by performing one of the attacks described above, and is thus able to generate valid hopping code messages. She sets the counter value to the maximum value inside the resynchronization window and sends two consecutive valid hopping codes. The receiver resynchronizes its counter to the new value. Consequently, the counter of the original transmitter is now considered to be outdated and the respective hopping codes are ignored by the receiver. Finally, the owner of the original transmitter needs to press the button very often, namely 2^{15} times, to increase the counter value back into the first window, where the transmitter produces valid hopping code messages.

Implications. This attack allows for locking out a legitimate user, leaving the impression that the KEELOQ RKE device is out of service. It can be performed by an unskilled adversary in the following scenario. Similarly to the eavesdropping device mentioned in Sect. 4.3, a spare transmitter enables a standard PC to transmit self-generated hopping codes. The program code for generating valid hopping codes could be provided by a skilled criminal, e.g., via the internet. Hence, this attack can have dramatic consequences, especially for the automotive industry, where reliability is of paramount importance. Apart from compromising the corporate image, the additional costs for increased customer support, e.g., fixing the spoofed devices, have to be considered.

5   Conclusion

We presented the first successful DPA and DEMA attacks on KEELOQ imple-mentations that can be applied to both IFF and code hopping devices. We demonstrated new techniques to reveal secret keys from commercial hardware and software implementations of KEELOQ. By applying these attacks, we illus-trated how to conduct a complete break of the KEELOQ code hopping scheme.

We revealed a manufacturer key from a receiver using a few 1000 power traces, and recovered the device key of a remote control with as few as 10 traces. We found that the investigated hardware implementations show an almost perfect leakage and hence constitute an easier target for SCA than the software implementations, whose data-dependent time variations made the DPA more difficult.

Analyzing real-world applications of KEELOQ and taking into account several key derivation schemes, we developed an eavesdropping attack that allows for cloning a transmitter from a distance, i.e., by intercepting at most two hopping code messages. This attack could be prevented if a 60 bit seed value, with good random properties, would be used for the key derivation. In addition, we introduced a denial of service attack for the KEELOQ code hopping mode. Both attacks can be performed by a completely unskilled person, as the technically challenging part, including the key extraction by means of SCA, needs to be conducted only once for each manufacturer and can thus be outsourced to criminal cryptographers.

This contribution shows that widespread commercial applications, claiming to be highly secure, can be practically broken with modest cost and efforts using SCA attacks. Thus, physical attacks must not be considered to be only relevant to the smart card industry or to be a mere academic exercise. Rather, effective countermeasures against side-channel attacks need to be implemented not only in high-value systems such as smart cards, but also in other embedded security applications.

Acknowledgements

We would like to thank Andrey Bogdanov for helpful discussions regarding the KEELOQ SCA attack, and for bringing cryptanalysis of KEELOQ to our attention.

References

1. Bogdanov, A.: Attacks on the KeeLoq Block Cipher and Authentication Systems. In: 3rd Conference on RFID Security 2007 (RFIDSec 2007), http://rfidsec07.etsit.uma.es/slides/papers/paper-22.pdf
2. Brier, E., Clavier, C., Olivier, F.: Correlation Power Analysis with a Leakage Model. In: Joye, M., Quisquater, J.-J. (eds.) CHES 2004. LNCS, vol. 3156, pp. 16–29. Springer, Heidelberg (2004)
3. Chari, S., Rao, J., Rohatgi, P.: Template Attacks. In: Cryptographic Hardware and Embedded Systems-Ches 2002: 4th International Workshop, Redwood Shores, CA, USA, August 13-15 (2002) (Revised Papers)
4. Courtois, N.T., Bard, G.V., Wagner, D.: Algebraic and Slide Attacks on KeeLoq. In: FSE 2008. LNCS. Springer, Heidelberg (2008)
5. Indesteege, S., Keller, N., Dunkelman, O., Biham, E., Preneel, B.: A Practical Attack on KeeLoq. In: Smart, N. (ed.) EUROCRYPT 2008. LNCS, vol. 4965. Springer, Heidelberg (2008)
6. Kocher, P.C., Jaffe, J., Jun, B.: Differential Power Analysis. In: Wiener, M. (ed.) CRYPTO 1999. LNCS, vol. 1666, pp. 388–397. Springer, Heidelberg (1999)

7. Kumar, S., Paar, C., Pelzl, J., Pfeiffer, G., Schimmler, M.: Breaking Ciphers with COPACOBANA - A Cost-Optimized Parallel Code Breaker. In: Goubin, L., Matsui, M. (eds.) CHES 2006. LNCS, vol. 4249, pp. 101–118. Springer, Heidelberg (2006)
8. Langer EMV-Technik. Details of Near Field Probe Set RF 2, http://www.langer-emv.de/en/produkte/prod_rf2.htm
9. Mangard, S., Pramstaller, N., Oswald, E.: Successfully Attacking Masked AES Hardware Implementations. In: Rao, J.R., Sunar, B. (eds.) CHES 2005. LNCS, vol. 3659, pp. 157–171. Springer, Heidelberg (2005)
10. Microchip. An Introduction to KeeLoq Code Hopping, http://ww1.microchip.com/downloads/en/AppNotes/91002a.pdf
11. Microchip. HCS200, KeeLoq Code Hopping Encoder, http://ww1.microchip.com/downloads/en/DeviceDoc/40138c.pdf
12. Microchip. HCS410, KeeLoq Code Hopping Encoder and Transponder, http://ww1.microchip.com/downloads/en/DeviceDoc/40158e.pdf
13. Microchip. HCS410/WM, KeeLoq Crypto Read/Write Transponder Module, http://ww1.microchip.com/downloads/en/DeviceDoc/41116b.pdf
14. Örs, S.B., Oswald, E., Preneel, B.: Power-Analysis Attacks on an FPGA - First Experimental Results. In: Walter, C.D., Koç, Ç.K., Paar, C. (eds.) CHES 2003. LNCS, vol. 2779, pp. 35–50. Springer, Heidelberg (2003)
15. Peeters, E., Standaert, F., Quisquater, J.: Power and Electromagnetic Analysis: Improved Model, Consequences and Comparisons. Integration, the VLSI Journal 40(1), 52–60 (2007)
16. Schramm, K., Leander, G., Felke, P., Paar, C.: A Collision-Attack on AES: Combining Side Channel- and Differential-Attack. In: Joye, M., Quisquater, J.-J. (eds.) CHES 2004. LNCS, vol. 3156, pp. 163–175. Springer, Heidelberg (2004)

Bug Attacks

Eli Biham[1], Yaniv Carmeli[1], and Adi Shamir[2]

[1] Computer Science Department,
Technion - Israel Institute of Technology,
Haifa 32000, Israel
{biham,yanivca}@cs.technion.ac.il
http://www.cs.technion.ac.il/~{biham,yanivca}
[2] Computer Science Department,
The Weizmann Institute of Science,
Rehovot 76100, Israel
adi.shamir@weizmann.ac.il

Abstract. In this paper we present a new kind of cryptanalytic attack which utilizes bugs in the hardware implementation of computer instructions. The best known example of such a bug is the Intel division bug, which resulted in slightly inaccurate results for extremely rare inputs. Whereas in most applications such bugs can be viewed as a minor nuisance, we show that in the case of RSA (even when protected by OAEP), Pohlig-Hellman, elliptic curve cryptography, and several other schemes, such bugs can be a security disaster: Decrypting ciphertexts on *any* computer which multiplies *even one pair of numbers* incorrectly can lead to full leakage of the secret key, sometimes with a single well-chosen ciphertext.

Keywords: Bug attack, Fault attack, RSA, Pohlig-Hellman, ECC.

1 Introduction

With the increasing word size and sophisticated optimizations of multiplication units in modern microprocessors, it becomes increasingly likely that they contain some undetected bugs. This was demonstrated by the accidental discovery of the Pentium division bug in the mid 1990's, by the less famous Intel 80286 `popf` bug (that set and then cleared the interrupt-enable bit during execution of the very simple `popf` instruction, when no change in the bit was necessary), by the recent discovery of a bug in the Intel Core 2 memory management unit (which allow memory corruptions outside of the range of permitted writing for a process), etc.

In this paper we show that if some intelligence organization discovers (or secretly plants) even one pair of single-word integers a and b whose product is computed incorrectly (even in a single low order bit) by a popular microprocessor, then any key in any RSA-based security program running on any one of the millions of PC's that contain this microprocessor can be easily broken, unless appropriate countermeasures are taken. In some cases, the full key can be retrieved with a single chosen ciphertext, while in other cases (such as RSA protected by the popular OAEP technique), a larger number of ciphertexts is required. The attack

D. Wagner (Ed.): CRYPTO 2008, LNCS 5157, pp. 221–240, 2008.
© International Association for Cryptologic Research 2008

is also applicable to other cryptographic schemes which are based on exponentiation modulo a prime or on point multiplication in elliptic curves, and thus almost all the presently deployed public key schemes are vulnerable to such an attack.

The new attack, which we call a *Bug Attack*, is related to the notion of fault attacks discovered by Boneh, Demillo and Lipton in 1996 [4], but seems to be much more dangerous in its implications. The original fault attack concentrated on soft errors that yield random results when induced at a particular point of time by the attacker (latent faults were briefly mentioned, but were never studied). They require physical possession of the computing device by the attacker, and the deliberate injection of a transient fault by operating this device in an unusual way (e.g., in a microwave oven, at high temperature, with high frequency clock, or with a sudden spike in the power supply). Such attacks are feasible against smart cards, but are much harder to carry out against PC's. In the new bug attack, the target PC's can be located at secure locations half a world away, and millions of PC's can be attacked simultaneously over the Internet, without having to manipulate the operating environment of each one of them individually. Unlike the case of fault attacks, in bug attacks the error is deterministic, and is triggered whenever a particular computation is carried out; the attacker cannot choose the timing or nature of the error, except by choosing the inputs of the computation.

Since the design of modern microprocessors is usually kept as a trade secret, there is no efficient method for the user to verify that a single multiplication bug does not exist. For example, there are 2^{128} pairs of inputs in a 64×64 bit multiplier, so we cannot try them all by exhaustive search. We can even expect that most of the 2^{128} pairs of inputs will never be multiplied on any processor. Even if we assume that Intel had learned its lesson and meticulously verified the correctness of its multipliers, there are many smaller manufacturers of microprocessors who may be less careful with their design, and less careful in testing the quality of the chips they produce. In addition, many PC's are sold with overclocked processors which are more likely to err when performing complex instructions such as 64-bit integer multiplication. The problem is not limited to microprocessors: Many cellular telephones are running RSA or elliptic curve computations on signal processors made by TI and others, FPGA or ASIC devices can embed in their design flawed multipliers from popular libraries of standard cell designs, and many security programs use optimized "bignum packages" written by others without being able to fully verify their correctness.

In addition to such innocent bugs, there is the issue of intentionally tampered hardware, which is a major security problem. In February 2005 the matter was addressed in a US Department of Defense (DoD) report [17], which warned about the risks of importing hardware from foreign countries to the US. Recently the F.B.I. reported that 3,500 counterfeit Cisco network components were discovered in the US, and some of them had even found their way into US military and government facilities [9]. Although in this case Cisco did not find any evidence of malicious modifications, it certainly demonstrates the feasibility of such a scenario. In [7], the open source design of the Leon3 processor was changed to exemplify that a hardware backdoor can be introduced into processors. The change (which affected only 0.05% of the logic gates) was shown to allow attackers to load a malicious firmware and take full control of the attacked machine. Even commercially sold bug-free processors can be made buggy by anyone along the supply chain who modifies their firmware via their built-in bug-fixing mechanisms.

What we show in this paper is that the innocent or intentional introduction of any bug into the multiplier of any processor (even when it affects only two specific inputs whose product contains a single erroneous low-order bit) can lead to a major security disaster, which can be secretly exploited in an essentially undetectable way by a sophisticated intelligence organization. Even though we are not aware of any such attacks being carried out in practice, hardware manufacturers and security experts should be aware of this possibility, and use appropriate countermeasures.

In this paper we present bug attacks against several widely deployed cryptosystems (such as Pohlig-Hellman [11], RSA [13], elliptic curve schemes, and some symmetric primitives), and against several implementations of these schemes. For all the discussed schemes, we show that the secret exponent can be retrieved by a chosen ciphertext attack, and in the case of Pohlig-Hellman, the secret exponent can also be retrieved by a chosen plaintext attack. In the case of RSA, we show that if decryption is performed using the Chinese remainder theorem (CRT) [10, Note 14.70] the public modulus n can be factored using a single chosen ciphertext. A particularly interesting observation is that even though RSA-OAEP [1] was designed to prevent chosen ciphertext attacks, we can actually use this protective mechanism as part of our bug attack in order to learn whether a bug was or was not encountered during the exponentiation process. This demonstrates that in spite of the similarity between bug attacks and fault attacks, their countermeasures can be very different. For example, just stopping an erroneous computation or recomputing the result with a different exponentiation algorithm may protect the scheme against fault attacks, but will leak the full key via a bug attack.

This paper is organized as follows: Section 2 gives an overview of the methods we use in most of our attacks, and describes the two most commonly used implementations of modular exponentiations: the left-to-right (LTOR) and right-to-left (RTOL) exponentiation algorithms. Section 3 presents the simplest bug attack on RSA when decryption is performed using the Chinese remainder theorem (CRT), using a single chosen ciphertext. Section 4 presents attacks on several cryptosystems when exponentiations are computed using the LTOR algorithm, and Section 5 presents attacks on the same schemes when the exponentiations are computed using the RTOL algorithm. In Section 6 we discuss bug attacks on elliptic curve schemes and some symmetric primitives. Section 7 summarizes the contributions of this paper, and presents the time and data complexities of all our attacks. Finally, Appendix A provides descriptions of the cryptosystems discussed in this paper.

2   Overview of Our Methods and Notations

We present several attacks which use multiplication bugs. We concentrate on these operations since multiplication and division are typically the most complex operations, their implementations are most aggressively optimized, and therefore bugs are more likely to exist in them than in simple operations like addition or XOR, and are less likely to be discovered by the manufacturers.

2.1 Multiplication of Big Numbers

In cryptography, we are often required to perform arithmetic operations on big numbers, which must be represented using more than a single 32-bit or 64-bit word. Arithmetic operations on such values must be broken down into arithmetic operations on the different words which comprise them. For example, when multiplying two very long integers x and y, each represented by ten words, each of the ten words of x is multiplied by each of the ten words of y, in some order, and the results are then summed up to the appropriate words of the product. If x contains a in the sense that one of the ten words of x is a, y contains b, and the processor produces an incorrect result when a and b are multiplied, then the result of multiplying $x \cdot y$ on that processor will typically be incorrect (unless there are multiple errors that exactly cancel each other during the computation, which is very unlikely when the other words in x and y are randomly chosen).

2.2 Notations

We use the notation $x \cdot y$ to denote the result of multiplying x by y on a bug-free processor, and $x \odot y$ to denote the result of the same computation when performed on a faulty processor. Similarly, the notation x^l denotes the value of x to the power l as computed on a bug-free processor, and $x^{\langle l \rangle}$ denotes the value of x to the power l as computed by a particular algorithm on a faulty processor (See Section 2.5 for details of popular exponentiation algorithms). Since we assume that faults are extremely rare, for most inputs we expect the result of the computation to be the same on both the faulty and the bug-free processors, and in these cases we use the notations $x \cdot y$ and x^l, even when referring to computations done on the faulty processor.

2.3 Methods

Our attacks request the decryptions of ciphertexts which may or may not invoke the execution of the faulty multiplications, depending on the bits of the secret exponent d. The results of those decryptions are used to retrieve the bits of the secret exponent d. We develop two methods for creating the conditions under which the buggy instructions are executed. The first method chooses a ciphertext C, such that an intermediate value x during the decryption process contains both a and b. If x is squared, then we expect that $x^2 \neq x^{\langle 2 \rangle}$, and thus the result of the entire decryption process is also expected to be incorrect. If x is multiplied by a different value y, which contains neither a nor b, then we expect that $x \cdot y = x \odot y$, and the decryption result is expected to be correct. The second method chooses C such that during decryption one intermediate value x

LTOR Exponentiation	RTOL Exponentiation

$z \leftarrow 1$
 For $k = \log n$ down to 0
    If $d_k = 1$ then $z \leftarrow z^2 \cdot x \bmod n$
    Otherwise, $z \leftarrow z^2 \bmod n$
 Output z
	$y \leftarrow x; z \leftarrow 1$
	 For $k = 0$ to $\log n$
	    If $d_k = 1$ then $z \leftarrow z \cdot y \bmod n$
	    $y \leftarrow y^2 \bmod n$
	 Output z

Fig. 1. The Two Basic Exponentiation Algorithms

contains a, while another value y contains b. If x and y are multiplied then it is expected that $x \cdot y \neq x \odot y$, and the result of decryption on the faulty processor is expected to be incorrect. If x and y are not multiplied by the decryption algorithm, we expect the decryption to be correct.

2.4   Complexity Analysis

Let w be the length (in bits) of the words of the processor. When analyzing complexities of our attacks throughout this paper we assume that numbers (both exponentiated values and exponents) are 1024-bit long, and that $w = 32$ (in the summary of the paper we also quote the complexities for $w = 64$). The standard representation of 1024-bit long numbers requires $\lceil 2^{10}/w \rceil$ words. Given a random 1024-bit value x, and a w-bit value a, the probability that x contains a (in any of its $2^{10}/w$ words) is about $2^{-w}2^{10}/w$. For $w = 32$ this probability is 2^{-27}, and for $w = 64$ it is 2^{-60}. Given two w-bit values a and b, the probability that x contains both a and b is about $\left(2^{-w}2^{10}/w\right)^2$. For $w = 32$ this probability is about 2^{-54}, and for $w = 64$ it is about 2^{-120}.

2.5   Exponentiation Algorithms

Given a value x and a secret exponent $d = d_{\log n}d_{\log n-1}\ldots d_1d_0$, the exponentiation $x \mapsto x^d \bmod n$ can be efficiently computed by several exponentiation algorithms [10, Chapter 14.6]. In this paper we present attacks against implementations that use the two basic exponentiation algorithms, LTOR (left-to-right) and RTOL (right-to-left), described in Figure 1. Our techniques can be easily adapted to attack implementations that use other exponentiation algorithms such as the sliding window algorithm, the k-ary exponentiation algorithm, etc.

2.6   Remarks

The following remarks apply to most of the attacks presented in this paper.

1. Microprocessors usually perform different sequences of microcode instructions when computing $a \cdot b$ and $b \cdot a$, and thus the bug is not expected to be

symmetric: for $a \cdot b$ the processor may give an incorrect result, while for $b \cdot a$ the result is correct. Therefore, the correctness of the result of multiplying two big numbers x and y, where x contains a and y contains b, depends on whether the implementation of $x \cdot y$ multiplies $a \cdot b$ or $b \cdot a$. We assume that such implementation details are known to the attacker.

2. Given a value w, the number of bits in the binary representation of w is $\lfloor \log_2 w \rfloor + 1$ (the indices of the bits of w are $0, \ldots, \lfloor \log w \rfloor$, where 0 is the index of the least significant bit, and $\lfloor \log w \rfloor$ is the index of the most significant bit). Throughout this paper we use $\log w$ (without the floor operator) as a shorthand for the index of the most significant bit of w.

3. It may be the case that more than one pair of buggy inputs $a \cdot b$ exist. In such cases, if γ multiplication bugs are known to the attacker, the complexities of some of the attacks we present can be decreased. In attacks where the attacker can control only one of the operands of the multiplication, and the other operand is expected to appear randomly, the time complexity can be decreased by a factor of $\min(\gamma, \lfloor \log n/w \rfloor)$. If some of the buggy pairs of operands share the same value for one of the operands, this factor can even get better (but it cannot be higher than γ). In attacks where both operands are expected to appear randomly, the time complexity can be decreased by a factor of γ. Note that symmetric bugs, where both the results of $a \cdot b$ and $b \cdot a$ are incorrect, are counted as two bugs.

4. If both operands of the buggy instruction are equal (i.e., $a = b$), the complexity of some of our attacks can be greatly reduced, while other attacks become impossible. The former case happens when attacks rely on faults in the squaring of values X, where X happens by chance to contain both a and b. In this case only one word (a) needs to appear in X, which makes the probability of this event much higher. On the other hand, attacks which use the existence of a bug in order to decide whether x and y were squared or multiplied together become impossible. When the attack requires that x contains a and that y contains b, our ability to distinguish between these cases depends on whether $a = b$.

3   Breaking CRT-RSA with One Chosen Ciphertext

We now describe a simple attack on RSA implementations in which decryptions are performed using the Chinese remainder theorem (CRT). Let $n = pq$ be the public modulus of RSA, where p and q are large primes, and assume without loss of generality that $p < q$. Knowing the target's public key n (but not its secret factors p and q), the attacker can easily compute a half size integer which is guaranteed to be between the two secret factors p and q of n. For example, $\lfloor \sqrt{n} \rfloor$ always satisfies $p \leq \lfloor \sqrt{n} \rfloor < q$, and any integer close to $\sqrt{n}$ is also likely to satisfy this condition. The attacker now chooses a ciphertext C which is the closest integer to $\sqrt{n}$, such that both a and b appear as low order words in C, and submits this "poisonous input" to the target PC.

The first step in the CRT-RSA computation is to reduce the input C modulo p and modulo q. Due to its choice, $C_p = C \bmod p$ is randomized modulo the smaller factor p, but $C_q = C \bmod q = C$ remains unchanged modulo the larger factor q. The next step in RSA-CRT is always to square the reduced inputs C_p and C_q, respectively. Since a and b are unlikely to remain in C_p, the computation mod p is likely to be correct. However, mod q the squaring operation will contain a step in which the word a is multiplied by the word b, and by our assumption the result will be incorrect. Assuming that the rest of the two computations mod p and q will be correct, the final result of the two exponentiations will be combined into a single output $\hat{M}$ which is likely to be correct mod p, but incorrect mod q. The attacker can then finish off his attack in the same way as the original fault attack, by computing the greatest common divisor (gcd) of n and $\hat{M}^e - C$, where e is the public exponent of the attacked RSA key. This gcd is the secret factor p of n.

Note that if such C cannot be found, then $q - p < 2^{2w}$. In this case, n can be easily factored by other methods (e.g., Fermat's factorization method, which will factor n in 2^w time without any calls to the decryption oracle).

4   Bug Attacks on LTOR Exponentiations

In this section we present bug attacks against several cryptosystems, where exponentiations are performed using the LTOR exponentiation algorithm. We first present chosen plaintext (or chosen ciphertext) attacks against the Pohlig-Hellman scheme, then present chosen ciphertext attacks against RSA, and finally discuss how to adapt our attacks on RSA to the case of RSA-OAEP.

4.1   Bug Attacks on Pohlig-Hellman

The Pohlig-Hellman cipher uses two secret exponents e and d: the former is used for encryption, and the latter for decryption. Given one of the secret exponents, the other can be computed by $d \equiv e^{-1} \pmod{p-1}$. We discuss adaptive and non-adaptive chosen ciphertext attacks which retrieve the bits of the decryption exponent d; similar chosen plaintext attacks can retrieve the encryption exponent e.

We start by presenting a simple adaptive attack, which demonstrates the basic idea of our technique. We later improve this attack with additional ideas.

4.1.1   Basic Adaptive Chosen Ciphertext

In this section, an attack which requires the decryption of $2 \log p$ chosen ciphertexts is presented. The attack retrieves the bits of the secret exponent one at a time, from $d_{\log p}$ to d_1 (d_0 is known to be one, as d is odd). Therefore, when the search for d_i is performed, we can assume that the bits $d_{i+1}, \ldots, d_{\log p}$ are already known.

The attack works as follows:

1. Choose a value X which contains the words a and b.
2. For $i = \log p$ down to 1 do

(a) Denote the value of the known bits of d by $d' = \sum_{k=i+1}^{\log p} 2^{k-(i+1)} d_k$.
(b) Compute $C = X^{1/d'} \bmod p$.
(c) Ask for the decryption $\hat{M} = C^{\langle d \rangle} \bmod p$ on the faulty processor.
(d) Obtain the correct decryption $M = C^d \bmod p$.
(e) If $M = \hat{M}$ conclude that $d_i = 1$, otherwise conclude that $d_i = 0$.

3. Set $d_0 = 1$.

The attack is based on the following observations. Since p is a known prime, the attacker can compute arbitrary roots modulo p. During the i'th iteration of the attack, the value of C is chosen such that when it is exponentiated to power d with LTOR, the intermediate value of the variable z after $\log p - i$ iterations is X. The next operation of the LTOR algorithm is either squaring z, or multiplying it by C, depending on the value of d_i. Since the intermediate value $z = X$ contains both a and b, we expect to get an incorrect decryption if z is squared (i.e., when $d_i = 0$), and a correct decryption if z is first multiplied by C (i.e., when $d_i = 1$).

Note that the bug-free decryption in Step 2d may be obtained on the same buggy microprocessor by using the multiplicative property of modular exponentiation. The attacker may request the decryption M' of $C' = C^3 \bmod p$ (or any other power of C which is not expected to cause the execution of the faulty instructions), and then check whether $\hat{M}^3 \equiv M' \pmod{p}$ to learn if an error had occurred. Thus, no calls to a bug-free decryption device that uses the same secret key (which is usually unavailable) is required. In fact, since the same value of X is used for each of the iterations, the correct decryption M can be computed from the value of the correct decryption in the previous iteration as: $M = \bar{M}^{d'/\bar{d}'} \bmod n$, where $\bar{M}$ and $\bar{d}'$ are the values of the corresponding variables in the previous iteration. Therefore, no additional decryption requests (beyond the first one) are needed in order to obtain all the correct decryption results throughout the attack.

The attack requires buggy decryption of $\log p + 1$ chosen ciphertexts to retrieve d, or buggy encryption of $\log p + 1$ chosen plaintexts to retrieve e. Each one of these values makes it easy to compute the other value since p is a known prime.

4.1.2 Improved Adaptive Chosen Ciphertext Attack

We observe that X can be selected such that both X and $X^{\langle 2 \rangle}$ contain a and b. A further improvement uses X's which contain a and b, such that when X is squared m times repeatedly on a faulty processor (for some $m > 0$), all the values $X^{\langle 2^j \rangle}$ contain a and b. Using such X, we can improve the expected complexity of the attack by a factor of $\alpha = 2 - 2^{-m}$. Further details on this improved attack will be presented in the full version of this paper.

4.1.3 Chosen Ciphertext Attack

The (non-adaptive) chosen ciphertext attack presented later in Section 4.2.2 is also applicable in the case of Pohlig-Hellman. The attack requires decryption of 2^{28} ciphertexts to retrieve the secret exponent d (the attack on RSA requires 2^{27} ciphertexts, but in the case of Pohlig-Hellman an additional decryption is required for each buggy decryption, in order to verify the correctness of the decryption). As in the previous attacks on Pohlig-Hellman, a similar chosen plaintext attack can retrieve the secret exponent e.

4.2   Bug Attacks on RSA

We describe several chosen ciphertext attacks on RSA, where the attacked implementation performs decryptions without using CRT. Instead, we assume that the decryption of a ciphertext C is performed by computing $C^d \bmod n$ using LTOR (where d is the secret exponent of RSA). We assume that the public exponent e and the public modulus n are known. The main difference between the case of RSA and the case of Pohlig-Hellman is that there is no known efficient algorithm to compute roots modulo a composite n, when the factorization of n is unknown.

Unlike the case of Pohlig-Hellman, in the case of RSA checking whether the decrypted message $\hat{M}$ is the correct decryption of a chosen ciphertext C can be easily done by checking whether $\hat{M}^e \equiv C \pmod{n}$. Thus, there is no need to request the decryptions of additional messages for this purpose.

4.2.1   Adaptive Chosen Ciphertext Attack

We describe an adaptive chosen ciphertext attack which requires the decryption of $\log n$ chosen ciphertexts by the target computer. The generation of each of the ciphertexts requires 2^{27} time on the attacker's (bug-free) computer, and thus the total time complexity of the attack is about 2^{37}.

Description of the attack:

1. For $i = \log n$ down to 1 do
   (a) Denote the value of the known bits of d by $d' = \sum_{k=i+1}^{\log n} 2^{k-(i+1)} d_k$.
   (b) Repeatedly choose random values C which contain b, until $C^{d'} \bmod n$ contains a.
   (c) Ask for the decryption $\hat{M} = C^{\langle d \rangle} \bmod n$ using the faulty processor.
   (d) Compute $\hat{C} = \hat{M}^e \bmod n$.
   (e) If $\hat{C} = C$ conclude that $d_i = 0$, otherwise conclude that $d_i = 1$.
2. Set $d_0 = 1$.

The attack is similar to the basic attack presented in Section 4.1.1, except that here only the word a is contained in the intermediate value of the exponentiation. The word b is contained in the ciphertext C, and therefore the roles of the correct and incorrect results are exchanged between $d_i = 0$ and $d_i = 1$.

During the execution of the LTOR algorithm, the intermediate value of the variable z after $\log n - i$ iterations contains a (due to the selection of C in Step 1b of the attack). If $d_i = 0$ then z is squared, and no errors in the computation are expected to occur, leading to $\hat{C} = C$ in Step 1e. If $d_i = 1$, then z is multiplied by C, which contains the word b, and due to the bug, the result of the exponentiation is expected to be incorrect, leading to $\hat{C} \neq C$ in Step 1e.

As explained in Section 2, the probability that the random number $C^{d'} \bmod n$ contains somewhere along it the word a is 2^{-27} (for our standard parameters).

Therefore, Step 1b takes an average time of 2^{27} exponentiations on the attacker's computer.

4.2.2   Chosen Ciphertext Attack

The previous adaptive attack on exponentiations using LTOR is the basis for the following non-adaptive chosen ciphertext attack. The attack requests the decryption of 2^{29} chosen ciphertexts, all of which contain the word b. It is expected that for every $0 \le i \le \log n$, there are about four ciphertexts for which the intermediate value of z after i rounds of the exponentiation algorithm contains the word a. The value of d_i can be determined by the correctness of the decryption of those ciphertexts, using considerations similar to the ones used in the attack of Section 4.2.1. If for some i there are no ciphertexts C_j for which $X_j = C_j^{d'} \bmod n$ contains a, there is no choice but to continue the attack recursively for both $d_i = 0$ and $d_i = 1$. However, when the wrong value is chosen, a contradiction may be encountered before retrieving the rest of the bits (i.e., more than one ciphertext C_j for which X_j contains a is found, and the decryption of some, but not all, of them is incorrect). By using standard results from the theory of branching processes, 2^{29} ciphertexts suffice to ensure that recursive calls which represent wrong bit values are quickly aborted.

Here are some details of this attack:

1. Choose 2^{29} random ciphertexts C_j $(1 \le j \le 2^{29})$ containing the word b, and ask for their decryptions $\hat{M}_j$ using the faulty processor.
2. For $i = \log n$ down to 1 do
   (a) Denote the value of the known bits of d by $d' = \sum_{k=i+1}^{\log n} 2^{k-(i+1)} d_k$.
   (b) For each ciphertext C_j compute $X_j = C_j^{d'} \bmod n$.
   (c) Consider all ciphertexts C_j such that X_j contains a:
      i. If for all such ciphertexts C_j it holds that $\hat{M}_j^e \bmod n = C_j$ then set $d_i = 0$.
      ii. If for all such ciphertexts C_j it holds that $\hat{M}_j^e \bmod n \ne C_j$ then set $d_i = 1$.
      iii. If there are no such ciphertexts try the rest of the attack for both $d_i = 0$ and $d_i = 1$.
      iv. If for some of these ciphertexts C_j, $\hat{M}_j^e \bmod n = C_j$ and for others $\hat{M}_j^e \bmod n \ne C_j$ (i.e., a previously set value of one of the bits is wrong) then backtrack.
3. Set $d_0 = 1$.

The data complexity may be increased in order to decrease the probability of not having ciphertexts C_j such that X_j contains a. Alternatively, it may be decreased, at the expense of more recursive guesses (Step 2(c)iii), with increased time complexity. If for every i there exists a j such that $C_j^{d'}$ contains b, the time complexity is equal to the data complexity (i.e., 2^{29}).

4.2.3   Known Plaintext Attack

The chosen ciphertext attack from Section 4.2.2 can be easily transformed into a known plaintext attack which requires 2^{56} known plaintexts. Among the 2^{56} plaintexts, only 2^{29} are expected to contain b. We can discard all the plaintexts which do not contain b, and use the rest as inputs for the attack described in Section 4.2.2.

Note that the known plaintexts must be the result of decrypting the corresponding ciphertexts on the faulty processor. The attack will not work if the given plaintext-ciphertext pairs are obtained by encrypting plaintexts (either on the attacker's computer or on the target computer).

4.3   Bug Attacks on OAEP

Since RSA has many mathematical properties such as multiplicativity, it is often used in modes of operation which protect it against attacks based on these properties. The most popular mode is OAEP [1], which is provably secure. We show here that although OAEP protects against "standard" attacks on RSA, it provides only limited protection against bug attacks, since it was not designed to deal with errors during the computation.

OAEP adds randomness and redundancy to messages before encrypting them with RSA, and rejects ciphertexts which do not display the expected redundancy when decrypted. Random ciphertexts are not expected to display such a redundancy, and are likely to be rejected by the receiver with overwhelming probability. To choose valid ciphertexts with certain desired characteristics, we choose random plaintexts and encrypt them using proper OAEP padding, until we get a ciphertext that has the desired structure by chance (since OAEP is a randomized cipher, we can also try to encrypt the same message with different random values, and thus can control the result of the decryption). Our main observation is that the structure we need in our attack (such as the existence of a certain word in the ciphertext) has a relatively high probability regardless of how much redundancy is added to the plaintext by OAEP, and the knowledge that a correctly constructed ciphertext was rejected suffices to conclude that some computational error occurred. We are thus exploiting the OAEP countermeasure itself in order to mount the new bug attack!

The attacks we present on RSA-OAEP are very similar to the attacks on RSA from Section 4.2, with some minor modifications. The same attacks are also applicable to OAEP+ [16].

4.3.1   Adaptive Chosen Ciphertext Attack

Unlike the attack of Section 4.2.1, OAEP stops us from choosing ciphertexts C which contain b, and thus in Step 1b we must choose random messages (on our own computer) until b "appears" in C at random. As explained in Section 2.4, the probability that this happens and in addition $C^{d'} \bmod n$ contains a is 2^{-54}. As mentioned above, computation errors are identified in Step 1e of the attack on OAEP by the mere rejection of the ciphertext, and there is no need to know the actual value which was rejected. The attack requires the decryption of $\log n$ chosen ciphertexts, and thus its total time complexity for 1024-bit n's is 2^{64}.

4.3.2   Chosen Ciphertext Attack

The (non-adaptive) chosen ciphertext attack on RSA from Section 4.2.2 can also be used in the case of OAEP. For a random message, the probability that the ciphertext contains b is 2^{-27}. In order to find 2^{29} messages with a ciphertext which contains b (as required by the attack), we have to try about 2^{56} random messages. Therefore, the attack requires the decryption of 2^{29} chosen ciphertexts, plus 2^{56} pre-computation time on the attacker's own computer. Once the decryptions of the chosen ciphertexts are available, the key can be retrieved in 2^{29} additional time.

5   Bug Attacks on RTOL Exponentiations

In this section we present attacks against Pohlig-Hellman, RSA, and RSA-OAEP, where exponentiations are performed using the RTOL exponentiation algorithm. In RTOL, the value of the variable y is squared in every iteration of the exponentiation algorithm, regardless of the bits of the secret exponent. Any error introduced into the value of y undergoes the squaring transformation in every subsequent iteration, and is propagated to the value of z if and only if the corresponding bit of the exponent is set. Consequently, every set bit in the binary representation of the exponent introduces a different error into the value of z, while zero bits do not introduce any errors. This allows us to mount efficient non-adaptive attacks, and to retrieve more than one bit from each chosen ciphertext, as described in the attacks presented in this section.

5.1   Bug Attacks on Pohlig-Hellman

We present a chosen ciphertext attack against Pohlig-Hellman, where exponentiations are performed using RTOL. The attack is aimed at retrieving the bits of the secret exponent d. As in Section 4.1, an identical chosen plaintext attack can retrieve the bits of the secret exponent e.

5.1.1   Chosen Ciphertext Attack

We present a (non-adaptive) chosen ciphertext attack which retrieves the secret key when the exponentiation is performed using RTOL. Let X be a value which contains the words a and b, and let $\beta = X^2/X^{\langle 2 \rangle}$. Unlike the improved attack on Pohlig-Hellman of Section 4.1.2, it does not help if $X^{\langle 2 \rangle}$ also contains a and b (on the contrary, it makes the analysis slightly more complicated). Each chosen ciphertext is used to retrieve r bits of the secret exponent d, where r is a parameter of the attack. The reader is advised to consider first the simplest case of $r = 1$.

The attack is carried out using the following steps:

1. For $i = \log p - (\log p \bmod r)$ down to 0 step $-r$
   (a) Compute $C = X^{1/2^{i-1}} \bmod p$.
   (b) Denote the value of the known bits of d by $d' = \sum_{k=i+r}^{\log p} 2^{k-(i+r)} d_k$.
   (c) Ask for the decryption $\hat{M} = C^{\langle d \rangle} \bmod p$ on the faulty processor.
   (d) Obtain the correct decryption $M = C^d \bmod p$.
   (e) Find an r-bit value u such that $M/\hat{M} = \beta^{2^r d' + u} \bmod p$ $(0 \le u < 2^r)$.
   (f) Denote the bits of u by $u_{r-1} u_{r-2} \dots u_1 u_0$.
   (g) Conclude that $d_{i+k} = u_k$, $\forall\, 0 \le k < r$.

Consider the decryption of C in Step 1c, for some i. Exponentiation by the RTOL algorithm sets $y = C$, and squares y repeatedly. After squaring it $i - 1$ times, the value of y becomes X, which contains both a and b. When y is squared again, a multiplicative error factor of β is introduced into its computed value (compared to its bug-free value). If $d_i = 1$ then z is multiplied by y, and thus the same multiplicative error factor of β is also propagated into the value of z. After the next squaring of y, it contains an error factor of β^2, which is propagated into the value of z when $d_{i+1} = 1$. In each additional iteration of the exponentiation the previous error in y is squared, and the error affects the result if and only if the corresponding bit of d is set. At the end of the exponentiation, the error factor in the final result is:

$$\frac{M}{\hat{M}} \equiv \prod_{k=i}^{\log p} \left(\beta^{2^{k-i}} \right)^{d_k} \equiv \beta^{\sum_{k=i}^{\log p} 2^{k-i} d_k} \pmod{p}.$$

Since only r bits of the exponent are unknown, they can be easily retrieved by performing $2^r - 1$ modular multiplications.

As in the attacks of Section 4.1, all the error-free decryption queries in Step 1c can be replaced by the decryption of one additional ciphertext on the faulty processor: The attacker can request the decryption M^3 of $C^3 \bmod p$ (or any other power of C which is not expected to cause a decryption error), and then in Step 1e can find an r-bit value u such that

$$\frac{M^3}{\hat{M}^3} \equiv \left[\prod_{k=i}^{\log p} \left(\beta^{2^{k-i}} \right)^{d_k} \right]^3 \equiv \beta^{3\left(2^r d' + u\right)} \pmod{p}.$$

The attack requires $2\lceil (\log p + 1)/r \rceil$ decryptions of chosen ciphertexts, and all of them can be pre-computed by $\log p$ modular square roots (Step 1a of the attack). Once the decryptions are available, each execution of Step 1e finds r bits of d using $2^r - 1$ multiplications, which is equivalent to about $2^r / \log p$ modular exponentiations. Since Step 1e is executed $\lceil (\log p + 1)/r \rceil$ times, the total time complexity is about $2^r / r$. For small values of r this time complexity is negligible compared to the time of the pre-computation. For $r \ge 12$, however, this computation takes longer, and there is a tradeoff between the time complexity and the data complexity.

5.2 Bug Attacks on RSA

Unlike the case of Pohlig-Hellman, there is no known efficient algorithm for extracting roots modulo a composite n with unknown factors. The chosen ciphertext attack presented in this section circumvents this problem by choosing random ciphertexts until a suitable ciphertext is found.

5.2.1 Chosen Ciphertext Attack

The attack in this case is similar to the attack on RTOL modulo a prime p (Section 5.1.1), except for some necessary adaptations to the case of RSA. The attack requires a pre-computation to find a value X which contains both a and b, and such that all the values $X^{1/2^{i-1}}$ for $1 \leq i \leq \log n$ are known (Step 2 in the following attack). The parameter r represents the number of bits retrieved in each iteration.

The detailed attack is as follows:

1. Choose a random ciphertext C_0, and let $t = 0$.
2. While $t \leq \log n$ or C_t does not contain both a and b do:
   (a) $t = t + 1$.
   (b) Compute $C_t = C_{t-1}^2 \bmod n$.
3. Let $X = C_t$ and let $X^{\langle 2 \rangle}$ be the result of squaring X on a faulty processor.
4. Let $\beta = X^2 / X^{\langle 2 \rangle} \bmod n$.
5. For $i = \log n - (\log n \bmod r)$ down to 0 step $-r$
   (a) Ask for the decryption $\hat{M}$ of $C = C_{t-i}$ using the faulty processor, $M = C_{t-i}^{\langle d \rangle} \bmod p$.
   (b) Denote the value of the known bits of d by $d' = \sum_{k=i+r}^{\log n} 2^{k-(i+1)} d_k$.
   (c) Compute $\hat{C} = \hat{M}^e \bmod n$.
   (d) Find an r-bit value u such that $C/\hat{C} \equiv \left(\beta^{2^r d' + u} \right)^e \pmod{n}$.
   (e) Denote the bits of u by $u_{r-1} u_{r-2} \ldots u_1 u_0$.
   (f) Conclude that $d_{i+k} = u_k$, $\forall\, 0 \leq k < r$.

A random ciphertext contains a and b with probability 2^{-54}, and therefore the pre-computation of Step 2 is expected to take time corresponding to 2^{54} modular multiplications (which is equivalent to 2^{44} modular exponentiations when $\log n = 1024$). In each iteration of the attack, r bits are retrieved by performing $2^r - 1$ modular multiplications, which are equivalent to about $(2^r - 1)/\log n$ modular exponentiations. Thus, once the decrypted ciphertexts are available, the attack requires a time equivalent to about

$$\left\lceil \frac{\log n}{r} \right\rceil \frac{2^r - 1}{\log n} \approx \frac{2^r - 1}{r}$$

modular multiplications. As in the attack of Section 5.1, this attack requires $\lceil \log n/r \rceil$ decryptions of pre-computed chosen ciphertexts. Step 5d finds r bits of the secret exponent d using $2^r - 1$ multiplications, and thus (as in the attack from Section 5.1.1) for large values of r there is a tradeoff between the time complexity and the data complexity.

5.3   Bug Attacks on OAEP Implementations That Use RTOL

5.3.1   Adaptive Chosen Ciphertext Attack

We present an adaptive chosen ciphertext attack for the case of RSA-OAEP when exponentiations are performed using RTOL. The presented attack resembles the attack from Section 4.3, but it identifies the bits of d starting from the *least* significant bit.

The description of the attack is as follows:

1. Set $d_0 = 1$.
2. For $i = 1$ to $\log n$
   (a) Denote the value of the known bits of d by $d' = \sum_{k=0}^{i-1} 2^k d_k \bmod n$.
   (b) Repeatedly encrypt random messages M until $C = E(M) = (\mathrm{OAEP}(M))^e$ satisfies that $C^{2^i} \bmod n$ contains a and $C^{d'} \bmod n$ contains b.
   (c) Ask for the decryption of C using the faulty processor.
   (d) If the decryption succeeds conclude that $d_i = 0$, otherwise conclude that $d_i = 1$.

After i iterations of the decryption exponentiation algorithm, the value of the variable z is $C^{d'} \bmod n$, and the value of the variable y is $C^{2^i} \bmod n$. The ciphertext C is chosen such that one of these values contains a and the other contains b. Therefore, if these values are multiplied ($d_i = 1$), then the result of the decryption is expected to be wrong, and the ciphertext is rejected, otherwise, no errors are expected to occur, and the decryption is expected to succeed ($d_i = 0$).

The complexity of finding the ciphertext in Step 2b is 2^{54}, and the complexity of the entire attack for 1024-bit n's is 2^{64} exponentiations on the attacker's computer. The attack requires $\log n$ chosen ciphertexts, which are decrypted on the target machine.

6   Bug Attacks on Other Schemes

6.1   Elliptic Curve Schemes

In cryptosystems based on elliptic curves, exponentiations are replaced by multiplying a point by a constant. It should be noted that the implementations of point addition (corresponding to multiplication in modular groups) and of point doubling (corresponding to squaring in modular groups) are different, but both of them use multiplications of large integers. Our bug attacks can be easily adapted in such a way that the bug is invoked only if two points are added (or alternatively, only if a point is doubled). The correctness or incorrectness of the result reveals the bits of the exponent.

6.2   Bug Attacks on Symmetric Primitives

Multiplication bugs can also be used to get information on the keys of symmetric ciphers which include multiplications, such as the block ciphers IDEA [8],

MARS [5], DFC [6], MultiSwap [14], Nimbus [18] and RC6 [12], the stream cipher Rabbit [3], the message authentication code UMAC [2], etc.

In IDEA, MARS, DFC, MultiSwap and Nimbus, subkeys are multiplied by intermediate values. If an encryption (or decryption) result is known to be incorrect, an attacker may assume that one of the subkeys used for these multiplications is a, and the corresponding intermediate value is b. For example, by selecting a plaintext which contains b in a word that is multiplied by a subkey, the attacker can easily check if the value of that subkey is a.

In Rabbit, a 32-bit value is squared to compute a 64-bit result used to update the internal state of the cipher. In faulty implementations with word size 8 or 16 (which are likely word sizes for smart card implementations), faults in the stream can give the attacker information about the internal state. Similarly, the block cipher RC6 uses multiplications of the form $A \cdot (2A + 1)$ for 32-bit values A, and thus multiplication bugs may cause errors in faulty implementations with word size 8 or 16. However, this is an unlikely scenario, since bugs in processors with small words are expected to cause frequent errors which can be easily discovered.

The MAC function UMAC uses multiplications of two words, both of which depend on the authenticated message. If an incorrect MAC is computed on a faulty processor, an attacker can gain information on the intermediate values of the computation.

Table 1. Summary of the Attacks Presented in This Paper

Scheme	Exp. Alg.	Attack	Sec.	Data	Pre-Comp. Time	Attack Time	Complexity for 32-bit Words*	Complexity for 64-bit Words*
Pohlig-Hellman	RTOL	CP/CC	5.1.1	$2\left\lceil\frac{\log p}{r}\right\rceil$**	$\log p$	$\frac{\log p + 2^r}{r}$**	$2^6/2^{10}/2^{27}$	$2^6/2^{10}/2^{27}$
	LTOR	ACP/ACC	4.1.1	$\log p$	–	$\log p$	$2^{10}/-/2^{10}$	$2^{10}/-/2^{10}$
	LTOR	CP/CC	4.1.3	$\frac{2\cdot 2^w \cdot w}{\log p}$	–	$\frac{2\cdot 2^w \cdot w}{\log p}$	$2^{28}/-/2^{28}$	$2^{61}/-/2^{61}$
RSA	CRT	CC	3	1	–	1	$1/-/1$	$1/-/1$
	RTOL	CC	5.2.1	$\left\lceil\frac{\log n}{r}\right\rceil$**	$\frac{2^{2w}w^2}{\log^2 n}$	$\frac{\log n + 2^r}{r}$**	$2^5/2^{54}/2^{27}$	$2^5/2^{120}/2^{27}$
	LTOR	ACC	4.2.1	$\log n$	–	$2^w \cdot w$	$2^{10}/-/2^{37}$	$2^{10}/-/2^{70}$
	LTOR	CC	4.2.2	$\frac{4\cdot 2^w \cdot w}{\log n}$	–	$\frac{4\cdot 2^w \cdot w}{\log n}$	$2^{29}/-/2^{29}$	$2^{62}/-/2^{62}$
	LTOR	KP	4.2.3	$\frac{4\cdot 2^{2w}\cdot w^2}{\log^2 n}$	–	$\frac{4\cdot 2^w \cdot w}{\log n}$	$2^{56}/-/2^{29}$	$2^{122}/-/2^{62}$
RSA with OAEP	RTOL	ACC	5.3.1	$\log n$	–	$\frac{2^{2w}\cdot w^2}{\log n}$	$2^{10}/-/2^{64}$	$2^{10}/-/2^{130}$
	LTOR	ACC	4.3.1	$\log n$	–	$\frac{2^{2w}\cdot w^2}{\log n}$	$2^{10}/-/2^{64}$	$2^{10}/-/2^{130}$
	LTOR	CC	4.3.2	$\frac{4\cdot 2^w \cdot w}{\log n}$	$\frac{4\cdot 2^{2w}\cdot w^2}{\log^2 n}$	$\frac{4\cdot 2^w \cdot w}{\log n}$	$2^{29}/2^{56}/2^{29}$	$2^{62}/2^{122}/2^{62}$

KP – Known Plaintext.
CP – Chosen Plaintext; ACP– Adaptive Chosen Plaintext.
CC – Chosen Ciphertext; ACC– Adaptive Chosen Ciphertext.
w is the word size (in bits) of the faulty processor.
* Complexity is described in terms of data/pre-computation time/attack time.
** r is a parameter of the attack. The presented numbers are for $r = 2^5$.

7   Summary and Countermeasures

We have presented several chosen ciphertext attacks against exponentiation based public-key and secret-key cryptosystems, including Pohlig-Hellman and RSA. We show such attacks for the two most common implementations of exponentiation. We also discuss the applicability of these techniques to elliptic curve cryptosystems and symmetric ciphers. The attacks and their complexities are summarized in Table 1.

There are various countermeasures against bug attacks. Many protection techniques against fault attacks are also applicable to bug attacks, but we stress that due to the differences between the techniques, most of them have to be adapted to the new environment. As shown in Sections 4.3 and 5.3, and unlike the case of fault attacks, the mere knowledge that an error has occurred suffices to mount an attack, even if the output of decryption is not available. Therefore, if a decryption is found to be incorrect, it can be dangerous to send out an error message, and the correct result must be computed by other means.

Possible ways to compute the correct result include using a different exponentiation algorithm, or relying on the multiplicative property of the discussed schemes to blind the computations. When blinding is used, an attacker has no control over the exponentiated values, and they are not made available to her. Thus, even if faults occur during the exponentiation, no information is leaked. However, this method renders the system vulnerable to timing attacks, as the decryption of ciphertexts which trigger the bug take longer than decryptions which succeed in the first attempt. In order to protect the implementation from timing attacks, the original exponentiations must be blinded, so that no unblinded exponentiations are performed. Another alternative is to exponentiate modulo $n \cdot r$, where r is a small (e.g., 32-bit) prime, rather than modulo n, and reduce mod n only at the last step.

Acknowledgments

The authors would like to thank Orr Dunkelman for his comments. The first two authors were supported in part by the Israel MOD Research and Technology Unit.

References

1. Bellare, M., Rogaway, P.: Optimal Asymmetric Encryption – How to encrypt with RSA (Extended Abstract). In: De Santis, A. (ed.) EUROCRYPT 1994. LNCS, vol. 950, pp. 92–111. Springer, Heidelberg (1995)
2. Black, J., Halevi, S., Krawczyk, H., Krovetz, T., Rogaway, P.: UMAC: Fast and Secure Message Authentication. In: Wiener, M. (ed.) CRYPTO 1999. LNCS, vol. 1666, pp. 215–233. Springer, Heidelberg (1999)
3. Boesgaard, M., Vesterager, M., Pedersen, T., Christiansen, J., Scavenius, O.: Rabbit: A New High Performance Stream Cipher. In: Johansson, T. (ed.) FSE 2003. LNCS, vol. 2887, pp. 307–329. Springer, Heidelberg (2003)

4. Boneh, D., DeMillo, R.A., Lipton, R.J.: On The Importance of Checking Cryptographic Protocols for Faults. In: Fumy, W. (ed.) EUROCRYPT 1997. LNCS, vol. 1233, pp. 37–51. Springer, Heidelberg (1997)
5. Burwick, C., Coppersmith, D., D'Avignon, E., Gennaro, R., Halevi, S., Jutla, C., Matyas Jr., S.M., O'Connor, L., Peyravian, M., Safford, D., Zunic, N.: MARS: A Candidate Cipher for AES. In: AES—The First Advanced Encryption Standard Candidate Conference, Conference Proceedings (1998)
6. Gilbert, H., Girault, M., Hoogvorst, P., Noilhan, F., Pornin, T., Poupard, G., Stern, J., Vaudenay, S.: Decorrelated Fast Cipher: An AES Candidate. In: AES—The First Advanced Encryption Standard Candidate Conference, Conference Proceedings (1998)
7. King, S.T., Tucek, J., Cozzie, A., Grier, C., Jiang, W., Zhou, Y.: Designing and Implementing Malicious Hardware, presented in LEET 08, http://www.usenix.org/events/leet08/tech/full_papers/king/king.pdf
8. Lai, X., Massey, J.L., Murphy, S.: Markov Ciphers and Differential Cryptanalysis. In: Davies, D.W. (ed.) EUROCRYPT 1991. LNCS, vol. 547, pp. 17–38. Springer, Heidelberg (1991)
9. Markoff, J.: F.B.I. Says the Military Had Bogus Computer Gear, New York Times (May 9, 2008), http://www.nytimes.com/2008/05/09/technology/09cisco.html
10. Menezes, A.J., van Oorschot, P.C., Vanstone, S.A.: Handbook of Applied Cryptography. CRC Press, Boca Raton (1996)
11. Pohlig, S.C., Hellman, M.E.: An Improved Algorithm for Computing Logarithms Over GF(p) and Its Cryptographic Significance. IEEE Transactions on Information Theory 24(1), 106–111 (1978)
12. Rivest, R.L., Robshaw, M.J.B., Sidney, R., Yin, Y.L.: The RC6 Block Cipher. In: AES—The First Advanced Encryption Standard Candidate Conference, Conference Proceedings (1998)
13. Rivest, R.L., Shamir, A., Adleman, L.: A Method for Obtaining Digital Signatures and Public-Key Cryptosystems. Communications of the ACM 21(2), 120–126 (1978)
14. Screamer, B.: Microsoft's Digital Rights Management Scheme – Technical Details (October 2001), http://cryptome.org/ms-drm.htm
15. Shamir, A., Rivest, R.L., Adleman, L.M.: Mental Poker. In: Klarner, D.A. (ed.) The Mathematical Gardner, pp. 37–43. Wadsworth (1981)
16. Shoup, V.: OAEP Reconsidered (Extended Abstract). In: Kilian, J. (ed.) CRYPTO 2001. LNCS, vol. 2139, pp. 239–259. Springer, Heidelberg (2001)
17. U.S.D. of Defense, Defense science board task force on high performance microchip supply (February 2005), http://www.acq.osd.mil/dsb/reports/2005-02-HPMS_Report_Final.pdf
18. Warner Machado, A.: The Nimbus Cipher: A Proposal for NESSIE, NESSIE Proposal (September 2000)

A   Brief Descriptions of Several Cryptosystems

A.1   The Pohlig-Hellman Cryptosystem and Pohlig-Hellman-Shamir Protocol

The Pohlig-Hellman cryptosystem [11] is a symmetric cipher. Let p be a large prime number. Alice and Bob share a secret key e, $1 \leq e \leq p-2$, $\gcd(e, p-1) = 1$.

When Alice wants to encrypt a message m, she computes $c = m^e \bmod p$. Bob can decrypt c by computing its e-th root modulo p. In practice, the decryption is performed by computing $c^d \bmod p$, where d is a decryption exponent such that $d \cdot e \equiv 1 \pmod{p-1}$. Note that given the encryption exponent e, the decryption exponent d can be easily computed, and thus e must be kept secret.

The Pohlig-Hellman-Shamir [15] keyless protocol allows encrypted communication between two parties that do not have shared secret keys. The protocol is based on the commutative properties of the Pohlig-Hellman cipher. Let p be a large prime number. Alice and Bob each has a secret encryption exponent (e_A and e_B, respectively) and a secret decryption exponent (d_A and d_B, respectively) such that $e_A \cdot d_A \equiv e_B \cdot d_B \equiv 1 \pmod{p-1}$. When Alice wishes to send Bob an encrypted message m, she sends $c_1 = m^{e_A} \bmod p$. Bob then computes $c_2 = c_1^{e_B} \bmod p$ and sends it back to Alice. Alice decrypts c_2 and sends the decryption $c_3 = c_2^{d_A} \bmod p$ to Bob. Finally, Bob decrypts c_3 to get the message $m = c_3^{d_B} \bmod p$. The protocol is secure under standard computational assumptions (The Diffie-Hellman assumption), but not against man in the middle attacks.

A.2   The RSA Cryptosystem

RSA [13] is a public-key cryptosystem. Let $n = pq$ be a product of two large prime integers. Bob has a public key (n, e) such that $\gcd(e, (p-1)(q-1)) = 1$, and a private key (n, d) such that $d \cdot e \equiv 1 \pmod{(p-1)(q-1)}$. When Alice wants to send bob an encrypted message m she computes $c = m^e \bmod n$. When Bob wants to decrypt the ciphertext he computes $c^d \equiv m^{de} \equiv m \pmod{n}$.

The security of RSA relies on the hardness of factoring n. If the factors of n are known, RSA can be easily broken.

A.3   RSA Decryption Using CRT

The modular exponentiations required by RSA are computationally expensive. Some implementations of RSA perform the decryption modulo p and q separately, and then use the Chinese remainder theorem (CRT) to compute the decryption $c^d \bmod n$. Such an implementation speeds up the decryption by a factor of 4 compared to naive implementations.

Given a ciphertext c, it is first reduced modulo p and modulo q. The two values are exponentiated modulo p and q separately: $m_p = c^{d_p} \bmod p$, and $m_q = c^{d_q} \bmod q$, where $d_p = d \bmod p - 1$ and $d_q = d \bmod q - 1$. Now m is computed using CRT, such that $m \equiv m_p \pmod{p}$ and $m \equiv m_q \pmod{q}$. This is done by computing $m = (x m_p + y m_q) \bmod n$, where x and y are pre-computed integers that satisfy:

$$\begin{cases} x \equiv 1 \pmod{p} \\ x \equiv 0 \pmod{q} \end{cases} \quad \text{and} \quad \begin{cases} y \equiv 0 \pmod{p} \\ y \equiv 1 \pmod{q} \end{cases}.$$

A.4   OAEP

Optimal Asymmetric Encryption Padding (OAEP) [1] and OAEP+ [16] are methods of encoding a plaintext before its encryption, with three major goals: adding randomization to deterministic encryption schemes (e.g., RSA), preventing the ciphertext from leaking information about the plaintexts, and preventing chosen ciphertext attacks. OAEP is based on two one-way functions G and H, which are used to create a two-round Feistel network, while OAEP+ uses three one-way functions. Only OAEP is described here.

Let $G : \{0,1\}^{k_0} \rightarrow \{0,1\}^{l+k_1}$, $H : \{0,1\}^{l+k_1} \rightarrow \{0,1\}^{k_0}$ be two one-way functions, where l is the length of the plaintext, and k_0, k_1 are security parameters. When Alice wants to compute the encryption C of a plaintext M, she chooses a random value $r \in \{0,1\}^{k_0}$ and computes

$$s = G(r) \oplus (M||0^{k_1}),$$
$$t = (H(s) \oplus r),$$
$$w = s||t,$$
$$C = E(w),$$

where $||$ denotes concatenation of binary vectors, and E denotes encryption with the underlying cipher. Decryption of c is performed by:

$$w = D(C),$$
$$s = w[0 \ldots l + k_1 - 1],$$
$$t = w[l + k_1 \ldots n - 1],$$
$$r = H(s) \oplus t,$$
$$y = G(r) \oplus s,$$
$$M = y[0 \ldots l - 1],$$
$$z = y[l \ldots l + k_1 - 1],$$

where D denotes decryption under the same cipher used in the encryption phase. If $z \neq 0^{k_0}$, then the ciphertext is rejected and no plaintext is provided. Otherwise, the decrypted plaintext is M.

Scalable Multiparty Computation with Nearly Optimal Work and Resilience

Ivan Damgård[1], Yuval Ishai[2,*], Mikkel Krøigaard[1],
Jesper Buus Nielsen[1], and Adam Smith[3]

[1] University of Aarhus, Denmark
{ivan,mk,buus}@daimi.au.dk
[2] Technion and UCLA
yuvali@cs.technion.ac.il
[3] Pennsylvania State University, USA
asmith@cse.psu.edu

Abstract. We present the first general protocol for secure multiparty computation in which the *total* amount of work required by n players to compute a function f grows only polylogarithmically with n (ignoring an additive term that depends on n but not on the complexity of f). Moreover, the protocol is also nearly optimal in terms of resilience, providing computational security against an active, adaptive adversary corrupting a $(1/2 - \epsilon)$ fraction of the players, for an arbitrary $\epsilon > 0$.

1 Introduction

Secure multiparty computation (MPC) allows n mutually distrustful players to perform a joint computation without compromising the privacy of their inputs or the correctness of the outputs. Following the seminal works of the 1980s which established the *feasibility* of MPC [4, 9, 22, 35], significant efforts have been invested into studying the *complexity* of MPC. When studying how well MPC scales to a large network, the most relevant goal minimizing the growth of complexity with the number of players, n. This is motivated not only by distributed computations involving inputs from many participants, but also by scenarios in which a (possibly small) number of "clients" wish to distribute a joint computation between a large number of untrusted "servers".

The above question has been the subject of a large body of work [2, 3, 11, 12, 14, 15, 19, 20, 21, 24, 26, 27, 28, 29]. In most of these works, the improvement over the previous state of the art consisted of either reducing the multiplicative overhead depending on n (say, from cubic to quadratic) or, alternatively, maintaining the same asymptotic overhead while increasing the fraction of players that can be corrupted (say, from one third to one half).

The current work completes this long sequence of works, at least from a crude asymptotic point of view: We present a general MPC protocol which is simultaneously optimal, up to lower-order terms, with respect to both efficiency and

---

* Supported in part by ISF grant 1310/06, BSF grant 2004361, and NSF grants 0430254, 0456717, 0627781.

D. Wagner (Ed.): CRYPTO 2008, LNCS 5157, pp. 241–261, 2008.
© International Association for Cryptologic Research 2008

resilience. More concretely, our protocol allows n players to evaluate an arbitrary circuit C on their joint inputs, with the following efficiency and security features.

COMPUTATION. The *total* amount of time spent by all players throughout the execution of the protocol is $\mathrm{poly}(k, \log n, \log |C|) \cdot |C| + \mathrm{poly}(k, n)$, where $|C|$ is the size of C and k is a cryptographic security parameter. Thus, the protocol is strongly *scalable* in the sense that the amount of work involving each player (amortized over the computation of a large circuit C) vanishes with the number of players. We write the above complexity as $\widetilde{\mathcal{O}}(|C|)$, hiding the low-order multiplicative $\mathrm{poly}(k, \log n, \log |C|)$ and additive $\mathrm{poly}(k, n)$ terms.[1]

COMMUNICATION. As follows from the bound on computation, the total number of bits communicated by all n players is also bounded by $\widetilde{\mathcal{O}}(|C|)$. This holds even in a communication model that includes only point-to-point channels and no broadcast. Barring a major breakthrough in the theory of secure computation, this is essentially the best one could hope for. However, unlike the case of computation, here a significant improvement cannot be completely ruled out.

RESILIENCE. Our protocol is computationally UC-secure [6] against an active, adaptive adversary corrupting at most a $(1/2 - \epsilon)$ fraction of the players, for an arbitrarily small constant $\epsilon > 0$. This parameter too is essentially optimal since robust protocols that guarantee output delivery require honest majority.

ROUNDS. The round complexity of the basic version of the protocol is $\mathrm{poly}(k, n)$. Using a pseudorandom generator that is "computationally simple" (e.g., computable in NC^1), the protocol can be modified to run in a constant number of rounds. Such a pseudorandom generator is implied by most standard concrete intractability assumptions in cryptography [1]. Unlike our main protocol, the constant-round variant only applies to functionalities that deliver outputs to a small (say, constant) number of players. Alternatively, it may apply to arbitrary functionalities but provide the weaker guarantee of "security with abort".

The most efficient previous MPC protocols from the literature [3, 12, 15, 28] have communication complexity of $\widetilde{\mathcal{O}}(n \cdot |C|)$, and no better complexity even in the semi-honest model. The protocols of Damgård and Nielsen [15] and Beerliova and Hirt [3] achieve this complexity with unconditional security. It should be noted that the protocol of Damgård and Ishai [12] has a variant that matches the asymptotic complexity of our protocol. However, this variant applies only to functionalities that receive inputs from and distribute outputs to a small number of players. Furthermore, it only tolerates a small fraction of corrupted players.

Techniques. Our protocol borrows ideas and techniques from several previous works in the area, especially [3, 12, 15, 28]. Similarly to [12], we combine the

---

[1] Such terms are to some extent unavoidable, and have also been ignored in previous works along this line. Note that the additive term becomes insignificant when considering complex computations (or even simple computations on large inputs), whereas the multiplicative term can be viewed as polylogarithmic under exponential security assumptions. The question of minimizing these lower order terms, which are significant in practice, is left for further study.

efficient secret sharing scheme of Franklin and Yung [20] with Yao's garbled circuit technique [35]. The scheme of Franklin and Yung generalizes Shamir's secret sharing scheme [33] to efficiently distribute a whole block of ℓ secrets, at the price of decreasing the security threshold. Yao's technique can be used to transform the circuit C into an equivalent, but very shallow, *randomized* circuit C_{Yao} of comparable size. The latter, in turn, can be evaluated "in parallel" on blocks of inputs and randomness that are secret-shared using the scheme of [20].

The main efficiency bottleneck in [12] is the need to distribute the blocks of randomness that serve as inputs for C_{Yao}. The difficulty stems from the fact that these blocks should be arranged in a way that reflects the structure of C. That is, each random secret bit may appear in several blocks according to a pattern determined by C. These blocks were generated in [12] by adding contributions from different players, which is not efficient enough for our purposes. More efficient methods for distributing many random secrets were used in [3, 15, 28]. However, while these methods can be applied to cheaply generate many blocks of the same pattern, the blocks we need to generate may have arbitrary patterns.

To get around this difficulty, we use a pseudorandom function (PRF) for reducing the problem of generating blocks of an arbitrary structure to the problem of generating independent random blocks. This is done by applying the PRF (with a key that is secret-shared between the servers) to a sequence of public labels that specifies the required replication pattern, where identical labels are used to generate copies of the same secret.

Another efficiency bottleneck we need to address is the cost of delivering the outputs. If many players should receive an output, we cannot afford to send the entire output of C_{Yao} to these players. To get around this difficulty, we propose a procedure for securely distributing the decoding process between the players without incurring too much extra work. This also has the desirable effect of dividing the work equally between the players.

Finally, to boost the fractional security threshold of our protocol from a small constant δ to a nearly optimal constant of $(1/2 - \epsilon)$, we adapt to our setting a technique that was introduced by Bracha [5] in the context of Byzantine Agreement. The idea is to compose our original protocol π_{out}, which is efficient but has a low security threshold ($t < n/c$), with another known protocol π_{in}, which is inefficient but has an optimal security threshold ($t < n/2$) in a way that will give us essentially the best of both worlds. The composition uses π_{in} to distribute the local computations of each player in π_{out} among a corresponding committee that includes a constant number of players. The committees are chosen such that any set including at most $1/2 - \epsilon$ of the players forms a majority in less than δn of the committees. Bracha's technique has been recently applied in the cryptographic contexts of secure message transmission [17] and establishing a network of OT channels [23]. We extend the generality of the technique by applying it as a method for boosting the security threshold of general MPC protocols with only a minor loss of efficiency.

2   Preliminaries

In this section we present some useful conventions.

Client-server model. Similarly to previous works, it will be convenient to slightly refine the usual MPC model as follows. We assume that the set of players consists of a set of *input clients* that hold the inputs to the desired computation, a set of n *servers*, $\mathcal{S} = \{S_1, \ldots, S_n\}$, that execute the computation, and a set of *output clients* that receive outputs. Since one player can play the role of both client(s) and a server, this is a generalization of the standard model. The number of clients is assumed to be at most linear in n, which allows us to ignore the exact number of clients when analyzing the asymptotic complexity of our protocols.

Complexity conventions. We will represent the functionality which we want to securely realize by a boolean circuit C with bounded fan-in, and denote by $|C|$ the number of gates in C. We adopt the convention that every input gate in C is labeled by the input client who should provide this input (alternatively, labeled by "random" in the case of a randomized functionality) and every output gate in C is labeled by a name of a single output client who should receive this output. In particular, distributing an output to several clients must be "paid for" by having a larger circuit. Without this rule, we could be asked to distribute the entire output $C(x)$ to all output clients, forcing the communication complexity to be more than we can afford. We denote by k a cryptographic security parameter, which is thought of as being much smaller than n (e.g., $k = O(n^\epsilon)$ for a small constant $\epsilon > 0$, or even $k = \text{polylog}(n)$).

Security conventions. By default, when we say that a protocol is "secure" we mean that it realizes in the UC model [6] the corresponding functionality with computational t-security against an *active* (malicious) and *adaptive* adversary, using synchronous communication over secure point-to-point secure channels. Here t denotes the maximal number of corrupted server; there is no restriction on the number of corrupted clients. (The threshold t will typically be of the form δn for some constant $0 < \delta < 1/2$.) The results can be extended to require only authenticated channels assuming the existence of public key encryption (even for adaptive corruptions, cf. [7]). We will sometimes make the simplifying assumption that outputs do not need to be kept private. This is formally captured by letting the ideal functionality leak $C(x)$ to the adversary. Privacy of outputs can be achieved in a standard way by having the functionality mask the output of each client with a corresponding input string picked randomly by this client.

3   Building Blocks

In this section, we will present some subprotocols that will later be put together in a protocol implementing a functionality F_{CP}, which allows to evaluate the same circuit in parallel on multiple inputs. We will argue that each subprotocol is *correct*: every secret-shared value that is produced as output is consistently shared, and *private*: the adversary learns nothing about secrets shared by uncorrupted parties. While correctness and privacy alone do not imply UC-security, when combined with standard simulation techniques for honest-majority MPC protocols they will imply that our implementation of F_{CP} is UC-secure.

Packed Secret-Sharing. We use a variant of the packed secret-sharing technique by Franklin and Yung [20]. We fix a finite field $\mathbb{F}$ of size $\mathcal{O}(\log(n)) = \widetilde{\mathcal{O}}(1)$ and share together a vector of field elements from $\mathbb{F}^\ell$, where ℓ is a constant fraction of n. We call $s = (s^1, \ldots, s^\ell) \in \mathbb{F}^\ell$ a *block*. Fix a generator α of the multiplicative group of $\mathbb{F}$ and let $\beta = \alpha^{-1}$. We assume that $|\mathbb{F}| > 2n$ such that $\beta^0, \ldots, \beta^{c-1}$ and $\alpha^1, \ldots, \alpha^n$ are distinct elements. Given $x = (x_0, \ldots, x_{c-1}) \in \mathbb{F}^c$, compute the unique polynomial $f(\mathbf{X}) \in \mathbb{F}[\mathbf{X}]$ of degree $\leq c-1$ for which $f(\beta^i) = x_i$ for $i = 0, \ldots, c-1$, and let $M_{c \to n}(x) = (y_1, \ldots, y_n) = (f(\alpha^1), \ldots, f(\alpha^n))$. This map is clearly linear, and we use $M_{c \to n}$ to denote both the mapping and its matrix. Let $M_{c \to r}$ consist of the top r rows of $M_{c \to n}$.

Since the mapping consists of a polynomial interpolation followed by a polynomial evaluation, one can use the fast Fourier transform (FFT) to compute the mapping in time $\widetilde{\mathcal{O}}(c) + \widetilde{\mathcal{O}}(n) = \widetilde{\mathcal{O}}(n)$. In [3] it is shown that $M_{c \to n}$ is *hyper-invertible*. A matrix M is hyper-invertible if the following holds: Let R be a subset of the rows, and let M_R denote the sub-matrix of M consisting of rows in R. Likewise, let C be a subset of columns and let M^C denote the sub-matrix consisting of columns in C. Then we require that M_R^C is invertible whenever $|R| = |C| > 0$. Note that from $M_{c \to n}$ being hyper-invertible and computable in $\widetilde{\mathcal{O}}(n)$ time, it follows that all $M_{c \to r}$ are hyper-invertible and computable in $\widetilde{\mathcal{O}}(n)$ time.

Protocol SHARE(D, d):

1. Input to dealer D: $(s^1, \ldots, s^\ell) \in \mathbb{F}^\ell$. Let $M = M_{\ell+t \to n}$, where $t = d - \ell + 1$.
2. D: Sample $r^1, \ldots, r^t \in_R \mathbb{F}$, let $(s_1, \ldots, s_n) = M(s^1, \ldots, s^\ell, r^1, \ldots, r^t)$, and send s_i to server S_i, for $i = 1, \ldots, n$.

The sharing protocol is given in Protocol SHARE(D, d). Note that $(s_1, \ldots, s_n)$ is just a t-private packed Shamir secret sharing of the secret block $(s^1, \ldots, s^\ell)$ using a polynomial of degree $\leq d$. We therefore call $(s_1, \ldots, s_n)$ a *d-sharing* and write $[s]_d = [s^1, \ldots, s^\ell]_d = (s_1, \ldots, s_n)$. In general we call a vector $(s_1, \ldots, s_n)$ a *consistent d-sharing* (over $\mathcal{S} \subseteq \{S_1, \ldots, S_n\}$) if the shares (of the servers in $\mathcal{S}$) are consistent with some d-sharing. For $a \in \mathbb{F}$ we let $a[s]_d = (as_1, \ldots, as_n)$ and for $[s]_d = (s_1, \ldots, s_n)$ and $[t]_d = (t_1, \ldots, t_n)$ we let $[s]_d + [t]_d = (s_1 + t_1, \ldots, s_n + t_n)$. Clearly, $a[s]_d + b[t]_d$ is a d-sharing of $as + bt$; We write $[as + bt]_d = a[s]_d + b[t]_d$. We let $[st]_{2d} = (s_1 t_1, \ldots, s_n t_n)$. This is a $2d$-sharing of the block $st = (s^1 t^1, \ldots, s^\ell t^\ell)$.

Below, when we instruct a server to check if $y = (y_1, \ldots, y_n)$ is d-consistent, it interpolates the polynomial $f(\alpha^i) = y_i$ and checks that the degree is $\leq d$. This can be done in $\widetilde{\mathcal{O}}(n)$ time using FFT.

To be able to reconstruct a sharing $[s]_{d_1}$ given t faulty shares, we need that $n \geq d_1 + 1 + 2t$. We will only need to handle up to $d_1 = 2d$, and therefore need $n = 2d + 1 + 2t$. Since $d = \ell + t - 1$ we need $n \geq 4t + 2\ell - 1$ servers. To get the efficiency we are after, we will need that ℓ, $n - 4t$ and t are $\Theta(n)$. Concretely we could choose, for instance, $t = n/8, \ell = n/4$.

Random Monochromatic Blocks. In the following, we will need a secure protocol for the following functionality:

Functionality MONOCHROM:

Takes no input.

Output: a uniformly random sharing $[b]_d$, where the block b is $(0, \ldots, 0)$ with probability $\frac{1}{2}$ and $(1, \ldots, 1)$ with probability $\frac{1}{2}$.

We only call the functionality k times in total, so the complexity of its implementation does not matter for the amortized complexity of our final protocol.

Semi-Robust VSS. To get a verifiable secret sharing protocol guaranteeing that the shares are d-consistent we adapt to our setting a VSS from [3].[2] Here and in the following subprotocols, several non-trivial modifications have to be made, however, due to our use of packed secret sharing, and also because directly using the protocol from [3] would lead to a higher complexity than we can afford.

Protocol SEMIROBUSTSHARE(d):

1. For each dealer D and each group of blocks $(x_1, \ldots, x_{n-3t}) \in (\mathbb{F}^\ell)^{n-3t}$ to be shared by D, the servers run the following in parallel:
   (a) D: Pick t uniformly random blocks $x_{n-3t+1}, \ldots, x_{n-2t}$ and deal $[x_i]_d$ for $i = 1, \ldots, n - 2t$, using SHARE(D, d).
   (b) All servers: Compute $([y_1]_d, \ldots, [y_n]_d) = M([x_1]_d, \ldots, [x_{n-2t}]_d)$ by locally applying M to the shares.
   (c) Each S_j: Send the share y_i^j of $[y_i]_d$ to S_i.
   (d) D: send the shares y_i^j of $[y_i]_d$ to S_i.
2. Now conflicts between the sent shares are reported. Let $\mathcal{C}$ be a set of subsets of $\mathcal{S}$, initialized to $\mathcal{C} := \emptyset$. Each S_i runs the following in parallel:
   (a) If S_i sees that D for some group sent shares which are not d-consistent, then S_i broadcasts (J'accuse, D), and all servers add $\{D, S_i\}$ to $\mathcal{C}$.
   (b) Otherwise, if S_i sees that there is some group dealt by D and some S_j which for this group sent y_i^j and D sent $y_i^{j\prime} \neq y_i^j$, then S_i broadcasts (J'accuse, $D, S_j, g, y_i^{j\prime}, y_i^j$) for all such S_j, where g identifies the group for which a conflict is claimed. At most one conflict is reported for each pair (D, S_j).
   (c) If D sees that $y_i^{j\prime}$ is not the share it sent to S_j for group g, then D broadcasts (J'accuse, S_j), and all servers add $\{D, S_j\}$ to $\mathcal{C}$.
   (d) At the same time, if S_i sees that y_i^j is not the share it sent to S_j for group g, then S_i broadcasts (J'accuse, S_j), and all servers add $\{S_i, S_j\}$ to $\mathcal{C}$.
   (e) If neither D nor S_i broadcast (J'accuse, S_j), they acknowledge to have sent different shares to S_j for group g, so one of them is corrupted. In this case all servers add $\{D, S_i\}$ to $\mathcal{C}$.
3. Now the conflicts are removed by *eliminating* some players:
   (a) As long as there exists $\{S_1, S_2\} \in \mathcal{C}$ such that $\{S_1, S_2\} \subseteq \mathcal{S}'$, let $\mathcal{S}' := \mathcal{S}' \setminus \{S_1, S_2\}$.
   (b) The protocol outputs the $[x_i]_d$ created by non-eliminated dealers.

---

[2] This protocol has an advantage over previous subprotocols with similar efficiency, e.g. from [12], in that it has *perfect* (rather than statistical) security. This makes it simpler to analyze its security in the presence of adaptive corruptions.

The protocol uses $M = M_{n-2t \to n}$ to check consistency of sharings. For efficiency, all players that are to act as dealers will deal at the same time. The protocol can be run with all servers acting as dealers. Each dealer D shares a *group* of $n - 3t = \Theta(n)$ blocks, and in fact, D handles a number of such groups in parallel. Details are given in Protocol SEMIROBUSTSHARE. Note that SEMIROBUSTSHARE(d) may not allow all dealers to successfully share their blocks, since some can be eliminated during the protocol. We handle this issue later in Protocol ROBUSTSHARE.

At any point in our protocol, $\mathcal{S}'$ will be the set of servers that still participate. We set $n' = |\mathcal{S}'|$ and $t' = t - e$ will be the maximal number of corrupted servers in $\mathcal{S}'$, where e is the number of pairs eliminated so far.

To argue correctness of the protocol, consider any surviving dealer $D \in \mathcal{S}'$. Clearly D has no conflict with any surviving server, i.e., there is no $\{D, S_i\} \in \mathcal{C}$ with $\{D, S_i\} \subset \mathcal{S}'$. In particular, all $S_i \in \mathcal{S}'$ saw D send only d-consistent sharings. Furthermore, each such S_i saw each $S_j \in \mathcal{S}'$ send the same share as D during the test, or one of $\{D, S_j\}, \{S_i, S_j\}$ or $\{D, S_i\}$ would be in $\mathcal{C}$, contradicting that they are all subsets of $\mathcal{S}'$.

Since each elimination step $\mathcal{S}' := \mathcal{S}' \setminus \{S_1, S_2\}$ removes at least one new corrupted server, it follows that at most t honest servers were removed from $\mathcal{S}'$. Therefore there exists $H \subset \mathcal{S}'$ of $n - 2t$ honest servers. Let $([y_i]_d)_{S_i \in H} = M_H([x_1]_d, \ldots, [x_{n-2t}]_d)$. By the way conflicts are removed, all $[y_i]_d$, $S_i \in H$ are d-consistent on $\mathcal{S}'$. Since M_H is invertible, it follows that all $([x_1]_d, \ldots, [x_{n-t}]_d) = M_H^{-1}([y_i]_d)_{S_i \in H}$ are d-consistent on $\mathcal{S}'$.

The efficiency follows from $n - 3t = \Theta(n)$, which implies a complexity of $\widetilde{\mathcal{O}}(\beta n) + \text{poly}(n)$ for sharing β blocks (here $\text{poly}(n)$ covers the $\mathcal{O}(n^3)$ broadcasts). Since each block contains $\Theta(n)$ field elements, we get a complexity of $\widetilde{\mathcal{O}}(\phi)$ for sharing ϕ field elements.

As for privacy, let $I = \{1, \ldots, n - 3t\}$ be the indices of the data blocks and let $R = \{n - 3t + 1, \ldots, n - 2t\}$ be the indices of the random blocks. Let $C \subset \{1, \ldots, n\}$, $|C| = t$ denote the corrupted servers. Then $([y_i]_d)_{i \in C} = M_C([x_1]_d, \ldots, [x_{n-2t}]_d) = M_C^I([x_i]_d)_{i \in I} + M_C^R([x_i]_d)_{i \in R}$. Since $|C| \nleq |R|$, M_C^R is invertible. So, for each $([x_i]_d)_{i \in D}$, exactly one choice of random blocks $([x_i]_d)_{i \in R} = (M_C^R)^{-1}(([y_i]_d)_{i \in C} - M_C^I([x_i]_d)_{i \in I})$ are consistent with this data, which implies perfect privacy.

Double Degree VSS. We also use a variant SEMIROBUSTSHARE(d_1, d_2), where each block x_i is shared both as $[x_i]_{d_1}$ and $[x_i]_{d_2}$ (for $d_1, d_2 \leq 2d$). The protocol executes SEMIROBUSTSHARE(d_1) and SEMIROBUSTSHARE(d_2), in parallel, and in Step 2a in SEMIROBUSTSHARE the servers also accuse D if the d_1-sharing and the d_2-sharing is not of the same value. It is easy to see that this guarantees that all $D \in \mathcal{S}'$ shared the same x_i in all $[x_i]_{d_1}$ and $[x_i]_{d_2}$.

Reconstruction. We use the following procedure for reconstruction towards a server R.

Protocol RECO(R, d_1):

1. The servers hold a sharing $[s]_{d_1}$ which is d_1-consistent over $\mathcal{S}'$ (and $d_1 \leq 2d$). The server R holds a set $\mathcal{C}_i$ of servers it knows are corrupted. Initially $\mathcal{C}_i = \emptyset$.

2. Each $S_i \in \mathcal{S}'$: Send the share s_i to R.

3. R: If the shares s_i are d_1-consistent over $\mathcal{S}' \setminus \mathcal{C}_i$, then compute s by interpolation. Otherwise, use error correction to compute the nearest sharing $[s']_{d_1}$ which is d_1-consistent on $\mathcal{S}' \setminus \mathcal{C}_i$, and compute s from this sharing using interpolation. Furthermore, add all S_j for which $s'_j \neq s_j$ to $\mathcal{C}_i$.

Computing the secret by interpolation can be done in time $\widetilde{\mathcal{O}}(n)$. For each invocation of the poly(n)-time error correction, at least one corrupted server is removed from $\mathcal{C}_i$, bounding the number of invocations by t. Therefore the complexity for reconstructing β blocks is $\widetilde{\mathcal{O}}(\beta n) + \text{poly}(n) = \widetilde{\mathcal{O}}(\phi)$, where ϕ is the number of field elements reconstructed.

At the time of reconstruction, some e eliminations have been performed to reach $\mathcal{S}'$. For the error correction to be possible, we need that $n' \geq d_1 + 1 + 2t'$. In the worst case one honest party is removed per elimination. So we can assume that $n' = n - 2e$ and $t' = t - e$. So, it is sufficient that $n \geq d_1 + 1 + 2t$, which follows from $n \geq 2d + 1 + 2t$ and $d_1 \leq 2d$.

Robust VSS. Protocol ROBUSTSHARE guarantees that all dealers can secret share their blocks, and can be used by input clients to share their inputs. Privacy follows as for SEMIROBUSTSHARE. Correctness is immediate. Efficiency follows directly from $n - 4t = \mathcal{O}(n)$, which guarantees a complexity of $\widetilde{\mathcal{O}}(\phi)$ for sharing ϕ field elements.

Protocol ROBUSTSHARE(d):

1. Each dealer D shares groups of $n - 4t$ blocks $x_1, \ldots, x_{n-4t}$. For each group it picks t random blocks $x_{n-4t+1}, \ldots, x_{n-3t}$, computes n blocks $(y_1, \ldots, y_n) = M(x_1, \ldots, x_{n-3t})$ and sends y_i to S_i. Here $M = M_{n-4t \rightarrow n}$.

2. The parties run SEMIROBUSTSHARE(d), and each S_i shares y_i.[a] This gives a reduced server set $\mathcal{S}'$ and a d-consistent sharing $[y'_i]_d$ for each $S_i \in \mathcal{S}'$.

3. The parties run RECO(D, d) on $[y'_i]_d$ for $S_i \in \mathcal{S}'$ to let D learn y'_i for $S_i \in \mathcal{S}'$.

4. D picks $H \subset \mathcal{S}'$ for which $|H| = n - 3t$ and $y'_i = y_i$ for $S_i \in H$, and broadcasts H, the indices of these parties.[b]

5. All parties compute $([x_1]_d, \ldots, [x_{n-3t}]_d) = M_H^{-1}([y_i]_d)_{i \in H}$. Output is $[x_1]_d, \ldots, [x_{n-4t}]_d$.

---

[a] In the main protocol, many copies of ROBUSTSHARE will be run in parallel, and S_i can handle the y_i's from all copies in parallel, putting them in groups of size $n - 3t$.

[b] $\mathcal{S}'$ has size at least $n - 2t$, and at most the t corrupted parties did not share the right value. When many copies of ROBUSTSHARE(d) are run in parallel, only one subset H is broadcast, which works for all copies.

Sharing Bits. We also use a variant ROBUSTSHAREBITS(d), where the parties are required to input bits, and where this is checked. First ROBUSTSHARE(d) is run to do the actual sharing. Then for each shared block $[x^1, \ldots, x^\ell]$ the parties compute $[y^1, \ldots, y^\ell]_{2d} = ([1, \ldots, 1]_d - [x^1, \ldots, x^\ell])[x^1, \ldots, x^\ell] = [(1 - x^1)x^1, \ldots, (1 - x^\ell)x^\ell]$. They generate $[1, \ldots, 1]_d$ by all picking the share 1. Note that $[y]_{2d} = [0, \ldots, 0]_{2d}$ if and only if all x^i were in $\{0, 1\}$.

For each dealer D all $[y]_{2d}$ are checked in parallel, in groups of $n' - 2t'$. For each group $[y_1]_{2d}, \ldots, [y_{n'-2t'}]_{2d}$, D makes sharings $[y_{n'-2t'+1}]_{2d}, \ldots, [y_{n'-t'}]_{2d}$ of $y_i = (0, \ldots, 0)$, using ROBUSTSHARE($2d$). Then all parties compute $([x_1]_{2d}, \ldots, [x_{n'}]_{2d}) = M([y_1]_{2d}, \ldots, [y_{n'-t'}]_{2d})$, where $M = M_{n'-t' \to n'}$. Then each $[x_i]_{2d}$ is reconstructed towards S_i. If all $x_i = (0, \ldots, 0)$, then S_i broadcasts ok. Otherwise S_i for each cheating D broadcasts (J'accuse, D, g), where D identifies the dealer and g identifies a group $([x_1]_{2d}, \ldots, [x_{n'}]_{2d})$ in which it is claimed that $x_i \neq (0, \ldots, 0)$. Then the servers publicly reconstruct $[x_i]_d$ (i.e., reconstruct it towards each server using RECO($2d, \cdot$)). If $x_i = (0, \ldots, 0)$, then S_i is removed from $\mathcal{S}'$; otherwise, D is removed from $\mathcal{S}'$, and the honest servers output the all-zero set of shares.

Let H denote the indices of $n' - t'$ honest servers. Then $([x_i]_{2d})_{i \in H} = M_H([y_1]_{2d}, \ldots, [y_{n'-t'}]_{2d})$. So, if $x_i = (0, \ldots, 0)$ for $i \in H$, it follows from $([y_1]_{2d}, \ldots, [y_{n'-t'}]_{2d}) = M_H^{-1}([x_i]_{2d})_{i \in H}$ that all $y_i = (0, \ldots, 0)$. Therefore D will pass the test if and only if it shared only bits. The privacy follows using the same argument as in the privacy analysis of Protocol SEMIROBUSTSHARE. The efficiency follows from $\Theta(n)$ blocks being handled in each group, and the number of broadcasts and public reconstructions being independent of the number of blocks being checked.

Resharing with a Different Degree. We need a protocol which given a d_1-consistent sharing $[x]_{d_1}$ produces a d_2-consistent sharing $[x]_{d_2}$ (here $d_1, d_2 \leq 2d$). For efficiency all servers R act as resharer, each handling a number of groups of $n' - 2t' = \Theta(n)$ blocks. The protocol is not required to keep the blocks x secret. We first present a version in which some R might fail.

Protocol SEMIROBUSTRESHARE(d_1, d_2):

- For each $R \in \mathcal{S}'$ and each group $[x_1]_{d_1}, \ldots, [x_{n'-2t'}]_{d_1}$ (all sharings are d_1-consistent on $\mathcal{S}'$) to be reshared by R, the servers proceed as follows:
- Run RECO(R, d_1) on $[x_1]_{d_1}, \ldots, [x_{n'-2t'}]_{d_1}$ to let R learn $x_1, \ldots, x_{n'-2t'}$.
- Run SEMIROBUSTSHARE(d_2), where each R inputs $x_1, \ldots, x_{n'-2t'}$ to produce $[x_1]_{d_2}, \ldots, [x_{n'-2t'}]_{d_2}$ (step 1a is omitted as we do not need privacy). At the same time, check that R reshared the same blocks, namely in Step 1b we also apply M to the $[x_1]_{d_1}, \ldots, [x_{n'-2t'}]_{d_1}$, in Step 2a open the results to the servers and check for equality. Conflicts are removed by elimination as in SEMIROBUSTSHARE.

Now all groups handled by $R \in \mathcal{S}'$ were correctly reshared with degree d_2. To deal with the fact that some blocks might not be reshared, we use the same idea as when we turned SEMIROBUSTSHARE into ROBUSTSHARE, namely the servers first apply $M_{n'-2t' \to n'}$ to each group of blocks to reshare, each of the resulting n'

sharings are assigned to a server. Then each server does SEMIROBUSTRESHARE on all his assigned sharings. Since a sufficient number of servers will complete this successfully, we can reconstruct d_2-sharings of the x_i's. This protocol is called ROBUSTRESHARE.

Random Double Sharings. We use the following protocol to produce double sharings of blocks which are uniformly random in the view of the adversary.

Protocol RANDOUSHA(d):

1. Each server S_i: Pick a uniformly random block $R_i \in_R \mathbb{F}^\ell$ and use SEMIROBUSTSHARE($d, 2d$) to deal $[R_i]_d$ and $[R_i]_{2d}$.
2. Let $M = M_{n' \to n' - t'}$ and let $([r_1]_d, \ldots, [r_{n'-t'}]_d) = M([R_i]_d)_{S_i \in \mathcal{S}'}$ and $([r_1]_{2d}, \ldots, [r_{n'-t'}]_{2d}) = M([R_i]_{2d})_{S_i \in \mathcal{S}'}$. The output is the pairs $([r_i]_d, [r_i]_{2d})$, $i = 1, \ldots, n' - t'$.

Security follows by observing that when $M = M_{n' \to n' - t'}$, then $M^H : \mathbb{F}^{n'-t'} \to \mathbb{F}^{n'-t'}$ is invertible when $|H| = n' - t'$. In particular, the sharings of the (at least) $n' - t'$ honest servers fully randomize the $n' - t'$ generated sharings in Step 2.

In the following, RANDOUSHA(d) is only run once, where a large number, β, of pairs $([r]_d, [r]_{2d})$ are generated in parallel. This gives a complexity of $\widetilde{\mathcal{O}}(\beta n) + \text{poly}(n) = \widetilde{\mathcal{O}}(\phi)$, where ϕ is the number of field elements in the blocks.

Functionality $F_{CP}(A)$

The functionality initially chooses a random bitstring $K_1, .., K_k$ where k is the security parameter. It uses gm blocks of input bits $z_1^1, \ldots, z_m^1, \ldots, z_1^g, \ldots z_m^g$. Each block z_u^v can be:

- owned by an input client. The client can send the bits in z_u^v to F_{CP}, but may instead send "refuse", in which case the functionality sets $z_i^j = (0, \ldots 0)$.
- Random, of type w, $1 \leq w \leq k$, then the functionality sets $z_u^v = (K_w, \ldots, K_w)$.
- Public, in which case some arbitrary (binary string) value for z_u^w is hard-wired into the functionality.

The functionality works as follows:

1. After all input clients have provided values for the blocks they own, compute $A(z_1^v, \ldots, z_m^v)$ for $v = 1..g$.
2. On input "open v to server S_a" from all honest servers, send $A(z_1^v, \ldots, z_m^v)$ to server S_a.

Parallel Circuit Evaluation. Let $A : \mathbb{F}^m \to \mathbb{F}$ be an arithmetic circuit over $\mathbb{F}$. For m blocks containing binary values $z_1 = (z_{1,1}, \ldots, z_{1,\ell}), \ldots, z_m = (z_{m,1}, \ldots, z_{m,\ell})$ we let $A(z_1, \ldots, z_m) = (A(z_{1,1}, \ldots, z_{m,1}), \ldots, A(z_{1,\ell}, \ldots, z_{m,\ell}))$. We define an ideal functionality F_{CP} which on input that consists of such a group of input blocks will compute $A(z_1, \ldots, z_m)$. To get an efficient implementation, we will handle g groups of input blocks, denoted $z_1^1, \ldots, z_m^1, \ldots, z_1^g, \ldots z_m^g$ in parallel. Some of these bits will be chosen by input clients, some will be random, and some are public values, hardwired into the functionality. See the figure for details. The subsequent protocol COMPPAR securely implements F_{CP}. As for its efficiency, let γ denote the number of gates in A, and let M denote the multiplicative depth of the circuit (the number of times Step 2b is executed). Assume that $M = \text{poly}(k)$, as will be the case later. Then the complexity is easily seen to be $\widetilde{\mathcal{O}}(\gamma gn) + M \text{poly}(n) = \widetilde{\mathcal{O}}(\gamma gn)$. Let μ denote the number of inputs on which A is being evaluated. Clearly $\mu = g\ell = \Theta(gn)$, giving a complexity of $\widetilde{\mathcal{O}}(\gamma \mu)$. If we assume that $\gamma = \text{poly}(k)$, as will be the case later, we get a complexity of $\widetilde{\mathcal{O}}(\gamma \mu) = \widetilde{\mathcal{O}}(\mu)$, and this also covers the cost of sharing the inputs initially.

Protocol COMPPAR(A):

1. The servers run RANDOUSHA(d) to generate a pair $([r]_d, [r]_{2d})$ for each multiplication to be performed in the following.
2. Input: for each input client D, run ROBUSTSHAREBITS(d) in parallel for all blocks owned by D. Run MONOCHROM k times to get $[K_t, \ldots, K_t]_d$, for $t = 1 \ldots k$, and let $[z_u^v]_d = [K_w, \ldots, K_w]_d$ if z_u^v is random of type w. Finally, for all public z_u^v, we assume that default sharings of these blocks are hardwired into the programs of the servers.
   The servers now hold packed sharings $[z_u^v]_d$, all of which are d-consistent on $\mathcal{S}'$. Now do the following, for each of the g groups, in parallel:
   (a) For all addition gates in A, where sharings $[x]_d$ and $[y]_d$ of the operands are ready, the servers compute $[x + y]_d = [x]_d + [y]_d$ by locally adding shares. This yields a d-consistent sharing on $\mathcal{S}'$.
   (b) Then for all multiplication gates in A, where sharings $[x]_d$ and $[y]_d$ of the operands are ready, the servers execute:
      i. Compute $[xy + r]_{2d} = [a]_d[b]_d + [r]_{2d}$, by local multiplication and addition of shares. This is a $2d$-consistent sharing of $xy + r$ on $\mathcal{S}'$.
      ii. Call ROBUSTRESHARE($2d, d$) to compute $[xy + r]_d$ from $[xy + r]_{2d}$. This is a d-consistent sharing of $xy + r$ on the reduced server set $\mathcal{S}'$. Note that all resharings are handled by one invocation of ROBUSTRESHARE. Finally compute $[xy]_d = [xy + r]_d - [r]_d$.
   (c) If there are still gates which were not handled, go to Step 2a.
3. Output: When all gates have been handled, the servers hold for each group a packed sharing $[A(z_1^v, \ldots, z_m^v)]_d$ which is d-consistent over the current reduced server set $\mathcal{S}'$. To open group v to server S_a, run RECO(S_a, d).

Lemma 1. *Protocol* COMPPAR *securely implements* F_{CP}.

Sketch of proof: The simulator will use standard techniques for protocols based on secret sharing, namely whenever an honest player secret-shares a new block, the simulator will hand random shares to the corrupt servers. When a corrupted player secret-shares a value, the simulator gets all shares intended for honest servers, and follows the honest servers' algorithm to compute their reaction to this. In some cases, a value is reconstructed towards a corrupted player as part of a subprotocol. Such values are always uniformly random and this is therefore trivial to simulate. The simulator keeps track of all messages exchanged with corrupt players in this way. The perfect correctness of all subprotocols guarantees that the simulator can compute, from its view of RobustShareBits, the bits shared by all corrupt input clients, it will send these to F_{CP}. When an input client or a server is corrupted, the simulator will get the actual inputs of the client, respectively the outputs received by the server. It will then construct a random, complete view of the corrupted player, consistent with the values it just learned, and whatever messages the new corrupted player has exchanged with already corrupted players. This is possible since all subprotocols have perfect privacy. Furthermore the construction can be done efficiently by solving a system of linear equations, since the secret sharing scheme is linear. Finally, to simulate an opening of an output towards a corrupted server, we get the correct value from the functionality, form a complete random set of shares consistent with the shares the adversary has already and the output value, and send the shares to the adversary. This matches what happens in a real execution: since all subprotocols have perfect correctness, a corrupted server would also in real life get consistent shares of the correct output value from all honest servers. It is straightforward but tedious to argue that this simulation is *perfect*.   □

4   Combining Yao Garbled Circuits and Authentication

To compute a circuit C securely, we will use a variant of Yao's garbled circuit construction [34, 35]. It can be viewed as building from an arbitrary circuit C together with a pseudorandom generator a new (randomized) circuit C_{Yao} whose depth is only $\mathrm{poly}(k)$ and whose size is $|C| \cdot \mathrm{poly}(k)$. The output of $C(x)$ is equivalent to the output of $C_{\mathsf{Yao}}(x, r)$, in the sense that given $C_{\mathsf{Yao}}(x, r)$ one can efficiently compute $C(x)$, and given $C(x)$ one can efficiently sample from the output distribution $C_{\mathsf{Yao}}(x, r)$ induced by a uniform choice of r (up to computational indistinguishability). Thus, the task of securely computing $C(x)$ can be reduced to the task of securely computing $C_{\mathsf{Yao}}(x, r)$, where the randomness r should be picked by the functionality and remain secret from the adversary.

In more detail, $C_{\mathsf{Yao}}(x, r)$ uses for each wire w in C two random encryption keys K_0^w, K_1^w and a random wire mask γ_w. We let $E_K()$ denote an encryption function using key K, based on the pseudorandom generator used. The construction works with an encrypted representation of bits, concretely $\mathrm{garble}_w(y) = (K_y^w, \gamma_w \oplus y)$ is called a *garbling of* y. Clearly, if no side information on keys or wire masks is known, $\mathrm{garble}_w(y)$ gives no information on y.

The circuit $C_{\mathsf{Yao}}(x, r)$ outputs for each gate in C a table with 4 entries, indexed by two bits (b_0, b_1). We can assume that each gate has two input wires l, r and output wire *out*. If we consider a circuit C made out of only NAND gates, $\barwedge$, a single entry in the table looks as follows:

$$(b_0, b_1): \quad E_{K_{b_0 \oplus \gamma_l}^l} \left(E_{K_{b_1 \oplus \gamma_r}^r} \left(\mathrm{garble}_{out} \left([b_0 \oplus \gamma_l] \barwedge [b_1 \oplus \gamma_r] \right) \right) \right) \ .$$

The tables for the output gates contain encryptions of the output bits without garbling, i.e., $[b_0 \oplus \gamma_l] \barwedge [b_1 \oplus \gamma_r]$ is encrypted. Finally, for each input wire w_i, carrying input bit x_i, the output of $C_{\mathsf{Yao}}(x, r)$ includes $\mathrm{garble}_{w_i}(x_i)$.

It is straightforward to see that the tables are designed such that given garble$_l(b_l)$, garble$_r(b_r)$, one can compute garble$_{out}(b_l \wedge b_r)$. One can therefore start from the garbled inputs, work through the circuit in the order one would normally visit the gates, and eventually learn (only) the bits in the output $C(x)$. We will refer to this as *decoding* the Yao garbled circuit.

In the following, we will need to share the work of decoding a Yao garbling among the servers, such that one server only handles a few gates and then passes the garbled bits it found to other servers. In order to prevent corrupt servers from passing incorrect information, we will augment the Yao construction with digital signatures in the following way.

The authenticated circuit $C_{\texttt{AutYao}}(x, r)$ uses a random input string r and will first generate a key pair $(sk, pk) = \text{gen}(r')$, for a digital signature scheme, from some part r' of r. It makes pk part of the output. Signing of message m is denoted $S_{sk}(m)$. It will then construct tables and encrypted inputs exactly as before, except that a table entry will now look as follows:

$$G(b_0, b_1) = E_{K^l_{b_0 \oplus \gamma_l}} \left(E_{K^r_{b_1 \oplus \gamma_r}} \left(\text{garble}_{out}([b_0 \oplus \gamma_l] \wedge [b_1 \oplus \gamma_r]), S_{sk}(e, b_0, b_1, L) \right) \right) ,$$

where $e = \text{garble}_{out}[b_0 \oplus \gamma_l] \wedge [b_1 \oplus \gamma_r]$ and L is a unique identifier of the gate. In other words, we sign exactly what was encrypted in the original construction, plus a unique label (b_0, b_1, L).

For each input wire w_i, it also signs garble$_{w_i}(x_i)$ along with some unique label, and makes garble$_{w_i}(x_i)$ and the signature σ_i part of the output. Since the gates in the Yao circuit are allowed to have fan-out,[3] we can assume that each input bit x_i to C appears on just one input wire w_i. Then the single occurrence of (garble$_{w_i}(x_i), \sigma_i$) is the only part of the output of $C_{\texttt{AutYao}}(x, r)$ which depends on x_i. We use this below.

5   Combining Authenticated Yao Garbling and a PRF

Towards using COMPPAR for generating $C_{\texttt{AutYao}}(x, r)$ we need to slightly modify it to make it more uniform.

The first step is to compute not $C_{\texttt{AutYao}}(x, r)$, but $C_{\texttt{AutYao}}(x, \text{prg}(K))$, where prg : $\{0, 1\}^k \to \{0, 1\}^{|r|}$ is a PRG and $K \in \{0, 1\}^k$ a uniformly random seed. The output distributions $C_{\texttt{AutYao}}(x, r)$ and $C_{\texttt{AutYao}}(x, \text{prg}(K))$ are of course computationally indistinguishable, so nothing is lost by this change. In fact, we use a very specific PRG: Let ϕ be a PRF with k-bit key and 1-bit output. We let $\text{prg}(K) = (\phi_K(1), \ldots, \phi_K(|r|))$, which is well known to be a PRG. Below we use $C_{\texttt{AutYao}}(x, K)$ as a short hand for $C_{\texttt{AutYao}}(x, \text{prg}(K))$ with this specific PRG.

The j's bit of $C_{\texttt{AutYao}}(x, K)$ depends on at most one input bit $x_{i(j)}$, where we choose $i(j)$ arbitrarily if the j'th bit does not depend on x. The uniform structure we obtain for the computation of $C_{\texttt{AutYao}}(x, K)$ is as follows.

---

[3] For technical reasons, explained below, we assume that no gate has fan-out higher than 3, which can be accomplished by at most a constant blow-up in circuit size.

Lemma 2. *There exists a circuit A of size* $\mathrm{poly}(k, \log |C|)$ *such that the j'th bit of* $C_{\mathtt{AutYao}}(x, K)$ *is* $A(j, x_{i(j)}, K)$.

This follows easily from the fact that Yao garbling treats all gates in C the same way and that gates can be handled in parallel. The proof can be found in [16].

It is now straightforward to see that we can set the parameters of the functionality F_{CP} defined earlier so that it will compute the values $A(j, x_{i(j)}, K)$ for all j. We will call F_{CP} with A as the circuit and we order the bits output by $C_{\mathtt{AutYao}}(x, K)$ into blocks of size ℓ. The number of such blocks will be the parameter g used in F_{CP}, and m will be the number of input bits to A. Blocks will be arranged such that the following holds for for any block given by its bit positions $(j_1, ..., j_\ell)$: either this block does nor depend on x or all input bits contributing to this output block, namely $(x_{i(j_1)}, \ldots, x_{i(j_\ell)})$, are given by one input client. This is possible as any input bit affects the same number of output bits, namely the bits in $\mathrm{garble}_{w_i}(x_i)$ and the corresponding signature σ_i.

We then just need to define how the functionality should treat each of the input blocks z_u^v that we need to define. Now, z_u^v corresponds to the v'th output block and to position u in the input to A. Suppose that the v'th output block has the bit positions $(j_1, .., j_\ell)$. Then if u points to a position in the representation of j, we set z_u^v to be the public value $(j_1^u, \ldots, j_\ell^u)$, namely the u'th bit in the binary representations of $j_1, ..., j_\ell$. If u points to the position where $x_{i(j)}$ is placed and block v depends on x, we define z_u^v to be owned by the client supplying $(x_{i(j_1)}, \ldots, x_{i(j_\ell)})$ as defined above. And finally if u points to position w in the key K, we define z_u^v to be random of type w.

This concrete instantiation of F_{CP} is called $F_{\mathtt{CompYao}}$, a secure implementation follows immediately from Lemma 1. From the discussion on CompPar, it follows that the complexity of the implementation is $\widetilde{\mathcal{O}}(|C|)$.

6   Delivering Outputs

Using $F_{\mathtt{CompYao}}$, we can have the string $C_{\mathtt{AutYao}}(x, K)$ output to the servers (ℓ bits at a time). We now need to use this to get the the results to the output clients efficiently. To this end, we divide the garbled inputs and encrypted gates into (small) subsets $\mathcal{G}_1, \ldots, \mathcal{G}_G$ and ask each server to handle only a fair share of the decoding of these.

We pick $G = n + (n - 2t)$ and pick the subsets such that no gate in $\mathcal{G}_g$ has an input wire w which is an output wire of a gate in $\mathcal{G}_{g'}$ for $g' > g$. We pick the subsets such that $|\mathcal{G}_g| = \widetilde{\mathcal{O}}(|C|/G)$, where $|\mathcal{G}_g|$ is the number of gates in $\mathcal{G}_g$. We further ensure that only the last $n - 2t$ subsets contain output wire carrying values that are to be sent to output clients. Furthermore, we ensure that all the L bits in the garbled inputs and encrypted gates for gates in $\mathcal{G}_g$ can be found in $\widetilde{\mathcal{O}}(L/\ell)$ blocks of $C_{\mathtt{AutYao}}(x, K)$. This is trivially achieved by ordering the bits in $C_{\mathtt{AutYao}}(x, K)$ appropriately during the run of CompPar.

We call a wire (name) w an *input wire* to $\mathcal{G}_g$ if there is a gate in $\mathcal{G}_g$ which has w as input wire, and the gate with output wire w (or the garbled input x_i for wire w) is not in $\mathcal{G}_g$. We call w an *output wire* from $\mathcal{G}_g$ if it is an output wire from a gate in $\mathcal{G}_g$ and is an input wire to another set $\mathcal{G}_{g'}$. We let the *weight* of $\mathcal{G}_g$, denoted $\|\mathcal{G}_g\|$, be the number of input wires to $\mathcal{G}_g$ plus the number of gates in $\mathcal{G}_g$ plus the number of output wires from $\mathcal{G}_g$. By the assumption that all gates have fan-out at most 3, $\|\mathcal{G}_g\| \leq 5|\mathcal{G}_g|$, where $|\mathcal{G}_g|$ is the number of gates in $\mathcal{G}_g$.

Protocol COMPOUTPUT:

1. All servers (in $\mathcal{S}'$): mark all $\mathcal{G}_g$ as unevaluated and let $c_i := 0$ for all S_i.[a]
2. All servers: let $\mathcal{G}_g$ be the lowest indexed set still marked as unevaluated, let $c = \min_{S_i \in \mathcal{S}'} c_i$ and let $S_i \in \mathcal{S}'$ be the lowest indexed server for which $c_i = c$.
3. All servers: execute open commands of F_{CompYao} such that S_i receives $\mathcal{G}_g$ and pk.
4. Each $S_j \in \mathcal{S}'$: for each input wire to $\mathcal{G}_g$, if it comes from a gate in a set handled by S_j, send the garbled wire value to S_i along with the signature.
5. S_i: If some S_j did not send the required values, then broadcast (J'accuse, S_j) for one such S_j. Otherwise, broadcast **ok** and compute from the garbled wire values and the encrypted gates for $\mathcal{G}_g$ the garbled wire values for all output wires from $\mathcal{G}_g$.
6. All servers: if S_i broadcasts (J'accuse, S_j), then mark all sets $\mathcal{G}_{g'}$ previously handled by S_i or S_j as unevaluated and remove S_i and S_j from $\mathcal{S}'$. Otherwise, mark $\mathcal{G}_g$ as evaluated and let $c_i := c_i + 1$.
7. If there are $\mathcal{G}_g$ still marked as unevaluated, then go to Step 2.
8. Now the ungarbled, authenticated wire values for all output wires from C are held by at least one server. All servers send pk to all output clients, which adopt the majority value pk. In addition all servers send the authenticated output wire values that they hold to the appropriate output clients, which authenticate them using pk.

---
[a] c_i is a count of how many $\mathcal{G}_g$ were handled by S_i.

The details are given in Protocol COMPOUTPUT. We call a run from Step 2 through Step 6 *successful* if $\mathcal{G}_g$ became marked as evaluated. Otherwise we call it *unsuccessful*. For each successful run one set is marked as evaluated. Initially G sets are marked as unevaluated, and for each unsuccessful run, at most $2\lceil G/n'\rceil$ sets are marked as unevaluated, where $n' = |\mathcal{S}'|$. Each unsuccessful run removes at least one corrupted party from $\mathcal{S}'$. So, it happens at most $G + t2\lceil G/n'\rceil$ times that a set is marked as evaluated, and since $n' \geq n - 2t \geq 2t$, there are at most $2G + 2t$ successful runs. There are clearly at most t unsuccessful runs, for a total of at most $2G + 4t \leq 2G + n \leq 3G$ runs. It is clear that the complexity of one run from Step 2 through Step 6 is $\|\mathcal{G}_g\| \cdot \text{poly}(k) + \text{poly}(n, k) = \widetilde{\mathcal{O}}(\|\mathcal{G}_g\|) = \widetilde{\mathcal{O}}(|\mathcal{G}_g|) = \widetilde{\mathcal{O}}(|C|/G)$. From this it is clear that the communication and computational complexities of COMPOUTPUT are $\widetilde{\mathcal{O}}(|C|)$.

The COMPOUTPUT protocol has the problem that t corrupted servers might not send the output values they hold. We handle this in a natural way by adding robustness to these output values, replacing the circuit C by a circuit C' derived from C as follows. For each output client, the output bits from C intended for this client are grouped into blocks, of size allowing a block to be represented as $n - 3t$ field elements $(x_1, \ldots, x_{n-3t})$. For each block, C' then computes $(y_1, \ldots, y_{n-2t}) = M(x_1, \ldots, x_{n-3t})$ for $M = M_{n-3t \to n-2t}$, and outputs the y-values instead of the x-values. The bits of $(y_1, \ldots, y_{n-2t})$ are still considered as output intended for the client in question. The output wires for the bits of $y_1, \ldots, y_{n-2t}$ are then added to the sets $\mathcal{G}_{n+1}, \ldots, \mathcal{G}_{n+n-2t}$, respectively. Since $|\mathcal{S}'| \geq n - 2t$ each of these $\mathcal{G}_g$ will be held by different servers at the end of COMPOUTPUT. So the output client will receive y_i-values from at least $n - 3t$ servers, say in set H, and can then compute $(x_1, \ldots, x_{n-3t}) = M_H^{-1}(y_i)_{S_i \in H}$. Since $|C'| = \tilde{\mathcal{O}}(|C|)$ and the interpolation can be done in time $\tilde{\mathcal{O}}(n)$ we maintain the required efficiency.

Our overall protocol π_{out} now consists of running (the implementation of) F_{CompYao} using C' as the underlying circuit, and then COMPOUTPUT. We already argued the complexity of these protocols.

A sketch of the proof of security: we want to show that π_{out} securely implements a functionality F_C that gets inputs for C from the input clients, leaks $C(x)$ to the adversary, and sends to each output client its part of $C(x)$.

We already argued that we have a secure implementation of F_{CompYao}, so it is enough to argue that we implement F_C securely by running F_{CompYao} and then COMPOUTPUT. First, by security of the PRG, we can replace F_{CompYao} by a functionality that computes an authenticated Yao-garbling $C_{\text{AutYao}}(x, r)$ using genuinely random bits, and otherwise behaves like F_{CompYao}. This will be indistinguishable from F_{CompYao} to any environment.

Now, based on $C(x)$ that we get from F_C, a simulator can construct a simulation of $C_{\text{AutYao}}(x, r)$ that will decode to $C'(x)$, by choosing some arbitrary x' and computing $C_{\text{AutYao}}(x', r)$, with the only exception that the correct bits of $C'(x)$ are encrypted in those entries of output-gate tables that will eventually be decrypted. By security of the encryption used for the garbling, this is indistinguishable from $C_{\text{AutYao}}(x, r)$.

The simulator then executes COMPOUTPUT with the corrupted servers and clients, playing the role of both the honest servers and F_{CompYao} (sending appropriate ℓ-bit blocks of the simulated $C_{\text{AutYao}}(x, r)$ when required). By security of the signature scheme, this simulated run of COMPOUTPUT will produce the correct values of $C'(x)$ and hence $C(x)$ as output for the clients, consistent with F_C sending $C(x)$ to the clients in the ideal process. Thus we have the following:

Lemma 3 (Outer Protocol). *Suppose one-way functions exist. Then there is a constant $0 < \delta < 1/2$ such that for any circuit C there is an n-server δn-secure protocol π_{out} for C which requires only $\text{poly}(k, \log n, \log |C|) \cdot |C| + \text{poly}(k, n)$ total computation (let alone communication) with security parameter k.*

We note that, assuming the existence of a PRG in NC^1, one can obtain a constant-round version of Lemma 3 for the case where there is only a constant number of output clients. The main relevant observation is that in such a case we can afford to directly deliver the outputs of C_{Yao} to the output clients, avoiding use of COMPOUTPUT. The round complexity of the resulting protocol is proportional to the *depth* of $C_{\text{Yao}}(x, K)$, which is poly(k).[4] To make the round complexity constant, we use the fact that a PRG in NC^1 allows to replace $C_{\text{Yao}}(x, K)$ by a similar randomized circuit $C'_{\text{Yao}}(x, K; \rho)$ whose depth is *constant* [1]. Applying COMPPAR to C'_{Yao} and delivering the outputs directly to the output clients yields the desired constant-round protocol. If one is content with a weaker form of security, namely "security with abort", then we can accommodate an arbitrary number of output clients by delivering all outputs to a single client, where the output of client i is encrypted and authenticated using a key only known to this client. The selected client then delivers the outputs to the remaining output clients, who broadcast an abort message if they detect tampering with their output.

7 Improving the Security Threshold Using Committees

In this section, we bootstrap the security of the protocol developed in the previous sections to resist coalitions of near-optimal size $(\frac{1}{2} - \epsilon)n$, for constant ϵ.

Theorem 1 (Main Result). *Suppose one-way functions exist. Then for every constant $\epsilon > 0$ and every circuit C there is an n-server $(\frac{1}{2} - \epsilon)n$-secure protocol Π for C, such that Π requires at most $\text{poly}(k, \log n, \log|C|) \cdot |C| + \text{poly}(k, n)$ total computation (and, hence, communication) with security parameter k.*

Moreover, if there exists a pseudorandom generator in NC^1 and the outputs of C are delivered to a constant number of clients, the round complexity of Π can be made constant with the same asymptotic complexity.

The main idea is to use *player virtualization* [5] to emulate a run of the previous sections' protocol among a group of n "virtual servers". Each virtual server is emulated by a committee of d real participants, for a constant d depending on ϵ, using a relatively inefficient SFE subprotocol that tolerates $\frac{d-1}{2}$ cheaters. The n (overlapping) committees are chosen so that an adversary corrupting $(\frac{1}{2} - \epsilon)n$ real players can control at most δn committees, where "controlling" a committee means corrupting at least $d/2$ of its members (and thus controlling the emulated server). As mentioned earlier (and by analogy which concatenated codes) we call the subprotocol used to emulate the servers the "inner" protocol, and the emulated protocol of the previous sections the "outer" protocol. For the inner protocol, we can use the protocol of Cramer, Damgård, Dziembowski, Hirt and Rabin [10] or a constant-round variant due to Damgård and Ishai [13].

The player virtualization technique was introduced by Bracha [5] in the context of Byzantine agreement to boost resiliency of a particular Byzantine agreement protocol to $(\frac{1}{3} - \epsilon)n$. It was subsequently used in several other contexts of distributed computing and cryptography, e.g. [17, 23, 25]. The construction of the committee sets below is explicit and implies an improvement on the parameters of the PSMT protocol of Fitzi *et al.* [17] for short messages.

---

[4] Note that $C_{\text{Yao}}(x, K)$ cannot have constant depth, as it requires the computation of a PRF to turn K into randomness for C_{Yao}.

We use three tools: the outer protocol from Lemma 3, the inner protocol and the construction of committee sets. The last two are encapsulated in the two lemmas below. The inner protocol will emulate an ideal, reactive functionality $\mathcal{F}$ which itself interacts with other entities in the protocol. For the general statement, we restrict $\mathcal{F}$ to be "adaptively well-formed" in the sense of Canetti et al. [8] (see Lindell [31, Sec. 4.4.3], for a definition). All the functionalities discussed in this paper are well-formed.

Lemma 4 (Inner Protocol, [10, 13]). *If one-way functions exist then, for every well-formed functionality $\mathcal{F}$, there exists a UC-secure protocol π_{in} among d players that tolerates any $t \leq \frac{d-1}{2}$ adaptive corruptions. For an interactive functionality $\mathcal{F}$, emulating a given round of $\mathcal{F}$ requires $\text{poly}(comp_{\mathcal{F}}, d, k)$ total computation, where $comp_{\mathcal{F}}$ is the computational complexity of $\mathcal{F}$ at that round, and a constant number of rounds.*

Strictly speaking, the protocols from [10, 13] are only for general secure function evaluation. To get from this the result above, we use a standard technique that represents the internal state of $\mathcal{F}$ as values that are shared among the players using verifiable secret sharing (VSS) Details can be found in [16].

Definition 1. *A collection S of subsets of $[n] = \{1, ..., n\}$ is a (d, ϵ, δ)-secure committee collection if all the sets in S (henceforth "committees") have size d and, for every set $B \subseteq [n]$ of size at most $(\frac{1}{2} - \epsilon)n$, at most a δ fraction of the committees overlap with B in $d/2$ or more points.*

Lemma 5 (Committees Construction). *For any $0 < \epsilon, \delta < 1$, there exists an efficient construction of a (d, ϵ, δ)-secure committee collection consisting of n subsets of $[n]$ of size $d = \mathcal{O}(\frac{1}{\delta\epsilon^2})$. Given an index i, one can compute the members of the i-th committee in time $\text{poly}(\log(n))$.*

The basic idea is to choose a sufficiently good expander graph on n nodes and let the members of the ith committee be the neighbors of vertex i in the graph. The lemma is proved in [16].

We note that the same construction improves the parameters of the perfectly secure message transmission protocol of Fitzi *et al.* [17] for short messages. To send a message of L bits over n wires while tolerating $t = (\frac{1}{2} - \epsilon)n$ corrupted wires, their protocol requires $\mathcal{O}(L) + n^{\Theta(1/\epsilon^2)}$ bits of communication. Plugging the committees construction above into their protocol reduces the communication to $\mathcal{O}(L + n/\epsilon^2)$. A similar construction to that of Lemma 5 was suggested to the authors of [17] by one of their reviewers ([17, Sec. 5]). This paper is, to our knowledge, the first work in which the construction appears explicitly.

The final, composed protocol Π will have the same input and output clients as π_{out} and n virtual servers, each emulated by a committee chosen from the n real servers. These virtual servers execute π_{out}. This is done in two steps:

First, we build a protocol Π' where we assume an ideal functionality $\mathcal{F}_i$ used by the i'th committee. $\mathcal{F}_i$ follows the algorithm of the i'th server in π_{out}. When π_{out} sends a message from server i to server j, $\mathcal{F}_i$ acts as dealer in the VSS to have members of the jth committee obtain shares of the message, members then give these as input to $\mathcal{F}_j$. See [16] for details on the VSS to be used. Clients exchange messages directly with the $\mathcal{F}_i$'s according to π_{out}. $\mathcal{F}_i$ follows its prescribed algorithm, unless a majority of the servers in the i'th committee are corrupted, in which case all its actions are controlled by the adversary, and it shows the adversary all messages it receives.

The second step is to obtain Π by using Lemma 4 to replace the $\mathcal{F}_i$'s by implementations via π_{in}.

The proof of security for Π' is a delicate hybrid argument, and we defer it to [16]. Assuming Π' is secure, the lemma below follows from Lemma 4 and the UC composition theorem:

Lemma 6. *The composed protocol Π is a computationally-secure SFE protocol that tolerates $t = (\frac{1}{2} - \epsilon)n$ adaptive corruptions.*

As for the computational and communication complexities of Π, we recall that these are both $\widetilde{\mathcal{O}}(|C|)$ for π_{out}. It is straightforward to see that the overhead of emulating players in π_{out} via committees amounts to a multiplicative factor of $O(\text{poly}(k, d))$, where d is the committee size, which is constant. This follows from the fact that the complexity of π_{in} is $\text{poly}(S, k, d)$ where S is the size of the computation done by the functionality emulated by π_{in}. Therefore the complexity of Π is also $\widetilde{\mathcal{O}}(|C|)$. This completes the proof of the main theorem.

References

1. Applebaum, B., Ishai, Y., Kushilevitz, E.: Computationally private randomizing polynomials and their applications. In: Proc. CCC 2005, pp. 260–274 (2005)
2. Beerliova-Trubiniova, Z., Hirt, M.: Efficient Multi-Party Computation with Dispute Control. In: Halevi, S., Rabin, T. (eds.) TCC 2006. LNCS, vol. 3876, pp. 305–328. Springer, Heidelberg (2006)
3. Beerliova-Trubiniova, Z., Hirt, M.: Perfectly-Secure MPC with Linear Communication Complexity. In: Canetti, R. (ed.) TCC 2008. LNCS, vol. 4948, Springer, Heidelberg (to appear, 2008)
4. Ben-Or, M., Goldwasser, S., Wigderson, A.: Completeness theorems for non-cryptographic fault-tolerant distributed computation. In: STOC 1988, pp. 1–10 (1988)
5. Bracha, G.: An $O(\log n)$ expected rounds randomized byzantine generals protocol. Journal of the ACM 34(4), 910–920 (1987)
6. Canetti, R.: Universally Composable Security: A New Paradigm for Cryptographic Protocols. In: Proc. FOCS 2001, pp. 136–145 (2001)
7. Canetti, R., Feige, U., Goldreich, O., Naor, M.: Adaptively Secure Multiparty Computation. In: Proc. STOC 1996, pp. 639–648 (1996)
8. Canetti, R., Lindell, Y., Ostrovsky, R., Sahai, A.: Universally composable two-party and multi-party secure computation. In: Proc. STOC 2002, pp. 494–503 (2002)
9. Chaum, D., Crépeau, C., Damgård, I.: Multiparty unconditionally secure protocols (extended abstract). In: Proc. STOC 1988, pp. 11–19 (1988)
10. Cramer, R., Damgård, I., Dziembowski, S., Hirt, M., Rabin, T.: Efficient Multiparty Computations Secure Against an Adaptive Adversary. In: Stern, J. (ed.) EUROCRYPT 1999. LNCS, vol. 1592, pp. 311–326. Springer, Heidelberg (1999)

11. Cramer, R., Damgård, I., Nielsen, J.: Multiparty computation from threshold homomorphic encryption. In: Pfitzmann, B. (ed.) EUROCRYPT 2001. LNCS, vol. 2045, pp. 280–299. Springer, Heidelberg (2001)

12. Damgård, I., Ishai, Y.: Scalable Secure Multiparty Computation. In: Dwork, C. (ed.) CRYPTO 2006. LNCS, vol. 4117, pp. 501–520. Springer, Heidelberg (2006)

13. Damgård, I., Ishai, Y.: Constant-Round Multiparty Computation Using a Black-Box Pseudorandom Generator. In: Shoup, V. (ed.) CRYPTO 2005, vol. 3621, pp. 378–394. Springer, Heidelberg (2005)

14. Damgård, I., Nielsen, J.: Universally Composable Efficient Multiparty Computation from Threshold Homomorphic Encryption. In: Boneh, D. (ed.) CRYPTO 2003. LNCS, vol. 2729, pp. 247–264. Springer, Heidelberg (2003)

15. Damgård, I., Nielsen, J.: Robust multiparty computation with linear communication complexity. In: Proc. Crypto 2007, pp. 572–590 (2007)

16. Damgård, I., Ishai, Y., Krøigaard, M., Nielsen, J., Smith, A.: Scalable Multiparty Computation with Nearly Optimal Work and Resilience (full version of this paper)

17. Fitzi, M., Franklin, M., Garay, J., Vardhan, H.: Towards optimal and efficient perfectly secure message transmission. In: Vadhan, S.P. (ed.) TCC 2007. LNCS, vol. 4392, pp. 311–322. Springer, Heidelberg (2007)

18. Fitzi, M., Hirt, M.: Optimally Efficient Multi-Valued Byzantine Agreement. In: Proc. PODC 2006, pp. 163–168 (2006)

19. Franklin, M.K., Haber, S.: Joint Encryption and Message-Efficient Secure Computation. In: Proc. Crypto 1993, pp. 266–277 (1993); Full version in Journal of Cyptoglogy 9(4), 217–232 (1996)

20. Franklin, M.K., Yung, M.: Communication Complexity of Secure Computation. In: Proc. STOC 1992, pp. 699–710 (1992)

21. Gennaro, R., Rabin, M.O., Rabin, T.: Simplified VSS and fast-track multiparty computations with applications to threshold cryptography. In: Proc. 17th PODC, pp. 101–111 (1998)

22. Goldreich, O., Micali, S., Wigderson, A.: How to play any mental game (extended abstract). In: Proc. STOC 1987, pp. 218–229 (1987)

23. Harnik, D., Ishai, Y., Kushilevitz, E.: How many oblivious transfers are needed for secure multiparty computation? In: Menezes, A. (ed.) CRYPTO 2007. LNCS, vol. 4622, pp. 284–302. Springer, Heidelberg (2007)

24. Hirt, M., Maurer, U.M.: Robustness for Free in Unconditional Multi-party Computation. In: Kilian, J. (ed.) CRYPTO 2001. LNCS, vol. 2139, pp. 101–118. Springer, Heidelberg (2001)

25. Hirt, M., Maurer, U.: Player simulation and general adversary structures in perfect multiparty computation. Journal of Cryptology 13(1), 31–60 (2000)

26. Hirt, M., Maurer, U.M., Przydatek, B.: Efficient Secure Multi-party Computation. In: Okamoto, T. (ed.) ASIACRYPT 2000. LNCS, vol. 1976, pp. 143–161. Springer, Heidelberg (2000)

27. Hirt, M., Nielsen, J.B.: Upper Bounds on the Communication Complexity of Optimally Resilient Cryptographic Multiparty Computation. In: Roy, B. (ed.) ASIACRYPT 2005. LNCS, vol. 3788, pp. 79–99. Springer, Heidelberg (2005)

28. Hirt, M., Nielsen, J.B.: Robust Multiparty Computation with Linear Communication Complexity. In: Menezes, A. (ed.) CRYPTO 2007. LNCS, vol. 4622, pp. 572–590. Springer, Heidelberg (2007)

29. Jakobsson, M., Juels, A.: Mix and Match: Secure Function Evaluation via Ciphertexts. In: Okamoto, T. (ed.) ASIACRYPT 2000. LNCS, vol. 1976, pp. 162–177. Springer, Heidelberg (2000)

30. Kushilevitz, E., Lindell, Y., Rabin, T.: Information theoretically secure protocols and security under composition. In: Proc. STOC 2006, pp. 109–118 (2006)
31. Lindell, Y.: Composition of Secure Multi-Party Protocols, A Comprehensive Study. Springer, Heidelberg (2003)
32. Lubotzky, A., Phillips, R., Sarnak, P.: Ramanujan graphs. Combinatorica 8(3), 261–277 (1988)
33. Shamir, A.: How to share a secret. Commun. ACM 22(6), 612–613 (1979)
34. Yao, A.C.: Theory and Applications of Trapdoor Functions (Extended Abstract). In: Proc. FOCS 1982, pp. 80–91 (1982)
35. Yao, A.C.: How to generate and exchange secrets. In: Proc. FOCS 1986, pp. 162–167 (1986)

Cryptographic Complexity of Multi-Party Computation Problems: Classifications and Separations

Manoj Prabhakaran⋆ and Mike Rosulek⋆

Department of Computer Science
University of Illinois, Urbana-Champaign
{mmp,rosulek}@uiuc.edu

Abstract. We develop new tools to study the relative complexities of secure multi-party computation tasks in the Universal Composition framework. When one task can be securely realized using another task as a black-box, we interpret this as a qualitative, complexity-theoretic reduction between the two tasks. Virtually all previous characterizations of MPC functionalities, in the UC model or otherwise, focus exclusively on secure function evaluation. In comparison, the tools we develop do not rely on any special internal structure of the functionality, thus applying to functionalities with arbitrary behavior. Our tools additionally apply uniformly to both the PPT and unbounded computation models.

Our first main tool is an exact characterization of realizability in the UC framework with respect to a large class of communication channel functionalities. Using this characterization, we can rederive all previously-known impossibility results as immediate and simple corollaries. We also complete the combinatorial characterization of 2-party secure function evaluation initiated by [12] and partially extend the combinatorial conditions to the multi-party setting. Our second main tool allows us to translate complexity separations in simpler MPC settings (such as the honest-but-curious corruption model) to the standard (malicious) setting. Using this tool, we demonstrate the existence of functionalities which are neither realizable nor complete, in the unbounded computation model.

1 Introduction

In this work, we seek to investigate the intrinsic "cryptographic complexity" of secure multiparty computation (MPC) functionalities. MPC functionalities can have a rich structure, being interactive, often randomized, computations involving more than one party. Clearly not all functionalities have equal cryptographic sophistication. For instance, one expects a task like oblivious transfer to be much more sophisticated than the mere task of communication or local computation. One could ask if the two-party task of commitment is any more complex than

⋆ Partially supported by NSF grant CNS 07-47027.

the task of two (mutually distrusting) parties generating unbiased coin-flips. We present a complexity-theoretic approach to asking and answering such questions.

At the heart of such an approach is identifying meaningful (or useful) notions of *reductions* between MPC functionalities, that would allow us to form "complexity classes" of functionalities with similar cryptographic complexity. The most natural notion of reduction for MPC functionalities is in terms of "secure realizability:" can one functionality $\mathcal{F}$ be securely realized given access to another functionality $\mathcal{G}$? Indeed, this notion of reduction has been extensively used in literature. Yet, the way this "reduction" was traditionally defined, it was *not transitive*. This severely restricted its usefulness as a reduction for studying cryptographic complexity. In the recently developed framework of Universal Composition (UC) [9], however, the *Universal Composition theorem* guarantees that the reduction based on secure realizability in that framework is indeed a transitive relation. It is in this framework that we ground our study.

Our results presented below can be viewed as relating to an abstract notion of complexity of MPC functionalities. More concretely, these can be interpreted as results on secure realizability in the UC framework.

Our Results. We introduce new techniques and tools to better understand and classify cryptographic complexity classes (as defined using secure realizability in the UC framework). We focus on tools that apply broadly to *arbitrary* functionalities, whereas most previous work either focused on secure function evaluation or involved *ad hoc* arguments specific to particular functionalities. Further, the main tools we develop apply in the standard UC model, as well as in the information theoretic (or computationally unbounded) variant.

We then apply our new tools to give more concrete results for specific functionalities and characterizations for important subclasses of functionalities. Our main results mostly involve showing *separations* in complexity among functionalities, as opposed to new protocol constructions. Our main results fall into two categories based on the techniques used:

Classifying Functionalities Using Splittability. We define a very general aspect of cryptographic complexity called *splittability*. We show that splittable functionalities are exactly the ones that have secure protocols in the plain model, with respect to static corruptions, using an idealized communication channel (Theorem 1). This is the first alternate characterization of realizability in the UC model.

Superficially, the definition of splittability is similar to the definition of realizability in the UC framework, and indeed, showing that a functionality is splittable is not much easier than directly showing that it is realizable. However, the main utility of the splittability characterization is that *it is often extremely easily to show that a functionality is unsplittable*. We rederive the impossibility of zero-knowledge proofs [9], bit commitment, coin-tossing, and oblivious transfer [11] as simple and easy consequences of this characterization. We also use splittability to complete the combinatorial characterization of 2-party secure function evaluation (SFE) initiated in [12, 13] (Theorem 5).

We generalize the notion of splittability as a transitive *binary relation* on functionalities, which we view as a complexity-theoretic reduction. Using this definition, we identify a class that includes all natural communication channels, and which we argue defines a natural class of "low cryptographic complexity." Then we show that for all $\mathcal{G}$ in this class, our exact characterization generalizes; that is, $\mathcal{F}$ is splittable with respect to $\mathcal{G}$ if and only if $\mathcal{F}$ has a secure protocol on the channel $\mathcal{G}$ (Theorem 3), with respect to static corruptions.

Furthemore, if a functionality is unsplittable according to the simpler, less general definition, then it has no secure protocol on any natural channel. Thus, splittability provides a powerful and easy way to separate the cryptographic complexities of many functionalities.

Our main technical results hold for multi-party functionalities, although the definitions become complicated and less intuitive for more than 2 parties. However, we show that the 2-party case yields some necessary conditions for multi-party functionalities (Theorem 7). We leave open the question of whether they are sufficient in general.

Passive Corruption and Deviation-Revealing Functionalities. A functionality's realizability depends crucially on the model of the adversary's corruption. For instance, in the unbounded computation model, functionalities like coin-flipping and commitment become trivial if the adversary is passive (honest-but-curious), while oblivious transfer still remains unrealizable. This motivates using alternate (and possibly unrealistic) corruption models to study the complexity of functionalities. We develop an effective technique to "lift" realizability separations from restricted corruption settings to the standard malicious corruption setting. While the techniques of splittability can give separations involving only relatively low-complexity functionalities, this second technique can yield separations among higher-complexity functionalities.

Translating separations in the restricted corruption settings to the standard setting is possible only for certain "well-behaved" functionalities. We identify such a well-behavedness property called *deviation revealing* and formulate an appropriate translation recipe (Theorem 4). As in the case of splittability, the deviation-revealing property is applicable to completely arbitrary functionalities.

Combining this recipe with known separations in various corruption models (as well as some easy observations), we show a sequence of four *natural* functionalities that form a hierarchy of *strictly increasing* complexity, in the unbounded computation model (Theorem 8). This implies that the two intermediate functionalities in this sequence are neither complete nor realizable (using any natural communication channel), and that there is more than one distinct intermediate level of complexity. Our result separating these two functionalities of intermediate complexity is perhaps unique since most previous works focused on only the extremes of complexity.

Related Work. Multiparty computation was introduced in the eighties, and secure protocols were given early on for realizing all functionalities [7, 15, 20, 26, 40]. However, the notion of security used was stand-alone security. MPC was also studied in an information theoretic setting, and with weaker models of adversarial behavior: honest-but-curious adversaries [18, 29, 30] and honest majority [7, 15]. In these models, much work has focused on developing alternate characterizations for the extremes of complexity: realizability [17, 18, 30, 31] and completeness [6, 22, 27, 28, 29, 32]; see the full version [37] for a more comprehensive survey of these results.

Canetti [8, 9] (and independently Pfitzmann and Waidner [35]) introduced the general framework of network-aware security used in this work, known as UC security. The first impossibility results in the UC framework were already given in [9], followed by more in [11, 12, 25]. Our splittability characterization is motivated by common techniques underlying these results. A somewhat similar technique appears in a different context in an impossibility proof by Dwork et al. [19]. Network-aware secure protocols for all functionalities were given under different variations of the model [4, 5, 10, 14, 24, 38]. Impossibility results were also shown for various settings with less standard notions of composability [3, 33, 34]. We remark that our results are meaningful in variations of the model which simply involve augmenting the UC model with a "set-up" functionality. However, some of our theory does not apply to the models in [5, 38], which effectively allow different computational powers for adversaries/environments and simulators.

2   Preliminaries

Some of our conventions differ from the original UC model. We now give an overview of the model while highlighting these (cosmetic) differences, which are motivated by our "complexity theoretic" view of MPC.

Modeling Conventions. The network-aware security framework for MPC includes four kinds of entities: an *environment*, multiple *parties*, an *adversary*, and a *functionality*. The functionality's program fully specifies an MPC problem, and as such, is the primary object we classify in this paper.

Emphasizing the generality of our theory, we do not specify any computational limitations on these network entities, but instead consider abstract classes of *admissible machines*. We only require that a machine that internally simulates several other admissible machines is itself admissible.[1] Our general results apply uniformly for any such system, the two most natural of which are *computationally unbounded systems* (which admit all probabilistic machines) and *PPT systems* (which admit all probabilistic, polynomial-time machines).

Unlike the original UC model, we model the communication among the environment, parties, and functionalities as an ideal, private, tamper-proof channel. In the UC model, an adversary can tamper with and delay such communications. Instead, we assume that functionalities themselves achieve the same effect by directly interacting with the adversary each time a party communicates with

---

[1] As such, our theory is *not* directly applicable to the network-aware security model introduced in [36, 38] and also used in [5], where an adversary can sometimes access extra computational power that an environment cannot.

the functionality. This difference is significant in defining *non-trivial protocols*, which we address later in this section. Furthermore, there is no built-in communication mechanism among the parties; all communication must be facilitated by a functionality. In this way, we are able to uniformly consider arbitrary channels.

We require that a protocol interact only with a *single instance* of some functionality. This is without loss of generality, since we can always consider a single "augmented" functionality that provides an interface to multiple independent sessions of simpler functionalities. This convention maintains the strict binary nature of our complexity reduction. Also, for simplicity, we assume that parties and communication ports of the functionality are numbered, and that a protocol which uses $\mathcal{F}$ must have the ith party interact only as the ith party to $\mathcal{F}$. Again, this is without loss of generality, as an "augmented" functionality could provide an interface to multiple different "port-mappings" of a simpler functionality. To emphasize a qualitative measure of cryptographic complexity, we generally (implicitly) consider reductions among such augmented functionalities. In the UC model, $\mathcal{F}$ and its augmented version $\mathcal{F}^+$ can be realized in terms of one another (though not following our notational conventions). Thus all of our results may be interpreted as being in terms of augmented or unaugmented functionalities, whichever is appropriate.

Notation. $\text{EXEC}[\mathcal{Z}, \mathcal{A}, \pi^{\mathcal{G}}]$ denotes the probability of the environment $\mathcal{Z}$ outputting 1 when it interacts with parties running the protocol $\pi^{\mathcal{G}}$ (i.e., π using $\mathcal{G}$ as the sole medium for interaction), in the presence of an adversary $\mathcal{A}$. We denote the "dummy protocol" used to access a functionality by ∂ (i.e., an ideal-world direct interaction with $\mathcal{F}$ will be denoted as running the protocol $\partial^{\mathcal{F}}$). We say π is a *secure realization* of $\mathcal{F}$ with respect to $\mathcal{G}$ if if for all adversaries $\mathcal{A}$, there exists a simulator $\mathcal{S}$ such that for all environments $\mathcal{Z}$, $\text{EXEC}[\mathcal{Z}, \mathcal{A}, \pi^{\mathcal{G}}]$ and $\text{EXEC}[\mathcal{Z}, \mathcal{S}, \partial^{\mathcal{F}}]$ are negligibly close. When there is a secure realization of $\mathcal{F}$ with respect to $\mathcal{G}$, we write $\mathcal{F} \sqsubseteq \mathcal{G}$. We define the natural complexity class $\text{REALIZ}^{\mathcal{G}} = \{\mathcal{F} \mid \mathcal{F} \sqsubseteq \mathcal{G}\}$, the class of functionalities that can be securely realized using $\mathcal{G}$. Our main results apply to both PPT and unbounded systems in a unified way, so our notation does not distinguish between them. To explicitly refer to PPT or unbounded systems, we write $\sqsubseteq_p, \text{REALIZ}_p$ and $\sqsubseteq_u, \text{REALIZ}_u$, respectively.

Non-trivial Protocols. In the standard UC model where an adversary can delay communications between functionality and parties, a protocol which does nothing is trivially a secure realization (since the same effect can be achieved in the ideal world by an adversary who indefinitely blocks all outputs). Thus it is necessary to restrict attention to *non-trivial* protocols [12, 14], which are secure even when the ideal-world adversary eventually delivers all messages.

In our model, all communication between parties and functionality is on an idealized channel that does not allow blocking, but we may consider functionalities that explicitly interact with the adversary, allowing it to block or delay outputs to honest parties. For such functionalities, we must also consider a definition of non-triviality for our results to be meaningful.

Definition 1. *Let* wrap($\mathcal{F}$) *be the functionality that runs* $\mathcal{F}$, *except that outputs generated by* $\mathcal{F}$ *are kept in an internal queue.* wrap($\mathcal{F}$) *informs the adversary of each such output (source, destination, and length) and delivers it only if/when the adversary instructs it to.*

Definition 2. *Let* π *be a secure realization of* wrap($\mathcal{F}$) *with respect to* wrap($\mathcal{G}$). *We say* π *is* non-trivial, *and write* wrap($\mathcal{F}$) $\sqsubseteq^{nt}$ wrap($\mathcal{G}$), *if* π *is also a realization of* $\mathcal{F}$ *with respect to* $\mathcal{G}$.

In other words, a secure realization is *non-trivial* if, in the optimistic case where the adversary delivers all messages on wrap($\mathcal{G}$), the protocol realizes $\mathcal{F}$ (which may guarantee delivery of outputs, for example).[2] Indeed, it is often the case that one would consider wrap($\mathcal{F}_{pvt}$) (or similar) as one's communication channel, and would be willing to settle for the security of wrap($\mathcal{F}$).

The important implication of Definition 2 is that $\mathcal{F} \not\sqsubseteq \mathcal{G}$ implies wrap($\mathcal{F}$) $\not\sqsubseteq^{nt}$ wrap($\mathcal{G}$). Thus the complexity separations we obtain (between more simply defined functionalities) also imply corresponding separations for the weaker, more realistic wrapped functionalities, with respect to non-trivial protocols.

3   Structural Results

In this section we present our two new tools for studying the realizability of functionalities. These tools apply to arbitrary functionalities and to both PPT and unbounded computational systems. We call this set of results our "structural results" to emphasize their generality. Later, in Section 4, we apply these structural results to specific settings and classes of functionalities to obtain concrete results.

3.1   Splittability of (Regular) 2-Party Functionalities

The main tool we develop to characterize classes REALIZ$^{\mathcal{G}}$ is a theory of *splittability*. For expositional clarity, we first present a special case of our splittability theory which captures the essential intuition and still has useful consequences in the more general setting. Then we remove these restrictions and present the general theory for 2-party functionalities. The general theory for the multi-party setting is more complicated and is defered to the full version.

In this section we restrict attention to a class of functionalities called 2REGULAR: the 2-party functionalities which do not directly interact with the adversary when no parties are corrupted, and whose behavior does not depend on which parties are corrupted. This class already includes all secure function evaluation (SFE) functionalities as they are typically defined. We also restrict

---

[2] This definition is slightly stronger than the non-triviality condition in [12, 14]. Their definition was arguably sufficient for secure function evaluation, but must be strengthened to be appropriate for more general functionalities. See the full version for more detailed justification.

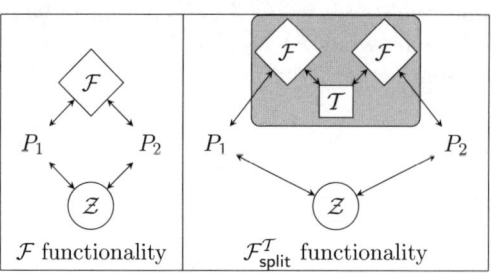

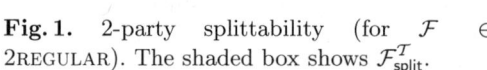

Fig. 1. 2-party splittability (for $\mathcal{F} \in$ 2REGULAR). The shaded box shows $\mathcal{F}_{\text{split}}^{\mathcal{T}}$.

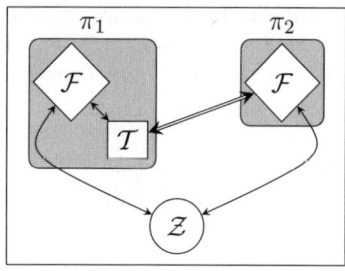

Fig. 2. Secure protocol π for a splittable functionality $\mathcal{F}$

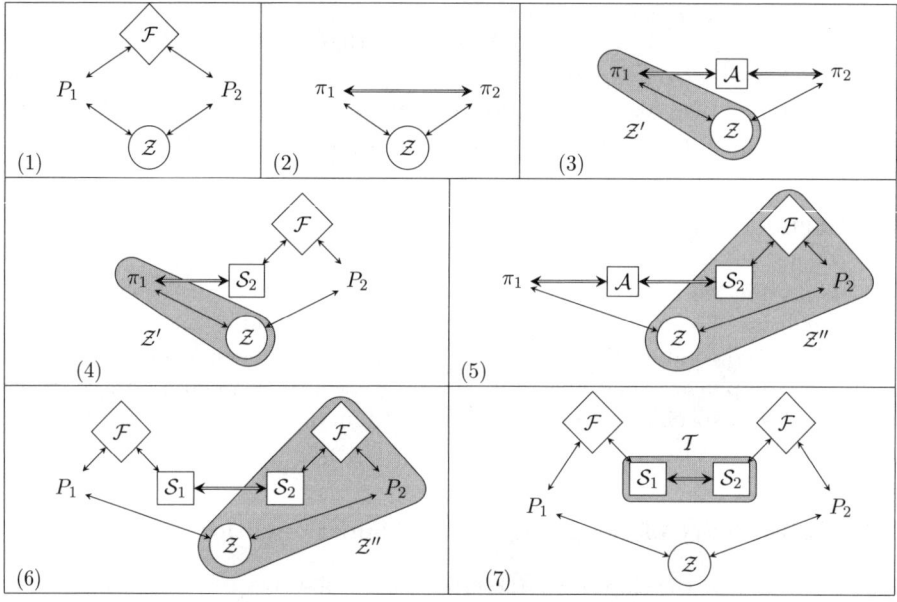

Fig. 3. Steps in the proof of Theorem 1. Given a protocol securely realizing $\mathcal{F}$, we apply the security guarantee three times: first with no parties corrupted (between boxes 1 and 2), then with a corrupt party P_1 which plays a man-in-the-middle between P_2 and an honest P_1 inside the environment (between boxes 3 and 4), and finally with a corrupt party P_2 which plays a man-in-the-middle between P_1 and the simulator from the previous step (between boxes 5 and 6), all with appropriately defined environments. The machine $\mathcal{T}$ required by the definition of splittability is derived from the simulators for the last two cases, by letting them simulate the protocol to each other.

attention to secure protocols that use an ideal communication channel. We let $\mathcal{F}_{\text{pvt}}$ denote the *completely private channel* functionality, which allows parties to privately send messages to other parties of their choice, and does not interact with the adversary at all (not even to notify that a message was sent).

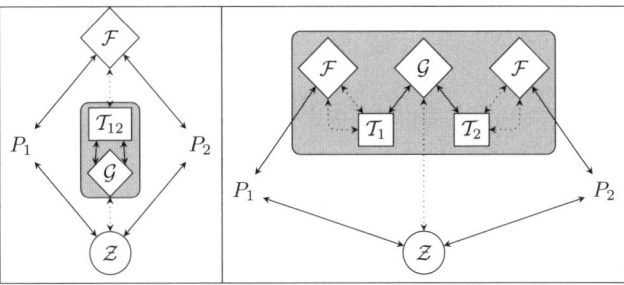

Fig. 4. General 2-party splittability (Definition 7). The shaded box in the far left shows $\mathcal{S}^{\mathcal{G},\mathcal{T}_{12}}$, and the other shaded box shows $\mathcal{F}_{\text{split}}^{\mathcal{G},\mathcal{T}_1,\mathcal{T}_2}$. Dotted lines indicate interaction as adversary or corrupted party.

Definition 3. *Let $\mathcal{F}$ be a 2-party functionality and $\mathcal{T}$ an admissible machine. Define $\mathcal{F}_{\text{split}}^{\mathcal{T}}$ as the compound functionality that does the following (See Fig. 1):*

$\mathcal{F}_{\text{split}}^{\mathcal{T}}$ internally simulates an instance of $\mathcal{T}$ and two independent instances of $\mathcal{F}$, which we call $\mathcal{F}_L$ and $\mathcal{F}_R$. The first party of $\mathcal{F}_{\text{split}}^{\mathcal{T}}$ directly interacts with $\mathcal{F}_L$ as its first party. The second party of $\mathcal{F}_{\text{split}}^{\mathcal{T}}$ directly interacts with $\mathcal{F}_R$ as its second party. The machine $\mathcal{T}$ interacts only with $\mathcal{F}_L$ and $\mathcal{F}_R$, as the other parties to these functionalities.

Definition 4 (Splittability). *$\mathcal{F} \in$ 2REGULAR is splittable if there exists a machine $\mathcal{T}$ such that $\mathcal{F}$ is indistinguishable from $\mathcal{F}_{\text{split}}^{\mathcal{T}}$. That is, for all environments $\mathcal{Z}$ and the dummy adversary $\mathcal{A}$ that corrupts no one, we have $\text{EXEC}[\mathcal{Z}, \mathcal{A}, \partial^{\mathcal{F}}] \approx \text{EXEC}[\mathcal{Z}, \mathcal{A}, \partial^{\mathcal{F}_{\text{split}}^{\mathcal{T}}}]$, where ∂ denotes the dummy protocol. We define 2REGSPLIT as the class of all splittable functionalities in 2REGULAR.*

At a very high level, $\mathcal{F}$ is *splittable* if there is a way to successfully mount an undetectable man-in-the-middle "attack" in the ideal world, between two independent instances of the functionality.

A crucial property of this definition is that it is often relatively easy to show that a functionality is unsplittable. See Section 4.1 for several examples involving some common functionalities.

Theorem 1. *$\mathcal{F} \in$ 2REGULAR is securely realizable using $\mathcal{F}_{\text{pvt}}$ if and only if $\mathcal{F}$ is splittable. That is, 2REGULAR $\cap$ REALIZ$^{\mathcal{F}_{\text{pvt}}}$ = 2REGSPLIT.*

Proof (Sketch). The easier direction is to see that 2REGSPLIT $\subseteq$ REALIZ$^{\mathcal{F}_{\text{pvt}}} \cap$ 2REGULAR. If $\mathcal{F}$ is splittable, then a protocol (π_1, π_2) can be derived as shown in Fig. 2. Note that the protocol uses a perfectly private channel for communication, which in our model is essentially the same kind of channel that the network entities use to communicate. Interestingly, the protocol at each end simulates a copy of the ideal functionality. Then the protocol's simulator can faithfully simulate the honest party's protocol merely by accessing the functionality in the ideal world.

The more interesting direction is showing that every realizable functionality is splittable. It generalizes the "split-adversary" technique used by Canetti,

Kushilevitz, and Lindell [12], and also has parallels with an impossibility proof by Dwork et al. [19].[3] A visual overview is given in Fig. 3.

3.2 General Theory of Splittability

Protocols in network-aware frameworks are usually considered to use less idealized channels than $\mathcal{F}_{\mathsf{pvt}}$. For instance, even the standard model of a private channel reveals to the adversary the fact that a message was sent, and often its length. It is also quite common to model functionalities which interact directly with the adversary, or whose behavior depends on which parties are corrupted.

In this section, we generalize the theory to apply to securely realizing *arbitrary* functionalities, using protocols which also use *arbitrary* functionalities as their "communication channels." Our theory also generalizes to multi-party functionalities; however, we present only the restriction of our results to 2-party functionalities. To state and prove our results for multi-party functionalities requires specialized notation, and is defered to the full version.

Definition 5. *Let $\mathcal{F}$ and $\mathcal{G}$ be 2-party functionalities and $\mathcal{T}_1$ and $\mathcal{T}_2$ be admissible machines. We define $\mathcal{F}_{\mathsf{split}}^{\mathcal{G},\mathcal{T}_1,\mathcal{T}_2}$ as the compound functionality that does the following, when interacting with honest parties P_1 and P_2 and an adversary (see Fig. 4):*

$\mathcal{F}_{\mathsf{split}}^{\mathcal{G},\mathcal{T}_1,\mathcal{T}_2}$ internally simulates instances of $\mathcal{T}_1$, $\mathcal{T}_2$, $\mathcal{G}$, and two independent instances of $\mathcal{F}$, which we call $\mathcal{F}_L$ and $\mathcal{F}_R$. P_1 directly interacts with $\mathcal{F}_L$ as its honest first party. P_2 directly interacts with $\mathcal{F}_R$ as its honest second party. The machine $\mathcal{T}_1$ interacts with $\mathcal{F}_L$ as its adversary, and as its corrupt second party. The machine $\mathcal{T}_2$ interacts with $\mathcal{F}_R$ as its adversary, and as its corrupt first party. Finally, $\mathcal{T}_1$ and $\mathcal{T}_2$ interact as honest first and second parties to $\mathcal{G}$, respectively. $\mathcal{F}_{\mathsf{split}}^{\mathcal{G},\mathcal{T}_1,\mathcal{T}_2}$'s adversary interacts directly as the adversary to $\mathcal{G}$.

Definition 6. *Let $\mathcal{G}$ be a functionality and $\mathcal{T}$ an admissible machine. We define $\mathcal{S}^{\mathcal{G},\mathcal{T}}$ as a simulator which internally runs a copy of $\mathcal{G}$ and $\mathcal{T}$, where $\mathcal{T}$ interacts as the adversary to the external functionality and as the two honest parties to $\mathcal{G}$. $\mathcal{S}^{\mathcal{G},\mathcal{T}}$ lets the external dummy adversary interact with $\mathcal{G}$ as its adversary (see Fig. 4).*

Definition 7 (General Splittability). *(See Fig. 4) For two 2-party functionalities $\mathcal{F}$ and $\mathcal{G}$, we say that $\mathcal{F}$ is splittable with respect to $\mathcal{G}$ (written $\mathcal{F} \prec \mathcal{G}$) if there exist machines $\mathcal{T}_{12}, \mathcal{T}_1, \mathcal{T}_2$ such that for all environments $\mathcal{Z}$, and the dummy adversaries $\mathcal{A}$ which corrupts no parties, we have $\mathrm{EXEC}[\mathcal{Z}, \mathcal{S}^{\mathcal{G},\mathcal{T}_{12}}, \partial^{\mathcal{F}}] \approx \mathrm{EXEC}[\mathcal{Z}, \mathcal{A}, \partial^{\mathcal{F}_{\mathsf{split}}^{\mathcal{G},\mathcal{T}_1,\mathcal{T}_2}}]$. We define $\mathrm{SPLIT}^{\mathcal{G}} = \{\mathcal{F} | \mathcal{F} \prec \mathcal{G}\}$ and $\mathrm{SPLIT}^* = \bigcup_{\mathcal{G}} \mathrm{SPLIT}^{\mathcal{G}}$.*

As in the previous section, our definitions (and results) apply to both PPT and computationally unbounded systems. We write $\mathrm{SPLIT}_u^{\mathcal{G}}$ or $\mathrm{SPLIT}_p^{\mathcal{G}}$, SPLIT_u^* or SPLIT_p^* and $\mathcal{F} \prec_u \mathcal{G}$ or $\mathcal{F} \prec_p \mathcal{G}$ to explicitly specify the type of the systems.

---

[3] The common thread in these proofs is to construct two separate corruption scenarios and consider a man-in-the-middle attack which pits the honest players in the two scenarios against each other.

As in Section 3.1, we aim to establish a relationship between splittability and realizability. Our main technical tools relating the two notions are given below:

Theorem 2. *For any 2-party functionalities $\mathcal{F}$, $\mathcal{G}$ and $\mathcal{H}$, the following hold:*

1. *If $\mathcal{F} \prec \mathcal{G}$ then $\mathcal{F} \sqsubseteq \mathcal{G}$.*   *[Splittability implies realizability]*
2. *If $\mathcal{F} \sqsubseteq \mathcal{G} \prec \mathcal{H}$, then $\mathcal{F} \prec \mathcal{H}$.*   *["Cross-transitivity"]*
3. *If $\mathcal{F} \prec \mathcal{G} \prec \mathcal{H}$, then $\mathcal{F} \prec \mathcal{H}$.*   *[Transitivity of $\prec$]*
4. *If $\mathcal{F} \sqsubseteq \mathcal{G} \sqsubseteq \mathcal{H}$, then $\mathcal{F} \sqsubseteq \mathcal{H}$.*   *[UC Theorem [9]]*

Proof (Sketch).

1. Analogous to the proof of Theorem 1, we can construct a protocol for $\mathcal{F}$ in the following way: The first party simulates $\mathcal{F}$ along with $\mathcal{T}_1$, while the second party simulates $\mathcal{F}$ along with $\mathcal{T}_2$ (as they interact in $\mathcal{F}_{\mathsf{split}}^{\mathcal{G},\mathcal{T}_1,\mathcal{T}_2}$). The simulator for a dummy adversary who corrupts no parties is $\mathcal{S}^{\mathcal{G},\mathcal{T}_{12}}$, and the simulator for a dummy adversary who corrupts party i can be constructed from $\mathcal{T}_i$ and $\mathcal{G}$.

2. This is the main technical tool. The proof is analogous to that of Theorem 1, but accounts for the fact that the communication channel used by a protocol for $\mathcal{F}$ is not $\mathcal{F}_{\mathsf{pvt}}$ but an arbitrary functionality $\mathcal{G}$, which in turn is splittable with respect to $\mathcal{H}$.

3. This is an immediate consequence of claims 1 and 2.

4. This is just a restatement of the universal composition theorem [9] in our notation. Though the original statement and proof of the UC theorem uses PPT systems, it is easy to see that it extends to the computationally unbounded systems as well.

In the simplified setting of Section 3.1, splittability provided an exact characterization of realizability with respect to completely private channels; namely, REALIZ$^{\mathcal{F}_{\mathsf{pvt}}}$ = SPLIT*. Ideally, we would like this characterization to generalize as REALIZ$^{\mathcal{F}}$ = SPLIT$^{\mathcal{F}}$ for all $\mathcal{F}$, but this is not the case. For instance $\mathcal{F} \in$ REALIZ$^{\mathcal{F}}$ for all $\mathcal{F}$, but $\mathcal{F} \notin$ SPLIT$^{\mathcal{F}}$ for several functionalities, e.g., commitment. However, the characterization does generalize for a certain class of functionalities.

Definition 8. *$\mathcal{F}$ is called* self-splittable *if $\mathcal{F} \prec \mathcal{F}$. We denote the class of all self-splittable functionalities as* SIMPLECHANNELS.

The class SIMPLECHANNELS can be viewed as a natural class of low cryptographic complexity in our landscape of complexity theory. Intuitively, $\mathcal{F} \prec \mathcal{F}$ means that $\mathcal{F}$ does not carry out any irreversible computation on its inputs. It can be easily seen that all typical communication channels (e.g., authenticated or unauthenticated, public or private, multicast or point-to-point, completely adversarially controlled), which are often implicitly incorporated into the network model, are in SIMPLECHANNELS.

Theorem 3. SPLIT$^{\mathcal{F}}$ = REALIZ$^{\mathcal{F}}$ *for all $\mathcal{F} \in$ SIMPLECHANNELS.*

In other words, functionalities which are realizable using a simple communication channel $\mathcal{F}$ are exactly those which are splittable with respect to $\mathcal{F}$. The proof follows as an easy consequence of Theorem 2. Interestingly, the simple communication channels are *exactly those functionalities for which this characterization holds.* That is, SIMPLECHANNELS $= \{\mathcal{F} \mid \text{SPLIT}^{\mathcal{F}} = \text{REALIZ}^{\mathcal{F}}\}$. As before, these statements hold for both PPT and computationally unbounded systems. However, note that SIMPLECHANNELS$_u$ and SIMPLECHANNELS$_p$ are different classes. For instance, a channel which applies a one-way permutation to its input is in SIMPLECHANNELS$_u \setminus$ SIMPLECHANNELS$_p$.

Relation to the Simplified Definition. The simplified definition of splittability (Definition 4) was elegant and easy to apply. We would still like to be able to use this simplified definition to say as much as possible about the complexity of functionalities, even in the general setting. The following lemma gives us a tool to do just that:

Lemma 1. SPLIT$^* =$ SPLIT$^{\mathcal{F}_{\text{pvt}}}$ $(=$ REALIZ$^{\mathcal{F}_{\text{pvt}}})$.

Intuitively, $\mathcal{F}_{\text{pvt}}$ is the "easiest" functionality to split with respect to. Equivalently, $\mathcal{F}_{\text{pvt}}$ is the most secure channel possible in the model. The special status of $\mathcal{F}_{\text{pvt}}$ is due to the fact that the network entities in our model communicate using essentially such a channel.

Most importantly, combining Lemma 1 with the characterization of Theorem 3, we see that if $\mathcal{F}$ is unsplittable according to the simple definition, then there is no secure $\mathcal{F}_{\text{pvt}}$-protocol for $\mathcal{F}$, and hence no secure protocol using *any* natural communication channel. As we shall see in Section 4, it is often very easy to show that a functionality is unsplittable according to the simpler definition. Thus, splittability gives us a convenient tool to easily show impossibility results of this kind.

3.3   Deviation Revealing Functionalities

Splittability provides a convenient way to give separations that involve the relatively low-complexity functionalities of SIMPLECHANNELS. However, splittability is virtually useless in distinguishing among higher complexity functionalities, which nevertheless exhibit a rich variety in their (intuitive) cryptographic complexities. For instance, one may ask whether $\mathcal{F}_{\text{OT}}$ (oblivious transfer) and $\mathcal{F}_{\text{com}}$ (commitment) have different cryptographic complexities or not. In this section we develop a tool to answer many of these questions.

We introduce a notion called *deviation revealing* functionalities, which will allow us to lift existing separations of functionalities derived in simpler settings (such as the honest-but-curious model) to the standard UC setting.

Relating Passive and Active Corruption. Consider the 2-party SFE functionality $\mathcal{F}_{\text{OR}}$ that evaluates the boolean OR of Alice and Bob's input bits and outputs it to only Bob. $\mathcal{F}_{\text{OR}}$ has a secure protocol in which Alice sends her input bit to Bob, and Bob locally computes the OR using that and his own input. This protocol is secure because if Bob wants to, he can learn Alice's bit even in the ideal world (by sending 0 to $\mathcal{F}_{\text{OR}}$). However, this is too much information for an "honest-but-curious" Bob when his input is 1. In fact it is known [18] that there is no secure protocol for $\mathcal{F}_{\mathsf{OR}}$ in the honest-but-curious, unbounded computation setting (where corruption in the ideal world must also be passive).

As such, in general, we cannot expect results about realizability in restricted corruption scenarios to imply anything about realizability in the unrestricted corruption model. However, several natural and important functionalities do not share this odd nature of $\mathcal{F}_{\mathsf{OR}}$. We formulate the deviation revealing condition to capture such "nicely behaved" functionalities.

Corruption Schemes. First we need to generalize the corruption model, to allow regular (active) corruption as well as the restricted passive (a.k.a honest-but-curious) corruption. For an m-party functionality a *corruption scheme* C is a subset of $\{\mathsf{none}, \mathsf{passive}, \mathsf{active}\}^m$. We say that a (static) adversary $\mathcal{A}$ C-*corrupts* (with respect to a protocol π) if the sequence of corruptions γ effected by $\mathcal{A}$ is in C. We will be interested in what we call *uniform corruption schemes*, wherein in each corruption sequence the corrupt parties either are all actively corrupted or are all passively corrupted: i.e., C is a uniform corruption scheme if it is a subset of $\{\mathsf{none}, \mathsf{passive}\}^m \cup \{\mathsf{none}, \mathsf{active}\}^m$.

For a corruption scheme C, we say that $\mathcal{F} \sqsubseteq^C \mathcal{G}$ if there exists a protocol π such that for all C-corrupting (with respect to π) $\mathcal{A}$, there exists a C-corrupting (with respect to the dummy protocol ∂) $\mathcal{S}$ such that for all environments $\mathcal{Z}$, $\mathrm{EXEC}[\mathcal{Z}, \mathcal{A}, \pi^{\mathcal{G}}] \approx \mathrm{EXEC}[\mathcal{Z}, \mathcal{S}, \partial^{\mathcal{F}}]$.

Deviation Revealing Functionalities. Intuitively, a deviation revealing functionality is one for which it is easy for the environment to detect whether an adversary is C-corrupting or not. However, to consider adversaries which deviate from C-corruption in benign ways, we use a more sophisticated definition. Note that, as with our other definitions, the following definition is given in terms of an ideal interaction with the functionality, and thus applies uniformly to arbitrary functionalities.

Definition 9. $\mathcal{F}$ *is* C-*deviation-revealing if for all adversaries* $\mathcal{A}$, *either:*

- *there exists a correctness environment* $\mathcal{Z}$ *such that* $\mathrm{EXEC}[\mathcal{Z}, \mathcal{A}, \partial^{\mathcal{F}}] \not\approx \mathrm{EXEC}[\mathcal{Z}, \widetilde{\mathcal{A}}, \partial^{\mathcal{F}}]$, *where* $\widetilde{\mathcal{A}}$ *is the dummy* C-*corrupting adversary;*
- *or, there exists a* C-*corrupting adversary* $\mathcal{A}'$ *such that for all environments* $\mathcal{Z}$, $\mathrm{EXEC}[\mathcal{Z}, \mathcal{A}, \partial^{\mathcal{F}}] \approx \mathrm{EXEC}[\mathcal{Z}, \mathcal{A}', \partial^{\mathcal{F}}]$.

A *correctness environment* is one which only interacts with the adversary by sending inputs to the corrupted parties and receiving their outputs.

Following is our toolkit for lifting relations in a C-corruption setting to the standard corruption setting

Theorem 4. *For any functionalities* $\mathcal{F}$, $\mathcal{G}$ *and* $\mathcal{H}$, *the following hold:*
1. *If* $\mathcal{F} \sqsubseteq^C \mathcal{G} \sqsubseteq^C \mathcal{H}$, *then* $\mathcal{F} \sqsubseteq^C \mathcal{H}$.   *[Universal Composition.]*
2. *If* $\mathcal{F}$ *is* C-*deviation revealing for a uniform* C, *then*
   a. $\mathcal{F} \sqsubseteq \mathcal{G} \implies \mathcal{F} \sqsubseteq^C \mathcal{G}$ *[C-realizability from realizability.]*
   b. $(\mathcal{F} \not\sqsubseteq^C \mathcal{H} \wedge \mathcal{G} \sqsubseteq^C \mathcal{H}) \implies \mathcal{F} \not\sqsubseteq \mathcal{G}$ *[Separation from C-separation.]*

4   Applications of the Theory

In this section, we apply the general theory developed in the previous section to specific settings and classes of functionalities, to obtain several new, concrete results, as easy consequences.

4.1   Simple Impossibility Results

A compelling aspect of our splittability characterization is that all previous impossibility results for the UC model can be obtained quite easily, because the splittability definition involves only interactions with ideal functionalities.

For instance, the *bit commitment* functionality ($\mathcal{F}_{\mathsf{com}}$) is unsplittable: Consider a simple environment which asks Alice to commit to a random bit, waits for Bob to receive acknowledgement of the commeitment, instructs Alice to reveal the bit, and finally checks whether Bob received the correct bit. In any potential split of $\mathcal{F}_{\mathsf{com}}$, $\mathcal{T}$ must at some point commit to a bit in one of the instances of $\mathcal{F}_{\mathsf{com}}$, but its view at that time is by definition independent of the environment's choice, and thus the bit that Bob eventually receives will be wrong with probability $1/2$.

Similarly, the coin-tossing functionality $\mathcal{F}_{\mathsf{coin}}$ is unsplittable, because any split of $\mathcal{F}_{\mathsf{coin}}$ simulates two *independent* copies of $\mathcal{F}_{\mathsf{coin}}$, and so (regardless of $\mathcal{T}$— it sends no input to either copy of $\mathcal{F}_{\mathsf{coin}}$) the two parties' outputs will disagree with probability $1/2$. Using similar arguments, it is a very easy exercise to see that several other important 2-party functionalities, such as *oblivious transfer* ($\mathcal{F}_{\mathsf{OT}}$) and, in PPT systems, *zero-knowledge proof* for languages in NP \ BPP, are unsplittable.

Applying Theorem 3 and Lemma 1, we can further see that these functionalities are unrealizable via protocols that use *any* simple communication channel (i.e., one from SIMPLECHANNELS). These impossibility results also rule out the possibility of *non-trivial protocols* for variants of these functionalities which allow the adversary to delay the honest parties' outputs.

4.2   Combinatorial Characterization for 2-Party SFE

We use the splittability characterization for 2REGULAR to give an explicit, *combinatorial* characterization for 2-party secure function evaluation. This subsumes and completes the characterizations initiated in [12, 13]. The impossibility results in [12, 13] were later extended in [25], to the setting where certain "trusted setup" functionalities $\mathcal{F}$ are also available for protocols to use. These extensions can also be easily derived in our framework by observing that these particular functionalities $\mathcal{F}$ are self-splittable, thus impossibility under $\mathcal{F}_{\mathsf{pvt}}$ implies impossibility under $\mathcal{F}$.

Definition 10. $\mathcal{F}$ *is a* 2-party secure function evaluation (SFE) *functionality if it waits for inputs x and y from the two parties, respectively, computes two (randomized) functions $f_1(x, y)$ and $f_2(x, y)$, and sends these values to the two parties, respectively. In this case, we write $\mathcal{F} = (f_1, f_2)$.*

Note that SFE functionalities are in the class 2REGULAR. We now define two properties of 2-party SFE functionalities which will be used in our characterization. We write $f(x, y) \approx g(x', y')$ to indicate that the two distributions are indistinguishable (computationally or statistically, depending on the system).

Definition 11. *We say that $\mathcal{F} = (f_1, f_2)$ has* unidirectional influence *if one party's output does not depend on the other party's input. That is, if $f_1(x, y) \approx f_1'(x)$ for some function f_1', or $f_2(x, y) \approx f_2'(y)$ for some function f_2'. Otherwise $\mathcal{F}$ has* bidirectional influence.

Definition 12. *Let $\mathcal{F} = (f_1, f_2)$ be a 2-party SFE functionality with unidirectional influence; say, $f_1(x, y) \approx f_1'(x)$. We say that $\mathcal{F}$ is* negligibly hiding *if there exists machines R_1, R_2 such that:*

$$\forall x, y : \Pr\left[(y^*, s) \leftarrow R_1; f_2\Big(R_2\big(s, f_2(x, y^*)\big), y\Big) \not\approx f_2(x, y)\right] \text{ is negligible}$$

The probability is over the randomness of f_1, f_2, R_1 and R_2.

If $\mathcal{F}$ is *deterministic* and has input domains of polynomial size (in the security parameter), then negligibly hiding is a simple combinatorial property: $\mathcal{F}$ is negligibly hiding if and only if there exists y such that $f_2(x, y) = f_2(x', y) \Rightarrow f_2(x, \cdot) \equiv f_2(x', \cdot)$. Our definition succinctly incorporates both the *completely revealing* and *efficiently invertible* properties of [12].

Theorem 5. *Let $\mathcal{F} = (f_1, f_2)$ be a 2-party SFE functionality. $\mathcal{F}$ is securely realizable (using $\mathcal{F}_{\mathsf{pvt}}$) if and only if $\mathcal{F}$ has unidirectional influence and is negligibly hiding.*

Proof (Sketch). We show that $\mathcal{F}$ is splittable if and only if it has unidirectional influence and is negligibly hiding. If it has bidirectional influence or is not negligibly hiding, then for every $\mathcal{T}$, it is straight-forward to construct an environment that distinguishes between $\mathcal{F}$ and $\mathcal{F}_{\mathsf{split}}^{\mathcal{T}}$.

On the other hand, if $\mathcal{F}$ has unidirectional influence (say, from the first party to the second) and is negligibly hiding, then a successful strategy for $\mathcal{T}$ is to choose its input to $\mathcal{F}_L$ according to R_1, then choose its input to $\mathcal{F}_R$ according to R_2. The definition of negligible hiding implies that the second party's output from such a $\mathcal{F}_{\mathsf{split}}^{\mathcal{T}}$ is correct with overwhelming probability. Unidirectional influence implies that the first party's output is correct, since (roughly) it depends only on the first party's input, not on anything provided by $\mathcal{T}$).

Again, we reiterate that Theorem 5 also characterizes the existence of *non-trivial protocols* for SFE functionalities in which the adversary can delay honest parties' outputs from the functionality.

4.3   Results for Multi-party Functionalities

For multi-party functionalities involving more than two parties, where the splittability definition is much more complicated, combinatorial characterizations like that of Theorem 5 seem difficult to come by. Nonetheless, we can use 2-party results to obtain some strong necessary conditions for the multi-party setting.

A well-known technique for studying m-party SFE functionalities is the *partitioning argument*: consider *2-party SFE functionalities* induced by partitioning of the m parties into two sets. If the original functionality is realizable, then clearly so is each induced 2-party functionality.

To exploit the partitioning argument, first we extend the notion of *influence* from Definition 11 to multi-party SFE: For $i \neq j$, if there exist inputs $x_1, \ldots, x_m$ and x_i' such that $f_j(x_1, \ldots, x_m) \not\approx f_j(x_1, \ldots, x_i', \ldots, x_m)$, then we say party i *influences* the output of party j, and write $i \overset{\mathcal{F}}{\rightsquigarrow} j$.

Corollary 6 (of Theorem 5) *If $\mathcal{F}$ is an m-party SFE functionality securely realizable using completely private channels, then in the directed graph induced by $\overset{\mathcal{F}}{\rightsquigarrow}$, either all edges have a common source, or all edges have a common destination.*

We see that there are only two simple kinds of securely realizable SFE functionalities. Let p be the common vertex in the graph induced by $\overset{\mathcal{F}}{\rightsquigarrow}$. If all edges are directed towards p, then we say that $\mathcal{F}$ is *aggregated* (via party p). If all edges are directed away from p, then we say that $\mathcal{F}$ is *disseminated via party p*.

3-party Characterization. We now show that in some instances, the above partitioning argument does lead to a sufficient condition for realizability of multiparty functionalities. We consider the class 3REGULAR of 3-party functionalities which do not interact with the adversary, and whose behavior does not depend on which parties are corrupted. We show that for functionalities in 3REGULAR that have a secure *honest-majority protocol*,[4] the partitioning argument along with our previous 2-party characterization suffices to characterize realizability.

Theorem 7. *If $\mathcal{F} \in$ 3REGULAR has an honest-majority protocol using channel $\mathcal{G} \in$ SIMPLECHANNELS, then $\mathcal{F} \sqsubseteq \mathcal{G}$ if and only if all 2-party restrictions of $\mathcal{F}$ are UC-realizable using $\mathcal{F}_{\mathsf{pvt}}$.*

Proof (Sketch). Informally, we construct a protocol for $\mathcal{F}$ by combining the honest-majority protocol for $\mathcal{F}$ and the machines guaranteed by each 2-party splitting of $\mathcal{F}$. For adversaries that corrupt only one party, a simulator is derived from the simulator for the honest-majority protocol. For adversaries that corrupt two parties, we use the splittability criterion to construct a simulator.

In particular, this implies that realizable 3-party SFE functionalities have a simple combinatorial characterization analogous to Theorem 5. Our protocol requires each player to internally simulate executions of another protocol with a weaker/different security guarantee (in this case, the 2-party restrictions and

---

[4] Honest majority protocols are known to exist for essentially all SFE functionalities using a broadcast channel. As outlined in [9], such protocols can be constructed by adapting the well-known information theoretically secure (stand-alone) protocols of [7, 15, 39].

the honest-majority protocol). This is somewhat comparable to the "MPC in the head" approach [21, 23], where significant efficiency gains are achieved in the standard corruption model by leveraging MPC protocols with security in the honest-majority settings. Our constructions indicate the possibility of extending this approach by having the parties carry out not a direct protocol execution, but a related simulation.

The same approach does not seem to apply for functionalities which interact with the adversary, whose behavior depends on which parties are corrupt, or which involve more than three parties (so that two parties do not form a strict majority). We leave it as an important open problem whether the partitioning argument along with our previous 2-party characterizations suffice for characterizing multi-party functionalities *in general*. Indeed, the analogous partitioning argument has been studied for the honest-but-curious setting and shown to be insufficient in this regard [16].

4.4   A Strict Hierarchy of Intermediate Complexities

Finally, we apply the main structural result of our deviation-revealing theory (Theorem 4) to identify a sequence of functionalities with strictly increasing complexities (in unbounded computation systems).

Theorem 8. $\text{REALIZ}_u^{\mathcal{F}_{\text{pvt}}} \subsetneq \text{REALIZ}_u^{\mathcal{F}_{\text{simex}}^+} \subsetneq \text{REALIZ}_u^{\mathcal{F}_{\text{com}}^+} \subsetneq \text{REALIZ}_u^{\mathcal{F}_{\text{OT}}^+}$.

Here, $\mathcal{F}_{\text{simex}}^+$, $\mathcal{F}_{\text{com}}^+$, and $\mathcal{F}_{\text{OT}}^+$ denote "augmented" versions of simultaneous exchange,[5] bit commitment, and oblivious transfer, respectively, in which the functionality provides multiple "sessions" instead of just one (as outlined in Section 2).

Proof (Sketch). The non-trivial observations in proving this are that $\mathcal{F}_{\text{OT}} \not\sqsubseteq_u \mathcal{F}_{\text{com}}^+$ and $\mathcal{F}_{\text{com}} \not\sqsubseteq_u \mathcal{F}_{\text{simex}}^+$. For the former, we consider the passive corruption scheme and for the latter we consider the corruption scheme where the sender (for $\mathcal{F}_{\text{com}}$) could be actively corrupt and the receiver could be passively corrupt. We exploit the fact that both $\mathcal{F}_{\text{OT}}$ and $\mathcal{F}_{\text{com}}$ are deviation revealing for these corruption schemes respectively; that $\mathcal{F}_{\text{com}}$ and $\mathcal{F}_{\text{simex}}$ respectively have trivial protocols (using, e.g., $\mathcal{F}_{\text{pvt}}$) in these settings; and that $\mathcal{F}_{\text{OT}}$ and $\mathcal{F}_{\text{com}}$ respectively do not have secure protocols (using $\mathcal{F}_{\text{pvt}}$) in these settings. Then by Theorem 4, $\mathcal{F}_{\text{OT}} \not\sqsubseteq_u \mathcal{F}_{\text{com}}^+$ and $\mathcal{F}_{\text{com}} \not\sqsubseteq_u \mathcal{F}_{\text{simex}}^+$.

The significance of Theorem 8 is to establish several distinct levels of *intermediate complexity* (i.e., distinct *degrees* of the $\sqsubseteq$ reduction). That is, $\mathcal{F}_{\text{simex}}$ and $\mathcal{F}_{\text{com}}$ are neither realizable nor complete for computationally unbounded systems. Incidentally, both of these functionalities *are complete* for PPT systems [14]. We leave it as an important open problem whether there is a zero-one law of complexity in PPT systems (i.e., whether all functionalities not in $\text{REALIZ}^{\mathcal{F}_{\text{pvt}}}$ are complete).

---

[5] The simulataneous exchange functionality $\mathcal{F}_{\text{simex}}$ takes two inputs bits x and y from the two parties, respectively, and outputs (x, y) to both. It is called simultaneous exchange because x must be chosen without knowledge of y, and vice-versa.

Acknowledgements

We would like to thank Ran Canetti, Cynthia Dwork, Yuval Ishai and Tal Malkin for useful discussions.

References

1. Proc. 20th STOC. ACM (1988)
2. Proc. 21st STOC. ACM (1989)
3. Barak, B., Prabhakaran, M., Sahai, A.: Concurrent non-malleable zero knowledge. In: Proc. 47th FOCS. IEEE, Los Alamitos (2006)
4. Barak, B., Canetti, R., Nielsen, J.B., Pass, R.: Universally composable protocols with relaxed set-up assumptions. In: FOCS, pp. 186–195. IEEE, Los Alamitos (2004)
5. Barak, B., Sahai, A.: How to play almost any mental game over the net - concurrent composition using super-polynomial simulation. In: Proc. 46th FOCS. IEEE, Los Alamitos (2005)
6. Beimel, A., Malkin, T., Micali, S.: The all-or-nothing nature of two-party secure computation. In: Wiener, M.J. (ed.) CRYPTO 1999. LNCS, vol. 1666, pp. 80–97. Springer, Heidelberg (1999)
7. Ben-Or, M., Goldwasser, S., Wigderson, A.: Completeness theorems for non-cryptographic fault-tolerant distributed computation. In: Proc. 20th STOC [1], pp. 1–10
8. Canetti, R.: Universally composable security: A new paradigm for cryptographic protocols. Cryptology ePrint Archive, Report 2000/067. Revised version of [9]
9. Canetti, R.: Universally composable security: A new paradigm for cryptographic protocols. Electronic Colloquium on Computational Complexity (ECCC) TR01-016 (2001). Previous version A unified framework for analyzing security of protocols, availabe at the ECCC archive TR01-016. Extended abstract in FOCS 2001 (2001)
10. Canetti, R., Dodis, Y., Pass, R., Walfish, S.: Universally composable security with global setup. In: Vadhan, S.P. (ed.) TCC 2007. LNCS, vol. 4392. Springer, Heidelberg (2007)
11. Canetti, R., Fischlin, M.: Universally composable commitments. Report 2001/055, Cryptology ePrint Archive, July 2001. Extended abstract appeared in CRYPTO 2001 (2001)
12. Canetti, R., Kushilevitz, E., Lindell, Y.: On the limitations of universally composable two-party computation without set-up assumptions. In: Biham, E. (ed.) EUROCRYPT 2003. LNCS, vol. 2656. Springer, Heidelberg (2003)
13. Canetti, R., Kushilevitz, E., Lindell, Y.: On the limitations of universally composable two-party computation without set-up assumptions. J. Cryptology 19(2), 135–167 (2006)
14. Canetti, R., Lindell, Y., Ostrovsky, R., Sahai, A.: Universally composable two-party computation. In: Proc. 34th STOC, pp. 494–503. ACM, New York (2002)
15. Chaum, D., Crépeau, C., Damgård, I.: Multiparty unconditionally secure protocols. In: Proc. 20th STOC [1], pp. 11–19
16. Chor, B., Ishai, Y.: On privacy and partition arguments. Information and Computation 167(1), 2–9 (2001)

17. Chor, B., Kushilevitz, E.: A zero-one law for boolean privacy (extended abstract). In: STOC [2], pp. 62–72
18. Chor, B., Kushilevitz, E.: A zero-one law for boolean privacy. SIAM J. Discrete Math. 4(1), 36–47 (1991)
19. Dwork, C., Lynch, N.A., Stockmeyer, L.J.: Consensus in the presence of partial synchrony. J. ACM 35(2), 288–323 (1988)
20. Goldreich, O., Micali, S., Wigderson, A.: How to play ANY mental game. In: ACM (ed.) Proc. 19th STOC, pp. 218–229. ACM, New York (1987)
21. Harnik, D., Ishai, Y., Kushilevitz, E., Nielsen, J.B.: Ot-combiners via secure computation. In: TCC 2008 (to appear, 2008)
22. Harnik, D., Naor, M., Reingold, O., Rosen, A.: Completeness in two-party secure computation: A computational view. J. Cryptology 19(4), 521–552 (2006)
23. Ishai, Y., Kushilevitz, E., Ostrovsky, R., Sahai, A.: Zero-knowledge from secure multiparty computation. In: STOC, pp. 21–30. ACM, New York (2007)
24. Kalai, Y.T., Lindell, Y., Prabhakaran, M.: Concurrent general composition of secure protocols in the timing model. In: STOC, pp. 644–653. ACM, New York (2005)
25. Kidron, D., Lindell, Y.: Impossibility results for universal composability in public-key models and with fixed inputs. Cryptology ePrint Archive, Report 2007/478 (2007), http://eprint.iacr.org/2007/478
26. Kilian, J.: Founding cryptography on oblivious transfer. In: STOC, pp. 20–31 (1988)
27. Kilian, J.: A general completeness theorem for two-party games. In: STOC, pp. 553–560. ACM, New York (1991)
28. Kilian, J.: More general completeness theorems for secure two-party computation. In: Proc. 32th STOC, pp. 316–324. ACM, New York (2000)
29. Kilian, J., Kushilevitz, E., Micali, S., Ostrovsky, R.: Reducibility and completeness in private computations. SIAM J. Comput. 29(4), 1189–1208 (2000)
30. Kushilevitz, E.: Privacy and communication complexity. In: FOCS, pp. 416–421. IEEE, Los Alamitos (1989)
31. Kushilevitz, E.: Privacy and communication complexity. SIAM J. Discrete Math. 5(2), 273–284 (1992)
32. Kushilevitz, E., Micali, S., Ostrovsky, R.: Reducibility and completeness in multiparty private computations. In: FOCS, pp. 478–489. IEEE, Los Alamitos (1994)
33. Lindell, Y.: General composition and universal composability in secure multi-party computation. In: Proc. 44th FOCS. IEEE, Los Alamitos (2003)
34. Lindell, Y.: Lower bounds for concurrent self composition. In: Naor, M. (ed.) TCC 2004. LNCS, vol. 2951, Springer, Heidelberg (2004)
35. Pfitzmann, B., Waidner, M.: Composition and integrity preservation of secure reactive systems. In: ACM Conference on Computer and Communications Security, pp. 245–254 (2000)
36. Prabhakaran, M.: New Notions of Security. PhD thesis, Department of Computer Science, Princeton University (2005)
37. Prabhakaran, M., Rosulek, M.: Cryptographic complexity of multi-party computation problems: Classifications and separations. Electronic Colloquium on Computational Complexity (ECCC) 15(50) (2008)
38. Prabhakaran, M., Sahai, A.: New notions of security: achieving universal composability without trusted setup. In: STOC, pp. 242–251. ACM, New York (2004)
39. Rabin, T., Ben-Or, M.: Verifiable secret sharing and multiparty protocols with honest majority. In: Proc. 21st STOC [2], pp. 73–85
40. Yao, A.C.: Protocols for secure computation. In: Proc. 23rd FOCS, pp. 160–164. IEEE, Los Alamitos (1982)

Cryptanalysis of MinRank

Jean-Charles Faugère[1], Françoise Levy-dit-Vehel[2],
and Ludovic Perret[1]

[1] SALSA Project
INRIA, Centre Paris-Rocquencourt
UPMC, Univ Paris 06, LIP6
CNRS, UMR 7606, LIP6
104, avenue du Président Kennedy
75016 Paris, France
jean-charles.faugere@inria.fr, ludovic.perret@lip6.fr
[2] ENSTA, 32 Boulevard Victor, 75739 Paris cedex 15
levy@ensta.fr

Abstract. In this paper, we investigate the difficulty of one of the most
relevant problems in multivariate cryptography – namely MinRank –
about which no real progress has been reported since [9, 19]. Our starting point is the Kipnis-Shamir attack [19]. We first show new properties
of the ideal generated by Kipnis-Shamir's equations. We then propose
a new modeling of the problem. Concerning the practical resolution, we
adopt a Gröbner basis approach that permitted us to actually solve challenges A and B proposed by Courtois in [8]. Using the multi-homogeneous
structure of the algebraic system, we have been able to provide a theoretical complexity bound reflecting the practical behavior of our approach.
Namely, when r' the dimension of the matrices minus the rank of the
target matrix in the MinRank problem is constant, then we have a poly-
nomial time attack $\mathcal{O}\left(\ln\left(q\right)n^3{}^{r'^2}\right)$. For the challenge C [8], we obtain
a theoretical bound of $2^{66.3}$ operations.

1 Introduction

The main purpose of this paper is the study of the MinRank (MR) problem.
MR was originally introduced in [26] as one of the natural questions in linear
algebra, and the authors there proved its NP-completeness. Later, it was restated
in [8] in the cryptographic context. Since then, it has been shown to be related
to several multivariate public key cryptosystems, for instance HFE [19, 23] and
TTM [9, 22]. MR is also the basis of an efficient zero-knowlege authentication
scheme [8]. According to the designer, this scheme is one of the most efficient
post-quantum (i.e. based on a NP-complete problem) authentication scheme.

We can consider that MinRank – with the Polynomial System Solving (PoSSo)
[17] and the Isomorphism of Polynomials (IP) [23] problems – is one of the main
problems in multivariate cryptography. Contrarily to PoSSo for which progresses
are reported continuously [12, 14], and IP which is now well mastered for most

D. Wagner (Ed.): CRYPTO 2008, LNCS 5157, pp. 280–296, 2008.
© International Association for Cryptologic Research 2008

of its cryptographic applications [15, 16], no advance has been reported on MR since [9, 19]. To this respect, it has to be noted that the paper of X. Jiang, J. Ding and L. Hu [18] deals with the particular context of HFE. They show that due to the particular structure of the equations the complexity of the Kipnis-Shamir attack is exponential. The theoretical argument underlying his observation does not apply to the context of the generic MR problem.

There exists two non-trivial general techniques for solving MinRank. A first technique, called *kernel attack*, consists in guessing some vectors of an appropriate kernel, and then solve a resulting linear system [8, 9]. Another technique, due to Kipnis and Shamir [8, 19], consists in modeling MR as a PoSSo problem, i.e. one in which the purpose is to solve a quadratic system of equations. It is a transposition of an original attack on HFE [8, 19]. Initially, the complexity of this attack was evaluated using relinearization.

The starting point of our study is Kipnis-Shamir (KS) attack on MR. We begin by proving an exact correspondence between the solutions found by KS attack and the solutions of MR; moreover, we show how Kipnis-Shamir's approach somehow include the so-called *minors attack* and *Schnorr's attack* on MR. We then propose a new method for solving MR, which can be viewed as an extension of Schnorr's attack. After that, we present our practical way of solving MR, namely by means of fast Gröbner bases algorithms (e.g. [12]). Our main practical result is the breaking of the challenges A and B proposed for the MR-authentication scheme [8]. The MinRank problem being NP-hard, we cannot expect to solve efficiently all the instances. Thus, there is a set of parameters which remains out of the scope of our approach. But it has to be noted that the challenges we break correspond to sets of parameters which are the most interesting for practical use. Consequently, the place of MR-authentication scheme in the hierarchy of post-quantum schemes should be re-evaluated.

1.1   Organization of the Paper. Main Results

In a first part of the paper, we recall known facts about the complexity of MR and its links with famous problems of coding theory. Then we detail two generic solving methods for MR, namely the kernel attack and Kipnis-Shamir attack. Section three is devoted to new properties satisfied by the equations generated by the Kipnis-Shamir attack (KS equations). In particular, we point out a bijection between the solutions of MR and the variety associated to the ideal generated by the KS equations. In the purpose of systematically studying and comparing all the modelings for MR, we show how the equations generated by other techniques – namely, the minors attack and Schnorr's attack – are included in the ideal of the KS equations. In section four, we describe our new modeling of the problem, that links the minors attack and Schnorr's attack. It appears that this new method is not the most efficient one, as quoted at the end of section 4. Thus, in order to evaluate the theoretical complexity of solving MR, we keep the best approach, that is, the one given by the KS equations. Section five provides such a theoretical complexity bound. It is obtained using multi-homogeneous properties of the equations. Our numerical results are presented in section six.

2   The MinRank Problem

First, let us recall the MinRank problem over a field $\mathbb{K}$:

MinRank (MR)
Input: positive integers N, n, r, k, and matrices $M_0; M_1, \ldots, M_k$ in $\mathcal{M}_{N \times n}(\mathbb{K})$
Question: is there a k-tuple $(\lambda_1, \ldots, \lambda_k)$ of elements of $\mathbb{K}$ such that:

$$\text{Rank}\left(\sum_{i=1}^{k} \lambda_i \cdot M_i - M_0 \right) \leq r.$$

We will in practice consider the search problem. If $N = n$, one gets a "square" instance of MR, that we call MR_s.

Property 1. MR_s and MR are poly-time many-one equivalent.

Proof. MR_s is a sub-problem of MR, in the sense that any instance of MR_s can be considered as an instance of MR.

Now, let g be a function from the set of instances of MR to the set of instances of MR_s, which maps a matrix M of size $N \times n$ to the square matrix of size $\max(N, n)$ obtained from M by adding $n - N$ rows (resp. $N - n$ columns) of zeroes (depending on whether $N < n$ or not). Then obviously $\text{Rank}(g(M)) = \text{Rank}(M)$, so that any yes-instance of MR is mapped to a yes-instance of MR_s by g; and conversely, any instance of MR which, by g, becomes a yes-instance of MR_s, is indeed a yes-instance of MR. $\square$

2.1   Complexity Considerations

Some complexity results have been proved for MR, linking it with other hard - or presumably hard - problems. The first one is a very simple reduction from Maximum Likelihood Decoding, proposed by Courtois in [9], which thus shows the NP-hardness of MR. Another less known fact is the link between MR and another problem in Coding Theory, namely Rank Decoding (RD). To this respect, the main result is that RD is poly-time many-one reducible to MR. The following lines are devoted to the RD problem and the proof of this result. Let us first recall what rank metric is.

Let $N, n \in \mathbb{N}^*$, and q be a power of a prime. We consider $\mathbb{F}_{q^N}$ as a vector space of dimension N over $\mathbb{F}_q$, and we fix a basis $\mathcal{B} = (b_1, \ldots, b_N)$ of it. For $\mathbf{x} = (x_1, \ldots, x_n) \in \mathbb{F}_{q^N}^n$, we denote by $\text{Rank}(\mathbf{x} \,|\, \mathbb{F}_q)$, the rank of the $(N \times n)$ matrix with entries in $\mathbb{F}_q$ given by $X = (x_{ij})_{ij}$, where, for each j, $1 \leq j \leq n$, $x_j = \sum_{i=1}^{N} x_{ij} b_i$. In other words, X is obtained from $\mathbf{x}$ by expressing each coordinate of x in $\mathcal{B}$. For any two vectors $\mathbf{x}, \mathbf{y} \in \mathbb{F}_{q^N}^n$, the quantity $d(\mathbf{x}, \mathbf{y}) = \text{Rank}(\mathbf{x} - \mathbf{y} \,|\, \mathbb{F}_q)$ defines a distance over $\mathbb{F}_{q^N}^n$, called *rank distance*. For codes over an extension field, rank distance is an analogue - although less discriminant - to the classical Hamming distance.

The Rank Decoding problem states as follows:

Rank Decoding (RD)

Input: positive integers N, n, r, k, a matrix G in $\mathcal{M}_{k \times n}(\mathbb{F}_{q^N})$ and a vector $c \in \mathbb{F}_{q^N}^n$.

Question: is there a vector $m \in \mathbb{F}_{q^N}^k$, such that $e = c - mG$ has rank $\mathrm{Rank}(e \,|\, \mathbb{F}_q) \leq r$?

If it is known a priori that $\mathrm{Rank}(e \,|\, \mathbb{F}_q) <= \lfloor (d-1)/2 \rfloor$, where d is the minimum (rank) distance of the considered code, then exactly one solution exists.

It is to be noted that MR can be seen as a *subfield subcode rank decoding* problem, where m has to be searched over the ground field $\mathbb{F}_q$. Indeed, let (N, n, r, k, G, c) be an instance of RD. Let $\mathcal{B} = (b_1, \ldots, b_N)$ be a basis of $\mathbb{F}_{q^N}$ over $\mathbb{F}_q$. Expressing each coordinate of c in this basis, we get an $N \times n$ matrix with entries in $\mathbb{F}_q$, that we call M_0. Analogously, for $1 \leq i \leq k$, expressing every entry of every row L_i of G in $\mathcal{B}$, we get an $N \times n$ matrix M_i over $\mathbb{F}_q$. Then $(N, n, r, k, M_0; M_1, \ldots, M_k)$ is an instance of MR that exactly corresponds to the instance (N, n, r, k, G, c) of RD in the sense that a solution $m = (\lambda_1, \ldots, \lambda_k)$ of this instance of MR will be a solution over $\mathbb{F}_q$ of the instance of RD, if ever such a solution exists.

As RD is in NP, we have that RD is poly-time many-one reducible to MR. A proof of this has been sketched in [7]. We give below a completely written proof of this property:

Proposition. [7] RD is poly-time many-one reducible to MR.

Proof. Fix a basis $\mathcal{B} = (b_1, \ldots, b_N)$ of $\mathbb{F}_{q^N}$, considered as an N-dimensional vector space over $\mathbb{F}_q$. Let f be the function that, to an instance $(N, n, r, k, G = (L_1, \ldots, L_k), c)$ of RD, associates the instance $(N, n, r, kN, M_0; M_1, \ldots, M_{kN})$ of MR, where M_0 is the $N \times n$ matrix with entries in $\mathbb{F}_q$ representing the vector $c \in \mathbb{F}_{q^N}^n$ in the basis $\mathcal{B}$, and $M_\ell = b_j L_i$ with $\ell = (i-1)N + j$, $1 \leq i \leq k$, $1 \leq j \leq N$, is the $N \times n$ matrix over $\mathbb{F}_q$ representing the product $b_j L_i$.

Let now $(N, n, r, k, G = (L_1, \ldots, L_k), c)$ be a yes-instance of RD, and let $m = (m_1, \ldots, m_k)$ be a vector of $\mathbb{F}_{q^N}^k$, solution to this instance, i.e. $\mathrm{Rk}((c-mG)|\mathbb{F}_q) = \mathrm{Rk}((c - \sum_{i=1}^k m_i L_i)|\mathbb{F}_q) \leq r$. For $1 \leq i \leq k$, let $m_i = \sum_{j=1}^N m_{ij} b_j$, $m_{ij} \in \mathbb{F}_q$, in the basis $\mathcal{B}$. Then, $m_i L_i = \sum_{j=1}^N m_{ij} b_j L_i$. We thus get

$$\mathrm{Rank}((c - mG)|\mathbb{F}_q) =$$

$$\mathrm{Rank}(M_0 - \sum_{i=1}^k \sum_{j=1}^N m_{ij} b_j L_i) = \mathrm{Rank}(M_0 - \sum_{i=1}^k \sum_{j=1}^N m_{ij} M_{(i-1)N+j}). \quad (1)$$

As this rank is $\leq r$, $\{m_{ij}\}_{1 \leq i \leq k, 1 \leq j \leq N}$ is a solution of the instance $(N, n, r, kN, M_0; M_1, \ldots, M_{kN}) = f(N, n, r, k, G, c)$ of MR.

Now let $(N, n, r, k, G = (L_1, \ldots, L_k), c)$ be an instance of RD, such that $f(N, n, r, k, G, c) = (N, n, r, kN, M_0; M_1, \ldots, M_{kN})$ is a yes-instance of MR. Let $m \in \mathbb{F}_q^{kN}$ be a solution of it. Then

$$\mathrm{Rk}(M_0 - \sum_{\ell=1}^{kN} m_\ell M_\ell) \leq r$$

Write $\ell = (i-1)N+j$, $1 \leq i \leq k$, $1 \leq j \leq N$. The equalities (1) then imply that the vector $(\sum_{j=1}^{N} m_j b_j, \sum_{j=1}^{N} m_{N+j} b_j, \ldots, \sum_{j=1}^{N} m_{(k-1)N+j} b_j)$, expressed in the basis $\mathcal{B}$ of $\mathbb{F}_{q^N}$, is a solution in $\mathbb{F}_{q^N}$ of the considered instance of RD.

2.2 Solving MinRank: Known Methods

We here consider a square instance $(n, r, M_0; M_1, \ldots, M_k)$ of MR. We are going to survey two methods to solve this problem. First, note that exhaustive search to find a tuple $(\lambda_1, \ldots, \lambda_k)$ of elements of $\mathbb{K}$ needs at most $(\#\mathbb{K}^k)n^3$ elementary operations on $n \times n$ matrices over $\mathbb{K}$.

The Kernel Attack. This first non-trivial attack on MR was proposed by Courtois and Goubin in citettmcrypt. It works as follows. First choose m vectors $\mathbf{x}^{(i)} \in \mathbb{F}_q^n$, $1 \leq i \leq m$ at random. Secondly, solve the system of mn equations for $(\mu_1, \ldots, \mu_k) \in \mathbb{F}_q^k : (M_0 - \sum_{j=1}^{k} \mu_j M_j)\mathbf{x}^{(i)} = \mathbf{0}_n$, $\forall 1 \leq i \leq m$. Note that if $m = \lceil \frac{k}{n} \rceil$, this system essentially has only one solution $\lambda = (\lambda_1, \ldots, \lambda_k)$.

Now set $E_\lambda = M_0 - \sum_{j=1}^{k} \lambda_j M_j$; we want E_λ to be of rank $\leq r$. If this were the case, then $\dim(\mathrm{Ker}E_\lambda) \geq n - r$ and so, for $\mathbf{x} \in \mathbb{F}_q^n$ chosen at random,

$$\Pr\{\mathbf{x} \in \mathrm{Ker}E_\lambda\} \geq q^{-r} \text{ and } \Pr\{\{\mathbf{x}^{(i)}, 1 \leq i \leq m\} \subseteq \mathrm{Ker}E_\lambda\} \geq q^{-mr}.$$

Thus, in order to find a λ such that E_λ has the desired rank, we have to run the above experiment (i.e. steps (1) and (2)) q^{mr} times on average. Taking the value of m as above, the complexity of this attack is thus $\mathcal{O}(q^{\lceil \frac{k}{n} \rceil r} k^3)$.

Kipnis-Shamir's Attack. In this attack, the MR problem is modeled as an MQ problem, i.e. one in which the purpose is to solve a quadratic system of equations. It is a transposition of an original attack on HFE due to Shamir and Kipnis [19]. In its principle, this attack is somehow dual to the previous one. The idea is to try to find a set of $n - r$ independent vectors of a special form in the kernel of : $E_\lambda = \sum_{i=1}^{k} \lambda_i \cdot M_i - M_0$. Putting the constraints into equations yields a quadratic system with unknowns a subset of coordinates of these vectors, together with the vector $\lambda = (\lambda_1, \ldots, \lambda_k)$.

In details, it works as follows: when λ is a solution of the considered MR instance, $\mathrm{rank}(E_\lambda) \leq r$. We want to express this rank condition as a large number of equations in a small number of variables. As $\dim(\mathrm{Ker}(E_\lambda)) \geq n - r$, there exists $n - r$ linearly independent vectors in $\mathrm{Ker}(E_\lambda)$. Name those vectors $x^{(1)}, \ldots, x^{(n-r)}$. Even if we fix the first $n - r$ coordinates of each vector to arbitrarily chosen values, we can still expect to get $n - r$ independent vectors. Each $x^{(i)}$ is thus of the form $x^{(i)} = (z_1, \ldots, z_{n-r}, x_1^{(i)}, \ldots, x_r^{(i)})$, where the z_is are chosen arbitrarily and $x_j^{(i)}$s are defined as new variables. The equalities:

$$\left(\sum_{i=1}^{k} \lambda_i \cdot M_i - M_0\right) x^{(i)} = \mathbf{0}_n, \text{ for all } i, 1 \leq i \leq n - r,$$

then yield a quadratic system of $(n-r)\cdot n$ equations in $r\cdot(n-r)+k$ unknowns. For $1 \leq i \leq n-r$ and $1 \leq j \leq n$, let $f_{i\,n+j}$ be the quadratic equation corresponding to j-th componant of $\left(\sum_{i=1}^{k} \lambda_i \cdot M_i - M_0\right) x^{(i)} = \mathbf{0}_n$. Here and in what follows, we shall denote by $\mathcal{I}_{KS} = \langle f_1, \ldots, f_{n\,(n-r)} \rangle$ the ideal generated by these quadratic equations. Here, and in what follows, we shall denote by $\mathcal{I}_{KS}$ the ideal generated by these quadratic equations.

3   A Fresh Look at Kipnis-Shamir's Attack

We here go deeper in the investigation of the properties of $\mathcal{I}_{KS}$. First, we precise the link between the variety associated to $\mathcal{I}_{KS}$ and the solutions of MR.

3.1   Properties of KS Equations

The next theorem shows that the modeling proposed by Kipnis and Shamir is somehow optimal, in the sense that the zeroes of the KS equations exactly correspond to the solutions of MR.

Theorem 1. *Let* $(n, k, M_0; M_1, \ldots, M_k, r) \in L_{MR}$[1]. *We shall denote by* $Sol(n, k, M_0; M_1, \ldots, M_k, r)$ *the set of solutions of MinRank on* $(n, k, M_0; M_1, \ldots, M_k, r)$. *There is a one-to-one correspondence between* $Sol(n, k, M_0; M_1, \ldots, M_k, r)$ *and the variety:*

$$V_{\mathbb{K}}(\mathcal{I}_{KS}) = \{\mathbf{z} \in \mathbb{K}^{r\cdot(n-r)+k} : f(\mathbf{z}) = 0, \text{for all } f \in \mathcal{I}_{KS}\},$$

$\mathcal{I}_{KS}$ *being the ideal defined from the considered instance of MR as in Section 2.*

Proof. Let $\mathbf{s} \in V_{\mathbb{K}}(\mathcal{I}_{KS}) \subset \mathbb{K}^{r\cdot(n-r)+k}$. We can suppose w.l.o.g. that the last $r\cdot(n-r)$ components of $\mathbf{s}$ correspond to the variables $x_j^{(i)}$s, i.e. the unknowns corresponding to the $n-r$ linearly independent vectors of a suitable kernel.

We can then write $\mathbf{s} = (\lambda, \mathbf{s}_1, \ldots, \mathbf{s}_{n-r}) \in \mathbb{K}^{r\cdot(n-r)+k}$, where each $\mathbf{s}_i \in \mathbb{K}^r$ and $\lambda = (\lambda_1, \ldots, \lambda_k) \in \mathbb{K}^k$. We then construct:

$$\mathbf{x}^{(i)} = (\ \underbrace{0, \ldots, 0}_{n-r \text{ zeroes}}, \mathbf{s}_i) \in \mathbb{K}^n.$$

By definition of $\mathcal{I}_{KS}$, it holds that $\left(\sum_{t=1}^{k} \lambda_t \cdot M_t - M_0\right) \mathbf{x}^{(i)} = \mathbf{0}_n$, for all $i, 1 \leq i \leq n-r$. The vectors $\mathbf{x}^{(i)}, 1 \leq i \leq n-r$ being independent vectors, we get:

$$\text{Rank}\left(\sum_{t=1}^{k} \lambda_t \cdot M_t - M_0\right) \leq r.$$

i.e. $\lambda = (\lambda_1, \ldots, \lambda_k)$ is a solution of MR on $(n, k, M_0; M_1, \ldots, M_k, r)$.

Conversely, the fact that any element of $Sol(n, k, M_0; M_1, \ldots, M_k, r)$ corresponds to a point of $V_{\mathbb{K}}(\mathcal{I}_{KS})$ has been explained in 2.2. $\qquad\square$

---

[1] The language associated to MinRank.

From this theorem, we can propose a classification of MinRank instances. As explained in the previous section, $\mathcal{I}_{KS}$ is generated by $(n - r) \cdot n$ equations in $r \cdot (n - r) + k$ variables. The system is underdefined as soon as the number of variables is greater than the number of equations. This implies that:

$$\Delta = r \cdot (n - r) + k - (n - r)n = (r - n) \cdot (n - r) + k = -(n - r)^2 + k$$

variables can take arbitrary values of $\mathbb{K}$. According to the previous theorem, there is a one-to-one correspondence between the zeroes of $\mathcal{I}_{KS}$ and the solutions of MR. It follows that an instance $(n, k, M_0; M_1, \ldots, M_k, r) \in L_{MR}$ has at least $\#\mathbb{K}^\Delta$ solutions if $\Delta > 0$. Our concern is to find only one solution. To this end, we can randomly fix Δ variables in the system corresponding to $\mathcal{I}_{KS}$. We will then get a system with the same number of variables as equations. It is then very likely that the corresponding variety will be reduced to a unique point (which will correspond to a solution of MR). This can be interpreted as follows:

Lemma 1. *Let* $(n, k, M_0; M_1, \ldots, M_k, r) \in L_{MR}$, *and set* $\Delta = -(n - r)^2 + k$. *If* $\Delta > 0$ *and* $(\lambda_{\Delta+1}, \ldots, \lambda_k) \in \mathcal{S}ol(n, k - \Delta, M_0; M_{\Delta+1}, \ldots, M_k, r)$ *then for all* $\mathbf{r} = (r_1, \ldots, r_\Delta) \in \mathbb{K}^\Delta$:

$$(r_1, \ldots, r_\Delta, \lambda_{\Delta+1}, \ldots, \lambda_k) \in \mathcal{S}ol(n, k, M_0; M_1, \ldots, M_k, r).$$

That is, all the solutions of $(n, k, M_0; M_1, \ldots, M_k, r) \in L_{MR}$ can be easily deduced from $\mathcal{S}ol(n, k - \Delta, M_0; M_{\Delta+1}, \ldots, M_k, r)$. This leads us to introduce the following terminology.

Definition 1. *Let* $(n, k, M_0; M_1, \ldots, M_k, r) \in L_{MR}$, *and set* $\Delta = -(n - r)^2 + k$.

- *If* $\Delta = 0$, *then we shall say that the instance* $(n, k, M_0; M_1, \ldots, M_k, r)$ *is* **well defined**.
- *If* $\Delta > 0$, *then we shall call* **normalization** *the process of deleting the first* Δ *matrices of* $(n, k, M_0; M_1, \ldots, M_k, r)$. *The result of this process is a* **normalized** *instance* $(n, k - \Delta, M_0; M_{\Delta+1}, \ldots, M_k, r)$.

In the sequel, we will always focus our attention to well defined instances. We would like to emphasize that this is not a restriction. Instances which are not well defined can be normalized, leading then to well defined instances. According to Lemma 1, it is sufficient to study such normalized instances, as a normalized instance will indeed permit to describe all the solutions of the initial instance.

We also would like to point out that Lemma 1 permits to classify instances of MR with respect to their difficulty. From the lemma, it is clear that if we are able to solve efficiently the well defined instance $(n, k, M_0; M_1, \ldots, M_k, r)$, then we will be able to solve efficiently any instance $(n, k', M_0; M_1, \ldots, M_{k'}, r)$ of MR with $k' \geq k$.

3.2 Relating KS to Other Algebraic Methods

We will here show another "optimality" feature of the KS equations, namely that the equations obtain via other algebraic methods are indeed included in the ideal generated by the KS equations.

In this section, we let $(n, k, M_0; M_1, \ldots, M_k, r) \in L_{MR}$ be an instance of Min-Rank and $\mathcal{I}_{KS}$ be the associated ideal. We will suppose that $\mathcal{I}_{KS}$ is radical, i.e.

$$\sqrt{\mathcal{I}_{KS}} = \{f \in \mathbb{K}[x_1, \ldots, x_m] : \exists r > 0 \text{ s. t. } f^r \in \mathcal{I}_{KS}\} = \mathcal{I}_{KS}.$$

In the cryptographic context, the ideals are usually radical. This is due to the fact that, for an ideal to be radical, it is sufficient that the field equations be included in it. In practice, we have not included the field equations in $\mathcal{I}_{KS}$. However, for proving the radicality of $\mathcal{I}_{KS}$, we can suppose w.l.o.g. that such equations are included in $\mathcal{I}_{KS}$.

The minors method. This method comes back to the very definition of MR, and expresses that $(\lambda_1, \ldots, \lambda_k)$ is a solution of MR if and only if all the minors of degree $r' > r$ of the matrix $\sum_{i=1}^{k} \lambda_i M_i - M_0$ are zero. In this context, we have the following

Proposition 1. Set $E(x_1, \ldots, x_k) = \sum_{i=1}^{k} x_i M_i - M_0$. Then all the minors of $E(x_1, \ldots, x_k)$ of degree $r' > r$ lie in $\mathcal{I}_{KS}$.

Proof. A solution of MR corresponds to a specialization of the variables $x_1, \ldots, x_k$ in $E(x_1, \ldots, x_k)$, leading to a matrix of rank $\leq r$. For such a specialization, all the minors $M_{r'}$ of rank $r' > r$ of the matrix $E(x_1, \ldots, x_k)$ must equal zero. A minor $M_{r'}$ is a polynomial of degree r' whose variables are $x_1, \ldots, x_k$.

It is clear that all the minors vanish on $V_{\mathbb{K}}(\mathcal{I}_{KS})$. Therefore, by Hilbert's Strong Nullstellensatz [1] (th. 2.2.5), we get that all the minors of rank $r' > r$ lie in the radical of $\mathcal{I}_{KS}$. This ideal being radical, it turns out that all those minors lie in $\mathcal{I}_{KS}$. □

Schnorr's method. This unpublished method due to Schnorr was quoted in [7]. The idea is to consider the multivariate polynomial

$$P(x_1, \ldots, x_k) = \text{Det}\left(\sum_{i=1}^{k} x_i M_i - M_0\right). \tag{2}$$

If $\lambda = (\lambda_1, \ldots, \lambda_k)$ is a solution of MR, then λ is a root of P of multiplicity greater than or equal to $n - r$. This means that such a λ is solution of $P(x_1, \ldots, x_k) = 0$ as well as $\left\{\frac{\partial^j P}{\partial x_i}(x_1, \ldots, x_k) = 0\right\}_{1 \leq i \leq k}$, for all j, $1 \leq j \leq n - r - 1$. This means that all these equations vanish on $V_{\mathbb{K}}(\mathcal{I}_{KS})$ and therefore belong to $\mathcal{I}_{KS}$.

To summarize, we have proved that the KS equations include the equations that one could obtain using minors or a basic property of the determinant. From a system solving point of view, this means that solving MR using KS equations is at least as efficient as solving MinRank using either of those alternative methods. We will see that KS equations lead to a more efficient solving. Before that, we present a new method for setting up a system of equations for MinRank.

4   A New Modeling

This new method will permit to link Schnorr and the minors methods. The starting point is similar to Schnorr's method, namely we will consider the polynomial given in (2). Remark that if $k = 1$, then $P(x_1) = \mathrm{Det}(x_1 \cdot I - M_0 M_1^{-1})$ is exactly the characteristic polynomial of $M_0 M_1^{-1}$. In this special context, MinRank corresponds to the problem of finding the eigenvalues of $M_0 M_1^{-1}$. It is well known that this can be done by computing the roots of the characteristic polynomial $P(x_1)$. Schnorr's method is a generalization of this technique.

We also would like to mention that MinRank is related to the so-called matrix pencils problem. The eigenvalue of a linear matrix pencil $(A, B) \in \mathcal{M}_{n \times n} \times \mathcal{M}_{n \times n}$ is a $\lambda \in \mathbb{K}$ such that $\mathrm{Det}(A - \lambda B) = 0$.

The generalized high order eigenvalue problem consists of finding the eigenvalues of for a matrix pencil $(M_1, \ldots, M_k) \in (\mathcal{M}_{n \times n})^k$; i.e. finding $\lambda \in \mathbb{K}$ such that $\mathrm{Det}\left(\sum_{i=1}^{k} \lambda^i M_i\right) = 0$. One can see that MinRank is a multivariate version of this problem.

We will now describe a new approach for modeling $\mathrm{MR}(n, k, r)$ as a set of algebraic polynomials. To do so, we remark that if $\lambda = (\lambda_1, \ldots, \lambda_k)$ is a solution of MinRank then λ is root of $P(x_1, \ldots, x_k)$ with "multiplicity" $n - r$. More precisely, the polynomial $P(x_1 + \lambda_1, \ldots, x_k + \lambda_k)$ has no terms of degree smaller or equal to $n - r$. Thus, similarly to the univariate case, the coefficients of the monomials of degree d are sums of the minors of degree $n - d$ of the matrix : $E(x_1, \ldots, x_k) = \sum_{i=1}^{k} x_i M_i - M_0$. In order to construct the system, we will introduce new variables $y_1, \ldots, y_k$ and consider the polynomial:

$$Q(y_1, \ldots, y_k) = P(x_1 + y_1, \ldots, x_k + y_k) \in \mathbb{K}[y_1, \ldots, y_k].$$

We can view this polynomial as a polynomial whose variables are $y_1, \ldots, y_k$ and coefficients are monomials in the variables $x_1, \ldots, x_k$. As explained, a solution of MinRank must vanish on all the monomials of degree smaller than $n - r$ in $Q(y_1, \ldots, y_k)$. The new system is then obtained by equating to zero the coefficients, in $Q(y_1, \ldots, y_k)$, of the monomials of degree d such that $0 < d < n - r$. Such coefficients are polynomials in the variables $x_1, \ldots, x_k$, of degree d, with $r < d < n$. Moreover, we can restrict our attention to the coefficients in $Q(y_1, \ldots, y_k)$ of degree $n - r - 1$. Thus we obtain a subset of linearly independent minors of $\sum_{i=1}^{k} x_i M_i - M_0$ which are polynomials in $x_1, \ldots, x_k$ of degree $r + 1$.

This then permits to establish a link between Schnorr and the minors methods. Let $\mathrm{M}_d(m)$ be the set monomials of degree d in m variables. We have $\#\mathrm{M}_d(m) = \binom{m+d-1}{d}$. We can also count precisely the number of minors of degree $r + 1$, i.e. $\#\mathrm{M}_{n-r-1}(k)$. We have obtained a system of $\#\mathrm{M}_{n-r-1}(k)$ algebraic equations of degree $r + 1$. Similarly to the previous section, one can prove that these new equations will also lie in $\mathcal{I}_{\mathrm{KS}}$, and more precisely in $\mathcal{I}_{\mathrm{KS}} \cap \mathbb{K}[x_1, \ldots, x_k]$. From a practical point of view, it turns out that the new approach is a little less efficient than the one of computing a Gröbner basis of $\mathcal{I}_{\mathrm{KS}}$. This is quite surprising since our new method will generate an overdefined system of equations.

5   Theoretical Bound on the Complexity

We will now try to explain such a behavior, and evaluate the complexity of computing a Gröbner basis of $\mathcal{I}_{\mathrm{KS}}$. To do so, we recall that the complexity of all known Gröbner bases algorithms depends on the so-called *degree of regularity* of the system [2, 3, 4, 10]. This corresponds to the maximal degree reached during a Gröbner basis computation. If d_{reg} is the degree of regularity of $\mathcal{I} \subset \mathbb{K}[x_1, \ldots, x_m]$, then the complexity of computing a Gröbner basis of $\mathcal{I}$ with F_5 [12] is:

$$\mathcal{O}\left(\left(\#M_{d_{\text{reg}}}(m) \right)^{\omega} \right),$$

with $\omega, 2 \leq \omega \leq 3$ the linear algebra constant.

In general, it is a difficult problem to know *a priori* the degree of regularity. For *regular and semi-regular systems* [2, 3, 4] (i.e. "ramdom" systems of algebraic equations), the behavior of the regularity is well mastered. For instance, if we suppose that KS equations are semi-regular, then we obtain a degree of regularity equal to $m + 1$, $m = r(n - r) + k$. Besides, we also know that the number of solutions is bounded from above by the Bézout bound, which is equal to 2^m for KS equations.

In our context, this is unsatisfactory. Indeed, this does not match with the experimental results that we will present in the next section. Typically, we have observed a degree of regularity which seems to be $\approx r + 2$ (see Section 6). Similarly, computing the degree of regularity of the systems obtained with one of the three other methods presented so far will not lead to a satisfactory bound.

To fill this gap between theory and practice, we have remarked that the ideal $\mathcal{I}_{\text{KS}}$ is *multi-homogeneous* (see for instance [21, 25]). Namely, the equations are homogeneous with respect to blocks of variables.

Definition 2. *Let* $S = \{f_1(x_1, \ldots, x_n) = 0, \ldots, f_m(x_1, \ldots, x_n) = 0\}$ *be an algebraic system of equations, and* $T = \{X^{(1)}, \ldots, X^{(k)}\}$ *be a partition of* $X = \{x_1, \ldots, x_n\}$ *s.t.:*

$$X^{(j)} = \{x_{j_1}, \ldots, x_{j_{k_j}}\}.$$

We shall say that S *is multi-homogeneous if the polynomials* f_i *are homogenous w.r.t. the* $X^{(j)}$*'s.*

For such systems, one can obtain new bounds for the degree of regularity and number of solutions [20, 21, 24].

Definition 3. *Let* $S = \{f_1(x_1, \ldots, x_n) = 0, \ldots, f_m(x_1, \ldots, x_n) = 0\}$ *be an algebraic system of equations, and* $T = \{X^{(1)}, \ldots, X^{(k)}\}$ *be a partition of* $X = \{x_1, \ldots, x_n\}$ *s.t.* $X^{(j)} = \{x_{j_1}, \ldots, x_{j_{k_j}}\}$. *We shall call partition vector of* T *the vector* $K = [k_1, \ldots, k_m]$. *Now, let* $d_{i,j}$ *be the degree of* f_i *restricted to the variables of* $X^{(j)}$. *We shall define the degree matrix under the partition* T *as:*

$$\begin{bmatrix} d_{1,1} & d_{1,2} & \cdots & d_{1,m} \\ d_{2,1} & d_{2,2} & \cdots & d_{2,m} \\ \vdots & \vdots & & \vdots \\ d_{n,1} & d_{n,2} & \cdots & d_{n,m} \end{bmatrix}.$$

The multi-homogeneous Bézout number associated to the partition T *is defined as the coefficient of* $z_1^{k_1} z_2^{k_2} \cdots z_m^{k_m}$ *in the following polynomial:*

$$(d_{1,1} z_1^{k_1} + d_{2,1} z_2^{k_2} + \cdots + d_{1,m} z_m^{k_m})(d_{2,1} z_1^{k_1} + \cdots d_{2,m} z_m^{k_m}) \cdots (d_{n,1} z_1^{k_1} + \cdots + d_{n,m} z_m^{k_m}).$$

This leads to the following result.

Theorem 2. *Let $r' = n - r$ be a constant, and we will consider instances of MinRank with parameters $(n, k = r'^2, r = n - r')$. If we denote by $\mathcal{S}ol$ the set of solutions of MinRank on such instances, it holds that:*

$$\# \mathcal{S}ol \leq \binom{n}{r'}^{r'}$$

For those particular instances, we can compute the variety of $\mathcal{I}_{KS}$ using Gröbner bases in:

$$\mathcal{O}\left(\ln(q)\, n^{3\, r'^2}\right),$$

where q is the size of the finite field $\mathbb{K}$.
In other words, the complexity of our attack is polynomial for instances of Min-Rank with $(n, k = r'^2, r = n - r')$.

Proof. First, we will assume an upper bound, say D, on the number of solutions of $\mathcal{I}_{KS}$. From such a D we can derive an upper bound on the complexity of computing a Lex-Gröbner basis from another (e.g. DRL) Gröbner basis using FGLM : D^3 (see [13, 27] for details).

Now to find such a D we exhibit a multi-homogeneous structure for the equations generating $\mathcal{I}_{KS}$. We can consider the following partition:

$$T = T_0 \cup \cdots \cup T_{n-r} = T_0 \cup \bigcup_{i=1}^{n-r} X^{(i)},$$

where $T_0 = [\lambda_1, \ldots, \lambda_k]$ and $T_i = X^{(i)} = [x_1^{(i)}, \ldots, x_r^{(i)}]$ (the $x_j^{(i)}$ are defined as in 2.2).

T is a partition of the set of variables. The degree of the polynomials $\{f_j\}_{1 \leq j \leq n \cdot (n-r)}$ with respect to T_0 (resp. $X^{(i)}$) will be denoted by $d_\ell^{(0)}$ (resp. $d_\ell^{(i)}$). The degree matrix corresponding to the partition T is:

$$\begin{bmatrix} d_1^{(0)} & \cdots & \cdots & d_1^{(n-r)} \\ d_2^{(0)} & \cdots & \cdots & d_2^{(n-r)} \\ \vdots & & & \vdots \\ d_{n \cdot (n-r)}^{(0)} & \cdots & \cdots & d_{n \cdot (n-r)}^{(n-r)} \end{bmatrix}.$$

Here, the partition vector is $K = [k, r, \ldots, r]$. As explained, the multi-homogeneous Bezout number is thus the coefficient of $z_1^k z_2^r \cdots z_{n-r+1}^r$ into the polynomial:

$$(z_1 + z_2)^n (z_1 + z_3)^n \cdots (z_1 + z_{n-r+1})^n.$$

Consequently, we can bound the number of solutions ($\#\mathcal{S}ol$) by $D = \binom{n}{r}^{n-r} \approx \left(\frac{n^{r'}}{r'!}\right)^{r'} = \frac{n^{r'^2}}{(r'!)^{r'}}$ assuming that $r' = n - r$ is constant when $n \longrightarrow \infty$. $\square$

Theorem 2 applies to challenges A, B, C. We obtain the following complexity bounds:

$$
\begin{array}{lccc}
(n, k, r) & (6, 9, 3) & (7, 9, 4) & (11, 9, 8) \\
\#\mathcal{S}ol \text{ (MH Bezout bound)} & 8000 & 42875 & 2^{22.1} \\
\text{Complexity bound } (\#\mathcal{S}ol)^3 & 2^{38.9} & 2^{46.2} & 2^{66.3}
\end{array}
$$

We would like to emphasize that such theoretical bounds are coherent with the results of the experiments that we are going to present.

6   Experimental Results

Initially, the complexity of the Kipnis-Shamir attack was evaluated using relinearization [19]. Here, we propose to use a more efficient tool for solving algebraic systems, namely fast Gröbner bases [5, 6] algorithms : F_5 [12] together with FGLM [13]. This choice permits to go one step further in the cryptanalysis of MinRank, especially for instances used in the ZK authentication scheme proposed in [8]. We have quoted below the set of parameters of the ZK authentication scheme recommended by the author of [8]. We have also given the number of equations and variables obtained using KS :

$A : \mathbb{F}_{65521}, k = 10, n = 6, r = 3$ (18 eq., and 19 variables)
$B : \mathbb{F}_{65521}, k = 10, n = 7, r = 4$ (21 eq., and 22 variables)
$C : \mathbb{F}_{65521}, k = 10, n = 11, r = 8$ (33 eq., and 35 variables)

One can remark that these instances of MR are not well defined. Thus, as explained in Lemma 1, we can fix $\Delta = 1$ variables for the challenges A, B (resp. $\Delta = 2$ variables for challenge C). This is then equivalent to solve MinRank on the following parameters :

$\mathbf{A} : \mathbb{F}_{65521}, k = 9, n = 6, r = 3$ (18 eq., and 18 variables)
$\mathbf{B} : \mathbb{F}_{65521}, k = 9, n = 7, r = 4$ (21 eq., and 21 variables)
$\mathbf{C} : \mathbb{F}_{65521}, k = 9, n = 11, r = 8$ (33 eq., and 33 variables)

The boldface letters $\mathbf{A}$, $\mathbf{B}$, $\mathbf{C}$ being the normalized of the challenges A, B, C respectively. Before presenting our practical results, we would like to explain the conditions of the experiments.

Generation of the instances
We have randomly generated k matrices $(M_1, \ldots, M_k) \in \mathcal{M}_{n,n}(\mathbb{K})^k$ and k coefficients $(\lambda_1, \ldots, \lambda_k) \in \mathbb{K}^k$ such that $\lambda = \sum_{i=1}^{k} \lambda_i \neq 0$. Finally, we have randomly selected a matrix $M \in \mathcal{M}_{n,n}(\mathbb{K})$ of rank r and set $M_0 = \sum_{i=1}^{k} \lambda_i (M_i - M)$. Thus we have $\text{Rank}\left(\sum_{i=1}^{k} \lambda_i M_i - M_0\right) = \text{Rank}(\lambda M) = r$.

Programming language – Workstation
The experimental results have been obtained with several Xeon bi-processors 3.2 Ghz, with 64 Gb of Ram. The instances of MinRank have been generated using the Maple software. The F_5 [12] and FGLM [13] algorithms have been implemented in C in the FGb software. We used this implementation for computing

Gröbner bases. From time to time, we use the last version of Magma (2.14) for obtaining these bases. This version of Magma includes efficient implementations of the F_4 [11] and FGLM algorithms. Hence, the reader can reproduce the experimental results. We were able to break the two challenges **A** and **B** using FGb or Magma. There is a huge gap between these challenges and challenge **C**, which seems intractable with the current implementation. However, we can estimate the complexity of our attack for the last challenge by:

- studying intermediate instances of the MinRank problem, i.e. $\mathrm{MR}(n, k, r)$ with $\mathbb{K} = \mathbb{F}_{65521}, n = r + 3$, $k = (n - r)^2 = 9$ and $r = 3, 4, 5, 6, 7, 8$.
- Since all the λ_i are in $\mathbb{K}$, we can perform an exhaustive search on some λ_i. Namely, we will suppose that we have $s > 0$ coefficients of a solution $(\lambda_1, \ldots, \lambda_k)$ of MinRank. This is equivalent to solve a $\mathrm{MR}(n - s, k, r)$ problem. From a system solving point of view, this means that we will solve $\#\mathbb{K}^s$ overdetermined systems. When $s > 0$ the number of solutions of the corresponding algebraic system is always 1 and any Gröbner basis for any monomial ordering gives the solution; consequently there is no need to apply the FGLM algorithm.

Table Notation
The following notation is used in the next table:

- T_{DRL} is the CPU time (in seconds) for computing a Gröbner basis for a total degree ordering.
- D is the number of solutions in the algebraic closure of $\mathbb{F}_{65521}$ ($D = 1$ when $s > 0$).
- T_{FGLM} is the CPU time (in seconds) for changing the basis to a lexicographic Gröbner basis using the FGLM algorithm. The complexity [13] is D^3.
- T is the time of our approach for finding a solution of MinRank; thus $T = T_{\mathrm{DRL}} + T_{\mathrm{FGLM}}$ when $s = 0$ and $T = T_{\mathrm{DRL}}$ when $s > 0$.
- d_{reg}, the maximum degree reached during the computation of a Gröbner basis with F_5.
- M, the maximum memory usage (in Mbytes) during a computation with F_5.
- $Log_2(N)$ is the log in base 2 of the number of arithmetic operations N for solving the MinRank problem. When $s = 0$, N is the total number of operations for the first Gröbner basis computation and FGLM.

Interpretation of the Results
Challenges **A** $(6, 9, 3)$ and **B** $(7, 9, 4)$ are completely broken. We emphasize that such sets of parameters were the most suited for a practical use of the ZK authentication scheme proposed in [8].

As explained in 3.1, we would be able to solve any instance $(n, k', M_0; M_1, \ldots, M_{k'}, r)$ of MR, with n, r as in the challenges and for all $k' > 9$. For example, all instances $(6, k', M_0; M_1, \ldots, M_{k'}, 3)$, with $k' > 9$ can be solved in one minute.

We have observed in our experiments that the maximum degree d_{reg} seems to be equal to $\max(r + 2, 4)$. We recall that the complexity of computing a Gröbner

		$\mathbb{K} = \mathbb{F}_{65521}$ MR(n,k,r)				
		Chall **A**	Chall **B**			Chall **C**
		$(6,9,3)$	$(7,9,4)$	$(8,9,5)$ $(9,9,6)$	$(10,9,7)$	$(11,9,8)$
$s=0$	$T_{\mathrm{DRL}} + T_{\mathrm{FGLM}}$ (FGb)	30.0+34.8	3794+2580	328233		
	$T_{\mathrm{DRL}} + T_{\mathrm{FGLM}}$ (Magma)	300+200	48745+∞	∞		
	memory : M	406.5	3113	58587		
	$Log_2(N)$	30.5	37.1	43,4		
	d_{reg}	5	6	7		
	Solutions: D	980	4116	14112		
$s=1$	T_{DRL} FGb	1.85	166.6	5649.7 590756		
	M	343.9	522.1	4548.7 43267		
	$Log_2(N)$	25,95	32.3	36.8 43.9		
	d_{reg}	4	5	6 6		
$s=2$	T_{DRL} FGb	0.5	5.5	632 14867		
	M	39.8	68.0	806,4 2510.3		
	$Log_2(N)$	24.1	27.5	34.1 38.7		
	d_{reg}	4	4	5 6		
$s=3$	T_{DRL} FGb	0.05	1.0	15.6 234.3	4248.4	56987
	M	35.5	44.9	75.4 888.6	2792.3	10539
	$Log_2(N)$	20.1	25.0	29.2 32.8	36.9	40.6
	d_{reg}	4	4	4 5	6	7

Fig. 1. Experimental results with FGb

basis with F_5 [12] is bounded by $\mathcal{O}\left(\#M_{d_{\mathrm{reg}}}(N)^3\right)$, where N is the number of variables. Here, $N = r(n-r) + k$. For MR$(r+3,9,r)$, we have $N = 3(r+3)$, yielding the bound:

$$\binom{N + d_{\mathrm{reg}} - 1}{d_{\mathrm{reg}}}^3 = \binom{3(r+3) + r+}{r+3}^3 \approx e^{3\,r+3\,(\frac{3}{2}+r)\ln(r)}.$$

For challenge **C**, we have $r = 8$ and we will obtain a complexity bound of 2^{120}. Of course this is a very pessimistic bound. We will now improve this bound.

Estimated Complexity of the attack

To obtain a better result, we use the following bound:

$$\#\mathrm{MR}(n,k,r) \leq (\#\mathbb{K})^s \times \#\mathrm{MR}(n-s,k,r),$$

where $\#\mathrm{MR}(n,k,r)$ is the number of operation to solve the corresponding min-rank problem MR(n,k,r). This bound is tight only when s is small. We can use our experimental results to derive new bounds for $\#\mathrm{MR}(r+3,9,r)$ and $\mathbb{K} = \mathbb{F}_{65521}$. For such parameters:

$$\log_2\left(\#\mathrm{MR}(r+3,9,r)\right) \leq 16s + \log_2\left(\#\mathrm{MR}(r+3-s,9,r)\right).$$

The following notation is used in the table below:
- $Nb(r, s) = 16s + \log_2\left(\#\text{MR}(r + 3 - s, 9, r)\right)$ is a logarithmic upper complexity bound for solving $\text{MR}(r + 3, 9, r)$.
- $Nb(r) = \log_2\left(\#\text{MR}(r + 3, 9, r)\right)$ is the exact number of oper. of our attack.
- "Security Bound" is the Log_2 of the complexity of the best approach known so far for solving MinRank. This is based on [8].
- "Estimated Bound" is an extrapolation of the complexity. This bound is not rigorous.
- $(\text{MHBezout})^3$ is the theoretical complexity bound obtained in the previous section.

	$\mathbb{K} = \mathbb{F}_{65521}$ $MR(n, k, r)$					
	Chall **A**	Chall **B**				Chall **C**
	(6,9,3)	(7,9,4)	(8,9,5)	(9,9,6)	(10,9,7)	(11,9,8)
$Nb(r)$	30,5	37,1	43,4			
$Nb(r, 1)$	42,0	48,3	52,8	59,9		
$Nb(r, 2)$	54,1	59,5	66,1	70,7		
$Nb(r, 3)$	68,2	72,9	77,2	80,8	84,9	88,6
Estimated Bound	30,5	37,1	43,4	50,4	57,4	64,4
Bezout^3	38,9	46,2	52,3	57,5	62,2	66,3
Security Bound	106	122				138

For challenge $(11, 9, 8)$ we obtain a complexity of 2^{88} for our attack. Clearly, this is not feasible in practice. However, this is much better than the previous security estimates (2^{138}) [8]. Still, this remains a pessimistic bound.

By using the estimated bound, which is less rigorous but more close to what we observed in practice (for the instances that we have been able to solve), we can evaluate the complexity of our attack to 2^{65}. We would like to emphasize that this is very close to the theoretical complexity bound $2^{66.3}$ obtained in the previous section using the particular structure of the algebraic system.

7   Conclusion

We have provided a unified view of the attacks known so far against the MinRank problem. We have also presented a new modeling of the problem that actually links the minors attack and Schnorr's method. From a practical point of view, our approach of solving the systems by means of fast Gröbner bases algorithms led to the breaking of the most practical challenges proposed for the MR-authentication scheme. On a more theoretical level, we showed that MinRank is polynomial when $n - r$ is constant. One line of research would now be to study the impact of our method on the solving of the Rank Decoding problem.

Acknowledgement. We wish to thank Mohab Safey El Din who brought multi-homogeneous papers to our attention. We also would like to thank the LIP6 for its cluster of computers that permitted us to conduct the experiments.

References

1. Adams, W.W., Loustaunau, P.: An Introduction to Gröbner Bases. Graduate Studies in Mathematics 3 (1994)
2. Bardet, M.: Étude des Systèmes Algébriques Surdéterminés. Applications aux Codes Correcteurs et à la Cryptographie. Thèse de doctorat, Université de Paris VI (2004)
3. Bardet, M., Faugère, J.-C., Salvy, B., Yang, B.-Y.: Asymptotic Behaviour of the Degree of Regularity of Semi-Regular Polynomial Systems. In: Proc. of MEGA 2005, Eighth International Symposium on Effective Methods in Algebraic Geometry (2005)
4. Bardet, M., Faugère, J.-C., Salvy, B.: On the Complexity of Gröbner Basis Computation of Semi-Regular Overdetermined Algebraic Equations. In: Proc. International Conference on Polynomial System Solving (ICPSS), pp. 71–75 (2004), http://www-calfor.lip6.fr/ICPSS/papers/43BF/43BF.htm
5. Buchberger, B., Collins, G.-E., Loos, R.: Computer Algebra Symbolic and Algebraic Computation, 2nd edn. Springer, Heidelberg (1982)
6. Buchberger, B.: Gröbner Bases : an Algorithmic Method in Polynomial Ideal Theory. Recent trends in multidimensional systems theory. Reider ed. Bose (1985)
7. Courtois, N.: Decoding Linear and Rank-Distance Codes, MinRank problem and Multivariate Cryptanalysis. In: CLC 2006, Darmstadt (September 2006)
8. Courtois, N.: Efficient Zero-knowledge Authentication Based on a Linear Algebra Problem MinRank. In: Boyd, C. (ed.) ASIACRYPT 2001. LNCS, vol. 2248, pp. 402–421. Springer, Heidelberg (2001)
9. Courtois, N., Goubin, L.: Cryptanalysis of the TTM Cryptosystem. In: Okamoto, T. (ed.) ASIACRYPT 2000. LNCS, vol. 1976, pp. 44–57. Springer, Heidelberg (2000)
10. Cox, D.A., Little, J.B., O'Shea, D.: Ideals, Varieties, and algorithms: an Introduction to Computational Algebraic Geometry and Commutative algebra. Undergraduate Texts in Mathematics. Springer, New York (1992)
11. Faugère, J.-C.: A New Efficient Algorithm for Computing Gröbner Basis: F_4. Journal of Pure and Applied Algebra 139, 61–68 (1999)
12. Faugère, J.-C.: A New Efficient Algorithm for Computing Gröbner Basis without Reduction to Zero: F_5. In: Proceedings of ISSAC, pp. 75–83. ACM press, New York (2002)
13. Faugère, J.-C., Gianni, P., Lazard, D., Mora, T.: Efficient Computation of Zero-Dimensional Gröbner Basis by Change of Ordering. Journal of Symbolic Computation 16(4), 329–344 (1993)
14. Faugère, J.-C., Joux, A.: Algebraic Cryptanalysis of Hidden Field Equations (HFE) Cryptosystems using Gröbner Bases. In: Boneh, D. (ed.) CRYPTO 2003. LNCS, vol. 2729, pp. 44–60. Springer, Heidelberg (2003)
15. Faugère, J.-C., Perret, L.: Polynomial Equivalence Problems: Algorithmic and Theoretical Aspects. In: Vaudenay, S. (ed.) EUROCRYPT 2006. LNCS, vol. 4004, pp. 30–47. Springer, Heidelberg (2006)
16. Fouque, P.-A., Macario-Rat, G., Stern, J.: Key Recovery on Hidden Monomial Multivariate Schemes. In: Smart, N. (ed.) EUROCRYPT 2008. LNCS, vol. 4965, pp. 19–30. Springer, Heidelberg (2008)
17. Garey, M.R., Johnson, D.B.: Computers and Intractability. A Guide to the Theory of NP-Completeness. W.H. Freeman, New York (1979)

18. Jiang, X., Ding, J., Hu, L.: Kipnis-Shamir's Attack on HFE Revisited. In: Proc. of Inscrypt 2007 (2007), http://eprint.iacr.org/2007/203
19. Kipnis, A., Shamir, A.: Cryptanalysis of the HFE Public Key Cryptosystem by Relinearization. In: Wiener, M. (ed.) CRYPTO 1999. LNCS, vol. 1666, pp. 19–30. Springer, Heidelberg (1999)
20. Li, T., Lin, Z., Bai, F.: Heuristic Methods for Computing the Minimal Multi-homogeneous Bézout Number. Applied Mathematics and Computation 146, 237–256 (2003)
21. Malajovich, G., Meer, K.: Computing Minimal Multi-homogeneous Bézout Numbers Is Hard. In: Diekert, V., Durand, B. (eds.) STACS 2005. LNCS, vol. 3404, pp. 244–255. Springer, Heidelberg (2005)
22. Moh, T.: A Public Key System with Signature and Master Key Functions. Communications in Algebra 27(5), 2207–2222 (1999)
23. Patarin, J.: Hidden Fields Equations (HFE) and Isomorphisms of Polynomials (IP): two new families of asymmetric algorithms. In: Maurer, U.M. (ed.) EUROCRYPT 1996. LNCS, vol. 1070, pp. 33–48. Springer, Heidelberg (1996)
24. Safey El Din, M., Trébuchet, P.: Strong bi-homogeneous Bezout theorem and its use in effective real algebraic geometry. INRIA Research Report RR, 46 pages (2006), http://hal.inria.fr/inria-00105204
25. Shafarevich, I.R.: Basic Algebraic Geometry. Springer Study edn. Springer, Berlin (1977)
26. Shallit, J.O., Frandsen, G.S., Buss, J.F.: The Computational Complexity of some Problems of Linear Algebra. BRICS series report, Aarhus, Denmark, RS-96-33, http://www.brics.dk/RS/96/33
27. Ha, H.T., Van Tuyl, A.: The regularity of points in multi-projective spaces. Journal of Pure and Applied Algebra 187(1-3), 153–167 (2004)

New State Recovery Attack on RC4

Alexander Maximov and Dmitry Khovratovich

Laboratory of Algorithmics, Cryptology and Security
University of Luxembourg
6, rue Richard Coudenhove-Kalergi, L-1359 Luxembourg
Alexander.Maximov@ericsson.com, Dmitry.Khovratovich@uni.lu

Abstract. The stream cipher RC4 was designed by R. Rivest in 1987, and it is a widely deployed cipher. In this paper we analyse the class RC4-N of RC4-like stream ciphers, where N is the modulus of operations, as well as the length of internal arrays. Our new attack is a state recovery attack which accepts the keystream of a certain length, and recovers the internal state. For the reduced RC4-100, our attack has total complexity of around 2^{93} operations, whereas the best previous attack (from Knudsen et al.) needs 2^{236} of time.

The complexity of the attack applied to the original RC4-256 depends on the parameters of specific states (patterns), which are in turn hard to discover. Extrapolated parameters from smaller patterns give us the attack of complexity about 2^{241}, and it is much smaller than the complexity of the best known previous attack 2^{779}. The algorithm of the new attack was implemented and verified on small cases.

Keywords: RC4, state recovery attack, key recovery attack.

1 Introduction

RC4 [Sch96] is a stream cipher designed by Ron Rivest in 1987, and since then it has been implemented in many various software applications to ensure privacy in communication. It is one of the most widely deployed stream ciphers and its most common application is to protect Internet traffic in the SSL protocol. Moreover, it has been implemented in Microsoft Lotus, Oracle Secure SQL, etc. The design of RC4 was kept secret until 1994 when it was anonymously leaked to the members of the Cypherpunk community. A bit later the correctness of the algorithm was confirmed.

In this paper we study a family RC4-N of RC4 like stream ciphers, where N is the modulus of operations. The internal state of RC4 is two registers $i, j \in \mathbb{Z}_N$ and a permutation S of all elements of $\mathbb{Z}_N$. Thus, RC4 has a huge state of $\log_2(N^2 N!)$ bits. For the original version, when $N = 256$, the size of the state is ≈ 1700 bits. This makes any time-memory trade-off attacks impractical. RC4-256 uses a variable length key from 1 to 256 bytes for its initialisation.

The initialisation procedure of RC4 has been thoroughly analysed in a large number of various papers, see e.g. [MS01,Man01,PP04]. These results show that the initialisation of RC4 is weak, and the secret key can be recovered with a small

D. Wagner (Ed.): CRYPTO 2008, LNCS 5157, pp. 297–316, 2008.
© International Association for Cryptologic Research 2008

portion of data/time. Because of these attacks, RC4 can be regarded as broken. However, if one would tweak the initialisation procedure, the cipher becomes secure again.

The simplicity of the keystream generating algorithm of RC4 has attracted many cryptanalysis efforts. In most analyses the scenario assumes that keystream of some length is given, and either a distinguishing ([Gol97, FM00, Max05, Man05]) or a state recovery ([KMP+98]) attack is of interest. A *state recovery attack* can be used to determine the actual security level of a cipher, if the initial internal state is considered as a secret key. The first state recovery attack was proposed by Knudsen et al in 1998 [KMP+98]. This had a computational complexity of 2^{779}. Some minor improvements were found in other literature, e.g. [MT98], but still, there is no attack even close to 2^{700}. One interesting attempt to improve the analysis was recently done in [Man05]. However, that attack is only a potential one [1], and the pretending time complexity claimed was around 2^{290}.

In this paper we propose a new state recovery attack on RC4-N. For the original design RC4-256 the total time complexity of the attack is less than 2^{579}, and under some realistic assumptions (see Section 6) a complexity would drop to 2^{241} (2^{272} under pessimistic extrapolations), requiring keystream of a similar length. This would mean that there is no additional gain in using a secret key longer than 30 bytes. We also show that in general if the secret key is of length N bits or longer the new attack is faster than exhaustive key search.

The idea of the new attack is as follows. The algorithm searches for a place in the keystream where the probability of a specific internal state, compliant with a chosen pattern, is high. Afterwards, the new state recovery algorithm is used together with a small portion of data (around $2N$ output words) in order to recover the internal state of the cipher in an iterative manner. This algorithm has been implemented and verified for **small values of** N, it has determined the correct internal state in *every* simulation run. The success rate of the full attack is shown to be at least 98%. For **large values of** N, where simulations were impossible, an upper bound for the average complexity of the attack is derived and calculated.

In the precomputation stage we search for a proper pattern to use in the attack. However, in this paper we skip a detailed analysis of that complexity since it is upper bounded by the time needed for the main stage of the attack (see Appendix B).

This paper is organized as follows. In Section 2 the new iterative *state recovery algorithm* is described in detail. Afterwards, Section 3 introduces various properties of a pattern that are needed for the recovering algorithm. An effective searching algorithm to find such patterns is also proposed in Appendix B (due to the page limitation and clarity of presentation). Section 4 describes several

---

[1] Mantin detects a large number of bytes of the state, and then applies Knudsen's attack given those bytes. However, this would reduce the complexity only if the knowns were located in a short window all together while this is not the case. This fact is confirmed in [Man05] (Section "State Recovery Attack").

techniques to detect specific states by observing the keystream, and also introduces additional properties of a pattern needed for detection purposes. Theoretical analysis of the state recovery algorithm and derivation of its complexity functions are performed in the full version of this paper [MK08]. All pieces of the attack are then combined in Section 5. Finally, we perform a set of simulations of the attack, summarize the results and conclude in Section 6. The paper ends with suggestions for further improvements and open problems in Section 7.

1.1   Notations

All internal variables of RC4 are over the ring $\mathbb{Z}_N$, where N is the size of the ring. To specify a particular instance of the cipher we denote it by RC4-N. Thus, the original design is RC4-256. Whenever applicable, $+$ and $-$ are performed in modulo N. At any time t the notation a_t denotes the value of a variable a at time t. The keystream is denoted by $\mathbf{z} = (z_1, z_2, \ldots)$, where z_i is a value $0 \leq z_i < N$. In all tables probabilities and complexities will be given in a logarithmical form with base 2.

1.2   Description of the Keystream Generator RC4-N

The new attack targets the keystream generation phase of RC4 and, thus, the initialisation procedure will not be described. We refer to, e.g., [Sch96] for a full description of RC4. After the initialisation procedure, the keystream generation algorithm of RC4 begins. Its description is given in Figure 1.

Internal variables:
$i,\ j$ – integers in $\mathbb{Z}_N$
$S[0 \ldots N-1]$ – a permutation of integers $0 \ldots N-1$
$S[\cdot]$ is initialised with the secret key
The keystream generator RC4-N
  $i = j = 0$
  Loop until we get enough symbols over $\mathbb{Z}_N$
    (A) $i = i + 1$
    (B) $j = j + S[i]$
    (C) swap$(S[i], S[j])$
    (D) $z_t = S[S[i] + S[j]]$

Fig. 1. The keystream generation algorithm of RC4-N

2   New State Recovery Algorithm

2.1   Previous Analysis: Knudsen's Attack

In [KMP$^+$98] Knudsen et al. have presented a basic recursive algorithm to recover the internal state of RC4. It starts at some point t in the keystream $\mathbf{z}$

given k known cells of the permutation S_t, which helps the recursion to cancel unlikely branches. The idea of the algorithm is simple. At every time t we have four unknowns:

$$j_t, \quad S_t[i_t], \quad S_t[j_t], \quad S_t^{-1}[z_t]. \tag{1}$$

One can simply simulate the pseudo random generation algorithm and, when necessary, guess these unknown values in order to continue the simulation. The recursion steps backward when a contradiction is reached due to previously wrong guesses. Additionally, it can be assumed that some k values are a priori known (guessed, given, or derived somehow), and this may reduce the complexity of the attack significantly. An important note is that the known k values should be located in a short window of the "working area" of the keystream, otherwise they cannot help to cancel hopeless branches.

The precise complexity of the attack was calculated in [KMP+98], and several tables for various values of N and k were given in Appendices D.1 and D.2 of [Man01]. As an example, the complete state recovery attack on RC4-256 would require time around 2^{779}.

2.2 Our Algorithm for State Recovery

In this section we propose an improved version of the state recovery algorithm. Assume that, at some time t in a window of length $w + 1$ of the keystream $\mathbf{z}$, all the values $j_t, j_{t+1}, j_{t+2}, \ldots, j_{t+w}$ are known. This means that for w steps the values $S_{t+1}[i_{t+1}], \ldots, S_{i+w}[i_{t+w}]$ are known as well, since they are derived as

$$S_{t+1}[i_{t+1}] = j_{t+1} - j_t, \quad \forall t. \tag{2}$$

Consequently, w equations of the following kind can be collected:

$$S_k^{-1}[z_k] = S_k[i_k] + S_k[j_k], \quad k = t+1, \ldots, t+w, \tag{3}$$

where only *two* variables are unknown,

$$S_k^{-1}[z_k], \quad S_k[j_k], \tag{4}$$

instead of *four* in Knudsen's attack, see (1). Let the set of consecutive w equations of the form (3) be called *a window of length w*.

Since all js in the window are known, then all swaps done during these w steps are known as well. This makes it possible to map the positions of the internal state S_t at any time t to the positions of some chosen *ground state* S_{t_0} at some ground time t_0 in the window. For simplicity, let us set $t_0 = 0$.

Our new state recovery algorithm is a recursive algorithm, shown in Figure 2. It starts with a collection of w equations, and attempts to solve them. A single equation is called *solved* or *processed* if its corresponding unknowns (4) have been explicitly derived or guessed. During the process, the window will dynamically increase and decrease. When the length of the window w is long enough (say, $w = 2N$), and all equations are solved, the ground state S_0 is likely to be fully recovered.

Now we give a more detailed description of the different parts of the algorithm.

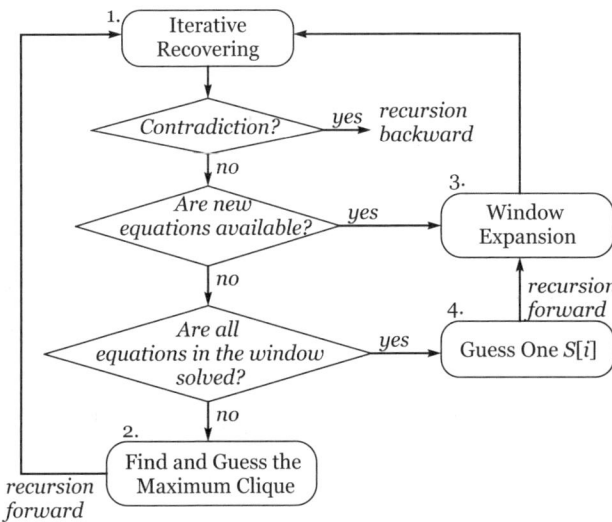

Fig. 2. New state recovery algorithm

Iterative Recovering (IR) Block. The *Iterative Recovering* block receives a number a of *active* equations (not yet processed) in the window of length w as input, and tries to derive the values of $S_t[j_t]$s and $S_t^{-1}[z_t]$s. To do that, the IR block goes through two steps iteratively, until no more new derivations are possible. If all previous guesses were correct, then all newly derived values (cells of the ground state) will be correct with probability 1. Otherwise, when the IR block finds a contradiction the recursion steps backward. The two steps are as follows.

A. Assume that, for one of the active equations its output symbol z_t is already allocated somewhere in the ground state. I.e., the value $S_t^{-1}[z_t]$ is known, and the second unknown $S_t[j_t]$ can explicitly be derived using (3).
   A contradiction arises if (a) $S_t[j_t]$ is already allocated and it is not equal to the derived value; (b) the derived value already exists in some other cell.
B. Already allocated values may give the value of $S_t[j_t]$ in another equation. Consequently, a new value $S_t^{-1}[z_t]$ can be derived via (3), which might possibly cause a contradiction.

Find and Guess the Maximum Clique (MC) Block. If no more active equations can explicitly be solved, $S_t^{-1}[z_t]$ for one t has to be guessed. The *Find and Guess the Maximum Clique* block analyses given active equations, and chooses the element that gives the maximum number of new derivations in consecutive recursive calls of the IR block. This element is then guessed.

The analysis is very simple. Let a active equations be vertices v_t in a graph representation. Two vertices $v_{t'}$ and $v_{t''}$ are connected if $z_{t'} = z_{t''}$ and/or $S_{t'}[j_{t'}]$

and $S_{t''}[j_{t''}]$ *refer* (like pointers) to the same cell of the ground state. Guessing any unknown variable in any connected subgraph solves all equations involved in that subgraph. Therefore, let us call these subgraphs *cliques*. The MC block searches for a maximum clique, and then guess *one* $S_t^{-1}[z_t]$ for one of the equations belonging to the clique. Afterwards, the IR block is called recursively.

Window Expansion (WE) Block. Obviously, the more equations we have the faster the algorithm works. Therefore, a new equation is added to the system as soon as the missing value $S[i]$ in the beginning or in the end of the window is derived. The *Window Expansion* block checks for this event and dynamically extends the window. Sometimes several equations are added at once, especially on the leafs of the recursion.

Guess One $S[i]$ (GSi) Block. If there are no active equations but the ground state S_0 is not yet fully determined, the window is then expanded by a direct guess of $S[i]$, in front or in back of the window. Then the WE, IR and MC blocks continue to work as usual. Additional heuristics can be applied for choosing which side of the window to be expanded for a larger success.

Appendix A provides an example that shows the steps of the outlined algorithm.

3   Precomputations: Finding Good Patterns

The algorithm presented in the previous section is used in the full state recovery attack as a part of it. Every time when the algorithm is running at some point of the keystream, its effectiveness depends on certain properties of the current internal state. Although these properties are not visible for the intruder, she may have a good guess about places in the keystream where the internal state has good properties (see Section 4), and apply the state recovery algorithm only at those places.

In this section we will define patterns (see Definition 1), they determine huge sets of internal states with common properties. If, for instance, a pattern has a large window then this certainly helps decreasing the complexity of the algorithm. However, the probability that the internal state is compliant with a certain pattern decreases with the number of conditions put on the pattern.

In this section we discuss properties of patterns that influence on the complexity of the attack, and also study their availability. We have also developed an efficient algorithm for finding these paterns, and it is located in Appendix B.

3.1   Generative States

Let us start with several definitions, some of which were previously defined in [MS01, Man01, Man05].

Definition 1 (d-order pattern). *A d-**order pattern** is a tuple*

$$A = \{i, j, P, V\}, \quad i, j \in \mathbb{Z}_N, \tag{5}$$

*where P and V are two vectors from $\mathbb{Z}_N^d$ with pairwise distinct elements. At a time t the internal state is said to be **compliant with** A if $i_t = i, j_t = j$, and d cells of the state S_t with indices from P contain corresponding values from V.* □

The example in Figure 4 in Appendix A illustrates how a 5-order pattern allows to receive a window of length 15. However, the higher the order, the less the probability of such a constraint to happen. Thus, we are interested in finding a low order pattern which generates a long window.

Definition 2 (w-generative pattern). *A pattern A is called w-**generative** if for any internal state compliant with A the next w clockings allow to derive w equations of the form (3), i.e., consecutive $w + 1$ values of js are known.* □

Table 1 demonstrates a 4-order 7-generative pattern $A=\{-7,-8,\{-6, -5, -4, 0\}, \{6, -1, 2, -2\}\}$, that supports the above definitions. Eight equations involve symbols of the keystream $z_{t+1}, \ldots, z_{t+8}$ associated with a certain time t. We say that the **keystream is true** if the internal state at time t is compliant with the pattern, otherwise we say the **keystream is random**.

Let another pattern B be derived from A as

$$B = A + \tau = \{i + \tau, j + \tau, P + \tau, V\}, \tag{6}$$

for some "shift" τ. The pattern B is likely to be w-generative as well. This happens when the properties of A are independent of N, which is the usual case.

3.2   Availability

We have done a set of simulations in order to find *maximum w-generative d*-order patterns, denoted by M_d. The results are given in Table 7(a) in Appendix C.

Table 1. An example of a 4-order 7-generative pattern

i_t	j_t	$S[i]$	$S[j]$	$S[i]+S[j]$	z_t	-6	-5	-4	-3	-2	-1	0	1	2	3	4	5
-7	-8	$-$	$-$	$-$	$-$	6	-1	2	x_1	x_2	x_3	-2	x_4	x_5	x_6	x_7	x_8
-6	-2	6	x_2	$6+x_2$	$*$	x_2	-1	2	x_1	6	x_3	-2	x_4	x_5	x_6	x_7	x_8
-5	-3	-1	x_1	$-1+x_1$	$*$	x_2	x_1	2	-1	6	x_3	-2	x_4	x_5	x_6	x_7	x_8
-4	-1	2	x_3	$2+x_3$	$*$	x_2	x_1	x_3	-1	6	2	-2	x_4	x_5	x_6	x_7	x_8
-3	-2	-1	6	5	x_8	x_2	x_1	x_3	6	-1	2	-2	x_4	x_5	x_6	x_7	x_8
-2	-3	-1	6	5	x_8	x_2	x_1	x_3	-1	6	2	-2	x_4	x_5	x_6	x_7	x_8
-1	-1	2	2	4	x_7	x_2	x_1	x_3	-1	6	2	-2	x_4	x_5	x_6	x_7	x_8
0	-3	-2	-1	-3	-2	x_2	x_1	x_3	-2	6	2	-1	x_4	x_5	x_6	x_7	x_8
1	$*$	x_4	$*$	$*$	$*$												

Table 2. Dependency of the maximum w from d, simulated and approximated values

	Real values from our simulations													Approximated values						
$d =$	2	3	4	5	6	7	8	9	10	11	12	13	14	15	16	17	18	19	20	21
$w_{\max} =$	6	10	15	21	27	31	37	42	50	55	61	68	76	82	88	94	100	106	112	118

Searching for a high order pattern is a challenging task since the computational complexity grows exponentially with d. The best result achieved in our work is a 14-order 76-generative pattern M_{14}.

Table 2 shows the dependency of a maximum achievable generativeness $w_{\max}$ from the order d. We can note that this dependency is almost linear, and it converges to $w_{\max} = 6d + \lambda$ as $d \to \infty$. We make the following conjecture.

Conjecture 1. The rate of $\frac{w_{\max}}{d} \approx 6$ as $d \to \infty$.[2]   □

That conjecture allows us to make a prediction about certain parameters for patterns with large d. These could not be found due to a very high precomputation complexity, but they are needed to analyse the attack for large N ($N = 128\ldots256$ in Table 3). However, given those parameters, d and w, we can derive theoretical complexities of the attack on average [3]. This has been done in [MK08]. An efficient search algorithm for patterns with desired properties is given in Appendix B.

4   Detection of Patterns in the Keystream

In the previous section we have studied properties of a pattern that are desirable for the state recovery algorithm to work fast and efficient. We have also shown (in Appendix B) how these patterns can be found, and introduced an efficient searching algorithm.

In this section we show how the internal state of RC4, compliant to a chosen pattern, can be detected by observing the keystream. If the detection is very good, then the state recovery algorithm might only have to be executed once, at the right location in the keystream.

The detection mechanism itself can be trivial (no detection at all), in which case the algorithm has to be run at every position of the keystream. On the other hand, a good detection may require a deep analysis of the keystream, where specific properties of the pattern can be used efficiently.

---

[2] Indeed, the "jump" of $w_{\max}$ as d increments by one is the sequence $\Gamma = \{4, 5, 6, 6, 4, 6, 5, 8, 5, 6, 7, 8, \ldots\}$. Obviously, for small d this "jump" is small, and it is notable that the "jump" increases for larger d. In our simulations heuristics were used (see Appendix B) when searching patterns for $d \geq 6$. This means that our "jumps" in the sequence Γ could possibly be larger if an optimal searching technique is applied, since our heuristic cannot guarantee that we get a pattern with the longest window. This suggests that the ratio $w \to 6d$ as $d \to \infty$ seems quite a fair conjecture.

[3] Because the relation $w = 6d + \lambda$ is a subject of discussions, we show in Table 4 that even more pessimistic conjectures do not affect the total complexity very much.

4.1   First Level of Analysis

The internal state of RC4 compliant to a d-order pattern A can be regarded as an *internal event* with probability

$$\Pr\{E_{\mathtt{int}}\} = N^{-d-1}. \tag{7}$$

When the internal event occurs, there is an *external event* $E_{\mathtt{ext}}$ observed in the keystream, which is associated with the pattern A, i.e., $\Pr\{E_{\mathtt{ext}}|E_{\mathtt{int}}\} = 1$. Applying Bayes' law we can derive the *detection probability* $\mathcal{P}_{\mathtt{det}}$ of the pattern A in the keystream as

$$\mathcal{P}_{\mathtt{det}} = \Pr\{E_{\mathtt{int}}|E_{\mathtt{ext}}\} = \frac{\Pr\{E_{\mathtt{int}}\}}{\Pr\{E_{\mathtt{ext}}\}}. \tag{8}$$

Our goal in this section is to study possible external events with high $\mathcal{P}_{\mathtt{det}}$ in order to increase the detection of the pattern.

Definition 3 (l-definitive pattern). *A w-generative pattern A is called l-**definitive** if there are exactly l out of w equations with determined $S[j]$s.*   □

It means that in l equations $S[i] + S[j]$ are known. If, additionally, $z' = S[S[i] + S[j]]$ is also known, then the correct value of $z_t = z'$ at the right position t of the keystream $\mathbf{z}$ detects the case *"the state at time t **is possibly** compliant to the pattern"*. Otherwise, when $z_t \neq z'$, it says that *"the state at time t **cannot be** compliant to the pattern"*.

For detection purposes a large l (up to d) is important. From our experiments we found that, however, a large l can be achieved via a slight reduction of the parameter w. This leads us to one more conjecture.

Conjecture 2. For any d and $w = w_{\mathtt{max}} - \lambda$ there exist a pattern with $l = d$, where λ is relatively small [4].   □

In the following definition we introduce other properties of a pattern that are important for its good detection via the keystream.

Definition 4 ($b_\alpha, b_\beta, b_\gamma$-$^{\alpha,\beta,\gamma}$predictive pattern). *Let us have an l-definitive pattern A and consider only those equations where $S[j]$s are determined. Then, the pattern A is called b_α-$^\alpha$**predictive** if for b_α of the l equations $S[S[i] + S[j]]$ is determined. For the remaining $l - b_\alpha$ equations two additional definitions are as follows. The pattern A is called b_β-$^\beta$**predictive** if for b_β pairs of the $l - b_\alpha$ equations the unknowns $S[S[i]+S[j]]$s must be the same. The set of b_β pairs must be of full rank. The pattern A is called b_γ-$^\gamma$**predictive** if the $l - b_\alpha$ equations contain exactly b_γ different variables of $S[S[i] + S[j]]$.*   □

---

[4] Table 6(a) in Appendix C contains patterns Xs with $l = d$ where w is still large, which supports the above conjecture. Indeed, Table 5 in Appendix B shows how the number of available patterns grows when relaxing the condition put on w. I.e., a slight reduction of w increases the chance of finding a pattern with $d = l$. This makes the conjecture fair.

These types of predictiveness are other properties of a pattern visible in the keystream. For example, it is not only necessary to search for known z' values (b_α of such), but one can also require that certain pairs of the keystream symbols (b_β of such) are equal $z_{t'} = z_{t''}$, which also helps to detect the pattern significantly.

The parameter b_α is usually quite moderate and to have it larger than 15 is quite difficult. However, the other criteria are more flexible and can be large. These new parameters follow the constraint

$$b_\alpha + b_\beta + b_\gamma = l \le d. \tag{9}$$

Consider the remaining $w - l$ equations of the pattern A where $S[j]$s are not determined. Let at time instances t_1 and t_2 one pair of these equations be such that the $S[i]$ values and the $S[j]$ pointers are equal. If the distance $\Delta_t = t_2 - t_1$ is small, it is likely that the output z_1 is the same as z_2. The probability of this event is

$$\Pr\{z_1 = z_2 | \Delta_t\} > \left(1 - \frac{\Delta_t}{N}\right) \cdot \left(1 - \frac{1}{N}\right)^{\Delta_t} \approx \exp\left(-\frac{2\Delta_t}{N}\right). \tag{10}$$

Definition 5 (b_θ-$^\theta$predictive pattern). *A pattern A is called b_θ-$^\theta$predictive if the number of such pairs (described above) is b_θ. Let the time distances of these pairs be $\Delta_1, \ldots, \Delta_{b_\theta}$, then the **cumulative distance** is the sum $\Pi_\theta = \Sigma_i \Delta_i$* □

These four types of predictiveness are direct external events for a pattern. One should observe the keystream and search for certain b_α symbols, check another b_β and b_θ pairs of symbols that they are equal, and also check that a group of b_γ symbols are different from the values of V and from each other. Thus, we have

$$\Pr\{E_{\text{ext}}\} = N^{-b_\alpha - b_\beta - b_\theta} \cdot \left[\frac{(N-d)!}{N^{b_\gamma}(N - d - b_\gamma)!}\right]$$

$$\Pr\{E_{\text{int}}\} \approx N^{-d-1} \cdot e^{-2\Pi_\theta/N}. \tag{11}$$

The example in Table 1 is a 4-definitive $b_\alpha = 1, b_\beta = 1, b_\gamma = 2, b_\theta = 0$-predictive pattern. For detection one has to test that $z_{t+6} = -2, z_{t+3} = z_{t+4}$, and z_{t+4}, z_{t+5} are different from the initial values at V and $z_{t+4} \ne z_{t+5}$. I.e., when, for example, $N = 64$, the detection probability is $64^{-5} \div (64^{-2} \cdot 60 \cdot 59/64^2) \approx 64^{-2.96}$ [5].

4.2   Second Level of Analysis

In fact, the first level of analysis allows to detect a pattern with probability at most N^{-1} (because j is not detectable), whereas with the second level of analysis it can be 1. Let us introduce a technique that we call a *chain of patterns*.

---

[5] Since γ-predictiveness has a minor influence on detection, we skip this parameter in future calculations.

Definition 6 (chain of patterns $A \to B$, distance, intersection). *Let us have two patterns $A = \{i_a, j_a, P_a, V_a\}$ and $B = \{i_b, j_b, P_b, V_b\}$. An event when two patterns appear in the keystream within the shortest possible time distance σ is called* **chain of patterns**, *and is denoted as $A \to B$ if B appears after A.*

The **chain distance** *σ between two patterns A and B is the shortest possible time between A's ending and B's beginning of their windows, i.e.,*

$$\sigma = i_b - (i_a + w_a) \mod N. \tag{12}$$

The **intersection** *of A and B is the number ξ of positions in A that are reused in B. These positions must not appear as $S[i]$ during σ clockings while the chain distance between A and B is approached.* $\square$

For example, let $A = \{0, 0, \{1, 3, 5, 6, 7, 8, 22, 23\}, \{2, 8, -3, -2, 1, 7, 4, -9\}\}$ and $B = \{34, 34, \{35, 36, 37, 38, 39, 44, 48, 52\}, \{8, -2, 1, 2, 4, -5, 5, 3\}\}$. After $w_a = 30$ clockings the first pattern becomes $A' = \{30, 28, \{15, 28, 30, 35, 36, 37, 38, 39\}, \{-3, -9, 7, 8, -2, 1, 2, 4\}\}$. Obviously, the last $\xi = 5$ positions can be reused in B, and after $\sigma = 4$ clockings a new pattern B ($w_b = 34$) can appear if $j_{t+34} = j_b$. The probability that the chain $A \to B$ appears is $N^{-9} \cdot N^{-4}$, multiplied by the probability that 5 elements from A' stay at the same locations during the next 4 clockings. This is much larger than the trivial $N^{-9} \cdot N^{-9}$. Thus, a more general theorem can be stated.

Theorem 1 (chain probability). *The probability of a chain $A \to B$ to appear is*

$$\mathcal{P}_{A \to B} = \Pr\{E_{\text{int}}\} \approx N^{-(d_a + d_b + 2 - \xi)} \cdot e^{-2(\Pi_{\theta a} + \Pi_{\theta b})/N} \cdot e^{-\xi}. \tag{13}$$

Proof. In [Man01] it has been shown that ξ elements stay in place during N clockings with an approximate probability $e^{-\xi}$. The remaining part comes from an assumption that the internal state is random, from where the proof follows.

$\square$

Obviously, the probability of the external event for the chain is

$$\Pr\{E_{\text{ext}}\} = N^{-(b_{\alpha a} + b_{\beta a} + b_{\theta a}) - (b_{\alpha b} + b_{\beta b} + b_{\theta b})}, \tag{14}$$

which can be smaller than $\Pr\{E_{\text{int}}\}$ (see Y_4 in Table 6 in Appendix C), confusing the equation (8). This happens since $\Pr\{E_{\text{ext}}\}$ is calculated assuming that the keystream is random. However, in RC4 only a portion of the observed external probability space can appear (which is another source for a distinguishing attack, but it is out of scope of this paper). Therefore, in the case when $\Pr\{E_{\text{ext}}\} < \Pr\{E_{\text{int}}\}$ we simply assume that the detection probability is 1.

Table 6 in Appendix C presents a few examples with a good trade-off (based on our intuition) between w and detectability for various d. Since the computation time for searching such patterns with multiple desired properties is really huge, only a few examples for small d were given. However, we believe that for large d it is possible to detect such patterns with a high probability, up to 1, applying the two proposed levels of analysis.

5   Complete State Recovery Attack on RC4

5.1   Attack Scenario and Total Complexity

Recall pattern detection techniques from Section 4. In the attack scenario an adversary analyses the keystream at every time t, and applies the state recovery algorithm if the desired internal event (pattern) is detected. In all cases except one the recovering algorithm deals with a random keystream.

Proposition 1 (Total Attack Complexities). *Let the detection probability be $\mathcal{P}_{\text{det}}$, then the total time C_T and data C_D complexities of the attack are*

$$C_T = \Pr\{E_{\text{int}}\}^{-1} + (\mathcal{P}_{\text{det}}^{-1} - 1) \cdot C_{\text{Rand}} + 1 \cdot C_{\text{True}}, \tag{15}$$
$$C_D = \Pr\{E_{\text{int}}\}^{-1}. \qquad\qquad \square$$

5.2   Success Rate of the Attack

The complexities C_{True} and C_{Random} are upper bounds for the *average* time the algorithm requires. It means that for some cases it could take more time than these bounds. In order to guarantee the upper bound of the total (not average) time complexity one can terminate the algorithm after, for example, C_{thr} operations. In this case the *success rate* of the attack can be determined.

Figure 3 shows density and cumulative functions for the time complexity of an example attack scenario. It shows that around 98% of all simulations of the attack have time smaller than the average $2^{29.28}$ (vertical line). When the keystream is random the termination makes the average time bound C_{Random} even smaller, since the random case is likely to be repeated very many times and the second term in (15) can only decrease.

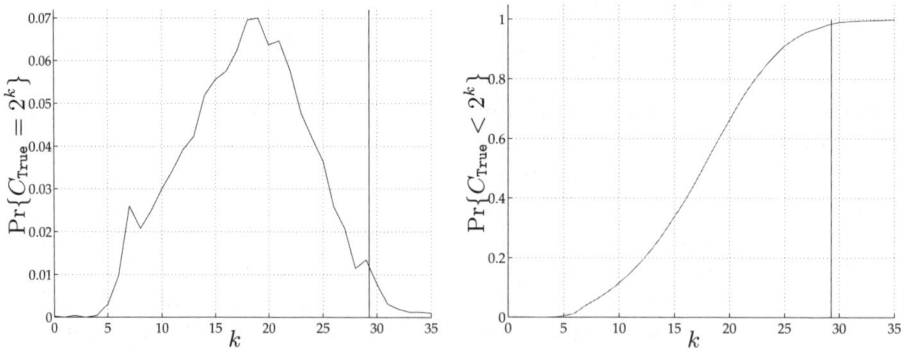

Fig. 3. Probability density (left) and cumulative (right) functions of the time C_{True} in logarithmical form ($k = \log_2 C_{\text{True}}$). The scenario is $N = 64$, M_8 and 2000 samples.

The plots in Figure 3 also show that even if the termination of the algorithm is done on the level $C_{\texttt{thr}} = \sqrt{C_{\texttt{True}}}$ ($\approx 2^{15}$), the success rate of the attack is still very high. I.e., the state recovery algorithm on RC4-64 can be done in time 2^{15} with success probability 35%! If a similar situation happens for large N (e.g., $N = 256$), then the full time complexity can be significantly decreased (perhaps, down to a square root of the estimated average complexity), and the success probability can still be very large.

6   Simulation Results and Conclusions

We have selected a set of test cases with various parameters and patters, and derived total data and time complexities of the new attack. Table 3 presents the results of this work. For example, when $N = 64$, the total complexity of the new attack is upper bounded by 2^{60}, if the pattern X_9 is used. This is much faster than, for example, Knudsen's attack, which complexity for this case is $2^{132.6}$. Even if $d = 9$ elements of the state are known, Knudsen's attack needs $2^{98.1}$ of time, which is still much higher. The complexity of a *potential* attack recently discussed by I. Mantin in [Man05] is also higher. As it was shown in Section 5.2, the success rate of the new attack is at least 98%.

Table 3 also contains intermediate probabilities and complexities for the attack, including theoretical ($\Delta = 0$) and attuned ($\Delta = 2$) values for $C_{\texttt{Rand}}$ and $C_{\texttt{True}}$. When it was possible, the real attack *on a true keystream* was simulated (real complexities for $C_{\texttt{True}}$ are shown in italic). In these simulations the complete state of RC4 was successfully recovered for every randomly generated keystream compliant with the corresponding pattern.

For larger N, patterns of a high order are needed to receive an attack of low complexity. The largest pattern that we could find in this work is M_{14}, and this was applied to attack RC4-N with $N = 128, 160, 200, 256$. These attack scenarios are those that we have in our hands already. However, the complexities received are not optimal, but they are still lower than in Knudsen's attack. Conjecture 1 and also discussions in Section 4 make it possible to approximate the parameters of a hypothetical pattern that is likely to exist ($\star$ – patterns). To be secure, we relate d and w as $w = 6d - 6$. The remaining parameters were chosen moderate as well. As the result, we obtained an attack on RC4-256 with the (upper bounded) total complexity of $2^{241.7}$, and this is the best state recovery attack known at the moment.

Since Conjecture 1 is discussible, we show in Table 4 that even pessimistic relations between w and d do not increase the attack complexity of approximated scenarious ($\star$) significantly. In general, we have noted the following tendency. For RC4-N with a secret key of length N bits or longer, the new attack can recover the internal state much faster than an exhaustive search. This observation can also be seen from the results in Table 3.

As the last point of the discussions we note that the key recovery attack can be easily converted from a state recovery attack. There are several papers dealing with recovering the secret key from a known internal state [MS01, Man01,

Table 3. Simulation results and comparisons with previous attacks

N	N = 64			N = 100		N = 128		N = 160		N = 200		N = 256	
Cases	I	II	III	IV	V	VI	VII	VIII	IX	X	XI	XII	XIII
Descriptions of the cases ($\star$ – are hypothetical cases)													
Pattern	$\mathcal{M}_8$	$\mathcal{V}_8$	$\mathcal{X}_9$	$\mathcal{X}_{11}$	$\mathcal{M}_{43}$	$\mathcal{M}_{44}$	$\star$	$\mathcal{M}_{44}$	$\star$	$\mathcal{M}_{44}$	$\star$	$\mathcal{M}_{44}$	$\star$
d	8	8	9	11	13	14	15	14	18	14	23	14	29
w	37	29	41	49	68	76	84	76	102	76	132	76	168
l	6	6	5	11	9	10	10	10	10	10	14	10	17
b_α	0	4	4	9	0	0	10	0	11	0	10	0	11
b_β	1	1	0	0	2	2	0	2	0	2	2	2	4
b_γ	5	1	1	2	7	8	0	8	0	8	2	8	2
b_θ	0	0	2	0	2	2	0	2	7	2	4	2	12
Π_θ	0	0	4	0	4	4	0	4	–	4	–	4	–
Internal/external/detection probabilities													
$\mathcal{P}_{\texttt{int}}$	-54.0	-65.8	-60.0	-79.7	-93.0	-105.0	-112.0	-109.8	-139.1	-114.7	-183.5	-120.0	-240.0
$\mathcal{P}_{\texttt{ext}}$	-6.0	-60.0	-36.0	-59.8	-26.6	-28.0	-70.0	-29.3	-131.8	-30.6	-122.3	-32.0	-216.0
$\mathcal{P}_{\texttt{det}}$	-48.0	-5.8	-24.0	-19.9	-66.4	-77.0	-42.0	-80.5	-7.3	-84.1	-61.2	-88.0	-24.0
Complexities of the state recovery algorithm when the keystream is true/random													
C_{Rand} Theor.	20.5	58.2	22.8	107.8	10.0	71.3	71.7	191.1	131.7	317.4	121.3	507.4	217.1
C_{Rand} Attun.	15.5	57.8	–	107.5	–	66.3	–	179.2	–	302.6	–	491.8	–
C_{True} Theor.	35.0	64.9	30.9	120.4	34.5	94.7	102.0	213.0	138.2	335.6	157.5	519.6	225.4
C_{True} Attun.	30.3	57.6	–	108.3	31.8	85.5	–	185.1	–	309.9	–	501.8	–
C_{True} Real	29.3	–	–	–	29.1	–	–	–	–	–	–	–	–
Total data/time complexity, and the comparison with previous attacks													
Knudsen's $C_K(0)$	132.6			236.6		324.8		431.4		572.0		779.7	
Knudsen's $C_K(d)$	101.7	101.7	98.1	189.3	181.0	261.3	256.9	364.6	346.1	501.9	458.2	705.9	629.3
Mantin's potential attack	73			114		147		186		243		290	
Our attack C_D	54.0	65.8	**60.0**	79.7	**93.0**	105.0	**112.0**	109.8	**139.1**	114.7	**183.4**	120.0	**240.0**
Our attack C_T	63.5	63.4	**60.0**	127.4	**93.1**	143.4	**113.7**	271.7	**140.4**	386.7	**184.0**	579.8	**241.7**

Table 4. Complexities of the attack on RC4-256 for various relations $w = \xi d + \lambda$. All scenarious show much better attack complexity than the best previous one 2^{779}.

	optimistic	realistic		pessimistic	
$N = 256$	$w = 6.5d - 17$	$w = 6d - 6$	$w = 6d - 12$	$w = 5d$	$w = 4d$
d, w	$d = 29, w = 171$	$d = 29, w = 168$	$d = 30, w = 168$	$d = 33, w = 165$	$d = 39, w = 156$
keystream	240.0	240.0	248.0	272.0	320.0
time	224.9	241.1	243.3	265.9	327.1

BC08]. However, this part works much faster than currently known state recovery attacks, and, therefore, we just refer to these papers without giving details.

7   Further Improvements and Open Problems

Pattern detection improvements. With a chain of patterns described in Section 4 one could reach a good detection. However, not only forward direction of chaining can be considered, but also backward one. Additionally, there is a possibility to analyse longer sequences of patterns in order to have a good detectability. Another idea is to use *unusual recyclable patterns* in a similar manner as in [Man05]. The difference is that these patterns are both recyclable and have a long window. For example, $A = \{0, -4, \{6, 4, 1, 5, 3\}, \{0, 1, 7, -2, -1\}\}$.

State recovery algorithm improvement. The GSi block can choose the corner (left or right) of the window to be extended by an additional heuristic analysis of the current situation during the process. Another improvement is achieved if the MC block could speculatively run the recursion for additional 1-3 extra forward steps for every possible guess, and, afterwards, make such a guess for which the number of sub branches is the minimum. The average time of the attack for this strategy is reduced.

Derivation and statistics. Our investigation showed that the derived theoretical upper bound gives a much larger complexity than the one received from the real simulations of the attack. Obviously, a better analysis of the algorithm's complexity is needed. This would allow a more accurate estimation of the total complexity, and it might improve the complexities in Table 3 significantly. Another interesting problem is to determine the density function of the recovering algorithm, likewise in Figure 3. This may allow us to decrease the complexity in square root times, maintaining a high success rate.

Other open problems. The search for patterns of a higher order with long windows is another challenging open question. We have shown that there are chains of patterns with short distances. The first pattern is used for the recovering algorithm, and the second one is for detection. However, another interesting question is whether or not the second pattern can also be used in the recovering algorithm.

Acknowledgements

We thank Martin Hell, Lars Knudsen, Matt Robshaw and also anonymous reviewers for their valuable comments and efforts which helped us to improve this paper significantly. This work was partly supported by University of Luxembourg and Ericsson AB.

References

[BC08]   Biham, E., Carmeli, Y.: Efficient reconstruction of rc4 keys from internal states. In: Fast Software Encryption 2008. Lecture Notes in Computer Science. Springer, Heidelberg (to appear, 2008)

[FM00]   Fluhrer, S.R., McGrew, D.A.: Statistical analysis of the alleged RC4 keystream generator. In: Schneier, B. (ed.) FSE 2000. LNCS, vol. 1978, pp. 19–30. Springer, Heidelberg (2001)

[Gol97]   Golić, J.D.: Linear statistical weakness of alleged RC4 keystream generator. In: Fumy, W. (ed.) EUROCRYPT 1997. LNCS, vol. 1233, pp. 226–238. Springer, Heidelberg (1997)

[KMP+98] Knudsen, L.R., Meier, W., Preneel, B., Rijmen, V., Verdoolaege, S.: Analysis methods for (alleged) RC4. In: Ohta, K., Pei, D. (eds.) ASIACRYPT 1998. LNCS, vol. 1514, pp. 327–341. Springer, Heidelberg (1998)

[Man01]   Mantin, I.: Analysis of the stream cipher RC4. Master's thesis, The Weizmann Institute of Science, Department of Applied Math and Computer Science, Rehovot 76100, Israel (2001)

[Man05]   Mantin, I.: Predicting and distinguishing attacks on RC4 keystream generator. In: Cramer, R. (ed.) EUROCRYPT 2005. LNCS, vol. 3494, pp. 491–506. Springer, Heidelberg (2005)

[Max05]   Maximov, A.: Two linear distinguishing attacks on VMPC and RC4A and weakness of RC4 family of stream ciphers. In: Gilbert, H., Handschuh, H. (eds.) FSE 2005. LNCS, vol. 3557, pp. 342–358. Springer, Heidelberg (2005)

[MK08]   Maximov, A., Khovratovich, D.: New state recovery attack on RC4 (accessed May 27, 2008) (2008), http://eprint.iacr.org/2008/017

[MS01]   Mantin, I., Shamir, A.: Practical attack on broadcast RC4. In: Matsui, M. (ed.) FSE 2001. LNCS, vol. 2355, pp. 152–164. Springer, Heidelberg (2002)

[MT98]   Mister, S., Tavares, S.E.: Cryptanalysis of RC4-like ciphers. In: Tavares, S., Meijer, H. (eds.) SAC 1998. LNCS, vol. 1556, pp. 131–143. Springer, Heidelberg (1999)

[PP04]   Paul, S., Preneel, B.: A new weakness in the RC4 keystream generator and an approach to improve the security of the cipher. In: Roy, B., Meier, W. (eds.) FSE 2004. LNCS, vol. 3017, pp. 245–259. Springer, Heidelberg (2004)

[Sch96]   Schneier, B.: Applied Cryptography: Protocols, Algorithms, and Source Code in C, 2nd edn. John Wiley&Sons, New York (1996)

A   Example Support for the State Recovery Algorithm

Figure 4 illustrates an example of the process of the IR block. In the example we start with specific values of i and j, and also $d = 5$ cells of the state S are filled with certain values, whereas the remaining cells are unknown. This constraint allows to collect $w = 15$ equations of the form (3). The keystream is given in the rightmost column of the table.

The first iteration, in Figure 4(b), finds that $z_6 = 4$ and $z_8 = -2$ are already allocated, thus solving equations 6 and 8 ($s_4 = 10, s_9 = 5$). Afterwards, given $s_9 = 5$, the IR block solves the equation 14 and successfully checks for a contradiction, in Figure 4(c). Finally, after the step (e) four additional cells of the state S were derived with probability 1.

When the IR block is processed, the input to the MC block is the maximum clique of size 4 equations with 5 unknowns, shown in Figure 4(f). It means that guessing only one unknown determines four other ones. Furthermore, the space of possible guesses is significantly reduced due to the higher probability of a contradiction to occur.

i_{t+1}	j_{t+1}	The part of the state S_t at time t, just before the swap-operation																			$S[i]$	$S[j]$	z
		1	2	3	4	5	6	7	8	9	10	11	12	13	14	15	16	17	18	19	20			
1	8	4	-2	1	8	-4	s_1	s_2	s_3	s_4	s_5	s_6	s_7	s_8	s_9	s_{10}	s_{11}	s_{12}	s_{13}	s_{14}	s_{15}	4	s_3	18
2	6	s_3	-2	1	8	-4	s_1	s_2	4	s_4	s_5	s_6	s_7	s_8	s_9	s_{10}	s_{11}	s_{12}	s_{13}	s_{14}	s_{15}	-2	s_1	29
3	7	s_3	s_1	1	8	-4	-2	s_2	4	s_4	s_5	s_6	s_7	s_8	s_9	s_{10}	s_{11}	s_{12}	s_{13}	s_{14}	s_{15}	1	s_2	6
4	15	s_3	s_1	s_2	8	-4	-2	1	4	s_4	s_5	s_6	s_7	s_8	s_9	s_{10}	s_{11}	s_{12}	s_{13}	s_{14}	s_{15}	8	s_{10}	16
5	11	s_3	s_1	s_2	s_{10}	-4	-2	1	4	s_4	s_5	s_6	s_7	s_8	s_9	8	s_{11}	s_{12}	s_{13}	s_{14}	s_{15}	-4	s_6	5
6	9	s_3	s_1	s_2	s_{10}	s_6	-2	1	4	s_4	s_5	-4	s_7	s_8	s_9	8	s_{11}	s_{12}	s_{13}	s_{14}	s_{15}	-2	s_4	4
7	10	s_3	s_1	s_2	s_{10}	s_6	s_4	1	4	-2	s_5	-4	s_7	s_8	s_9	8	s_{11}	s_{12}	s_{13}	s_{14}	s_{15}	1	s_5	12
8	14	s_3	s_1	s_2	s_{10}	s_6	s_4	s_5	4	-2	1	-4	s_7	s_8	s_9	8	s_{11}	s_{12}	s_{13}	s_{14}	s_{15}	4	s_9	-2
9	12	s_3	s_1	s_2	s_{10}	s_6	s_4	s_5	s_9	-2	1	-4	s_7	s_8	4	8	s_{11}	s_{12}	s_{13}	s_{14}	s_{15}	-2	s_7	21
10	13	s_3	s_1	s_2	s_{10}	s_6	s_4	s_5	s_9	s_7	1	-4	-2	s_8	4	8	s_{11}	s_{12}	s_{13}	s_{14}	s_{15}	1	s_8	6
11	9	s_3	s_1	s_2	s_{10}	s_6	s_4	s_5	s_9	s_7	s_8	-4	-2	1	4	8	s_{11}	s_{12}	s_{13}	s_{14}	s_{15}	-4	s_7	9
12	7	s_3	s_1	s_2	s_{10}	s_6	s_4	s_5	s_9	-4	s_8	s_7	-2	1	4	8	s_{11}	s_{12}	s_{13}	s_{14}	s_{15}	-2	s_5	1
13	8	s_3	s_1	s_2	s_{10}	s_6	s_4	-2	s_9	-4	s_8	s_7	s_5	1	4	8	s_{11}	s_{12}	s_{13}	s_{14}	s_{15}	1	s_9	10
14	12	s_3	s_1	s_2	s_{10}	s_6	s_4	-2	1	-4	s_8	s_7	s_5	s_9	4	8	s_{11}	s_{12}	s_{13}	s_{14}	s_{15}	4	s_5	16
15	20	s_3	s_1	s_2	s_{10}	s_6	s_4	-2	1	-4	s_8	s_7	4	s_9	s_5	8	s_{11}	s_{12}	s_{13}	s_{14}	s_{15}	8	s_{15}	17
16	?																							

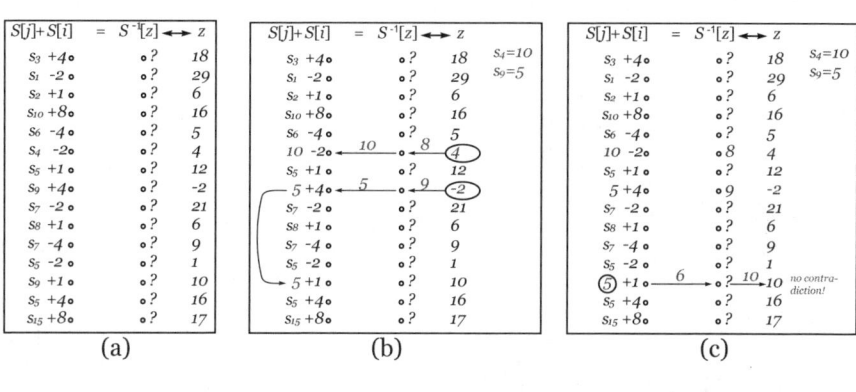

Fig. 4. Example of the iterative reconstruction process

B   Searching Technique

Since the search space for a d-order pattern grows exponentially with d, only patterns of order $d \leq 6$ were analysed before in various literature, e.g., in [Man05]. In this section we suggest a few techniques that accelerate this search significantly, and allow to search and analyse patterns of order up to $d \leq 15$, approximately, on a usual desktop PC.

First, we need to make some observations on the construction of patterns. Afterwards, several ideas based on the observation for improving the algorithm follow.

All "good" patterns found have Vs with values from a short interval $I_\delta = [-\delta \ldots + \delta]$, where $\delta \approx 10 \ldots 25$ is quite conservative. From this we make the following conjecture.

Conjecture 3. A pattern with the largest w is likely found among all possible combinations for $i = 0, j \in I_\delta, V \in I_\delta^d$, with a moderate value of $\delta \ll N$.   □

This conjecture will be used as the basis for a *significant* improvement in the *searching technique* of such patterns.

Table 5 provides the number of patterns for $\delta = 15$, and various values of d and w. When d and δ are fixed, the amount of desired patterns can be exponentially increased by letting w be slightly less than $w_{\max}$. This approach can help finding patterns with additional properties which are introduced in Section 4.

The first idea is to set $i = 0$ due to (6), and for the remaining variables only a small set of values I_δ for some δ should be tested due to Conjecture 3.

A straightforward approach would be to allocate d values in a vector S and then to check the desired properties of the pattern. The time complexity of this approach is $O\left(\binom{N}{d}\binom{|I_\delta|}{d}|I_\delta|\right)$, which is still very large. Our *second idea* is to allocate a new element in S only when it is necessary. This will significantly decrease the time complexity.

The diagram of a recursive algorithm exploiting the first two ideas is shown in Figure 5, but it can be improved with the following *heuristic*. The *third idea* is to start searching for a desired pattern somewhere in the middle of its future window. Let us split d as $d = d_{\mathtt{fwd}} + d_{\mathtt{back}}$ and then start the algorithm in Figure 5 allowing to allocate exactly $d_{\mathtt{fwd}}$ cells of S. At the point ($*$) the current length of the window w is compared with some threshold $w_{\mathtt{thr}}$. If $w \geq w_{\mathtt{thr}}$, then a similar recursive algorithm starts, but it goes backward and allocates remaining $d_{\mathtt{back}}$ cells of S. This double-recursion results in a pattern with w *likely* to be close to the maximum possible length of the window.

Searching of a d-order pattern is a precomputation stage of the attack.

Theorem 2. *The complexity of the precomputation stage is less than the total complexity of the attack.*

Table 5. The number of different constraints for specific d and w, when $\delta = 15$

d		The number of patterns A_d when $\delta = 15$.								
$\downarrow$	$w \rightarrow$	15	14	13	12	11	10	9	8	7	6
4	$\#\{A_4\} \rightarrow$	1	3	10	26	226	863	5234	21702	114563	853012
	$w \rightarrow$	21	20	19	18	17	16	15	14	13	12
5	$\#\{A_5\} \rightarrow$	1	4	6	15	66	252	652	1879	6832	27202
	$w \rightarrow$	27	26	25	24	23	22	21	20	19	18
6	$\#\{A_6\} \rightarrow$	1	2	7	42	81	177	371	799	2646	10159

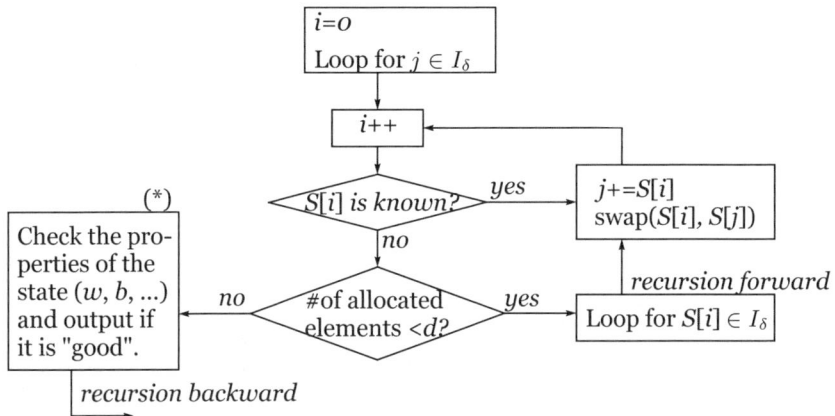

Fig. 5. Recursive algorithm for searching patterns with large w

Proof. Assume we are interested a d-order pattern. To start with, one should loop j of N values. Afterwards, the algorithm tries to allocate the first value in $V[]$ at some first location in $P[]$, which is another inner loop of N values, and so on. At the end we got $d+1$ inner loops, each of N values. Thus, the complexity of this non-heuristic and non-optimized searching algorithm is $O(N^{d+1})$. The attack requires a keystream of the same size, thus, it proofs the statement. $\square$

C   Patterns Used in This Paper

Table 6. Various patterns that were achieved by our simulations (part I)

Reference	level of analysis	i, j	Pattern description	order	generative	definitive	α-predictive	β-predictive	θ-predictive	cumulative distance	chain distance	intersection	new guesses	Pr{E_{int}}	Pr{E_{ext}}
Ref.	A.l.	i, j	P, V	d	w	l	b_α	b_β	b_θ	Π_θ	σ	ξ	ψ	Int	Ext
(a) Trade-off between w and l, the first level of analysis															
$\mathcal{X}_8$	1st	-18,-5	P ={8 9 -17 -16 -15 -14 -13 -9}	8	33	5	5	0	2	4	–	–	–	$N^{-9}e^{-8/N}$	N^{-7}
			V ={2, -1, -5, -2, 1, 4, 5, 3}												
$\mathcal{X}_9$	1st	-20, -23	P ={0, 1, 5, -19, -17, -16, -15, -14, -7}	9	41	5	4	0	2	4	–	–	–	$N^{-10}e^{-8/N}$	N^{-6}
			V ={-5, 8, 3, 5, 4, -2, -1, 2, 1}												
$\mathcal{X}_{10}$	1st	-25, -25	P ={6, 8, -24, -22, -20, -19, -18, -17, -3, -2 }	10	47	6	5	0	2	4	–	–	–	$N^{-11}e^{-8/N}$	N^{-7}
			V ={3, 4, 2, 8, -3, -2, 1, 7, 0, -5}												
$\mathcal{X}_{11}$	1st	-37, -37	P ={-36, -35, -34, -33, -30, -29, -28, -15, -13, -10, 5}	11	49	11	9	0	.	.	–	–	–	N^{-12}	N^{-9}
			V ={10, -4, -1, 11, 3, -2, 1, 9, -3, -7, 2}												
(b) Good detection through the second level of analysis															
$\mathcal{Y}_4$	1st	-7, -7	P ={-6, -5, -3, -1}, V ={3, 2, -1, 0}	4	9	4	4	0	.	.	–	–	–	N^{-5}	N^{-4}
	2nd	-2, -1	P' ={0, 2, -1}, V' ={0, -1, 2}	3	4	3	3	0	.	.	-4	3	0	$N^{-6}e^{-3}$	N^{-7}
$\mathcal{Y}_7$	1st	-24, -19	P ={1, -23, -22, -20, -18, -10, -3}	7	26	6	5	0	.	.	–	–	–	N^{-8}	N^{-5}
			V ={-3, -2, 4, 5, 1, 0, -1}												
	2nd	-5, -2	P' ={0, 1, -4, -3, -2}, V' ={2, -1, 1, 0, -2}	5	6	5	5	0	.	.	-7	4	1	$N^{-10}e^{-4}$	N^{-10}
$\mathcal{Y}_8$	1st	-26, -27	P ={-25, -24, -23, -20, -19, -18, -16, -4}	8	29	6	4	1	0	0	–	–	–	N^{-9}	N^{-5}
			V ={5, 1, 4, -3, -1, 2, 3, -2}												
	2nd	-7, 2	P' ={0, 3, -6, -5, -4, -3, -2}, V' ={-2, 3, 0, -1, 1, -3, 2}	7	10	7	7	0	0	0	-10	6	1	$N^{-12}e^{-6}$	N^{-12}

Table 7. Various patterns that were achieved by our simulations (part II)

Ref.	i, j	P, V	d	w	l	b_α	b_β	b_γ	b_θ	Π_θ
$\mathcal{W}_2$	0, -1	P ={1, 3}, V ={3, -1}	2	6	0	0	0	0	1	1
$\mathcal{W}_3$	0, -1	P ={1, 3, 4}, V={3, 2, -1}	3	10	3	0	1	2	0	0
$\mathcal{W}_4$	0, -2	P ={1, 3, 4, 5}, V={4, 3, -2, 1}	4	15	1	0	0	1	1	2
$\mathcal{W}_5$	0, -2	P ={1, 2, 4, 6, 8}, V={5, 2, -3, 6, -1}	5	21	0	0	0	0	0	0
$\mathcal{W}_6$	0, 0	P ={1, 2, 3, 4, 5, 20}, V={7, -1, 5, -3, 2, -9}	6	27	3	0	1	2	0	0
$\mathcal{W}_7$	0, 5	P ={1, 2, 4, 6, 8, 9, 16}, V={-2, 4, 7, 1, 3, -3, 8}	7	31	4	0	0	4	1	2
$\mathcal{W}_8$	0, 5	P ={1, 2, 4, 6, 14, 18, 19, 25}	8	37	6	0	1	5	0	0
		V ={-2, 4, 5, 1, 3, -3, 2, -1}								
$\mathcal{W}_9$	0, 9	P ={1, 2, 3, 6, 7, 8, 11, 20, 24}	9	42	6	0	1	5	1	2
		V ={-4, -1, 10, 3, -2, 11, 1, 4, -6}								
$\mathcal{W}_{10}$	0, 3	P ={1, 2, 3, 5, 8, 10, 18, 21, 22, 23}	10	50	4	1	1	2	1	2
		V ={1, 5, -3, 8, -7, 3, -2, -5, 9, -1}								
$\mathcal{W}_{11}$	0, -1	P ={1, 2, 3, 4, 6, 9, 11, 13, 21, 30, 33}	11	55	10	0	1	9	0	0
		V ={6, 5, -3, 1, 4, -4, 7, -1, 2, -9, 8}								
$\mathcal{W}_{12}$	0, 6	P ={1, 2, 3, 4, 5, 9, 15, 17, 34, 35, 43, 45 }	12	59	8	1	0	7	2	4
		V ={2, -2, 1, 12, -7, 7, 8, -3, 0, -5, 3, 4}								
$\mathcal{W}_{13}$	0, 0	P ={1, 3, 5, 6, 7, 8, 22, 23, 31, 32, 34, 44, 52}	13	68	9	0	2	7	2	4
		V ={2, 8, -3, -2, 1, 7, 4, -9, 5, 10, -14, -5, 3}								
$\mathcal{W}_{14}$	0, 15	P ={1, 2, 3, 4, 5, 11, 13, 30, 31, 39, 40, 42, 52, 60}	14	76	10	0	2	8	2	4
		V ={-7, -2, 1, 2, 7, 8, -3, 4, -9, 5, 10, -14, -5, 3}								

Dynamic Threshold Public-Key Encryption

Cécile Delerablée[1,2] and David Pointcheval[2]

[1] Orange Labs
cecile.delerablee@orange-ftgroup.com
[2] ENS-CNRS-INRIA
david.pointcheval@ens.fr

Abstract. This paper deals with *threshold public-key encryption* which allows a pool of players to decrypt a ciphertext if a given threshold of authorized players cooperate. We generalize this primitive to the dynamic setting, where any user can *dynamically* join the system, as a possible recipient; the sender can *dynamically* choose the authorized set of recipients, for each ciphertext; and the sender can *dynamically* set the threshold t for decryption capability among the authorized set. We first give a formal security model, which includes strong robustness notions, and then we propose a candidate achieving all the above dynamic properties, that is semantically secure in the standard model, under a new non-interactive assumption, that fits into the general Diffie-Hellman exponent framework on groups with a bilinear map. It furthermore compares favorably with previous proposals, *a.k.a.* threshold broadcast encryption, since this is the first threshold public-key encryption, with dynamic authorized set of recipients and dynamic threshold that provides constant-size ciphertexts.

1 Introduction

In a threshold public-key encryption (in short, TPKE) system [6, 9, 12, 15, 18, 35], the decryption key corresponding to a public key is shared among a set of n users (or servers). In such a system, a ciphertext can be decrypted only if at least t users cooperate. Below this threshold, no information about the plaintext is leaked, which is crucial in all the applications where one cannot fully trust a unique person, but possibly a pool of individuals.

Electronic voting is a classical example for such a threshold encryption primitive: only a pool of bodies is trusted not to cooperate for decrypting individual ballots, but for opening the final result only. The scrutineers cannot be individually trusted, but they globally are, since they control each other. Key-escrow is another application where the distribution of trust is a requirement, or in identity-based cryptography for the secret key extraction [6], as well as any decryption procedure requiring a judge decision.

However, one of the main limitation of standard TPKE is that authorized sets (the public keys) and the threshold t are often fixed during the setup, or at least t is fixed during the key generation phase: the threshold is intrinsic to the public key, and thus cannot be tuned at the encryption time. Additional

D. Wagner (Ed.): CRYPTO 2008, LNCS 5157, pp. 317–334, 2008.
© International Association for Cryptologic Research 2008

flexibility could be useful in many applications in order to avoid the generation of multiple keys for the same purpose, but with different properties only, such as different partners in the authorized set or different thresholds.

Related Work. To this aim, different notions were proposed, like identity-based threshold encryption (decryption) [1, 26], or threshold broadcast encryption (or dynamic threshold encryption) [9, 10, 11, 20].

The scheme proposed in [11] appears to be the closest to what we are dealing with is this paper, and the most efficient. Indeed, this is the first (threshold broadcast encryption) scheme that provides ciphertexts of length smaller than $O(s)$ (actually, this is $O(s-t)$), where s is the size of the authorized set, and t is the threshold, and they are not fixed during the setup, but at the encryption time only. They proposed a scheme in the PKI scenario, and then in the identity-based case. But they let as an open problem to find a scheme with smaller ciphertexts.

Note that such a threshold encryption primitive is close to broadcast encryption [8, 14, 16, 22, 24, 27]: A ciphertext c sent to an authorized set $\mathcal{S}$, under a threshold $t = 1$, actually allows any player in $\mathcal{S}$ to individually decrypt c. However, for a larger threshold, such a primitive does not seem related to broadcast anymore, hence the name of *(dynamic) threshold public-key encryption*. A quite recent primitive, the so-called *attribute-based encryption* [23], is also related to this threshold decryption capability, according to the number of common attributes owned by the recipient. However, the first constructions were not dynamic, since the required attributes were decided at the key-generation phase (key-policy). Ciphertext-policy for attribute-based encryption is more dynamic [3], since it allows to decide about the threshold at the encryption time. However, no join-computation is required/possible for the decryption, contrarily to the usual notion of *threshold cryptography*, where a pool of players are needed to cooperate in order to perform the private computation. For practical reasons, it is preferable when the private computation needs the cooperation of several players, but in a non-interactive way. In the following, we are thus interested in *non-interactive threshold public-key encryption systems*. The latter feature of non-interactivity is also considered as an important one in [19], which deals with dynamic threshold cryptography too, but for signatures only.

Our Contributions. In this paper, to capture previous notions, we propose a generalization of *threshold public-key encryption* (TPKE) to the dynamic setting, where any user can *dynamically* join the system, as a possible recipient; the sender can *dynamically* choose the authorized set of recipients, for each ciphertext; the sender can *dynamically* set the threshold t for decryption capability among the authorized set.

We first formalize this notion, and propose a security model, which deals with all the usual notions of secrecy, but also of robustness, which is important in group-oriented protocols. For our security model we start from [11]. Then, we enhance it with algorithms able to check the validity of all the objects: first a ValidateCT algorithm that (publicly) checks whether a ciphertext is valid with respect to the authorized set and the threshold; and a ShareVerify algorithm that (publicly) checks whether the players honestly computed their partial decryptions. We then present a new scheme, which is fully dynamic, and secure in the standard model. Our scheme is the first one with constant-size ciphertexts, which answers positively to the above problem, for the non-adaptive case. For the security analysis, we introduce a new assumption. Whereas it is a non-interactive assumption and thus easily falsifiable [28], it is non-standard. Since it falls under the Boneh, Boyen and Goh [5] paradigm, in the generic group model, we can have some confidence into the actual intractability, but relying on a more standard assumption remains an interesting open problem. For the robustness, for efficiency reasons, we achieve it in the random oracle model [2].

2   Definitions

This section is dedicated to the definition of the new primitive, and the security model.

2.1   Dynamic Threshold Public-Key Encryption

Our goal is to generalize the notion of *threshold public-key encryption* to the dynamic setting, where

- any user can *dynamically* join the system (the Join algorithm), as a possible recipient,
- the sender can *dynamically* choose the authorized set $\mathcal{S}$ of recipients, for each ciphertext,
- the sender can *dynamically* set the threshold t for decryption capability among the authorized set $\mathcal{S}$.

A (robust) dynamic threshold public-key encryption scheme is a tuple of algorithms $\mathcal{DTPKE} = $ (Setup, Join, Encrypt, ValidateCT, ShareDecrypt, ShareVerify, Combine) described as follows:

Setup(λ). Takes as input a security parameter λ. It outputs a set of parameters Param $= $ (MK, EK, DK, VK, CK), where MK is the master secret key, EK is the encryption key, DK is the decryption key, VK is the verification key, and CK the combining key. The master secret key MK is kept secret by the issuer, whereas the four other keys are public parameters.

Join(MK, ID). Takes as input the master secret key MK and the identity ID of a new user who wants to join the system. It outputs the user's keys (usk, upk, uvk), the private key usk, for decryption; the public key upk, for encryption; and the verification key uvk. The private key usk is privately given to the user, whereas upk and uvk are widely distributed, with an authentic link to ID.

Encrypt(EK, $\mathcal{S}, t, M$). Takes as input the encryption key EK, the authorized set $\mathcal{S}$ (or the public keys upk of users lying in $\mathcal{S}$), a threshold t, and a message M to be encrypted. It outputs a ciphertext.

ValidateCT(EK, $\mathcal{S}$, t, C). Takes as input the encryption key EK, the authorized set $\mathcal{S}$ (or the users' public keys upk), a threshold t, and a ciphertext C. It checks whether C is a valid ciphertext with respect to EK, $\mathcal{S}$ and t.

ShareDecrypt(DK, ID, usk, C). Takes as input the decryption key DK, a user ID and his private key usk, as well as a ciphertext C. It outputs a decryption share σ or $\perp$.

ShareVerify(VK, ID, uvk, C, σ). Takes as input the verification key VK, a user ID and his verification key uvk, as well as a ciphertext C and a decryption share σ. It checks whether σ is a valid decryption share with respect to uvk. This algorihtm is crucial if robustness is required.

Combine(CK, $\mathcal{S}$, t, C, T, Σ). Takes as input the combining key CK, a ciphertext C, a subset $T \subseteq \mathcal{S}$ of t authorized users, and $\Sigma = (\sigma_1, \ldots, \sigma_t)$ a list of t decryption shares. It outputs a cleartext M or $\perp$.

Remark 1. As already explained, for practical efficiency, we focus on non-interactive ShareDecrypt algorithms, and public Combine algorithm.

2.2 Key Encapsulation Method

For content distribution, or any encryption of a large plaintext, the by-now classical technique is the KEM-DEM methodology [34], where an ephemeral secret key is first generated, and used with an appropriate symmetric mechanism to encrypt the data. In such a case, we modify the above algorithms:

Encrypt(EK, $\mathcal{S}$, t). Takes as input the encryption key EK, the authorized set $\mathcal{S}$ (or the users' public keys upk) and a threshold t. It outputs an ephemeral key K, and the key encapsulation material, called the header Hdr. The key K will be later used with the message to be encrypted with the DEM;

The header Hdr is thus the encryption of the ephemeral key, whereas the full header will denote the concatenation of the header and the authorized set $\mathcal{S}$, with the threshold t. The ciphertext will denote the concatenation of all the data: the full header and the DEM part (the data encrypted with the ephemeral key).

ValidateCT(EK, $\mathcal{S}$, t, Hdr). Takes as input the encryption key EK and a full header $(\mathcal{S}, t, \text{Hdr})$. It checks whether Hdr is a valid header with respect to EK, $\mathcal{S}$ and t.

In ShareDecrypt and ShareVerify, the header Hdr only is given, instead of C;

Combine(CK, $\mathcal{S}$, t, Hdr, T, Σ). Takes as input the combining key CK, a full header $(\mathcal{S}, t, \text{Hdr})$, a subset T of t authorized users in $\mathcal{S}$, and $\Sigma = (\sigma_1, \ldots, \sigma_t)$ a list of t decryption shares. It outputs the ephemeral key K or $\perp$. The key K will be later used with the ciphertext to be decrypted with the DEM;

In the following, we thus focus on this KEM-DEM methodology, and thus use the header Hdr only.

2.3   Security Model

Such a scheme must satisfy the following informal properties. They will be for-malized later.

Correctness. For any header Hdr associated to an ephemeral key K during an encryption with a set S of registered users and a threshold t, if t users from this authorized set correctly produced the partial decryptions σ_i,
- the ShareVerify algorithm on any $(\mathsf{VK}, \mathsf{ID}_i, \mathsf{uvk}_i, C, \sigma_i)$ accepts;
- the Combine algorithm on set $\Sigma = \{\sigma_i, i = 1, \ldots, t\}$ outputs K;

Robustness. For any header Hdr associated to an ephemeral key K during an encryption with a set S of registered users and a threshold t, if t users (assumed to be) from this authorized set produce partial decryptions σ_i that are all accepted by the ShareVerify algorithm, then the Combine algorithm outputs K;

Privacy. For any header Hdr encrypted for a set S of registered users with a threshold t, any collusion that contains less than t users from this authorized set cannot learn any information about the ephemeral key.

Following [35] and [6], we can more formally define the above privacy notion, under the classical semantic-security notion [21], under various attacks [29, 31], using a game between an adversary $\mathcal{A}$ and a challenger. Both the adversary and the challenger are given as input a security parameter λ.

Setup: The challenger runs $\mathsf{Setup}(\lambda)$ to obtain the set of parameters $\mathsf{Param} = (\mathsf{MK}, \mathsf{EK}, \mathsf{DK}, \mathsf{VK}, \mathsf{CK})$. The public parameters $(\mathsf{EK}, \mathsf{DK}, \mathsf{VK}, \mathsf{CK})$ are given to the adversary.

Query phase 1:   The adversary $\mathcal{A}$ adaptively issues queries:
- Join query, on input an ID: The challenger runs the Join algorithm on input $(\mathsf{MK}, \mathsf{ID})$, to create a new user in the system.
- Corrupt query, on input an identity ID: The challenger forwards the cor-responding private key to the adversary.
- ShareDecrypt query, on input an ID and a header Hdr: The challenger runs the ShareDecrypt algorithm on Hdr, using the corresponding secret keys, and forwards the resulting partial decryption to the adversary.

Challenge: $\mathcal{A}$ outputs a target set of users $S^\star$ and a threshold $t^\star$. The chal-lenger randomly selects $b \leftarrow \{0, 1\}$ and runs algorithm Encrypt to obtain $(K_0, \mathsf{Hdr}^\star) = \mathsf{Encrypt}(\mathsf{EK}, S^\star, t^\star)$, and randomly chooses an ephemeral key K_1. The challenger then returns $(K_b, \mathsf{Hdr}^\star)$ to $\mathcal{A}$.
There is the natural constraint that $S^\star$ contains at most $t^\star - 1$ corrupted ID's.

Query phase 2:   The adversary continues to adaptively issue Join, Corrupt and ShareDecrypt queries, as in phase 1, but with the constraint that the number of identities ID such that $\mathsf{Corrupt}(\mathsf{ID})$ or $\mathsf{ShareDecrypt}(\mathsf{ID}, \mathsf{Hdr}^\star)$ queries have been asked is less than $t^\star - 1$.

Guess: Finally, the adversary $\mathcal{A}$ outputs a guess $b' \in \{0, 1\}$ and wins the game if $b = b'$.

The advantage of $\mathcal{A}$ is defined as $\mathsf{Adv}_{\mathcal{A}}^{\mathsf{cca}}(\lambda) = \left| \Pr[b' = b] - \frac{1}{2} \right|$. As usual, we denote by $\mathsf{Adv}_{T,n,s,t,q_C,q_D}^{\mathsf{cca}}(\lambda)$ the maximal value of $\mathsf{Adv}_{\mathcal{A}}^{\mathsf{cca}}(\lambda)$, over the adversaries $\mathcal{A}$ that run within time T, and where n, s, t, q_C and q_D are upper-bounds for the numbers of Join-queries, the size of $\mathcal{S}^\star$, $t^\star$, the number of Corrupt and ShareDecrypt queries respectively.

Non-Adaptive Adversary (NAA). We can restrict the adversary to decide before the setup which set $\mathcal{S}^\star$ as well as the threshold $t^\star$ will be sent to the challenger.

Non-Adaptive Corruption (NAC). We can also restrict the adversary to decide before the setup which identities will be corrupted.

Chosen-Plaintext Adversary (CPA). As usual, we can also prevent the adversary from issuing share decryption queries ($q_D = 0$).

Of course, the more adaptive the adversary is in the security analysis, the more secure the scheme is. But as a first step, in the following, we will focus on a basic security level:

Definition 2. (n, s, t, q_C)-**IND-NAA-NAC-CPA** security *(non-adaptive adversary, non-adaptive corruption, chosen-plaintext attacks). At initialization time, the attacker outputs the set $\mathcal{S}^\star$ of size s and a set C of identities that it wants to corrupt, of size q_C. The threshold $t^\star$ is set to t. Then the attacker does not have access to the **ShareDecrypt**-oracle.*

2.4   Extensions

Our threshold public-key encryption definition can be extended in various ways: first, in the ID-based setting, and then with improved access structures.

ID-Based Threshold Encryption. In the ID-based setting, the Join algorithm is replaced by the Extract algorithm that just generates the user's decryption key from the identity, in a similar way as done in [13], and let the public key and the verification key to be this identity.

More General Access Structure. To any identity ID, one can virtually associate several sub-identities $\mathsf{ID}\|1, \ldots, \mathsf{ID}\|k$, and then derive several sets of keys $(\mathsf{usk}_1, \mathsf{upk}_1, \mathsf{uvk}_1), \ldots, (\mathsf{usk}_k, \mathsf{upk}_k, \mathsf{uvk}_k)$. By including several sub-identities of the same user in an authorized set $\mathcal{S}$, one can give different weights for each user in the decryption capability.

This description is quite general and covers all the classical cases, but also quite sophisticated access structures, according to the way the private keys of the sub-identities are distributed:

- If the private keys of a given identity are all given to the same party, by including several sub-ID of the same party in the set $\mathcal{S}$ one gives a bigger weight to this party (and even the possibility for him to decrypt alone, whereas two other parties need to cooperate, etc).

– If the private keys (related to the sub-identities of ID) are distributed to distinct users, a threshold among these users will be needed to decrypt a message sent to ID (implicitly using $\mathcal{S} = (\text{ID}\|1, \ldots, \text{ID}\|w)$). This is the classical t-out-of-n threshold decryption scheme.

3 Computational Assumptions

Our construction will make use of groups with bilinear maps [7, 25], and a new computational assumption, that fits into the general Diffie-Hellman exponent framework proposed by Boneh, Boyen and Goh [5]. This framework does not provide a definite answer about the real intractability, but is a starting point for getting confidence.

3.1 Bilinear Maps

Let $\mathbb{G}_1$, $\mathbb{G}_2$ and $\mathbb{G}_T$ be three cyclic groups of prime order p. A bilinear map $e(\cdot, \cdot)$ is a map $\mathbb{G}_1 \times \mathbb{G}_2 \to \mathbb{G}_T$ such that for any generators $g_1 \in \mathbb{G}_1$, $g_2 \in \mathbb{G}_2$ and $a, b \in \mathbb{Z}_p$,

– $e\left(g_1{}^a, g_2{}^b\right) = e\left(g_1, g_2\right)^{ab}$ (Bilinearity)
– $e\left(g_1, g_2\right) \neq 1$ (Non-degeneracy).

A bilinear map group system is a tuple $\mathcal{B} = (p, \mathbb{G}_1, \mathbb{G}_2, \mathbb{G}_T, e(\cdot, \cdot))$, composed of objects as described above. $\mathcal{B}$ may also include a group generator. We impose all group operations as well as the bilinear map $e(\cdot, \cdot)$ to be efficiently computable, i.e. in time $\text{poly}(|p|)$.

Note that our construction just makes use of an arbitrary bilinear map group system, without any particular additional property. In particular, we do not need $\mathbb{G}_1$ and $\mathbb{G}_2$ to be distinct or equal. Neither do we require the existence of an efficient isomorphism going either way between $\mathbb{G}_1$ and $\mathbb{G}_2$, as it is the case for some pairing-based systems.

3.2 The Multi-sequence of Exponents Diffie-Hellman Assumption

As in [14], our security proof uses the general Diffie-Hellman exponent theorem due to Boneh, Boyen and Goh [5]. They indeed introduced a class of assumptions which includes a lot of (by-now familiar) assumptions, that appeared in the past with new pairing-based schemes. It includes for example DDH (in $\mathbb{G}_T$), BDH, q–BDHI, and q–BDHE assumptions. Even if group systems equipped with bilinear maps are far from being generic, an intractability result in this framework is a first step for getting some confidence in the actual intractability. In our case, we assume the intractability of the following decisional problem (ℓ, m, t)-MSE–DDH:

Let $\mathcal{B} = (p, \mathbb{G}_1, \mathbb{G}_2, \mathbb{G}_T, e(\cdot, \cdot))$ be a bilinear map group system and let ℓ, m and t be three integers. Let g_0 be a generator of $\mathbb{G}_1$ and h_0 a generator of $\mathbb{G}_2$. Given two random coprime polynomials f and g, of respective orders ℓ and m, with pairwise distinct roots $x_1, \ldots, x_\ell$ and $y_1, \ldots, y_m$ respectively, as well as several sequences of exponentiations

$$
\begin{array}{ll}
x_1, \ldots, x_\ell, & y_1, \ldots, y_m \\
g_0, g_0{}^\gamma, \ldots, g_0{}^{\gamma^{\ell+t-2}}, & g_0{}^{k \cdot \gamma \cdot f(\gamma)}, \\
g_0{}^\alpha, g_0{}^{\alpha \cdot \gamma}, \ldots, g_0{}^{\alpha \cdot \gamma^{\ell+t}}, & \\
h_0, h_0{}^\gamma, \ldots, h_0{}^{\gamma^{m-2}}, & \\
h_0{}^\alpha, h_0{}^{\alpha \cdot \gamma}, \ldots, h_0{}^{\alpha \cdot \gamma^{2m-1}}, & h_0{}^{k \cdot g(\gamma)},
\end{array}
$$

and $T \in \mathbb{G}_T$, decide whether T is equal to $e\left(g_0, h_0\right)^{k \cdot f(\gamma)}$ or to some random element of $\mathbb{G}_T$.

The following statement is a corollary of Theorem 7 [5] which can be found in section 6. It provides an intractability bound in the generic model [33], but in groups equipped with pairings. We emphasize on the fact that, whereas the assumption has several parameters, it is non-interactive, and thus easily falsifiable [28].

Corollary 3 (Generic Security). *For any probabilistic algorithm $\mathcal{A}$ that totalizes of at most q_G queries to the oracles performing the group operations in $\mathbb{G}_1, \mathbb{G}_2, \mathbb{G}_T$ and the bilinear map $e\left(\cdot, \cdot\right)$,*

$$
\mathsf{Adv}^{\mathsf{mse\text{-}ddh}}(\ell, m, t, \mathcal{A}) \leq \frac{(q_G + 4(\ell + t) + 6m + 4)^2 \cdot d}{2p}
$$

with $d = 4(\ell + t) + 6m + 2$.

4   Our Construction

4.1   Description

In this section we present our new dynamic threshold public-key encryption ($\mathcal{DTPKE}$), with constant size ciphertexts. Basically, the encryption algorithm specifies the authorized-user set with an inclusion technique as in the broadcast encryption schemes [8, 13]. Moreover this authorized set is combined with a set of dummy users, in order to be consistent with the value of the threshold (this is a well-known technique in threshold encryption). We make use of the Aggregate algorithm (over $\mathbb{G}_T$) described in [14] to combine the decryption shares. The Aggregate algorithm simply exploits the fact that a product of inverses of coprime polynomials can be written as a sum of inverses of affine polynomials. Thus given some elements in $\mathbb{G}_T$ of the right form, one can combine the exponents using some group operations. We provide below a description of the case which interests us and refer to [14] for more details.

Setup(λ). Given the security parameter λ, a system with groups and a bilinear map $\mathcal{B} = (p, \mathbb{G}_1, \mathbb{G}_2, \mathbb{G}_T, e\left(\cdot, \cdot\right))$ is constructed such that $|p| = \lambda$. Also, two generators $g \in \mathbb{G}_1$ and $h \in \mathbb{G}_2$ are randomly selected as well as two secret values γ and $\alpha \in \mathbb{Z}_p^\star$. Finally, a set $\mathcal{D} = \{d_i\}_{i=1}^{m-1}$ of values in $\mathbb{Z}_p$ is randomly selected, where m is the maximal size of an authorized set. This corresponds to a set of dummy users, that will be used to complete a set of authorized users.

$\mathcal{B}$ constitutes the system parameters. The master secret key is defined as $\mathsf{MK} = (g, \gamma, \alpha)$. The encryption key is $\mathsf{EK} = \left(m, u, v, h^\alpha, \{h^{\alpha \cdot \gamma^i}\}_{i=1}^{2m-1}, \mathcal{D} \right)$, and the combining key is $\mathsf{CK} = \left(m, h, \{h^{\gamma^i}\}_{i=1}^{m-2}, \mathcal{D} \right)$, where $u = g^{\alpha \cdot \gamma}$, and $v = e(g, h)^\alpha$. In the following, we denote by $\mathcal{D}_i$ the i first elements of $\mathcal{D}$. Note that $\mathsf{DK} = \emptyset$, since no general data are needed for partial decryption. Furthermore, this version of the scheme does not provide robustness, we thus do not define VK yet. Robustness will be studied later.

$\mathsf{Join}(\mathsf{MK}, \mathsf{ID})$. Given $\mathsf{MK} = (g, \gamma, \alpha)$, and an identity ID, it randomly chooses $x \in \mathbb{Z}_p^\star$ (different from all previous ones, included dummy users data in $\mathcal{D}$), and outputs the user's keys $(\mathsf{usk}, \mathsf{upk})$ with:

$$\mathsf{upk} = x , \quad \mathsf{usk} = g^{\frac{1}{\gamma + x}}.$$

The private key usk is privately given to the user, whereas upk is widely published, in an authentic way (again, since robustness is not dealt with here, we do not set uvk yet).

$\mathsf{Encrypt}(\mathsf{EK}, \mathcal{S}, t)$. Given the encryption key EK, a set $\mathcal{S}$ of users, which is identified to $\mathcal{S} = \{\mathsf{upk}_1 = x_1, \ldots, \mathsf{upk}_s = x_s\}$ and a threshold t (with $t \le s = |\mathcal{S}| \le m$), $\mathsf{Encrypt}$ randomly picks $k \in \mathbb{Z}_p^\star$, and computes $\mathsf{Hdr} = (C_1, C_2)$ and K, where

$$C_1 = u^{-k} , \qquad C_2 = h^{k \cdot \alpha \cdot \prod_{x_i \in \mathcal{S}} (\gamma + x_i) \cdot \prod_{x \in \mathcal{D}_{m+t-s-1}} (\gamma + x)} , \qquad K = v^k .$$

$\mathsf{Encrypt}$ then outputs the full header $(\mathcal{S}, t, \mathsf{Hdr} = (C_1, C_2))$ and the secret key K, which will be used to encrypt the message. The crucial point is that $\mathsf{Encrypt}$ includes a set of $m + t - s - 1$ dummy users, in order to obtain a polynomial of degree exactly $m + t - 1$ in the exponent of h. This way, exploiting the cooperation of t authorized users together with a combining key that contains $\left(h, \{h^{\gamma^i}\}_{i=1}^{m-2} \right)$ is sufficient to decrypt a ciphertext (see the $\mathsf{Combine}$ algorithm).

$\mathsf{ValidateCT}(\mathsf{EK}, \mathcal{S}, t, \mathsf{Hdr})$. Given the encryption key EK and a full header $(\mathcal{S}, t)$ and $\mathsf{Hdr} = (C_1, C_2))$, as above, one can compute

$$C_1' = u^{-1} , \qquad C_2' = h^{\alpha \cdot \prod_{x \in \mathcal{S} \cup \mathcal{D}_{m+t-s-1}} (\gamma + x)}.$$

One should notice that a header $\mathsf{Hdr} = (C_1, C_2)$ is valid with respect to $\mathcal{S}$ if and only if there exists a scalar k such that $C_1 = {C_1'}^k$ and $C_2 = {C_2'}^k$. Moreover, one can note that in such a header, a correct $\mathcal{S}$ contains at least t keys of some users. As a consequence, $\mathsf{ValidateCT}$ simply checks whether $e(C_1, C_2') = e(C_1', C_2)$ and $\mathcal{S}$ is correct, or not.

ShareDecrypt(ID, usk, Hdr). In order to retrieve a share σ of a decryption key encapsulated in the header $\mathsf{Hdr} = (C_1, C_2)$, user with identity ID and the corresponding public key upk and private key $\mathsf{usk} = g^{\frac{1}{\gamma+x}}$ computes

$$\sigma = e\,(\mathsf{usk}, C_2) = e\,(g, h)^{\frac{k\cdot\alpha\cdot\prod_{x_i\in\mathcal{S}\cup\mathcal{D}_{m+t-s-1}}(\gamma+x_i)}{\gamma+x}}.$$

Combine(CK, C, T, Σ). Given $\mathcal{S}$, t, $\mathsf{Hdr} = (C_1, C_2)$, CK, a subset T of t users ($T \subseteq \mathcal{S}$) and Σ the corresponding decryption shares, outputs

$$K = \left(e\left(C_1, h^{p_{(T,\mathcal{S})}(\gamma)}\right) \cdot \mathsf{Aggregate}(\mathbb{G}_T, \Sigma)\right)^{\frac{1}{c_{(T,\mathcal{S})}}},$$

with $c_{(T,\mathcal{S})}$ a constant in $\mathbb{Z}_p$ and $p_{(T,\mathcal{S})}$ a polynomial of degree $m - 2$, that both allow to cancel a part corresponding to the $m - 1$ decryption shares (over $m+t-1$) that are not in the input. Note that since $p_{(T,\mathcal{S})}$ is of degree $m - 2$, $h^{p_{(T,\mathcal{S})}(\gamma)}$ is computable from CK. More precisely, we have:

$$p_{(T,\mathcal{S})}(\gamma) = \frac{1}{\gamma} \cdot \left(\prod_{x\in\mathcal{S}\cup\mathcal{D}_{m+t-s-1}-T}(\gamma + x) - c_{(T,\mathcal{S})}\right),$$

$$c_{(T,\mathcal{S})} = \prod_{x\in\mathcal{S}\cup\mathcal{D}_{m+t-s-1}-T} x,$$

$$\mathsf{Aggregate}(\mathbb{G}_T, \Sigma) = \mathsf{Aggregate}\left(\mathbb{G}_T, \left\{e\,(g, C_2)^{\frac{1}{\gamma+x}}\right\}_{x\in T}\right)$$

$$= e\,(g, C_2)^{\frac{1}{\prod_{x\in T}(\gamma+x)}}$$

$$= e\,(g, h)^{k\cdot\alpha\cdot\prod_{x_i\in\mathcal{S}\cup\mathcal{D}_{m+t-s-1}-T}(\gamma+x_i)}$$

Correctness. Assuming C is well-formed, and Σ is correct:

$$K' = e\left(C_1, h^{p_{(T,\mathcal{S})}(\gamma)}\right) \cdot \mathsf{Aggregate}(\mathbb{G}_T, \Sigma)$$

$$= e\left(g^{-k\cdot\alpha\cdot\gamma}, h^{p_{(T,\mathcal{S})}(\gamma)}\right) \cdot e\,(g, C_2)^{\frac{1}{\prod_{x\in T}(\gamma+x)}}$$

$$= e\,(g, h)^{-k\cdot\alpha\cdot\gamma\cdot_{(T,\mathcal{S})}(\gamma)} \cdot e\,(g, h)^{k\cdot\alpha\cdot\prod_{x\in\mathcal{S}\cup\mathcal{D}_{m+t-s-1}-T}(\gamma+x)}$$

$$= e\,(g, h)^{k\cdot\alpha\cdot c_{(T,\mathcal{S})}} = K^{c_{(T,\mathcal{S})}}.$$

Thus $K'^{\frac{1}{c_{(T,\mathcal{S})}}} = K$.

Efficiency. In our construction, ciphertexts remain constant (plus the authorized set $\mathcal{S}$ that contains the x_i's of the authorized users only, which is unavoidable and thus optimal). Moreover, our Encrypt algorithm is very efficient, since it does not need any pairing computation, whereas in [11], $3(s-t)$ pairing computations are needed, with s the size of the authorized set. Furthermore, any additional encryption for the same target set only require 3 exponentiations.

4.2   Aggregation of 1-Degree Terms: **Aggregate**

The Combine algorithm requires the computation of

$$L = e\left(g, C_2\right)^{\frac{1}{(\gamma + x_1) \ldots (\gamma + x_t)}} \in \mathbb{G}_T$$

given $\Sigma = \{\sigma_j = e\left(g, C_2\right)^{\frac{1}{\gamma + x_j}}\}_{j=1}^t$ where the x_j's are pairwise distinct. We recall how $\mathsf{Aggregate}(\mathbb{G}_T, \cdots)$ allows to compute L from the x_j's and the σ_j's, as described in [14].

Description. Given $x_1, \ldots, x_t$ and σ_j for $1 \leq j \leq t$, let us define for any (j, ℓ) such that $1 \leq j < \ell \leq r$,

$$L_{j,\ell} = \sigma_\ell^{\frac{1}{\prod_{\kappa=1}^{j}(\gamma + x_\kappa)}} = e\left(g, C_2\right)^{\frac{1}{(\gamma + x_\ell)} \cdot \frac{1}{\prod_{\kappa=1}^{j}(\gamma + x_\kappa)}} .$$

The Aggregate algorithm consists in computing sequentially $L_{j,\ell}$ for $j = 1, \ldots, t-1$ and $\ell = j + 1, \ldots, t$ using the induction

$$L_{j,\ell} = \left(\frac{L_{j-1,j}}{L_{j-1,\ell}}\right)^{\frac{1}{x_\ell - x_j}}$$

and posing $L_{0,\ell} = \sigma_\ell$ for $\ell = 1, \ldots, t$. The algorithm finally outputs $L_t = L_{t-1,t}$.

4.3   Security Analysis

This section is devoted to the proof of the IND-NAA-NAC-CPA security level for our system, under our new MSE–DDH assumption.

Security Result. Let $\mathcal{DTPKE}$ denote our construction, described above, Section 4.1. We can state the following security result.

Theorem 4. *For any ℓ, m, t, $Adv^{\mathsf{ind}}_{\mathcal{DTPKE}}(\ell, m, t) \leq 2 \cdot Adv^{\mathsf{mse-ddh}}(\ell, m, t)$.*

The rest of this section is dedicated to proving Theorem 4. To establish the semantic security of $\mathcal{DTPKE}$ against static adversaries, we assume an adversary $\mathcal{A}$ that breaks the scheme under an (ℓ, m, t)-collusion and we build an algorithm $\mathcal{R}$ that distinguishes the two distributions of the (ℓ, m, t)-MSE–DDH problem.

Both the adversary and the challenger are given as input m, the maximal size of a set of authorized users $\mathcal{S}$, ℓ the total number of Join queries that can be issued by the adversary, and a threshold t.

Algorithm $\mathcal{R}$ is given as input a group system $\mathcal{B} = (p, \mathbb{G}_1, \mathbb{G}_2, \mathbb{G}_T, e\left(\cdot, \cdot\right))$, and an (ℓ, m, t)-MSE–DDH instance in $\mathcal{B}$ (as described in Section 3.2). We thus have two coprime polynomials f and g, of respective orders ℓ and m, with their pairwise distinct roots $(x_1, \ldots, x_\ell)$ and $(x_{\ell+t}, \ldots, x_{\ell+t+m-1})$, and $\mathcal{R}$ is furthermore given

$$g_0, g_0^{\gamma}, \ldots, g_0^{\gamma^{\ell+t-2}}, \qquad\qquad g_0^{k \cdot \gamma \cdot f(\gamma)},$$
$$g_0^{\alpha}, g_0^{\alpha \cdot \gamma}, \ldots, g_0^{\alpha \cdot \gamma^{\ell+t}},$$
$$h_0, h_0^{\gamma}, \ldots, h_0^{\gamma^{m-2}},$$
$$h_0^{\alpha}, h_0^{\alpha \cdot \gamma}, \ldots, h_0^{\alpha \cdot \gamma^{2m-1}}, \qquad\qquad h_0^{k \cdot g(\gamma)},$$

as well as $T \in \mathbb{G}_T$ which is either equal to $e\left(g_0, h_0\right)^{k \cdot f(\gamma)}$ or to some random element of $\mathbb{G}_T$. For the sake of simplicity, we state that f and g are unitary polynomials, but this is not a mandatory requirement:

$$f(X) = \prod_{i=1}^{\ell}(X + x_i), \quad q(X) = \prod_{i=\ell+1}^{\ell+t-1}(X + x_i), \quad g(X) = \prod_{i=\ell+t}^{\ell+t+m-1}(X + x_i).$$

The polynomial f corresponds to a set of ℓ users not in the target set, that can be corrupted. The polynomial q corresponds to a set of $t - 1$ users of the target set that can be corrupted. The polynomial g corresponds to the m users of the target set that cannot be corrupted. We will thus be able to simulate $\ell + t - 1$ decryption keys (corruptions), with $t - 1$ of them, only, in the target set.

For $i \in [1, \ell + t - 1]$, we set

$$f_i(x) = \frac{f(x) \cdot q(\gamma)}{x + x_i},$$

which is a polynomial of degree $\ell + t - 2$.

Init: The adversary $\mathcal{A}$ outputs a set $\mathcal{S}^\star = \{ \mathsf{ID}_1^\star, \ldots, \mathsf{ID}_{s^\star}^\star \}$ of identities that he wants to attack (the target authorized set), and a set $\overline{\mathcal{C}} = \{ \overline{\mathsf{ID}}_1, \ldots, \overline{\mathsf{ID}}_c \}$ of identities that he wants to corrupt, with $c \leq \ell$ and $|\mathcal{S}^\star \cap \overline{\mathcal{C}}| \leq t - 1$;

Setup: To generate the system parameters, $\mathcal{R}$ formally sets $g = g_0^{f(\gamma) \cdot q(\gamma)}$ (but without computing it, since it does not need to publish it) and sets

$$h = h_0, \qquad u = g_0^{\alpha \cdot \gamma \cdot f(\gamma) \cdot q(\gamma)} = g^{\alpha \cdot \gamma},$$

$$v = e\left(g_0, h_0\right)^{\alpha \cdot f(\gamma) \cdot q(\gamma)} = e(g, h)^\alpha.$$

The two latter formulae can be computed from the instance input, since $f \cdot q$ is of degree $\ell + t - 1$;

$\mathcal{R}$ then sets the set $\mathcal{D} = \{ d_i \}_{i=1}^{m-1}$ corresponding to dummy users:

- $\mathcal{D}_{m+t-s^\star-1} = \{ d_i \}_{i=1}^{m+t-s^\star-1}$ is a subset of $\{ x_j \}_{j=\ell+t}^{\ell+t+m-1}$. This subset corresponds to the dummy users included to complete the target set in the challenge.
- $\{ d_i \}_{i=m+t-s^\star}^{m-1}$ is a set of random elements in $\mathbb{Z}_p$

Finally, $\mathcal{R}$ defines the encryption key as $\mathsf{EK} = \left(m, u, v, h^\alpha, \{ h^{\alpha \cdot \gamma^i} \}_{i=1}^{2m-1}, \mathcal{D} \right)$, and the combining key as $\mathsf{CK} = \left(h, \{ h^{\gamma^i} \}_{i=1}^{m-2}, \mathcal{D} \right)$. Note that $\mathcal{R}$ can by no means compute the value of g. But we do not need it.

Generation of users' keys:

- For each $\overline{\mathsf{ID}} \in \overline{\mathcal{C}}$, $\mathcal{R}$ computes and sends $(\overline{\mathsf{usk}}, \overline{\mathsf{upk}})$ to $\mathcal{A}$ with

$$\overline{\mathsf{upk}} = x_i,$$

$$\overline{\mathsf{usk}} = g_0^{f_i(\gamma)} = g^{\frac{1}{\gamma + x_i}},$$

with the following constraint: if $\mathsf{ID} \in \mathcal{S}^\star$, then the corresponding x_i must be taken in $\{ x_j \}_{j=\ell+1}^{\ell+t-1}$. Otherwise x_i must be taken in $\{ x_j \}_{j=1}^{\ell}$.

- For each $\mathsf{ID} \in \mathcal{S}^\star - \mathcal{S}^\star \cap \overline{\mathcal{C}}$, $\mathcal{R}$ sends $\mathsf{upk} = x_i$ to $\mathcal{A}$, with the following constraint: x_i must be taken in $\{x_j\}_{j=\ell+t}^{\ell+t+m-1} - \mathcal{D}_{m+t-s^\star-1}$.
- For each $\mathsf{ID} \notin \mathcal{S}^\star \cup \overline{\mathcal{C}}$, $\mathcal{R}$ sends $\mathsf{upk} = x$ to $\mathcal{A}$, with $x \notin \{x_j\}_{j=1}^{\ell+t+m-1}$.

$\mathcal{R}$ runs $\mathcal{A}$ on the system parameters $\mathcal{B}$ and $(\mathsf{EK}, \mathsf{CK})$, and on the target set $\mathcal{S}^\star$.

Challenge: Algorithm $\mathcal{R}$ computes $\mathsf{Encrypt}$ to obtain

$$(\mathsf{Hdr}^\star, \mathcal{S}^\star, t, K) = \mathsf{Encrypt}(\mathsf{Param}, \mathsf{EK}, \mathcal{S}^\star, t), \text{ with}$$

$$C_1 = g_0^{-k \cdot \gamma \cdot f(\gamma)}, \qquad C_2 = h_0^{k \cdot g(\gamma)}, \qquad K = T,$$

$$|\mathcal{S}| = s^\star, \qquad \mathcal{S}^\star \subseteq \{x_i\}_{i=\ell+1}^{\ell+m+t-1} - \mathcal{D}_{m+t-s^\star-1}.$$

One can verify that, if we set $k' := \frac{k}{\alpha \cdot q(\gamma)}$, then

$$C_1 = u^{-k'}, \qquad C_2 = h^{k' \cdot \alpha \cdot \prod_{x_i \in \mathcal{S}^\star}(\gamma + x_i) \cdot \prod_{x \in \mathcal{D}_{m+t-s^\star-1}}(\gamma + x)}.$$

Note that if $T = e(g_0, h_0)^{k \cdot f(\gamma)}$, then $K = v^{k'}$.

The challenger then randomly selects $b \leftarrow \{0,1\}$, sets $K_b = K$, and sets K_{1-b} to a random value in $\mathcal{K}$. The challenger returns $(\mathsf{Hdr}^\star, K_0, K_1)$ to $\mathcal{A}$.

Guess: Finally, the adversary $\mathcal{A}$ outputs a guess $b' \in \{0,1\}$ and wins the game if $b = b'$.

One has

$$\mathsf{Adv}^{\mathsf{mse\text{-}ddh}}(\mathcal{R}) = \Pr[b' = b|\mathsf{real}] - \Pr[b' = b|\mathsf{random}]$$

$$= \frac{1}{2} \times (\Pr[b' = 1|b = 1 \wedge \mathsf{real}] - \Pr[b' = 1|b = 0 \wedge \mathsf{real}])$$

$$- \frac{1}{2} \times (\Pr[b' = 1|b = 1 \wedge \mathsf{random}] + \Pr[b' = 1|b = 0 \wedge \mathsf{random}]).$$

Now in the random case, the distribution of b is independent from the adversary's view wherefrom

$$\Pr[b' = 1|b = 1 \wedge \mathsf{random}] = \Pr[b' = 1|b = 0 \wedge \mathsf{random}].$$

In the real case however, the distributions of all variables defined by $\mathcal{R}$ perfectly comply with the semantic security game since all simulations are perfect. Therefore

$$\mathsf{Adv}^{\mathsf{ind}}_{\mathcal{DTPKE}}(\mathcal{A}) = \Pr[b' = 1|b = 1 \wedge \mathsf{real}] - \Pr[b' = 1|b = 0 \wedge \mathsf{real}].$$

Putting it altogether, we get the conclusion.

5   Extensions in the Random Oracle Model

We insist on the fact that the previous construction is in the standard model, without any additional non-standard setup assumption. However, some improvements can be achieved in the random oracle model [2].

5.1   Robustness

First, note that in our security model, we defined the robustness, as a very interesting feature (ShareVerify). Such a verification seems hard to do in the standard model, in our previous scheme. It was not available in [11] either. However, in the random oracle model, we can use proofs of equality of discrete logarithms for providing it, at almost no additional cost in our scheme above:

- when decrypting (C_1, C_2), using usk, one can generate $\text{usk}' = \text{usk}^\delta$, for a random δ, together with $\sigma = e(\text{usk}, C_2)$.
- the validity can be checked by the existence of a common value δ such that

$$e\left(\text{usk}', (h^{\alpha\gamma}) \times (h^\alpha)^{\text{upk}}\right) = v^\delta \quad e\left(\text{usk}', C_2\right) = \sigma^\delta.$$

The latter can be a usual Schnorr-like proof π of equality of discrete logarithms [32] (the existence of a common exponent δ), and its non-interactive version using the Fiat-Shamir paradigm [17, 30].

The verification key is thus simply $\text{uvk} = \text{upk} = x$, the partial decryption consists of the triple $(\sigma, \text{usk}', \pi)$, and the ShareVerify algorithm checks the validity of π.

5.2   Identity-Based

It is also simple to get an ID-based version in the random oracle model, as in [11] and [13], by simply taking $\text{upk} = x = \mathcal{H}(\text{ID})$ as in [4].

6   Intractability of (ℓ, m, t)-MSE–DDH

6.1   Notations

For the sake of simplicity, we focus to the symmetric case ($\mathbb{G}_1 = \mathbb{G}_2 = \mathbb{G}$). Let then $\mathcal{B} = (p, \mathbb{G}, \mathbb{G}, \mathbb{G}_T, e(\cdot, \cdot))$ be a bilinear map group system. Let $g_0 \in \mathbb{G}$ be a generator of $\mathbb{G}$, and set $g = e(g_0, g_0) \in \mathbb{G}_T$. Let s, n be positive integers and $P, Q \in \mathbb{F}_p[X_1, \ldots, X_n]^s$ be two s-tuples of n-variate polynomials over $\mathbb{F}_p$. Thus, P and Q are just two lists containing s multivariate polynomials each: we write $P = (p_1, p_2, \ldots, p_s)$ and $Q = (q_1, q_2, \ldots, q_s)$ and impose that $p_1 = q_1 = 1$. For any function $h : \mathbb{F}_p \to \Omega$ and vector $(x_1, \ldots, x_n) \in \mathbb{F}_p^n$, the notation $h(P(x_1, \ldots, x_n))$ stands for

$$(h(p_1(x_1, \ldots, x_n)), \ldots, h(p_s(x_1, \ldots, x_n))) \in \Omega^s.$$

We use a similar notation for the s-tuple Q. Let $f \in \mathbb{F}_p[X_1, \ldots, X_n]$. It is said that f depends on (P, Q), which we denote by $f \in \langle P, Q \rangle$, when there exists a linear decomposition

$$f = \sum_{1 \leq i, j \leq s} a_{i,j} \cdot p_i \cdot p_j + \sum_{1 \leq i \leq s} b_i \cdot q_i, \qquad a_{i,j}, b_i \in \mathbb{Z}_p.$$

Let P, Q be as above and $f \in \mathbb{F}_p[X_1, \ldots, X_n]$. The (P, Q, f)-General Diffie-Hellman Exponent problems are defined as follows.

Definition 5 ((P, Q, f)-**GDHE**). *Given the tuple*

$$H(x_1, \ldots, x_n) = \left(g_0{}^{P(x_1,\ldots,x_n)}, g^{Q(x_1,\ldots,x_n)}\right) \in \mathbb{G}^s \times \mathbb{G}_T^s ,$$

compute $g^{f(x_1,\ldots,x_n)}$.

Definition 6 ((P, Q, f)-**GDDHE**). *Given* $H(x_1, \ldots, x_n) \in \mathbb{G}^s \times \mathbb{G}_T^s$ *as above and* $T \in \mathbb{G}_T$, *decide whether* $T = g^{f(x_1,\ldots,x_n)}$.

We refer to [5] for a proof that (P, Q, f)-GDHE and (P, Q, f)-GDDHE have generic security when $f \notin \langle P, Q \rangle$. We will prove that our construction is secure by first exhibiting the polynomials P, Q and f involved in the security proofs, and then by showing that $f \notin \langle P, Q \rangle$.

6.2   (ℓ, m, t)-MSE–DDH

In this section, we prove the intractability of distinguishing the two distributions involved in the (ℓ, m, t)-MSE–DDH problem (cf. Corollary 3, section 4.3). We first review some results on the General Diffie-Hellman Exponent Problem, from [5]. In order to be the most general, we assume the easiest case for the adversary: when $\mathbb{G}_1 = \mathbb{G}_2$, or at least that an isomorphism that can be easily computed in either one or both ways is available.

Theorem 7 ([5]). *Let* $P, Q \in \mathbb{F}_p[X_1, \ldots, X_n]$ *be two s-tuples of n-variate polynomials over* $\mathbb{F}_p$ *and let* $F \in \mathbb{F}_p[X_1, \ldots, X_n]$. *Let* d_P *(resp.* d_Q, d_F*) denote the maximal degree of elements of* P *(resp. of* Q, F*) and pose* $d = \max(2d_P, d_Q, d_F)$. *If* $F \notin \langle P, Q \rangle$ *then for any generic-model adversary* $\mathcal{A}$ *totalizing at most* q_G *queries to the oracles (group operations in* $\mathbb{G}, \mathbb{G}_T$ *and evaluations of e) which is given* $H(x_1, \ldots, x_n)$ *as input and tries to distinguish* $g^{F(x_1,\ldots,x_n)}$ *from a random value in* $\mathbb{G}_T$, *one has*

$$\mathsf{Adv}(\mathcal{A}) \leq \frac{(q_G + 2s + 2)^2 \cdot d}{2p} .$$

Proof (of Corollary 3). In order to conclude with Corollary 3, we need to prove that our problem lies in the scope of Theorem 7. As already said, we consider the weakest case $\mathbb{G}_1 = \mathbb{G}_2 = \mathbb{G}$ and thus pose $h_0 = g_0{}^\beta$. Our problem can be reformulated as (P, Q, F)-GDHE where

$$P = \begin{pmatrix} 1, \gamma, \gamma^2, \ldots, \gamma^{\ell+t-2}, & k \cdot \gamma \cdot f(\gamma) \\ \alpha, \alpha \cdot \gamma, \alpha \cdot \gamma^2, \ldots, \alpha \cdot \gamma^{\ell+t}, & \\ \beta, \beta \cdot \gamma, \beta \cdot \gamma^2, \ldots, \beta \cdot \gamma^{m-2} & \\ \alpha \cdot \beta, \alpha \cdot \beta \cdot \gamma, \alpha \cdot \beta \cdot \gamma^2, \ldots, \alpha \cdot \beta \cdot \gamma^{2m-1}, & k \cdot \alpha \cdot \beta \cdot g(\gamma) \cdot q(\gamma) \end{pmatrix}$$

$$Q = 1$$

$$F = k \cdot \beta \cdot f(\gamma),$$

and thus $n = 4$ and $s = 2(\ell + t) + 3m + 1$. We have to show that F is independent of (P, Q), i.e. that no coefficients $\{a_{i,j}\}_{i,j=1}^{s}$ and b_1 exist such that

$F = \sum_{i,j=1}^{s} a_{i,j} p_i p_j + b_1$ where the polynomials p_i are the one listed in P above. By making all possible products of two polynomials from P which are multiples of $k \cdot \beta$, we want to prove that no linear combination among the polynomials from the list R below leads to F:

$$R = \begin{pmatrix} k \cdot \beta \cdot g(\gamma), \ k \cdot \beta \cdot \gamma \cdot g(\gamma), \ldots, \ k \cdot \beta \cdot \gamma^{\ell+t-2} \cdot g(\gamma), \\ k \cdot \beta \cdot \gamma \cdot f(\gamma), \ k \cdot \beta \cdot \gamma^2 \cdot f(\gamma), \ldots, \ k \cdot \beta \cdot \gamma^{m-1} \cdot f(\gamma) \end{pmatrix} .$$

We simplify the task to refuting a linear combination of elements of the list R' below which leads to $f(\gamma)$:

$$R' = \begin{pmatrix} g(\gamma), \ \gamma \cdot g(\gamma), \ldots, \ \gamma^{\ell+t-2} \cdot g(\gamma), \\ \gamma \cdot f(\gamma), \ \gamma^2 \cdot f(\gamma), \ldots, \ \gamma^{m-1} \cdot f(\gamma) \end{pmatrix} .$$

Any such linear combination can be written as

$$f(\gamma) = A(\gamma) \cdot f(\gamma) + B(\gamma) \cdot g(\gamma)$$
$$\Leftrightarrow f(\gamma) \cdot (1 - A(\gamma)) = B(\gamma) \cdot g(\gamma)$$

where A and B are polynomials such that $A(0) = 0$, $\deg A \leq m - 1$ and $\deg B \leq \ell + t - 2$. Since f and g are coprime by assumption, we must have $g \mid 1 - A$. Since $\deg g = m$ and $\deg A \leq m - 1$ this implies $A = 1$, which contradicts $A(0) = 0$. $\square$

7 Conclusion

We presented a generalization of threshold public-key encryption to the dynamic setting. We first proposed a security model and then a new scheme, which is non-interactive, fully dynamic, and which is the first one to achieve constant-size ciphertexts. However, our scheme can be viewed as a first step toward the problem, since it still presents a few restrictions: our security proof relies on a new and non-standard assumption, and does not prevent adaptive adversaries, nor chosen-ciphertext attacks. However, it applies in the standard model.

Acknowledgements

We would like to thank the anonymous referees for their fruitful comments.

The second author was supported in part by the European Commission through the IST Program under Contract IST-2002-507932 ECRYPT, and by the CELAR.

References

1. Baek, J., Zheng, Y.: Identity-based threshold decryption. In: Bao, F., Deng, R., Zhou, J. (eds.) PKC 2004. LNCS, vol. 2947, pp. 262–276. Springer, Heidelberg (2004)
2. Bellare, M., Rogaway, P.: Random oracles are practical: A paradigm for designing efficient protocols. In: ACM CCS 1993, pp. 62–73. ACM Press, New York (1993)

3. Bethencourt, J., Sahai, A., Waters, B.: Ciphertext-policy attribute-based encryption. In: IEEE Symposium on Security and Privacy, pp. 321–334. IEEE Computer Society, Los Alamitos (2007)

4. Boneh, D., Boyen, X.: Efficient selective-ID secure identity based encryption without random oracles. In: Cachin, C., Camenisch, J.L. (eds.) EUROCRYPT 2004. LNCS, vol. 3027, pp. 223–238. Springer, Heidelberg (2004)

5. Boneh, D., Boyen, X., Goh, E.-J.: Hierarchical identity based encryption with constant size ciphertext. In: Cramer, R. (ed.) EUROCRYPT 2005. LNCS, vol. 3494, pp. 440–456. Springer, Heidelberg (2005)

6. Boneh, D., Boyen, X., Halevi, S.: Chosen ciphertext secure public key threshold encryption without random oracles. In: Pointcheval, D. (ed.) CT-RSA 2006. LNCS, vol. 3860, pp. 226–243. Springer, Heidelberg (2006)

7. Boneh, D., Franklin, M.K.: Identity-based encryption from the Weil pairing. In: Kilian, J. (ed.) CRYPTO 2001. LNCS, vol. 2139, pp. 213–229. Springer, Heidelberg (2001)

8. Boneh, D., Gentry, C., Waters, B.: Collusion resistant broadcast encryption with short ciphertexts and private keys. In: Shoup, V. (ed.) CRYPTO 2005. LNCS, vol. 3621, pp. 258–275. Springer, Heidelberg (2005)

9. Canetti, R., Goldwasser, S.: An efficient threshold public key cryptosystem secure against adaptive chosen ciphertext attack. In: Stern, J. (ed.) EUROCRYPT 1999. LNCS, vol. 1592, pp. 90–106. Springer, Heidelberg (1999)

10. Chai, Z., Cao, Z., Zhou, Y.: Efficient id-based broadcast threshold decryption in ad hoc network. In: IMSCCS (2), pp. 148–154. IEEE Computer Society, Los Alamitos (2006)

11. Daza, V., Herranz, J., Morillo, P., Ràfols, C.: CCA2-secure threshold broadcast encryption with shorter ciphertexts. In: Susilo, W., Liu, J.K., Mu, Y. (eds.) ProvSec 2007. LNCS, vol. 4784, pp. 35–50. Springer, Heidelberg (2007)

12. De Santis, A., Desmedt, Y., Frankel, Y., Yung, M.: How to share a function securely. In: 26th ACM STOC, pp. 522–533. ACM Press, New York (1994)

13. Delerablée, C.: Identity-based broadcast encryption with constant size ciphertexts and private keys. In: Kurosawa, K. (ed.) ASIACRYPT 2007. LNCS, vol. 4833, pp. 200–215. Springer, Heidelberg (2007)

14. Delerablée, C., Paillier, P., Pointcheval, D.: Fully collusion secure dynamic broadcast encryption with constant-size ciphertexts or decryption keys. In: Takagi, T., Okamoto, T., Okamoto, E., Okamoto, T. (eds.) Pairing 2007. LNCS, vol. 4575, pp. 39–59. Springer, Heidelberg (2007)

15. Desmedt, Y., Frankel, Y.: Threshold cryptosystems. In: Brassard, G. (ed.) CRYPTO 1989. LNCS, vol. 435, pp. 307–315. Springer, Heidelberg (1990)

16. Fiat, A., Naor, M.: Broadcast encryption. In: Stinson, D.R. (ed.) CRYPTO 1993. LNCS, vol. 773, pp. 480–491. Springer, Heidelberg (1994)

17. Fiat, A., Shamir, A.: How to prove yourself: Practical solutions to identification and signature problems. In: Odlyzko, A.M. (ed.) CRYPTO 1986. LNCS, vol. 263, pp. 186–194. Springer, Heidelberg (1987)

18. Frankel, Y.: A practical protocol for large group oriented networks. In: Quisquater, J.-J., Vandewalle, J. (eds.) EUROCRYPT 1989. LNCS, vol. 434, pp. 56–61. Springer, Heidelberg (1990)

19. Gennaro, R., Halevi, S., Krawczyk, H., Rabin, T.: Threshold RSA for dynamic and ad-hoc groups. In: Smart, N. (ed.) EUROCRYPT 2008. LNCS, vol. 4965, pp. 88–107. Springer, Heidelberg (2008)

20. Ghodosi, H., Pieprzyk, J., Safavi-Naini, R.: Dynamic threshold cryptosystems: A new scheme in group oriented cryptography. In: PRAGOCRYPT 1996, pp. 370–379 (1996)
21. Goldwasser, S., Micali, S.: Probabilistic encryption. Journal of Computer and System Sciences 28(2), 270–299 (1984)
22. Goodrich, M.T., Sun, J.Z., Tamassia, R.: Efficient tree-based revocation in groups of low-state devices. In: Franklin, M. (ed.) CRYPTO 2004. LNCS, vol. 3152, pp. 511–527. Springer, Heidelberg (2004)
23. Goyal, V., Pandey, O., Sahai, A., Waters, B.: Attribute-based encryption for fine-grained access control of encrypted data. In: ACM CCS 2006, pp. 89–98. ACM Press, New York (2006)
24. Halevy, D., Shamir, A.: The LSD broadcast encryption scheme. In: Yung, M. (ed.) CRYPTO 2002. LNCS, vol. 2442, pp. 47–60. Springer, Heidelberg (2002)
25. Joux, A.: A one-round protocol for tripartite diffie-hellman. In: Bosma, W. (ed.) ANTS 2000. LNCS, vol. 1838, pp. 385–394. Springer, Heidelberg (2000)
26. Libert, B., Quisquater, J.-J.: Efficient revocation and threshold pairing based cryptosystems. In: 22nd ACM PODC, pp. 163–171. ACM Press, New York (2003)
27. Naor, D., Naor, M., Lotspiech, J.: Revocation and tracing schemes for stateless receivers. In: Kilian, J. (ed.) CRYPTO 2001. LNCS, vol. 2139, pp. 41–62. Springer, Heidelberg (2001)
28. Naor, M.: On cryptographic assumptions and challenges (invited talk). In: Boneh, D. (ed.) CRYPTO 2003. LNCS, vol. 2729, pp. 96–109. Springer, Heidelberg (2003)
29. Naor, M., Yung, M.: Public-key cryptosystems provably secure against chosen ciphertext attacks. In: 22nd ACM STOC. ACM Press, New York (1990)
30. Pointcheval, D., Stern, J.: Security arguments for digital signatures and blind signatures. Journal of Cryptology 13(3), 361–396 (2000)
31. Rackoff, C., Simon, D.R.: Non-interactive zero-knowledge proof of knowledge and chosen ciphertext attack. In: Feigenbaum, J. (ed.) CRYPTO 1991. LNCS, vol. 576, pp. 433–444. Springer, Heidelberg (1992)
32. Schnorr, C.-P.: Efficient identification and signatures for smart cards. In: Brassard, G. (ed.) CRYPTO 1989. LNCS, vol. 435, pp. 239–252. Springer, Heidelberg (1990)
33. Shoup, V.: Lower bounds for discrete logarithms and related problems. In: Fumy, W. (ed.) EUROCRYPT 1997. LNCS, vol. 1233, pp. 256–266. Springer, Heidelberg (1997)
34. Shoup, V.: ISO 18033-2: An emerging standard for public-key encryption, Final Committee Draft (December 2004)
35. Shoup, V., Gennaro, R.: Securing threshold cryptosystems against chosen ciphertext attack. In: Nyberg, K. (ed.) EUROCRYPT 1998. LNCS, vol. 1403, pp. 1–16. Springer, Heidelberg (1998)

On Notions of Security for Deterministic Encryption, and Efficient Constructions without Random Oracles

Alexandra Boldyreva[1], Serge Fehr[2], and Adam O'Neill[1]

[1] Georgia Institute of Technology, Atlanta, GA, USA
{aboldyre,amoneill}@cc.gatech.edu
[2] CWI, Amsterdam, Netherlands
Serge.Fehr@cwi.nl

Abstract. The study of deterministic public-key encryption was initiated by Bellare et al. (CRYPTO '07), who provided the "strongest possible" notion of security for this primitive (called PRIV) and constructions in the random oracle (RO) model. We focus on constructing efficient deterministic encryption schemes *without* random oracles. To do so, we propose a slightly weaker notion of security, saying that no partial information about encrypted messages should be leaked as long as each message is a-priori hard-to-guess *given the others* (while PRIV did not have the latter restriction). Nevertheless, we argue that this version seems adequate for many practical applications. We show equivalence of this definition to single-message and indistinguishability-based ones, which are easier to work with. Then we give general constructions of both chosen-plaintext (CPA) and chosen-ciphertext-attack (CCA) secure deterministic encryption schemes, as well as efficient instantiations of them under standard number-theoretic assumptions. Our constructions build on the recently-introduced framework of Peikert and Waters (STOC '08) for constructing CCA-secure *probabilistic* encryption schemes, extending it to the deterministic-encryption setting as well.

1 Introduction

1.1 Background and Overview

MOTIVATION. Deterministic public-key encryption (where the encryption algorithm is deterministic) was studied by Bellare, Boldyreva and O'Neill [1]. They proposed a semantic-security-style definition of privacy for it, called PRIV, which requires that no partial information about multiple, possibly-dependent messages is leaked from their encryptions, while appropriately taking into account two inherent limitations of deterministic encryption: privacy is only possible for messages that are a-priori hard-to-guess by the adversary, and some information about a message leaks unavoidably, namely its encryption. Both the chosen-plaintext (CPA) and chosen-ciphertext-attack (CCA) cases were considered, and the authors designed several constructions meeting them.

D. Wagner (Ed.): CRYPTO 2008, LNCS 5157, pp. 335–359, 2008.
© International Association for Cryptologic Research 2008

Deterministic encryption seems interesting and useful. As discussed in [1], it allows for fast searching on encrypted data; moreover, deterministic encryption can be length-preserving, which can be needed for securing legacy code or in bandwidth-critical applications. Finally, we find that the study of deterministic encryption can have applications to normal (randomized) encryption as well.

However, the constructions of [1] are only proven secure in the random oracle (RO) model [4]. Of course, finding alternative schemes secure in the standard model (i.e. without random oracles) is desirable, as a growing number of papers have raised concerns about the "soundness" of the RO model (e.g. [8,22,2] to name a few). Finding deterministic encryption schemes secure in the standard model was left as an important open problem in [1].

THIS PAPER. We construct efficient deterministic encryption schemes without random oracles, secure under standard number-theoretic assumptions. The notion of security we use, however, is slightly weaker than that of [1], in that it considers the encryption of *block-sources*. That is, it guarantees no partial information about encrypted messages is leaked, as long as each message is a-priori hard-to-guess *given the other messages*. We believe this notion to nevertheless be suitable for a variety of practical applications, for example the encryption high-entropy data containing social security or phone numbers. In such examples, messages can depend on one another, e.g. share a common prefix, yet the foregoing condition is satisfied.

RELATED WORK. The encryption of high-entropy sources was first considered in the information-theoretic, symmetric-key setting by Russell and Wang [28], and the problem was studied in greater generality (under the name "entropic security") by Dodis and Smith [20,19]. Entropic security was later studied in the quantum setting by Desrosiers and Dupuis [16,17].

1.2   Main Results

EQUIVALENT DEFINITIONS. We show that PRIV-security for block-sources is equivalent to PRIV-security for a *single* hard-to-guess message. The latter was briefly introduced (using a slightly different formulation) in [1] under the name PRIV1, where it was shown *strictly weaker* than PRIV, but beyond that this notion remained unstudied. We also show equivalence of PRIV1 to a single-message, indistinguishability-based notion, which is handier to work with. The proof is non-trivial and employs ideas from [20] and [16,17], used for showing the equivalence between entropic security for information-theoretic symmetric-key (quantum) encryption schemes and an indistinguishability-based notion. All our results about the definitions extend to the CCA setting as well.

GENERAL CONSTRUCTIONS. We present general constructions of both CPA- and CCA-secure deterministic encryption schemes, building on the recently-introduced framework of Peikert and Waters [26] for constructing (randomized) IND-CCA encryption schemes in the standard model. Recall that [26] introduces a framework of "lossy" trapdoor functions (TDFs) — TDFs that operate in one two possible "modes," an injective one and an un-invertible lossy one, for which the outputs are indistinguishable. We observe that if the lossy mode also acts as a *universal hash function* [9,10] (in which case we say it has a *universal hash mode*), then the lossy TDF in injective mode is in fact a secure deterministic encryption scheme in our sense. Indeed, this follows straightforwardly under our indistinguishability-based security notion by the Leftover-Hash Lemma (LHL) [23,6]. We extend the connection between lossy TDFs and deterministic encryption schemes to the CCA setting as well: our general CCA-secure construction can be viewed as a "deterministic" version of the general IND-CCA scheme of [26]. Unlike the latter it does not use a one-time signature scheme but rather a hash function H that is both target-collision resistant (TCR) [24,5] and universal. It also uses a lossy TDF F and an all-but-one (ABO) TDF G (the latter is a generalization of the former introduced in [26] whose first input is drawn from a set of *branches*, one of which is lossy), where as before lossiness must be strengthened to universality. The encryption of message m under our scheme has the form $(H(m), F(m), G(H(m), m))$.

DDH-BASED INSTANTIATIONS. We obtain instantiations of our general constructions based on the decisional Diffie-Hellman assumption (DDH) rather straightforwardly. In fact, we show that the DDH-based lossy and ABO TDF constructs of [26] already suffice; that is, they indeed have "universal" lossy modes. To construct an appropriate hash function for our CCA-secure scheme, we use the discrete-log-based, collision-resistant (and thus TCR) construct of [11] and show that it is also universal. However, some care needs to be taken about its choice of parameters, because the range of the hash must be "compatible" with the ABO TDF in our construction. Nevertheless, we demonstrate ways to achieve compatibility for two popular choices of groups where DDH is believed hard.

EXTENDING OUR GENERAL CONSTRUCTIONS. While our DDH-based instantiations fit neatly into a conceptual framework of "deterministic encryption with universal hash mode," they are not particularly efficient. Moreover, the other instantiations of lossy and ABO TDFs in [26] do *not* (at least immediately) give universal lossy modes. Our solution to this problem is to extend our general constructions in an efficient way such that the extra universality requirement on the underlying primitives is eliminated. These extensions derive from a novel application of a "crooked" version of the LHL due to Dodis and Smith [19], which tells us that if one applies an invertible, pairwise-independent hash function (e.g. the usual $H_{a,b}(x) = ax + b$ construct over a finite field) to a message before encrypting it under our general constructions, then "lossiness" of the underlying primitives (in addition to TCR for hash function H in the CCA case) alone suffices for security.

EFFICIENT PAILLIER-BASED SCHEMES. In particular, the above extensions allow us to instantiate our schemes using the "more-efficient" Paillier-based [25] lossy and ABO TDFs of [26]. However, these constructs are still far from optimal.

Borrowing a technique of Damgård and Nielsen [14,15], we devise new Paillier-based constructs of lossy and ABO TDFs having public-key size on the order of (instead of quadratic in) the message length and essentially no ciphertext expansion; moreover, they compare to standard Paillier encryption computationally. In order to encrypt messages with potentially-small min-entropy (relative to the length of a message), our constructs actually use a generalization of Paillier's scheme due to Damgård and Jurik [13]. Under this generalization, we also construct a hash function for H in the extended CCA-secure construction that is provably TCR based on the same assumption (decisional composite residuosity), and whose range is compatible with the ABO scheme. However, for practical efficiency one can instead use a TCR cryptographic hash function such as SHA256 or the constructs of [5,29] for H. This is in fact another pleasing consequence of extending our general constructions, since before H was required to be both TCR and *universal*, which seems to preclude using a cryptographic hash function.

1.3 Concurrent Work

Concurrently and independently, Bellare, Fischlin, O'Neill and Ristenpart [3] define several multi-message, semantic-security-style definitions for deterministic encryption and prove them equivalent to PRIV definition of [1]. They also propose and prove equivalent an indistinguishability-based definition, but their proof techniques are different from ours. Namely, they consider an "intermediate" definitional variant that we do not. Also, they propose a new deterministic encryption scheme based on general assumptions, whereas our constructions are based on number-theoretic assumptions and are efficient. No constructions secure against chosen-ciphertext attacks are given in [3].

Our efficient Paillier-based instantiations of lossy and ABO TDFs were independently discovered by [27].

2   Preliminaries

ALGORITHMS, PROBABILITIES AND SOURCES. Algorithms implicitly take as additional input the unary encoding 1^k of the security parameter k; they may be randomized and must run in poly-time in k unless indicated otherwise. Integer parameters are also implicitly polynomial functions of k. Adversaries are *non-uniform* and as such receive an auxiliary input of polynomial-size in k, which we also usually leave implicit. For a random variable Y, we write $y \xleftarrow{\$} Y$ to denote that y is sampled according to Y's distribution; furthermore, for an algorithm A, by $y \xleftarrow{\$} A(x)$ we mean that A is executed on input x and the output is assigned to y. (In the case that A gets no input we slightly abuse notation and write $y \xleftarrow{\$} A$ instead of $y \xleftarrow{\$} A()$.) We denote by $\Pr[A(x) = y : x \xleftarrow{\$} X]$ the probability that A outputs y on input x when x is sampled according to X. We say that an adversary A *has advantage ϵ in distinguishing X from Y* if $\Pr[A(x) = 1 : x \xleftarrow{\$} X]$ and $\Pr[A(y) = 1 : y \xleftarrow{\$} Y]$ differ by at most ϵ.

When more convenient, we use the following probability-theoretic notation instead. We write P_X for the distribution of random variable X and $P_X(x)$ for the probability that X puts on value x, i.e. $P_X(x) = \Pr[X = x]$. Similarly, we write $P_{X|\mathcal{E}}$ for the probability distribution of X conditioned on event $\mathcal{E}$, and P_{XY} for the joint distribution of random variables X, Y. The *statistical distance* between X and Y is given by $\Delta(X, Y) = \frac{1}{2}\sum_x |P_X(x) - P_Y(x)|$. It is well-known that if $\Delta(X, Y)$ is at most ϵ then any (even computationally unbounded) adversary A has advantage at most ϵ in distinguishing X from Y.

The *min-entropy* of a random variable X is $\mathrm{H}_\infty(X) = -\log(\max_x P_X(x))$. The *worst-case conditional* min-entropy of X given Y is defined as $\mathrm{H}_\infty(X|Y) = -\log(\max_{x,y} P_{X|Y=y}(x))$, and the *average conditional* min-entropy of X given Y as $\tilde{\mathrm{H}}_\infty(X|Y) = -\log(\sum_y P_Y(y) \max_x P_{X|Y=y}(x))$. A random variable X over $\{0,1\}^\ell$ is called a (t, ℓ)-*source* if $\mathrm{H}_\infty(X) \geq t$, and a list $\boldsymbol{X} = (X_1, \ldots, X_n)$ of random variables over $\{0,1\}^\ell$ is called a (t, ℓ)-*block-source* of length n if $\mathrm{H}_\infty(X_i | X_1 \ldots X_{i-1}) \geq t$ for all $i \in \{1, \ldots, n\}$.

A value $\nu \in \mathbb{R}$ depending on k is called *negligible* if its absolute value goes to 0 faster than any polynomial in k, i.e. $\forall c > 0\ \exists k_\circ \in \mathbb{N}\ \forall k \geq k_\circ : |\nu| < 1/k^c$.

PUBLIC-KEY ENCRYPTION. An encryption scheme is a triple of algorithms $\mathcal{AE} = (\mathcal{K}, \mathcal{E}, \mathcal{D})$, satisfying the usual syntax except that for convenience we also give the decryption algorithm $\mathcal{D}$ the public key. For simplicity, we only consider a message-space of $\{0,1\}^\ell$, and we say that $\mathcal{AE}$ is an ℓ-*bit encryption scheme* if for all messages in the message space $\{0,1\}^\ell$

$$\Pr\big[\mathcal{D}(sk, \mathcal{E}(pk, m)) \neq m : (pk, sk) \xleftarrow{\$} \mathcal{K}\big]$$

is negligible. We say that $\mathcal{AE}$ is *deterministic* if $\mathcal{E}$ is deterministic. Note that we require the message-space to depend only on the security parameter and not on the specific public key; as in [1] this is somewhat crucial to our security definitions.

HASHING. An ℓ-*bit hash function* $\mathcal{H} = (\mathcal{K}, H)$ with domain $\{0,1\}^\ell$ consists of a key-generation algorithm and a hash algorithm.[1] Again, we omit the well-known syntax and restrict to domain $\{0,1\}^\ell$ for simplicity. We say $\mathcal{H}$ has a 2^r-*bounded hash range* if its range $R = \{H(K, x) \mid K \in \mathcal{K}, x \in D\}$ is bounded by $|R| \leq 2^r$ in size. We say that $\mathcal{H}$ with range R is *universal* if for all $x_1 \neq x_2 \in \{0,1\}^\ell$

$$\Pr\big[H(K, x_1) = H(K, x_2) : K \xleftarrow{\$} \mathcal{K}\big] \leq \frac{1}{|R|},$$

and we say it is *pairwise-independent* if for all $x_1 \neq x_2 \in \{0,1\}^\ell$ and all $y_1, y_2 \in R$

$$\Pr\big[H(K, x_1) = y_1 \wedge H(K, x_2) = y_2 : K \xleftarrow{\$} \mathcal{K}\big] \leq \frac{1}{|R|^2}.$$

We say $\mathcal{H}$ is *collision-resistant* (CR) if for every poly-time A the *CR-advantage*

$$\mathbf{Adv}_{\mathcal{H}}^{\mathrm{cr}}(A) = \Pr\big[H(K, x_1) = H(K, x_2) : K \xleftarrow{\$} \mathcal{K}; (x_1, x_2) \xleftarrow{\$} A(K)\big]$$

---

[1] Note that we are not only interested in "compressing" hash functions, e.g. images and pre-images might have the same bit-length.

is negligible. Similarly, we say $\mathcal{H}$ is *target-collision resistant* (TCR) if for every poly-time A the *TCR-advantage*

$$\mathbf{Adv}_{\mathcal{H}}^{\mathrm{tcr}}(A) = \Pr\left[H(K, x_1) = H(K, x_2) : (x_1, \mathrm{st}) \xleftarrow{\$} A \,;\, K \xleftarrow{\$} \mathcal{K} \,;\, x_2 \xleftarrow{\$} A(K, \mathrm{st}) \right]$$

is negligible. As discussed in [5] TCR has some potential benefits over CR, such as being easier to achieve and allowing for shorter output lengths.

3   Security Definitions

The PRIV notion of security for deterministic encryption introduced in [1] asks that it be hard to guess any partial information[2] of a list of messages given their encryptions, as long as the list has component-wise high (super-logarithmic) min-entropy. We introduce a slight weakening of this notion where each message must have high min-entropy *conditioned on values of the other messages*. This notion seems to nevertheless suffice in some practical applications, for example in the encryption of high-entropy data containing phone or social security numbers that can share prefixes but are otherwise uncorrelated. We then consider two other security definitions in order of increasing simplicity and ease-of-use; in the next section we prove that they are all equivalent.

PRIV FOR BLOCK-SOURCES. The following is a semantic-security-style definition that considers the encryption of multiple messages under the same public-key. For an ℓ-bit encryption scheme $\mathcal{AE} = (\mathcal{K}, \mathcal{E}, \mathcal{D})$ and list $\boldsymbol{m} = (m_1, \dots, m_n)$ of messages, we write $\boldsymbol{\mathcal{E}}(pk, \boldsymbol{m})$ below as shorthand for $(\mathcal{E}(pk, m_1), \dots, \mathcal{E}(pk, m_n))$.

Definition 1. *An ℓ-bit encryption scheme $\mathcal{AE} = (\mathcal{K}, \mathcal{E}, \mathcal{D})$ is PRIV-secure for (t, ℓ)-block-sources if for any (t, ℓ)-block-source $\boldsymbol{M} = (M_1, \dots, M_n)$ of polynomial length n, any function $f : \{0,1\}^{n\ell} \to \{0,1\}^*$ and all poly-time adversaries A, the PRIV-advantage*

$$\mathbf{Adv}_{\mathcal{AE}}^{\mathrm{priv}}(A, f, \boldsymbol{M}) = \mathbf{Real}_{\mathcal{AE}}(A, f, \boldsymbol{M}) - \mathbf{Ideal}_{\mathcal{AE}}(A, f, \boldsymbol{M})$$

is negligible, where

$$\mathbf{Real}_{\mathcal{AE}}(A, f, \boldsymbol{M}) = \Pr\left[A(pk, \boldsymbol{\mathcal{E}}(pk, \boldsymbol{m})) = f(\boldsymbol{m}) : (pk, sk) \xleftarrow{\$} \mathcal{K} \,;\, \boldsymbol{m} \xleftarrow{\$} \boldsymbol{M}\right] \; and$$

$$\mathbf{Ideal}_{\mathcal{AE}}(A, f, \boldsymbol{M}) = \Pr\left[A(pk, \boldsymbol{\mathcal{E}}(pk, \boldsymbol{m}')) = f(\boldsymbol{m}) : (pk, sk) \xleftarrow{\$} \mathcal{K} \,;\, \boldsymbol{m}, \boldsymbol{m}' \xleftarrow{\$} \boldsymbol{M}\right]$$

A SINGLE-MESSAGE DEFINITION. Consider Definition 1 with the restriction that only (t, ℓ)-block-sources of length $n = 1$ are allowed; that is, a (t, ℓ)-source M replaces block-source $\boldsymbol{M}$ in the definition. Call the resulting notion *PRIV1-security* for (t, ℓ)-*sources*, where we define $\mathbf{Real}_{\mathcal{AE}}(A, f, M)$ and $\mathbf{Ideal}_{\mathcal{AE}}(A, f, M)$ as well as the PRIV1-advantage $\mathbf{Adv}_{\mathcal{AE}}^{\mathrm{priv1}}(A, f, M)$ accordingly.

We note that (an alternative formulation of) PRIV1 was already considered in [1], and it was shown to be strictly weaker than their multi-message notion

---

[2] To make the definition achievable, the partial information must not depend on the public key. This is reasonable since real data does not depend on any public key.

PRIV. We will show that in the setting of *block*-sources the single- and multi-message definitions are equivalent.

AN INDISTINGUISHABILITY-BASED FORMULATION. We also consider the following indistinguishability-based formulation of PRIV1 inspired by [20], which is handier to work with. It asks that it be hard to distinguish the encryptions of two plaintexts, each drawn from a different (public-key-independent) high-entropy distribution on the message-space.

Definition 2. *An ℓ-bit encryption scheme $\mathcal{AE} = (\mathcal{K}, \mathcal{E}, \mathcal{D})$ is* PRIV1-IND *-secure for (t, ℓ)-sources if for any (t, ℓ)-sources M_0 and M_1 and all poly-time adversaries A, the* PRIV1-IND*-advantage*

$$\mathbf{Adv}_{\mathcal{AE}}^{\mathrm{priv1\text{-}ind}}(A, M_0, M_1) = \mathbf{Guess}_{\mathcal{AE}}(A, M_0) - \mathbf{Guess}_{\mathcal{AE}}(A, M_1)$$

is negligible, where for $b \in \{0, 1\}$

$$\mathbf{Guess}_{\mathcal{AE}}(A, M_b) = \Pr\left[A(pk, \mathcal{E}(pk, m_b)) = 1 : (pk, sk) \xleftarrow{\$} \mathcal{K} ; m_b \xleftarrow{\$} M_b\right].$$

We note that concurrently and independently, [3] gives an indistinguishability-based formulation of the multi-message PRIV definition from [1] (that does not restrict to block-sources).

EXTENSION TO CHOSEN-CIPHERTEXT ATTACKS (CCA). For simplicity, the presented definitions only consider the case of chosen-plaintext attacks (CPA).[3] To extend the definitions to the *chosen-ciphertext-attack* (CCA) setting, we can additionally provide the adversary A in each definition with access to decryption oracle $\mathcal{D}(pk, sk, \cdot)$, which it may query on any ciphertext not appearing in its input. We denote the resulting notions with "-CCA" (e.g. PRIV-CCA for block-sources). Our equivalence results in the following also hold in the CCA setting.

Remark 1. The PRIV definition (and similarly the PRIV1 definition) in [1] requires the pair $(\boldsymbol{m}, s)$ of message-list $\boldsymbol{m}$ and partial-information s on $\boldsymbol{m}$ to be *poly-time samplable*. We do not have such restrictions in our definitions. On the other hand, we ask s to be a *deterministic* function $s = f(\boldsymbol{m})$ of $\boldsymbol{m}$; this latter restriction, however, is without loss of generality, as we argue in Remark 1 below (as long as we allow f to be unbounded). Thus, our definitions remain at least as strong as their corresponding formulations in the style of [1]. The reason for omitting samplability restrictions is for generality and to simplify our results and proofs, and because they are actually not required for the security of our constructions. Furthermore, this strengthening of the definitions is not crucial for our equivalence results; see Remark 4.

Remark 2. PRIV1 (similarly PRIV for block-sources) remains equivalent if we allow f to be *randomized*; i.e., on input m the function f is evaluated as $f(m; r)$

---

[3] Actually, the plaintexts themselves in the definitions are not chosen by the adversary. This is a minor semantic point that we ignore.

for r chosen independently according to some fixed probability distribution (typically uniform) on a finite domain. This equivalence holds for both the "private seed" model, where adversary A does not learn r, and the "public coin" model, where r is given to A (or in a combination of the two). Indeed, if for some adversary, randomized function and block-source, the advantage of A is in absolute value lower-bounded by ε *on average* over the random choice of r, then the same lower-bound holds for some specific choice of r. (The other direction is trivial.)

Note that the "private seed" model covers the case, similar to [1], where a message-and-partial-info pair (m, s) is chosen according to an *arbitrary joint probability distribution* P_{MS} (with $H_\infty(M) \geq t$ and a finite domain for s), as we can always understand the message m as instead sampled according to its distribution P_M and then the partial-information s computed with conditional distribution $P_{S|M=m}$ by means of a randomized function (which can always be done since we do not require f to be efficient[4]). Thus, if in the "private seed" model we restrict the message-and-partial-info pair to be poly-time samplable, then our PRIV1 definition is equivalent to that from [1].

Remark 3. It also suffices in the above definitions to consider *predicates f*, i.e., binary functions to $\{0, 1\}$. This actually follows from Lemma 3 of [16] (and verifying that their proof also works in our poly-time-adversary setting). The idea is to consider the Goldreich-Levin (i.e. inner-product) predicate of the partial information with a random string, and use Remark 2. The resulting adversary loses a factor 2 in its advantage and its running-time increases by $O(n\ell)$. (The technique also works for definitions in the style of [1]; i.e., it suffices to consider partial information of length 1 there.)

4   Equivalence of the Definitions

We show that all three definitions, namely PRIV for block-sources, PRIV1 and PRIV1-IND, are equivalent. Our strategy is as follows. We take PRIV1 as our starting point, and we first show that it is equivalent to PRIV1-IND. Later we show that it is also equivalent to PRIV for block-sources.

Theorem 1. *Let $\mathcal{AE}$ be an ℓ-bit encryption scheme. Then for any (t, ℓ)-sources M_0, M_1 and any adversary A, there exists a (t, ℓ)-source M, an adversary B and a function f such that*

$$\mathbf{Adv}_{\mathcal{AE}}^{\mathrm{priv1\text{-}ind}}(A, M_0, M_1) \leq 2 \cdot \mathbf{Adv}_{\mathcal{AE}}^{\mathrm{priv1}}(B, f, M),$$

and the running-time of B is that of A. And, for any $(t + 1, \ell)$-source M, any function f and any adversary A, there exists an adversary B and (t, ℓ)-sources M_0, M_1 such that

---

[4] E.g., r could consist of a list of suitable choices for s, one choice for each possible m, and f would select and output the right entry.

$$\mathbf{Adv}_{\mathcal{AE}}^{\mathrm{priv1}}(A, f, M) \leq 2 \cdot \mathbf{Adv}_{\mathcal{AE}}^{\mathrm{priv1\text{-}ind}}(B, M_0, M_1),$$

and the running-time of B is that of A plus $O(\ell)$.   □

The proof borrows and combines ideas from [20] and [16,17], used for showing the equivalence between entropical security for information-theoretic symmetric (quantum) encryption schemes and an indistinguishability-based notion.[5]

The proof of the first claim relies on Remark 2 and is straightforward, since distinguishing M_0, M_1 given their encryptions is equivalent to guessing b from the encryption of M_b where b is a random bit. For the second claim, note that if $f(M)$ is easy-to-guess given the encryption of M, then M conditioned on $f(M) = 0$ and M conditioned on $f(M) = 1$ are easy to distinguish. However, one of these distributions may have much smaller min-entropy than M (if f is unbalanced). To avoid (almost all of) this entropy loss we can "mix" them appropriately with M. Moreover, the resulting distributions become poly-time samplable if the pair $(M, f(M))$ is (see Remark 4).

Proof. We start with the first claim. Let M_0, M_1 and A be as given. Let M to be the balanced "mixture" of M_0 and M_1, and f be the corresponding "indicator function;" i.e., M is sampled by choosing a random bit b and then outputting m sampled according to M_b, and the partial information $f(m)$ is defined as b. Such a joint probability distribution on m and b is allowed by Remark 2. Let B be the PRIV1-adversary that on inputs pk, c runs A on the same inputs and outputs the result. Then $\mathrm{H}_\infty(M) \geq t$ and we have

$$
\begin{aligned}
\mathbf{Adv}_{\mathcal{AE}}^{\mathrm{priv1}}(B, f, M) &= \mathbf{Real}_{\mathcal{AE}}(B, f, M) - \mathbf{Ideal}_{\mathcal{AE}}(B, f, M) \\
&= \left(\frac{1}{2}(1 - \mathbf{Guess}_{\mathcal{AE}}(A, M_0)) + \frac{1}{2}\mathbf{Guess}_{\mathcal{AE}}(A, M_1) \right) - \frac{1}{2} \\
&= \frac{1}{2}(\mathbf{Guess}_{\mathcal{AE}}(A, M_1) - \mathbf{Guess}_{\mathcal{AE}}(A, M_0)) \\
&= \frac{1}{2}\mathbf{Adv}_{\mathcal{AE}}^{\mathrm{priv1\text{-}ind}}(A, M_0, M_1);
\end{aligned}
$$

this proves the first claim.

For the second claim, let A, f, M be as given. We first note that by Remark 3, we may assume that $f : \{0,1\}^\ell \to \{0,1\}$, at the cost of losing at most a factor 2 in A's advantage and increasing its running-time by $O(\ell)$. Consider the independent random variables M_0 and M_1, with respective distributions

$$P_{M_0} = r_0 P_{M|f(M)=0} + r_1 P_M \qquad \text{and} \qquad P_{M_1} = r_1 P_{M|f(M)=1} + r_0 P_M,$$

---

[5] Note that the definition of entropic security may come in different flavors, named *ordinary* and *strong* in [16]. The (ordinary) notion used in [20] makes their proof much more cumbersome since Remark 3 does not apply (directly). Our definition of PRIV corresponds to the *strong* flavor.

where $r_0 = P_{f(M)}(0)$ and $r_1 = P_{f(M)}(1)$. Then for any $m \in \{0,1\}^\ell$

$$P_{M_0}(m) = r_0 P_{M|f(M)=0}(m) + r_1 P_M(m) = P_{Mf(M)}(m,0) + r_1 P_M(m)$$
$$\leq 2^{-t-1} + r_1 2^{-t-1} \leq 2^{-t} ,$$

and similarly $P_{M_1}(m) \leq 2^{-t}$, so that $H_\infty(M_0), H_\infty(M_1) \geq t$ as required. Let B be the PRIV1-IND adversary that runs the same code as A. It remains to argue that B can distinguish M_0 and M_1. In order to simplify notation, we let Y, Y_0 and Y_1 be the random variables defined by $Y = A(PK, \mathcal{E}(PK, M))$, $Y_0 = A(PK, \mathcal{E}(PK, M_0))$ and $Y_1 = A(PK, \mathcal{E}(PK, M_1))$, where PK describes a public key generated by $\mathcal{K}$.[6] We have

$$\mathbf{Adv}_{\mathcal{AE}}^{\text{priv1-ind}}(B, M_0, M_1) = \mathbf{Guess}_{\mathcal{AE}}(B, M_1) - \mathbf{Guess}_{\mathcal{AE}}(B, M_0)$$
$$= P_{Y_1}(1) - P_{Y_0}(1) = P_{Y_1}(1) - (1 - P_{Y_0}(0)) = P_{Y_1}(1) + P_{Y_0}(0) - 1 , \quad (1)$$

where the second equality is by construction. Note that $P_{Y_0} = r_0 P_{Y|f(M)=0} + r_1 P_Y$ and similarly for P_{Y_1}. It follows that

$$P_{Y_0}(0) + P_{Y_1}(1)$$
$$= \big(r_0 P_{Y|f(M)=0}(0) + r_1 P_Y(0)\big) + \big(r_1 P_{Y|f(M)=1}(1) + r_0 P_Y(1)\big)$$
$$= \big(r_0 P_{Y|f(M)=0}(0) + r_1 P_{Y|f(M)=1}(1)\big) + \big(r_0 P_Y(1) + r_1 P_Y(0)\big)$$
$$= \big(r_0 P_{Y|f(M)=0}(0) + r_1 P_{Y|f(M)=1}(1)\big) + 1 - \big(r_0 P_Y(0) + r_1 P_Y(1)\big)$$
$$= \big(P_{Yf(M)}(0,0) + P_{Yf(M)}(1,1)\big) + 1 - \big(P_{f(M)}(0)P_Y(0) + P_{f(M)}(1)P_Y(1)\big)$$
$$= \Pr[Y = f(M)] - \Pr[Y = f(M')] + 1$$
$$= \mathbf{Real}_{\mathcal{AE}}(A, f, M) - \mathbf{Ideal}_{\mathcal{AE}}(A, f, M) + 1 = \mathbf{Adv}_{\mathcal{AE}}^{\text{priv1}}(A, f, M) + 1 ,$$

where M' is an independent identically-distributed copy of M. Note that we use $r_0 + r_1 = 1$ and $P_Y(0) + P_Y(1) = 1$ in the third equality and in the second-to-last we use that we can switch the roles of m and m' in the definition of $\mathbf{Ideal}_{\mathcal{AE}}(A, f, M)$. Substituting into equation (1), we obtain

$$\mathbf{Adv}_{\mathcal{AE}}^{\text{priv1-ind}}(B, M_0, M_1) = \mathbf{Adv}_{\mathcal{AE}}^{\text{priv1}}(A, f, M) .$$

Taking into account the factor-2 loss, this proves the second claim. $\qquad \square$

Next, we show that PRIV1 for (t, ℓ)-sources implies PRIV for (t, ℓ)-block-sources; the reverse implication holds trivially.

Theorem 2. Let $\mathcal{AE} = (\mathcal{K}, \mathcal{E}, \mathcal{D})$ be an ℓ-bit encryption scheme. For any (t, ℓ)-block-source M of length n, any function $f : \{0,1\}^{n\ell} \to \{0,1\}^*$ and any adversary A, there exists a (t, ℓ)-source M, a function g and an adversary B such that

$$\mathbf{Adv}_{\mathcal{AE}}^{\text{priv1}}(A, M, f) \leq 10n \cdot \mathbf{Adv}_{\mathcal{AE}}^{\text{priv}}(B, M, g) .$$

---

[6] It makes no difference for the upcoming argument whether we consider the same or a fresh public key for Y, Y_0 and Y_1.

Furthermore, the running-time of B is at most that of A plus $O(n\ell)$.   □

Interestingly, the proof is not a straightforward hybrid argument, but makes intensive use of Theorem 1. The idea is to consider the probability of the adversary A in guessing $f(\boldsymbol{M})$ when given the encryption of a list of *independent* and *uniformly* distributed messages and compare this both to $\mathbf{Ideal}_{\mathcal{AE}}(A, f, \boldsymbol{M})$ and to $\mathbf{Real}_{\mathcal{AE}}(A, f, \boldsymbol{M})$, making use of hybrid arguments and the PRIV1-IND-security of $\mathcal{AE}$ (which follows from its assumed PRIV1-security).

Proof. Let $A, \boldsymbol{M}, f$ be as given. By Remark 3, we may assume that f is *binary*, at the cost of losing a factor 2 in A's advantage and increasing its running-time by $O(n\ell)$. Furthermore, we may assume the PRIV1-advantage to be non-negative (otherwise we flip A's output bit). To simplify notation, we write $\mathbf{Adv}(A)$ below as shorthand for $\mathbf{Adv}_{\mathcal{AE}}^{\mathrm{priv1}}(A, \boldsymbol{M}, f)$. Consider the probability

$$\mathbf{u}_{\mathcal{AE}}(A, \boldsymbol{M}, f) = \Pr\big[A(pk, \boldsymbol{\mathcal{E}}(pk, \boldsymbol{u})) = f(\boldsymbol{m}) : (pk, sk) \leftarrow \mathcal{K} \, ; \, \boldsymbol{m} \leftarrow \boldsymbol{M} \, ; \, \boldsymbol{u} \leftarrow \boldsymbol{U}\big]$$

with $\boldsymbol{U} = (U_1, \ldots, U_n)$ being n independent copies of the uniform distribution on $\{0, 1\}^{\ell}$. Note that we can re-write $\mathbf{Adv}(A)$ as

$$\big(\mathbf{Real}_{\mathcal{AE}}(A, f, \boldsymbol{M}) - \mathbf{u}_{\mathcal{AE}}(A, f, \boldsymbol{M})\big) + \big(\mathbf{u}_{\mathcal{AE}}(A, f, \boldsymbol{M}) - \mathbf{Ideal}_{\mathcal{AE}}(A, f, \boldsymbol{M})\big).$$

Intuitively, this implies that if $\mathbf{Adv}(A)$ is "large" then one of the above two summands must be as well. We show that in either case we can construct a (t, ℓ)-source M, a function g and an adversary B as claimed. We start with the latter case. Specifically, suppose that

$$\mathbf{u}_{\mathcal{AE}}(A, f, \boldsymbol{M}) - \mathbf{Ideal}_{\mathcal{AE}}(A, f, \boldsymbol{M}) \geq \frac{2}{5}\mathbf{Adv}(A) \, .$$

We construct a PRIV1-IND adversary B with running-time that of A plus $O(n\ell)$ and two (t, ℓ)-sources with resulting PRIV1-IND advantage lower bounded by $2\mathbf{Adv}(A)/5n$; Theorem 1 then implies the claim (taking into account the factor-2 loss by our initial assumption that f is binary). We use a hybrid argument. For $i \in \{0, \ldots, n\}$ consider the probability

$$\mathbf{h}_{\mathcal{AE}}^{1,i}(A, f, \boldsymbol{M}) = \Pr\left[\begin{array}{c} A(pk, \boldsymbol{\mathcal{E}}(pk, (m'_1, \ldots, m'_i, u_{i+1}, \ldots, u_n))) = f(\boldsymbol{m}) : \\ (pk, sk) \xleftarrow{\$} \mathcal{K} \, ; \, \boldsymbol{m}, \boldsymbol{m}' \xleftarrow{\$} \boldsymbol{M}, \boldsymbol{u} \xleftarrow{\$} \boldsymbol{U} \end{array}\right] \, .$$

It follows that there exists a j such that $\mathbf{h}_{\mathcal{AE}}^{1,j}(A, f, \boldsymbol{M}) - \mathbf{h}_{\mathcal{AE}}^{1,j+1}(A, f, \boldsymbol{M})$ is at least $2\mathbf{Adv}(A)/5n$. Furthermore, this lower-bound holds for some specific choices $\dot{m}'_1, \ldots, \dot{m}'_j$ of $m'_1, \ldots m'_j$ and some specific choice $\dot{m}$ of $\boldsymbol{m}$. We assume for simplicity that $f(\dot{m}) = 1$; if it is 0 the argument is similar. This implies that there exists an adversary B, which on inputs pk, c samples $\boldsymbol{u} \xleftarrow{\$} \boldsymbol{U}$ and returns

$$A(pk, \mathcal{E}(pk, \dot{m}'_1), \ldots, \mathcal{E}(pk, \dot{m}'_j), c, \mathcal{E}(pk, u_{j+2}), \ldots, \mathcal{E}(pk, u_n)) \, ,$$

and two (t, ℓ)-sources, namely M_{j+1} conditioned on $M_1 = \dot{m}'_1, \ldots, M_j = \dot{m}'_j$ and U_{j+1}, such that the resulting PRIV1-IND advantage is lower bounded by $2\mathbf{Adv}(A)/5n$, as required.

We move to the other case, where we have

$$\mathbf{Real}_{\mathcal{AE}}(A, f, \boldsymbol{M}) - \mathbf{u}_{\mathcal{AE}}(A, f, \boldsymbol{M}) \geq \frac{3}{5}\mathbf{Adv}(A) .$$

We use another hybrid argument. Specifically, for $i \in \{0, \ldots, n\}$ consider the probability

$$\mathbf{h}_{\mathcal{AE}}^{2,i}(A, f, \boldsymbol{M}) = \Pr\left[\begin{array}{c} A(pk, \mathcal{E}(pk, (m_1, \ldots, m_i, u_{i+1}, \ldots, u_n))) = f(\boldsymbol{m}) : \\ (pk, sk) \xleftarrow{\$} \mathcal{K} ; \ \boldsymbol{m} \xleftarrow{\$} \boldsymbol{M}, \boldsymbol{u} \xleftarrow{\$} \boldsymbol{U} \end{array}\right].$$

Again it follows that there exists a j such that $\mathbf{h}_{\mathcal{AE}}^{2,j+1}(A, f, \boldsymbol{M}) - \mathbf{h}_{\mathcal{AE}}^{2,j}(A, f, \boldsymbol{M})$ is at least $3\mathbf{Adv}(A)/5n$, and that this lower-bound holds for some specific choices $\dot{m}_1, \ldots, \dot{m}_j$ of $m_1, \ldots, m_j$. Let us denote the corresponding probabilities with these choices by $\dot{\mathbf{h}}_{\mathcal{AE}}^{2,j+1}(A, f, \boldsymbol{M})$ and $\dot{\mathbf{h}}_{\mathcal{AE}}^{2,j}(A, f, \boldsymbol{M})$. Consider now the (t, ℓ)-source M with distribution $P_M = P_{M_{j+1}|M_1=\dot{m}_1, \ldots, M_j=\dot{m}_j}$. By assumption we have $\mathrm{H}_\infty(M) \geq t$. Also, consider the "randomized" function g (in the "private seed" model) defined as

$$g(m; m_{j+2}, \ldots, m_n) = f(\dot{m}_1, \ldots, \dot{m}_j, m, m_{j+2}, \ldots, m_n) ,$$

with $m_{j+2}, \ldots, m_n$ chosen according to the distribution of $M_{j+2}, \ldots, M_n$, conditioned on $M_1 = \dot{m}_1, \ldots, M_j = \dot{m}_j$ and $M_{j+1} = m$. By Remark 2, it indeed suffices to consider such a function. Let B be the PRIV1 adversary that on input pk, c, samples $\boldsymbol{u} \xleftarrow{\$} \boldsymbol{U}$ and outputs

$$A\big(pk, \mathcal{E}(pk, \dot{m}_1), \ldots, \mathcal{E}(pk, \dot{m}_j), c, \mathcal{E}(pk, u_{j+2}), \ldots, \mathcal{E}(pk, u_k)\big).$$

Now by construction, $\mathbf{Real}_{\mathcal{AE}}(B, g, M)$ coincides with $\dot{\mathbf{h}}_{\mathcal{AE}}^{2,j+1}(A, f, \boldsymbol{M})$ and thus $\mathbf{Real}_{\mathcal{AE}}(B, g, M) - \dot{\mathbf{h}}_{\mathcal{AE}}^{2,j}(A, f, \boldsymbol{M}) \geq 3\mathbf{Adv}(A)/5n$. We consider two cases. If $\dot{\mathbf{h}}_{\mathcal{AE}}^{2,j}(A, f, \boldsymbol{M}) - \mathbf{Ideal}_{\mathcal{AE}}(B, g, M) \geq \mathbf{Adv}(A)/5n$ then the claim follows. Otherwise, $\dot{\mathbf{h}}_{\mathcal{AE}}^{2,j}(A, f, \boldsymbol{M}) - \mathbf{Ideal}_{\mathcal{AE}}(B, g, M)$ is at least $2\mathbf{Adv}(A)/5n$. Then this lower-bound also holds for some particular choices $\dot{m}_{j+1}, \ldots, \dot{m}_n$ of $m_{j+1}, m_{j+2}, \ldots, m_n$ in the definition of $\dot{\mathbf{h}}_{\mathcal{AE}}^{2,j}(A, f, \boldsymbol{M})$ and the same choices of $m, m_{j+2}, \ldots, m_n$ in the definition of $\mathbf{Ideal}_{\mathcal{AE}}(B, g, M)$. Let us denote the corresponding probabilities with these choices by $\ddot{\mathbf{h}}_{\mathcal{AE}}^{2,j}(A, f, \boldsymbol{M})$ and $\mathbf{Id\ddot{e}al}_{\mathcal{AE}}(B, g, M)$. Furthermore, let us assume for simplicity that $f(\dot{m}_1, \ldots, \dot{m}_n) = 1$. Then re-using B as a PRIV1-IND adversary, by construction $\mathbf{Guess}_{\mathcal{AE}}(B, U_{j+1}) = \ddot{\mathbf{h}}_{\mathcal{AE}}^{2,j}(A, f, \boldsymbol{M})$ and $\mathbf{Guess}_{\mathcal{AE}}(B, M) = \mathbf{Id\ddot{e}al}_{\mathcal{AE}}(B, g, M)$, so the claim follows by Theorem 1 (though now with different choices of B, g, M in the statement). $\qquad\square$

Remark 4. Our proof of Theorem 1 also works if as in [1] we require message-and-partial-info pairs (M, S) in the PRIV1 definition, and message-sources M_0 and M_1 in the PRIV1-IND definition to be *poly-time samplable* (allowing S to depend probabilistically on M). Indeed, in the proof of the first claim, note that if M_0 and M_1 are poly-time samplable then so is the pair (M_B, B) where B is a random bit. In the second, note that if the message-and-partial-info pair (M, S), where S is a bit, is poly-time samplable then the following is a poly-time sampler for M_0 (the sampler for M_1 is symmetric): Sample (m, s) and output m if $s = 0$; else, sample (m', s') and output m'. (Specifically the running-time of the sampler is at most twice that of the original one in this case.) As such, essentially the same proof can be used to obtain equivalence between the multi-message PRIV and IND definitions shown in [3] as well.

Similarly, our proof of Theorem 2 also works when restricting to such poly-time samplable message-and-partial-info pairs, where though in the PRIV definition for block-sources we need that $(\boldsymbol{M}, S)$ can be sampled by a poly-time algorithm *conditioned on any fixed choice for* $M_1, \ldots, M_j$ and for any j. Indeed, in the reduction we fix a particular choice $\dot{m}_1, \ldots, \dot{m}_j$ for $M_1, \ldots, M_j$ (for some j) and construct a PRIV1 adversary based upon the message-and-partial-info pair (M_{j+1}, S) conditioned on $(M_1, \ldots, M_j) = (\dot{m}_1, \ldots, \dot{m}_j)$. This is poly-time samplable under the above samplability condition on $(\boldsymbol{M}, S)$.

5   General CPA- and CCA-Secure Constructions

We propose general constructions of deterministic encryption that are CPA- and CCA-secure under our security notions. The constructions derive from an interesting connection between deterministic encryption and "lossy" trapdoor functions introduced by Peikert and Waters [26]. These are trapdoor functions with a (un-invertible) "lossy" mode in which the function loses information about its input, and for which the outputs of the "normal" and "lossy" modes are (computationally) indistinguishable. Viewing trapdoor functions as deterministic encryption schemes in our context, we develop a similar framework of *deterministic encryption with hidden hash mode*.

5.1   CPA-Secure Construction

For our CPA-secure construction, we introduce the following notion.

DETERMINISTIC ENCRYPTION WITH HIDDEN HASH MODE. We say that $\mathcal{AE} = (\mathcal{K}, \tilde{\mathcal{K}}, \mathcal{E}, \mathcal{D})$ is a deterministic ℓ-bit encryption scheme with *hidden hash mode* (HHM), or simply HHM deterministic encryption scheme, with a 2^r-*bounded hash range* if $(\mathcal{K}, \mathcal{E}, \mathcal{D})$ is an ℓ-bit deterministic encryption scheme, and the following conditions are satisfied:

- (*Algorithm* $\tilde{\mathcal{K}}$ *induces a hash.*) There is an induced hash function $\mathcal{H}_{\mathcal{E}} = (\tilde{\mathcal{K}}, H_{\mathcal{E}})$ with domain $\{0, 1\}^\ell$ and a 2^r-bounded hash range, where algorithm $\tilde{\mathcal{K}}$ outputs $\tilde{pk}$, and $H_{\mathcal{E}}$ on inputs $\tilde{pk}, m$ returns $\mathcal{E}(\tilde{pk}, m)$. (Typically $r \ll \ell$.)
- (*Hard to tell* $\tilde{pk}$ *from* pk.) Any poly-time adversary A has negligible advantage, denoted $\mathbf{Adv}_{\mathcal{AE}}^{\mathrm{hhm}}(A)$, in distinguishing the first outputs of $\tilde{\mathcal{K}}$ and $\mathcal{K}$.

The "alternate" key-generation algorithm $\tilde{\mathcal{K}}$ is used only for security proofs; we assume it produces only a public key and no secret key. In the case that the induced encryption scheme $\mathcal{H}_{\mathcal{E}}$ in the first property is universal, we say that scheme $\mathcal{AE}$ has a *hidden universal-hash mode* (HUHM).

HUHM IMPLIES CPA-SECURITY. We show that a deterministic encryption scheme with hidden universal-hash mode is in fact PRIV-secure for block-sources. In other words, if the lossy mode of a lossy trapdoor function is universal, then it is a CPA-secure deterministic encryption scheme in our sense.

Theorem 3. *Let $\mathcal{AE} = (\mathcal{K}, \tilde{\mathcal{K}}, \mathcal{E}, \mathcal{D})$ be an ℓ-bit deterministic encryption scheme with a HUHM and a 2^r-bounded hash range. Then for any adversary A, any (t, ℓ)-sources M_0, M_1 and any $\epsilon > 0$ such that $t \geq r + 2\log(1/\epsilon)$, there exists an adversary B such that*

$$\mathbf{Adv}_{\mathcal{AE}}^{\mathrm{priv1\text{-}ind}}(A, M_0, M_1) \leq 2 \cdot \left(\mathbf{Adv}_{\mathcal{AE}}^{\mathrm{hhm}}(B) + \epsilon \right) .$$

Furthermore, the running-time of B is that of A.   □

The idea of the proof is simple: in the experiments for the PRIV1-IND adversary A, the alternate key generation algorithm $\tilde{\mathcal{K}}$ of $\mathcal{AE}$ may be used instead of $\mathcal{K}$ without A being able to tell the difference; then, the Leftover Hash Lemma (LHL) [23,6] implies that "encryptions" are essentially uniform, so it is impossible for A to guess from which source the encrypted message originated. (Note that it is not crucial here that the output distribution be *uniform*, but merely independent of the input distribution.)

Proof. For $b \in \{0, 1\}$, by definition of $\mathbf{Adv}_{\mathcal{AE}}^{\mathrm{hhm}}$, the probability

$$\mathbf{Guess}_{\mathcal{AE}}(A, M_b) \ = \ \Pr\left[A(pk, \mathcal{E}(pk, m_b)) = 1 : (pk, sk) \stackrel{\$}{\leftarrow} \mathcal{K} \,;\, m_b \stackrel{\$}{\leftarrow} M_b\right]$$

differs from the probability

$$\Pr\left[A(\tilde{pk}, \mathcal{E}(\tilde{pk}, m_b)) = 1 : \tilde{pk} \stackrel{\$}{\leftarrow} \tilde{\mathcal{K}} \,;\, m_b \stackrel{\$}{\leftarrow} M_b\right]$$

by at most $\sum_m P_{M_b}(m) \, \mathbf{Adv}_{\mathcal{AE}}^{\mathrm{hhm}}(B_m)$, where B_m on any input pk simply runs and outputs $A(pk, \mathcal{E}(pk, m))$. By the universal property of the hash mode and applying the LHL, it follows that the above probability is within ϵ of

$$\Pr\left[A(\tilde{pk}, c) = 1 : \tilde{pk} \stackrel{\$}{\leftarrow} \tilde{\mathcal{K}} \,;\, c \stackrel{\$}{\leftarrow} R\right]$$

where R denotes the range of the induced hash function $\mathcal{H}_\mathcal{E}$. But now, this probability does not depend on b anymore, and thus

$$\mathbf{Adv}_{\mathcal{AE}}^{\mathrm{priv1\text{-}ind}}(A, M_0, M_1) \ \leq \ \sum_m \left(P_{M_0}(m) + P_{M_1}(m)\right) \mathbf{Adv}_{\mathcal{AE}}^{\mathrm{hhm}}(B_m) + 2\epsilon$$

from which the claim follows by a suitable choice of m.   □

5.2   CCA-Secure Construction

In order to extend the connection between lossy TDFs and deterministic encryption to the CCA setting, we first generalize our notion of deterministic encryption with HHM in a similar way to the all-but-one (ABO) TDF primitive defined in [26].

ALL-BUT-ONE DETERMINISTIC ENCRYPTION. An *all-but-one* (ABO) deterministic encryption scheme $\mathcal{AE} = (\mathcal{K}, \mathcal{E}, \mathcal{D})$ with a 2^r-bounded hash range is such that each of $\mathcal{K}, \mathcal{E}, \mathcal{D}$ takes an additional input b from an associated *branch-set* $\mathcal{B}$. (For $\mathcal{E}$ it is given as the second input.) In particular, each $b^* \in \mathcal{B}$ yields particular algorithms $\mathcal{K}_{b^*}, \mathcal{E}_{b^*}, \mathcal{D}_{b^*}$. *If no branch input is specified, it is assumed to be a fixed "default" branch.* The following conditions must hold:

- (*One branch induces a hash.*) For any $b \in \mathcal{B}$, there is an induced hash function $\mathcal{H}_{\mathcal{E}_b} = (\mathcal{K}_b, H_{\mathcal{E}_b})$ with a 2^r-bounded hash range, where algorithm $\mathcal{K}_b$ returns pk_b, and $H_{\mathcal{E}_b}$ on inputs pk_b, x returns $\mathcal{E}(pk, b, x)$.
- (*Other branches encrypt.*) For any $b_1 \neq b_2 \in \mathcal{B}$, the triple $(\mathcal{K}_{b_1}, \mathcal{E}_{b_2}, \mathcal{D}_{b_2})$ is a deterministic encryption scheme.
- (*Hash branch is hidden.*) For any $b_1, b_2 \in \mathcal{B}$, any adversary A has negligible advantage, denoted $\mathbf{Adv}_{\mathcal{AE}}^{abo}(A)$, in distinguishing the first outputs of $\mathcal{K}_{b_1}$ and $\mathcal{K}_{b_2}$.

In the case that for all $b \in \mathcal{B}$ the induced hash function $\mathcal{H}_{\mathcal{E}_b}$ in the first condition is universal, we say that scheme $\mathcal{AE}$ is *universal-ABO*.

THE CONSTRUCTION. For our general CCA-secure construction, we show how to adapt the IND-CCA *probabilistic* encryption scheme of [26] to the deterministic-encryption setting. In particular, our construction does not use a one-time signature scheme as in [26] but rather a TCR hash function.

Let $\mathcal{AE}_{hmm} = (\mathcal{K}_{hmm}, \mathcal{E}_{hmm}, \mathcal{D}_{hmm})$ be an ℓ-bit deterministic encryption scheme with a HHM and a $2^{r_{hmm}}$-bounded hash range, let $\mathcal{AE}_{abo} = (\mathcal{K}_{abo}, \mathcal{E}_{abo}, \mathcal{D}_{abo})$ be an ℓ-bit ABO deterministic encryption scheme with branch set $\mathcal{B}$ and a $2^{r_{abo}}$-bounded hash range, and let $\mathcal{H}_{tcr} = (\mathcal{K}_{tcr}, H_{tcr})$ be a ℓ-bit hash function with a $2^{r_{tcr}}$-bounded hash range $R \subseteq \mathcal{B}$. Key-generation algorithm $\mathcal{K}_{cca}$ of the associated deterministic encryption scheme $\mathcal{AE}_{cca} = (\mathcal{K}_{cca}, \mathcal{E}_{cca}, \mathcal{D}_{cca})$ runs $\mathcal{K}_{tcr}$, $\mathcal{K}_{hhm}$, and $\mathcal{K}_{abo}$ to obtain outputs $K_{tcr}, (pk_{hhm}, sk_{hhm}), (pk_{abo}, sk_{abo})$, respectively; it then returns $(K_{tcr}, pk_{hhm}, pk_{abo})$ as public key pk and sk_{hhm} as secret key sk. The encryption and decryption algorithms of are defined as follows:

Algorithm $\mathcal{E}_{cca}(pk, m)$	**Algorithm** $\mathcal{D}_{cca}(pk, sk, h\|c_1\|c_2)$
$h \leftarrow H_{tcr}(K_{tcr}, m)$	$m' \leftarrow \mathcal{D}_{hhm}(sk_{hhm}, c_1)$
$c_1 \leftarrow \mathcal{E}_{hhm}(pk_{hhm}, m)$	$c' \leftarrow \mathcal{E}_{cca}(pk, m')$
$c_2 \leftarrow \mathcal{E}_{abo}(pk_{abo}, h, m)$	If $c' = h\|c_1\|c_2$ then return m'
Return $h\|c_1\|c_2$	Else return $\perp$

We show that if the HHM and ABO schemes in fact induce universal hash functions, and hash function $\mathcal{H}_{tcr}$ is universal as well, then the construction indeed achieves PRIV-CCA-security for block-sources.

Theorem 4. *Let $\mathcal{AE}_{cca} = (\mathcal{K}_{cca}, \mathcal{E}_{cca}, \mathcal{D}_{cca})$ be as above, and suppose that $\mathcal{AE}_{hhm}$ has a HUHB, $\mathcal{AE}_{abo}$ is universal-ABO, and that $\mathcal{H}_{tcr}$ is universal. Then for any adversary A, any (t, ℓ)-block-sources M_0, M_1, and all $\epsilon > 0$ such that $t \geq r_{tcr} + r_{hhm} + r_{abo} + 2\log(1/\epsilon)$, there exists adversaries $B_{tcr}, B_{hhm}, B_{abo}$ such that*

$$\mathbf{Adv}_{\mathcal{AE}_{\mathrm{cca}}}^{\mathrm{priv1\text{-}ind\text{-}cca}}(A, M_0, M_1)$$
$$\leq 2 \cdot \left(\mathbf{Adv}_{\mathcal{H}_{\mathrm{tcr}}}^{\mathrm{tcr}}(B_{\mathrm{tcr}}) + \mathbf{Adv}_{\mathcal{AE}_{\mathrm{hhm}}}^{\mathrm{hhm}}(B_{\mathrm{hhm}}) + \mathbf{Adv}_{\mathcal{AE}_{\mathrm{abo}}}^{\mathrm{abo}}(B_{\mathrm{abo}}) + 3\epsilon \right).$$

Furthermore, the running-times of $B_{\mathrm{tcr}}, B_{\mathrm{hhm}}, B_{\mathrm{abo}}$ are essentially that of A. $\square$

The formal proof is in the full paper [7]. The idea is that, in the experiments for the PRIV1-IND-CCA adversary A, we may first replace the input branch to $\mathcal{AE}_{\mathrm{abo}}$ by the hash (under $\mathcal{H}_{\mathrm{TCR}}$) of "challenge message" m; then, using the secret key of $\mathcal{AE}_{\mathrm{abo}}$ to answer A's decryption queries, we may replace $\mathcal{K}_{\mathrm{hhb}}$ by the hash-inducing generator $\tilde{\mathcal{K}}_{\mathrm{hhb}}$. Crucial to this is that A cannot produce a valid decryption query that contains a hash h' colliding with the hash h of m, but this is guaranteed by the TCR property of $\mathcal{H}_{\mathrm{tcr}}$ and the fact that each message has exactly one possible encryption. Now, the only information A sees on m are universal hashes of it. If m has enough min-entropy, then, intuitively, the LHL implies that each of these hashes are close to uniform, independent of the specific distribution of m, bounding A's advantage to be small.

One technical subtlety is that although the concatenation of independent instances of universal hash functions is again universal, in our case the universal hash function $\mathcal{H}_{\mathcal{E}_{\mathrm{abo}}}$ coming from the ABO scheme depends (via the branch) on the outcome of the universal hash function $\mathcal{H}_{\mathrm{tcr}}$. We overcome this by using the Generalized Leftover Hash Lemma and several observations from [18].

APPLICATION TO WITNESS-RECOVERING DECRYPTION. We remark that our construction (as well as for that in Section 7), when converted into an IND-CCA probabilistic encryption scheme using the KEM-DEM-style conversion of [3],[7] yields, to the best of our knowledge, the first such scheme without ROs that is truly *witness-recovering*; that is, via the decryption process the receiver is able to recover *all* of the randomness used by a sender to encrypt the message. The constructs of [26] technically do not achieve this since, as the authors note, in their IND-CCA scheme the receiver does not recover the randomness used by the sender to generate a key-pair for the one-time signature scheme.

6   Schemes Based on DDH

In this section, we give instantiations of our general CPA- and CCA-secure constructions based the well-known decisional Diffie-Hellman assumption (DDH) in the corresponding groups. The "hidden branches" of the presented HHM and ABO schemes, as well as the TCR hash function in the instantiations are indeed universal, so Theorem 4 applies to show that they are PRIV-CCA-secure for block-sources. (Our CCA-secure construction uses the CPA one as a building-block, so we focus on instantiation of the former here.)

---

[7] Note that security of the resulting probabilistic scheme only requires the base deterministic scheme to be secure for the encryption of a single high-entropy message.

HHM AND ABO SCHEMES. In fact, the deterministic encryption scheme with HUHB and the universal-ABO deterministic encryption schemes are precisely the corresponding DDH-based constructs from [26] of lossy and ABO (with branch-set $\mathbb{Z}_p$ where prime p is the order of group $\mathbb{G}$ in which DDH holds) trapdoor functions with 2^k-bounded hash ranges, where k is the bit-size of p. It suffices to observe that the "lossy branches" of these functions are in fact universal. These constructs are recalled in Appendix A, where this observation is justified. Our results demonstrate that these constructs have stronger security properties than were previously known.

UNIVERSAL-TCR HASH. To fully instantiate our CCA-secure construction, it remains to design an ℓ-bit hash function whose range is contained in the branch-set $\mathbb{Z}_p$ of the DDH-based ABO scheme, and which is both universal and TCR (we will call such hashes "universal-TCR"). We accomplish this slightly differently for two popular choices of group $\mathbb{G}$ in which DDH is believed to hold, giving rise to two possible concrete instantiations of the construction.

Instantiation 1. Let $\mathbb{G}$ be a group of prime order $p = 2q+1$, where q is also prime (i.e. p is a so-called *safe* prime). Let p have size k. This covers the case of $\mathbb{G}$ as an appropriate elliptic-curve group where DDH is hard. Let $QR(\mathbb{Z}_p^*) = \{x^2 \mid x \in \mathbb{Z}_p^*\}$ be the subgroup of quadratic residues modulo p. Note that $QR(\mathbb{Z}_p^*)$ has order $(p-1)/2 = q$. (Also note that we can sample from $QR(\mathbb{Z}_p^*)$ by choosing a random $x \in \mathbb{Z}_p^*$ and returning x^2.) In this case we can use the following hash function, based on the general construct from [11]. Define the key-generation and hash algorithms of ℓ-bit hash function $\mathcal{H}_1 = (\mathcal{K}_1, H_1)$ as follows:

Algorithm $\mathcal{K}_1$	Algorithm $H_1((R_1, \ldots, R_l), x)$
$R_1, \ldots, R_l \overset{\$}{\leftarrow} QR(\mathbb{Z}_p^*)$	$\pi \leftarrow \prod_{i=1}^{l} R_i^{x_i}$
Return $(R_1, \ldots, R_l)$	Return π

Above, x_i denotes the i-th bit of string x.

Proposition 1. *Hash function $\mathcal{H}_1$ defined above is CR assuming the discrete-logarithm problem (DLP) is hard in $QR(\mathbb{Z}_p^*)$, and is universal with a 2^{k-2}-bounded hash range $QR(\mathbb{Z}_p^*)$ contained in $\mathbb{Z}_p$.*

The proof is in the full paper [7]. Note that the hardness of the DLP is a weaker assumption than DDH (although it is made in a different group).

Instantiation 2. Now let $\mathbb{G}$ be $QR(\mathbb{Z}_{p'}^*)$, where $p' = 2p + 1$ is as before a safe prime, so that $|\mathbb{G}| = p$ is also prime. This is another popular class of groups where DDH is believed hard. To instantiate the universal-TCR hash function in this case, we would like to use $\mathcal{H}_1$ from Instantiation 1, but this cannot (immediately) work since $QR(\mathbb{Z}_{p'}^*)$ is not a subset of $\mathbb{Z}_p$. However, we can modify hash algorithm H_1 to output $\mathsf{encode}(\pi)$ instead of π, where encode is an bijection from $QR(\mathbb{Z}_{p'}^*)$ to $\mathbb{Z}_p$. Namely, we can use the "square-root coding" function from [12]: $\mathsf{encode}(\pi) = \min\{\pm\pi^{(p'+1)/4}\}$. Here $\pm\pi^{(p'+1)/4}$ are the two square-roots of π, using the fact that for any safe prime $p' > 5$ we have p' is congruent to 3 mod 4.

While our DDH-based schemes are a definite proof of concept that secure deterministic encryption can be constructed from a widely-accepted number-theoretic assumption, they are rather inefficient. In particular, the constructs of [26] follow a "matrix encryption" approach and have public keys with order ℓ^2 group elements and ciphertexts with order ℓ group elements. We seek more efficient schemes. However, the other instantiations of lossy and ABO TDFs given in [26] do not (immediately) give universal lossy branches. To overcome this difficulty, we first show how to extend our general constructions in an efficient way to provide security when *any* lossy and ABO TDFs are used.

7   Extended General Constructions

GENERALIZED "CROOKED" LHL. In our security proofs we used the fact that the "lossy modes" of the underlying primitives, unlike those defined in [26], act as universal hash functions, allowing us to apply the LHL. However, the conclusion of the LHL was actually stronger than we needed, telling us that output of the lossy modes are *uniform* (and not merely input-independent). We show that the extra universality requirement can actually be avoided, not only for the HHB and ABO schemes but also the TCR hash function, by slightly extending our constructions. The extensions derive from a variant of the LHL due to Dodis and Smith [19, Lemma 12]. We actually need the following generalization of it analogous to the generalization of the standard LHL in [18].

Lemma 1. (Generalized "Crooked" LHL) *Let* $\mathcal{H} = (\mathcal{K}, H)$ *be an ℓ-bit pairwise-independent hash function with range R, and let $f : R \to S$ be a function to a set S. Let the random variable K describe the key generated by $\mathcal{K}$, and U the uniform distribution over R. Then for any random variable X over $\{0,1\}^\ell$ and any random variable Z such that $\tilde{H}_\infty(X|Z) \geq \log|S| + 2\log(1/\epsilon) - 2$, we have $\Delta\big((f(H(K,X)), Z, K), (f(U), Z, K)\big) \leq \epsilon$.*

The proof is the full version [7]. Intuitively, the lemma says that if we compose a pairwise-independent hash function with *any* lossy function, the output of the composition is essentially input-independent (but not necessarily uniform), as long as the input has enough (average conditional) min-entropy. This suggests the following extension to our CCA-secure construction. (We treat the CCA case here, since the extension to our CPA-secure construction is evident from it.)

EXTENDED CCA-SECURE CONSTRUCTION. Let $\mathcal{E}_{cca} = (\mathcal{K}_{cca}, \mathcal{E}_{cca}, \mathcal{D}_{cca})$ be as defined in Section 5.2, and let $\mathcal{H}_{pi} = (\mathcal{K}_{pi}, H_{pi})$ be an ℓ-bit invertible pairwise-independent hash function with range $\{0,1\}^\ell$. Invertibility of $\mathcal{H}_{pi}$ means that there is a polynomial-time algorithm I such that for all K_{pi} that can be output by $\mathcal{K}_{pi}$ and all $m \in \{0,1\}^\ell$ we have $I(K_{pi}, H_{pi}(K_{pi}, m))$ outputs m. The key-generation algorithm $\mathcal{K}_{cca}^+$ of the associated composite scheme $\mathcal{AE}_{cca}^+ = (\mathcal{K}_{cca}^+, \mathcal{E}_{cca}^+, \mathcal{D}_{cca}^+)$ is the same as $\mathcal{K}_{cca}$ except it also generates three independent hash keys $K_{pi,1}, K_{pi,2}, K_{pi,3}$ via $\mathcal{K}_{pi}$ which are included in the public key pk. The encryption and decryption algorithms are defined as follows:

Alg $\mathcal{E}_{\text{cca}}^+((\{K_{\text{pi, i}}\}, pk_{\mathcal{AE}}), m)$
For $i = 1$ to 3 do $h_i \leftarrow H_{\text{pi}}(K_{\text{pi},i}, m)$
$h \leftarrow H(K_{\text{tcr}}, h_1)$
$c_1 \leftarrow \mathcal{E}_{\text{hhm}}(pk_{\text{hhm}}, h_2)$
$c_2 \leftarrow \mathcal{E}_{\text{abo}}(pk_{\text{abo}}, h_3)$
Return $h\|c_1\|c_2$

Alg $\mathcal{D}_{\text{cca}}^+((\{K_{\text{pi, i}}\}, pk_{\mathcal{AE}}), sk_{\mathcal{AE}}, c)$
Parse c as $h\|c_1\|c_2$
$h_1' \leftarrow \mathcal{D}_{\text{hhm}}(sk_{\text{hhm}}, c_1)$
$m' \leftarrow I(K_{\text{pi},2}, h_1')$
$c' \leftarrow \mathcal{E}_{\text{cca}}^+((\{K_{\text{pi, i}}\}, pk_{\mathcal{AE}}), m')$
If $c' = h\|c_1\|c_2$ then return m'
Else return $\bot$

Concretely, viewing ℓ-bit strings as elements of the finite field $\mathbb{F}_{2^\ell}$, we can use for $\mathcal{H}_{\text{pi}}$ the standard construct $\mathcal{H}_\ell = (\mathcal{K}_\ell, H_\ell)$ where $\mathcal{K}_\ell$ outputs a random $a, b \in \mathbb{F}_{2^\ell}$ and H_ℓ on inputs $(a, b), x$ returns $ax + b$, which is clearly invertible.[8] Using Lemma 1, we obtain the following result.

Theorem 5. *Let $\mathcal{AE}_{\text{cca}}^+ = (\mathcal{K}_{\text{cca}}^+, \mathcal{E}_{\text{cca}}^+, \mathcal{D}_{\text{cca}}^+)$ be as defined above. Then for any adversary A, any $(t; \ell)$-block-sources M_0, M_1 and any $\epsilon > 0$ such that $t \geq r_{\text{tcr}} + r_{\text{hhm}} + r_{\text{abo}} + 2\log(1/\epsilon) + 2$, there exist adversaries $B_{\text{tcr}}, B_{\text{hhm}}, B_{\text{abo}}$ such that*

$$\mathbf{Adv}_{\mathcal{AE}_{\text{cca}}^+}^{\text{priv1-ind-cca}}(A, M_0, M_1)$$
$$\leq 2 \cdot \left(\mathbf{Adv}_{\mathcal{H}_{\text{tcr}}}^{\text{tcr}}(B_{\text{tcr}}) + \mathbf{Adv}_{\mathcal{AE}_{\text{hhm}}}^{\text{hhm}}(B_{\text{hhm}}) + \mathbf{Adv}_{\mathcal{AE}_{\text{abo}}}^{\text{abo}}(B_{\text{abo}}) + 3\epsilon \right).$$

Furthermore, the running-times of $B_{\text{tcr}}, B_{\text{hhm}}, B_{\text{abo}}$ are essentially that of A. $\square$

DECREASING THE KEY SIZE. A potential drawback of the extended CCA-secure scheme is its public-key size, due to including three hash keys for $\mathcal{H}_{\text{pi}}$. But in fact we can usually re-use the same key, i.e. take $K_{\text{pi},1} = K_{\text{pi},2} = K_{\text{pi},3}$. For the security proof to go through, we just need the minor technical condition that the range of the hash function $\mathcal{H}_{\mathcal{E}_{\text{abo}}}$[9] induced by the lossy branch of the ABO scheme be independent of the particular branch b. This condition is met by all known instantiations. In the proof, this condition allows us to apply Lemma 1 to the "concatenation" of the hash functions $\mathcal{H}_{\text{tcr}}, \mathcal{H}_{\text{hhb}}$ and $\mathcal{H}_{\text{abo}}$. Details are provided in the full paper [7].

ADVANTAGES. In our extended CCA-secure construction, we can use *any* lossy and ABO TDF, as defined in [26]. In particular, we can use the Paillier- and lattice-based constructs of [26], although we obtain even more efficient Paillier-based schemes in the next section. Also, since $\mathcal{H}_{\text{tcr}}$ in the extended scheme need only be TCR and "lossy," it can be a cryptographic hash function such as SHA256 or the efficient TCR constructs of [5,29] in practice. (Security of the basic scheme required $\mathcal{H}_{\text{tcr}}$ to be both TCR and *universal*, whereas cryptographic hash functions fail to meet the latter.) Such instantiation of $\mathcal{H}_{\text{tcr}}$ is compatible

---

[8] Note that $\mathcal{K}_\ell$ must compute a representation of $\mathbb{F}_{2^\ell}$, which can be done in expected polynomial-time. Alternatively, a less-efficient, matrix-based instantiation of $\mathcal{H}_{\text{pi}}$ runs in strict polynomial time and is invertible with high probability (over the choice of the hash key).

[9] Technically, $\mathcal{H}_{\mathcal{E}_{\text{abo}}}$ should have a third subscript "b," but we drop it here and in similar instances for ease of notation.

with any ABO scheme with branch-set $\mathcal{B}$ for which $\{0,1\}^n \subseteq \mathcal{B}$ for n sufficiently large (so that the probability of hashing to the "default" lossy branch of the ABO scheme is small). Again, this is satisfied for all known instantiations.

8   Efficient Schemes Based on Paillier's DCR Assumption

To improve on efficiency over their DDH-based constructs, [26] suggests basing their matrix-encryption approach on Paillier encryption [25] (which uses the group $\mathbb{Z}_{N^2}$ for an RSA modulus N) instead. One then obtains HHM (or lossy TDF) and ABO schemes with an N-bounded hash range and offering roughly a factor $\log N$ savings in public-key and ciphertext size, namely public keys contain order $(\ell/\log N)^2$ group elements and ciphertexts order $\ell/\log N$ group elements for ℓ-bit messages, and for which encryption requires the latter amount of exponentiations. Based on a technique introduced by Damgård and Nielsen [14,15], we propose new Paillier-based schemes that use an entirely different (i.e. non-matrix-encryption-based) approach and have even better efficiency: they are essentially length-preserving, have about ℓ-bit public keys, and compare to standard Paillier encryption computationally.

SETTING. Let $\mathcal{K}$ be an algorithm that outputs $(N, (p,q))$ where $N = pq$ and p, q are random $k/2$-bit primes. Paillier's *decisional composite residuosity* (DCR) *assumption* [25] states that any poly-time adversary A has negligible advantage in distinguishing a from $a^N \bmod N^2$ for random $(N, (p,q))$ output by $\mathcal{K}$ and random $a \in \mathbb{Z}_{N^2}^*$. Let $s \geq 1$ be polynomial in k. Our schemes actually use a generalization of Paillier encryption, based on the same assumption, to the group $\mathbb{Z}_{N^{s+1}}$ due to Damgård and Jurik [13], with some modifications in the spirit of [14,15]. The schemes have message-space $\{0,1\}^{(s+1)(k-1)}$ (i.e. $\ell = (s+1)(k-1)$), where we regard messages as elements of $\{0, \ldots, 2^{s(k-1)}\} \times \{1, \ldots, 2^{k-1}+1\}$, chosen so that it is contained in the "usual" message-space $\mathbb{Z}_{N^s} \times \mathbb{Z}_N$ for any possible N output by $\mathcal{K}$.

THE NEW DETERMINISTIC ENCRYPTION SCHEME WITH HHM. Define scheme $\mathcal{AE}_{\mathrm{hhm}} = (\mathcal{K}_{\mathrm{hhm}}, \tilde{\mathcal{K}}_{\mathrm{hhm}}, \mathcal{E}_{\mathrm{hhm}}, \mathcal{D}_{\mathrm{hhm}})$ as follows (decryption is specified below):

Alg $\mathcal{K}_{\mathrm{hhm}}$	**Alg $\tilde{\mathcal{K}}_{\mathrm{hhm}}$**	**Alg $\mathcal{E}_{\mathrm{hhm}}((g,N),(x,y))$**
$(N,(p,q)) \xleftarrow{\$} \mathcal{K}$	$(N,(p,q)) \xleftarrow{\$} \mathcal{K}$	If $\gcd(y,N) \neq 1$
$a \xleftarrow{\$} \mathbb{Z}_N^*$	$a \xleftarrow{\$} \mathbb{Z}_N^*$	Then abort
$g \leftarrow (1+N)a^{N^s} \bmod N^{s+1}$	$\tilde{g} \leftarrow a^{N^s} \bmod N^{s+1}$	$c \leftarrow g^x y^{N^s} \bmod N^{s+1}$
Return $((g,N),(p,q))$	Return $(\tilde{g},N)$	Return c

Decryption algorithm $\mathcal{D}_{\mathrm{hhm}}$ on inputs $(g,N),(p,q),c$ uses standard Paillier-decryption as in [13] to recover x, then computes y by taking the N^s-th root of c/g^x (which can be done efficiently given p,q) and returns (x,y). The fact that the scheme indeed has a HHM, i.e. that the first outputs of $\mathcal{K}_{\mathrm{hhm}}$ and $\tilde{\mathcal{K}}_{\mathrm{hhm}}$ above are indistinguishable, follows under DCR by security of the underlying "randomized" encryption scheme of [13]: g output by $\mathcal{K}_{\mathrm{hhm}}$ is an encryption of 1 under this scheme and $\tilde{g}$ output by $\tilde{\mathcal{K}}_{\mathrm{hhm}}$ is an encryption of 0.

Note that the hash range is isomorphic to $\mathbb{Z}_N^*$, hence the scheme has a 2^k-bounded hash range. Also note that the size of this range does *not* depend on parameter s; in hidden hash mode the encryption function "looses" a $1-1/(s+1)$ fraction of the information on the plaintext, so by increasing s we can make the scheme arbitrarily (i.e. $1 - o(1)$) lossy as defined in [26]. This has some useful consequences. First, it allows to securely encrypt long messages with small min-entropy relative to the length of the message. Second, it permits a purely black-box construction of an ABO scheme with many branches having the same amount of lossiness, via the reduction in [26, Section 3.3]. (The latter applies in the lossy TDF context as well.) However, we obtain a much more efficient ABO scheme directly in the following.

THE NEW ABO DETERMINISTIC ENCRYPTION SCHEME. Define scheme $\mathcal{AE}_{abo} = (\mathcal{K}_{abo}, \mathcal{E}_{abo}, \mathcal{D}_{abo})$ with branch-set Z_{N^s} as follows:

Algorithm $\mathcal{K}_{abo}(b^*)$	**Algorithm** $\mathcal{E}_{abo}((g, N), b, (x, y))$
$(N, (p, q)) \xleftarrow{\$} \mathcal{K}$	If $\gcd(y, N) \neq 1$ then abort
$a \xleftarrow{\$} \mathbb{Z}_N^*$	$h \leftarrow g/(1 + N)^b \bmod N^{s+1}$
$g \leftarrow (1 + N)^{b^*} a^{N^s} \bmod N^{s+1}$	Else $c \leftarrow h^x y^{N^s} \bmod N^{s+1}$
Return $((g, N), (p, q))$	Return c

where decryption works essentially as in the previous scheme. A similar analysis shows that under DCR it is indeed an ABO scheme with 2^k-bounded hash range.

TCR HASH. To instantiate our extended CCA-secure construction, it remains to specify a TCR hash function with range the branch-set Z_{N^s} of the above ABO scheme. One way is to use a "heuristic" cryptographic hash function, as discussed in Section 7. This approach also yields, via the KEM-DEM-style conversion of [3], a quite efficient, witness-recovering IND-CCA (probabilistic) encryption scheme. However, for completeness, we give below an alternative construction of a provably CR hash function based on the computational analogue of DCR, which dovetails nicely with our ABO scheme for $s \geq 2$.

We now regard the $(s + 1)(k - 1)$-bit messages as elements of $(x_1, \ldots, x_s, y) \in \{0, \ldots, 2^{k-1}\}^s \times \{1, \ldots, 2^{k-1} + 1\}$ and define hash function $\mathcal{H}_2 = (\mathcal{K}_2, H_2)$ as:

Algorithm $\mathcal{K}_2$	**Algorithm** $H_2((g_1, \ldots, g_s), (x_1, \ldots x_s, y))$
$(N, (p, q)) \xleftarrow{\$} \mathcal{K}$	$\pi \leftarrow g_1^{x_1} \cdots g_s^{x_s} y^N \bmod N^2$
For $i = 1$ to s do:	Return π
$\quad a_i \xleftarrow{\$} Z_N^*$; $g_i \leftarrow a_i^N \bmod N^2$	
Return $(g_1, \ldots, g_s)$	

Proposition 2. *Hash function $\mathcal{H}_2$ defined above is CR assuming the computational composite residuosity assumption [25] holds (relative to $\mathcal{K}$).*

Proof. Given an adversary A that produces a collision, we construct an adversary A' which computes an N-th root of a random N-th power $h = a^N$ in $\mathbb{Z}_{N^2}$. On input h, A' chooses a random index $i^* \in \{1, \ldots, s\}$ and runs $\mathcal{K}_2$ but replaces g_{i^*} by $g_{i^*} \leftarrow h$. Then, it runs A on inpt $(g_1, \ldots, g_s, y)$ and obtains a collision with probability $\mathbf{Adv}^{\mathrm{CR}}(A)$, i.e., $(x_1, \ldots, x_s, y) \neq (x_1, \ldots, x_s, y)$ such that $g_1^{x_1} \cdots g_s^{x_s} y^N = g_1^{x_1'} \cdots g_s^{x_s'} y'^N$. In this case, note that $x_j \neq x_j'$ for some j, as otherwise $y^N = y'^N$ modulo N^2 which implies that also $y = y'$ modulo N, and with probability $1/s$: $i^* = j$. Furthermore, note that we may assume that $x_{i^*} - x_{i^*}'$ is co-prime with N, as otherwise A' can immediately factor N. It follows that if indeed $i^* = j$ then A' can efficiently compute integers σ and τ such that $2\sigma(x_{i^*} - x_{i^*}') + \tau N = 1$. Raising both sides of

$$g_{i^*}^{x_{i^*} - x_{i^*}'} = \prod_{\substack{i=1 \\ i \neq i^*}}^{s} g_i^{x_i' - x_i} \cdot \left(\frac{y'}{y}\right)^N$$

to the power σ and multiplying both sides with $g^{\tau N}$ results in

$$g_{i^*} = \left(\prod_{\substack{i=1 \\ i \neq i^*}}^{s} a_i^{x_i' - x_i} \cdot \frac{y'}{y} g^{\tau} \right)^N.$$

Thus, with probability $\mathbf{Adv}^{\mathrm{CR}}(A)/s$, A' obtains a N-th root of $g_{i^*} = h$.   □

Note that the hash function is "compatible" with the above ABO deterministic encryption scheme in that a hash value it produces lies in Z_{N^s} as long as $s \geq 2$ and the N from the hash function is not larger than the N from the ABO scheme; in fact, it is not too hard to verify that the hash function and the ABO scheme may safely use the same N, so that the latter condition is trivially satisfied.

Acknowledgements

We thank Eike Kiltz, Chris Peikert, and Yevgeniy Dodis for helpful discussions and comments, and the anonymous reviewers of Crypto 2008 for their suggestions. Alexandra Boldyreva was supported in part by NSF CAREER award 0545659. Serge Fehr was supported by a VENI grant from the Dutch Organization for Scientific Research (NWO). Adam O'Neill was supported in part by the above-mentioned grant of the first author.

References

1. Bellare, M., Boldyreva, A., O'Neill, A.: Deterministic and efficiently searchable encryption. In: Menezes, A. (ed.) CRYPTO 2007. LNCS, vol. 4622. Springer, Heidelberg (2007)
2. Bellare, M., Boldyreva, A., Palacio, A.: An uninstantiable random-oracle-model scheme for a hybrid-encryption problem. In: Cachin, C., Camenisch, J.L. (eds.) EUROCRYPT 2004. LNCS, vol. 3027. Springer, Heidelberg (2004)
3. Bellare, M., Fischlin, M., O'Neill, A., Ristenpart, T.: Deterministic encryption: Definitional equivalences and constructions without random oracles. In: CRYPTO 2008. LNCS. Springer, Heidelberg (2008)
4. Bellare, M., Rogaway, P.: Random oracles are practical: a paradigm for designing efficient protocols. In: CCS 1993. ACM, New York (1993)

5. Bellare, M., Rogaway, P.: Collision-resistant hashing: Towards making UOWHFs practical. In: Kaliski Jr., B.S. (ed.) CRYPTO 1997. LNCS, vol. 1294. Springer, Heidelberg (1997)

6. Bennett, C., Brassard, G., Crepeau, C., Maurer, U.: Generalized privacy amplification. Transactions on Information Theory 41(6) (1995)

7. Boldyreva, A., Fehr, S., O'Neill, A.: On notions of security for deterministic encryption, and efficient constructions without random oracles. Full version of this paper (2008), http://eprint.iacr.org/2008/

8. Canetti, R., Goldreich, O., Halevi, S.: The random oracle methodology, revisited. In: STOC 1998. ACM, New York (1998)

9. Carter, J.L., Wegman, M.N.: Universal classes of hash functions. Journal of Computer and System Sciences 18 (1979)

10. Carter, J.L., Wegman, M.N.: New hash functions and their use in authentication and set equality. Journal of Computer and System Sciences 22 (1981)

11. Chaum, D., van Heijst, E., Pfitzmann, B.: Cryptographically strong undeniable signatures, unconditionally secure for the signer. In: Feigenbaum, J. (ed.) CRYPTO 1991. LNCS, vol. 576. Springer, Heidelberg (1992)

12. Cramer, R., Shoup, V.: A practical public key cryptosystem provably secure against adaptive chosen ciphertext attack. In: Krawczyk, H. (ed.) CRYPTO 1998. LNCS, vol. 1462. Springer, Heidelberg (1998)

13. Damgård, I., Jurik, M.: A generalisation, a simplification and some applications of paillier's probabilistic public-key system. In: Kim, K.-c. (ed.) PKC 2001. LNCS, vol. 1992. Springer, Heidelberg (2001)

14. Damgård, I., Nielsen, J.-B.: Perfect hiding and perfect binding universally composable commitment schemes with constant expansion factor. In: Yung, M. (ed.) CRYPTO 2002. LNCS, vol. 2442. Springer, Heidelberg (2002)

15. Damgård, I., Nielsen, J.-B.: Universally composable efficient multiparty computation from threshold homomorphic encryption. In: Boneh, D. (ed.) CRYPTO 2003. LNCS, vol. 2729. Springer, Heidelberg (2003)

16. Desrosiers, S.: Entropic security in quantum cryptography. ArXiv e-Print quant-ph/0703046 (2007), http://arxiv.org/abs/quant-ph/0703046

17. Desrosiers, S., Dupuis, F.: Quantum entropic security and approximate quantum encryption. arXiv e-Print quant-ph/0707.0691 (2007), http://arxiv.org/abs/0707.0691

18. Dodis, Y., Ostrovsky, R., Reyzin, L., Smith, A.: Fuzzy extractors: How to generate strong keys from biometrics and other noisy data, http://eprint.iacr.org/2003/235; Preliminary version appeared in: EUROCRYPT 2004. LNCS, vol. 3027. Springer, Heidelberg (2004)

19. Dodis, Y., Smith, A.: Correcting errors without leaking partial information. In: STOC 2005. ACM, New York (2005)

20. Dodis, Y., Smith, A.: Entropic security and the encryption of high entropy messages. In: Kilian, J. (ed.) TCC 2005. LNCS, vol. 3378. Springer, Heidelberg (2005)

21. ElGamal, T.: A public key cryptosystem and signature scheme based on discrete logarithms. In: Transactions on Information Theory, vol. 31. IEEE, Los Alamitos (1985)

22. Goldwasser, S., Tauman Kalai, Y.: On the (in)security of the Fiat-Shamir paradigm. In: FOCS 2003. IEEE, Los Alamitos (2003)

23. Hastad, J., Impagliazzo, R., Levin, L., Luby, M.: A pseudorandom generator from any one-way function. Journal of Computing 28(4) (1999)

24. Naor, M., Yung, M.: Universal one-way hash functions and their cryptographic applications. In: STOC 1989. ACM, New York (1989)
25. Paillier, P.: Public-key cryptosystems based on composite degree residuosity classes. In: Stern, J. (ed.) EUROCRYPT 1999. LNCS, vol. 1592. Springer, Heidelberg (1999)
26. Peikert, C., Waters, B.: Lossy trapdoor functions and their applications. In: STOC 2008. ACM, New York (2008)
27. Rosen, A., Segev, G.: Efficient lossy trapdoor functions based on the composite residuosity assumption. In: Cryptology ePrint Archive: Report 2008/134 (2008)
28. Russell, A., Wang, H.: How to fool an unbounded adversary with a short key. In: Knudsen, L.R. (ed.) EUROCRYPT 2002. LNCS, vol. 2332. Springer, Heidelberg (2002)
29. Shoup, V.: A composition theorem for universal one-way hash functions. In: Preneel, B. (ed.) EUROCRYPT 2000. LNCS, vol. 1807. Springer, Heidelberg (2000)

A   DDH-Based Lossy and ABO TDFs of Peikert-Waters

In [26] the authors introduce a form of "matrix encryption" that they use to realize lossy and ABO TDFs based on encryption schemes allowing some linear-algebraic operations to be performed on ciphertexts. We briefly recall this and the resulting schemes here (using our terminology of HHM and ABO deterministic encryption schemes rather than lossy and ABO TDFs). For concreteness we describe the schemes based on DDH. Moreover, although this was not shown in [26], the "lossy branches" of the DDH-based schemes are *universal*, so we can use them towards instantiating our basic CPA- and CCA-secure constructions. Throughout the description we fix a group $\mathbb{G}$ of prime order p with generator g in which DDH is believed to hold.

ElGamal-based matrix encryption. We first review the ElGamal-based method of [26] for encrypting $\ell \times \ell$ boolean matrices. The public key is $(g^{s_1}, \ldots, g^{s_\ell})$, where $s_1, \ldots, s_\ell \in \mathbb{Z}_p$ are random, and $(s_1, \ldots s_\ell)$ is the secret key. The encryption of an $\ell \times \ell$ boolean matrix $A = (a_{ij})$ is the matrix $C = (c_{ij})$ of pairs of elements in $\mathbb{G}$, where $c_{ij} = (g^{a_{ij}} g^{s_i \cdot r_i}, g^{r_i})$ for random $r_1, \ldots, r_\ell \in \mathbb{Z}_p$. Note that the same randomness is re-used for elements in the same row and the same component of the public key is re-used for elements in the same column. Under the DDH assumption, the encryption of any matrix using this scheme is indistinguishable from the encryption of any other one [26, Lemma 5.1].

The schemes. We briefly describe the DDH-based deterministic encryption scheme with HHM from [26]. The (normal) key-generation algorithm of the scheme outputs an encryption of the $(\ell \times \ell)$ identity-matrix I under the above scheme as the public key, and the s_j's as the secret key. To encrypt a message $\boldsymbol{x} = (x_1, \ldots, x_\ell) \in \{0,1\}^\ell$ one multiplies $\boldsymbol{x}$ (from the left) into the encrypted public-key matrix by using the homomorphic property of ElGamal: ciphertext $\boldsymbol{c} = (c_1, \ldots, c_\ell)$ is computed as

$$c_j = \left(\prod_i u_{ij}^{x_i}, \prod_i v_{ij}^{x_i} \right).$$

It is easy to verify that $c_j = \left(g^\rho, g^{x_j} h_j^\rho\right)$ with $\rho = \sum_i r_i x_i \in \mathbb{Z}_p$, so that standard ElGamal decryption allows to recover x_j when given s_j (using the fact that $x_j \in \{0,1\}$). The alternate key-generation algorithm of the scheme outputs an encryption of the $(\ell \times \ell)$ all-zero matrix rather than of the identity matrix, so that the encryption of a message x results in the ciphertext c with $c_j = \left(g^\rho, h_j^\rho\right)$ where, as before, $\rho = \sum_i r_i x_i$. Thus, c only contains limited information on x, namely $\rho = \sum_i r_i x_i \in \mathbb{Z}_p$. This makes the encryption function *lossy*, as required in [26], but it is also easy to see that it also makes the encryption function a universal hash function. Indeed, the encryptions c and c' of two distinct messages x and x' collide if and only if the corresponding $\rho = \sum_i r_i x_i$ and $\rho' = \sum_i r_i x_i'$ collide, which happens with probability $1/q$ (over the choices of the r_i's). Thus, for any ℓ, we obtain an ℓ-bit deterministic encryption scheme with HHM having 2^k-bounded hash range, where k is the bit-size of p. We omit the description of the corresponding DDH-based ℓ-bit ABO scheme with 2^k-bounded hash range obtained from [26]. Essentially the same analysis applies to show that its lossy branch is universal as well.

Deterministic Encryption: Definitional Equivalences and Constructions without Random Oracles

Mihir Bellare[1], Marc Fischlin[2], Adam O'Neill[3], and Thomas Ristenpart[1]

[1] Dept. of Computer Science & Engineering, University of California at San Diego
9500 Gilman Drive, La Jolla, CA 92093-0404, USA
{mihir,tristenp}@cs.ucsd.edu
http://www-cse.ucsd.edu/~{mihir,tristenp}
[2] Dept. of Computer Science, Darmstadt University of Technology
Hochschulstrasse 10, 64289 Darmstadt, Germany
fischlin@informatik.tu-darmstadt.de
http://www.fischlin.de/
[3] College of Computing, Georgia Institute of Technology
801 Atlantic Drive, Atlanta, GA 30332, USA
amoneill@cc.gatech.edu
http://www.cc.gatech.edu/~amoneill

Abstract. We strengthen the foundations of deterministic public-key encryption via definitional equivalences and standard-model constructs based on general assumptions. Specifically we consider seven notions of privacy for deterministic encryption, including six forms of semantic security and an indistinguishability notion, and show them all equivalent. We then present a deterministic scheme for the secure encryption of uniformly and independently distributed messages based solely on the existence of trapdoor one-way permutations. We show a generalization of the construction that allows secure deterministic encryption of independent high-entropy messages. Finally we show relations between deterministic and standard (randomized) encryption.

1 Introduction

The foundations of public-key encryption, as laid by Goldwasser and Micali [23] and their successors, involve two central threads. The first is definitional equivalences, which aim not only to increase our confidence that we have the "right" notion of privacy but also to give us definitions that are as easy to use in applications as possible. (Easy-to-use indistinguishability is equivalent to the more intuitive, but also more complex, semantic security [21, 23, 24, 28].) The second (of the two threads) is to obtain schemes achieving the definitions under assumptions as minimal as possible. In this paper we pursue these same two threads for *deterministic* encryption [3], proving definitional equivalences and providing constructions based on general assumptions.

D. Wagner (Ed.): CRYPTO 2008, LNCS 5157, pp. 360–378, 2008.
© International Association for Cryptologic Research 2008

DETERMINISTIC ENCRYPTION. A public-key encryption scheme is said to be deterministic if its encryption algorithm is deterministic. Deterministic encryption was introduced by Bellare, Boldyreva, and O'Neill [3]. The motivating application they gave is efficiently searchable encryption. Deterministic encryption permits logarithmic time search on encrypted data, while randomized encryption only allows linear time search [13, 27], meaning a search requires scanning the whole database. This difference is crucial for large outsourced databases which cannot afford to slow down search. Of course deterministic encryption cannot achieve the classical notions of security of randomized encryption, but [3] formalize a semantic security style notion PRIV that captures the "best possible" privacy achievable when encryption is deterministic, namely that an adversary provided with encryptions of plaintexts drawn from a message-space of high (super-logarithmic) min-entropy should have negligible advantage in computing any public-key independent *partial information function* of the plaintexts. The authors provide some schemes in the random-oracle (RO) model [5] meeting this definition but leave open the problem of finding standard model schemes.

The PRIV definition captures intuition well but is hard to work with. We would like to find simpler, alternative definitions of privacy for deterministic encryption —restricted forms of semantic security as well as an indistinguishablility style definition— that are equivalent to PRIV. We would also like to find schemes not only in the standard model but based on general assumptions.

NOTIONS CONSIDERED. We define seven notions of privacy for deterministic encryption inspired by the work of [3, 19]. These include a notion IND in the indistinguishability style and six notions —A-CSS, B-CSS, BB-CSS, A-SSS, B-SSS, BB-SSS— in the semantic-security style. The IND definition —adapted from [19]— asks that the adversary be unable to distinguish encryptions of plaintexts drawn from two, adversary-specified, high-entropy message spaces, and is simple and easy to use. The semantic security notions are organized along two dimensions. The first dimension is the class of partial information functions considered, and we look at three choices, namely arbitrary (A), boolean (B), or balanced boolean (BB). (A boolean function is balanced if the probabilities that it returns 0 or 1 are nearly the same.) The second dimension is whether the formalization is simulation (S) based or comparison (C) based.[1] The PRIV notion of [3] is A-CSS in our taxonomy. Low-end notions —think of BB as the lowest, then B then A and similarly C then S in the other dimension— are simpler and easier to use in applications, while high end ones are more intuitively correct. The question is whether the simplifications come at the price of power.

DEFINITIONAL EQUIVALENCES. We show that all seven notions discussed above are equivalent. The results are summarized in Figure 1. These results not only

---

[1] In the first case, A's success in computing partial information about plaintexts from ciphertexts is measured relative to that of a simulator, while in the second it is measured relative to A's own success when it is given the encryption of plaintexts independent of the challenge ones. The terminology is from [8] who prove equivalence between simulation and comparison based notions of non-malleability.

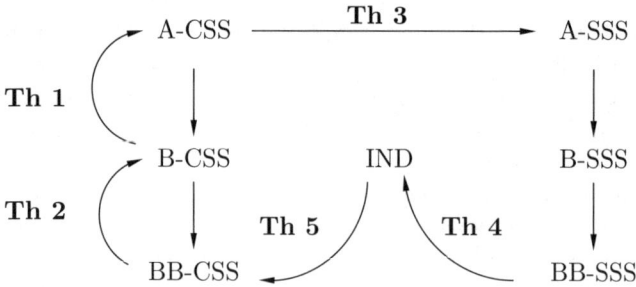

Fig. 1. Notions of security for deterministic encryption schemes and implications showing that all seven notions are equivalent. An arrow $X \to Y$ means that every scheme secure under X is also secure under Y. Unlabeled implications are trivial.

show that semantic security for boolean functions (predicates) is as powerful as semantic security for arbitrary functions, but (perhaps surprisingly) that one can even restrict attention to boolean functions that are balanced, meaning semantic security for balanced boolean functions is as powerful as semantic security for arbitrary functions. We note that balance in this context originates with [19] but they only use it as a tool. We explicitly define and consider the notions BB-CSS and BB-SSS because they bear a natural and intuitive relation to IND and because we feel that the use made of balance by [19] indicates it is important. The proofs of our results rely on new techniques compared to [17, 18, 19].

DEFINITIONAL FRAMEWORK. We believe that an important and useful contribution of our paper is its definitional framework. Rather than an experiment per notion, we have a few core experiments and then use the approach of [6], capturing different notions via different adversary classes. Advantages of this approach are its easy extendability —for example we can capture the notions of [12] by simply introducing a couple of new adversary classes— and the ability to capture many definitional variants in a way that is unified, concise and yet precise.

A CONSTRUCTION FOR UNIFORM MESSAGES. Constructing a non-RO model deterministic encryption scheme meeting our strong notions of security seems like a very challenging problem. We are however able to make progress on certain special cases. We present a deterministic encryption scheme DE1 for the secure encryption of independent, uniformly distributed messages. The scheme is not only without random oracles but based on general trapdoor one-way permutations. To encrypt a random message x one iterates a trapdoor permutation f on x a number of times to get a point y. Let r denote the sequence of Goldreich-Levin [22] hardcore bits obtained in the process. Then one uses a standard IND-CPA scheme —which exists assuming trapdoor one-way permutations— to encrypt y with coins r. The interesting aspect of the scheme, and the source of the difficulty in analyzing it, is its cyclic nature, namely that the coins used for the IND-CPA encryption depend on the plaintext y that is IND-CPA encrypted. The proof manages to show that an adversary who, given y, can distinguish r from random can recover x *even though* this adversary may have partial information about the underlying seed x. The proof exploits in a crucial way that the equivalence between A-CSS and B-CSS holds even for uniformly and independently distributed messages.

ANOTHER PERSPECTIVE. A deterministic encryption scheme is (syntactically) the same thing as a family of injective trapdoor functions. Our notions can then be seen as an extension of the usual notion of one-wayness. Our construction is then a family of injective trapdoor functions which hides all (possible) partial information about its (randomly chosen) input. We believe this is a natural and useful strengthening of the usual notion of a trapdoor function that is fully achieved under standard assumptions in our work.

EFFICIENCY. The general assumption notwithstanding, our scheme admits efficient instantiations. For example with squaring as the trapdoor permutation [9] and Blum-Goldwasser [10] as the bare IND-CPA scheme, encryption and decryption come in at about double that of Blum-Goldwasser with no increase in ciphertext size. See Section 5.

A GENERALIZATION. We generalize our construction to obtain a non-RO model deterministic scheme DE2 for the encryption of independent, high min-entropy (but not necessarily uniform) plaintexts. The assumption used is that one has a trapdoor permutation that is one-way for high min-entropy distributions on its input. This increase in assumption strength is in some sense necessary, since deterministic encryption secure for some distribution trivially provides a one-way injective trapdoor function for that distribution.

FROM DETERMINISTIC TO RANDOMIZED ENCRYPTION. Another central foundational theme is relations between primitives, meaning determining which primitives imply which others. From this perspective we consider how to build IND-CPA-secure standard (randomized) encryption from PRIV-secure deterministic encryption. The obvious approach would be to use the deterministic encryption scheme as a trapdoor one-way function within some well-known general construction [22]. However, this approach leads to large ciphertexts, and we would hope to achieve better efficiency when using a primitive that provides more than one-wayness. We provide a much more efficient construction using a hybrid encryption-style approach, in which the deterministic scheme encrypts a fresh session key padded with extra randomness and the session key is used to encrypt the message. See [4] for the details.

CCA. Lifting our notions and equivalences to the CCA setting is straightforward; see [4]. Our above-mentioned construction of a randomized encryption scheme from a deterministic one works even in the CCA setting. This means, in particular, that we can generically build witness-recovering IND-CCA encryption schemes [25] from arbitrary CCA-secure deterministic schemes.

(Witness-recovering encryption allows, during decryption, recovery of all randomness used to generate a ciphertext.) CCA-secure witness-recovering encryption is of use in further applications [16], and only very recently was a (not very efficient) standard-model construction produced [25]. Our construction shows that building CCA-secure deterministic schemes is at least as hard as building witness-recovering probabilistic encryption.

RELATED WORK. Dodis and Smith's work on entropic security [19] has in common with ours the consideration of privacy for messages of high min-entropy. But there are important differences in the settings, namely that theirs is information-theoretic and symmetric while ours is computational and public-key. Dodis and Smith [19] introduce definitions that in our framework are IND, B-SSS, and BB-SSS, to complement the A-SSS-like information-theoretic notion originally proposed by Russell and Wang [26]. Also, Desrosiers [17] and Desrosiers and Dupuis [18] subsequently treat quantum entropic security, providing notions similar to our framework's B-CSS and A-CSS. These works provide some relations between the notions they define. While some of their techniques and implications lift to our setting, others do not. The salient fact that emerges is that prior work *does not* imply equivalence of all seven notions we consider. In particular, the BB-SSS and BB-CSS notions are not considered in [17, 18] and Dodis and Smith [19] only provide reductions for BB-SSS implying A-SSS that result in inefficient or restricted adversaries. See [4] for more information.

Another setting that deals with high min-entropy messages is that of perfectly one-way hash functions (POWHF), introduced by Canetti [14] and further studied by Canetti, Micciancio, and Reingold [15]. These are randomized hash functions that produce publically-verifiable outputs. Our definitions and equivalences can be adapted to the POWHF setting.

INDEPENDENT WORK. In concurrent and independent work, Boldyreva, Fehr, and O'Neill [12] consider a relaxation of PRIV in which message sequences need to not merely have high entropy but each message must have high entropy even given the others. They prove some relations between their notions using techniques of [17, 18, 19] but do not consider as many notions as us and in particular do not consider balance. Their schemes achieve stronger notions of security then our DE1 but at the cost of specific algebraic assumptions as opposed to our general one. Combining their results with ours shows that our DE2 achieves their notion of security while using a general (even though non-standard) assumption.

2   Preliminaries

NOTATION AND CONVENTIONS. If x is a string then $|x|$ denotes its length; if x is a number then $|x|$ denotes its absolute value; if S is a set then $|S|$ denotes its size. We denote by λ the empty string. If S is a set then $X \leftarrow_\$ S$ denotes that X is selected uniformly at random from S. We let $x[i \ldots j]$ denote bits i through j of string x, for $1 \leq i \leq j \leq |x|$. By $x_1 \parallel \cdots \parallel x_n$ we denote the concatenation of $x_1, \ldots, x_n$. Vectors are denoted in boldface, e.g. $\mathbf{x}$. If $\mathbf{x}$ is a vector then $|\mathbf{x}|$

denotes the number of components of $\mathbf{x}$ and $\mathbf{x}[i]$ denotes its ith component for $1 \leq i \leq |\mathbf{x}|$. If $i \geq 1$ is an integer, we use B_i as shorthand for $\{0,1\}^i$. By $\langle a, b \rangle$ we denote the inner product modulo 2 of equal-length strings a, b.

We write $\alpha \leftarrow_{\$} X(x, y, \ldots)$ to denote running X on inputs $(x, y, \ldots)$ with fresh random coins and assigning the result to α. We let $[X(x, y, \ldots)]$ denote the set of possible outputs of X when run on $x, y, \ldots \in \{0,1\}^*$. An algorithm X is *non-uniform* if its first input is 1^k and there is a collection $\{C_k\}_{k \in \mathbb{N}}$ of (randomized) circuits such that C_k computes $X(1^k, \ldots)$. The running time is the circuit size. A function f is called *negligible* if it approaches zero faster than the reciprocal of any polynomial, that is, for any polynomial p, there exists $n_p \in \mathbb{N}$ such that $f(n) \leq 1/p(n)$ for all $n \geq n_p$. "PT" stands for polynomial time. We denote by Λ the algorithm that on any inputs returns λ.

PUBLIC-KEY ENCRYPTION. A *public-key encryption (PKE)* scheme $\Pi = (\mathcal{K}, \mathcal{E}, \mathcal{D})$ is a triple of PT algorithms. The key generation algorithm $\mathcal{K}$ takes input 1^k, where $k \in \mathbb{N}$ is the security parameter, and outputs a public-key, secret-key pair (pk, sk). The encryption algorithm $\mathcal{E}$ takes inputs 1^k, pk, and plaintext $x \in \{0,1\}^*$ and outputs a ciphertext. The deterministic decryption algorithm $\mathcal{D}$ takes inputs 1^k, sk, and ciphertext y and outputs either a plaintext x or $\perp$. We say that Π is *deterministic* if $\mathcal{E}$ is deterministic. If $\mathbf{x}$ is a vector of plaintexts, then we write $\mathbf{y} \leftarrow_{\$} \mathcal{E}(1^k, pk, \mathbf{x})$ to denote component-wise encryption of $\mathbf{x}$, i.e. $\mathbf{y}[i] \leftarrow_{\$} \mathcal{E}(1^k, pk, \mathbf{x}[i])$ for all $1 \leq i \leq |\mathbf{x}|$.

3   Security Notions for Deterministic PKE

We first provide formal definitions and then discuss them.

SEMANTIC SECURITY. An *SS-adversary* $A = (A_c, A_m, A_g)$ is a tuple of non-uniform algorithms. A_c takes as input a unary encoding 1^k of the security parameter $k \in \mathbb{N}$ and returns a string st representing some state information. A_m takes input 1^k and st, and returns a vector of challenge messages $\mathbf{x}$ together with a test string t that represents some information about $\mathbf{x}$. A_g takes 1^k, a public key and the component-wise encryption of $\mathbf{x}$ under this key, and tries to compute t. The running time of A is defined as the sum of the running times of A_c, A_m, A_g, so that A is PT if A_c, A_m, A_g are all PT.

Let $\Pi = (\mathcal{K}, \mathcal{E}, \mathcal{D})$ be a PKE scheme, $A = (A_c, A_m, A_g)$ an SS-adversary, and S a simulator (a non-uniform algorithm). Let $k \in \mathbb{N}$. Figure 2 displays the css (comparison-based semantic security) and sss (simulation-based semantic security) experiments. We define the css advantage and sss advantage of A by

$$\mathbf{Adv}_{\Pi,A}^{css}(k) = 2 \cdot \Pr\left[\mathbf{Exp}_{\Pi,A}^{css}(k) \Rightarrow \mathsf{true}\right] - 1, \text{ and} \tag{1}$$

$$\mathbf{Adv}_{\Pi,A,S}^{sss}(k) = 2 \cdot \Pr\left[\mathbf{Exp}_{\Pi,A,S}^{sss}(k) \Rightarrow \mathsf{true}\right] - 1. \tag{2}$$

Our approach to defining the six notions of semantic security of interest to us is to associate to each a corresponding class of adversaries and ask that the

$\mathbf{Exp}_{\Pi,A}^{\mathrm{css}}(k)$	$\mathbf{Exp}_{\Pi,A,S}^{\mathrm{sss}}(k)$	$\mathbf{Exp}_{\Pi,I}^{\mathrm{ind}}(k)$
$b \leftarrow_\$ \{0,1\}$; $st \leftarrow_\$ A_\mathrm{c}(1^k)$	$b \leftarrow_\$ \{0,1\}$; $st \leftarrow_\$ A_\mathrm{c}(1^k)$	$b \leftarrow_\$ \{0,1\}$; $st \leftarrow_\$ I_\mathrm{c}(1^k)$
$(\mathbf{x}_0, t_0) \leftarrow_\$ A_\mathrm{m}(1^k, st)$	$(\mathbf{x}, t) \leftarrow_\$ A_\mathrm{m}(1^k, st)$	$\mathbf{x}_b \leftarrow_\$ I_\mathrm{m}(1^k, b, st)$
$(\mathbf{x}_1, t_1) \leftarrow_\$ A_\mathrm{m}(1^k, st)$	$(pk, sk) \leftarrow_\$ \mathcal{K}(1^k)$	$(pk, sk) \leftarrow_\$ \mathcal{K}(1^k)$
$(pk, sk) \leftarrow_\$ \mathcal{K}(1^k)$	$\mathbf{c} \leftarrow_\$ \mathcal{E}(1^k, pk, \mathbf{x})$	$\mathbf{c} \leftarrow_\$ \mathcal{E}(1^k, pk, \mathbf{x}_b)$
$\mathbf{c} \leftarrow_\$ \mathcal{E}(1^k, pk, \mathbf{x}_b)$	If $b = 1$ then	$b' \leftarrow_\$ I_\mathrm{g}(1^k, pk, \mathbf{c}, st)$
$g \leftarrow_\$ A_\mathrm{g}(1^k, pk, \mathbf{c}, st)$	$\quad g \leftarrow_\$ A_\mathrm{g}(1^k, pk, \mathbf{c}, st)$	Ret $(b' = b)$
If $g = t_1$ then $b' \leftarrow 1$	Else $g \leftarrow_\$ S(1^k, pk, st)$	
Else $b' \leftarrow 0$	If $g = t$ then $b' \leftarrow 1$	
Ret $(b' = b)$	Else $b' \leftarrow 0$	
	Ret $(b' = b)$	

Fig. 2. Three experiments for defining security of encryption schemes

advantage of any adversary in this class be negligible. We proceed to define the relevant classes.

An SS-adversary $A = (A_\mathrm{c}, A_\mathrm{m}, A_\mathrm{g})$ is *legitimate* if there exists a function $v(\cdot)$, called the number of messages, and a collection $\{\mathbf{y}_k\}_{k \in \mathbb{N}}$ of *reference* message-vectors such that the following three conditions hold. First, $|\mathbf{x}| = v(k)$ for all $(\mathbf{x}, t) \in [A_\mathrm{m}(1^k, st)]$ and all $st \in \{0,1\}^*$. Second, $|\mathbf{x}[i]| = |\mathbf{y}_k[i]|$ for all $(\mathbf{x}, t) \in [A_\mathrm{m}(1^k, st)]$, all $st \in \{0,1\}^*$, and all $1 \le i \le v(k)$. Third, the function

$$\nu(k) = \Pr\left[\, \mathrm{eq}(\mathbf{x}, \mathbf{y}_k) = 0 \,:\, st \leftarrow_\$ A_\mathrm{c}(1^k) \,;\, (\mathbf{x}, t) \leftarrow_\$ A_\mathrm{m}(1^k, st) \,\right]$$

is negligible, where

$$\mathrm{eq}(\mathbf{x}, \mathbf{y}_k) = \begin{cases} 1 \text{ if } \forall i, j: \ \mathbf{x}[i] = \mathbf{x}[j] \text{ iff } \mathbf{y}_k[i] = \mathbf{y}_k[j] \\ 0 \text{ otherwise.} \end{cases} \tag{3}$$

(The third condition reflects that every deterministic scheme leaks plaintext equality.) Let $\mathcal{A}_\mathrm{SS}$ be the set of all legitimate, PT SS-adversaries. We say that A has *trivial state function* if $A_\mathrm{c} = \Lambda$. Let $\mathcal{A}_\lambda$ be the set of all SS-adversaries with trivial state functions.

Without loss of generality (through suitable padding) we can assume there is a function $\ell(\cdot)$ such that the output of $A_\mathrm{g}(1^k, \cdot, \cdot)$ and any test string t output by $A_\mathrm{m}(1^k, \cdot)$ always have length $\ell(k)$. We call ℓ the *information length* of A. An SS-adversary $A = (A_\mathrm{c}, A_\mathrm{m}, A_\mathrm{g}) \in \mathcal{A}_\mathrm{SS}$ is *boolean* if it has information length $\ell(\cdot) = 1$. Let $\mathcal{A}_\mathrm{B} \subseteq \mathcal{A}_\mathrm{SS}$ be the class of all boolean SS-adversaries. A boolean SS-adversary $A = (A_\mathrm{c}, A_\mathrm{m}, A_\mathrm{g}) \in \mathcal{A}_\mathrm{B}$ is δ-*balanced* if for every st we have

$$\left| \Pr\left[\, t = 0 \,:\, (\mathbf{x}, t) \leftarrow_\$ A_\mathrm{m}(1^k, st) \,\right] - \frac{1}{2} \right| \le \delta \,. \tag{4}$$

When $\delta = 0$ we say that A is *perfectly balanced*. We say that A is *balanced* if it is δ-balanced for some $\delta < 1/2$. Let $\mathcal{A}_\mathrm{BB}^\delta \subseteq \mathcal{A}_\mathrm{B}$ be the class of all δ-balanced boolean SS-adversaries. An SS-adversary $A = (A_\mathrm{c}, A_\mathrm{m}, A_\mathrm{g}) \in \mathcal{A}_\mathrm{SS}$ has *min-entropy* μ if

$$\Pr\left[\, \mathbf{x}[i] = x \,:\, (\mathbf{x}, t) \leftarrow_\$ A_\mathrm{m}(1^k, st) \,\right] \le 2^{-\mu(k)}$$

for all $k \in \mathbb{N}$, all $1 \leq i \leq v(k)$, all $x \in \{0,1\}^*$, and all $st \in \{0,1\}^*$. Let $\mathcal{A}_{\mathrm{ME}}^{\mu} \subseteq \mathcal{A}_{\mathrm{SS}}$ be the class of all SS-adversaries with min-entropy μ. We say that A has *high min-entropy* if it is in $\mathcal{A}_{\mathrm{ME}}^{\mu}$ for some $\mu(k) \in \omega(\log k)$. Let $\mathcal{A}_{\mathrm{HE}} \subseteq \mathcal{A}_{\mathrm{SS}}$ be the class of all SS-adversaries that have high min-entropy.

Let Π be a PKE scheme. We say that Π is A-CSS secure if $\mathbf{Adv}_{\Pi,A}^{\mathrm{css}}(\cdot)$ is negligible for all $A \in \mathcal{A}_{\mathrm{HE}} \cap \mathcal{A}_\lambda$; Π is B-CSS-secure if $\mathbf{Adv}_{\Pi,A}^{\mathrm{css}}(\cdot)$ is negligible for all $A \in \mathcal{A}_{\mathrm{HE}} \cap \mathcal{A}_\lambda \cap \mathcal{A}_{\mathrm{B}}$; and Π is BB-CSS-secure if there exists $\delta < 1/2$ such that $\mathbf{Adv}_{\Pi,A}^{\mathrm{css}}(\cdot)$ is negligible for all $A \in \mathcal{A}_{\mathrm{HE}} \cap \mathcal{A}_\lambda \cap \mathcal{A}_{\mathrm{BB}}^{\delta}$.

Similarly, we say that Π is A-SSS-secure if for all $A \in \mathcal{A}_{\mathrm{HE}} \cap \mathcal{A}_\lambda$ there exists a PT simulator S such that $\mathbf{Adv}_{\Pi,A,S}^{\mathrm{sss}}(\cdot)$ is negligible; Π is B-SSS-secure if for all $A \in \mathcal{A}_{\mathrm{HE}} \cap \mathcal{A}_\lambda \cap \mathcal{A}_{\mathrm{B}}$ there exists a PT simulator S such that $\mathbf{Adv}_{\Pi,A,S}^{\mathrm{sss}}(\cdot)$ is negligible; and Π is BB-SSS-secure if there exists $\delta < 1/2$ such that for all $A \in \mathcal{A}_{\mathrm{HE}} \cap \mathcal{A}_\lambda \cap \mathcal{A}_{\mathrm{BB}}^{\delta}$ there exists a PT simulator S such that $\mathbf{Adv}_{\Pi,A,S}^{\mathrm{sss}}(\cdot)$ is negligible.

The *message space* of an SS-adversary $A = (A_{\mathrm{c}}, A_{\mathrm{m}}, A_{\mathrm{g}})$ is the algorithm A_{d} that on input 1^k, st lets $(\mathbf{x}, t) \leftarrow_\$ A_{\mathrm{m}}(1^k, st)$ and returns $\mathbf{x}$. An SS-adversary is said to produce *independent messages* if the coordinates of $\mathbf{x}$ are independently distributed when $\mathbf{x} \leftarrow_\$ A_{\mathrm{d}}(1^k, st)$ for all k, st. Let $\mathcal{A}_\times$ be the class of all SS-adversaries which produce independent messages.

For each $d \in \{0,1\}$, we let $\mathbf{Exp}_{\Pi,A}^{\mathrm{css\text{-}}d}(k)$ be the same as $\mathbf{Exp}_{\Pi,A}^{\mathrm{css}}(k)$ except that the first line sets $b \leftarrow d$ rather than picking b at random. We similarly define $\mathbf{Exp}_{\Pi,A,S}^{\mathrm{sss\text{-}}d}(k)$. A standard argument gives

$$\mathbf{Adv}_{\Pi,A}^{\mathrm{css}}(k) = \Pr\left[\, \mathbf{Exp}_{\Pi,A}^{\mathrm{css\text{-}}1}(k) \Rightarrow \mathsf{true} \,\right] - \Pr\left[\, \mathbf{Exp}_{\Pi,A}^{\mathrm{css\text{-}}0}(k) \Rightarrow \mathsf{false} \,\right] \quad \text{and} \quad (5)$$

$$\mathbf{Adv}_{\Pi,A,S}^{\mathrm{sss}}(k) = \Pr\left[\, \mathbf{Exp}_{\Pi,A,S}^{\mathrm{sss\text{-}}1}(k) \Rightarrow \mathsf{true} \,\right] - \Pr\left[\, \mathbf{Exp}_{\Pi,A,S}^{\mathrm{sss\text{-}}0}(k) \Rightarrow \mathsf{false} \,\right] . \quad (6)$$

INDISTINGUISHABILITY. An *IND-adversary* $I = (I_{\mathrm{c}}, I_{\mathrm{m}}, I_{\mathrm{g}})$ is a tuple of non-uniform algorithms. I_{c} takes as input 1^k and returns a string st representing some state information. I_{m} takes input 1^k, a bit b, and st, and returns a vector of messages $\mathbf{x}$. I_{g} takes 1^k, a public key, the component-wise encryption of $\mathbf{x}$ under this key, and st and tries to compute the bit b. The running time of I is defined as the sum of the running times of $I_{\mathrm{c}}, I_{\mathrm{m}}, I_{\mathrm{g}}$, so that I is PT if $I_{\mathrm{c}}, I_{\mathrm{m}}, I_{\mathrm{g}}$ are all PT.

Let $\Pi = (\mathcal{K}, \mathcal{E}, \mathcal{D})$ be a PKE scheme, $I = (I_{\mathrm{c}}, I_{\mathrm{m}}, I_{\mathrm{g}})$ an IND-adversary and $k \in \mathbb{N}$. Figure 2 displays the ind experiment. We define the ind advantage of I by

$$\mathbf{Adv}_{\Pi,I}^{\mathrm{ind}}(k) = 2 \cdot \Pr\left[\, \mathbf{Exp}_{\Pi,I}^{\mathrm{ind}}(k) \Rightarrow \mathsf{true} \,\right] - 1 . \quad (7)$$

We next define classes of IND-adversaries. An IND-adversary $I = (I_{\mathrm{c}}, I_{\mathrm{m}}, I_{\mathrm{g}})$ is *legitimate* if there exists a function $v(\cdot)$, called the number of messages, and a collection $\{\mathbf{y}_k\}_{k\in\mathbb{N}}$ of *reference* message-vectors such that the following three conditions hold. First, $|\mathbf{x}| = v(k)$ for all $(\mathbf{x}, t) \in [I_{\mathrm{m}}(1^k, b, st)]$, all $b \in \{0,1\}$, and all $st \in \{0,1\}^*$. Second, $|\mathbf{x}[i]| = |\mathbf{y}_k[i]|$ for all $(\mathbf{x}, t) \in [I_{\mathrm{m}}(1^k, b, st)]$, all $b \in \{0,1\}$, all $st \in \{0,1\}^*$, and all $1 \leq i \leq v(k)$. Third, the function

$$\nu(k) = \Pr\left[\, \mathrm{eq}(\mathbf{x}, \mathbf{y}_k) = 0 \,:\, st \leftarrow_\$ I_{\mathrm{c}}(1^k) \,;\, b \leftarrow_\$ \{0,1\} \,;\, (\mathbf{x}, t) \leftarrow_\$ I_{\mathrm{m}}(1^k, b, st) \,\right]$$

is negligible, where $\mathrm{eq}(\mathbf{x}, \mathbf{y}_k)$ was defined by (3). Let $\mathcal{I}$ be the set of all legitimate, polynomial time IND-adversaries. We say that I has *trivial state function* if $I_\mathrm{c} = \Lambda$. Let $\mathcal{I}_\lambda \subseteq \mathcal{I}$ be the set of all IND-adversaries with trivial state functions. An IND-adversary $I = (I_\mathrm{c}, I_\mathrm{m}, I_\mathrm{g}) \in \mathcal{I}$ has *min-entropy* μ if

$$\Pr\left[\mathbf{x}[i] = x \ : \ \mathbf{x} \leftarrow_{\!\$} I_\mathrm{m}(1^k, b, st) \right] \ \leq \ 2^{-\mu(k)}$$

for all $k \in \mathbb{N}$, all $b \in \{0,1\}$, all $1 \leq i \leq v(k)$, all $x \in \{0,1\}^*$, and all $st \in \{0,1\}^*$. Let $\mathcal{I}_\mathrm{ME}^\mu \subseteq \mathcal{I}$ be the class of all IND-adversaries with min-entropy μ. We say I has *high min-entropy* if it is in $\mathcal{I}_\mathrm{ME}^\mu$ for some $\mu(k) \in \omega(\log k)$. Let $\mathcal{I}_\mathrm{HE}$ be the class of all IND-adversaries that have high min-entropy. We say that Π is IND-secure if $\mathbf{Adv}_{\Pi,I}^{\mathrm{ind}}(\cdot)$ is negligible for all $I \in \mathcal{I}_\mathrm{HE} \cap \mathcal{I}_\lambda$.

For each $d \in \{0,1\}$, we let $\mathbf{Exp}_{\Pi,I}^{\mathrm{ind}\text{-}d}(k)$ be the same as $\mathbf{Exp}_{\Pi,I}^{\mathrm{ind}}(k)$ except that the first line sets $b \leftarrow d$ rather than picking b at random. A standard argument gives

$$\mathbf{Adv}_{\Pi,A}^{\mathrm{ind}}(k) \ = \ \Pr\left[\mathbf{Exp}_{\Pi,I}^{\mathrm{ind}\text{-}1}(k) \Rightarrow \mathsf{true} \right] - \Pr\left[\mathbf{Exp}_{\Pi,I}^{\mathrm{ind}\text{-}0}(k) \Rightarrow \mathsf{false} \right] \ . \quad (8)$$

DISCUSSION. A-CSS is exactly the PRIV definition of [3]. As discussed in [3], it is important that A_m does not take input the public key, and this carries over to I_m. In the classical setting a standard hybrid argument [2] shows that the security of encrypting one message implies the security of encrypting multiple messages. In the deterministic encryption setting this is not true in general, which is why $A_\mathrm{m}, I_\mathrm{m}$ output vectors of messages.

Following [3], message spaces are not explicit but rather implicitly defined by their PT sampling algorithms A_m and I_m. As a consequence, message spaces are PT sampleable.

Following [3], the partial information function is not explicit. Think of t as its value on $\mathbf{x}$. This is more general because t is allowed to depend on coins underlying the generation of $\mathbf{x}$ rather than merely on $\mathbf{x}$ itself. (This is stronger than merely allowing the function to be randomized, which is standard.) It allows us in particular to capture "history." However, we show in [4] that this formulation is equivalent to one where the partial information is computed as a function of the message. Note that the (implicit or explicit) partial information functions are PT.

Our security definitions quantify only over adversaries with trivial state functions. We do this for compatibility with [3, 19]. So why introduce the common state function at all? The reason is that it is useful in proofs. Indeed, [19] use such a function implicitly in many places. We believe making it explicit increases clarity. In the end we can always hardwire a "best" state and thereby end up with an adversary in $\mathcal{A}_\lambda$.

4   Relating the Security Notions

In this section we justify the implications summarized by Figure 1. The implications given by the unlabeled arrows are trivial and can be justified by the fact that $X \to Y$ whenever the adversary class corresponding to Y is a subset of the one corresponding to X. We focus on the implications: A-CSS $\Rightarrow$ A-SSS; BB-SSS $\Rightarrow$ IND; IND $\Rightarrow$ BB-CSS; BB-CSS $\Rightarrow$ B-CSS; and B-CSS $\Rightarrow$ A-CSS.

Theorem 1. *[B-CSS $\Rightarrow$ A-CSS] Let $\Pi = (\mathcal{K}, \mathcal{E}, \mathcal{D})$ be a PKE scheme. Let $A = (A_{\mathrm{c}}, A_{\mathrm{m}}, A_{\mathrm{g}}) \in \mathcal{A}_{\mathrm{ME}}^{\mu} \cap \mathcal{A}_{\lambda}$ be an SS-adversary having information length $\ell(\cdot)$. Then there exists a boolean SS-adversary $A' = (A_{\mathrm{c}}', A_{\mathrm{m}}', A_{\mathrm{g}}') \in \mathcal{A}_{\mathrm{ME}}^{\mu} \cap \mathcal{A}_{\lambda} \cap \mathcal{A}_{\mathrm{B}}$ such that for all $k \in \mathbb{N}$*

$$\mathbf{Adv}_{\Pi,A}^{\mathrm{css}}(k) \le 2 \cdot \mathbf{Adv}_{\Pi,A'}^{\mathrm{css}}(k) . \tag{9}$$

A' has the same message space as A and its running time is that of A plus $\mathcal{O}(\ell)$. $\quad\square$

Proof. The proof is from [19] and repeated here in order to provide intuition for Theorem 2. Below we write ℓ for $\ell(k)$. Then let

alg. $A_{\mathrm{c}}^{*}(1^{k})$:	**alg. $A_{\mathrm{m}}^{*}(1^{k}, (r, s))$:**	**alg. $A_{\mathrm{g}}^{*}(1^{k}, pk, \mathbf{c}, (r, s))$:**
$r \leftarrow_{\$} \{0,1\}^{\ell}$	$(\mathbf{x}, t) \leftarrow_{\$} A_{\mathrm{m}}(1^{k}, \lambda)$	$g \leftarrow_{\$} A_{\mathrm{g}}(1^{k}, pk, \mathbf{c}, \lambda)$
$s \leftarrow_{\$} \{0,1\}$	Ret $(\mathbf{x}, \langle r, t \rangle \oplus s))$	Ret $\langle r, g \rangle \oplus s$
Ret (r, s)		

Then $A^{*} = (A_{\mathrm{c}}^{*}, A_{\mathrm{m}}^{*}, A_{\mathrm{g}}^{*})$ is certainly boolean, and

$$P_{A^{*}}(k) = P_{A}(k) + \frac{1}{2}[1 - P_{A}(k)]$$

$$Q_{A^{*}}(k) = Q_{A}(k) + \frac{1}{2}[1 - Q_{A}(k)]$$

where $P_{X}(k) = \Pr\left[\mathbf{Exp}_{\Pi,X}^{\mathrm{css}\text{-}1}(k) \Rightarrow \mathsf{true}\right]$ and $Q_{X}(k) = \Pr\left[\mathbf{Exp}_{\Pi,X}^{\mathrm{css}\text{-}0}(k) \Rightarrow \mathsf{false}\right]$. Subtracting, we get $\mathbf{Adv}_{\Pi,A^{*}}^{\mathrm{css}}(k) = \frac{1}{2} \cdot \mathbf{Adv}_{\Pi,A}^{\mathrm{css}}(k)$. We are not done yet because A^{*} does not have trivial state function. Let A' be obtained from A^{*} by hardwiring in a "best" choice of r, s and we are done. $\quad\blacksquare$

Now we wish to show that BB-CSS $\Rightarrow$ B-CSS. Note that if the adversary A' constructed in the proof of Theorem 1 were balanced, we would be done. But, A' need not be balanced. Dodis and Smith [19] give a partial solution to this problem, showing that it is in fact possible to find an r that, when hardwired into A^{*}, results in a balanced adversary, as long as $p \le \epsilon^{2}/4$, where p is the maximum probability of any t being output by A_{m} and $\epsilon = \mathbf{Adv}_{\Pi,A}^{\mathrm{css}}(\cdot)$.

We will remove this restriction by proceeding as follows. Let A be a given SS-adversary, which from Theorem 1 we can assume is boolean (but not balanced). We again construct an A^{*} with non-trivial state, but this will consist of n independently chosen keys $\mathbf{K}[1], \ldots, \mathbf{K}[n]$ for a family of pairwise independent hash functions H. Then $A_{\mathrm{m}}^{*}(1^{k}, \mathbf{K})$ first runs $(\mathbf{x}, t) \leftarrow_{\$} A_{\mathrm{m}}(1^{k}, \lambda)$ and then returns $(\mathbf{x}, H(\mathbf{K}[i], t))$ for random $i \in \{1, \ldots, n\}$, while $A_{\mathrm{g}}^{*}(1^{k}, pk, \mathbf{c}, \mathbf{K})$ picks its own independent random j and returns $H(\mathbf{K}[j], A_{\mathrm{g}}(1^{k}, pk, \mathbf{c}, \lambda))$. Our analysis will show that for a suitable choice of n there exists a choice of the vector $\mathbf{K}$ which, when hardwired into A^{*}, yields an adversary A' having all the claimed properties. The theorem is below and the proof is in the full version [4].

Theorem 2. [BB-CSS $\Rightarrow$ B-CSS] *Let* $\Pi = (\mathcal{K}, \mathcal{E}, \mathcal{D})$ *be a PKE scheme. Let* $A = (A_c, A_m, A_g) \in \mathcal{A}^\mu_{\mathrm{ME}} \cap \mathcal{A}_\lambda \cap \mathcal{A}_{\mathrm{B}}$ *be a boolean SS-adversary. Let* $\epsilon(\cdot) = \mathbf{Adv}^{\mathrm{css}}_{\Pi,A}(\cdot) > 0$ *and let* $\delta = 1/4$. *Then there exists an SS-adversary* $A' = (A'_c, A'_m, A'_g) \in \mathcal{A}^\mu_{\mathrm{ME}} \cap \mathcal{A}_\lambda \cap \mathcal{A}^\delta_{\mathrm{BB}}$ *such that for all* $k \in \mathbb{N}$

$$\mathbf{Adv}^{\mathrm{css}}_{\Pi,A}(k) \leq 4n(k) \cdot \mathbf{Adv}^{\mathrm{css}}_{\Pi,A'}(k) \,,$$

where $n(k) = \max\{485\,,\ \lceil 64 \cdot \ln(1/\epsilon(k)) + 64\ln 4\rceil\}$. A' *has the same message space as* A *and its running time is that of* A *plus* $\mathcal{O}(\log(1/\epsilon(k)) + k)$. □

Below are theorem statements for the other three implications. Proofs are given in the full version [4].

Theorem 3. [A-CSS $\Rightarrow$ A-SSS] *Let* $\Pi = (\mathcal{K}, \mathcal{E}, \mathcal{D})$ *be a PKE scheme. Let* $A = (A_c, A_m, A_g) \in \mathcal{A}^\mu_{\mathrm{ME}} \cap \mathcal{A}_\lambda$ *be an SS-adversary outputting at most* v *messages. Then there exists a simulator* S *such that for all* $k \in \mathbb{N}$

$$\mathbf{Adv}^{\mathrm{sss}}_{\Pi,A,S}(k) \leq \mathbf{Adv}^{\mathrm{css}}_{\Pi,A}(k) \,.$$

The running time of S *is that of* A *plus the time to perform* v *encryptions.* □

Theorem 4. [BB-SSS $\Rightarrow$ IND] *Let* $\Pi = (\mathcal{K}, \mathcal{E}, \mathcal{D})$ *be a PKE scheme. Let* $I = (I_c, I_m, I_g) \in \mathcal{I}^\mu_{\mathrm{ME}} \cap \mathcal{I}_\lambda$ *be an IND-adversary. Let* $\delta = 0$. *Then there exists an SS-adversary* $A = (A_c, A_m, A_g) \in \mathcal{A}^\mu_{\mathrm{ME}} \cap \mathcal{A}_\lambda \cap \mathcal{A}^\delta_{\mathrm{BB}}$ *such that for any simulator* S *and all* $k \in \mathbb{N}$

$$\mathbf{Adv}^{\mathrm{ind}}_{\Pi,I}(k) \leq 2 \cdot \mathbf{Adv}^{\mathrm{sss}}_{\Pi,A,S}(k) \,.$$

The running time of A *is that of* I. □

Theorem 5. [IND $\Rightarrow$ BB-CSS] *Let* $\Pi = (\mathcal{K}, \mathcal{E}, \mathcal{D})$ *be a PKE scheme. Let* $0 \leq \delta < 1/2$ *and let* $A = (A_c, A_m, A_g) \in \mathcal{A}^\mu_{\mathrm{ME}} \cap \mathcal{A}_\lambda \cap \mathcal{A}^\delta_{\mathrm{BB}}$ *be an SS-adversary. Then there exists an ind-adversary* $I = (I_c, I_m, I_g) \in \mathcal{I}^\nu_{\mathrm{ME}} \cap \mathcal{I}_\lambda$ *such that for all* $k \in \mathbb{N}$

$$\mathbf{Adv}^{\mathrm{css}}_{\Pi,A}(k) \leq 2 \cdot \mathbf{Adv}^{\mathrm{ind}}_{\Pi,I}(k) + 2^{-k} \,.$$

I *has min-entropy* $\nu(k) = \mu(k) - 1 + \log(1 - 2\delta)$ *and its running time is that of* A *plus the time for* $\lceil -(\log(2/(1+2\delta)))^{-1}\rceil(k+3)+1$ *executions of* A_m. □

5   Deterministic Encryption from Trapdoor Permutations

We construct a deterministic encryption scheme, without ROs, that meets our definitions in the case that the messages being encrypted are uniformly and independently distributed. It is based on the existence of trapdoor permutations. In [4] we generalize the construction to independently distributed messages of high min-entropy μ, but under the (stronger and non-standard) assumption of the existence of trapdoor permutations that are one-way under all input distributions of min entropy μ.

PRIMITIVES. A family of trapdoor permutations $\mathcal{TP} = (G, F, \overline{F})$ is a triple of PT algorithms, with the last two being deterministic. On input 1^k, the key

alg. $\mathcal{K}(1^k)$:	alg. $\mathcal{E}(1^k, pk, x)$:	alg. $\mathcal{D}(1^k, sk, c)$:
$(\phi, \tau) \leftarrow\!\!{}^\$ G(1^k)$	$(\phi, \overline{pk}, s) \leftarrow pk$	$(\tau, \overline{sk}) \leftarrow sk$
$s \leftarrow\!\!{}^\$ \{0,1\}^k$	$y \leftarrow F_\phi^{n(k)}(x)$	$y \leftarrow \overline{\mathcal{D}}(1^k, \overline{sk}, c)$
$(\overline{pk}, \overline{sk}) \leftarrow\!\!{}^\$ \overline{\mathcal{K}}(1^k)$	$\omega \leftarrow \mathcal{G}(1^k, 1^{n(k)}, \phi, x, s)$	$x \leftarrow \overline{F}_\tau^{n(k)}(y)$
$pk \leftarrow (\phi, \overline{pk}, s)$	$c \leftarrow \overline{\mathcal{E}}(1^k, \overline{pk}, y ; \omega)$	Ret x
$sk \leftarrow (\tau, \overline{sk})$	Ret c	
Ret (pk, sk)		

Fig. 3. Algorithms defining our deterministic encryption scheme $\Pi = (\mathcal{K}, \mathcal{E}, \mathcal{D})$

generation algorithm G returns a pair (ϕ, τ) of strings such that $F_\phi(\cdot) = F(\phi, \cdot)$ is a permutation on $\{0,1\}^k$ and $\overline{F}_\tau(\cdot) = \overline{F}(\tau, \cdot)$ is its inverse. If $f \colon \{0,1\}^k \to \{0,1\}^k$ then $f^i \colon \{0,1\}^k \to \{0,1\}^k$ is defined inductively by $f^0(x) = x$ and $f^{i+1}(x) = f(f^i(x))$ for $i \geq 0$ and $x \in \{0,1\}^k$. The Blum-Micali-Yao [11, 28], Goldreich-Levin [22] generator $\mathcal{G}_{\mathcal{TP}}$ takes input $1^k, 1^n, \phi$ and $x, s \in B_k$ and returns

$$\langle F_\phi^0(x), s \rangle \ \| \ \langle F_\phi^1(x), s \rangle \ \| \ \cdots \ \| \ \langle F_\phi^{n-1}(x), s \rangle \ .$$

To discuss the security of our scheme, we say that an SS-adversary is uniform if for every k and every st the components of $\mathbf{x}$ are uniformly and independently distributed over $\{0,1\}^k$ when $(\mathbf{x}, t) \leftarrow\!\!{}^\$ A_m(1^k, st)$. We let $\mathcal{A}_{UN}$ be the class of all uniform SS-adversaries. If $f : B_k \to B_k$ then $f(\mathbf{x})$ denotes the vector whose i^{th} component is $f(\mathbf{x}[i])$. We let $\mathcal{G}_{\mathcal{TP}}(1^k, 1^n, \phi, \mathbf{x}, s)$ be the vector whose i^{th} component is $\mathcal{G}_{\mathcal{TP}}(1^k, 1^n, \phi, \mathbf{x}[i], s)$.

THE CONSTRUCTION. We fix a (randomized) encryption scheme $\overline{\Pi} = (\overline{\mathcal{K}}, \overline{\mathcal{E}}, \overline{\mathcal{D}})$. Assume that $\overline{\mathcal{E}}(1^k, \cdot, \cdot)$ draws its coins from $\{0,1\}^{n(k)}$, and write $\overline{\mathcal{E}}(1^k, pk, x ; \omega)$ for the execution of $\overline{\mathcal{E}}$ on inputs $1^k, pk, x$ and coins ω. Let $\mathcal{TP} = (G, F, \overline{F})$ be a family of trapdoor permutations and $\mathcal{G}_{\mathcal{TP}}$ the associated generator. Our deterministic encryption scheme $\Pi = (\mathcal{K}, \mathcal{E}, \mathcal{D})$ is defined as shown in Figure 3. We refer to it as DE1.

INTUITION. A weird aspect of our scheme is that one is encrypting, under the standard scheme $\overline{\mathcal{E}}$, a message y under coins ω that are related to y. The challenge is to show that this works assuming $\mathcal{TP}$ is one-way and $\overline{\Pi}$ is IND-CPA. So let $A = (A_c, A_m, A_g) \in \mathcal{A}_{UN} \cap \mathcal{A}_\lambda$ be an adversary with associated information length $\ell(\cdot)$ and number of messages $v(\cdot)$ that is successful in violating the A-CSS security of Π. It is not hard to see that the assumed security of $\overline{\Pi}$ allows us to reduce our task to showing that it is hard for a PT adversary D to have a non-negligible advantage in computing the challenge bit b in the following distinguishing game. The game generates $\phi, \tau, \overline{pk}, \overline{sk}, s$ as done by $\mathcal{K}(1^k)$ and lets $(\mathbf{x}, t) \leftarrow\!\!{}^\$ A_m(1^k, \lambda)$. It lets

$$\boldsymbol{\omega}_1 \leftarrow \mathcal{G}_{\mathcal{TP}}(1^k, 1^{n(k)}, \phi, \mathbf{x}, s) \quad \text{and} \quad \boldsymbol{\omega}_0 \leftarrow\!\!{}^\$ B_{n(k)}^{v(k)} \ ,$$

picks a random challenge bit b, and provides the adversary D with ϕ, s, $F_\phi^{n(k)}(\mathbf{x})$, $\boldsymbol{\omega}_b$, and t. Now, D's task would be merely the standard (and known to be hard) one of breaking the pseudorandomness of $\mathcal{G}_{T\mathcal{P}}$ (meaning, we would be done) but for one catch, namely that D has "help" information t about the seed(s) $\mathbf{x}$. If we could somehow remove it we would be done, but this seems hard to do directly. Instead, we first produce from D an adversary I' that solves (although still with help) a computational (rather than decision) problem, namely that of inverting F_ϕ: given ϕ, $F_\phi(x)$, and $\ell(\cdot)$ bits of information about x, our adversary computes x. This is obtained by noting that the Goldreich-Levin [22] and Blum-Micali-Yao [11, 28] proof of pseudorandomness of $\mathcal{G}_{T\mathcal{P}}$ based on the one-wayness of $T\mathcal{P}$ generalizes to say that $\mathcal{G}_{T\mathcal{P}}$ remains pseudorandom in the presence of $\ell(\cdot)$ bits of help information about the seed assuming $T\mathcal{P}$ is one-way in the presence of $\ell(\cdot)$ bits of help information about the input. Now we need to turn I' into an adversary succeeding at the same task, but without help. We appeal to Theorem 1, which allows us to assume our starting adversary A was boolean, meaning $\ell(\cdot) = 1$. In this case it is easy to dispense with the help provided to I because we can try both values of it and lower our success probability by at most a factor of 2.

We remark that we have made crucial use of the fact that the adversary constructed by Theorem 1 has the same message space as the original one. This means that if the latter is in $\mathcal{A}_{UN}$ then so is the former, so that B-CSS for uniform adversaries implies A-CSS for uniform adversaries. We now proceed to the full proof.

OWPs AND PRGs WITH HELP. For our proof, we will need to extend the usual frameworks of one-wayness and pseudorandomness to adversaries with "help." An inversion adversary $J = (J_c, J_p, J_s)$ is a triple of non-uniform algorithms. If $T\mathcal{P} = (G, F, \overline{F})$ is a family of trapdoor permutations we let

$$\mathbf{Adv}_{T\mathcal{P},J}^{\mathrm{owf}}(k) = \Pr\left[\mathbf{Exp}_{T\mathcal{P},J}^{\mathrm{owf}}(k) \Rightarrow \mathsf{true}\right]$$

where the experiment is shown in Figure 4. The running time of J is defined as the sum of the running times of J_c and J_s, so that J is PT if J_c, J_s are PT. (J_p is not required to be PT.) We say that J has help-length $\ell(\cdot)$ if the output of $J_p(1^k, \cdot, \cdot, \cdot)$ is always of length $\ell(k)$. We say that J is unaided if it has help length $\ell(\cdot) = 0$. We let $\mathcal{J}_\ell$ denote the class of all PT inversion adversaries with help length $\ell(\cdot)$. We say $T\mathcal{P}$ is one-way for help-length $\ell(\cdot)$ if $\mathbf{Adv}_{T\mathcal{P},J}^{\mathrm{owf}}(\cdot)$ is negligible for all $J \in \mathcal{J}_\ell$. We say that $T\mathcal{P}$ is one-way if it is one-way for help-length $\ell(\cdot) = 0$. The following, although trivial, will be very useful.

Proposition 1. *Let $T\mathcal{P}$ be a family of trapdoor permutations and J an inversion adversary with help-length $\ell(\cdot)$. Then there is an inversion adversary J' with help-length 0 such that*

$$\mathbf{Adv}_{T\mathcal{P},J}^{\mathrm{owf}}(k) \leq 2^{\ell(k)} \cdot \mathbf{Adv}_{T\mathcal{P},J'}^{\mathrm{owf}}(k)$$

for all k, and the running time of J' is that of J plus $\mathcal{O}(\ell)$.   □

$\mathbf{Exp}^{\mathrm{owf}}_{\mathcal{TP},J}(k)$	$\mathbf{Exp}^{\mathrm{prg}\text{-}v}_{\mathcal{TP},D,n}(k)$
$(\phi,\tau) \leftarrow_\$ G(1^k)\,;\ st \leftarrow_\$ J_{\mathrm{c}}(1^k,\phi)$	$(\phi,\tau) \leftarrow_\$ G(1^k)\,;\ st \leftarrow_\$ D_{\mathrm{c}}(1^k,\phi)$
$x \leftarrow_\$ \{0,1\}^k\,;\ t \leftarrow_\$ J_{\mathrm{p}}(1^k,x,\phi,st)$	$\mathbf{x} \leftarrow_\$ B_k^{v(k)}\,;\ s \leftarrow_\$ \{0,1\}^k\,;\ d \leftarrow_\$ \{0,1\}$
$y \leftarrow F_\phi(x)\,;\ x' \leftarrow_\$ J_{\mathrm{s}}(1^k,\phi,st,y,t)$	$t \leftarrow_\$ D_{\mathrm{p}}(1^k,\mathbf{x},\phi,st)$
Ret $(x = x')$	$\boldsymbol{\omega}_1 \leftarrow \mathcal{G}_{\mathcal{TP}}(1^k,1^{n(k)},\phi,\mathbf{x},s)$
	$\boldsymbol{\omega}_0 \leftarrow_\$ B_{n(k)}^{v(k)}$
	$d' \leftarrow_\$ D_{\mathrm{g}}(1^k,\phi,st,F_\phi^{n(k)}(\mathbf{x}),\boldsymbol{\omega}_d,s,t)$
	Ret $(d = d')$

Fig. 4. (Left) Experiment defining one-wayness of $\mathcal{TP} = (G,F,\overline{F})$. **(Right)** Experiment defining pseudorandomness of $\mathcal{G}_{\mathcal{TP}}$.

Proof. Let $J = (J_{\mathrm{c}}, J_{\mathrm{p}}, J_{\mathrm{s}})$ and $J' = (J_{\mathrm{c}}, \Lambda, J'_{\mathrm{s}})$ where $J'_{\mathrm{s}}(1^k, \phi, st, y, \lambda)$ lets $t \leftarrow_\$ \{0,1\}^{\ell(k)}$ and returns $J_{\mathrm{s}}(1^k, \phi, st, y, t)$. ∎

A PRG adversary $D = (D_{\mathrm{c}}, D_{\mathrm{p}}, D_{\mathrm{g}})$ is a triple of non-uniform algorithms. If $\mathcal{TP} = (G, F, \overline{F})$ is a family of trapdoor permutations and $\mathcal{G}_{\mathcal{TP}}$ is the corresponding generator we let

$$\mathbf{Adv}^{\mathrm{prg}\text{-}v}_{\mathcal{TP},D,n}(k) = 2 \cdot \Pr\left[\ \mathbf{Exp}^{\mathrm{prg}\text{-}v}_{\mathcal{TP},D,n}(k) \Rightarrow \mathsf{true}\ \right] - 1$$

where the experiment is shown in Figure 4 and $v(\cdot), n(\cdot) \colon \mathbb{N} \to \mathbb{N}$. The running time of D is defined as the sum of the running times of D_{c} and D_{g}, so that D is PT if $D_{\mathrm{c}}, D_{\mathrm{g}}$ are PT. (D_{p} is not required to be PT.) We say that D has help-length $\ell(\cdot)$ if the output of $D_{\mathrm{p}}(1^k,\cdot,\cdot,\cdot)$ is always of length $\ell(k)$. We let $\mathcal{D}_\ell$ denote the class of all PT PRG-adversaries with help length $\ell(\cdot)$. We say $\mathcal{G}_{\mathcal{TP}}$ is pseudorandom for help-length $\ell(\cdot)$ if $\mathbf{Adv}^{\mathrm{prg}\text{-}v}_{\mathcal{TP},D,n}(\cdot)$ is negligible for all $D \in \mathcal{D}_\ell$ and all polynomials v, n. We say that $\mathcal{G}_{\mathcal{TP}}$ is pseudorandom if it is pseudorandom for help-length $\ell(\cdot) = 0$. We remark that it is important that D_{p} does not get s as input, meaning the help information is only about x. The following says that if $\mathcal{TP}$ is one-way for help-length $\ell(\cdot)$ then $\mathcal{G}_{\mathcal{TP}}$ is pseudorandom for help-length $\ell(\cdot)$. The case $\ell(\cdot) = 0$ is the standard result [11, 22, 28] saying that $\mathcal{G}_{\mathcal{TP}}$ is pseudorandom if $\mathcal{TP}$ is one-way. The proof of the following is in the full version [4].

Lemma 1. *Let $\mathcal{TP} = (G, F, \overline{F})$ be a family of trapdoor permutations. Let $v(\cdot)$, $n(\cdot)$ be polynomials. Let D be a PRG-adversary with help-length $\ell(\cdot)$ and let $\epsilon(\cdot) = \mathbf{Adv}^{\mathrm{prg}\text{-}v}_{\mathcal{TP},D,n}(\cdot) > 0$. Then there is an inversion adversary J with help-length $\ell(\cdot)$ such that*

$$\epsilon(k) \le 4n(k)v(k) \cdot \mathbf{Adv}^{\mathrm{owf}}_{\mathcal{TP},J}(k)$$

and the running time of J is

$$T_J = \mathcal{O}(k^3 n^4 v^4 \epsilon^{-4}) + \mathcal{O}(T_D + nvT_F)k^2 n^2 v^2 \epsilon^{-2}\ ,$$

where T_X is the running time of X. □

IND-CPA. Associate to (randomized) encryption scheme $\overline{\Pi} = (\overline{\mathcal{K}}, \overline{\mathcal{E}}, \overline{\mathcal{D}})$ and adversary B the experiment $\mathbf{Exp}^{\text{ind-cpa}}_{\overline{\Pi}, B}(k)$ defined by

$$b \leftarrow_\$ \{0, 1\} \,;\, (\overline{pk}, \overline{sk}) \leftarrow_\$ \overline{\mathcal{K}}(1^k) \,;\, b' \leftarrow_\$ B^{\overline{\mathcal{E}}_{\overline{pk}}(\text{LR}(\cdot, \cdot, b))}(\overline{pk}) \,;\, \text{Ret } (b = b')$$

where $\text{LR}(M_0, M_1, b) = M_b$. B is an IND-CPA adversary if all its oracle queries consist of equal length strings. Let

$$\mathbf{Adv}^{\text{ind-cpa}}_{\overline{\Pi}, B}(k) = 2 \cdot \Pr\left[\mathbf{Exp}^{\text{ind-cpa}}_{\overline{\Pi}, B}(k) \Rightarrow \text{true}\right] - 1 \,.$$

We say that $\overline{\Pi}$ is IND-CPA secure if $\mathbf{Adv}^{\text{ind-cpa}}_{\overline{\Pi}, B}(\cdot)$ is negligible for all PT IND-CPA adversaries B.

SECURITY OF OUR SCHEME. The following says that our scheme is B-CSS secure for uniform adversaries assuming $\mathcal{TP}$ is one-way and $\overline{\Pi}$ is IND-CPA secure. By Theorem 1 it is A-CSS secure for uniform adversaries under the same assumptions and a constant factor loss in security. Since the existence of one-way trapdoor permutations implies the existence of IND-CPA secure encryption schemes we obtain the results under the sole assumption of the existence of one-way trapdoor permutations.

Theorem 6. *Let $\mathcal{TP} = (G, F, \overline{F})$ be a family of trapdoor permutations and $\overline{\Pi} = (\overline{\mathcal{K}}, \overline{\mathcal{E}}, \overline{\mathcal{D}})$ an encryption scheme. Let $\Pi = (\mathcal{K}, \mathcal{E}, \mathcal{D})$ be the associated deterministic encryption scheme as per our construction above. Let $A = (A_c, A_m, A_c) \in \mathcal{A}_B \cap \mathcal{A}_\lambda \cap \mathcal{A}_{UN}$ be an SS-adversary against Π with advantage $\epsilon(\cdot) = \mathbf{Adv}^{\text{css}}_{\Pi, A}(\cdot) > 0$ and number of messages $v(\cdot)$. Then there is an unaided inversion adversary J and an IND-CPA adversary B such that for all $k \in \mathbb{N}$*

$$\epsilon(k) \leq 2 \cdot \mathbf{Adv}^{\text{ind-cpa}}_{\overline{\Pi}, B}(k) + 16n(k)v(k) \cdot \mathbf{Adv}^{\text{owf}}_{\mathcal{TP}, J}(k) \,. \tag{10}$$

The running time of B is that of A plus $\mathcal{O}(nT_F + T_G)$ and it makes $v(k)$ oracle queries. The running time of J is

$$\mathcal{O}(k^3 n^4 v^4 \epsilon^{-4}) + \mathcal{O}(T_A + T_{\overline{\mathcal{E}}} + T_{\overline{\mathcal{K}}} + nvT_F) \cdot k^2 n^2 v^2 \epsilon^{-2} \tag{11}$$

where T_X is the running time of X.   □

Proof. Consider the experiments of Figure 5. There $\overline{\mathcal{E}}(1^k, \overline{pk}, \mathbf{y} \,;\, \boldsymbol{\omega})$ is the vector whose i^{th} component is $\overline{\mathcal{E}}(1^k, \overline{pk}, \mathbf{y}[i] \,;\, \boldsymbol{\omega}[i])$. Let

$$P_a = \Pr\left[\mathbf{Exp}^{d\text{-}a}_{\Pi, A}(k) \Rightarrow \text{true}\right]$$

for $a \in \{0, 1\}$. Then

$$\mathbf{Adv}^{\text{css}}_{\Pi, A}(k) = 2P_1 - 1 = 2(P_1 - P_0) + (2P_0 - 1) \,.$$

Adversary B is shown in Figure 5, and we omit the (easy) analysis establishing that

$$2P_0 - 1 \leq \mathbf{Adv}^{\text{ind-cpa}}_{\overline{\Pi}, B}(k) \,.$$

$\mathbf{Exp}_{\Pi,A}^{d\text{-}1}(k)$ / $\boxed{\mathbf{Exp}_{\Pi,A}^{d\text{-}0}(k)}$	$\text{adversary } B^{\overline{\mathcal{E}_{\overline{pk}}}(\mathrm{LR}(\cdot,\cdot,b))}(\overline{pk}):$
$b \leftarrow_{\$} \{0,1\}$	$(\mathbf{x}_0,t_0),(\mathbf{x}_1,t_1) \leftarrow_{\$} A_m(1^k,\lambda)$
$(\mathbf{x}_0,t_0),(\mathbf{x}_1,t_1) \leftarrow_{\$} A_m(1^k,\lambda)$	$(\phi,\tau) \leftarrow_{\$} G(1^k)\;;\; s \leftarrow_{\$} \{0,1\}^k$
$(\phi,\tau) \leftarrow_{\$} G(1^k)\;;\; s \leftarrow_{\$} \{0,1\}^k$	$pk \leftarrow (\phi,\overline{pk},s)$
$(\overline{pk},\overline{sk}) \leftarrow_{\$} \overline{\mathcal{K}}(1^k)\;;\; pk \leftarrow (\phi,\overline{pk},s)$	$\mathbf{y}_0 \leftarrow F_\phi^{n(k)}(\mathbf{x}_0)\;;\; \mathbf{y}_1 \leftarrow F_\phi^{n(k)}(\mathbf{x}_1)$
$\boldsymbol{\omega} \leftarrow \mathcal{G}_{\mathcal{TP}}(1^k,1^{n(k)},\phi,\mathbf{x}_b,s)$	For $i = 1,\dots,v(k)$ do
$\boxed{\boldsymbol{\omega} \leftarrow_{\$} B_{n(k)}^{v(k)}}$	$\quad \mathbf{c}[i] \leftarrow_{\$} \overline{\mathcal{E}_{\overline{pk}}}(\mathrm{LR}(\mathbf{y}_0[i],\mathbf{y}_1[i],b))$
$\mathbf{y} \leftarrow F_\phi^{n(k)}(\mathbf{x}_b)\;;\; \mathbf{c} \leftarrow \overline{\mathcal{E}}(1^k,\overline{pk},\mathbf{y}\,;\,\boldsymbol{\omega})$	$g \leftarrow_{\$} A_g(1^k,pk,\mathbf{c},\lambda)$
$g \leftarrow_{\$} A_g(1^k,pk,\mathbf{c},\lambda)$	If $g = t_1$ then Ret 1 else Ret 0
If $g = t_1$ then $b' \leftarrow 1$ else $b' \leftarrow 0$	
Ret $(b = b')$	

Fig. 5. (Left) Experiments used in the proof of Theorem 6. The experiment d-0 includes the boxed statement while d-1 does not. **(Right)** IND-CPA adversary for proof of Theorem 6.

alg. $D_{\mathrm{p}}(1^k,\mathbf{x},\phi,\lambda)$:	**alg.** $D_{\mathrm{g}}(1^k,\phi,\lambda,\mathbf{y},\boldsymbol{\omega},s,t)$:
Repeat	$c \leftarrow_{\$} \{0,1\}\;;\; \mathbf{y}_1 \leftarrow \mathbf{y}\;;\; t_1 \leftarrow t\;;\; \boldsymbol{\omega}_1 \leftarrow \boldsymbol{\omega}$
$\quad (\mathbf{x}',t') \leftarrow_{\$} A_m(1^k,\lambda)$	$(\mathbf{x}_0,t_0) \leftarrow_{\$} A_m(1^k,\lambda)$
Until $(\mathbf{x}' = \mathbf{x})$	$(\overline{pk},\overline{sk}) \leftarrow_{\$} \overline{\mathcal{K}}(1^k)\;;\; pk \leftarrow (\phi,\overline{pk},s)$
$t \leftarrow t'$	$\boldsymbol{\omega}_0 \leftarrow \mathcal{G}_{\mathcal{TP}}(1^k,1^{n(k)},\phi,\mathbf{x}_0,s)\;;\; \mathbf{y}_0 \leftarrow F_\phi^{n(k)}(\mathbf{x}_0)$
Ret t	$\mathbf{c} \leftarrow \overline{\mathcal{E}}(1^k,\overline{pk},\mathbf{y}_c\,;\,\boldsymbol{\omega}_c)$
	$g \leftarrow_{\$} A_g(1^k,pk,\mathbf{c},\lambda)$
	If $(g = t_c)$ then $c' \leftarrow 1$ else $c' \leftarrow 0$
	Ret $c \oplus c' \oplus 1$

Fig. 6. PRG adversary for proof of Theorem 6

Next we define PRG-adversary $D = (\Lambda, D_{\mathrm{p}}, D_{\mathrm{g}})$ with help length $\ell(\cdot)$ as shown in Figure 6 and claim that

$$P_1 - P_0 \le 2 \cdot \mathbf{Adv}_{\mathcal{TP},D,n}^{\mathrm{prg}\text{-}v}(k)\,. \tag{12}$$

Let J' be the inversion adversary obtained from D by Lemma 1. It also has help-length $\ell(\cdot)$. Now apply Proposition 1 to get inversion adversary J with help-length 0. In [4] we justify (12), (10) and (11) to conclude the proof.   ∎

INSTANTIATIONS. DE1 admits quite efficient instantiations. Say we want to encrypt a 1024 bit (random) message. Let the trapdoor one-way permutation be squaring modulo a 1024-bit composite number N [9] that is part of the public key. Then the PRG requires n squarings, where n is the number of bits of randomness required by the (randomized) encryption scheme $\overline{\Pi}$. Let $\overline{\Pi}$ be the

Blum-Goldwasser scheme [10], also using a 1024-bit modulus. (This modulus, also part of the public key, must be different from N.) Then encryption cost of DE1 is that of Blum-Goldwasser (1024 squarings) plus $n = 1024$ squarings for the PRG to get coins for $\overline{\Pi}$. (We assume here, and below, an efficient mapping from bits to group elements, otherwise n increases by a small amount.) Decryption time also doubles, coming in at about 4 exponentiations modulo 512 bit numbers (less than one 1024 bit exponentiation!) using Chinese remainders. The ciphertext size is that of Blum-Goldwasser, namely 2048 bits, and security rests solely on factoring. Alternatively, let $\overline{\Pi}$ be El Gamal hybrid encryption using a 160-bit group. (A universal hash of the DH key is used to one-time symmetrically encrypt the data.) Encryption time for DE1 is that of hybrid El Gamal plus the time for $n = 320$ squarings modulo N, decryption time is 2 exponentiations modulo 512 bit numbers plus one 160-bit exponentiation. and the ciphertext size is only 1344 bits. The security assumption is now factoring + DDH.

DISCUSSION. One might ask why we did not work with IND rather than with CSS notions. The reason is that it is unclear how to meaningfully capture the case of uniformly and independently distributed messages with IND. We could certainly say that an IND-adversary $I = (I_c, I_m, I_g)$ is uniform if for every k and every st, b the components of $\mathbf{x}$ are uniformly distributed over $\{0, 1\}^k$ when $\mathbf{x} \leftarrow_\$ I_m(1^k, b, st)$. But such an adversary would always have zero advantage.

Acknowledgments

Mihir Bellare was supported in part by NSF grants CNS 0524765 and CNS 0627779 and a gift from Intel Corporation. Marc Fischlin was supported in part by the Emmy Noether Program Fi 940/2-1 of the German Research Foundation (DFG). Adam O'Neill was supported in part by Alexandra Boldyreva's NSF CAREER award 0545659. Thomas Ristenpart was supported in part by the above-mentioned grants of the first author.

References

1. Bellare, M.: The Goldreich-Levin Theorem (manuscript),
   http://www-cse.ucsd.edu/users/mihir/papers/gl.pdf
2. Bellare, M., Boldyreva, A., Micali, S.: Public-key encryption in a multi-user setting: Security proofs and improvements. In: Preneel, B. (ed.) EUROCRYPT 2000. LNCS, vol. 1807, pp. 259–274. Springer, Heidelberg (2000)
3. Bellare, M., Boldyreva, A., O'Neill, A.: Deterministic and efficiently searchable encryption. In: Menezes, A. (ed.) CRYPTO 2007. LNCS, vol. 4622, pp. 535–552. Springer, Heidelberg (2007)
4. Bellare, M., Fischlin, M., O'Neill, A., Ristenpart, T.: Deterministic Encryption: Definitional Equivalences and Constructions without Random Oracles. Full version of this paper. IACR ePrint archive (2008) http://eprint.iacr.org/
5. Bellare, M., Rogaway, P.: Random oracles are practical: A paradigm for designing efficient protocols. In: Conference on Computer and Communications Security – CCS 1993, pp. 62–73. ACM, New York (1993)

6. Bellare, M., Rogaway, P.: Robust computational secret sharing and a unified account of classical secret-sharing goals. In: Conference on Computer and Communications Security – CCS 2007, pp. 172–184. ACM, New York (2007)
7. Bellare, M., Rogaway, P.: The security of triple encryption and a framework for code-based game-playing proofs. In: Vaudenay, S. (ed.) EUROCRYPT 2006. LNCS, vol. 4004, pp. 409–426. Springer, Heidelberg (2006)
8. Bellare, M., Sahai, A.: Non-malleable encryption: Equivalence between two notions, and an indistinguishability-based characterization. In: Wiener, M. (ed.) CRYPTO 1999. LNCS, vol. 1666, pp. 519–536. Springer, Heidelberg (1999)
9. Blum, L., Blum, M., Shub, M.: A simple unpredictable pseudo-random number generator. SIAM Journal on Computing 15, 364–383 (1986)
10. Blum, M., Goldwasser, S.: An efficient probabilistic public-key encryption scheme which hides all partial information. In: Blakely, G.R., Chaum, D. (eds.) CRYPTO 1984. LNCS, vol. 196, pp. 289–302. Springer, Heidelberg (1984)
11. Blum, M., Micali, S.: How to generate cryptographically strong sequences of pseudorandom bits. SIAM Journal on Computing 13, 850–864 (1984)
12. Boldyreva, A., Fehr, S., O'Neill, A.: On notions of security for deterministic encryption, and efficient constructions without random oracles. In: Wagner, D. (ed.) CRYPTO 2008. LNCS, vol. 5157, pp. 335–359. Springer, Heidelberg (2008)
13. Boneh, D., Di Crescenzo, G., Ostrovsky, R., Persiano, G.: Public key encryption with keyword search. In: Cachin, C., Camenisch, J.L. (eds.) EUROCRYPT 2004. LNCS, vol. 3027, pp. 506–522. Springer, Heidelberg (2004)
14. Canetti, R.: Towards realizing random oracles: Hash functions that hide all partial information. In: Kaliski Jr., B.S. (ed.) CRYPTO 1997. LNCS, vol. 1294, pp. 455–469. Springer, Heidelberg (1997)
15. Canetti, R., Micciancio, D., Reingold, O.: Perfectly one-way probabilistic hash functions (Preliminary version). In: Symposium on the Theory of Computation – STOC 1998, pp. 131–141 (1998)
16. Damgaard, I., Hofheinz, D., Kiltz, E., Thorbek, R.: Public-key encryption with non-interactive opening. In: Malkin, T. (ed.) CT-RSA 2008. LNCS, vol. 4964, pp. 239–255. Springer, Heidelberg (2008)
17. Desrosiers, S.: Entropic security in quantum cryptography. arXiv e-Print quant-ph/0703046 (2007), http://arxiv.org/abs/quant-ph/0703046
18. Desrosiers, S., Dupuis, F.: Quantum entropic security and approximate quantum encryption. arXiv e-Print quant-ph/0707.0691 (2007), http://arxiv.org/abs/0707.0691
19. Dodis, Y., Smith, A.: Entropic security and the encryption of high entropy messages. In: Kilian, J. (ed.) TCC 2005. LNCS, vol. 3378, pp. 556–577. Springer, Heidelberg (2005)
20. El Gamal, T.: A public-key cryptosystem and a signature scheme based on discrete logarithms. In: Blakely, G.R., Chaum, D. (eds.) CRYPTO 1984. LNCS, vol. 196, pp. 10–18. Springer, Heidelberg (1985)
21. Goldreich, O.: A uniform complexity treatment of encryption and zero-knowledge. Journal of Cryptology 6, 21–53 (1993)
22. Goldreich, O., Levin, L.: A hard-core predicate for all one-way functions. In: Symposium on the Theory of Computation – STOC 1989, pp. 25–32. ACM, New York (1989)
23. Goldwasser, S., Micali, S.: Probabilistic encryption. Journal of Computer and Systems Sciences 28(2), 412–426 (1984)
24. Micali, S., Rackoff, C., Sloan, R.: The notion of security for probabilistic cryptosystems. SIAM Journal on Computing 17(2), 412–426 (1988)

25. Peikert, C., Waters, B.: Lossy trapdoor functions and their applications. In: Symposium on the Theory of Computing – STOC 2008, pp. 187–196. ACM, New York (2008)
26. Russell, A., Wang, H.: How to fool an unbounded adversary with a short key. In: Knudsen, L.R. (ed.) EUROCRYPT 2002. LNCS, vol. 2332, pp. 133–148. Springer, Heidelberg (2002)
27. Song, D., Wagner, D., Perrig, A.: Practical techniques for searches on encrypted data. In: Symposium on Security and Privacy, pp. 44–55. IEEE, Los Alamitos (2000)
28. Yao, A.: Theory and applications of trapdoor functions. In: Symposium on Foundations of Computer Science – FOCS 1982, pp. 80–91. IEEE, Los Alamitos (1982)

Communication Complexity in Algebraic Two-Party Protocols

Rafail Ostrovsky[*] and William E. Skeith III[**]

Department of Computer Science and Department of Mathematics,
University of California, Los Angeles
rafail@cs.ucla.edu,wskeith@math.ucla.edu

Abstract. In cryptography, there has been tremendous success in building various two-party protocols with small communication complexity out of homomorphic semantically-secure encryption schemes, using their homomorphic properties in a black-box way. A few notable examples of such primitives include items like single database Private Information Retrieval (PIR) schemes (introduced in [15]) and private database update with small communication (introduced in [5]). In this paper, we illustrate a general methodology for determining what types of protocols can and cannot be implemented with small communication by using homomorphic encryption in a black-box way.

We hope that this work will provide a simple "litmus test" of feasibility for black-box use of known homomorphic encryption schemes by other cryptographic researchers attempting to develop new protocols with low communication. Additionally, a precise mathematical language for reasoning about such problems is developed in this work, which may be of independent interest. We stress that the class of algebraic structures for which we prove communication complexity lower bounds is large, and covers practically all known semantically-secure homomorphic cryptosystems (including those based upon bilinear maps).

Finally, we show the following equivalence which relates group homomorphic encryption and a major open question of designing a so-called fully-homomorphic cryptosystem: a fully homomorphic encryption scheme (over a non-zero ring) exists if and only if there exists homomorphic encryption over any finite non-abelian simple group. This result somewhat generalizes results of Barrington [1] (to any group containing a finite non-abelian simple subgroup) and of Maurer and Rhodes [18], and in fact gives a *constructive* proof of the 1974 result Werner [28]. (This also answers an open question posed by Rappe in [23], who in 2004 proved a special case of this result.)

Keywords: homomorphic encryption, fully homomorphic encryption, private information retrieval, PIR writing, keyword search, communication complexity, algebraic lower bounds.

[*] Supported in part by IBM Faculty Award, Xerox Innovation Group Award, NSF grants 0430254, 0716835, 0716389 and U.C. MICRO grant.
[**] Supported in part by U.C. Chancellor's Presidential Dissertation Fellowship 2006-2007 and by NSF grant 0430254.

© International Association for Cryptologic Research 2008

1   Introduction

One of the central problems in cryptography is that of finding a public key encryption scheme that would allow "computation on encrypted data". In its full generality the problem could be simply stated as follows: to find a public key encryption scheme such that given encryptions of arbitrary plaintexts $\mathcal{E}(x_1), \ldots, \mathcal{E}(x_n)$ it is possible *without the decryption key* to compute $\mathcal{E}(f(x_1, \ldots, x_n))$ for any polynomial-time computable function f. Naturally, if one can find a public-key cryptosystem that is "fully homomorphic", i.e. allows operations on ciphertext that preserve the structure of a ring, and hence allows computation of the ubiquitous *"NAND"* operation on the underlying plaintext, it would give a general solution to the above problem. Indeed, the reason this is such a central problem is that it would create a powerful mechanism to arbitrarily manipulate encrypted data without sacrificing privacy. This problem was posed nearly 30 years ago by Rivest, Adelman and Dertouzos [24]. We do not know if such an encryption scheme exists in its full generality, though various partial answers are known: One partial answer is abelian group-homomorphic encryption: given $\mathcal{E}(x)$ and $\mathcal{E}(y)$, where x and y come from some abelian group, there exist cryptosystems that can compute $\mathcal{E}(x*y)$, where $*$ is the group operation. Examples include ElGamal [9], where the group operation is multiplication, Goldwasser and Micali [10] where the operation is addition modulo 2, and Pallier [22] where the group operation is addition modulo a large composite. Recent progress by Boneh, Goh and Nissim [3] has shown that more is possible: they designed a cryptosystem that allows an arbitrary number of additions and a single multiplication (of the underlying plaintext) by manipulating ciphertexts only. I.e., polynomials of total degree 2 can be computed on ciphertext. Another approach at building fully-homomorphic encryption was considered by Sander, Young, and Yung [26], but only applied to Boolean operations that doubled the ciphertext size at every step. As a result, one could only perform a few Boolean operations before the ciphertext size became impractical. A partial negative result was given by Boneh and Lipton [4].

Many useful protocols and primitives have been derived from such homomorphic schemes in a "black box" way, essentially just manipulating the homomorphic properties to construct various systems. Prominent examples include single-database private information retrieval (PIR), originally introduced by Kushilevitz and Ostrovsky [15] and collision-resistant hashing as shown by Ishai, Kushilevitz, and Ostrovsky [14]. (For more details regarding this approach to PIR, see the survey of Ostrovsky and Skeith [20].) In this work, we show a variety of communication complexity lower bounds for natural tasks when constructed in a similar, but somewhat more restricted manner (to further improve communication complexity, the aforementioned protocols often use repeated encryption, destroying the algebraic value of the resulting ciphertext). More accurately we'll illustrate a single basic task that cannot be algebraically accomplished (with small communication) in various structures (e.g., that of *any* abelian group). This result will give us a simple criterion or "litmus test" for determining the feasibility of constructing communication-efficient protocols in general, and a rule out the possibility for constructing many communication-efficient protocols based on the black box use of homomorphic encryption. Along the way, we'll also develop a mathematical language and technique for reasoning about such questions, which may be of independent interest. A lot of effort has been put into designing new cryptosystems that allow the structure to be as rich as possible, but our lower bounds capture an even larger class of algebraic structures than what current homomorphic encryption schemes provide.

1.1   Our Results

A few of the main results in this work are as follows, where n represents the database size in a PIR-writing scheme:

Theorem 1 (informal). We prove $\Omega(n)$ bound for algebraic **PIR-writing** based on **any** abelian group homomorphic encryption.

Theorem 2 (informal). We prove $\Omega(\sqrt{n})$ bound for algebraic PIR-writing based on the cryptosystem of Boneh, Goh and Nissim [3]. We note that the work of Boneh, Kushilevitz, Ostrovsky and Skeith [5] shows a matching upper bound for PIR-writing using [3] in a black-box way. Thus, we prove a *matching* black-box lower bound for [5].

Theorem 3 (informal). We prove $\Omega(\sqrt[t]{n})$ bound for algebraic PIR-writing based on homomorphic encryption that allows evaluation of total degree t multivariate polynomials on ciphertext. (We stress that cryptosystems for such structures are not known today beyond polynomials of total degree 2.)

Theorem 4 (informal). We show a constructive proof of a 1974 theorem of Werner [28] demonstrating the existence of a fully homomorphic encryption scheme (over a non-zero ring) if and only if there exists homomorphic encryption over any finite non-abelian simple group. (In the full version [21], we also show an explicit construction to implement a composable "NAND" gate from a group formula in any non-abelian simple group.) This also generalizes the result of Barrington [1] to all groups containing a finite non-abelian simple subgroup, as well as generalizing a result from Rappe [23].

A central element of this paper, from which we will derive a number of results, is an algebraic lower bound for a certain task- that of specifying "characteristic vectors" over a group. For a group G, we call a vector $(v_1, ..., v_n) \in G^n$ "characteristic" for a set $S \subset [n]$ if $v_i \neq \mathbf{0}_G$ if and only if $i \in S$, where $\mathbf{0}_G$ is the identity of G. We'll show that

> **Theorem (informal):** For *any abelian group*, communication complexity $\Omega(n)$ is required to "algebraically" specify characteristic vectors of arbitrary singleton subsets of $[n]$.

A formal statement of this idea appears as Theorem 2.

We stress that this statement holds for *all* abelian groups. For intuition, one may consider the case of linear algebra, in which the group G is of prime order, and has a field structure which could be put in place. It is a relatively simple exercise to prove this special case of our theorem, just arguing about the degree of vector spaces. However, note that this technique does not get very far. As the reader will see from Example 1 below, these ideas don't apply to general abelian groups G, even in the special case of cyclic groups. (Note that there is not even a well-defined notion of degree in this setting.) A "degree-based" argument could likely be carried out via free-module analysis, but it will substantially complicate and obfuscate matters, and furthermore it will yield a weaker version of the theorem. The abstract approach taken here will yield a strong algebraic result which will be of great utility later on, when we generalize to other structures.

Additionally, we prove a smooth trade-off in communication complexity as the number of non-identity elements in the characteristic vectors increases, and as mentioned, we also generalize to other algebraic structures, which contain virtually all structures that are preserved by known homomorphic encryption schemes. In particular, we prove results for *any abelian group* as well as results for arbitrary rings, in a setting restricted to polynomials of total degree t. (For an example of the case $t = 2$, see the cryptosystem of Boneh, Goh, and Nissim [3].) Finally, we'll show a number of natural cryptographic protocols that would imply the functionality of generating characteristic vectors, and hence derive algebraic lower bounds for the communication involved in these protocols as well.

As one will see after an examination of our algebraic results, they are in fact quite general. Since the results for abelian groups apply to all affine maps, this rules out many possibilities which do not necessarily come from group formulas. (For example, arbitrary endomorphisms may now be included in the class of "formulas" even though there is generally no way to compute all endomorphisms via an abelian group formula.) In particular, even if one changes their representation of data to be not just one group element, but many, and furthermore manipulates each of these elements independently, our results still apply (this is a simple consequence of Corollary 2).

Finally, regarding the equivalence of ring and group homomorphic encryption, we demonstrate that with *any* simple non-abelian group structure one can (constructively) compute all finite functions via group formulas and thus, the existence of any cryptosystem homomorphic over a simple non-abelian group implies a fully-homomorphic encryption scheme. This work can be found in the later sections, and somewhat generalizes results of Barrington [1] and Rhodes [18], however it is essentially a new and constructive version of the results of Werner [28] and may be of independent interest. This also answers an open question posed by Rappe in [23], who in 2004 proved a special case of this result.

1.2   Related Work

The lower bounds that we consider are most closely related to computational lower bounds on number theoretic problems when algorithms are restricted only to underlying group operations. For example, Boneh and Lipton [4] examine the computational difficulty breaking any algebraically homomorphic (over a field) cryptosystem. In contrast, our lower bounds are on communication complexity and apply to a wide variety of algebraic structures. Other related works are that of Shoup [27] and Maurer and Wolf [17], which consider computational difficulty of the discrete logarithm problem, and other number-theoretic problems in cyclic groups, provided that the algorithms do not exploit any specific properties of the representation of group elements.

Again, our lower bounds are geared towards communication complexity and program size, rather than computational complexity, but similar to these works, we focus only on algorithms that utilize nothing other than the underlying algebraic structures. However, we consider a greater variety of structures in our work (including arbitrary abelian groups and bounded degree multivariate polynomials over rings).

1.3   Overview, Motivation and Intuition

Often times, novel cryptographic protocols are developed using homomorphic encryption as building block (and often it is the only necessary ingredient). Many basic protocols can be constructed in this way, for example, private information retrieval, oblivious transfer, and collision-resistant hashing, to name a few. Indeed, such methods have accomplished much in the past, and continue to prove themselves as fruitful techniques. However, the types of algebraic structures available in homomorphic encryption schemes are limited. Not much beyond the structure of an abelian group can be preserved under an encryption scheme. Quite clearly, abelian groups have limited computing power. If one simply examines the number of distinct m-variable "formulas" in a finite abelian group G of order k in comparison to the number of G-valued functions (as set maps) that depend on m variables, one can't help but notice a great discrepancy in cardinality, so indeed, there is much that cannot be computed using only abelian group formulas. But what are these functions? Furthermore, in what sense can they not be computed or represented?

Using a black box model for homomorphic encryption, one is limited to only computing such formulas. However, there are a vast number of other types of "algebraic" maps which cannot necessarily be derived from any such formula that we study as well. We'll show that these maps also do not suffice for our tasks. As a somewhat trivial example, consider the endomorphism on $G = \mathbb{Z}_p \times \mathbb{Z}_p$ obtained by switching the coordinates. If one has only black box access to the group operation, then this endomorphism is not computable, however, if elements are represented as coordinate pairs, then computing this map is trivial. This is by no means the most complex example, but it does illustrate the benefits of an abstract approach (which will naturally cover all endomorphisms).

As mentioned before, there have been many protocols of great utility derived from homomorphic encryption over abelian groups (e.g. [15,6,14]). However, as the authors believe, for every such useful protocol in the literature, there are many dead ends, lying at the bottom of stacks of paper upon researchers' desks. But until now, there has not been much formal proof that these dead ends are actually just that. This work provides some basic proofs of lower bounds for a few simple protocols, based on these algebraic assumptions. More importantly, any task that can be reduced to our basic task is also immediately impossible to accomplish in an algebraic way with small communication complexity. We illustrate the power of these reductions on a number of examples below.

1.4   Implications of the Results

As we have mentioned, the applications of these results as lower bounds for cryptographic protocols are limited to an algebraic setting and are not absolutes. However, in many situations the bounds are quite practical. We'd like to take a moment to better illustrate and clarify where these results apply and where they do not. Additionally, we'll demonstrate the algebraic strength of the results, which are quite complete in the algebraic context.

The practical cryptographic significance of the results primarily deals with building protocols for computing on encrypted data. Let us consider single-database PIR, introduced by Kushilevitz and Ostrovsky [15]. PIR schemes are often based upon homomorphic encryption (e.g., [15],[16]). In the most efficient versions of these schemes, note that the answers to queries can be viewed as encryptions of the appropriate database elements- however, due to the repeated segmentation and application used to achieve better efficiency, these encryptions have *no algebraic value* after the second iteration (i.e. recursive calls in [15] scheme.) Roughly what is meant is that there is no way to combine two or more of these results (without the decryption key) to obtain an encryption of some other meaningful combination of the original elements. Looking at PIR alone, in its own context, this is not much of a problem. However, if PIR were to be used as a subroutine for some larger computation on encrypted data, this lack of algebraic value of the PIR "answer" could be very inconvenient. For example, the keyword search of [19] could be improved greatly if an efficient *algebraic* PIR protocol existed (see Section 3 for more details). To summarize very briefly, these results apply to situations where it is necessary to preserve (in the ciphertext) algebraic value of the results of underlying computation on encrypted data. In situations where algebraic value can safely be destroyed, there are often much more communication-efficient solutions.

We obtain results that hold *for all abelian groups*, as well as several other structures. We believe that this level of generality is a necessity. Details can be found in [21] (the full version of this work), but we summarize here. First of all, just using the simple structure of abelian groups, and the general abundance of cryptosystems that are homomorphic over cyclic groups, it is not hard to imagine constructing a homomorphic cryptosystem over virtually any abelian group. (For a more formal approach to this idea, see [11].) So, to make the results have any significance at all, a study at this level of generality is necessary, even though it requires additional machinery. In the case say of prime order cyclic groups, linear algebra suffices to solve the problem. However, this already breaks down for a general cyclic group (see Section 2). In addition to being insufficient from the start, a less general approach to the problem will also interfere the generalizations to other structures (see 3), as well as weakening the basic applications. For example, in addition to all algebraic formulas, the algebraic results as stated here cover the entire ring of endomorphisms (which in general may have no algebraic formula at all, much less a linear one). This would greatly complicate the set of functions to consider, and makes an elementary approach difficult. However, the abstract approach eliminates these issues[1].

2 Preliminaries and Basic Results

Most notations used are standard, and the algebraic notation used is typically consistent with [13]. However, a more comprehensive list of the notation used in this work can be found in [21], the full version.

2.1 Equivalence of Homomorphic Encryption over Non-abelian Simple Groups and Rings

We'll begin by stating a positive result regarding the equivalence of homomorphic encryption over non-abelian simple groups and rings. For more thorough formalizations, please see [21], the full version of this work.

Theorem 1. *Let G be a finite non-abelian simple group. Then any function $f : \{0,1\}^m \longrightarrow \{0,1\}^n$ can be represented solely in terms of the group operation of G.*

First, we'll prove a few elementary lemmas, and then the theorem (which again, uses only basic techniques from algebra). To begin, recall that from the Feit-Thompson theorem and Cauchy's theorem, we have that every non-abelian simple group of finite order has an element of order 2.

Lemma 1. *Let G be a finite group and suppose that $S \subset G$ is conjugation invariant (i.e., $\forall s \in S, g \in G$ we have $gsg^{-1} \in S$). Then $\langle S \rangle \lhd G$.*

The proof is straightforward, but can be found in its entirety in the full version [21].

Consider for a moment, the conjugacy classes. For an element $x \in G$, we will denote the conjugacy class by $\mathrm{Cl}_G(x)$. I.e.,

$$\mathrm{Cl}_G(x) = \{y \in G \mid y = gxg^{-1} \text{ some } g \in G\}$$

Recall that we can define a natural action of G on $\mathrm{Cl}_G(x)$ for any $x \in G$: for all $s \in \mathrm{Cl}_G(x)$, simply define $g \cdot s = gsg^{-1}$. Now, let G be a non-abelian simple group of finite order. From Cauchy's theorem, we know that there exists $x \in G$ such that x has order 2. Consider $\mathrm{Cl}_G(x)$. Let $|\mathrm{Cl}_G(x)| = k$. It must be the case that $k > 1$. If not, then every element of G conjugates x to itself, and hence we

---

[1] Again, consider the simple example of $G = \mathbb{Z}_p \times \mathbb{Z}_p$ as a black box and as a direct product coordinate representation. The endomorphism $\varphi \in \mathrm{Hom}_{\mathbb{Z}}(G, G)$ by $(a, b) \mapsto (b, a)$ is not computable as a formula, but clearly is computable if given a coordinate representation.

have $x \in \mathbf{Z}(G)$, the center of G. But of course this is impossible since the center of a group is always normal and we assumed that G is simple. So, the conjugacy class of x has at least two elements. Recall next, that whenever a group acts on a set S of size k, there is an induced homomorphism,

$$\varphi : G \longrightarrow S_k$$

Since the action of G on $\mathrm{Cl}_G(x)$ is obviously transitive, and since the size k of the class of x is greater than 1, we see that φ cannot be the trivial homomorphism which sends all elements to the identity, and hence $\ker(\varphi) \neq G$. But, since G is simple, we in fact know that $\ker(\varphi)$ must be the trivial subgroup $\{e\}$, since the kernel is always normal. *Therefore, every element of G acts non-trivially on the set $\mathrm{Cl}_G(x)$.*

We will extract the useful information into the following lemma which we have just now proved.

Lemma 2. *Let G be a finite, non-abelian simple group, and let $x \in G$ be an element of order 2. Then there exists an element $y \in \mathrm{Cl}_G(x)$ such that $yxy^{-1} \neq x$, and hence, such that $[x, y] \neq e$.*

Using these facts, we can now prove Theorem 1.

Proof. We will simply show that the function $\mathrm{NAND}(a, b)$ is computable in this way, which suffices to prove the theorem since any such function $f : \{0, 1\}^m \longrightarrow \{0, 1\}^n$ can be written in terms of compositions of NAND alone. More precisely, we will show that for an element x of order 2, the set $\{e, x\}$ can be identified with $\{0, 1\}$ respectively, and the operation NAND can be computed solely in terms of the group operation of G.

So, to begin, let $x \in G$ be of order 2, which as we discussed exists by Cauchy's theorem. Define $C = \mathrm{Cl}_G(x)$. As discussed, $|C| > 1$. Consider $S = [C, C]$, the set of *commutators in C*. Note that the subset S is conjugation invariant since it is generated by $C = \mathrm{Cl}_G(x)$, which is quite clearly conjugation-invariant. Hence by Lemma 1, the subgroup generated by *these specific commutators*, is a normal subgroup: $\langle S \rangle = \langle [C, C] \rangle \triangleleft G$ However, by Lemma 2, we know that $|S| > 1$, as there are at least 2 non-commuting elements. But, we assumed that G was simple. Therefore, we have in fact that $\langle S \rangle = G$. So, in particular, there exists some product, $s_1 s_2 \cdots s_k$ of commutators in C such that $s_1 s_2 \cdots s_k = x$ So, each $s_i = [r_i, t_i]$ where r_i and t_i are both conjugate to x. Therefore we have sequences of group elements, $\{g_i\}_{i=1}^k$ and $\{h_i\}_{i=1}^k$ such that $[g_i x g_i^{-1}, h_i x h_i^{-1}] = s_i$ We are now ready to define our $\mathrm{NAND}(a, b)$. First, define the function $\mathrm{AND}(a, b)$ as follows:

$$\mathrm{AND}(a, b) = \prod_{i=1}^{k} [g_i a g_i^{-1}, h_i b h_i^{-1}]$$

It is now easy to observe that it performs the appropriate function on our inputs from $\{e, x\}^2$. Whenever a or b is set to the identity, every commutator will of course be the identity since all elements commute with e. However, if both a and

b are set to the group element x, the by our design, we will have $\text{AND}(x, x) = x$, exactly as desired. Now, since x has order 2, we can simply define $\text{NAND}(a, b) = \text{AND}(a, b)x$. This completes the proof.

Corollary 1. *Constructing a fully homomorphic encryption scheme over a ring with identity is equivalent to constructing a group homomorphic encryption over any finite non-abelian simple group. In particular, it is equivalent to constructing a homomorphic encryption scheme over A_5, the smallest such group.*

Proof: This is almost immediate, but see the full version [21], or [23] for more detail.

As mentioned, detailed examples and formalizations can be found in the full version [21] of this work.

2.2   Generating Encryptions of Characteristic Vectors: Motivation

This example provides a simple description of a protocol that can't be non-trivially implemented with abelian group algebra. Later, we'll show a variety of problems (usually related to PIR or PIR-writing) which would imply a protocol like this. Hence, these too cannot be implemented with abelian group algebra.

We could, at this point, formalize a cryptographic protocol about generating characteristic-type vectors over a group, but it may be convenient to postpone such a definition and instead get right to the main algebraic point. So for the moment, we'll just explain in simple terms the algebraic task we are trying to accomplish.

Consider the following problem: Let $n, m \in \mathbb{Z}^+$, and let G be an abelian group. Define the following elements $v_i \in G^n$:

$$v_i = (\mathbf{0}_G, ..., \mathbf{0}_G, x_i, \mathbf{0}_G, ..., \mathbf{0}_G)$$

where $x_i \neq \mathbf{0}_G$ appears in the i-th position.[2] Let $\{\mathbf{m}_i\}_{i=1}^n \subset G^m$ and let f be an arbitrary affine group map in m variables from $G^m \longrightarrow G^n$, i.e., $f = f_m + c$ where $f_m : G^m \longrightarrow G^n$ is linear and $c \in G^n$. Note that these affine maps can express all possible abelian group formulas on a set of variables (see the full version [21] for complete formalization and definitions). The question is

Question 1. **(Informal)** If $f(\mathbf{m}_i) = v_i$ for all $i \in [n]$, what can be said about $|G^m|$? In particular, how small can it be?

We will soon answer this question in a variety of contexts, but first we'll give an example to help motivate the question and our lower bound. The phrasing used regarding the size estimation was deliberate: we don't isolate or bound m alone, because we cannot bound m in a non-trivial way. It is in fact possible to accomplish the above result with $m = 1$, even for a cyclic group. However, as we'll show in our lower bound, this comes at the cost of increasing the size of G.

---

[2] We give x an index i simply to show that it need not be uniform across all vectors.

Example 1. Let $n \in \mathbb{Z}^+$, and let $N = \prod_{i=1}^{n} p_i$, where p_i is the i-th prime number. Define $G = \mathbb{Z}_N$. Define integers $\{z_i\}_{i=1}^{n}$ as follows:

$$z_i = \prod_{j \neq i} p_j$$

Then, since all the primes were distinct, it is easy to verify that

$$(z_i z_j \neq 0 \bmod N) \iff (i = j)$$

So, we could define a linear function $f = (f_1, ..., f_n)$ from $G \longrightarrow G^n$ by $f_i(x) = z_i \cdot x$, and we would have $f(z_i) = v_i$, for some elements $v_i \in G^n$ which fit the above description of a complete set of characteristic vectors.

However, in the preceding example, notice that n different primes had to divide the order of G. Hence, $|G| > 2^n$ is of exponential size in n. We will show that even using affine maps, this is always the case: to generate n orthogonal-type characteristic vectors with m group elements always requires a group G such that G^m has exponential size in n, although the statement we prove has a more abstract setting.

2.3   A Basic Algebraic Result

Here, we will make precise the relationship regarding n and the size of an abelian group that can algebraically generate a complete set of n characteristic vectors over an abelian group G. Again, to conserve space, we direct the reader to the full version [21] for most of the proofs.

Theorem 2. *Let $n \in \mathbb{Z}^+$ and let G, A be abelian groups. Let $V = \{v_i\}_{i=1}^{n} \subset G^n$ be any collection of elements so that the j-th position of v_i is $\mathbf{0}_G$ if and only if $i \neq j$. Then if $F = f + c$ is an affine map from $A \longrightarrow G^n$ such that $V \subset F(A)$ then we have $\log(|A|) \in \Omega(n)$. More specifically, if $A \subset G^m$, we have that*

$$\log(|G|) \geq \frac{n}{m+1}$$

The proof of this theorem is given in the full version [21]. The pieces used are outlined below, and their proofs can also be found in [21] as well. To begin, we'll prove the following lemma which will help us analyze affine maps and translated characteristic vectors.

Lemma 3. *Let R be a finite ring with identity, and let M be a (unitary) R-module. Let $\Omega = \{\omega_i\}_{i=1}^{k} \subset M$ be a finite collection of elements. Let $\Omega' = \{(\omega_i + c)\}_{i=1}^{k}$ for some fixed element $c \in M$. Then $\langle \Omega' \rangle$, the module generated by Ω', increases in size by at most a factor of $|R|$ over the size of $\langle \Omega \rangle$. I.e.,*

$$\frac{|\langle \Omega' \rangle|}{|\langle \Omega \rangle|} \leq |R|$$

Proof: See the full version [21].

In light of Lemma 3, we need only to analyze "un-translated" characteristic-type vectors. If they generate a large module, then so will the translated vectors. It is quite clear any such module generated by elements like those in V will be exponential in size, however to be complete, we provide a formal proof.

Observation 3. *Let G be a finite abelian group. Let $n \in \mathbb{Z}^+$. Define elements $v_i \in G^n$ by $v_{ij} = \delta_{ij} \cdot \alpha_i$ for some $\alpha_i \neq 0 \in G$, and $\delta_{ij} \in \mathbb{Z}$ with $\delta_{ii} = 1$ for all i and $\delta_{ij} = 0$ for $i \neq j$. Let $H = \langle \{v_i\}_{i=1}^n \rangle$, the subgroup of G^n generated by the v_i. Then $|H| \geq 2^n$.*

Proof: See the full version [21].

We'll also make use of a few very elementary observations from group theory. As elementary as they may be, proofs can none the less be found in the full version [21].

Observation 4. *Let G be an abelian group and let $a, b \in G$ with $x = \operatorname{ord}(a), y = \operatorname{ord}(b)$. Then $\operatorname{ord}(ab) \mid \operatorname{lcm}(x, y)$.*

Observation 5. *Let G, H be groups, and let $f : G \longrightarrow H$ be a homomorphism. Then for all $g \in G$, we have that $\operatorname{ord}(f(g)) \mid \operatorname{ord}(g)$.*

Observation 6. *Let G be a group, and let $(a, b) \in G \times G$. Then $\operatorname{ord}((a, b)) = \operatorname{lcm}(\operatorname{ord}(a), \operatorname{ord}(b))$.*

Observation 7. *Let G be an abelian group, and suppose that there exists $N \in \mathbb{Z}^+$ such that $N \cdot g = \mathbf{0}_G$ for all $g \in G$, where $\cdot$ denotes $\mathbb{Z}$-module action. Then, G is a $\mathbb{Z}_N$-module, where the action is inherited from that of $\mathbb{Z}$.*

These basic observations and lemmas are enough for the proof of Theorem 2 (which again, can be found in the full version [21]).

2.4   Functions That Change Multiple Values

We can also generalize this algebraic result to include other types of vectors, where $F(\mathfrak{m}_i)$ has the i-th component non-identity, but possibly some other number of positions are non-identity elements as well. If the function F has the ability to change arbitrary subsets of c elements for a constant c, then our original results clearly apply, as you could re-organize G^n as a product $G^c \times \cdots \times G^c$ with n/c components. (Without loss of generality, we assume $c|n$.) However, the bounds still apply for less powerful classes of functions. We will show that *any* function that produces vectors with $c(n)$ or fewer non-identity positions at a time has communication complexity $\Omega(n/c(n))$, provided only that it is complete- i.e., for every position, it has the ability to produce a vector that is non-identity in that position. Here, $c(n)$ is any positive function of n, and note also that the number of non-identity positions per $\mathfrak{m}_i$ need not be uniform- we only ask that it is bounded by $c(n)$. We'll prove this by showing that we can always re-organize

G^n into a product of larger components (of size $c(n)$) so that the original function F produces orthogonal characteristic-type vectors in the original sense, only over $(G^{c(n)})^{n/c(n)}$. Then, the proof follows immediately from the original result. Consider the following lemma.

Lemma 4. *Let $c \in \mathbb{Z}^+$. Let $\{S_k\}_{k \in \Gamma}$ be a collection of sets such that $S_k \subseteq [n]$, $|S_k| \leq c$ for all $k \in [n]$ and such that the $\{S_k\}$ form a cover of $[n]$, i.e., $\bigcup_{k \in \Gamma} S_k = [n]$. Then there exists $X \subseteq [n]$ and a sub-collection of sets $\{S_{k_j}\}_{k_j \in \Lambda \subseteq \Gamma}$ such that $S_{k_j} \cap S_{k_{j'}} \cap X = \varnothing$ whenever $j \neq j'$ yet $S_{k_j} \cap X \neq \varnothing$ for at least $\lceil n/c \rceil$ of the sets S_{k_j}.*

Proof: See the full version [21].

Corollary 2. *Let $n \in \mathbb{Z}^+$ and let G, A be abelian groups. Let $w(x)$ be a positive valued function and let $V = \{v_i\}_{i=1}^n \subset G^n$ be any collection of elements so that the i-th position of v_i is not equal to $\mathbf{0}_G$, and at most $w(n)$ total positions of v_i are non-identity for all $i \in [n]$. Then if $F = f + c$ is an affine map from $A \longrightarrow G^n$ such that $V \subset F(A)$ then we have $\log(|A|) \in \Omega(n/w(n))$.*

Proof: See the full version [21].

2.5   Polynomials of Bounded Total Degree

Recently, new cryptosystems have been developed with additional homomorphic properties (see [3]), which provide the ability to compute on ciphertext, polynomials of total degree at most 2. Here, we will generalize our original algebraic result to apply to algebraic functions of the form of any polynomial of total degree t, over a ring R. Although the following result will apply to the ring of polynomials over any ring R (it need not have an identity or be commutative), this result has the most meaning in the case of commutative rings with identity, since in this case the ring of multivariate polynomials coincides precisely with our notion of "algebraic formula", which is formalized in [21], the full version of this work. (For a non-commutative ring, there's a more general structure that serves as the set of all formulas.)

Corollary 3. *Let $n \in \mathbb{Z}^+$ and let R be any ring. Let $V = \{v_i\}_{i=1}^n \subset R^n$ be any collection of elements so that the j-th position of v_i is not equal to $\mathbf{0}_R$ precisely when $j = i$, for all $i, j \in [n]$. Then if $F : R^m \longrightarrow R^n$ is such that $F = (F_1, ..., F_n)$ with each $F_i \in R[X_1, ..., X_m]$ of total degree less than or equal to t (a constant) and has $V \subset F(R^m)$ then we have $\left(\sqrt[t]{\log(|R|)}\right) m \in \Omega(\sqrt[t]{n})$. In particular, if $|R|$ is independent of n, then $m \in \Omega(\sqrt[t]{n})$.*

Proof: See the full version [21].

3   Applications of Algebraic Results

We will discuss here a number of protocols which are both easy to state, and would provide desirable functionalities, yet under algebraic assumptions, they cannot be very well implemented with existing technology.

3.1   Private Database Modification (PIR Writing)

As seen in [5], the ability to privately modify an encrypted database in a communication efficient way could provide a valuable tool for private computation. One very natural approach to such a problem, is to proceed in a manner analogous to many PIR protocols, and use homomorphic encryption as a building block (as was done in [5]).

The protocol would then communicate encrypted values which encode the modification to take place, and then the database owner would execute some algebraic operations on the encrypted database and the description given by the user to update the database contents. Since all of the communication consisted only of encrypted values, CPA-type security comes easily from a hybrid type argument. Unfortunately, we have very limited structures available to homomorphic encryption schemes. Almost always, what is preserved is the operation of an abelian group. At best, the ability to evaluate polynomials of total degree 2 is provided (see [3]). It will follow from our preliminary algebraic results, that these types of algebraic protocols cannot be very well implemented with existing encryption schemes. We'll often speak of "algebraic" maps, which will usually mean functions that are obtainable from some type of formula involving only the operations of the algebraic structure. A precise, formal, and detailed exposition of this idea is given in [21], the full version of this work. Also, we've omitted some of the formal protocol-type definitions to improve readability, since there isn't much surprising about them, and most readers of this paper could likely re-invent them in a few minutes. We'll instead give an informal description of the protocol here. For precise statements, again we direct the reader to the full version [21].

Let **U** be a user that wishes to update the database, and denote by **DB** the database owner. We'll summarize a protocol for algebraic database modification between **U** and **DB** via the following steps, in which we assume that G is an abelian group. Below, we'll just describe the algebra involved. In an actual privacy preserving protocol, everything will of course be computed on ciphertext in some homomorphic encryption scheme over G.

1. **U** selects $\mathfrak{m}_i = (g_1, ..., g_m)$ to modify position i and sends $\mathfrak{m}_i$ to **DB**.
2. **DB** computes an algebraic function $F(X, \mathfrak{m}_i, H)$ of the database $X \in G^n$, the modification description $\mathfrak{m}_i$, and other inputs of his own, $H \in G^\epsilon$.
3. **DB** replaces X by $X' = F(X, \mathfrak{m}_i, H)$

Clearly the algebra involved in this protocol implies the ability to algebraically generate complete sets of characteristic vectors:

Claim. An algebraic protocol for database modification over an abelian group implies an algebraic function (affine map) with a complete set of characteristic vectors in the image.

Proof sketch: Define a database $X = \{0_G\}_{i=1}^n$, which is the identity in all positions. Apply **DB**'s function to obtain $X' = F(X, \mathfrak{m}_i, H)$ where $\mathfrak{m}_i$ describes a modification for position i. Then clearly $X' = v_i$, a characteristic vector in G^n, non-identity at position i.   □

Therefore, by Theorem 2, if we build such a protocol based on a homomorphic cryptosystem over *any* abelian group, it will necessarily have linear communication complexity. Note the strong sense in which this is true: abelian group formulas always correspond to affine maps, but certainly not every affine map comes from such a formula.[3] So, we've shown that even if we allow **DB** to somehow compute arbitrary affine maps on the ciphertext values, it still does not suffice to accomplish this task. Furthermore, Theorem 2 did not even assume that the groups were the same. So, even if the database elements are encrypted in some other cryptosystem which is homomorphic over a group different than that of the descriptions, and even if we were provided the ability to compute all algebraic maps from one to the other on ciphertext, we still couldn't produce a non-trivial protocol over abelian groups. We'll summarize these ideas as

Corollary 4. *There are no non-trivial Algebraic Oblivious Database Modifiers over an abelian group. I.e., any oblivious database modifier based on the operations of an abelian group has communication complexity $\Omega(n)$.*

3.2 Algebraic and Homomorphic Protocols for PIR

As a second corollary, we consider "algebraic", or "homomorphic" protocols for private information retrieval. One may have observed, as the authors have, that the query results for PIR protocols usually fall into one of two categories: either (a) they have no (or very limited) algebraic value[4] or homomorphic properties, or (b) the server side communication is non-constant, i.e., the results of a query return many items, not just an encryption of one value in the database[5]. A protocol for private information retrieval that returns encryptions of single values which retain algebraic and homomorphic properties could be a very useful tool in private computation[6], especially in non-interactive settings. In what follows, we present evidence that the absence of such protocols is perhaps to be expected.

We'll try to establish a basic definition that captures the properties that we desire, and encapsulates most existing work possessing these properties. Suppose that the values in a database have some algebraic structure. For now, say that of an abelian group which we'll denote $(G, \cdot)$. We will denote the return value of a PIR query for the i-th position of a database by $\mathrm{PIR}(i)$, which consists of one or more encrypted database elements. Let $S_i = \{s_j\}_{j=1}^k$ denote the set of values from the database that are returned by a PIR query for position i.

---

[3] Again, consider $G = \mathbb{Z}_p \times \mathbb{Z}_p$ and $\varphi \in \mathrm{Hom}_{\mathbb{Z}}(G, G)$ by $(a, b) \mapsto (b, a)$.

[4] See the work of [7] for an example of such a PIR protocol having "limited" algebraic value.

[5] See [15] for such an example, but many PIR protocols based on homomorphic encryption (over an abelian group) have this property.

[6] For example, in the keyword search of [19], the dictionary size could be reduced.

Suppose for a moment that the domain from which PIR query returns reside has the algebraic structure of a group as well, say $(G', \star)^7$. To name just a few examples, one can see that the PIR protocols of [15], [6], [7] all fit this description. We could then make the following definition:

Definition 1. *Using the notation established above, we say that a PIR protocol is* **homomorphic** *if for a given database $X \in G^n$, we have that $\mathcal{D}(\text{PIR}(i) \star \text{PIR}(j)) = S_i \cdot S_j$ where $\mathcal{D}$ is the function from the PIR protocol that decrypts the query results.*

Note also that for such a PIR protocol to be of much utility as a subroutine in some non-interactive private computation, it is almost essential to have $|S_i| = 1$, or at least bounded by a small constant. If not, then the party which is to perform a computation on the return values of a homomorphic PIR query will likely not have any information about where the relevant element is in the query results. Hence, if such a party wishes to perform a computation on t variables obtained via homomorphic PIR queries, it would in general require repeatedly performing the computation on all $|S_i|^t$ possible sequences to ensure that the right variables were involved at least once. Furthermore, it may not be possible for any party to distinguish which of the resulting outputs in fact corresponds to the desired computation, even after decryption.

Finally note that from the definition of homomorphic PIR, we see that the results of queries are in fact encryptions of elements in some homomorphic cryptosystem. To create such a PIR protocol, a very natural approach is to manipulate the algebraic structure of some such homomorphic cryptosystem. This motivates the following definition.

Definition 2. *We say that a PIR protocol is* **algebraic** *if the following hold:*

1. *A query consists of an ordered sequence of ciphertexts in some cryptosystem where the plaintext set A has some algebraic structure.*
2. *To process a query, the database owner computes on ciphertext some algebraic function of the query's array, this function being determined by the contents of the database to obtain an array of ciphertext which will be the results of the query.*

For precise definitions of the term "algebraic function", we again direct the reader to [21], the full version of this work. In the case of abelian groups, these definitions yield affine maps as our model of algebraic functions.

Corollary 5. *Consider an abelian group algebraic PIR protocol with sender-side communication complexity $g(n)$ and server-side communication complexity $h(n)$.*

---

[7] There is no assumption that the group representing the query returns are the same as the database elements, or even that they are encryptions of database elements, exactly. It could be the case that as a part of the encryption, the group that the database elements come from is first homomorphically transformed, and then transformed back as a part of decryption. The general way that we've stated our algebraic results will be useful for such a definition.

Then $g(n)h(n) = \Omega(n)$. More specifically, if $k(n)$ is any positive integer-valued function and if the server's response consists of $k(n)$ encrypted values, then the sender-side communication complexity is $\Omega(n/k(n))$.

Proof: See the full version [21].

Using Corollary 3, we can generalize this result to cryptosystems that may have additional homomorphic properties (see [3]), showing $\Omega(\sqrt[t]{n})$ bounds if total degree t polynomials over a ring R can be computed on ciphertext.

For example, if given a cryptosystem that allows polynomials of fixed total degree t to be computed on ciphertext over some ring R, we can easily construct an algebraic PIR protocol with sender-side communication $\Theta(\sqrt[t]{n})$ and server-side complexity $\Theta(1)$ (see [3], or [21] for details of a simple example). However, this is in fact meets a lower bound: In general, if such a protocol has sender-side complexity $g(n)$ and server-side complexity $h(n)$, then we can show that $g(n)h(n) = \Omega(\sqrt[t]{n})$, which is a simple consequence of Corollary 3.

3.3   Private Keyword Searching [19]

As another relatively simple corollary, we resolve (under our algebraic assumptions) an open problem posed by Ostrovsky and Skeith [19] regarding extending the query semantics for private searching on streaming data. We show that without new homomorphic encryption schemes with additional properties, their methods cannot be extended to perform conjunctive queries.

Corollary 6. *The problem of private keyword search on streaming data as proposed in [19], has no non-trivial algebraic solution for a conjunctive query of two or more terms if the underlying cryptosystem is only group homomorphic over an abelian group.*

Remark: We will assume the same basic framework as developed in [19] for a solution and show that there is no such solution that performs conjunctive queries. Specifically, we assume that a dictionary with an associated array of ciphertexts is used to conditionally encrypt documents as in [19].

Proof. First note that the protocol inherently gives rise to an algebraic method for generating complete sets of characteristic vectors: Suppose that the dictionary D has size m. Each word has its role in the query encoded via an encrypted group element, say in some group G. Look at the encoded dictionary (un-encrypted) as the set G^m. Suppose we have a protocol as described in [19] for some query that involves k variables. Running this protocol on m^k documents which run over all unique k-tuples from the dictionary gives us a set of characteristic vectors inside of $G^{(m^k)}$. So, we can think of this as an algebraic map from $G^m \longrightarrow G^{(m^k)}$, which (unless the query is somewhat trivial) contains a complete set of characteristic vectors in the image. But, now the question is how many positions are non-identity in each vector? This of course depends on the query. Suppose that the query is a disjunction of terms. Each vector in $G^{(m^k)}$ will have at least $m^{k-1}k$ positions that are non-identity, since $k-1$ entries could be arbitrary as long as one contains a keyword. So, the ratio of total positions to non-identity positions is less than m and our algebraic lower bounds give no contradiction (which of course should be the case since [19] gives such a construction). But now consider a conjunctive query, just of two terms. In the same way as described above, this gives rise to an algebraic function for characteristic vectors from $G^m \longrightarrow G^{(m^2)}$, however this time we have $\mathcal{O}(1)$ positions of each vector are non-identity. So, applying Corollary 2, we see that no such protocol can exist based on an abelian group. More generally, from Corollary 3, we see that if given the ability to compute total degree t polynomials, we can construct a protocol that executes a conjunction of *at most t* terms.

We believe that this example illustrates particularly well a situation in which the bounds proved in this work are especially useful. The entire method of [19] critically depends on the ability to generate these types of characteristic vectors so that the final representation is an encryption in a homomorphic scheme. This is the case since the functionality of characteristic vectors is used as a subroutine for the larger procedure, and so to continue the computation (i.e., writing to the buffer, etc.) it is necessary that the output have algebraic value. So, since we have proven that this subroutine is impossible to implement in the required manner, it seems that improving the work of [19] would require either a completely new approach, or new designs of homomorphic encryption schemes, such as fully-homomorphic encryption.

It is this type of information that we hope will save researchers time and effort in the future. Applying these bounds may not give an absolute impossibility, but it can quickly eliminate a very large space of what might otherwise seem to be feasible approaches to the problem.

References

1. Barrington, D.: Bounded-Width Polynomial-Size Branching Programs Recognize Exactly Those Languages in NC. In: STOC 1986, pp.1–5 (1986)
2. Boneh, D., Crescenzo, G., Ostrovsky, R., Persiano, G.: Public Key Encryption with Keyword Search. In: Cachin, C., Camenisch, J.L. (eds.) EUROCRYPT 2004. LNCS, vol. 3027, pp. 506–522. Springer, Heidelberg (2004)
3. Boneh, D., Goh, E., Nissim, K.: Evaluating 2-DNF Formulas on Ciphertexts. In: Kilian, J. (ed.) TCC 2005. LNCS, vol. 3378, pp. 325–341. Springer, Heidelberg (2005)
4. Boneh, D., Lipton, R.: Searching for Elements in Black Box Fields and Applications. In: Koblitz, N. (ed.) CRYPTO 1996. LNCS, vol. 1109, pp. 283–297. Springer, Heidelberg (1996)
5. Boneh, D., Kushilevitz, E., Ostrovsky, R., Skeith, W.: Public Key Encryption that Allows PIR Queries. In: Menezes, A. (ed.) CRYPTO 2007. LNCS, vol. 4622. Springer, Heidelberg (2007)
6. Chang, Y.C.: Single Database Private Information Retrieval with Logarithmic Communication. In: Wang, H., Pieprzyk, J., Varadharajan, V. (eds.) ACISP 2004. LNCS, vol. 3108. Springer, Heidelberg (2004)

7. Cachin, C., Micali, S., Stadler, M.: Computationally private information retrieval with polylogarithmic communication. In: Stern, J. (ed.) EUROCRYPT 1999. LNCS, vol. 1592, pp. 402–414. Springer, Heidelberg (1999)
8. Chor, B., Goldreich, O., Kushilevitz, E., Sudan, M.: Private information retrieval. J. of the ACM 45, 965–981 (1998)
9. ElGamal, T.: A Public-Key Cryptosystem and a Signature Scheme Based on Discrete Logarithms. IEEE Transactions on Information Theory IT-31(4), 469–472 (1985)
10. Goldwasser, S., Micali, S.: Probabilistic encryption. J. Comp. Sys. Sci. 28(1), 270–299 (1984)
11. Grigoriev, D., Ponomarenko, I.: Homomorphic public-key cryptosystems over groups and rings. Quaderni di Mathematica 13, 305–325 (2004)
12. Herstein, I.N.: Abstract Algebra. Prentice-Hall, Englewood Cliffs (1986, 1990, 1996)
13. Hungerford, T.W.: Algebra. Springer, Berlin (1984)
14. Ishai, Y., Kushilevitz, E., Ostrovsky, R.: Sufficient Conditions for Collision-Resistant Hashing. In: Kilian, J. (ed.) TCC 2005. LNCS, vol. 3378. Springer, Heidelberg (2005)
15. Kushilevitz, E., Ostrovsky, R.: Replication is not needed: Single database, computationally-private information retrieval. In: FOCS 1997, pp. 364–373 (1997)
16. Lipmaa, H.: An Oblivious Transfer Protocol with Log-Squared Communication. IACR ePrint Cryptology Archive 2004/063
17. Maurer, U., Wolf, S.: Lower bounds on generic algorithms in groups. In: Nyberg, K. (ed.) EUROCRYPT 1998. LNCS, vol. 1403, pp. 72–84. Springer, Heidelberg (1998)
18. Maurer, W., Rhodes, J.: A property of finite non-Abelian simple groups. Proc. Am. Math. Soc. 16, 522–554 (1965)
19. Ostrovsky, R., Skeith, W.: Private Searching on Streaming Data. In: Shoup, V. (ed.) CRYPTO 2005. LNCS, vol. 3621, pp. 397–430. Springer, Heidelberg (2005); Journal of Cryptology 20(4), 397–430 (October 2007)
20. Ostrovsky, R., Skeith, W.: A Survey of Single Database PIR: Techniques and Applications. In: Okamoto, T., Wang, X. (eds.) PKC 2007. LNCS, vol. 4450. Springer, Heidelberg (2007)
21. Ostrovsky, R., Skeith, W.: Algebraic Lower Bounds for Computing on Encrypted Data. Electronic Colloquium on Computational Complexity report TR07-22
22. Paillier, P.: Public Key Cryptosystems based on CompositeDegree Residue Classes. In: Stern, J. (ed.) EUROCRYPT 1999. LNCS, vol. 1592, pp. 223–238. Springer, Heidelberg (1999)
23. Rappe, D.K.: Homomorphic Cryptosystems and their Applications Ph.D. Thesis, under E. Becker and J. Patarin (2004)
24. Rivest, R.L., Adleman, L., Dertouzos, M.L.: On data banks and privacy homomorphisms. In: DeMillo, R.A., et al. (eds.) Foundations of Secure Computation, pp. 169–179. Academic Press, London (1978)
25. Rivest, R.L., Shamir, A., Adleman, L.: A method for obtaining digital signatures and public key cryptosystems. Commun. ACM 21, 120–126 (1978)
26. Sander, T., Young, A., Yung, M.: Non-Interactive CryptoComputing For NC1. In: FOCS 1999, pp. 554–567 (1999)
27. Shoup, V.: Lower Bounds for Discrete Logarithms and Related Problems. In: Fumy, W. (ed.) EUROCRYPT 1997. LNCS, vol. 1233, pp. 256–266. Springer, Heidelberg (1997)
28. Werner, H.: Finite simple non-Abelian groups are functionally complete. Bull. Soc. Roy. Sci. Liège 43, 400 (1974)

Beyond Uniformity: Better Security/Efficiency Tradeoffs for Compression Functions

Martijn Stam

EPFL, Switzerland
martijn.stam@epfl.ch

Abstract. Suppose we are given a perfect $n + c$-to-n bit compression function f and we want to construct a larger $m + s$-to-s bit compression function H instead. What level of security, in particular collision resistance, can we expect from H if it makes r calls to f? We conjecture that typically collisions can be found in $2^{(nr+cr-m)/(r+1)}$ queries. This bound is also relevant for building a $m + s$-to-s bit compression function based on a blockcipher with k-bit keys and n-bit blocks: simply set $c = k$, or $c = 0$ in case of fixed keys.

We also exhibit a number of (conceptual) compression functions whose collision resistance is close to this bound. In particular, we consider the following four scenarios:

1. A $2n$-to-n bit compression function making two calls to an n-to-n bit primitive, providing collision resistance up to $2^{n/3}/n$ queries. This beats a recent bound by Rogaway and Steinberger that $2^{n/4}$ queries to the underlying random n-to-n bit function suffice to find collisions in any rate-1/2 compression function. In particular, this shows that Rogaway and Steinberger's recent bound of $2^{(nr-m-s/2)/r)}$ queries (for $c = 0$) crucially relies upon a uniformity assumption; a blanket generalization to arbitrary compression functions would be incorrect.

2. A $3n$-to-$2n$ bit compression function making a single call to a $3n$-to-n bit primitive, providing collision resistance up to 2^n queries.

3. A $3n$-to-$2n$ bit compression function making two calls to a $2n$-to-n bit primitive, providing collision resistance up to 2^n queries.

4. A single call compression function with parameters satisfying $m \leq n + c, n \leq s, c \leq m$. This result provides a tradeoff between how many bits you can compress for what level of security given a single call to an $n + c$-to-n bit random function.

1 Introduction

Hash function design based on idealized primitives has recently undergone a surge in popularity. One of the earliest approaches is Merkle's use of the ideal cipher model to argue the collision resistance of his double length construction [10]. The use of the ideal cipher model has also been instrumental in proving security properties of single call blockcipher based compression functions by Black, Rogaway and Shrimpton [3]. These 1-call blockcipher based constructions have the disadvantage of rekeying every round, which is expensive.

D. Wagner (Ed.): CRYPTO 2008, LNCS 5157, pp. 397–412, 2008.
© International Association for Cryptologic Research 2008

An alternative is the use of a blockcipher with its key fixed or, slightly relaxed, simply a a random n-to-n bit function. Black, Cochran and Shrimpton [2] show that no compression function can exist making only a single call to a fixed key ideal cipher yet still achieving collision resistance. Indeed, two queries suffice to find a collision with certainty.

Rogaway and Steinberger [16] have recently generalized this result considerably. They consider a compression function that maps $m + s$ bits to s bits using r calls to n-to-n bit random functions.[1] Central to their results is the yield of an adversary, that is the number of compression function evaluations that can be made after q queries. It can be shown that if $q = 2^{(nr-n-s/2)/r}$, a greedy adversary can evaluate the compression function on at least $2^{s/2}$ different inputs. Assuming the corresponding evaluations are uniformly distributed implies a collision can be expected (birthday paradox). Consequently [16, Theorem 2], for any compression function satisfying the uniformity assumption, $2^{(nr-n-s/2)/r}$ queries suffice to find a collision with high probability.

One could argue that any good compression function ought to be 'collision-uniform'. But what happens if the compression function is somehow 'bad' and the assumption does not hold? In the case of standard birthday attacks on compression functions [1], deviation from uniformity only reduces collision resistance and it is tempting to generalize to the current scenario. In any case, in the original[2] interpretation of their results, Rogaway and Steinberger silently drop any mention of the uniformity assumption and seemingly claim that, for *any* compression function, an adversary will be able to find collisions with high probability after only $2^{(nr-n-s/2)/r}$ queries. In particular, this would imply that around $2^{n/4}$ queries would typically suffice to find collisions in a 2-call $2n$-to-n bit compression function (a result also alluded to by Shrimpton and Stam [19]).

We show that this interpretation is *incorrect*. In particular, we demonstrate a 2-call compression function that provably requires around $2^{n/3}/n$ queries to find a collision. A first impression why this might be possible is already contained in the bound on the number of queries required under the uniformity assumption. Indeed, if $2^{(nr-n-s/2)/r}$ queries were required, this would indicate that enlarging the state size s would actually *reduce* the collision resistance.[3] This is clearly incorrect, since one can always just expand the state by keeping part of it fixed, a measure that will not influence collision resistance. Nonetheless, the bound $2^{(nr-n-s/2)/r}$ is useful, since it provides us with a means to identify the ideal state size for a given rate. Using an ordinary birthday attack would require $2^{s/2}$ queries, intuitively the optimal state size is that for which the yield-based bound coincides with the standard birthday bound. This crossover occurs for $s = 2(nr - m)/(r + 1)$, heuristically yielding collision resistance up to $q = 2^{(nr-m)/(r+1)}$ queries, or $q = 2^{n(r-1)/(r+1)}$ assuming $m = n$.

---

[1] Our notation deviates from theirs; we emphasize the size s of the chaining variable, or state, and the size m of message material to be hashed when the compression function would be Merkle-Damgård iterated.

[2] In a response to an early manuscript of this paper, the phrasing is more accurate in an updated version [14].

[3] This problem is actually less clear from Rogaway and Steinberger's formulation of the bound.

For the aforementioned 2-call compression function, the optimal state size is $2n/3$, implying we could expect collision resistance up to $2^{n/3}$ for a 2-call $2n$-to-n bit compression function. We give a surprisingly simple compression function almost achieving this bound. For 3-call schemes, the optimal state size is n, yielding collision resistance $2^{n/2}$, coinciding with the Rogaway-Steinberger bound. Shrimpton and Stam [18] and Rogaway and Steinberger [15] already gave distinct 3-call $2n$-to-n bit compression functions achieving collision resistance up to almost $2^{n/2}$ queries. For 4-call schemes, the optimal state size is $6n/5$, yielding collision resistance $2^{3n/5}$. Thus a 4-call double length $3n$-to-$2n$ bit function can be expected to be broken within $2^{3n/5}$ queries, not the $2^{n/4}$ queries as reported by Rogaway and Steinberger. In particular, this indicates that one might already achieve more security with a 4-call double length function than what could be achieved with a single length function. However, we do not yet have a construction matching this bound.

One could object to compression functions that are not as uniform when it comes to their collision behaviour. We agree, but only up to a point. At the core of our 2-call $2n$-to-n bit construction is a 2-call $\frac{5}{3}n$-to-$\frac{2}{3}n$ bit compression function, also with collision resistance up to about $2^{n/3}$ queries. This smaller-state compression function is expected to behave collision-uniform. Thus, in this particular case, the choice really is between a collision-uniform compression function outputting n bits and being collision resistant up to $2^{n/4}$ queries on the one hand, and a collision-uniform compression function outputting only $\frac{2}{3}n$ bits yet being collision resistant up to $2^{n/3}$ queries. We believe the latter option is more desirable in practice.

We stress that our work does not contradict or invalidate [16, Theorem 2] in any way; we do show that by dropping uniformity, or rather state size, one can do better. We also point out that many of the bounds obtained by Rogaway and Steinberger do not have uniformity as a premise: for any compression function just slightly over $2^{(nr-m)/r}$ queries are guaranteed to give a collision [16, Theorem 1] and similarly for any hash function that makes on average r calls per message block, $n2^{(nr-n)/r}$ queries will suffice [16, Theorem 3].

We then ask ourselves the question what happens if the underlying primitive already compresses, that is, if we use idealized $n+c$-to-n bit functions instead of n-to-n bit ones as underlying primitive. This question is most interesting if the compression function to be constructed has a larger input size than the idealized primitive (i.e., $m + s > n + c$, cf. the examples above) or outputs more bits (i.e., $s > n$), which is relevant for instance for the construction of double length compression functions. We show that if we build a $m + s$-to-s bit compression function H making r calls to an $n + c$-to-n bit primitive, we can expect to find collision after $2^{(nr+cr-m)/(r+1)}$ queries. We believe this bound to be tight up to some pathological cases, namely when 2^{nr} or $2^{s/2}$ is smaller than said bound. We also prove an upper bound on indifferentiability.

Assuming our conjectured bound can be achieved has interesting implications for the construction of double-length hash functions, where $m = n$ and $s = 2n$. In particular, one call to a $3n$-to-n bit primitive or two calls to a $2n$-to-n bit primitive would suffice to obtain optimal collision resistance in the compression function. This contrasts starkly with earlier approaches, where either more calls needed to be made, or the collision resistance can only (partially) be proven in the iteration [5, 7, 8, 11, 12, 13, 17, 20]. For both scenarios we give a construction that we believe offers the required collision resistance, up to a small factor. Against non-adaptive adversaries the proof is surprisingly straightforward; against adaptive adversaries we need a reasonable assumption. Although our compression functions sport impressive collision resistance, they do have some obvious shortcomings when other properties are taken into account. As such, we consider them more a proof of concept—the setting of a bar—than actual proposals to be implemented and used as is. We leave open the problem of designing cryptographically satisfactory compression functions reaching the collision resistance bound provided in this paper.

Finally, we present a general single call construction for the case $m \leq n + c, n \leq s$, and $c \leq m$, achieving collision resistance up to $2^{(n+c-m)/2}$ queries. This provides a tradeoff of how much message bits you can hash and what collision resistance you can expect. It also fills the gap between the impossibility result of Black et al. [2] for $c = 0$ and $m = n$ and the trivial optimally collision resistant solution when $c = n$ and $m = n$.

Notwithstanding the emphasis in this paper on random $n + c$-to-n bit functions, the bounds are also indicative for ideal ciphers with k bit keys and n bit blocks, by setting $c = k$. Fixed-key ideal ciphers correspond to $c = 0$. No constructions in this scenario are presented, although our constructions with the public random functions replaced with ideal ciphers in Davies-Meyer mode are obvious candidates.

Our paper is organized as follows. In Section 2 we introduce notation and recall some relevant results. In Section 3, we discuss upper bounding the probability of finding collisions and, to a lesser extent, preimages in the compression function. Finally, Section 4 consists of four parts, each detailing a construction that (almost) meets the upper bound from its preceding section.

2 Background

2.1 General Notation

For a positive integer n, we write $\{0, 1\}^n$ for the set of all bitstrings of length n. When X and Y are strings we write $X \| Y$ to mean their concatenation and $X \oplus Y$ to mean their bitwise exclusive-or (xor). Unless specified otherwise, we will consider bitstrings as elements in the group $(\{0, 1\}^n, \oplus)$.

For positive integers m and n, we let $\mathrm{Func}(m, n)$ denote the set of all functions mapping $\{0, 1\}^m$ into $\{0, 1\}^n$. We write $f \xleftarrow{\$} \mathrm{Func}(m, n)$ to denote random sampling from the set $\mathrm{Func}(m, n)$ and assignment to f. Unless otherwise specified, all finite sets are equipped with a uniform distribution.

2.2 Compression Functions

A *compression function* is a mapping from $\{0, 1\}^m \times \{0, 1\}^s$ to $\{0, 1\}^s$ for some $m, s > 0$. For us, a compression function H must be given by a program that, given (M, V), computes $H^{f_1, \ldots, f_r}(M, V)$ via access to a finite number of specified oracles $f_1, \ldots, f_r$, where we use the convention to write oracles that are provided to an algorithm as superscripts.

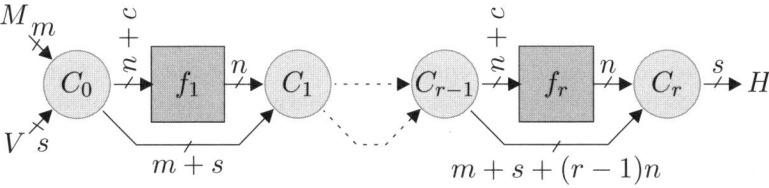

Fig. 1. General form of a $m + s$-to-s bit compression function based on r calls to underlying $n + c$-to-n bit primitive

Let $f_1, \ldots, f_r$ be random functions from $\{0,1\}^{n+c} \to \{0,1\}^n$. Let $C_i : \{0,1\}^s \times \{0,1\}^m \times (\{0,1\}^n)^i \to \{0,1\}^{n+c}$, for $i = 0, \ldots, r-1$, and $C_r : \{0,1\}^s \times \{0,1\}^m \times (\{0,1\}^n)^r \to \{0,1\}^s$, be processing functions. Compression of a message block then proceeds as follows: Given an s-bit state V and m bit message M, compute output $H = H^{f_1, \ldots, f_r}(M, V)$ by

$$Y_1 \leftarrow f_1(C_0(M, V))$$
$$Y_2 \leftarrow f_2(C_1(M, V, Y_1))$$
$$\vdots$$
$$Y_r \leftarrow f_r(C_{r-1}(M, V, Y_1, \ldots, Y_{r-1}))$$
$$H \leftarrow C_r(M, V, Y_1, \ldots, Y_r)$$

as illustrated by Figure 1. In particular, we assume the functions $f_1, \ldots, f_r$ are always called in the same, fixed order. Dynamic, input-dependent ordering can to some extent be modelled by only considering the combined function $\tilde{f}(i, x) = f_i(x)$, thus increasing c by $\lg r$.

Normally one defines the rate as the reciprocal of the number of calls made to the underlying primitive. For primitives have differing input sizes this could skew the comparison: we suggest to use $R = m/(rn + rc)$ as rate, that is the number of bits compressed divided by the total number of input bits taken by the underlying primitives. However, most of the time we will concentrate on the number of calls made, and simply talk of an r-call primitive.

A compression function can be made into a hash function by iterating it. We briefly recall the standard Merkle-Damgård iteration [4, 10], where we assume that there is already some injective padding from $\{0,1\}^* \to (\{0,1\}^m)^*$ in place. Given an initial vector $V_0 \in \{0,1\}^s$ define $\mathcal{H}^H : (\{0,1\}^m)^* \to \{0,1\}^s$ as follows for $\mathbf{M} = (M_1, \ldots, M_\ell)$:

1. Set $V_i \leftarrow H^{f_1, \ldots, f_r}(M_i, V_{i-1})$ for $i = 1, \ldots, \ell$.
2. Output $\mathcal{H}^H(\mathbf{M}) = V_\ell$.

In particular, the hash of the empty message $\mathbf{M} = \emptyset$ corresponds to $\ell = 0$, so $\mathcal{H}^H(\emptyset) = V_0$, the initial vector. Baring this iteration in mind, given a compression function $H : \{0,1\}^m \times \{0,1\}^s \to \{0,1\}^s$ we will refer to the $\{0,1\}^m$ part of the input as 'message' and the $\{0,1\}^s$ as the state. In particular, we refer to s as the state size; increasing the state size will reflect the size of both the input and output of the compression function.

A *collision-finding adversary* is an algorithm with access to one or more oracles, whose goal it is to find collisions in some specified compression or hash function. It is standard practice to consider information-theoretic adversaries only. Currently this seems to provide the only handle to get any provable results. Information-theoretic adversaries are computationally unbounded and their complexity is measured only by the number of queries made to their oracles. Without loss of generality, such adversaries are assumed not to repeat queries to oracles nor to query an oracle outside of its specified domain.

Definition 1. *Let $n, c, m, s > 0$ be integer parameters, and fix an integer $r > 0$. Let $H \colon \{0,1\}^m \times \{0,1\}^s \to \{0,1\}^s$ be a compression function taking r oracles $f_1, \ldots, f_r \colon \{0,1\}^{n+c} \to \{0,1\}^n$. Let $\mathcal{A}$ be a collision-finding adversary for H that takes r oracles. The collision-finding advantage of $\mathcal{A}$ is defined to be*

$$\mathbf{Adv}^{\mathrm{coll}}_{H(n)}(\mathcal{A}) = \Pr\left[f_1..f_r \xleftarrow{\$} \mathrm{Func}(n+c, n), (M, V), (M', V') \leftarrow \mathcal{A}^{f_1 \cdot \cdot f_r} : \right.$$
$$\left. (M, V) \neq (M', V') \text{ and } H^{f_1 \cdot \cdot f_r}(M, V) = H^{f_1 \cdot \cdot f_r}(M', V')\right]$$

Define $\mathbf{Adv}^{\mathrm{coll}}_{H(n)}(q)$ as the maximum advantage over all adversaries making at most q queries to each of their oracles.

Definition 2. *Let $n, c, m, s > 0$ be integer parameters, and fix an integer $r > 0$. Let $H \colon \{0,1\}^m \times \{0,1\}^s \to \{0,1\}^s$ be a compression function taking r oracles $f_1, \ldots, f_r \colon \{0,1\}^{n+c} \to \{0,1\}^n$. Let $\mathcal{A}$ be a preimage-finding adversary for H that takes r oracles. The preimage-finding advantage of $\mathcal{A}$ is defined to be*

$$\mathbf{Adv}^{\mathrm{preim}}_{H(n)}(\mathcal{A}) = \Pr\left[f_1..f_r \xleftarrow{\$} \mathrm{Func}(n+c, n), (M, V) \xleftarrow{\$} \{0,1\}^{m+s}, \right.$$
$$\left. H \leftarrow H^{f_1 \cdot \cdot f_r}(M, V), (M', V') \leftarrow \mathcal{A}^{f_1 \cdot \cdot f_r}(H) : H = H^{f_1 \cdot \cdot f_r}(M', V')\right]$$

Define $\mathbf{Adv}^{\mathrm{preim}}_{H(n)}(q)$ as the maximum advantage over all adversaries making at most q queries in total to their oracles.

For future reference we offer the following little lemma, which basically states that one can increase the state size of a compression function by simply forwarding the extra state bits untouched, without aversely affecting preimage or collision resistance. The lemma is mainly of theoretical use, since simply outputting part of the input is counter to practical hash design.

Lemma 3. *Let m, s, s' be positive integers with $s' > s$. Let $H \colon \{0,1\}^m \times \{0,1\}^s \to \{0,1\}^s$ be a hash function. Define $H' \colon \{0,1\}^m \times \{0,1\}^{s'} \to \{0,1\}^{s'}$ by $H'(M, V\|V') = (H(M, V)\|V')$ where $V \in \{0,1\}^s$ and $V' \in \{0,1\}^{s'-s}$. Then H' inherits its collision resistance and preimage resistance from H, that is, for any adversary $\mathcal{A}'$ on H' there is an adversary $\mathcal{A}$ on H with essentially the same complexity and advantage.*

Proof: We first prove the statement for collision resistance. Let an adversary $\mathcal{A}'$ on the collision resistance of H' be given. Then $\mathcal{A}$ runs $\mathcal{A}'$ and, supposing $\mathcal{A}'$ outputs a collision (M, V, V') and $(\tilde{M}, \tilde{V}, \tilde{V}')$, outputs (M, V) and $(\tilde{M}, \tilde{V})$. Then $H(M, V) = H(\tilde{M}, \tilde{V})$ and $\tilde{V} = \tilde{V}'$. Because $(M, V, V') \neq (\tilde{M}, \tilde{V}, \tilde{V}')$ this implies that $(M, V) \neq (\tilde{M}, \tilde{V})$, making it a collision on H.

The proof for preimage resistance is similar. Let an adversary $\mathcal{A}'$ on the preimage resistance of H' be given and suppose $\mathcal{A}$ needs to find a preimage of $Z \in \{0, 1\}^s$, where Z is distributed by applying H to the uniform distribution over $\{0, 1\}^{m+s}$. Then $\mathcal{A}$ randomly selects $V' \in \{0, 1\}^{s'-s}$ and runs $\mathcal{A}'$ on input $Z' = (Z || V')$. By construction, Z' is distributed correctly (as if applying H' to the uniform distribution over $\{0, 1\}^{m+s'}$), so suppose $\mathcal{A}'$ outputs a preimage (M, V, V'). Then $\mathcal{A}$ outputs (M, V) as preimage of Z under H.   Q.E.D.

2.3 Collisions in Uniform Samples

With $(\{0, 1\}^n)^q$ we denote the set of q-element vectors, or q-*vectors*, in which each element is an n-bit string. When $\mathbf{a} \in (\{0, 1\}^n)^q$, we will write $\mathbf{a} = (a_1, \ldots, a_q)$ when we wish to stress its components. We will use U to denote the uniform distribution over $(\{0, 1\}^n)^q$ (where n and q will often follow from the context). Thus U corresponds to sampling q strings from $\{0, 1\}^n$ uniformly and independently *with* replacement.

If in a random sample some value appears *exactly* k times, we say there is a k-*way collision* in that sample. Let $M_U(k)$ be the random variable describing the number of k-way collisions when the samples are drawn according to the distribution U. We recall the following well known "urns and balls" result [6]. The expected number of k-way collisions is $\mathbb{E}[M_U(k)] = N \binom{q}{k} \left(\frac{1}{N}\right)^k \left(1 - \frac{1}{N}\right)^{q-k}$, where $N = 2^n$. Thus, $\mathbb{E}[M_U(k)]$ follows a (scaled) binomial distribution with parameters $1/N$ and q. Asymptotically, this would correspond to a scaled Poisson distribution with parameter q/N (provided q/N remains bounded).

The probability of finding any sort of collision is at most $\frac{q^2}{2N}$. We can also bound the probability of finding a k-way collision, see Lemma 4 below. In particular, for $q = 2^n/n$ and $k = n$ the probability $\Pr[M_U(n) > 0] < (2/n)^n$ tends to zero for increasing n and with a little bit more work $\sum_{k \geq n} \Pr[M_U(n) > 0] \leq 2(2/n)^n$, also tending to zero.

Lemma 4. *Let q, n, and k be positive integers. Then $\Pr[M_U(k) > 0] \leq 2^n (q/2^n)^k$.*

3 Upper Bounding Collision and Preimage Resistance

3.1 Introduction

Let us consider an $m + s$-to-s bit compression function that uses one call to each of r independent $n + c$-to-n bit random functions $f_1, \ldots, f_r$. We are interested in what kind of collision respectively preimage resistance we can expect. Before we discuss the main line of attack, we mention two other attacks.

Firstly, since the compression function maps to s-bit strings, we know that collisions can be expected after $2^{s/2}$ queries, whereas preimages will be found after just under 2^s queries. Note that these complexities depend only on the size of the compression functions output and not on the dimensions of the underlying primitive, how often it is called, or how many bits are compressed.

Collisions (and in many cases preimages as well) can often also be found using 2^{nr} queries, essentially by guessing the output of the queries corresponding to a certain H-input. Consider C_r, the final function mapping $m + s + rn$ bits to s bits. To find a collision, evaluate the compression function for a random value. If C_r is balanced, every possible output has 2^{m+rn} preimages, each of $m + s + rn$ bits. Parse into $M \times V \times Y_1 \times \cdots \times Y_r$ and evaluate the compression function on input (M, V). With probability 2^{-nr} the Y_i values will correspond with that of the chosen C_r preimage, resulting in a collision. Since there are 2^{m+rn} preimages we can hope that these all have distinct (M, V), so a collision can be found in 2^{nr} queries. If n is relatively small compared to c, this attack might beat the other two.

For the final and main attack the adversary tries to maximize the number of compression function evaluations it can make given his queries. Shrimpton and Stam [18] call this the yield.

Definition 5. *Let $H^{f_1, \ldots, f_r}$ be a compression function based on a primitives $f_1, \ldots, f_r$. The yield of an adversary after a set of queries to $f_1, \ldots, f_r$, is the number of inputs to H for which he can compute $H^{f_1, \ldots, f_r}$ given the answers to his queries. With $\mathrm{yield}(q)$ we denote the maximum expected yield given q queries to each of the oracles $f_1, \ldots, f_r$.*

The central theorem is the following generalization of a result by Rogaway and Steinberger [16], who give the result for $c = 0$ only (Shrimpton and Stam [19] give the result for $c = 0$ and $r = 2$ only).

Theorem 6. *Let $H^{f_1, \ldots, f_r}$ be an $m + s$-to-s bit compression function making one call to each of the $n + c$-to-n bit primitives $f_1, \ldots, f_r$. Then $\mathrm{yield}(q) \geq 2^{m+s}(q/2^{n+c})^r$.*

Proof: Consider the following greedy adversary. Let $0 < i < r$. Suppose that after q queries to each of the oracles $f_1, \ldots, f_i$ the adversary can compute $Y_0, \ldots, Y_i$ for Q_i input pairs $(M, V) \in \{0, 1\}^m \times \{0, 1\}^s$. Since there are 2^{n+c} possible inputs to f_{i+1}, on average for each possible input to f_{i+1} there are $Q_i/2^{n+c}$ inputs to the compression function for which the adversary can compute all intermediate chaining values $Y_0, \ldots, Y_i$. If the adversary queries f_{i+1} on the q values for which he knows most intermediate chaining paths, he will be able to compute Y_{i+1} for at least $Q_i q/2^{n+c}$ values (by the pigeon hole principle). With finite induction and using that $Q_0 = 2^{m+s}$ it follows that this adversary can compute Y_r, and hence the compression output, for at least $2^{m+s}(q/2^{n+c})^r$ values.   *Q.E.D.*

3.2   Rogaway and Steinberger's Bounds (Generalized)

Rogaway and Steinberger [16] observe that if $\mathrm{yield}(q) > 2^s$ a collision is guaranteed. Moreover, if $\mathrm{yield}(q) > 2^s$ one expects to be able to find preimages, provided the compression function has sufficiently uniform behaviour.[4] Similarly, if $\mathrm{yield}(q) > 2^{s/2}$ one

---

[4] The preimage result erroneously omits a uniformity assumption [16, Theorem 4], corrected in [14, Theorem 4].

would expect a collision, again provided sufficiently uniform behaviour. The following formulation captures the loosely stated uniformity assumption, taking into account preimages as well. See [14] for a finegrained description (also of Theorem 8 with $c = 0$).

Assumption 7. *Let H^f be an $m + s$-to-s bit compression function making r calls to $n + c$-to-n bit primitive f. Let $\mathcal{A}$ be the adversary that optimizes its yield according to the proof of Theorem 6. Then the spread and occurence of collisions for the adversary's compression function evaluations behave as if these $\text{yield}(q)$ elements were drawn at random.*

Theorem 8. *(Case $c = 0$ corresponds to [16, Theorem 2]) Let H^f be an $m + s$-to-s bit compression function making r calls to $n + c$-to-n bit primitive f.*

1. *If $q \geq 2^{(nr+cr-m-s/2)/r}$ then $\text{yield}(q) \geq 2^{s/2}$ and, under Assumption 7 a collision in H^f can be found with high probability.*
2. *If $q \geq 2^{(nr+cr-m)/r}$ then $\text{yield}(q) \geq 2^s$ and a collision in H^f can be found with certainty.*
3. *If $q \geq 2^{(nr+cr-m)/r}$ then $\text{yield}(q) \geq 2^s$ and, under Assumption 7 preimages in H^f can be found with high probability.*

Proof: Given the lower bound on the yield (Theorem 6) it suffices to determine those q for which $2^{m+s}(q/2^{n+c})^r \geq 2^{s/2}$ respectively $2^{m+s}(q/2^{n+c})^r \geq 2^s$ holds.   Q.E.D.

3.3   New Bounds

It is easy to see that Assumption 7 cannot be true for all possible H and examples, for which the greedy adversary does not find collisions within the required amount of queries suggested by Theorem 8, are easy to find (another more efficient adversary might exist). Moreover, upon closer inspection the bound of Lemma 8 has a very strange consequence: increasing the state size reduces the security! (Note that this problem does not exist for the bound on preimage resistance.) This is counterintuitive; indeed, given a $m+s$-to-s bit compression function one can easily increase the state size without affecting collision or preimage resistance by simply forwarding the extra state bits untouched (Lemma 3).

The solution to this problem presents itself naturally: first determine the optimal state size. For this we need to determine for which state size the direct yield-based bound and the generic birthday bound coincide. That is, for which s do we have $2^{s/2} = 2^{(nr+cr-m-s/2)/r}$. Taking logarithms and some simple formula manipulation leads to $s = 2(nr + cr - m)/(r + 1)$, corresponding to collision resistance $q = 2^{(nr+cr-m)/(r+1)}$. All in all this leads us to the following conjecture.

Conjecture 9. *Let H^f be an $m + s$-to-s bit compression function making r calls to $n + c$-to-n bit primitive f. Then collisions can be found for $q \leq 2^{(nr+cr-m)/(r+1)}$.*

The yield can also be used to obtain bounds on the indifferentiability of a construction (we refer to [9] for an introduction to hash function indifferentiability).

Table 1. Security Bounds for Single-Length Constructions ($s = n$). Listed are the approximate number of queries after which a certain property will be broken.

	Collision Conjecture 9 $2^{n(1-2/(r+1))}$	Collision [16, Theorem 2] $2^{n(1-3/(2r))}$	Preimages [16, Theorem 4] $2^{n(1-1/r)}$	Indifferentiable Theorem 10 $\approx 2^{n(1-1/(r-1))}$
$r = 2$	$2^{n/3}$	$2^{n/4}$	$2^{n/2}$	2
$r = 3$	$2^{n/2}$	$2^{n/2}$	$2^{2n/3}$	$2^{n/2}$
$r = 4$			$2^{3n/4}$	$2^{2n/3}$
$r = 5$			$2^{4n/5}$	$2^{3n/4}$

Table 2. Security Bounds for Double-Length Constructions ($s = 2n$). Listed are the approximate number after which a certain property will be broken.

	Collision Conjecture 9 $2^{n(1-2/(r+1))}$	Collision [16, Theorem 2] $2^{n(1-2/r)}$	Preimages [16, Theorem 4] $2^{n(1-1/r)}$	Indifferentiable Theorem 10 $\approx 2^{n(1-2/(r-1))}$
$r = 3$	$2^{n/2}$	$2^{n/3}$	$2^{2n/3}$	2
$r = 4$	$2^{3n/5}$	$2^{n/2}$	$2^{3n/4}$	$2^{n/3}$
$r = 5$	$2^{2n/3}$	$2^{3n/5}$	$2^{4n/5}$	$2^{n/2}$

Theorem 10. *Let H^f be an $m+s$-to-s bit compression function making r calls to $n+c$-to-n bit primitive f. Then H^f is differentiable with high probability from a random $m + s$-to-s bit function after $q > 2^{n+c}(\frac{nr}{s}2^{n+c-m-s})^{1/(r-1)}$ queries.*

Proof: (Sketch) We claim that if $\mathrm{yield}(q) > nqr/s$ the construction cannot possibly be indifferentiable. Suppose the adversary is communicating with a real $m + s$-to-s bit public random function H and simulated $f_1, \ldots, f_r$. After q calls to each of $f_1, \ldots, f_r$, the adversary has received a total of qrn bits in answers. Yet he can now predict the outcome of H for $\mathrm{yield}(q)$ values, i.e., a total of $\mathrm{yield}(q)s$ completely random bits. The claim follows from incompressibility of completely random bitstrings.

Using Theorem 6 we get differentiability for $2^{m+s}(q/2^{n+c})^r > nqr/s$ or $q > 2^{n+c}(\frac{nr}{s}2^{n+c-m-s})^{1/(r-1)}$.   *Q.E.D.*

3.4 Interpretation

In Tables 1 and 2 we look at the maximal attainable security under our conjecture and theorem. Both tables are for non-compressing n-to-n bit underlying public random functions (so $c = 0$), similar to the random permutation setting studied by Rogaway and Steinberger. Their bounds for compression functions satisfying the uniformity assumption are included in the tables. Table 1 focuses on single length compression functions, that is $s = n$, and Table 2 focuses on double length compression functions, so $s = 2n$. In both cases $m = n$ bits of message are compressed.

In the interesting cases where our upper bound on collision resistance is higher than Rogaway and Steinbergers, notably $r = 2$ for Table 1 and all of Table 2 we suggest to actually reduce the state size to match the provided collision resistance (e.g., $s = 2n/3$ for $r = 2$ in Table 1). Increasing the state size would either reduce collision resistance or introduce questionable behaviour invalidating the uniformity assumption (cf. Lemma 3).

One can also look at the maximum number of message bits one can hope to compress given a targeted leved of collision resistance and number of calls to the underlying public random function. For a collision resistance level of $2^{n/2}$ queries, Conjecture 9 implies one can hash at most $m \leq (\frac{n}{2} + c)r - \frac{n}{2}$ message bits. For double-length constructions and corresponding target of 2^n queries the number of bits increases to $m \leq cr - n$.

Finally, for $c = 0$ and writing rate $R = m/nr$, the bound can be rewritten as $q \leq 2^{n(1-1/R)(r+1)/r}$ indicating that asymptotically (in r) one can get collision resistance up to $2^{n(1-1/R)}$ queries. Up to constants this is the same as the bound by Rogaway and Steinberger [16, Theorem 3], but an important difference is that their bound is rigorously proven, whereas ours follows from a conjecture.

4 Matching Collision Resistant Constructions

4.1 Case I: A Rate-1/2 Single Length Compression Function

Our main result in this section is a compression function with state size $s = 2n/3$ and almost optimal collision resistance, making only 2 calls to a n-to-n bit public random function. Let $M \in \{0,1\}^n$ and $V \in \{0,1\}^{2n/3}$. Define (Figure 2)

$$H^{f_1,f_2}(M, V) = V \oplus \mathrm{msb}_{2n/3}(f_2((V\|0^{n/3}) \oplus f_1(M)))$$

The state size can be expanded by forwarding the extra chaining bits untouched (Lemma 3), giving rise to a $2n$-to-n bit compression function beating the upper bound of $2^{n/4}$ queries for compression functions satisfying the uniformity assumption.

Theorem 11. *Let H^{f_1,f_2} be as given above. Then*

$$\mathbf{Adv}_{H(n)}^{\mathrm{coll}}(q) \leq q^2/2^{n+1} + 2^{n/3}(q/2^{n/3})^n + q(q-1)n^2/2^{2n/3}.$$

Since the third term is initially dominant, an adversary needs to asks roughly $2^{n/3}/n$ queries for a reasonable advantage.

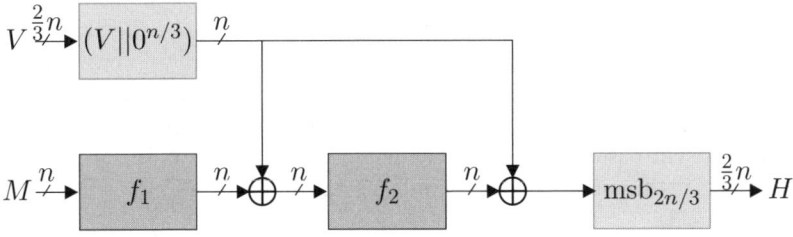

Fig. 2. The 2-call compression function H based on public random n-to-n bit functions f_1 and f_2

Proof: Let an adversary $\mathcal{A}$ be given. We need to bound its success probability. Suppose that, whenever $\mathcal{A}$ makes a query to f_1, we do not answer with f_1's answer, but with a randomly drawn value in $\{0,1\}^n$. From the adversary's point of view, this is the same distribution. What's more, we can even decide upon a list of answers to the adversary's f_1 queries ahead of time. Since the input to f_1 is not used elsewhere in the compression function and f_2 is independent of f_1, we could even give this list to the adversary in advance of any query.

Thus we can replace the q queries to f_1 by a list $\mathbf{a}$ of q values drawn at random with replacement from $\{0,1\}^n$, given to the adversary before any queries to f_2 are made. Let us consider the probability that the adversary finds a collision on the i-th query to f_2. Let X_i be the value queried by the adversary. This will allow the adversary to evaluate H for those $a \in \mathbf{a}$ whose first $n/3$ bits coincide with X_i. Let's call this number k_i. Moreover, unless two values $a, a' \in \mathbf{a}$ are identical, this cannot lead to a collision based on the i'th f_2-query alone. Note that collisions in $\mathbf{a}$ occur with probability $\leq q^2/2^{n+1}$.

Suppose that after $i-1$ queries the adversary can evaluate H for Q_{i-1} different values. Then the probability of a collision is on the i-th query is at most $k_i Q_{i-1}/2^{2n/3}$. Note that $Q_i = \sum_{j=1}^{i-1} k_j$. With probability at most $2^{n/3}(q/2^{n/3})^n$ some n-way collision occurs (Lemma 4), otherwise all $k_i < n$ and $Q_i < (i-1)n$. Thus the probability of a collision (provided no n-way collisions occur in the $n/3$ upper most bits of $\mathbf{a}$) is upper bounded by $\sum_{i=1}^{q}(i-1)n^2/2^{2n/3} = q(q-1)n^2/2^{2n/3}$.   Q.E.D.

As an aside, our construction shares some of the disadvantages of the rate-1/3 construction by Shrimpton and Stam [19]. In particular, finding a collision in f_1 leads to many collisions in H (although the gap between finding a single collision in H and one in f_1 is significantly bigger this time). Implementing f_1 using a fixed key ideal cipher in Davies-Meyer mode does not affect the security (PPP Switching Lemma [18, Lemma 6]), but the effect of replacing f_2 with a fixed-key ideal cipher is less clear.

4.2   Case II: A Single-Call Double Length Compression Function

We will now consider a $3n$-to-$2n$ bit compression function based on a single call to a random $3n$-to-n bit function. We show that there exists a compression function for which the number of random function queries to find a collision is of the order 2^n for non-adaptive adversaries, that need to commit to all their queries in advance of receiving any answers. Subsequently we indicate why we expect the advantage not to drop significantly when taking into account adaptive adversaries and discuss a variant based on two random $2n$-to-n bit functions.

Let us first define the hash function. For ease of exposition, we consider the hash function to have three n-bit inputs U, V, and W. Moreover, we will interpret n-bit strings as elements in $\mathbb{F}_{2^n}$. The input U, V, W is then used to define a quadratic polynomial over $\mathbb{F}_{2^n}$. The hash consists of the output of f (on input U, V, W) and the polynomial evaluated in this output. In other words, to compute $H^f(U, V, W)$ do the following (and see Figure 3):

1. Set $Y \leftarrow f(U, V, W)$
2. Output $H^f(U, V, W) = (Y \| WY^2 + VY + U)$.

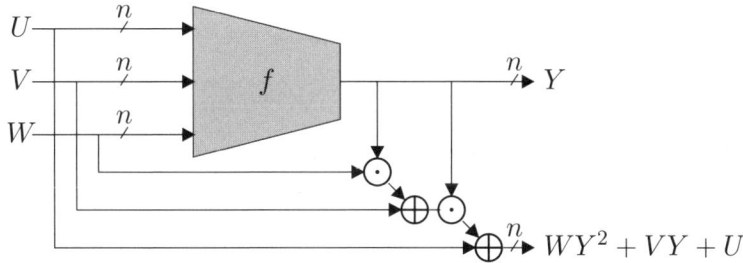

Fig. 3. A single call double length compression function with close to optimal collision resistance. Arithmetic over $\mathbb{F}_{2^n}$.

Theorem 12. *Let H^f be as given above. Then*

$$\mathbf{Adv}^{\mathrm{coll}}_{H(n)}(q) \leq q(q-1)/2^{2n} ,$$

for non-adaptive adversaries. Hence a non-adaptive adversary needs to ask roughly 2^n queries for a reasonable advantage.

Proof: Suppose an adversary wants to find a collision (U, V, W) and (U', V', W'). Let $Y = f(U, V, W)$ and $Y' = f(U', V', W')$. Then we need that $Y = Y'$ and $(W - W')Y^2 + (V - V')Y + (U - U') = 0$. Since $(U, V, W) \neq (U', V', W')$ for a collision, this means that Y needs to be a root of a non-zero quadratic polynomial. There are at most two roots over $\mathbb{F}_{2^n}$. The probability that both $f(U, V, W)$ and $f(U', V', W')$ evaluate to the same root is at most $2/2^{2n}$. Since we assume a non-adaptive adversary, we can use a union bound over all pairs (U, V, W) and (U', V', W') to obtain an upper bound on the adversary finding a collision of $\binom{q}{2}2/2^{2n} = q(q-1)/2^{2n}$.   *Q.E.D.*

The question that remains to be answered is what happens if the adversary is adaptive. In this case, given a list of queries (U_j, V_j, W_j) with answers Y_j for $j = 1, \ldots, i-1$ with $i < q$, the adversary will already know when he makes a new query (U_i, V_i, W_i) for which $0 < j < i$ it holds that Y_j satisfies $(U_i - U_j)Y_j^2 + (V_i - V_j)Y_j + (W_i - W_j) = 0$. He finds a collision if Y_i will equal one of those Y_j, so he optimizes his probability querying (U_i, V_i, W_i) that maximizes the number of $0 < j < i$ for which this query could result in a collision. Suppose there are several j's we target, let's say $j_1, \ldots, j_\ell$. Then one can express W_i as a function of $(U_j, V_j, W_j, Y_j), U_i$, and V_i for any of the j's. Because W_i is unique, one can subsequently express V_i in U_i and any two of the (U_j, V_j, W_j, Y_j)'s. Taking this one step further leads to U_i expressed in any triple of the (U_j, V_j, W_j, Y_j). For $\ell > 3$ the corresponding (U_j, V_j, W_j, Y_j) already need to satisfy an increasing (in ℓ) number of conditions. Since the Y_j are random, we believe one can upper bound the probability p_ℓ of an ℓ-tuple occurring satisfying these conditions, leading to an overall bound on the collision probability of $p_\ell + \sum_{i=1}^{q} \ell/2^n = p_\ell + q\ell/2^n$.

In the full version we show that if one instantiates the $3n$-to-n bit function f with a cascade of two random $2n$-to-n bit functions f_1 and f_2 (i.e., $f(U, V, W) = f_2(f_1(U, V), W)$) the construction remains secure. For any quasi-adaptive adversary that provides a list of queries to f_1 and, only after receiving the answers to this list, will prepare a list of queries to f_2, we bound the advantage

$$\mathbf{Adv}_{H(n)}^{\mathrm{coll}}(q) \leq n^2 q^2/2^{2n} + (2q/2^n)^n \;,$$

so again $2^n/n$ queries are needed for a reasonable advantage.

4.3   Case III: Single Call Efficiency/Security Tradeoff

In this section we show that our conjectured upper bound can be achieved for single call schemes when $m \leq n + c$, $n \leq s$ and $c \leq m$, achieving collision resistance up to $2^{(n-c+m)/2}$ queries. This parameter set might seem a bit artificial at first, but it neatly fills the gap between Black et al.'s impossibility result of creating a $2n$-to-n bit compression function using a single n-to-n bit random function and the trivial construction of making a $2n$-to-n bit compression function based on a single call to a random $2n$-to-n bit function. Indeed, we have $n + c \leq m + s$ in this case, so unless the equality holds, we need to compress before calling f.

For the moment we concentrate on the case $s \leq n + c$, we deal with $s > n + c$ at the end of this section. Recall that the hash function $H^f(M, V)$ has two inputs, M and V and a single oracle f. Split $V = V_0 || V_1$ in two where $|V_0| = n + c - m$ and $|V_1| = m + s - n - c$. Split $M = M_0 || M_1$ in two where $|M_0| = n + c - s$ and $|M_1| = m + s - n - c$. Finally split f's output $F = F_0 || F_1$ in two where $|F_0| = n + c - m$ and $|F_1| = m - c$. Then define

$$C_0(M, V) = (M_0 || M_1 \oplus V_1 || V_0)$$
$$C_1(M, V, F) = (F_0 || (F_1 || 0^{s-n}) \oplus V_1).$$

Theorem 13. *Let H^f be as given above. Then*

$$\mathbf{Adv}_{H(n)}^{\mathrm{coll}}(q) \leq q^2/2^{n+c-m+1} \;.$$

Hence an adversary needs to ask roughly $2^{(n+c-m)/2}$ queries for a reasonable advantage.

Proof: Suppose an adversary wants to find a collision (M, V) and (M', V') on H^f. Let $X = C_0(M, V)$, $F = f(X)$, and $H = C_1(M, V, F)$. Similar for X', F', and H'. A collision means $(M, V) \neq (M', V')$ yet $H = H'$. A priori, there are two possibilities in such a case. Either $X = X'$ or not. If $X = X'$ then also $F = F'$ and in particular $F_1 = F_1'$. Since $H = H'$ implies $F_1 \oplus V_1 = F_1' \oplus V_1'$, we get $V_1 = V_1'$. This combined with $X = X'$ would already imply $(M, V) = (M', V')$, hence no collision.

Thus we are left with the case $X \neq X'$. In that case $H = H'$ requires $F_0 = F_0'$. This is outside the adversary's control, his advantage after q queries follows the birthday bound based on F_0's length, so is upper bounded by $\frac{1}{2}q^2/|F_0| = q^2/2^{n+c+1-m}$, so roughly $|F_0|^{1/2}$ queries are needed to find a collision for F_0.   Q.E.D.

Note that, in the proof above, once an F_0-collision has been obtained, one can pick V_1 freely and (uniquely) complete the collision given F_1, F_1', X and X'. Thus the adversary needs about $2^{(n+c-m)/2}$ queries to find his first collision, but it will immediately be a $2^{m+s-n-c}$-way collision.

If $s > n + c$ this multicollision behaviour can be changed slightly. In that case one can simply feed forward $s - n - c$ bits of the state to the next without any processing (and apply the construction above on the $s' = n + c$ remaining bits). Finding a collision will still take time $2^{(n+c-m)/2}$, but this time the adversary will find 2^{s-n-c} collisions that are 2^m-way.

5   Conclusion

Thanks to Phil Rogaway, John Steinberger, Stefano Tessaro and the anonymous Crypto'08 referees for their valuable feedback. Thanks to the folks in Bristol for their hospitality during a vital phase in the writing of this paper and special thanks to Tom Shrimpton for great discussions and feedback on this work from the initial stages to the end.

References

1. Bellare, M., Kohno, T.: Hash function balance and its impact on birthday attacks. In: Cachin, C., Camenisch, J.L. (eds.) EUROCRYPT 2004. LNCS, vol. 3027, pp. 401–418. Springer, Heidelberg (2004)
2. Black, J., Cochran, M., Shrimpton, T.: On the impossibility of highly efficient blockcipher-based hash functions. In: Cramer, R. (ed.) EUROCRYPT 2005. LNCS, vol. 3494, pp. 526–541. Springer, Heidelberg (2005)
3. Black, J., Rogaway, P., Shrimpton, T.: Black-box analysis of the block-cipher-based hash-function constructions from PGV. In: Yung, M. (ed.) CRYPTO 2002. LNCS, vol. 2442. Springer, Heidelberg (2002)
4. Damgård, I.: A design principle for hash functions. In: Brassard, G. (ed.) CRYPTO 1989. LNCS, vol. 435. Springer, Heidelberg (1990)
5. Hirose, S.: Some plausible constructions of double-length hash functions. In: Robshaw, M. (ed.) FSE 2006. LNCS, vol. 4047, pp. 210–225. Springer, Heidelberg (2006)
6. Johnson, N.L., Kotz, S.: Urn Models and Their Applications. John Wiley and Sons, Inc., Chichester (1977)
7. Knudsen, L., Muller, F.: Some attacks against a double length hash proposal. In: Lai, X., Chen, K. (eds.) ASIACRYPT 2006. LNCS, vol. 4284, pp. 462–473. Springer, Heidelberg (2006)
8. Lucks, S.: A collision-resistant rate-1 double-block-length hash function. In: Biham, E., Handschuh, H., Lucks, S., Rijmen, V. (eds.) Symmetric Cryptography, number 07021 in Dagstuhl Seminar Proceedings, Dagstuhl, Germany, 2007, Schloss Dagstuhl, Germany. Internationales Begegnungs- und Forschungszentrum für Informatik (IBFI) (2007)
9. Maurer, U., Tessaro, S.: Domain extension of public random functions: Beyond the birthday barrier. In: Menezes, A. (ed.) CRYPTO 2007. LNCS, vol. 4622, pp. 187–204. Springer, Heidelberg (2007)
10. Merkle, R.: One way hash functions and DES. In: Brassard, G. (ed.) CRYPTO 1989. LNCS, vol. 435, pp. 428–466. Springer, Heidelberg (1990)
11. Mironov, I., Narayanan, A.: Domain extension for random oracles: Beyond the birthday-paradox bound. In: ECRYPT Hash Workshop 2007, Barcelona, May 24–25 (2007)

12. Nandi, M., Lee, W., Sakurai, K., Lee, S.: Security analysis of a 2/3-rate double length compression function in black-box model. In: Gilbert, H., Handschuh, H. (eds.) FSE 2005. LNCS, vol. 3557, pp. 243–254. Springer, Heidelberg (2005)

13. Peyrin, T., Gilbert, H., Muller, F., Robshaw, M.: Combining compression functions and block cipher-based hash functions. In: Lai, X., Chen, K. (eds.) ASIACRYPT 2006. LNCS, vol. 4284, pp. 315–331. Springer, Heidelberg (2006)

14. Rogaway, P., Steinberger, J.: Security/efficiency tradeoffs for permutation-based hashing. Full version of [16] available through authors' website

15. Rogaway, P., Steinberger, J.: Constructing cryptographic hash functions from fixed-key blockciphers. In: Wagner, D. (ed.) CRYPTO 2008. LNCS, vol. 5157, pp. 433–450. Springer, Heidelberg (2008)

16. Rogaway, P., Steinberger, J.: Security/efficiency tradeoffs for permutation-based hashing. In: Smart, N. (ed.) EUROCRYPT 2008. LNCS, vol. 4965, pp. 220–236. Springer, Heidelberg (2008)

17. Seurin, Y., Peyrin, T.: Security analysis of constructions combining FIL random oracles. In: Biryukov, A. (ed.) FSE 2007. LNCS, vol. 4593, pp. 119–136. Springer, Heidelberg (2007)

18. Shrimpton, T., Stam, M.: Efficient collision-resistant hashing from fixed-length random oracles. In: ECRYPT Hash Workshop 2007, Barcelona, May 24–25 (2007)

19. Shrimpton, T., Stam, M.: Building a collision-resistant compression function from non-compressing primitives. In: ICALP 2008, Part II, vol. 5126, pp. 643–654. Springer, Heidelberg (2008); Supersedes [18]

20. Steinberger, J.: The collision intractability of MDC-2 in the ideal-cipher model. In: Naor, M. (ed.) EUROCRYPT 2007. LNCS, vol. 4515, pp. 34–51. Springer, Heidelberg (2007)

Compression from Collisions,
or Why CRHF Combiners Have a Long Output

Krzysztof Pietrzak

CWI Amsterdam, The Netherlands

Abstract. A black-box combiner for collision resistant hash functions (CRHF) is a construction which given black-box access to two hash functions is collision resistant if at least one of the components is collision resistant.

In this paper we prove a lower bound on the output length of black-box combiners for CRHFs. The bound we prove is basically tight as it is achieved by a recent construction of Canetti et al [Crypto'07]. The best previously known lower bounds only ruled out a very restricted class of combiners having a very strong security reduction: the reduction was required to output collisions for both underlying candidate hash-functions given a single collision for the combiner (Canetti et al [Crypto'07] building on Boneh and Boyen [Crypto'06] and Pietrzak [Eurocrypt'07]).

Our proof uses a lemma similar to the elegant "reconstruction lemma" of Gennaro and Trevisan [FOCS'00], which states that any function which is not one-way is compressible (and thus uniformly random function must be one-way). In a similar vein we show that a function which is not collision resistant is compressible. We also borrow ideas from recent work by Haitner et al. [FOCS'07], who show that one can prove the reconstruction lemma even relative to some very powerful oracles (in our case this will be an exponential time collision-finding oracle).

1 Introduction

Combiners. A robust black-box $(1,2)$-combiner for some cryptographic primitive α is a construction, which given black-box access to two components, securely implements α if *either* of the two components securely implements α. More generally, for $k \leq \ell$, one can consider black-box (k,ℓ)-combiners which securely implement α, if at least k of the ℓ components the combiner has access to securely implement α. In this introduction, we will mostly talk about $(1,2)$-combiners (and simply call them combiners), but the results in the paper are stated for general (k,ℓ)-combiners.

Combiners for CRHFs. Combiners can be used as a hedge against the failure of a concrete construction: the combiner remains secure as long as at least one of the two combined constructions is not broken. In light of the many recent attacks on popular collision resistant hash functions [20,21], combiners for CRHFs are of particular practical interest. A function $H : \{0,1\}^* \rightarrow \{0,1\}^v$

D. Wagner (Ed.): CRYPTO 2008, LNCS 5157, pp. 413–432, 2008.
© International Association for Cryptologic Research 2008

is collision-resistant, if no efficient algorithm can find two inputs $M \neq M'$ where $H(M) = H(M')$, such a pair (M, M') is called a collision for H.[1]

One trivially gets a $(1, 2)$-combiner for CRHF by simply concatenating the outputs of the components:

$$C^{H_1, H_2}(X) = H_1(X) \| H_2(X). \tag{1}$$

This is a robust combiner for any reasonable definition of "robust", as a collision for $C^{H_1, H_2}(.)$ is also a collision for $H_1(.)$ and $H_2(.)$. Unfortunately the output length ℓ_{out}^C of C^{H_1, H_2} is twice the output length ℓ_{out}^H of its components, which makes this combiner not very useful for practical applications, where the output length is usually a crucial parameter, and doubling it is not an option. The existence of black-box combiners for CRHFs with "short" output length has first been investigated by Boneh and Boyen [2] who showed that no "highly efficient" robust combiner with output length $\ell_{out}^C < 2\ell_{out}^H$ exists. Here "highly efficient" means that the combiner is allowed only one query to each of its components (thus the combiner (1) who achieves $\ell_{out}^C = 2\ell_{out}^H$ is the best "highly efficient" black-box combiner one can hope for). Subsequently, for the more general case where one allows the combiner C any number q_C of oracle gates, a lower bound of $\ell_{oul}^C \geq 2\ell_{out}^H - O(\log q_C)$ was proven [16].

In [2,16] a combiner is defined as a pair (C, P), where the oracle circuit C defines the construction of the combiner, and the oracle PPTM P is the "security proof" for C. The security definition requires that for any hash functions H_1, H_2, and any collision M, M' (i.e. $M \neq M'$ and $C^{H_1, H_2}(M) = C^{H_1, H_2}(M')$), we have that $P^{H_1, H_2}(M, M')$ finds a collision for H_1 and H_2 (here we say that a collision for H_i is found if P makes two queries $X \neq X'$ to H_i where $H_i(X) = H_i(X')$). This is a good definition, as such a (C, P) clearly achieves what one intuitively would require from a robust combiner. But when proving impossibility results, one should try to use a definition which is as general as possible, and ideally should cover (and thus rule out) any black-box construction which would satisfy what one intuitively would consider a robust combiner. In this paper we consider what is arguably the most general definition of black-box combiners for CRHFs.

A General Definition. Informally, we define a randomized black-box combiner for CRHFs as a pair (C, P), where C is a *randomized* oracle circuit, and the oracle PPTM P is the security reduction. For $0 \leq \rho \leq 1$, we say that an oracle $\mathcal{B}$ ρ-breaks C^{H_1, H_2}, if on input some randomness R the oracle $\mathcal{B}$ outputs a collision for $C^{H_1, H_2}(R, .)$ for at least a ρ fraction of the R's. The combiner (C, P) is ρ-robust if for any H_1, H_2 and any $\mathcal{B}$ which ρ-breaks C^{H_1, H_2} the PPTM P in the random experiment $P^{\mathcal{B}, H_1, H_2}$ (let us stress that P can query its oracles $\mathcal{B}, H_1, H_2$ *adaptively*) finds a collision for H_1 and a collision for H_2 with high probability.

Thus if (C, P) is ρ-robust, by picking the randomness for C^{H_1, H_2} uniformly at random, with probability at least $1 - \rho$ we will get a construction which is

---

[1] This definition is intentionally kept informal as there are some issues which make it tricky to have a definition for collision-resistant hash-functions which is theoretically and practically satisfying, see e.g. [18] for a discussion.

secure if either H_1 or H_2 is. A combiner (C, P) is *efficient*, if C and P make a polynomial number of oracle queries, and *robust* if it is ρ-robust with $\rho \in o(1)$.[2]

Remark 1 (on the definition). In practice it's not enough to require that C and P make a polynomial number of queries, one would require that their total running time is polynomial. One would also require ρ-security where ρ is negligible, not only $\rho \in o(1)$. But keep in mind that we want to prove an impossibility result, so using such an "undemanding" definition makes the impossibility result actually stronger.

The Main Result. Theorem 1 in this paper states that no black-box combiners exist whose output length ℓ_{out}^C is significantly smaller than what can be achieved by concatenation. For the special case of $(1, 2)$-combiners, where concatenation achieves a length of $2\ell_{out}^H$, this means that no efficient and robust combiner exists whose output length satisfies $\ell_{out}^C = 2\ell_{out}^H - \omega(\log \ell_{out}^H)$. This result is tight because $\ell_{out}^C = 2\ell_{out}^H - \Theta(\log \ell_{out}^H)$ is achievable as we'll explain in Section 2.

Combining CRHF *Families*. In our definition, the (randomized) combiner is instantiated with two hash-functions H_1, H_2. A seemingly more general definition would allow an instantiation of the combiner with two *families* $\mathcal{H}_1, \mathcal{H}_2$ of hash functions, and only require that the reduction $P^{\mathcal{B}, \mathcal{H}_1, \mathcal{H}_2}$ outputs a collision for some $h_1 \in \mathcal{H}_1$ and some $h_2 \in \mathcal{H}_2$. Here the combiner $C^{\mathcal{H}_1, \mathcal{H}_2}(R, M)$ can query different, adaptively chosen functions from $\mathcal{H}_1$ and $\mathcal{H}_2$. Our impossibility also rules out the case where one considers combiners for families as just described, the reason is that we can always view a single hash function $H_b : \{0,1\}^* \to \{0,1\}^v$ as a family $\mathcal{H}_b = \{0,1\}^k \times \{0,1\}^* \to \{0,1\}^v$ (where the first k bits $K \in \{0,1\}^k$ of the input define the hash function $h_b^K \in \mathcal{H}_b$ as $h_b^K(M) = H_b(K \| M)$).

Note that a collision M, M' for any $h_b^K(.) = H_b(K\|.) \in \mathcal{H}_b$ directly gives a collision $K\|M, K\|M'$ for H_b. Thus if (C, P) is not a black-box combiner in our sense, which we prove by showing that there exist $H_1, H_2, \mathcal{B}$ where $P^{\mathcal{B}, H_1, H_2}$ does not output collisions for both, H_1 and H_2 (except with negligible probability), it directly follows that $P^{\mathcal{B}, \mathcal{H}_1, \mathcal{H}_2}$ will not be able to output collisions for some $h_1^K \in \mathcal{H}_1$ and some $h_2^{K'} \in \mathcal{H}_2$ either.

The Canetti et al Lower Bound. *Randomized* combiners for CRHFs have recently been considered by Canetti et al.[3],[3] who proved a basically tight $\ell_{out}^C \geq 2\ell_{out}^H - O(\log \ell_{out}^H)$ lower bound on the output length for randomized combiners using a definition which is basically equivalent to the one in this paper, but with the restriction, that the reduction P is only allowed a single query to the breaking oracle $\mathcal{B}$. We see no good reason to motivate this restriction on the reduction, except for the fact that the existing combiners are of this form. In particular,

---

[2] By $\rho \in o(1)$ we mean that $\rho \in o(\ell_{out}^H)/\ell_{out}^H$, i.e. ρ drops below any constant for a sufficiently large security parameter which w.l.o.g. will be the output length ℓ_{out}^H of the components H_i.

[3] Let us mention that the main topic of their paper is security amplification, not combiners, for CRHFs.

a reduction which needs, say, any two different collisions[4] for the combiner in order to break the components, would still be perfectly convincing.

Related Work. We only consider black-box combiners, and not general combiners, where the combiner gets a full description (e.g. as a circuit) of the underlying primitives and thus is not limited to access the primitives in a black-box manner. This can be justified by noticing that most known cryptographic constructions (not only of combiners) are black-box, which means that the construction only uses the input/output behaviour of the underlying components, and moreover the security proof is black-box, which means that the reduction only uses a successful adversary against the construction in a black-box fashion in order to break the security assumption on the underlying component, such constructions are called "fully black-box" in the taxonomy of [17]. The few exceptions of non black-box constructions (notably the GMW construction of zero-knowledge proofs for NP [6] and Barak's work [1]), are very inefficient. Thus even if non black-box combiners with short output should exist, it would be very surprising if they actually were of any practical relevance. The motivation to restrict oneself to black-box constructions is that it is often feasible to *rule out* such constructions, by using the fact that a black-box reduction is relativizing, i.e. it holds relative to any oracle. Thus a way to rule out the existence of a black-box construction of some primitive α from some primitive β, is to come up with a hypothetical (usually non-efficient) oracle, such that relative to this oracle β exists, but α does not. This technique was first used in the context of cryptographic reductions by Impagliazzo and Rudich, who in their seminal paper [10] prove the impossibility of a black-box construction of key-agreement from one-way functions. Another classical result along this line due to Simon [19] proves the impossibility of constructing CRHFs from one way functions, the breaking oracle in this paper is inspired by this work.

Kim et al. [11] were the first to consider lower bound on the *efficiency* (as opposed to mere feasibility) of black-box constructions. They prove a lower bound on the number of calls to a one-way permutation needed to implement a pseudorandom generator (a thigh bound was subsequently proven in [5]).

The concept of a combiners has first been explicitly considered by Herzberg [9] (who called them "tolerant constructions") and later by Harnik et al. [8], who coined the term "robust combiner". For many natural cryptographic primitives like one-way functions, PRGs or CRHFs $(1, 2)$-combiners are trivially seen to exist. For other primitives like commitments and oblivious transfer the question is open [8,9,13,14].

As mentioned already in the introduction, combiners for CRHFs have been investigated by [2,3,16]. Fischlin and Lehmann [4] consider CRHFs combiners in an ideal setting, and in this setting are able to give a construction which is *more secure* than any of its components. Fischlin, Lehmann and Pietrzak recently constructed a robust $(1, 2)$-combiner for hash-functions with output

---

[4] Here "different collision" can either mean two different collisions for $C^{H_1,H_2}(R,.)$ and any randomness R, or a collision for $C^{H_1,H_2}(R,.)$ and (a not necessarily different one) for $C^{H_1,H_2}(R',.)$ where $R \neq R'$.

length $2\ell_{out}^H$, which simultaneously combines several properties, namely collision-resistance, target collision-resistance, message authentication, pseudorandomness, one-wayness and – at the price of a slightly longer output – indifferentiability from a random oracle [12].

2   Combiners for CRHFs: Definition and Constructions

Notation and some Basic Definitions. For $X, Y \in \{0,1\}^*$ we denote with $X\|Y$ the concatenation of X and Y. For $a \in \mathbb{N}$ we denote with $[a]$ the set $\{0, 1, \ldots, a-1\}$ and $\langle a \rangle_b$ denotes the binary representation of a, padded with 0's to length b, e.g. $\langle 5 \rangle_6 = 000101$. A pair M, M' is a *collision* for a function F if $F(M) = F(M')$ and $M \neq M'$. We call M, M' a *pseudocollision* for F if $F(M) = F(M')$ (but not necessarily $M \neq M'$). With $X \overset{*}{\leftarrow} \mathcal{X}$ we denote that X is assigned a value chosen uniformly at random from the set $\mathcal{X}$.

PPTM stands for Probabilistic Polynomial time Turing Machine. An *oracle* PPTM A (oPPTM for short) is a PPTM with an extra oracle tape, where $A^{\mathcal{O}_1,\ldots,\mathcal{O}_z}$ denotes the random experiment where A runs having access to the oracles $\mathcal{O}_1, \ldots, \mathcal{O}_z$ via its oracle tape: A can write a query (i, X) on the tape, and in the next step the value $\mathcal{O}_i(X)$ is written on the tape. Let

$$\mathsf{qry}^{\mathcal{O}_i}(A^{\mathcal{O}_1,\ldots,\mathcal{O}_z})$$

denote the queries that A makes to the oracle $\mathcal{O}_i$. In this paper, the oracles will always be hash functions $H_1, H_2, \ldots$ and possibly a breaking oracle $\mathcal{B}$. The collision predicate col^{H_i} is defined for the random experiment $A^{H_1,\ldots,H_\ell}$ and holds if A finds a collision for H_i, i.e. A makes two distinct queries X, X' to H_i where $H_i(X) = H_i(X')$, formally[5]

$$\mathsf{col}^{H_i}(A^{H_1,\ldots,H_z}) \iff \exists X, X' \in \mathsf{qry}^{H_i}(A^{H_1,\ldots,H_z}) : X \neq X' \wedge H_i(X) = H_i(X')$$

More generally, for $\mathcal{H} \subseteq \{H_1, \ldots, H_\ell\}$ we define the predicate

$$\mathsf{col}^{\mathcal{H}}(A^{H_1,\ldots,H_z}) \iff \forall H_i \in \mathcal{H} : \mathsf{col}^{H_i}(A^{H_1,\ldots,H_z})$$

which holds if A finds a collisions for all H_i in $\mathcal{H}$. Finally

$$\mathsf{col}_t(A^{H_1,\ldots,H_z}) \iff \exists \mathcal{H} \subseteq \{H_1, \ldots, H_\ell\}, |\mathcal{H}| = t : \mathsf{col}^{\mathcal{H}}(A^{H_1,\ldots,H_z})$$

holds if A finds collisions for at least t of the H_i's.

Definition 1 (Randomized Black-Box Combiner For CRHFs). CONSTRUCTION: *A randomized (k, ℓ)-combiner for CRHFs is a pair (C, P) where C is an oracle circuit $C : \mathcal{R} \times \{0,1\}^m \to \{0,1\}^n$ and P is an oracle PPTM.*
REDUCTION: *An oracle $\mathcal{B}$ ρ-breaks $C^{H_1,\ldots,H_\ell}$ if $\mathcal{B}(R)$ outputs a collision for $C^{H_1,\ldots,H_\ell}(R,.)$ for at least a ρ-fraction of the possible choices of the randomness $R \in \mathcal{R}$, and $\perp$ on the remaining inputs.*

---

[5] Note that we e.g. write simply $\mathsf{col}^H(A^H)$ to denote "the predicate col^H is satisfied in the random experiment A^H", with $\neg\mathsf{col}^H(A^H)$ we denote the complementary event.

(C, P) *is ρ-robust (where ρ can be a function of $v \in \mathbb{N}$) if for all $H_1, \ldots, H_\ell$: $\{0,1\}^* \rightarrow \{0,1\}^v$ and any oracle $\mathcal{B}$ which ρ-breaks $C^{H_1, \ldots, H_\ell}$ the PPTM P in the random experiment $P^{\mathcal{B}, H_1, \ldots, H_\ell}$ finds collisions for at least $\ell - k + 1$ of the H_i's with probability at least .9, i.e.*

$$\Pr_{H_1, \ldots, H_\ell, P's \ coins} [\mathsf{col}_{\ell-k+1}(P^{\mathcal{B}, H_1, \ldots, H_\ell})] \geq .9 \qquad (2)$$

EFFICIENCY: *Let q_C denote the number of oracle gates in C, and q_P be an upper bound on the number of oracle queries made by P, where we do not count oracle queries to $\mathcal{B}$ where the answer is $\perp$ (as we want to prove a negative result, not accounting for such queries makes the negative result stronger). Then the combiner (C, P) is efficient if q_C and q_P are polynomial in v.*

SECURITY: *An efficient combiner (C, P) is robust, if it is ρ-robust where $\rho = \rho(v)$ is smaller than any positive constant for sufficiently large v.*

Remark 2 (on the constant .9). The probability .9 in (2) is over the random coins of P. We chose to fix this probability to the arbitrary constant .9 instead of adding an additional parameter in the security definition, as the constant .9 can be replaced with any value ϵ where ϵ is noticeable[6] and bounded away from 1 by some exponentially small amount, by changing the running time of P only by a polynomial factor. The reason is that if some efficient combiner (C, P) satisfies (2) for some ϵ (instead .9), then for any $z = poly(v)$, we get an efficient combiner (C, P_z) which satisfies (2) with probability $1 - (1 - \epsilon)^z$, where P_z simply simulates P z times using fresh random coins for each run.

Concatenation Combiner. We trivially get a robust and very efficient (k, ℓ)-combiner, by concatenating the output of any $\ell - k + 1$ of the components.

$$C^{H_1, \ldots, H_\ell}(R, M) = H_1(M) \| H_2(M) \| \ldots \| H_{\ell-k+1}(M). \qquad (3)$$

This combiner is an ρ-robust (k, ℓ)-combiner for any $\rho > 0$, where

$$n = (\ell - k + 1)v \qquad q_C = \ell - k + 1 \qquad q_P = 1$$

The reduction P achieving the above parameters, simply queries the oracle $\mathcal{B}$ on distinct $R \in \mathcal{R}$ until it gets a collision (as $\rho > 0$, there will be at least one).

Random Concatenation Combiner. As a generalization of the previous combiner, we can consider the combiner $C : [\binom{\ell}{c}] \times \{0,1\}^m \rightarrow \{0,1\}^n$ where we concatenate the output of c randomly chosen components. For $c < \ell - k + 1$ this combiner has shorter output than (3), but also is only ρ-robust for a ρ which is bounded away from 0 by a constant, and thus is not a "robust combiner". The only reason we mention this construction here is to make clear, that the upper bound on ρ which we will need in our impossibility result is necessary.

Below each R in the randomness space $[\binom{\ell}{c}]$ is parsed as a c element subset $1 \leq R_1 < R_2 < \ldots < R_c \leq \ell$ of $[\ell]$.

$$C^{H_1, \ldots, H_\ell}(R, M) = H_{R_1}(M) \| H_{R_2}(M) \| \ldots \| H_{R_c}(M)$$

---

[6] I.e. at least $1/poly(v)$ for some positive polynomial poly.

In the full version of the paper we prove that this combiner is a ρ-robust (k, ℓ)-combiner for any $\rho > \binom{\ell-k}{c} / \binom{\ell}{c}$ with parameters

$$n = cv \qquad q_C = c \qquad q_P = \ell - k + 2 - c$$

Thus efficient ρ-robust (k, ℓ)-combiners with output length $(\ell - k + 1)v$ exists for any $\rho > 0$, on the other extreme, we can get (by setting $c = 1$ in the above construction) ρ-robust combiners for any $\rho > 1 - k/\ell$ with an output length of only v. This can be slightly improved as we'll describe now.

The Canetti et al $(1,1)$-Combiner. A remarkable construction of Canetti et al [3] is a $(1,1)$ black-box Combiner S which, from any CRHF H with range v, constructs a CRHF S^H with range $v - \Delta$. Unfortunately, for *efficient* combiners, Δ must be logarithmic, as the running time of S increases exponentially in Δ.

We will shortly sketch the idea of the Canetti et al. combiner, for the detailed construction of that combiner we refer the reader to the original paper [3]. Let $H : \{0,1\}^w \to \{0,1\}^v$ be a hash function. First one finds a string $\gamma \in \{0,1\}^\Delta$ where for a random z, the prefix of $H(z)$ is γ with good probability.[7] Let $\tilde{H}(z)$ denote $H(z)$ but with the first Δ bits deleted, and let

$$Z := \{z \in \{0,1\}^w : \text{ the prefix of } H(z) \text{ is } \gamma\}$$

Note that any collision z, z' for $\tilde{H}$ where $z, z' \in Z$ is also a collision for H, as

$$H(z) = \gamma \| \tilde{H}(z) = \gamma \| \tilde{H}(z') = H(z')$$

Thus we have constructed a CRHF $\tilde{H} : Z \to \{0,1\}^{v-\Delta}$ from a CRHF $H : \{0,1\}^w \to \{0,1\}^v$. This is almost a $(1,1)$-combiner with output length $v - \Delta$, except that the domain is some strange set Z. We must somehow map $\{0,1\}^{w'}$ where $w' > v$ injectively to a (subset) of Z in order to get a CRHF $\{0,1\}^{w'} \to \{0,1\}^v$. As shown in [3] this can be achieved, albeit inefficiently in time 2^Δ.

One can replace the H_i's with S^{H_i} in the combiners considered before in order to get shorter output, e.g. for the concatenation combiner (3) we get

"Shrinked" Concatenation Combiner. The combiner (with S as above)

$$C^{H_1,\dots,H_\ell}(R, M) = S^{H_1}(M) \| S^{H_2}(M) \| \dots \| S^{H_{\ell-k+1}}(M) \qquad (4)$$

satisfies for any $\rho > 0$

$$n = (\ell - k + 1)(v - \Delta) \qquad q_C = 2^{O(\Delta)}(\ell - k + 1) \qquad q_P = O(2^\Delta)$$

Main Theorem. In this paper we'll prove that the bound achieved by the combiners (4) is basically tight.

---

[7] The expected probability for a random γ is $2^{-\Delta}$, we're fine with anything not much smaller than that, say $2^{-\Delta-1}$, such a good γ can be found by sampling.

Theorem 1 (Main). *If (C, P), where*

$$C : \{0,1\}^m \to \{0,1\}^n$$

is an efficient and robust randomized (k, ℓ)-combiner for CRHFs with range $\{0,1\}^v$, then

$$n \geq (\ell - k + 1)v - O(\log(q_C)).$$

This theorem is stated in asymptotic terms so it is easy to parse, but we prove a quantitative statement. The quantitative statements are given by Proposition 3 for the special case of $(1,1)$-combiners, and in Proposition 4 for general (k, ℓ)-combiner. In particular, the exact meaning of "efficient" in the theorem above is given by equation (30), where $q_P^{\mathcal{B}}$ and q_P^H denote an upper bound on the number of oracle queries the reduction P makes to the breaking oracle and to the candidate hash functions respectively, so $q_P = q_P^{\mathcal{B}} + q_P^H$. Throughout the paper we assume w.l.o.g. that $q_P^{\mathcal{B}}, q_P^H$ and q_C are at least one.

Lower Bounds for Restricted Combiners. A result analogous to the statement of Theorem 1 has been proven for restricted cases of combiners. Starting with [2], who proved it for *deterministic* combiners (i.e. where $\mathcal{R}$ in Definition 1 is empty), and where the construction C was only allowed to query each H_i exactly once. A simpler proof without the latter restriction (but still deterministic) was subsequently given in [16]. The proof was further simplified in [3], who also for the first time considered the randomized case, but under the restriction that the reduction P queries the breaking oracle at most once. This special case seems much easier to prove than the general one. As the main idea behind the proof of the special case, which is a probabilistic argument, is also used in the proof of the general case, we give the full statement and proof of the special case below.

Proposition 1 (following [3]). *For some n, m, v with $m > n$, assume that (C, P) where*

$$C : \{0,1\}^m \to \{0,1\}^n$$

is a 1-robust (k, ℓ)-combiner for CRHFs with range $\{0,1\}^v$, with the additional constraint that P is querying the breaking oracle only once. Let ϵ denote the success probability (over P's random coins) of P, i.e. for any breaking oracle $\mathcal{B}$ which on input R outputs a collision for $C^{H_1,\ldots,H_\ell}(R, .)$ [8]

$$\forall H_1, \ldots, H_\ell : \Pr_{P's \text{ coins}}[\text{col}_{k+1}(P^{\mathcal{B}, H_1, \ldots, H_\ell})] \geq \epsilon$$

Then the output length n of C satisfies

$$n \geq (\ell - k + 1)(v + 1 - 2\log q_P) - \log\left(\binom{\ell}{\ell - k + 1}\right) + \log(\epsilon) + 1 \qquad (5)$$

---

[8] Here Remark 2 (after Def.1) does not apply, as now we can't run P several times to amplify ϵ as we're only allowed one query to $\mathcal{B}$. So unlike in the general case where we arbitrarily set $\epsilon = .9$, here it is necessary to keep ϵ as a parameter.

Before we prove the proposition, let us remark that for the practically relevant case where P is efficient and ϵ is noticeable, (5) can be written as

$$n \geq (\ell - k + 1)(v - O(\log v))$$

which, up to the constant hidden in the O term, matches parameters of the combiner (4).

Proof. We will only prove the case for $k = 1$ and $\ell = 2$ and explain at the end how to adapt the proof for the general k and ℓ.

Let A be any oracle PPTM making at most q_A oracle queries and $H : \{0,1\}^* \to \{0,1\}^v$ be uniformly random. The probability that any two (distinct) queries made by A to H give a collision for H is $1/2^v$, taking the union bound over all $q_A(q_A - 1)/2$ possible pairs of queries

$$\Pr_{H, A's\ coins}[\mathsf{col}^H(A^H)] \leq q_A(q_A - 1)/2^{v+1} < q_A^2/2^{v+1}. \tag{6}$$

Now consider an oracle PPTM A which expects two oracles, making at most q_A queries to each of them. Let $H_1, H_2 : \{0,1\}^* \to \{0,1\}^v$ be uniformly random and independent. As the H_i's are independent, the probability that P will find collisions for both is the product of what we had in eq.(6).

$$\Pr_{H_1, H_2, A's\ coins}[\mathsf{col}^{H_1, H_2}(A^{H_1, H_2})] \leq (q_A^2/2^{v+1})^2 \tag{7}$$

Now let (C, P) be a combiner as in the statement of the proposition. Let A be an oracle PPTM where A^{H_1, H_2} simulates $P^{\mathcal{B}, H_1, H_2}$, but answers the (single) $\mathcal{B}$ query R made by P with random $M \xleftarrow{*} \{0,1\}^m, M' \xleftarrow{*} \{0,1\}^m$. Note that P will output collisions for H_1, H_2 with probability ϵ conditioned on the event that M, M' is a collision for $C^{H_1, H_2}(R, .)$.

$$\Pr_{H_1, H_2, A's\ coins}[\mathsf{col}^{H_1, H_2}(A^{H_1, H_2})]$$
$$\geq \Pr[\mathsf{col}^{H_1, H_2}(P^{\mathcal{B}, H_1, H_2})] \cdot \Pr[M \neq M' \wedge C^{H_1, H_2}(R, M) = C^{H_1, H_2}(R, M')]$$
$$\geq \epsilon \cdot (2^{-n} - 2^{-m}) \geq \epsilon \cdot 2^{-n+1} \tag{8}$$

Where in the last step we used $m > n$ which holds as C is shrinking. Now by (7) and (8) we must have $\epsilon \cdot 2^{-n+1} \leq (q_P^2/2^{v+1})^2$, solving for n gives

$$n \geq 2(v + 1 - 2\log q_P) + \log(\epsilon) + 1$$

which is (5) for the case where $k = 1, \ell = 2$. For the general case of (k, ℓ)-combiners, we can similarly upper and lower bound the probability of a PPTM A in finding collision for at least $\ell - k + 1$ of its ℓ oracles as

$$\epsilon \cdot 2^{-n+1} \leq \Pr_{H_1, \ldots, H_\ell, A's\ coins}[\mathsf{col}_{\ell-k+1}(A^{H_1, \ldots, H_\ell})] \leq \binom{\ell}{\ell - k + 1}(q_P^2/2^{v+1})^{\ell-k+1}$$

Solving this inequality for n then gives

$$n - 1 \geq (\ell - k + 1)(v + 1 - 2\log q_P) - \log\left(\binom{\ell}{\ell - k + 1}\right) + \log(\epsilon). \qquad \square$$

3   Proof Outline

We will prove our main result gradually, introducing new techniques and ideas in each step. First, in Lemma 1 we show that a uniformly random function is collision resistant, using the fact that such a function cannot be compressed. Based on this technique, we then prove Proposition 3 which implies Theorem 1 for the special case $k = \ell = 1$. Finally, Proposition 4 proves the general case. Due to space reasons, the proof of Proposition 4 is only given in the full version of the paper [15].

Collisions imply Compressibility, Section 4. Gennaro and Trevisan [5] give a very elegant proof that a uniformly random permutation $\pi : \{0,1\}^v \to \{0,1\}^v$ is one-way against poly-size, non-uniform adversaries. On a high level, they show that if P is an efficient[9] adversary which inverts π on many inputs, i.e. for many x we have $A^\pi(\pi(x)) = x$, then π has a "short" description relative to P. This is impossible as a uniformly random π is incompressible, and thus such an P cannot exist (i.e. π is one-way).

We adapt this proof in order to show that a uniformly random *function* $H : \{0,1\}^w \to \{0,1\}^v$ is *collision resistant*. This has been independently discovered by the authors of [7], the proof given in this paper is due to Thomas Holenstein (via personal communication with Iftach Haitner), and is much simpler than the one we had originally.

Lower Bounds for Black-Box Combiners via Incompressibility. The just sketched proof is by no means the easiest way to show that a uniformly random function is collisions resistant.[10]

The advantage of such a "incompressibility based" proof is that it extends to the case where P additionally gets access to a carefully defined "combiner breaking" oracle $\mathcal{B}$, which itself can make much more queries to the hash function(s) than what is needed to find collisions for uniformly random functions with output length v bits (which means roughly $2^{v/2}$ queries), as we'll explain below. This approach is inspired by a recent work of Haitner et al [7], the Gennaro-Trevisan reconstruction lemma [5] and Simon's breaking oracle [19].

Lower bound for $(1,1)$-combiners, Section 5. In order to rule out the existence of an efficient ρ-robust black-box combiner (C, P) with output length $n = v - \omega(\log v)$, one must come up with oracles $H, \mathcal{B}$ such that

- $C^H : \{0,1\}^r \times \{0,1\}^m \to \{0,1\}^n$ is not collision resistant, in the sense that $\mathcal{B}(R)$ outputs a collision for $C^H(R, .)$ on at least a ρ-fraction of the $R \in \{0,1\}^r$.

---

[9] Here efficient means that the number of oracle queries made by P must be much smaller than what would be required to invert π by brute force search (but can still be exponential).

[10] The straight forward way to prove this, is to argue that for any two distinct queries X_a, X_b made by P we have $\Pr[H(X_a) = H(X_b)] = 2^{-v}$, and thus by taking the union bound over all $q(q-1)/2$ pairs of queries, the probability that there exist any X_a, X_b where $H(X_a) = H(X_b)$ is at most $q(q-1)/2^{v+1}$.

– $H : \{0,1\}^w \to \{0,1\}^v$ is collision resistant (even relative to $\mathcal{B}$), in the sense that the probability that $P^{H,\mathcal{B}}$ finds a collision (where the probability is over the random coins of P) is small, which means < 0.9 (cf. Remark 2).

The oracle hash function $H : \{0,1\}^w \to \{0,1\}^v$ is chosen uniformly at random. The breaking oracle $\mathcal{B}$ samples, for each possible input $R \in \{0,1\}^r$, a random pseudocollision Z_R, Z_R' for $C^H(R, .)$. On input R the oracle $\mathcal{B}$ outputs Z_R, Z_R' if this is a "safe" collision, by which we mean that the H queries needed in order to evaluate $C^H(R, Z_R)$ and $C^H(R, Z_R')$ do not contain a collision for H. If the collision is not safe, then $\mathcal{B}(R)$ outputs $\bot$.

Using the fact that the output length n of C^H is by $\omega(\log v)$ bits shorter than the output length of H, one can show (using a probabilistic argument like in the proof of Proposition 1), that with high probability most collisions Z_R, Z_R' will be safe, and thus $\mathcal{B}$ will ρ-break C^H for a ρ which is exponentially close to 1. This is the only part of the proof where we use the fact that C has short output.

It remains to prove that P cannot find collisions for H, even with the powerful combiner breaking oracle $\mathcal{B}$. Intuitively, $\mathcal{B}$ should not be of much help in finding collision for H, as it only returns random collisions for $C^H(R, .)$ which are "safe" (as described above), and thus do not (at least trivially) give collisions for H. To actually prove this, we show that if $P^{H,\mathcal{B}}$ finds collisions with high probability, then we can use P to compress H, which is impossible as H is uniformly random, thus such a P cannot exist.

Lower bound for (k, ℓ)-combiners, Section 5. To rule out the existence of an efficient ρ-robust (k, ℓ)-black-box combiner (C, P) with output length $n = (\ell - k + 1)v - \omega(\log v)$, we will construct ℓ hash functions $H_1, \ldots, H_\ell \stackrel{\text{def}}{=} H^\ell$ and a breaking oracle $\mathcal{B}$ which ρ-breaks C^{H^ℓ}, but at least k of the H_i's are collision resistant even relative to $\mathcal{B}$. The ρ we achieve will be exponentially close to $1/\binom{\ell}{k}$, which is tight because (as explained in the last section) for $\rho > 1/\binom{\ell}{k}$ combiners with output length only $(\ell - k + 1)v$ exist. The $H^\ell = H_1, \ldots, H_\ell : \{0,1\}^w \to \{0,1\}^v$ are chosen uniformly at random. The breaking oracle $\mathcal{B}$ samples, for each $R \in \{0,1\}^r$ a collision Z_R, Z_R' for $C^{H^\ell}(R, .)$ (or, a pseudocollision to be precise, as there's a tiny 2^{-m} probability that $Z_R = Z_R'$). We say that Z_R, Z_R' is a safe collision for H_i, if the evaluation of $C^{H^\ell}(R, .)$ on inputs Z_R, Z_R' does not contain a collision of H_i. By a probabilistic argument, one can show that with high probability a random collision will be safe for at least k of the H_i's (here we need the fact that the output length of C is short). This again implies that there exists a subset $\Gamma \subset \{1, \ldots, \ell\}$ of size k, such that for (almost) a $1/\binom{\ell}{k}$ fraction of the R's, let's call it $\mathcal{R}_\Gamma$, the collision Z_R, Z_R' is safe for all the H_i with $i \in \Gamma$. Now $\mathcal{B}$ on input R outputs Z_R, Z_R' if $R \in \mathcal{R}_\Gamma$, and $\bot$ otherwise. Intuitively, the H_i where $i \in \Gamma$ should be still be collision resistant even relative to $\mathcal{B}$. To prove this we show that if an efficient P exists where $P^{\mathcal{B},H^\ell}$ finds a collision for any H_i where $i \in \Gamma$ with high probability, then this H_i can be compressed, which is impossible as H_i is uniformly random, and thus such a P cannot exist.

4   Collisions Imply Compressibility

For a function $H : \{0,1\}^w \to \{0,1\}^v$, we denote with $\widetilde{H} \in \{0,1\}^{2^w v}$ the function table of H, which is a binary $2^w \times v$ matrix. We number the rows from 0 to $2^w - 1$, thus the i'th row contains the value $H(i)$. Such a function table can be uniquely encoded by a bit-string of length $2^w v$.

A random variable H can be compressed to s bits, if there exists a pair com, dec of functions (possibly probabilistic using joint randomness) such that for any $t \in \mathbb{N}$ and $\widetilde{H}_1, \ldots, \widetilde{H}_t$ being independent instantiations of H, we have

$$\mathop{\mathrm{E}}_{\widetilde{H}_1, \ldots, \widetilde{H}_t} [|\mathsf{com}(\widetilde{H}_1, \ldots, \widetilde{H}_t)|] \leq t \cdot s \tag{9}$$

$$\mathop{\Pr}_{\widetilde{H}_1, \ldots, \widetilde{H}_t} [\mathsf{dec}(\mathsf{com}(\widetilde{H}_1, \ldots, \widetilde{H}_t)) = \widetilde{H}_1, \ldots, \widetilde{H}_t] = 1 \tag{10}$$

As already proved by Shannon, a function table which is chosen uniformly at random, cannot be compressed, i.e.

Proposition 2. *A uniformly random function $H : \{0,1\}^w \to \{0,1\}^v$ cannot be compressed to less than $2^w v$ bits.*

By the following proposition, any function H for which there exists an efficient collision finding algorithm P, can be compressed.

Lemma 1. *Let P be an oracle PPTM which makes at most q_P oracle queries. Let H be a random variable taking as value functions $\{0,1\}^w \to \{0,1\}^v$. For $0 \leq \delta \leq 1$, if P finds a collision with probability δ:*

$$\mathop{\Pr}_{H, P's \ coins} [\mathsf{col}^H(P^H)] = \delta \tag{11}$$

then H can be compressed to

$$1 + 2^w v - \delta(v - 2\log(q_P)) \qquad bits. \tag{12}$$

Using Proposition 2 we get the following Corollary

Corollary 1. *Let $H : \{0,1\}^w \to \{0,1\}^v$ be uniformly random, then any P which for some $\delta > 0$ satisfies eq. (11) makes at least $q_P \geq 2^{v/2 - 1/2\delta}$ oracle queries.*

Proof (of Corollary). If H is uniformly random, then by Proposition 2 expression (12) is at least $2^w v$ which means $1 \geq \delta(v - 2\log(q_P))$, or equivalently $v/2 - 1/2\delta \leq \log q_P$ which implies $q_P \geq 2^{v/2 - 1/2\delta}$ by exponentiating on both sides.   □

Proof (of Lemma 1). Consider a variable H taking as values functions $\{0,1\}^w \to \{0,1\}^v$ and any PPTM P making at most q_P oracle queries. If P^H does not find a collision for H, then we do not compress at all, in this case $\mathsf{col}(\widetilde{H})$ is simply a 0 followed by $\widetilde{H}$. Otherwise let $X_1, X_2, \ldots$ denote the oracle queries made by P^H and let X_{c_1}, X_{c_2} where $c_1 < c_2$ denote the collision found. Let $\widetilde{H}^- \in \{0,1\}^{(2^w - c_2)v}$ denote $\widetilde{H}$ with the rows $X_1, \ldots, X_{c_2}$ (containing the value

$H(X_1), \ldots, H(X_{c_2}))$ deleted. Now $\mathsf{com}(\widetilde{H})$ is a 1 followed by an encoding of the indices c_1, c_2 followed by the first $c_2 - 1$ oracle answers $H(X_1), \ldots, H(X_{c_2-1})$ and finally $\widetilde{H}^-$, i.e.

$$\mathsf{com}(\widetilde{H}) = \begin{cases} 0\|\widetilde{H} & \text{if } \neg\mathsf{col}(P^H) \\ 1\|\langle c_1 \rangle_{\log_{q_P}}\|\langle c_2 \rangle_{\log_{q_P}}\|H(X_1)\|\ldots\|H(X_{c_2-1})\|\widetilde{H}^- & \text{if } \mathsf{col}(P^H) \end{cases}$$

On input more than one function table, com simply compresses each function table separately, and then concatenates the outputs, i.e.

$$\mathsf{com}(\widetilde{H}_1, \ldots, \widetilde{H}_t) = \mathsf{com}(\widetilde{H}_1)\|\ldots\|\mathsf{com}(\widetilde{H}_t)$$

Before we describe the decompression algorithm, let us check that this compression really achieves the length as claimed in eq.(12). The output length of $\mathsf{com}(\widetilde{H})$ is $1 + 2^w v$ if P does not find a collision for H, which by assumption happens with probability $1 - \delta$. Otherwise the length is $1 + (c_2 - 1)v + (2^w v - c_2)v + 2\log q_P$, which gives an expected length of

$$\mathsf{E}[|\mathsf{com}\widetilde{H}|] = 1 + (1-\delta)2^w v + \delta\left((2^w - 1)v + 2\log q_P\right) = 1 + 2^w v - \delta(v - 2\log q_P)$$

as claimed. The decompression algorithm dec, on input $T = \mathsf{com}(\widetilde{H}_1, \ldots, \widetilde{H}_t)$ first parses T into $\mathsf{com}(H_1)$ to $\mathsf{com}(H_t)$ which can be done as the length (there are only 2 possibilities) of $\mathsf{com}(H_1)$ can be uniquely determined reading only the first bit. We can then strip off $\mathsf{com}(H_1)$, the first bit of the remaining string determines the length of $\mathsf{com}(H_2)$, and so on. We thus must only show how to decompress a single compressed function table $T = \mathsf{com}(\widetilde{H})$. On input $T = \mathsf{com}(\widetilde{H})$, dec parses T as $b\|T'$, where $b \in \{0,1\}$. If $b = 0$ the output is T' and we are done. Otherwise parse T' as

$$\langle c_1 \rangle_{\log_{q_P}}\|\langle c_2 \rangle_{\log_{q_P}}\|H(X_1)\|\ldots\|H(X_{c_2-1})\|\widetilde{H}^-$$

Now simulate P^H up to the point where P asks the c_2'th oracle query X_{c_2}.[11] Note that we can answer the first $c_2 - 1$ oracle queries made by P as we know $H(X_1), \ldots, H(X_{c_2-1})$. Now, by construction we also know $H(X_{c_2})$, as it is equal to $H(X_{c_1})$. We can now reconstruct (and output) $\widetilde{H}$ from the reduced table $\widetilde{H}^-$ as we know all missing values $H(X_1)$ to $H(X_{c_2})$ and also the positions X_1 to X_{c_2} where to insert them in $\widetilde{H}^-$ in order to get $\widetilde{H}$.

Before we continue proving Theorem 1, we need a few more definitions.

Definition 2 (safe collisions, the predicate safeCol). *Let $H^\ell = H_1, \ldots, H_\ell$ be ℓ hash functions and A be an oPPTM. We say that Z, Z' is a safe collision for H_i (with respect to A^{H^ℓ})*

1. $A^{H^\ell}(Z) = A^{H^\ell}(Z')$ (but not necessarily $Z \neq Z'$)

---

[11] As P can be probabilistic, we need com and dec to use the same random coins for P. Alternatively, we can just fix the randomness of P as to maximize $\Pr[\mathsf{col}^H(P^H)]$.

2. *during the evaluation of $A^{H^\ell}(.)$ on inputs Z and Z', there are no two queries $X \neq X'$ to H_i where $H_i(X) = H_i(X')$.*

We have $\mathsf{safeCol}_{H_i}^{A^{H^\ell}}(Z, Z')$ if Z, Z' is a safe collision. For any $1 \leq k \leq \ell$, $\mathsf{safeCol}_k^{A^{H^\ell}}(Z, Z')$ holds if for at least k different i's, $\mathsf{safeCol}_{H_i}^{A^{H^\ell}}(Z, Z')$ holds.

Intuitively, when given $Z \neq Z'$ where $\mathsf{safeCol}_{H_i}^{A^{H^\ell}}(Z, Z')$, one learns a collision for A^{H^ℓ}, but this collision does not (at least not trivially) give us a collision for H_i.

Definition 3 ($\prec$). *If we consider a random experiment where some oPPTM runs making queries to its oracle(s). Then for two queries X, Y (not necessarily to the same oracle) we denote by $X \prec Y$ that the query X is made before the query Y is made.*

5   Lower Bounds

Lower bound for $(1, 1)$-combiners. In this section we prove a Proposition which implies Theorem 1 for the special case $k = \ell = 1$. The word "combiner" is a bit misleading in this case, "shrinker" would be more appropriate, as we ask for a construction which given access to a hash function H with range $\{0, 1\}^v$, gives a hash function whose output length n is "significantly" shorter than v.

Proposition 3 (implies Thm.1 for the special case $k = \ell = 1$). *Let $C : \{0, 1\}^r \times \{0, 1\}^m \to \{0, 1\}^n$ be an oracle circuit with input range $m := v + 1$ bits and with q_C oracle gates, where for some $t > 0$*

$$n := v - 2\log(q_C) - t \tag{13}$$

then, if for some oracle PPTM P (which makes $q_P^\mathcal{B}$ oracle calls to the breaking oracle and q_P^H oracle calls to the components) it is the case that (C, P) is a ρ-robust $(1, 1)$-combiner with $\rho := 1 - 2^{-t+3}$, then for some constant $\alpha > 0$

$$v \leq \log q_P^\mathcal{B} + \log q_C + 2(\log(q_P^H + \alpha q_C q_P^\mathcal{B})) + 6 \tag{14}$$

or equivalently,

$$2^v \leq q_P^\mathcal{B} \cdot q_C \cdot (q_P^H \cdot \alpha q_C q_P^\mathcal{B})^2 \cdot 64$$

in particular, (C, P) is not efficient, as by the above, either C or P must make an exponential number of queries.

Remark 3 (on the constant α). A concrete bound on α in (14) can be determined from the proof of Lemma 4 given in the full version of the paper. A rough estimate suggests that setting $\alpha = 1000$ is far on the safe side. This seems large, but note that only the logarithm of α appears in the expression.

Remark 4 (on the input length). Proposition 3 only rules out combiners which hash their input down by $m - n = t + 2\log q_P + 1$ bits. This implies impossibility for the general case, where the input length can be arbitrary as long as the combiner is shrinking. The reason is that using the Merkle-Damgård construction, one can get a CRHF with any input length from a CRHF which hashes down only one bit.

The Oracle. We now define the oracles, which consist of the hash function H and the breaking oracle $\mathcal{B}$. The oracle H is sampled uniformly at random from all functions $\{0,1\}^w \to \{0,1\}^v$. The oracle $\mathcal{B}$ will be completely defined by a function $\phi : \{0,1\}^* \to \{0,1\}^m$ which we sample uniformly at random. This ϕ defines for each randomness $R \in \{0,1\}^r$ a pseudocollision[12] Z_R, Z'_R for $C^H(R,.)$ as follows: $Z_R := \phi(R)$ and $Z'_R := \phi(R\|\langle i \rangle)$, where i is the smallest integer such that $C^H(R, Z_R) = C^H(R, Z'_R)$. The input/output behavior of oracle $\mathcal{B}$ is now defined as

$$\mathcal{B}(R) = \begin{cases} Z_R, Z'_R & \text{if } \mathsf{safeCol}_H^{C^H(R,.)}(Z_R, Z'_R) \\ \perp & \text{otherwise} \end{cases}$$

So $\mathcal{B}(R)$ outputs Z_R, Z'_R only if this is a safe collision.

To prove that $\mathcal{B}$ breaks the security of any combiner, we'll need the following technical lemma (for the special case $\ell = 1$), which states that a randomly sampled collision for a combiner C^{H^ℓ} will be safe for many of the $H_\ell = H_1, \ldots, H_\ell$. For how many exactly of course depends on the output length of C. For space reasons we only prove this lemma in the full version.

Lemma 2. *For any oracle circuit $C : \{0,1\}^m \to \{0,1\}^n$ with q_C oracle gates, and ℓ independent uniformly random functions $H^\ell = H_1, \ldots, H_\ell : \{0,1\}^* \to \{0,1\}^v$. For X, X', sampled as $X \xleftarrow{*} \{0,1\}^m$ and $X' \xleftarrow{*} C^{H^\ell}(X)^{-1}$, then for $k \leq \ell$*

$$\Pr[\mathsf{safeCol}_k^{C^{H^\ell}}(X, X')] \geq 1 - 2^{n-m} - (q_C(q_C - 1))^{\ell-k+1} \cdot \binom{\ell}{\ell-k+1} 2^{n-(\ell-k+1)\cdot v}$$

$\mathcal{B}$ $1 - 2^{-t+3}$ **breaks** C^H. Let $I_R = 1$ if $\mathcal{B}(R) \neq \perp$ and $I_R = 0$ otherwise. From Lemma 2 (for $\ell = 1$) if follows that (recall that ϕ is the randomness used by $\mathcal{B}$)

$$\Pr_{H,\phi}[I_R = 0] \leq 2^{n-v-1} + q_C(q_C - 1) \cdot 2^{n-v+1} < q_C^2 \cdot 2^{n-v+1} \qquad (15)$$

Note that $\mathcal{B}$ ρ-breaks C^H, where ρ is the fraction of R's for which $\mathcal{B}(R) \neq \perp$. By (15) ρ is a random variable with expectation

$$\mathsf{E}_{H,\phi}[\rho] = 2^{-r} \sum_{R \in \{0,1\}^r} \Pr_{H,\phi}[I_R = 1] > 1 - q_C^2 \cdot 2^{n-v+1} = 1 - 2^{-t+1}$$

where in the last step we used (13). Applying the Markov inequality,[13] we get $\Pr[\rho < 1 - \gamma 2^{-t+1}] \leq 1/\gamma$ for any $\gamma > 0$, we will use this bound with $\gamma = 4$, i.e.

$$\Pr_{H,\phi}[\rho < 1 - 2^{-t+3}] \leq 1/4 \qquad (16)$$

---

[12] Recall that X, X' is a pseudocollision for F if $F(X) = F(X')$ but (unlike for collisions) we must not necessarily have $X \neq X'$.

[13] Unfortunately the I_R's are not independent, thus Chernoff is not an option here.

Hard to find collisions for H relative to $\mathcal{B}$. We will now show that one cannot find collisions in H even relative to the powerful oracle $\mathcal{B}$.

Lemma 3. *Let (C, P) be as in the statement of Proposition 3 where*

$$v > \log q_P^{\mathcal{B}} + \log q_C + 2(\log(q_P^H + \alpha q_C q_P^{\mathcal{B}})) + 6 \tag{17}$$

and

$$\Pr_{H,\phi,P's\ coins}[\mathsf{col}^H(P^{H,\mathcal{B}})] \geq .675 \tag{18}$$

then H can be compressed below $2^w v$ bits.

Before we prove this lemma, we first how it implies Proposition 3.

Proof (of Proposition 3). let $\mathcal{E}$ denote the event that $\mathcal{B}$ ρ-breaks C^H with $\rho \geq 1 - 2^{-t+3}$, using (16) and the $1 - 2^{-t+3}$ security of (C, P) in the last step

$$\Pr_{H,\phi,P's\ coins}[\mathsf{col}^H(P^{\mathcal{B},H})] \geq \Pr_{H,\phi,P's\ coins}[\mathcal{E}] \cdot \Pr_{H,\phi,P's\ coins}[\mathsf{col}^H(P^{\mathcal{B},H})|\mathcal{E}] \geq \frac{3}{4} \cdot 0.9$$

Assume H is uniformly random, then by Lemma 3 the function table of H can be compressed below $2^w v$ bits, which contradicts Proposition 2, thus (17) must be wrong.   □

We split the proof of Lemma 3 into two parts. First, Lemma 4 below states that from an oPPTM which finds collisions with high probability as required by eq.(18), we can construct another oPPTM which finds collisions of a special kind, called "very good collisions".[14] Second, Lemma 5 below states that any oPPTM which finds very good collisions for H, implies that H can be compressed.

Very Good Collisions. We now define the "very good collisions" predicate vgCol just mentioned. This predicate has a quite intuitive meaning: $\mathsf{vgCol}(Q^{H,\mathcal{B}})$ if there's a collision, and the H query leading to the collision is fresh, in the sense that it is not in qry_R for some $\mathcal{B}$ query R. More formally, for an oPPTM Q consider the random experiment $Q^{H,\mathcal{B}}$, where $X_1, X_2, \ldots, X_{c_2}$ denotes the H queries, and $R_1, R_2, \ldots, R_j$ denotes the $\mathcal{B}$ queries made by Q. If $\mathsf{col}(Q^{H,\mathcal{B}})$, let X_{c_1}, X_{c_2} denote the collision found by Q. Let qry_R denote all the H queries one must make in order to evaluate $C^H(R, .)$ on the pseudocollision Z_R, Z'_R as sampled by $\mathcal{B}$, i.e.

$$\mathsf{qry}_R := \mathsf{qry}^H(C^H(R, Z_R)) \cup \mathsf{qry}^H(C^H(R, Z'_R)) \tag{19}$$

Then the very good collisions predicate $\mathsf{vgCol}(Q^{H,\mathcal{B}})$ holds if

$$\mathsf{col}(Q^{H,\mathcal{B}}) \text{ and the collision } X_{c_1}, X_{c_2} \text{ satisfies } \forall R \prec X_{c_2} : X_{c_2} \notin \mathsf{qry}_R \tag{20}$$

---

[14] We leave the term "good collision" for an intermediate kind of collision which will only come up in the proof.

From Good Collisions to Very Good Collisions

Lemma 4. *If for a PPTM P*

$$\Pr_{H,\phi,P's \ coins}[\mathsf{col}(P^{H,\mathcal{B}})] \geq .675 \tag{21}$$

then there exists a PPTM Q where

$$\Pr_{H,\phi,Q's \ coins}[\mathsf{vgCol}(Q^{H,\mathcal{B}})] \geq .5 \tag{22}$$

and for a constant α

$$q_Q^{\mathcal{B}} = q_P^{\mathcal{B}} \qquad q_Q^H = q_P^H + \alpha q_C q_P^{\mathcal{B}} \tag{23}$$

We omit the proof of this lemma for space reasons. The basic idea of the proof is to let $Q^{H,\mathcal{B}}$ simply simulate $P^{H,\mathcal{B}}$, but whenever P is about to make a $\mathcal{B}$ query R, Q will additionally sample some random $V_1, \ldots, V_\alpha$ and make all the H queries needed to compute $C^H(R, V_i)$. One can show that if the output P gets on his $\mathcal{B}$ query R is likely to contain a collision (which will not be a very good collision), then the H queries Q makes, will also be likely to contain a very good collision. It is the proof of this lemma where we need the fact that $\mathcal{B}(R)$ will output the collision Z_R, Z'_R only if this is a safe collision.

Very Good Collisions Imply Compressibility

Lemma 5. *Let $\mathcal{B}$ be an oracle (sampled as described earlier in this section) and let H be a random variable taking as values functions $\{0,1\}^w \to \{0,1\}^v$. If a PPTM Q satisfies*

$$\Pr_{H,\phi,Q's \ coins}[\mathsf{vgCol}^H(Q^{\mathcal{B},H})] \geq .5 \tag{24}$$

then for any $0 \leq \gamma \leq 1$, H can be compressed to

$$1 + 2^w v - (1-p)(v - \gamma - 2\log q_Q^H) \tag{25}$$

bits, where $p := 0.5 + q_Q^{\mathcal{B}} \cdot q_C \cdot 2^{-\gamma v}$.

Before we prove this Lemma, let us show how Lemma 4 and 5 imply Lemma 3.

Proof (of Lemma 3). A P as in (18) implies by Lemma 4 a Q as in (22), which by Lemma 5 implies that H can be compressed to (25) bits. This expression is less than $2^w v$ (as required by the lemma) if

$$(0.5 - q_Q^{\mathcal{B}} \cdot q_C \cdot 2^{-\gamma v})(v - \gamma v - 2\log q_Q^H) > 1 \tag{26}$$

By setting $\gamma v := \log q_Q^{\mathcal{B}} + \log q_C + 2$ the first bracket on the left side of (26) becomes $1/4$, if now $v > \log q_Q^{\mathcal{B}} + \log q_C + 2\log q_Q^H + 6 + \log(\ell)$, which by (23) is exactly the requirement (17) from the lemma, then the second bracket in (26) is > 4, thus as $1/4 \cdot 4 = 1$ (26) holds. □

Proof (of Lemma 5). The proof is similar to the proof of Lemma 1, except that now we must additionally handle the breaking oracle $\mathcal{B}$. For this we will additionally need some shared randomness for com and dec, namely a pairwise independent function $\tau : \{0,1\}^w \to \{0,1\}^{\gamma v}$.

If we don't have a very good collision, the compression com simply outputs the whole function table

$$\mathsf{com}(\widetilde{H}) = 0\|\widetilde{H} \quad \text{if} \quad \neg\mathsf{vgCol}^H(Q^{\mathcal{B},H})$$

Otherwise let $X_1, X_2, \ldots, X_{c_1}, \ldots, X_{c_2}$ denote the H queries, and $R_1, R_2, \ldots, R_j$ denote the $\mathcal{B}$ queries made by $Q^{\mathcal{B},H}$, where X_{c_1}, X_{c_2} denotes the collision found by Q. With qry_R as defined in (19), let

$$\mathcal{X} := \mathsf{qry}_{R_1} \cup \ldots \cup \mathsf{qry}_{R_j}$$

We define the predicate miss as $\mathsf{miss} \iff \exists X \in \mathcal{X} : \tau(X) = \tau(X_{c_2})$.

The size of $\mathcal{X}$ is upper bounded by $2 \cdot j \cdot q_C \le 2 \cdot q_Q^{\mathcal{B}} \cdot q_C$. As we now consider the case where $\mathsf{vgCol}^H(Q^{\mathcal{B},H})$, we have $X_{c_2} \notin \mathcal{X}$ (cf. (20)). Further, because τ is pairwise independent, for any $X \ne X_{c_2}$ we have $\Pr[\tau(X) = \tau(X_{c_2})] < 2^{-\gamma v}$. Taking the union bound over all $X \in \mathcal{X}$

$$\Pr[\mathsf{miss}] \le \Pr[\exists X \in \mathcal{X} : \tau(X_{c_2}) = \tau(X)]$$
$$\le \sum_{X \in \mathcal{X}} \Pr[\tau(X_{c_2}) = \tau(X)] \le |\mathcal{X}| \cdot 2^{-\gamma v} \le 2 \cdot q_Q^{\mathcal{B}} \cdot q_C \cdot 2^{-\gamma v} \quad (27)$$

In the case where we have miss, com again simply outputs the whole table.

$$\mathsf{com}(\widetilde{H}) = 0\|\widetilde{H} \quad \text{if} \quad \mathsf{vgCol}^H(Q^{\mathcal{B},H}) \wedge \mathsf{miss}$$

We will now define an oracle $\mathcal{B}_\tau$, which almost behaves as $\mathcal{B}$, and in particular, whenever we have $\neg\mathsf{miss}$ in $Q^{\mathcal{B},H}$, then the oracle answers of $\mathcal{B}$ in $Q^{\mathcal{B}}$ are identical to the answers of $\mathcal{B}_\tau$ in $Q^{\mathcal{B}_\tau,H}$. Recall that $\mathcal{B}$ on input R samples a pseudocollision Z_R, Z'_R by setting $Z_R := \phi(R)$ and then computes, for $i = 1, 2, \ldots$, the value $Z_i = C^H(R\|\langle i\rangle)$ until $C^H(R, Z_R) = C^H(R, Z_i)$, it then assigns $Z'_R := Z_i$. The oracle $\mathcal{B}_\tau$ does exactly the same, but if the evaluation of $C^H(R, Z_R)$ requires to make an H query X here $\tau(X) = \tau(X_{c_2})$, then $\mathcal{B}_\tau$ does not make this query, but stops and outputs $\bot$. Also, whenever the evaluation of $C^H(R, Z_i)$ requires to make an H query X here $\tau(X) = \tau(X_{c_2})$, then $\mathcal{B}_\tau$ does not make this query, but proceeds with Z_{i+1}. Note that $\mathcal{B}(R)$ and $\mathcal{B}_\tau(R)$ will find the same pseudocollision, iff $\tau(X_{c_2}) \notin \mathsf{qry}_R$.

Recall that we now only consider the case where $\mathsf{vgCol}^H(Q^{\mathcal{B},H})$ and $\neg\mathsf{miss}$. Consider the random experiment $Q^{\mathcal{B}_\tau,H}$, and let $A_1, A_2, \ldots, A_\sigma$, where $A_\sigma = X_{c_2}$ denote all the H queries done by Q plus the H queries made by $\mathcal{B}_\tau$ (in the order as they are made in the random experiment $Q^{\mathcal{B}_\tau,H}$ up to the "collision finding" H query X_{c_2}, but without repetitions). So each H query by Q increases the sequence $A_1, A_2, \ldots$ at most by one, whereas a $\mathcal{B}_\tau$ query by Q can increase it by arbitrary many values. Let $\widetilde{H}^-$ denote the function table of H, but with the rows $A_1, A_2, \ldots, A_\sigma$ deleted. The compression algorithm com for the remaining cases is now defined as

$$\mathsf{com}(\widetilde{H}) = 1\|\tau(X_{c_2})\|\langle c_1\rangle_{\log q_Q}\|\langle c_2\rangle_{\log q_Q}\|H(A_1)\|\ldots\|H(A_{\sigma-1})\|\widetilde{H}^- \quad (28)$$

$$\text{if} \quad \mathsf{vgCol}^H(Q^{\mathcal{B},H}) \wedge \neg\mathsf{miss} \quad (29)$$

Let us check that this compression really compresses as claimed by eq. (25). If $\neg\mathsf{vgCol}^H(Q^{\mathcal{B},H})$, or $\mathsf{vgCol}^H(Q^{\mathcal{B},H}) \wedge \mathsf{miss}$ then then $|\mathsf{com}(\widetilde{H})| = 2^w v + 1$. By (24),(27) this happens with probability at most $p := 0.5 + 0.5(2 \cdot q_Q^{\mathcal{B}} \cdot q_C \cdot 2^{-\gamma v})$. Otherwise by (29) $|\mathsf{com}(\widetilde{H})|$ has length only $1 + (2^w - 1 + \gamma)v + 2\log q_Q^H$. Thus

$$\mathsf{E}[|\mathsf{com}(\widetilde{H})|] \leq 1 + 2^w v - (1 - p)(v - \gamma - 2\log q_Q^H)$$

as required by (25). Decompression is straight forward and omitted for space reasons. □

Lower bound for (k, ℓ)-combiners. In the full version of this paper [15] we prove the following Proposition which implies Theorem 1 for general (k, ℓ)-combiners.

Proposition 4. *Let $C : \{0,1\}^r \times \{0,1\}^m \to \{0,1\}^n$ be an oracle circuit with input range[15] $m := \ell \cdot (v+1)$ bits and q_C oracle gates, where for some $t > 0$*

$$n := (\ell - k + 1) \cdot (v - 2\log(q_C)) - t$$

then, if for some oracle PPTM P which makes $q_P^{\mathcal{B}}$ oracle calls to the breaking oracle and q_P^H oracle calls to its components it is the case that (C, P) is a ρ-robust (k, ℓ)-combiner with $\rho := 1/\binom{\ell}{k} - 2^{-t+\ell+2}$, then

$$v \leq \log q_P^{\mathcal{B}} + \log q_C + 2(\log(q_P^H + \alpha q_C q_P^{\mathcal{B}})) + 6 + \log(\ell) \quad (30)$$

in particular, (C, P) is not efficient.

Acknowledgements

I'd like to thank the anonymous reviewers from Crypto'08 for their many helpful comments and suggestions.

References

1. Barak, B.: How to go beyond the black-box simulation barrier. In: 42nd FOCS, pp. 106–115. IEEE Computer Society Press, Los Alamitos (2001)
2. Boneh, D., Boyen, X.: On the impossibility of efficiently combining collision resistant hash functions. In: Dwork, C. (ed.) CRYPTO 2006. LNCS, vol. 4117, pp. 570–583. Springer, Heidelberg (2006)

---

[15] Remark 4 applies here too.

3. Canetti, R., Rivest, R.L., Sudan, M., Trevisan, L., Vadhan, S.P., Wee, H.: Amplifying collision resistance: A complexity-theoretic treatment. In: Menezes, A. (ed.) CRYPTO 2007. LNCS, vol. 4622, pp. 264–283. Springer, Heidelberg (2007)
4. Fischlin, M., Lehmann, A.: Security-amplifying combiners for collision-resistant hash functions. In: Menezes, A. (ed.) CRYPTO 2007. LNCS, vol. 4622, pp. 224–243. Springer, Heidelberg (2007)
5. Gennaro, R., Trevisan, L.: Lower bounds on the efficiency of generic cryptographic constructions. In: 41st FOCS, pp. 305–313. IEEE Computer Society Press, Los Alamitos (2000)
6. Goldreich, O., Micali, S., Wigderson, A.: Proofs that yield nothing but their validity for all languages in NP have zero-knowledge proof systems. Journal of the. ACM 38(3), 691–729 (1991)
7. Haitner, I., Hoch, J.J., Reingold, O., Segev, G.: Finding collisions in interactive protocols: a tight lower bound on the round complexity of statistically-hiding commitments. In: 48th FOCS, pp. 669–679. IEEE Computer Society Press, Los Alamitos (2007)
8. Harnik, D., Kilian, J., Naor, M., Reingold, O., Rosen, A.: On robust combiners for oblivious transfer and other primitives. In: Cramer, R. (ed.) EUROCRYPT 2005. LNCS, vol. 3494, pp. 96–113. Springer, Heidelberg (2005)
9. Herzberg, A.: On tolerant cryptographic constructions. In: Menezes, A. (ed.) CT-RSA 2005. LNCS, vol. 3376, pp. 172–190. Springer, Heidelberg (2005)
10. Impagliazzo, R., Rudich, S.: Limits on the Provable Consequences of One-way Permutations. In: Proc. 21st ACM STOC, pp. 44–61. ACM Press, New York (1989)
11. Kim, J.H., Simon, D.R., Tetali, P.: Limits on the efficiency of one-way permutation-based hash functions. In: 40th FOCS, pp. 535–542. IEEE Computer Society Press, Los Alamitos (1999)
12. Lehmann, A., Fischlin, M., Pietrzak, K.: Robust multi-property combiners for hash functions revisited. In: ICALP 2008. LNCS. Springer, Heidelberg (2008)
13. Meier, R., Przydatek, B.: On robust combiners for private information retrieval and other primitives. In: Dwork, C. (ed.) CRYPTO 2006. LNCS, vol. 4117, pp. 555–569. Springer, Heidelberg (2006)
14. Meier, R., Przydatek, B., Wullschleger, J.: Robuster combiners for oblivious transfer. In: Vadhan, S.P. (ed.) TCC 2007. LNCS, vol. 4392, pp. 404–418. Springer, Heidelberg (2007)
15. Pietrzak, K.: www.cwi.nl/~pietrzak/publications.html
16. Pietrzak, K.: Non-trivial black-box combiners for collision-resistant hash-functions don't exist. In: Naor, M. (ed.) EUROCRYPT 2007. LNCS, vol. 4515, pp. 23–33. Springer, Heidelberg (2007)
17. Reingold, O., Trevisan, L., Vadhan, S.P.: Notions of reducibility between cryptographic primitives. In: Naor, M. (ed.) TCC 2004. LNCS, vol. 2951, pp. 1–20. Springer, Heidelberg (2004)
18. Rogaway, P.: Formalizing human ignorance. In: Nguyên, P.Q. (ed.) VIETCRYPT 2006. LNCS, vol. 4341, pp. 211–228. Springer, Heidelberg (2006)
19. Simon, D.R.: Finding collisions on a one-way street: Can secure hash functions be based on general assumptions? In: Nyberg, K. (ed.) EUROCRYPT 1998. LNCS, vol. 1403, pp. 334–345. Springer, Heidelberg (1998)
20. Wang, X., Yin, Y.L., Yu, H.: Finding collisions in the full SHA-1. In: Shoup, V. (ed.) CRYPTO 2005. LNCS, vol. 3621, pp. 17–36. Springer, Heidelberg (2005)
21. Wang, X., Yu, H.: How to break MD5 and other hash functions. In: Cramer, R. (ed.) EUROCRYPT 2005. LNCS, vol. 3494, pp. 19–35. Springer, Heidelberg (2005)

Constructing Cryptographic Hash Functions from Fixed-Key Blockciphers

Phillip Rogaway[1] and John Steinberger[2]

[1] Department of Computer Science, University of California, Davis, USA
[2] Department of Mathematics, University of British Columbia, Canada

Abstract. We propose a family of compression functions built from fixed-key blockciphers and investigate their collision and preimage security in the ideal-cipher model. The constructions have security approaching and in many cases equaling the security upper bounds found in previous work of the authors [24]. In particular, we describe a $2n$-bit to n-bit compression function using three n-bit permutation calls that has collision security $N^{0.5}$, where $N = 2^n$, and we describe $3n$-bit to $2n$-bit compression functions using five and six permutation calls and having collision security of at least $N^{0.55}$ and $N^{0.63}$.

Keywords: blockcipher-based hashing, collision-resistant hashing, compression functions, cryptographic hash functions, ideal-cipher model.

1 Introduction

This paper is about fixed-key constructions for turning blockciphers into compression functions. When we say that a blockcipher-based construction is *fixed key* we mean that just a handful of constants are used as keys, so that our starting point is actually a small collection of permutations. The idea of doing cryptographic hashing from such a starting point was introduced by Preneel, Govaerts, and Vandewalle some 15 years ago [20], but the approach did not catch on.

For years, the customary starting point for building cryptographic hash functions has been (non-fixed-key) blockciphers, even if this hasn't always been made explicit. But a fixed-key design has some definite advantages. For one thing, blockciphers usually have significant key-setup costs, so fixing the keys can be good for efficiency. More than that, blockcipher designs are typically not very conservative with respect to related-key attacks—but this and more is needed when a blockcipher is used in any standard way to build a collision-resistant hash function. A fixed-key design effectively addresses this concern, banishing the need for related-key-attack security on a large space of keys. Finally, a fixed-key design concentrates the cryptographic work into a highly compact primitive, say a map $\pi\colon \{0,1\}^{128} \to \{0,1\}^{128}$ instead of a map $E\colon \{0,1\}^{512} \times \{0,1\}^{160} \to \{0,1\}^{160}$. Cryptographically processing fewer than half as many bits, a fixed-key design embodies an aspiration for minimalism.

So far, it has not been clear that it is *possible* to turn an n-bit permutation into a hash function that outputs n or more bits and has desirable collision-resistance bounds. Nobody has ever demonstrated such a design, and, three

D. Wagner (Ed.): CRYPTO 2008, LNCS 5157, pp. 433–450, 2008.
© International Association for Cryptologic Research 2008

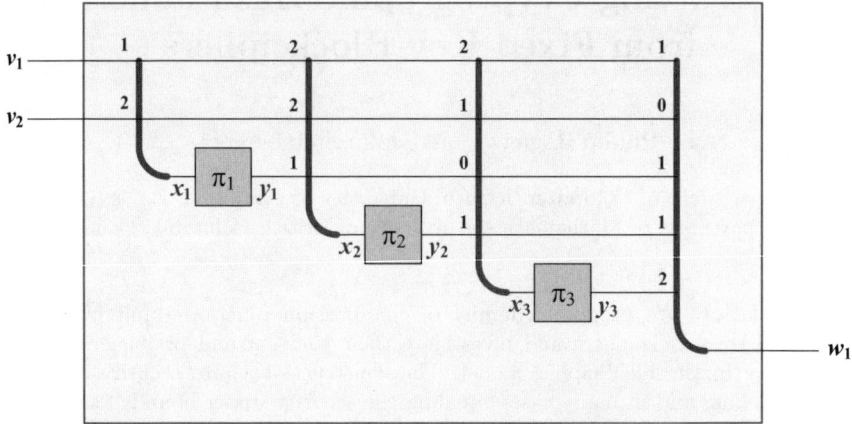

$$
\begin{array}{ll}
\textbf{algorithm } \mathrm{LP}^{A}_{231}\,(v_1\,v_2) & \textbf{algorithm } \mathrm{LP}^{A}_{mkr}\,(v_1\cdots v_m) \\
x_1 \leftarrow v_1 + \mathbf{2}v_2 & \textbf{for } i \leftarrow 1 \textbf{ to } k \textbf{ do} \\
y_1 \leftarrow \pi_1(x_1) & \quad x_i \leftarrow a_i \cdot (v_1,\ldots,v_m,\,y_1,\ldots,y_{i-1}) \\
x_2 \leftarrow \mathbf{2}\,v_1 + \mathbf{2}\,v_2 + y_1 & \quad y_i \leftarrow \pi_i(x_i) \\
y_2 \leftarrow \pi_2(x_2) & \textbf{for } i \leftarrow 1 \textbf{ to } r \textbf{ do} \\
x_3 \leftarrow \mathbf{2}\,v_1 + v_2 + y_2 & \quad w_i \leftarrow a_{k+i} \cdot (v_1,\ldots,v_m,\,y_1,\ldots,y_k) \\
y_3 \leftarrow \pi_3(x_2) & \textbf{return } w_1 \cdots w_r \\
w_1 \leftarrow v_1 + y_1 + y_2 + \mathbf{2}\,y_3 & \\
\textbf{return } w_1 &
\end{array}
$$

Fig. 1. Top & bottom left: Compression function LP^{A}_{231} (usually denoted LP231) for a suitable matrix $A = (a_{ij})$. Horizontal lines carry $n = 128$ bit strings and x_1, x_2, x_3, w_1 are computed as in $x_3 = \mathbf{2}\,v_1 + v_2 + y_2$ with arithmetic and constants $\mathbf{0}, \mathbf{1}$, and $\mathbf{2} = 0^{126}10$ all in $\mathbb{F}_{2^{128}}$. Permutations $\pi_1, \pi_2, \pi_3 \colon \{0,1\}^n \to \{0,1\}^n$ are modeled as independent random permutations to which the adversary can make forward and backwards queries. **Bottom right:** Defining LP^{A}_{mkr} for an $(k+r) \times (k+m)$ matrix A over $\mathbb{F}_{2^n}$. Function lp^{A}_{mkr} is identical but uses a single permutation π.

years ago, Black, Cochran, and Shrimpton seemed to cast a shadow on the possibility [6]. They showed that a prior construction in the literature was wrong, in the sense of having a query-efficient attack, and that, in fact, so will *any* iterated hash function whose underlying compression function maps $2n$ bits to n bits using a single call to an n-bit permutation. But Black *et al.* never said that provably-sound permutation-based hashing is impossible—only that such a scheme couldn't be *extremely* efficient.

In earlier work, the authors extended the Black *et al.* findings [24]. We showed that one expects to find a collision in an $m \xrightarrow{k} r$ compression function—meaning one that maps mn bits to rn bits by calling a sequence of k n-bit permutations— with about $N^{1-(m-0.5r)/k}$ queries, where $N = 2^n$. The result is in the ideal-cipher model and assumes a random-looking compression-function, formalized by a notion of *collision-uniformity*. The result suggests that a desirable-in-practice $2 \xrightarrow{2} 1$ construction can't deliver acceptable security so a $2 \xrightarrow{3} 1$ design is the

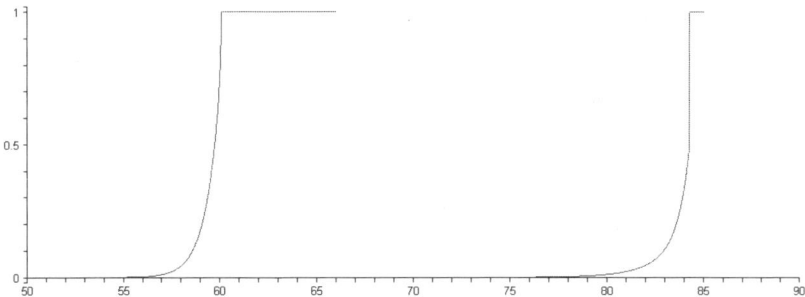

Fig. 2. Proven collision resistance (left curve) and preimage resistance (right curve) for LP231 with $n = 128$ and a suitable matrix A, as given by Theorems 1 and 2. In the ideal-permutation model, no adversary asking $q = 2^x$ queries (the x-axis) can have find a collision (left) or preimage (right) with probability exceeding the value shown.

best one can hope for in a good single-length construction. Similarly, a collision-uniform $3 \xrightarrow{4} 2$ construction can't have a satisfying provably-good bound, and even a (collision-uniform) $3 \xrightarrow{5} 2$ construction will fail at around $N^{3/5}$ queries, and a $3 \xrightarrow{6} 2$ design at around $N^{2/3}$ queries. The same paper also proves that the preimage resistance of a $m \xrightarrow{k} r$ scheme is limited to about $N^{1-(m-r)/k}$ queries, again assuming random-like behavior, now formalized as *preimage uniformity*. Stam has recently shown that the collision-uniformity assumption cannot be removed [29].

RESULTS. In this paper we give practical constructions that approach the limits described above for uniform permutation-based compression functions. Given numbers n, m, k, and r, and given an appropriate matrix A, we define an $m \xrightarrow{k} r$ compression function LP^A_{mkr}. The matrix A is $(k+r) \times (k+m)$ entries of n-bit strings, which we regard as points in the field $\mathbb{F}_{2^n}$. We will often omit mention of A and, when we do, move the subscripts up, say writing LP231 in lieu of LP^A_{231}. By varying A the LP^A_{mkr} schemes encompass *all* permutation-based $m \xrightarrow{k} r$ constructions where each permutation's input as well as each n-bit block of output is linearly determined from all that's come before. See Fig. 1. We call a function from this schema an LP compression-function (linearly-determined, permutation-based).

We first study LP231, the smallest potentially "good" LP scheme. We exhibit a condition on the matrix A for which the scheme demonstrably achieves security to about $N^{1/2}$ queries, which is of course optimal for any hash function that outputs n bits. Our analysis is in terms of concrete security, and for $n = 128$, it turns out that the adversary must ask more than $2^{59.72}$ queries to get a 0.5 chance to find a collision. See the left-hand curve of Fig. 2.

Besides collision resistance we also consider the preimage resistance of LP231. Under the same conditions on its matrix, the scheme has asymptotic security of $N^{2/3}$ queries, which is optimal for any preimage-uniform $2 \xrightarrow{3} 1$ scheme [24].

scheme	maps	collision resistance			preimage resistance		
		security	attack	tight?	security	attack	tight?
LP231, lp231	$2 \xrightarrow{3} 1$	$N^{0.50}$	$N^{0.50}$	✓	$N^{0.67}$	$N^{0.67}$	✓
LP241, lp241	$2 \xrightarrow{4} 1$	$N^{0.50}$	$N^{0.50}$	✓	$N^{0.75}$	$N^{0.75}$	✓
LP352, lp352	$3 \xrightarrow{5} 2$	$N^{0.55}$	$N^{0.60}$		$N^{0.80}$	$N^{0.80}$	✓
LP362, lp362	$3 \xrightarrow{6} 2$	$N^{0.63}$	$N^{0.66}$		$N^{0.80}$	$N^{0.83}$	
$\text{LP}_{231}^{\text{SS}}$	$2 \xrightarrow{3} 1$	$N^{0.50}$	$N^{0.50}$	✓	$N^{0.50}$	$N^{0.67}$	
$\text{lp}_{231}^{\text{SS}}$	$2 \xrightarrow{3} 1$	N^{0}	N^{0}		N^{0}	N^{0}	

Fig. 3. Rows 1–4: Automated analyses of our schemes instantiated with an appropriate sequence of matrices. The attacks are from prior work [24]. **Rows 5–6:** Automated analysis of the SS-scheme [28] and its single-permutation variant.

Numerically, for $n = 128$ one must ask more than $2^{84.25}$ queries to get a 0.5 chance to find a given preimage. See the right curve of Fig. 2.

Next we look at LP352 and LP362. Such double-length schemes are important because, in practice, the source of our permutations is likely to be AES, and its blocklength of $n = 128$ is below what people want in a hash function's output. We also look at the LP scheme just like LP_{mkr}^{A} except that a single permutation is used throughout, instead of k independent ones. This is desirable because it yields a more minimalist scheme, one needing less hardware or a smaller memory footprint. Let lp_{mkr}^{A} be defined as in Fig. 1 but with all permutations the same. When we don't want to specify A we write lp231, lp352, and so on.

For the mechanisms named in the last paragraph, doing an analysis like we did for LP231 seems infeasible, generating too many cases. We therefore developed a computer program that is capable of carrying out these analyses. It is fed a matrix A and computes asymptotic upper bounds for collision resistance and preimage resistance. For LP352 our program establishes collision resistance to at least $N^{0.55}$ queries; for LP362, at least $N^{0.63}$. The same program finds bounds for the single-permutation constructions lp231, lp352, and lp362 that are the same as for the corresponding multiple-permutation schemes. See Fig. 3. All of these results assume an appropriate matrix A (see Section 4).

While collision security of $N^{0.55}$ or $N^{0.63}$ may not seem excessive for a compression function of output length $2n$ (whose collision security could ideally reach N), one should bear in mind that the attacks described in [24] assume information-theoretic adversaries. We do not know any *actual* (implementable) attacks for LP352 or LP362 that have running time less than N. Also, collision and preimage security may increase when compression functions are used in an iterated construction.

FURTHER RELATED WORK. Shrimpton and Stam describe a $2 \xrightarrow{3} 1$ compression function based on an n-bit to n-bit random *function* [28]. While cryptographic practice does not seem to directly provide such objects, their construction is the first to obtain a good collision-resistance bound starting from a non-compressing primitive. To make a permutation-based scheme they suggest to instantiate each random function as a permutation with a feed-forward xor. They go on to explain that it's equivalent to modify just the first two, resulting in a scheme, call it $\text{LP}_{231}^{\text{SS}}$, that is LP_{231}^{A} where A has rows 10000, 01000, 11110, and 10101. This matrix does not satisfy the criterion associated to our concrete security bound, but our computer-based technique can handle it, proving asymptotic collision-resistance of $N^{0.5}$ queries. Preimage resistance is also $N^{0.5}$, shy of the desired $N^{2/3}$. This is not a deficiency in the analysis, as it matches a known attack by Joux [28]. We point out that $\text{lp}_{231}^{\text{SS}}$, the single-permutation variant of $\text{LP}_{231}^{\text{SS}}$, is completely insecure, admitting a two-query preimage-finding attack and a our-query collision-finding one. See Fig. 3.

An interesting recent hash function is the *sponge construction* of Bertoni, Daemen, Peeters, and Van Assche [4, 5]. The mechanism turns an n-bit permutation (or function) into an arbitrary-output-length hash function that is indifferentiable from a random oracle [8] (which is stronger than collision and preimage resistance). But the concrete security bounds shown do not enable its use with a 128-bit permutation; security is never as high as $N^{1/2}$, and approaching that is paid for with the scheme's rate.

2   Preliminaries

THE MODEL. A *compression function* is a map $H\colon \{0,1\}^{mn} \to \{0,1\}^{rn}$ for a given n, m, and r, with $m > r$. It is *permutation-based* if it is given by a (deterministic) program with oracle access to permutations $\pi_1, \pi_2, \ldots, \pi_k\colon \{0,1\}^n \to \{0,1\}^n$. Numbers k and n are constants associated to H. An *adversary* $\mathcal{A}$ for finding collisions in H is a probabilistic Turing machine with oracle access to the permutations used by H and their inverses. When $\mathcal{A}$ asks $(+1, i, x)$ it receives $y = \pi_i(x)$ (a *forward query*), and when it asks $(-1, i, y)$ it gets $x = \pi_i^{-1}(y)$ (a *backwards query*). We assume $\mathcal{A}$ never asks a *pointless* query, defined as a repeated query, asking $\pi_i(x)$ and later $\pi_i^{-1}(\pi_i(x))$, or asking $\pi_i^{-1}(y)$ and later $\pi_i(\pi_i^{-1}(y))$.

Now run the adversary $\mathcal{A}$ with its collection of oracles, instantiating each permutation by a uniformly chosen one. By the convention just given, the i-th query made by the adversary is answered by a uniform point from a set of size at least $N' = N - i + 1 > N - q$ where $N = 2^n$ and q is the total number of queries asked. Record the queries the adversary makes into a *query history* $\mathcal{Q}$ where the elements of $\mathcal{Q}$ are triplets (i, x, y). A triplet (i, x, y) is in the query history if the adversary asks $\pi_i(x)$ and gets back y, or it asks $\pi_i^{-1}(y)$ and gets back x. We usually specify the value of i in a triplet implicitly by way of the names of the other two arguments, like (x_1, y_1) for $(1, x_1, y_1)$, or (x_i', y_i') for (i, x_i', y_i').

The adversary *wins* if it outputs values v, v' with $v \neq v'$ such that $H(v) = H(v')$. We assume that the adversary makes the queries necessary to compute $H(v)$ and $H(v')$. As a result, one can tell from looking at the adversary's query history $\mathcal{Q}$ whether it has the information necessary to construct a collision. We thus dispense with the adversary having to output anything, giving the adversary the win if its query history contains a queries from which a collision in H can be constructed. Let $\mathbf{Adv}_H^{\text{coll}}(\mathcal{A})$ be the probability that A wins. The probability is taken over the random permutations $\pi_1, \ldots, \pi_k$ and $\mathcal{A}$'s coins (if any). Let $\mathbf{Adv}_H^{\mathrm{coll}}(q)$ be the maximal value of $\mathbf{Adv}_H^{\mathrm{coll}}(q)$ over all adversaries $\mathcal{A}$ that ask at most q queries, in all, to its oracles. Let $\mathsf{Coll}(\mathcal{Q})$ be the event that a collision can be constructed from queries in $\mathcal{Q}$. Note that $\mathbf{Adv}_H^{\mathrm{coll}}(\mathcal{A}) = \Pr[\mathsf{Coll}(\mathcal{Q})]$.

We can similarly define preimage resistance. The adversary's collection of oracles and the rules under which it operates are the same as for collision resistance, but now the adversary is said to win if it finds the preimage for a particular point $w \in \{0,1\}^{rn}$. Rather than asking the adversary to output the actual preimage, we can again look at the adversary's query history $\mathcal{Q}$ to see if it holds the information necessary to determine a preimage for w under H. We denote this predicate $\mathsf{Preim}_{H,w}(\mathcal{Q})$. We will usually omit the subscripts. We let $\mathbf{Adv}_H^{\mathrm{pre}}(\mathcal{A}) = \max_w\{\Pr[\mathcal{A} \text{ asks queries } \mathcal{Q} \text{ for which } \mathsf{Preim}_{H,w}(\mathcal{Q})]\}$. Informally, a compression function is preimage resistant if the adversary can't invert a given point without asking an unreasonable number of queries.

THE SUBJECT SCHEMES. Identify n-bit strings and points in the finite field $\mathbb{F}_{2^n}$. In this way n-bit strings inherit addition and multiplication. Given vectors of n-bit strings $x = (x_1, \ldots, x_a) \in (\{0,1\}^n)^a$ and $y = (y_1, \ldots, y_b) \in (\{0,1\}^n)^b$ let $x \cdot y$ be the n-bit string $\sum_{i=1}^{\min(a,b)} x_i y_i$ that is the inner product of the shorter vector and the truncated longer one. The i-th row of a matrix A is the vector a_i and its row-i, column-j element is a_{ij}.

Fix positive integers n, k, m and r and a $(k+r) \times (m+k)$ matrix A of n-bit strings and permutations $\pi_1, \ldots, \pi_k \colon \{0,1\}^n \to \{0,1\}^n$. We call these variables the *parameters* of our scheme. They induce a hash function $\mathrm{LP}_{mkr}^A \colon \{0,1\}^{mn} \to \{0,1\}^{rn}$ that is defined in Fig. 1. The number n is called the *blocksize*; m is the *input length*; r is the *output length*; k is the *number of permutation calls*; A is the scheme's *matrix*, and $(\pi_1, \ldots, \pi_k)$ are its *permutations*. If they are all different then we're in the *multiple-permutation* setting and the adversary has forwards and backwards access to each. If a single permutation π is used throughout then we're in the *single-permutation* setting and the adversary has forwards and backwards access to it. Compression functions LP_{mkr}^A (multiple permutations) and lp_{mkr}^A (a single permutation) differ only in this way.

3   Concrete Security Bounds for LP231

Our theorem statements refer to a function $\beta(q, p, b, B)$. It is defined as the probability that the sum of q independent random variables that are b with probability p and value 0 with probability $1-p$ exceeds B. It is easy to get good upper bounds on β. For example if $B < b$ then $\beta(q, p, b, B) \leq pq$, and if $b \neq 0$ then, letting $x = B/b$ and $t = \lfloor x \rfloor + 1$, observe that, by the sum bound,

$$\beta(q, p, b, B) = \beta(q, p, 1, x) \leq p^t \binom{q}{t}. \tag{1}$$

This "binomial bound" is sharp enough for our purposes except for code associated to computing Corollary 4. There we bound β by combining the binomial bound and a standard Chernoff bound, selecting the smallest upper bound.

We will prove that LP231 achieves good collision resistance and preimage resistance if its matrix A satisfies a certain "independence criterion" that will be defined within the proof. A random matrix will satisfy the criterion with high probability, while a sample small-entry matrix A that works is

$$A = \begin{bmatrix} 1 & 2 & 0 & 0 & 0 \\ 2 & 2 & 1 & 0 & 0 \\ 2 & 1 & 0 & 1 & 0 \\ 1 & 0 & 1 & 1 & 2 \end{bmatrix}. \tag{2}$$

The numbers represent points in $\mathbb{F}_{2^{128}}$ by identifying their binary representation with coefficient vectors of a polynomial (eg, $\mathbf{3} = \mathbf{x}+\mathbf{1}$). We use $\mathbf{x^{128} + x^7 + x^2 + x + 1}$ as our irreducible polynomial. We are now ready to state our main result on the security of LP231.

Theorem 1. *Fix $n, q \geq 1$ and let $H = \mathrm{LP}_{231}^{A}$ where $A \in \mathbb{F}_{2^n}^{4 \times 5}$ satisfies the independence criterion. Let $N = 2^n$ and $N' = N - q$. Then, for any positive integers b_1, b_2, B_1, and B_2,*

$$\begin{aligned} \mathbf{Adv}_H^{\mathrm{coll}}(q) \leq\; & 12N\,\beta(q, 1/N', 1, b_1) + 4N\,\beta(q, 1/N', 1, b_2) \\ & + 12N\,\beta(q, q/N', b_1, B_1) + 2N\,\beta(q, q/N', b_2, B_2) \\ & + 4N\,\beta(q, q/N', b_1, B_2) + 3\,\beta(q, qB_1/N', 1, 0) + \beta(q, qB_2^2/N', 1, 0)\,. \end{aligned}$$

The content of Theorem 1 is a bit opaque, both because of the non-standard function β and the universally quantified b_1, b_2, B_1, B_2. Indeed for a given q and n one must optimize the constants b_1, b_2, B_1, B_2, likely by computer, to get the best possible bound. To clarify the strength of Theorem 1, the following two corollaries will help.

Corollary 1. *Let $H_n = \mathrm{LP}_{231}^{A_n}$ where $A_n \in \mathbb{F}_{2^n}^{4 \times 5}$ satisfies the independence criterion. Let $\epsilon > 0$. Then $\lim\limits_{n \to \infty} \mathbf{Adv}_{H_n}^{\mathrm{coll}}(2^{n/2 - \epsilon}) = 0$.* □

Corollary 2. *Let $H = \mathrm{LP}_{231}^{A}$ where $A \in \mathbb{F}_{2^{128}}^{4 \times 5}$ satisfies the independence criterion. Then $\mathbf{Adv}_H^{\mathrm{coll}}(2^{59.72}) < 0.5$.* □

The first corollary captures the asymptotic behavior of the formula in Theorem 1. The proof is in the full version of this paper [23]. There one chooses reasonable but non-optimal values of b_1, b_2, B_1, B_2 and the bound falls out. As commented on earlier, the asymptotic statement is the best one can hope for in a $2 \to 1$ construction because of the always-present birthday attack. The second corollary is obtained by computer-aided optimization of b_1, b_2, B_1, B_2 and q for $n = 128$. The selected constants are $(b_1, b_2, B_1, B_2) = (1, 1, 12, 12)$. In Fig. 2 we show a graph of our security bound for the case of $n = 128$ (the left-hand curve) with the choice of constants just named. The birthday attack (elided for clarity) would appear just to the right of that curve, rising from 0 to 1 with a midpoint at $q = 2^{65.7}$. The space between such curves is the "gap" in the proven collision-resistance of LP231. It is about a factor of 60.

For preimage resistance we have analogous results, with the bound pushed well to the right. The proof of the following is in the full version of this paper [23].

Theorem 2. *Fix $n, q \geq 1$ and let $H_n = \text{LP}_{231}^A$ where matrix $A \in \mathbb{F}_{2^n}^{4 \times 5}$ satisfies the independence criterion. Let $N = 2^n$ and $N' = N - q$. Then, for any positive integers b_1, b_2 and B_2,*

$$\mathbf{Adv}_H^{\text{pre}}(q) \leq 12N\,\beta(q, 1/N', 1, b_1) + 4N\,\beta(q, 1/N', 1, b_2)$$
$$+ 2N\,\beta(q, q/N', b_2, B_2) + 4N\,\beta(q, q/N', b_1, B_2)$$
$$+ \beta(q, B_2/N', 1, 0)\,. \qquad \qquad \square$$

We once again clarify the strength of the theorem through a couple of corollaries.

Corollary 3. *Let $H_n = \text{LP}_{231}^{A_n}$ where $A_n \in \mathbb{F}_{2^n}^{4 \times 5}$ satisfies the independence criterion. Let $\epsilon > 0$. Then $\lim_{n \to \infty} \mathbf{Adv}_{H_n}^{\text{pre}}(2^{2n/3 - \epsilon}) = 0$.* $\qquad \square$

Corollary 4. *Let $H = \text{LP}_{231}^{A_n}$ where $A \in \mathbb{F}_{2^{128}}^{4 \times 5}$ satisfies the independence criterion. Then $\mathbf{Adv}_H^{\text{pre}}(2^{84.25}) < 0.5$.* $\qquad \square$

The proof of Corollary 3 is in the full version of this paper [23]. For Corollary 4, we select $b_1 = b_2 = 2$ and $B_2 = 2^{41.51}$, determined by computer optimization. In Fig. 2 we illustrate the preimage-resistance security bound for $n = 128$ as the right-hand curve.

While collision-resistance security is asymptotically as good as anything that can be achieved by an n-bit output compression function, preimage resistance is limited by the fact that this is a $2 \xrightarrow{3} 1$ schemes [24]. For $n = 128$, the known attack has at least a 0.5 chance of success with $2^{86.92}$ queries. This implies a "gap" in the proven preimage-resistance of LP231 of about a factor of 6.

OVERVIEW OF THE PROOF OF THEOREM 1. The proofs of the two theorems are similar. We will give a *very* high-level sketch of Theorem 1; due to space limitation, for the actual proof, we must direct the reader to the full version of this paper [23].

The problem of finding a collision is first reformulated in terms of solving a set of linear equations using variables from the query history. To upper bound the probability of the adversary obtaining a solution to the equations, we first upper bound the probability adversary getting lucky in certain ways that could be helpful in finding a solution, and then upper bound the probability of obtaining a solution assuming the adversary has not been "lucky". One obtains a final bound by adding the probability of getting lucky to the probability of obtaining a solution without being lucky. (Being "lucky" may mean, for example, finding a large number of solutions to a certain subsystem of equations.) In terms of events, one defines an event $\mathsf{Lucky}(\mathcal{Q})$ and then one gives separate upper bounds for $\Pr[\mathsf{Lucky}(\mathcal{Q})]$ and $\Pr[\neg\mathsf{Lucky}(\mathcal{Q}) \wedge \mathsf{Coll}(\mathcal{Q})]$; the sum of the two bounds is an upper bound for $\Pr[\mathsf{Coll}(\mathcal{Q})]$. In fact there are really two levels or "stages" of luckiness: one defines events $\mathsf{Lucky1}(\mathcal{Q})$ and $\mathsf{Lucky2}(\mathcal{Q})$, and then upper bounds the three probabilities $\Pr[\mathsf{Lucky1}(\mathcal{Q})]$, $\Pr[\neg\mathsf{Lucky1}(\mathcal{Q}) \wedge \mathsf{Lucky2}(\mathcal{Q})]$ and $\Pr[\neg\mathsf{Lucky2}(\mathcal{Q}) \wedge \mathsf{Coll}(\mathcal{Q})]$, and takes the sum of the three terms. These three terms correspond to the three lines in the bound of Theorem 1 (in that order). The event $\mathsf{Lucky1}$ is defined in terms of parameters b_1 and b_2 and the event $\mathsf{Lucky2}$ is defined in terms of parameters B_1 and B_2, hence the appearance of these variables in the statement of Theorem 1.

The above description is still rather simplified; for example event $\mathsf{Lucky1}(\mathcal{Q})$ is really the disjunction of two events, "$\mathsf{WinA}(\mathcal{Q})$" and "$\mathsf{WinB}(\mathcal{Q})$", which are themselves the disjunction of 12 and 4 different events. Similarly $\mathsf{Lucky2}(\mathcal{Q})$ is the disjunction of three events—"$\mathsf{WinC}(\mathcal{Q})$", "$\mathsf{WinD}(\mathcal{Q})$" and "$\mathsf{WinE}(\mathcal{Q})$"—which are themselves the disjunction of 12, 2 and 4 different events each. Finally $\mathsf{Coll}(\mathcal{Q})$ is shown equivalent to an event $\mathsf{WinF}(\mathcal{Q})$ which is the disjunction of 4 different events. All in all, approximately 40 different named events are considered. This kind of multilevel decomposition of the collision event into sub-events is similar to the technique employed in the analysis of MDC-2 [30].

4   Automated Analyses of LP Compression Functions

OVERVIEW. We now describe the theory underlying a computer program we designed to get asymptotic security bounds like those of Corollaries 1 and 3 but for constructions larger than LP231, where the analysis gets too complex to do by hand. The method is applicable both to collision-resistance and preimage-resistance, and can be applied to both the multiple-permutation and single-permutation settings. (For the latter, our intuition has always been that, with a suitable matrix, lp231 should offer comparable security to LP231. But the amount of casework is much larger; analyzing lp231 by hand would require about 30 times as much paper as LP231.) The security lower bounds produced by our program are always *correct*, but are not always *tight*. For example, we suspect that a good realization of LP362 has collision security of $N^{0.6}$, but our automated analysis only proves a bound of $N^{0.55}$.

While the program is mainly designed to prove security lower bounds, it can also be used to find attacks. For example, the Joux preimage attack on $\mathrm{LP}^{\mathrm{SS}}_{231}$ (Shrimpton-Stam with permutations and feed-forward xors [28]) was located by our program before we knew of it. The program likewise discovered constant-query collision and preimage attacks on $\mathrm{lp}^{\mathrm{SS}}_{231}$. Of course it is not too hard to reconstruct such attacks with foreknowledge of their existence, but it is useful to have a program that can locate such attacks automatically. Note that because of correctness, an easily-attacked scheme is always given a poor bound by the program (at least as bad as the attack shows), but a poor bound does not necessarily indicate the program has located an attack, because the weak bound could be caused by the program's own limitations.

At this point our program can only obtain asymptotic results, not concrete numerical bounds like those in Corollaries 2 and 4. One could, theoretically, trace back through the computation and replace asymptotic bounds by numerical ones, thereby obtaining concrete numerical values. We may try to do so in the future, but it is harder than it sounds.

At least one scheme to which we applied our automated analysis, $\mathrm{LP}^{\mathrm{SS}}_{231}$, is small enough that one should be able to get numerical results by carrying out a hand analysis of the type made for LP231. For most schemes this would not be feasible, as the number of cases searched by the code is too large.

REDUCTION AND BASIC IDEAS. The problem of finding a collision or preimage can be recast as the problem of finding a solution to the system $M\mathbf{x} = b$ where M is a matrix with constant entries in $\mathbb{F}_{2^n}$, where b is a constant vector with entries in $\mathbb{F}_{2^n}$, and where $\mathbf{x}$ is a vector to be filled with the adversary's queries and possibly containing some free variables which the adversary can arbitrarily set. For example, for LP_{231}^A with a matrix $A = (a_{ij})$, finding a collision is equivalent to solving the system

$$\begin{bmatrix} a_{11} & a_{12} & -1 & 0 & 0 & 0 & 0 & 0 & 0 & 0 & 0 & 0 & 0 & 0 & 0 \\ a_{21} & a_{22} & 0 & a_{23} & -1 & 0 & 0 & 0 & 0 & 0 & 0 & 0 & 0 & 0 & 0 \\ a_{31} & a_{32} & 0 & a_{33} & 0 & a_{34} & -1 & 0 & 0 & 0 & 0 & 0 & 0 & 0 & 0 \\ 0 & 0 & 0 & 0 & 0 & 0 & 0 & 0 & a_{11} & a_{12} & -1 & 0 & 0 & 0 & 0 \\ 0 & 0 & 0 & 0 & 0 & 0 & 0 & 0 & a_{21} & a_{22} & 0 & a_{23} & -1 & 0 & 0 \\ 0 & 0 & 0 & 0 & 0 & 0 & 0 & 0 & a_{31} & a_{32} & 0 & a_{33} & 0 & a_{34} & -1 & 0 \\ a_{41} & a_{42} & 0 & a_{43} & 0 & a_{44} & 0 & a_{45} & -a_{41} & -a_{42} & 0 & -a_{43} & 0 & -a_{44} & 0 & -a_{45} \end{bmatrix} \mathbf{x} = \begin{bmatrix} 0 \\ 0 \\ 0 \\ 0 \\ 0 \\ 0 \\ 0 \end{bmatrix}$$

where $\mathbf{x} = (v_1, v_2, x_1, y_1, x_2, y_2, x_3, y_3, v_1', v_2', x_1', y_1', x_2', y_2', x_3', y_3')^\mathsf{T}$. (Our matrix has minus signs for readability, but these have no effect in a field of characteristic two.) Here v_1, v_2, v_1', v_2' are variables the adversary can choose arbitrarily, but each pair (x_i, y_i) or (x_i', y_i') must come from the adversary's query history. The first three rows stipulate the linear relationships between the inputs of the first word and the inputs/outputs of the permutations in that word; the next three rows do the same for the second word; and the last row stipulates a collision. Note that the adversary can obtain a trivial solution by setting $(x_i, y_i) = (x_i', y_i')$ for $i = 1, 2, 3$, which corresponds to a "collision" between two equal words. Since this is of course uninteresting, we require that $(x_i, y_i) \neq (x_i', y_i')$ for some i, or more precisely that (x_i, y_i) is a *distinct query* from (x_i', y_i') for some i, as two queries could have the same inputs and outputs and still be distinct (if they come from different permutations).

Since the latter condition is a bit cumbersome to work with, it is easier to assume that *all* queries in $\mathbf{x}$ are distinct, and to analyze with separate linear systems the special cases in which the queries are not all distinct. For example, for the adversary to obtain an attack with $(x_3, y_3) = (x_3', y_3')$ but with all other queries distinct, the adversary must solve the modified system

$$\begin{bmatrix} a_{11} & a_{12} & -1 & 0 & 0 & 0 & 0 & 0 & 0 & 0 & 0 & 0 & 0 & 0 \\ a_{21} & a_{22} & 0 & a_{23} & -1 & 0 & 0 & 0 & 0 & 0 & 0 & 0 & 0 & 0 \\ a_{31} & a_{32} & 0 & a_{33} & 0 & a_{34} & -1 & 0 & 0 & 0 & 0 & 0 & 0 & 0 \\ 0 & 0 & 0 & 0 & 0 & 0 & 0 & 0 & a_{11} & a_{12} & -1 & 0 & 0 & 0 \\ 0 & 0 & 0 & 0 & 0 & 0 & 0 & 0 & a_{21} & a_{22} & 0 & a_{23} & -1 & 0 \\ 0 & 0 & 0 & 0 & 0 & 0 & -1 & 0 & a_{31} & a_{32} & 0 & a_{33} & 0 & a_{34} \\ a_{41} & a_{42} & 0 & a_{43} & 0 & a_{44} & 0 & 0 & -a_{41} & -a_{42} & 0 & -a_{43} & 0 & -a_{44} \end{bmatrix} \mathbf{x}' = \begin{bmatrix} 0 \\ 0 \\ 0 \\ 0 \\ 0 \\ 0 \\ 0 \end{bmatrix}$$

obtained by adding the x_3'-column to the x_3-column, adding the y_3'-column to the y_3-column, and dropping the x_3' and y_3'-columns: this accounts for setting

$x_3 = x_3'$ and $y_3 = y_3'$. Here $\mathbf{x}' = (v_1, v_2, x_1, y_1, x_2, y_2, x_3, y_3, v_1', v_2', x_1', y_1', x_2', y_2')^\mathsf{T}$ is $\mathbf{x}$ with x_3' and y_3' dropped. Thus by analyzing a separate system for each way of identifying queries, the problem reduces to analyzing the probability of solving systems of the form $M\mathbf{x} = b$ where the queries in $\mathbf{x}$ are all distinct.

Our program operates in the limit, where the entries of M and b, as well as the query coordinates themselves, come from larger and larger fields $\mathbb{F}_{2^n}$. This requires some additional explanations. Say that two matrices of M and M' of the same dimensions but over possibly different fields are *column similar* if the rank of each subset of columns of M is equal to the rank of the corresponding subset in M'. What our program effectively does is this: given a sequence of column similar matrices $(M_n)_{n \geq h}$ where each M_i is a matrix with entries in $\mathbb{F}_{2^i}$, and given an arbitrary infinite sequence $(b_n)_{n \geq h}$ and a number $\alpha \in (0, 1)$, it upper bounds the limit as $n \to \infty$ of the probability of the adversary obtaining a solution to $M_n \mathbf{x} = b_n$ if it uses no more than $q = N^\alpha$ queries. Of course we cannot input an infinite sequence $(M_n)_{n \geq h}$ or $(b_n)_{n \geq h}$; all the program ever knows about $(M_n)_{n \geq h}$ is the rank of each subset of columns of M_h, which it knows because it is given M_h itself. As for the sequence of right-hand sides $(b_n)_{n \geq h}$, its values are never used to compute the upper bound, so we do not care.

The result of the computation is obviously dependent on M_h, so we must explain how this matrix is chosen. Say that one wishes to instantiate LP^A_{mkr} with $n = 128$, so the entries of A will be points in $\mathbb{F}_{2^{128}}$. Then $M_h = M_{128}$ is a function of A, and A is simply chosen at random. One can choose the entries of A uniformly in $\mathbb{F}_{2^{128}}$, or one can limit the entries of A to numbers with few bits in order to increase the efficiency of LP^A_{mkr} (if the entries are chosen from too small a set, however, M_h can have poor linear independence properties leading to a sub-optimal bound; then A must be re-chosen).

One can significantly increase the speed of the computation by giving the program a matrix $M_{h'}$ that is column-similar to M_h but where h' is smaller than $h = 128$, such as, say, $h' = 7$, because field operations over $\mathbb{F}_{2^7}$ can be implemented more efficiently than over $\mathbb{F}_{2^{128}}$. Finding such a matrix $M_{h'}$ requires trial and error, but is generally not too hard. If $M_{h'}$ cannot be found, a second method is to generate $M_{h'}$ directly by choosing a random matrix A' with entries in $\mathbb{F}_{2^{h'}}$ and then to use a field embedding of $\mathbb{F}_{2^{h'}}$ into $\mathbb{F}_{2^{128}}$ to recover A. The only disadvantage of this method is that one does not obtain a matrix A with small entries. Sample matrices A and $M_{h'}$ used for our computations are given in the full version of this paper [23].

SYSTEM CONSTRAINTS. The problem thus reduces to upper bounding the probability of the adversary obtaining a solution to a system $M_n \mathbf{x} = b_n$ where M_n is an unknown matrix over $\mathbb{F}_{2^n}$ column similar to a given matrix M_h with entries in $\mathbb{F}_{2^h}$ and b_n is an arbitrary vector with entries in $\mathbb{F}_{2^n}$. But, beyond this, there are also stipulations on the solution $\mathbf{x}$, certain entries of which must come from the adversary's query history. We now make this part precise.

A *system constraint* ρ on a matrix M is a partition of M's columns into a set of *free* columns, a set of *null* columns, and zero or more sets of two columns each called *query pairs*. Each query pair of has a *first* and *second* column. When we multiply M by a column vector $\mathbf{x}$ its entries inherit ρ's partition of the columns of M, so we can speak of the "free" or "null" entries of $\mathbf{x}$, and of pairs of entries of $\mathbf{x}$ that form query pairs. If M_n is a matrix with entries in $\mathbb{F}_{2^n}$ and b_n is a vector with entries in $\mathbb{F}_{2^n}$ and $\mathcal{Q} = \mathcal{Q}(A)$ is the query history of an adversary interacting with ideal-permutation oracles of domain (and range) $\mathbb{F}_{2^n}$, then we say that $\mathcal{Q}$ *solves* $M_n\mathbf{x} = b_n$ *with respect to* ρ, or that $\mathcal{Q}$ *solves* $(M_n\mathbf{x} = b_n, \rho)$, if there exists a vector $\mathbf{x}$ with entries in $\mathbb{F}_{2^n}$ such that: (i) every null entry of $\mathbf{x}$ is 0, (ii) if (x, y) is a query pair of $\mathbf{x}$ with first element x and second element y then (i, x, y) is a query in $\mathcal{Q}$ for some π_i, and (iii) no two query pairs of $\mathbf{x}$ correspond to the *same* query in $\mathcal{Q}$. If ρ is understood we simply say that $\mathcal{Q}$ solves $M_n\mathbf{x} = b_n$.

Note that in the above definition we do not attach any importance to which query pairs are mapped to which kind of queries in $\mathcal{Q}$; in fact, it does not even matter for the definition how many oracles the adversary is interacting with. All that matters is that when the adversary makes a query to one of its oracles, the resulting output comes uniformly at random from a pool of size at least $N' = N-q$. As always, the adversary can make both forward and backward queries.

From here on it will simplify the discussion if we assume there is a single matrix M and vector b under discussion, which implicitly depend on n. However one should remember that all one knows about M is the rank of any subset of columns of M, and that b is arbitrary.

INDUCTION. Note that it is trivial to tell whether the adversary can solve $(M\mathbf{x} = b, \rho)$ if ρ has no query pairs, since in this case the problem reduces to the feasibility of a standard system of linear equations. Determining the probability of the adversary solving $(M\mathbf{x} = b, \rho)$ becomes increasingly complicated as ρ has more query pairs. To upper bound the probability of the adversary solving $(M\mathbf{x} = b, \rho)$ the program works by induction on the number of query pairs in ρ, using bounds on the probability of solving systems $(M\mathbf{x} = b, \rho')$ where ρ' has fewer query pairs than ρ.

What is ultimately of interest is whether for a given $\alpha \in (0, 1)$ the adversary has zero probability, in the limit, of finding *some* solution to the system in $q = N^\alpha$ queries (consider α as fixed throughout). But to get the induction to work we also need to know, if the adversary has nonzero chance of solving the system in the limit, *how many* solutions the adversary can reasonably expect to get. Here the "number" of solutions of a system $(M\mathbf{x} = b, \rho)$ with respect to a query history $\mathcal{Q}$ is counted in a special way: solutions $\mathbf{x}$ and $\mathbf{x}'$ only count as distinct if the values of some query pair are different in them (meaning they are mapped to distinct queries of $\mathcal{Q}$). If ρ has no query pairs at all, then the number of solutions is defined as 1.

The following definition formalizes the idea of "how many solutions the adversary can reasonably expect to get". It is central to the analysis.

Definition 1. *Let $\alpha \in (0, 1)$. A number $\beta \geq 0$ is an α-**threshold** of $(M\mathbf{x} = b, \rho)$ if for any $\epsilon > 0$, the probability that there exists some c for which the adversary obtains at least $N^{\beta+\epsilon}$ solutions to $(M\mathbf{x} = c, \rho)$ in $q = N^\alpha$ queries goes to zero as $n \to \infty$.*

Observe that b is immaterial in the definition. Also note the α-threshold is not unique since every number greater than an α-threshold is also an α-threshold. On the other hand, it is easy to see that the set of α-thresholds is a half-closed interval: if $\beta + \epsilon$ is an α-threshold for every $\epsilon > 0$, then β is also an α-threshold.

As an example, consider the system $M\mathbf{x} = b$ where M consists of a single row and $\mathbf{x} = (x, y)^{\mathsf{T}}$ has two variables, so that the system can be written

$$m_{11}x + m_{12}y = b_1 \tag{3}$$

where b_1 is the unique entry of b and where $[m_{11}\ m_{12}] = M$. Take ρ to be the system constraint on M with a unique query pair consisting of (m_{11}, m_{12}) (so ρ has no free columns or null columns), so that solving $(M\mathbf{x} = b, \rho)$ means getting a query (i, x, y) such that (x, y) solves (3). If $m_{11}, m_{12} \neq 0$ one can show that the probability of finding such a query in $q = N^\alpha$ queries where $\alpha \in (0, 1)$ goes to 0 as $n \to \infty$ (each query made has chance $1/N' \approx 1/N$ of solving the system, so a sum bound shows the chance of success is at most N^α/N, which is $\ll 1$ for large n). Furthermore, one can show the probability of there being a $c \in \mathbb{F}_{2^n}$ such that $(M\mathbf{x} = c, \rho)$ has many solutions—say more than N^ϵ— approaches zero as $n \to \infty$, for fixed $\epsilon > 0$ (otherwise put, the set of values $\{m_{11}x + m_{12}y : (i, x, y) \in \mathcal{Q}(A)\}$ is "well spread-out" in $\mathbb{F}_{2^n}$). By definition, this means that $(M\mathbf{x} = b, \rho)$ has zero α-threshold.

In the above example the α-threshold of $(M\mathbf{x} = b, \rho)$ does not depend on α. As a second example, consider a system $M\mathbf{x} = b$ where M is a 1×4 matrix, written

$$m_{11}x_1 + m_{12}y_1 + m_{13}x_2 + m_{14}y_2 = b_1. \tag{4}$$

We assume each m_{1j} is nonzero. Let ρ partition the columns of M into two query pairs, (m_{11}, m_{12}) and (m_{13}, m_{14}). Then one can show that $\max(0, 2\alpha - 1)$ is an α-threshold of $(M\mathbf{x} = b, \rho)$. (Intuitively, each query made, whether forward or backward, gives a random value of $m_{11}x_1 + m_{12}y_1$ and a random value of $m_{13}x_2 + m_{14}y_2$; thus q queries give q^2 pairs of random values; for any $c \in \mathbb{F}_{2^n}$ the expected number of such pairs that sum to c is $q^2/N = N^{2\alpha-1}$; one can finally show that the probability of there being a c with more than $N^\epsilon N^{\max(0, 2\alpha-1)}$ pairs summing to it approaches zero as $n \to \infty$ for fixed $\epsilon > 0$, which by definition means that $\max(0, 2\alpha - 1)$ is an α-threshold of $(M\mathbf{x} = b, \rho)$.) Thus, for example, if an adversary makes $q = N^\alpha = N^{3/4}$ queries there is negligible chance there will be some c for which $(M\mathbf{x} = c, \rho)$ has more than $N^{1/2+\epsilon}$ solutions, for fixed $\epsilon > 0$ and as $n \to \infty$.

The crux of the problem lies in knowing how to compute an α-threshold for a system $(M\mathbf{x} = b, \rho)$ given α-thresholds for every system $(M\mathbf{x} = b, \rho')$ where ρ' has fewer query pairs than ρ. There are a few ways for doing this, possibly giving different numbers (the method giving the smallest result is the better, obviously). We will sketch some of the methods. Note that if ρ has no query pairs then $(M\mathbf{x} = b, \rho)$ has a zero α-threshold, by our earlier convention on how to count the number of solutions to a system.

THE INDUCTION STEP. Take a system $(M\mathbf{x} = b, \rho)$. Say each system $(M\mathbf{x} = b, \rho')$ where ρ' has fewer query pairs than ρ has a (previously computed)

α-threshold $\beta_{\rho'}$. We can assume the adversary has found at most $N^{\beta_{\rho'}+\epsilon}$ solutions for each system $(M\mathbf{x} = b, \rho')$. Here $\epsilon > 0$ can be as small as desired. Making these assumptions costs us a small additive constant for each ρ', because it *is* possible for the adversary to obtain more than $N^{\beta_{\rho'}+\epsilon}$ solutions to $(M\mathbf{x} = c, \rho')$ for some c and ρ', if it is lucky, but, by the definition of an α-threshold, the sum of these additive constants goes to zero as $n \to \infty$.

We introduce some more notation. We write $(M\mathbf{x} = b, \rho, \mathcal{Q})$ to emphasize that the set of solutions of $(M\mathbf{x} = b, \rho)$ depends on $\mathcal{Q}$. Also recall that two solutions of $(M\mathbf{x} = b, \rho, \mathcal{Q})$ are counted as distinct only if the two solutions map some query pair of ρ to different queries in $\mathcal{Q}$; a good way to think of the set of solutions is that each solution only specifies the values for the query pairs of ρ, the free variables being left unspecified. We write the elements of $\mathcal{Q}$ as tuples (x, y) rather than as triplets (i, x, y) since the index of the permutation queried does not matter in the model.

One can first observe that, for any solution $\mathbf{x}$ of $(M\mathbf{x} = b, \rho, \mathcal{Q})$, one of the queries in $\mathcal{Q}$ used by $\mathbf{x}$ comes last in chronological order in $\mathcal{Q}$. We say this query *creates* the solution. A natural way to bound the probability of the adversary obtaining many solutions to $(M\mathbf{x} = b, \rho)$ is to compute the probability that any given query creates a solution, and to consider that only q queries are made in total, so that only so many solutions can be expected. To further break down the problem one can evaluate separately, for each query pair in ρ, the probability that a given query creates a solution where the query is matched *to that pair*.

So take a query pair (C_1, C_2) in ρ, where C_1, C_2 are the two columns of the query pair. Say the adversary has already made a certain sequence $\mathcal{Q}'$ of queries, and is now making, say, a new forward query $\pi(x)$ to one of its oracles; we want to know the probability that the resulting output $y = \pi(x)$ creates a solution $\mathbf{x}$ of $(M\mathbf{x} = b, \rho, \mathcal{Q})$ where $\mathcal{Q} = \mathcal{Q}' \cup \{(x, y)\}$ and where the coefficients x, y are used for the columns C_1, C_2 respectively (to denote this, we will say that the query (x, y) is *used in position* (C_1, C_2)). We distinguish between the case when C_2 is linearly independent of the free columns of ρ and the case when it is not.

Say first that C_2 is linearly independent of the free columns of ρ. Let ρ' be the system constraint obtained from ρ by changing C_2 into a free column and changing C_1 into a null column. Also let $b' = b - xC_1$. Then for each solution of $(M\mathbf{x} = b, \rho, \mathcal{Q})$ where (x, y) is used in position (C_1, C_2) there is exactly one solution of $(M\mathbf{x} = b', \rho', \mathcal{Q}')$. Moreover, for every solution $\mathbf{x}$ in $(M\mathbf{x} = b', \rho', \mathcal{Q}')$ there is *at most one* value of y for which the query (x, y) extends the solution $\mathbf{x}$ to a solution of $(M\mathbf{x} = b, \rho, \mathcal{Q})$ where (x, y) is used in position (C_1, C_2). This is because C_2 is linearly independent from the free columns of ρ, so that when the coefficients of all columns of M have been fixed except for the coefficient of C_2 and the coefficients of the free columns of ρ, there is at most one coefficient for C_2 for which the remaining linear system will have a solution for a given right-hand side. So the number of successful outputs y is at most the number of solutions of $(M\mathbf{x} = b', \rho', \mathcal{Q}')$.

If ρ' has an α-threshold β, then the probability the returned value y will give a solution of $(M\mathbf{x} = b, \rho, \mathcal{Q})$ where (x, y) is used in position (C_1, C_2) is thus at most $N^{\beta+\epsilon}/N'$, since $(M\mathbf{x} = b', \rho', \mathcal{Q}')$ has at most $N^{\beta+\epsilon}$ solutions and since y is returned uniformly at random from a set of size at least N'. Since the adversary makes $q = N^\alpha$ queries in total the "expected" number of solutions is $N^{\alpha+\beta+\epsilon}/N'$. There are two cases to distinguish: $\alpha + \beta \leq 1$ and $\alpha + \beta > 1$. For the first case, the expected number of solutions is $\leq N^\epsilon$ and it isn't hard to show that the probability that there exists a c for which the adversary obtains more than $N^{2\epsilon}$ solutions to $(M\mathbf{x} = c, \rho)$ in this way goes to zero as $n \to \infty$. If $\alpha + \beta > 1$, a similar argument shows that the probability of the adversary obtaining more than $N^{\alpha+\beta+2\epsilon-1}$ solutions to $(M\mathbf{x} = c, \rho)$ for some c in this way goes to 0 as $n \to \infty$. If the adversary's only means of constructing solutions to $(M\mathbf{x} = b, \rho)$ was to fill in last the values of the query pair (C_1, C_2) with a forward query, one would thus obtain an α-threshold of $\max(0, \alpha + \beta - 1)$ for $(M\mathbf{x} = b, \rho)$, because $\epsilon > 0$ is arbitrary. But of course there are the other possibilities to consider: backward queries and solutions were the last query pair filled in is not (C_1, C_2). (In the end, the maximum α-threshold is retained unless some query pair column of ρ is linearly dependent on the free columns of ρ; see the next case.)

In the second case the column C_2 is linearly dependent on the free columns of ρ. In this case let ρ' be obtained from ρ by setting to "null" both C_1 and C_2, and let $b' = b - xC_1$. Then the number of solutions of $(M\mathbf{x} = b, \rho, \mathcal{Q})$ where (x, y) is in position (C_1, C_2) is equal to the number of solutions of $(M\mathbf{x} = b', \rho', \mathcal{Q}')$. Let β be the α-threshold of $(M\mathbf{x} = b, \rho')$. Since there are at most $N^{\beta+\epsilon}$ solutions to $(M\mathbf{x} = b', \rho')$, and since there are at most $q = N^\alpha$ queries (x, y) in the query history, the total number of solutions of $(M\mathbf{x} = b, \rho)$ obtained in q queries is at most $N^{\alpha+\beta+\epsilon}$. Because $\epsilon > 0$ is arbitrary, $\alpha + \beta$ is therefore an α-threshold of $(M\mathbf{x} = b, \rho)$. Note that unlike in the first case, we do not need to examine any other query pairs of ρ to establish this threshold. This benefits the program's speed, as having the computation of ρ's α-threshold depend only on one other α-threshold instead of depending on several other α-thresholds helps stave off a combinatorial explosion of cases.

A few other techniques for computing α-thresholds recursively, which sometimes give better bounds, are used by the program. These are discussed in the paper's full version [23]

FINAL PROBABILITY OF SUCCESS. What ultimately interests us is the adversary's probability of obtaining *some* solution to the system $(M\mathbf{x} = b, \rho)$ where ρ is the original "root" system constraint with the maximum number of query pairs and b is the original right-hand side. If, after applying the recursion, the system $(M\mathbf{x} = b, \rho)$ is found to have a nonzero α-threshold, then we cannot conclude anything; the program has effectively failed to show the scheme requires at least $q = N^\alpha$ queries to break. But if $(M\mathbf{x} = b, \rho)$ has a zero α-threshold, then for any $\epsilon > 0$ the probability of obtaining at least one solution of $(M\mathbf{x} = b, \rho)$ in $q = N^{\alpha-\epsilon}$ queries goes to 0 as $n \to \infty$. Seeing so requires revisiting how the α-thresholds are obtained. From the recursion process, there is only one way that a system constraint can have zero α-threshold: when, at every query, the adversary has probability at most $N^{\beta+\epsilon'}/N'$ of obtaining a solution for some β with $\alpha+\beta \leq 1$ (see the first case considered for the computation of α-thresholds). By choosing $\epsilon' < \epsilon$, then, we see that the adversary's final probability of success is bounded by the probability of solving the original system in $q = N^{\alpha-\epsilon}$ queries where each query has chance at most $N^{\beta+\epsilon'}/N'$ of giving a solution; but since $\alpha-\epsilon+\beta+\epsilon' < 1$, a sum bound directly shows that this probability of success goes to 0 as $n \to \infty$. Thus the scheme asymptotically requires at least N^α queries to be broken. To find the best α the program simply does a binary search.

RESULTS. Findings produced using our program are presented in Fig. 3. Some of the cases are solved quickly, but others take over an hour and involve the exploration of millions of system constraints.

5   Discussion

It may be interesting to compare the efficiency of LP compression functions and conventional blockcipher-based ones. It is conventional to use *rate* as a rough measure of efficiency, but the rate of a blockcipher-based construction, as conventionally defined [17, p. 340], doesn't even attend to the number of key bits employed. The simplest way to correct this is to say that the *adjusted rate* of a blockcipher-based hash-function is the number of message bits processed per blockcipher input bits, the latter including plaintext bits and key bits (for simplicity, equally weighted). Then SHA-1 would have an adjusted rate of 0.76; Davies-Meyer [16], 0.5; MDC-2 [16], 0.27; Hirose's double-length construction [11], 0.17; and MDC-4, 0.13. From this vantage, the adjusted rate of LP231, 0.33, and LP362, 0.17, are competitive. Regardless, adjusted rate is a coarse measure of efficiency, and the current work aims only to probe the security and feasibility of LP compression functions, not to vanquish any other design.

This paper has only dealt with making a compression function, not a full-fledged hash function. Of course you can always turn the former into the latter using Merkle-Damgård [9, 18] or any of the other techniques that have emerged in recent years [1, 3, 10, 22], but the "best" approach remains to be seen. Also, we have considered only collision and preimage resistance. Certainly there are other desirable properties one should aim for in a contemporary construction, like being indifferentiable from a random oracle [8].

Acknowledgments

Most of the work on this paper was carried out while the second author was in the Department of Mathematics at UC Davis. Both authors received funding from NSF grant CCR-0208842 and a gift from Intel; many thanks to Intel, particularly Jesse Walker, and to the NSF, for their kind support.

References

1. Andreeva, E., Neven, G., Preneel, B., Shrimpton, T.: Seven-property preserving iterated hashing: ROX. In: Kurosawa, K. (ed.) ASIACRYPT 2007. LNCS, vol. 4833, pp. 130–146. Springer, Heidelberg (2007)
2. Bellare, M., Ristenpart, T.: Hash functions in the dedicated-key setting: design choices and MPP transforms. In: Arge, L., Cachin, C., Jurdziński, T., Tarlecki, A. (eds.) ICALP 2007. LNCS, vol. 4596, pp. 399–410. Springer, Heidelberg (2007)
3. Bellare, M., Ristenpart, T.: Multi-property-preserving hash domain extension and the EMD transform. In: Lai, X., Chen, K. (eds.) ASIACRYPT 2006. LNCS, vol. 4284, pp. 299–314. Springer, Heidelberg (2006)
4. Bertoni, G., Daemen, J., Peeters, M., Van Assche, G.: On the indifferentiability of the sponge construction. In: Smart, N. (ed.) EUROCRYPT 2008. LNCS, vol. 4965, pp. 181–197. Springer, Heidelberg (2008)
5. Bertoni, G., Daemen, J., Peeters, M., Van Assche, G.: Sponge functions. In: Ecrypt Hash Workshop (2007), http://sponge.noekeon.org/
6. Black, J., Cochran, M., Shrimpton, T.: On the impossibility of highly-efficient blockcipher-based hash functions. In: Cramer, R. (ed.) EUROCRYPT 2005. LNCS, vol. 3494, pp. 526–541. Springer, Heidelberg (2005)
7. Black, J., Rogaway, P., Shrimpton, T.: Black-box analysis of the block-cipher-based hash function constructions from PGV. In: Yung, M. (ed.) CRYPTO 2002. LNCS, vol. 2442, pp. 320–335. Springer, Heidelberg (2002)
8. Coron, J., Dodis, Y., Malinaud, C., Puniya, P.: Merkle-Damgård revisited: how to construct a hash function. In: Shoup, V. (ed.) CRYPTO 2005. LNCS, vol. 3621, pp. 430–448. Springer, Heidelberg (2005)
9. Damgård, I.: A design principle for hash functions. In: Brassard, G. (ed.) CRYPTO 1989. LNCS, vol. 435, pp. 416–427. Springer, Heidelberg (1990)
10. Dodis, Y., Pietrzak, K., Puniya, P.: A new mode of operation for block ciphers and length-preserving MACs. In: Smart, N. (ed.) EUROCRYPT 2008. LNCS, vol. 4965, pp. 198–219. Springer, Heidelberg (2008)
11. Hirose, S.: Some plausible construction of double-block-length hash functions. In: Robshaw, M. (ed.) FSE 2006. LNCS, vol. 4047, pp. 210–225. Springer, Heidelberg (2006)
12. Hattori, M., Hirose, S., Yoshida, S.: Analysis of double block length hash functions. In: Paterson, K.G. (ed.) Cryptography and Coding 2003. LNCS, vol. 2898, pp. 290–302. Springer, Heidelberg (2003)
13. Joux, A.: Multicollisions in iterated hash functions. In: Franklin, M. (ed.) CRYPTO 2004. LNCS, vol. 3152, pp. 306–316. Springer, Heidelberg (2004)
14. Knudsen, L., Lai, X., Preneel, B.: Attacks on fast double block length hash functions. Journal of Cryptology 11(1), 59–72 (1998)
15. Lucks, S.: A failure-friendly design principle for hash functions. In: Roy, B. (ed.) ASIACRYPT 2005. LNCS, vol. 3788, pp. 474–494. Springer, Heidelberg (2005)
16. Matyas, S., Meyer, C., Oseas, J.: Generating strong one-way functions with cryptographic algorithm. IBM Tech. Disclosure Bulletin 27, 5658–5659 (1985)
17. Menezes, A., van Oorschot, P., Vanstone, S.: Handbook of Applied Cryptography. CRC Press, Boca Raton (1996)
18. Merkle, R.: One way hash functions and DES. In: Brassard, G. (ed.) CRYPTO 1989. LNCS, vol. 435, pp. 428–446. Springer, Heidelberg (1990)
19. Nandi, M.: Designs of efficient secure large hash values. Cryptology ePrint report 2005/296

20. Preneel, B., Govaerts, R., Vandewalle, J.: Hash functions based on block ciphers: a synthetic approach. In: Stinson, D.R. (ed.) CRYPTO 1993. LNCS, vol. 773, pp. 368–378. Springer, Heidelberg (1994)
21. Preneel, B., Govaerts, R., Vandewalle, J.: On the power of memory in the design of collision resistant hash functions. In: AUSCRYPT 1992. LNCS, vol. 718, pp. 105–121. Springer, Heidelberg (1993)
22. Ristenpart, T., Shrimpton, T.: How to build a hash function from any collision-resistant function. In: Kurosawa, K. (ed.) ASIACRYPT 2007. LNCS, vol. 4833, pp. 147–163. Springer, Heidelberg (2007)
23. Rogaway, P., Steinberger, J.: Constructing cryptographic hash functions from fixed-key blockciphers. Full version of this paper. Available from either author's web page (manuscript, 2008)
24. Rogaway, P., Steinberger, J.: Security/efficiency tradeoffs for permutation-based hashing. In: Smart, N. (ed.) EUROCRYPT 2008. LNCS, vol. 4965, pp. 220–236. Springer, Heidelberg (2008)
25. Peyrin, T., Gilbert, H., Matthew, F., Robshaw, J.: Combining compression functions and block cipher-based hash functions. In: Lai, X., Chen, K. (eds.) ASIACRYPT 2006. LNCS, vol. 4284, pp. 315–331. Springer, Heidelberg (2006)
26. Satoh, T., Haga, M., Kurosawa, K.: Towards secure and fast hash functions. TIE-ICE: IEICE Transactions on Communications/Electronics/Information and Systems, 55–62 (1999)
27. Shannon, C.: A mathematical theory of communication. Bell System Technical Journal 27, 379–423, 623–656 (1948)
28. Shrimpton, T., Stam, M.: Building a collision-resistant compression function from non-compressing primitives. In: Aceto, L., Damgård, I., Goldberg, L.A., Halldorsson, M.M., Ingólfsdóttir, A., Walukiewicz, I. (eds.) ICALP 2008. LNCS, vol. 5126. Springer, Heidelberg (2008)
29. Stam, M.: Beyond uniformity: better security/efficiency tradeoffs for compression function security. In: Wagner, D. (ed.) CRYPTO 2008. LNCS, vol. 5157, pp. 397–412. Springer, Heidelberg (2008)
30. Steinberger, J.: The collision intractability of MDC-2 in the ideal-cipher model. In: Naor, M. (ed.) EUROCRYPT 2007. LNCS, vol. 4515, pp. 34–51. Springer, Heidelberg (2007)

Distributed Private Data Analysis: Simultaneously Solving How and What

Amos Beimel, Kobbi Nissim, and Eran Omri

Department of Computer Science, Ben Gurion University, Be'er Sheva, Israel
{beimel,kobbi,omrier}@cs.bgu.ac.il

Abstract. We examine the combination of two directions in the field of privacy concerning computations over distributed private inputs – *secure function evaluation* (SFE) and *differential privacy*. While in both the goal is to privately evaluate some function of the individual inputs, the privacy requirements are significantly different. The general feasibility results for SFE suggest a natural paradigm for implementing differentially private analyses distributively: First choose *what* to compute, i.e., a differentially private analysis; Then decide *how* to compute it, i.e., construct an SFE protocol for this analysis. We initiate an examination whether there are advantages to a paradigm where both decisions are made simultaneously. In particular, we investigate under which accuracy requirements it is beneficial to adapt this paradigm for computing a collection of functions including Binary Sum, Gap Threshold, and Approximate Median queries. Our results yield new separations between the local and global models of computations for private data analysis.

1 Introduction

We examine the combination of two directions in the field of privacy concerning distributed private inputs – secure function evaluation [1, 3, 13, 18] and differential privacy [7, 9]. While in both the goal is to privately evaluate some function of individual inputs, the privacy requirements are significantly different.

Secure function evaluation (SFE) allows n parties $p_1, \ldots, p_n$, sharing a common interest in distributively computing a function $f(\cdot)$ of their inputs $\mathbf{x} = (x_i, \ldots, x_n)$, to compute $f(\mathbf{x})$ while making sure that no coalition of t or less curious parties learns anymore than the outcome of $f(\mathbf{x})$. I.e., for every such coalition, executing the SFE protocol is equivalent to communicating with a trusted party that is given the private inputs $\mathbf{x}$ and releases $f(\mathbf{x})$. SFE has been the subject of extensive cryptographic research (initiated in [1, 3, 13, 18]), and SFE protocols exist for any feasible function $f(\cdot)$ in a variety of general settings.

SFE is an important tool for achieving privacy of individual entries – no information about these entries is leaked beyond the outcome $f(\mathbf{x})$. However this guarantee is insufficient in many applications, and care must be taken in choosing the function $f(\cdot)$ to be computed – any implementation, no matter how secure, of a function $f(\cdot)$ that leaks individual information would not preserve individual privacy. A criterion for functions that preserve privacy of individual

D. Wagner (Ed.): CRYPTO 2008, LNCS 5157, pp. 451–468, 2008.
© International Association for Cryptologic Research 2008

entries, *differential privacy*, has evolved in a sequence of recent works [2, 6–9, 11, 12]. Alongside, techniques have been developed for constructing a differentially private analysis $\hat{f}(\cdot)$ approximating a desired analysis $f(\cdot)$, by means of adding carefully chosen random noise that conceals any single individual's contribution to $f(\cdot)$ [2, 9, 15, 16].

Combining these two lines of research – SFE and differential privacy – we get a very natural paradigm for constructing protocols that preserve differential privacy, making use of the generality of SFE:

1. Decide on *what* to compute, i.e., a differentially private analysis $\hat{f}(\cdot)$ that approximates a desired analysis $f(\cdot)$. This can be done while abstracting out all implementation issues, assuming the computation is performed by a trusted party that only announces the outcome of the analysis.
2. Decide on *how* to compute $\hat{f}(\cdot)$, i.e., construct an SFE protocol for computing $\hat{f}(\mathbf{x})$ either by using one of the generic transformations of the feasibility results mentioned above, or by crafting an efficient protocol that utilizes the properties of $\hat{f}(\cdot)$.

This natural paradigm yields a conceptually simple recipe for constructing distributed analyses preserving differential privacy, and, furthermore, allows a valuable separation of our examinations of the *what* and *how* questions. However, comparing the privacy requirements from SFE protocols with differential privacy suggests that this combination may result in sub-optimal protocols. For example, differential privacy is only concerned with how the view of a coalition changes when one (or only few) of the inputs are changed, whereas SFE protocols are required to keep these views indistinguishable even when significant changes occur, if these changes do not affect the function's outcome. Hence, it is interesting to learn whether there are advantages to a paradigm where the analysis to be computed and the protocol for computing it are chosen simultaneously.

The main model of distribution we consider is of semi-honest parties $p_1, \ldots, p_n$ that perform a computation over their private inputs $x_1, \ldots, x_n$, while maintaining differential privacy with respect to coalitions of size up to t (see Definition 2 below). This model has been examined thoroughly in cryptography, and was shown to enable SFE in a variety of settings [1, 3, 13, 18]. We note that while it is probably most natural to consider a setting where the players are computationally limited, we present our results in an information theoretic setting, as this setting allows us to prove lowerbounds on protocols, and hence demonstrate rigorously when constructing differentially private protocols is better than using the natural paradigm.

The second model we consider is the *local model*[1]. Protocols executing in the local model have a very simple communication structure, where each party p_i can only communicate with a designated semi-honest party C, referred to as a *curator*. The communication can either be *non-interactive*, where each party

---

[1] Also referred to in the literature as *randomized response* and *input perturbation*. This model was originally introduced by Warner [17] to encourage survey responders to answer truthfully, and has been studied extensively since.

sends a single message to the curator which replies with the protocol's outcome, or *interactive*, where several rounds of communication may take place.

1.1   Our Results

We initiate an examination of the paradigm where an analysis and the protocol for computing it are chosen simultaneously. We begin with two examples that present the potential benefits of using this paradigm: it can lead to simpler protocols, and more importantly it can lead to more efficient protocols. The latter example is of the Binary Sum function, $\text{SUM}(x_1, \ldots, x_n) = \sum_{i=1}^{n} x_i$ for $x_i \in \{0, 1\}$.

The major part of this work examines whether constructing protocols for computing an approximation $\hat{f}(\cdot)$ to $\text{SUM}(\cdot)$, that are not SFE protocols for $\hat{f}(\cdot)$, yields an efficiency gain[2]. Ignoring the dependency on the privacy parameter, our first observation is that for approximations with additive error $\approx \sqrt{n}$ there is a gain – for a natural class of *symmetric* approximation functions (informally, functions where the outcome does not depend on the order of inputs), it is possible to construct differentially private protocols that are much more efficient than any SFE protocol for a function in this class. Moreover, these differentially private protocols are secure against coalitions of size up to $t = n - 1$, and need not rely on secure channels.

The picture changes when we consider additive error smaller than $\sqrt{n}$. This follows from a sequence of results. We prove first that no such local non-interactive protocols exist (by itself, this contribution is not new, see below). Furthermore, no local protocols with $\ell \leq \sqrt{n}$ rounds and additive error $\sqrt{n}/\tilde{O}(\ell)$ exist. In particular, no local interactive protocol with $o(\sqrt{n/\log(n)})$ rounds exists for computing $\text{SUM}(\cdot)$ within constant additive error[3]. Finally, the bounds on local protocols imply that no distributed protocols exist that use $o(nt)$ messages, and approximates $\text{SUM}(\cdot)$ within additive error $\sqrt{n}/\tilde{O}(\ell)$ in ℓ rounds. Considering the natural paradigm, i.e., computing a differentially-private approximation to $\text{SUM}(\cdot)$ using SFE, we get a protocol for approximating $\text{SUM}(\cdot)$ with $O(1)$ additive error, and sending $O(nt)$ messages.

1.2   Techniques

We prove our lowerbound in sequence of reductions. We begin with a simple reduction from any differentially private protocol for SUM to a gap version of the threshold function GAP-TR. Henceforth, it is enough to prove our lowerbound for GAP-TR.

In the heart of our lowerbound for GAP-TR is a transformation from efficient distributed protocols into local interactive protocols, showing that if there are

---

[2] We only consider *oblivious protocols* where the communication pattern is independent of input and randomness (see Section 2.2).

[3] This is in contrast to the centralized setup where $\text{SUM}(\cdot)$ can be computed within $O(1)$ additive error.

distributed differentially-private protocols for GAP-TR$(\cdot)$ in which half of the parties interact with less than $t+1$ other parties, then there exist differentially-private protocol for GAP-TR$(\cdot)$ in the local interactive model. This allows us to prove our impossibility results in the local model, a model which is considerably simpler to analyze.

In analyzing the local non-interactive model, we prove lowerbounds borrowing from analyses in [6, 11]. The main technical difference is that our analysis holds for general protocols, whereas the work in [6, 11] was concerned with proving feasibility of privacy preserving computations, and hence the analysis of very specific protocols.

To extend our lowerbounds from the local non-interactive to interactive protocols, we decompose an ℓ-round interactive protocol to ℓ one-round protocols, analyze the ℓ protocols, and use composition to obtain the lowerbound.

1.3   Related Work

Secure function evaluation and private data analysis were first tied together in the *Our Data, Ourselves (ODO)* protocols [8]. Their constructions – distributed SFE protocols for generating shares of random noise used in private data analyses – follow the natural paradigm discussed above. They do, however, avoid utilizing generic SFE feasibility results to gain on efficiency. We note that a point of difference between the protocols in [8] and the discussion herein is that ODO protocols are secure against malicious parties, in a computational setup, whereas we deal with semi-honest parties in an information theoretic setup.

Lowerbounds on the local non-interactive model were previously presented implicitly in [9, 14], and explicitly in [6, 10]. The two latter works are mainly concerned with what is called the global (or centralized) interactive setup, but have also implications to approximation to SUM in the local non-interactive model, namely, that it is impossible to approximate it within additive error $o(\sqrt{n})$, a similar consequence to our analysis of local non-interactive protocols. However (to the best of our understanding), these implications of [6, 10] do not imply the lowerbounds we get for local interactive protocols.

Chor and Kushilevitz [4] consider the problem of securely computing modular sum when the inputs are distributed. They show that this task can be done while sending roughly $n(t+1)/2$ messages. Furthermore, they prove that this number of messages is optimal for a family of protocols that they call oblivious. These are protocols where the communication pattern is fixed and does not depend on the inputs or random inputs. In our work we also only prove lowerbounds for oblivious protocols.

2   Preliminaries

Notation. D denotes an arbitrary domain. A *vector* $\mathbf{x} = (x_1, \ldots, x_n)$ is an ordered sequence of n elements of D. Vectors $\mathbf{x}, \mathbf{x}'$ are *neighboring* if they differ on exactly one entry, and are *T-neighboring* if they differ on a single entry, whose index is *not* in $T \subset [n]$. The *Laplace distribution*, $\mathrm{Lap}(\lambda)$, is the continuous probability distribution with probability density function $h(y) = \frac{\exp(-|y|/\lambda)}{2\lambda}$ (hence, $\mathbb{E}[Y] = 0$, $\mathrm{Var}[Y] = 2\lambda^2$, and $\Pr[|Y| > k\lambda] = e^{-k}$).

2.1 Differential Privacy

Our privacy definition (Definition 2 below) can be viewed as a distributed variant of ϵ-differential privacy (a.k.a. ϵ-indistinguishability). Informally, a computation is differentially private if any change in a single private input may only induce a small change in the distribution on its outcomes.

Definition 1 ([9]). *Let* $\hat{f} : \mathcal{D}^n \rightarrow R$ *be a randomized function from domain* $\mathcal{D}^n$ *to range* R. *We say that* $\hat{f}(\cdot)$ *is* ϵ-differentially private *if for all neighboring vectors* $\mathbf{x}, \mathbf{x}'$, *and for all possible sets of outcomes* $\mathcal{V} \subseteq R$ *it holds that* $\Pr[\hat{f}(\mathbf{x}) \in \mathcal{V}] \leq e^{\epsilon} \cdot \Pr[\hat{f}(\mathbf{x}') \in \mathcal{V}]$. *The probability is taken over the randomness of* $\hat{f}(\cdot)$.

Several frameworks for constructing differentially private functions by means of perturbation are presented in the literature (see [2, 9, 15, 16]). The most basic transformation on a function f that yields a differentially private function is via the framework of *global sensitivity* [9], where the outcome $f(\mathbf{x})$ is modified by the addition of noise sampled from the Laplace distribution, calibrated to the global sensitivity of f,

$$\hat{f}(\mathbf{x}) = f(\mathbf{x}) + Y, \tag{1}$$

where $Y \sim \mathrm{Lap}(\mathrm{GS}_f/\epsilon)$, and $\mathrm{GS}_f = \max |f(\mathbf{x}) - f(\mathbf{x}')|$, with the maximum taken over neighboring $\mathbf{x}, \mathbf{x}'$.

Example 1. The binary sum function $\mathrm{SUM} : \{0,1\}^n \rightarrow \mathbb{R}$ is defined as $\mathrm{SUM}(\mathbf{x}) = \sum_{i=1}^{n} x_i$. For every two neighboring $\mathbf{x}, \mathbf{x}' \in \{0,1\}^n$ we have that $|\mathrm{SUM}(\mathbf{x}) - \mathrm{SUM}(\mathbf{x}')| = 1$ and hence $\mathrm{GS}_{\mathrm{SUM}} = 1$. Applying (1), we get an ϵ-differentially private approximation, $\hat{f}(\mathbf{x}) = \mathrm{SUM}(\mathbf{x}) + Y$, where $Y \sim \mathrm{Lap}(1/\epsilon)$.

2.2 Differentially Private Protocols

We consider a distributed setting, where semi-honest parties $p_1, \ldots, p_n$ hold private inputs $x_1, \ldots, x_n$ respectively and engage in a protocol Π in order to compute (or approximate) a function $f(\cdot)$ of their joint inputs. The protocol Π is executed in a synchronous environment with point-to-point secure (untappable) communication channels, and is required to preserve privacy with respect to coalitions of size up to t. Following [4], we only consider a *fixed-communication* protocol Π (also called an oblivious protocol) where every channel is either (i) active in every run of Π (i.e., at least one bit is sent over the channel), or (ii) never used[4]. Parties that are adjacent to at least $t + 1$ active channels are called *popular* other parties are called *lonely*.

---

[4] Our proofs also work in a relaxed setting where every channel is either (i) used in at least a constant fraction of the runs of Π (where the probability is taken over the coins of Π), or (ii) is never used.

The main definition we will work with is an extension of Definition 1 to a distributed setting. Informally, we require that differential privacy is preserved with respect to any coalition of size up to t.

Definition 2. *Let Π be a protocol between n (semi-honest) parties. For a set $T \subseteq [n]$, let $\text{View}_T(x_1, \ldots, x_n)$ be the random variable containing the inputs of the parties in T (i.e. $\{x_i\}_{i \in T}$), the random coins of the parties in T, and the messages that the parties in T received during the execution of the protocol with private inputs $\mathbf{x} = (x_1, \ldots, x_n)$.*

We say that Π is (t, ϵ)-differentially private if for all $T \subset [n]$, where $|T| \le t$, for all T-neighboring $\mathbf{x}, \mathbf{x}'$, and for all possible sets $\mathcal{V}_T$ of views of parties in T:

$$\Pr[\text{View}_T(\mathbf{x}) \in \mathcal{V}_T] \le e^\epsilon \cdot \Pr[\text{View}_T(\mathbf{x}') \in \mathcal{V}_T], \tag{2}$$

where the probability is taken over the randomness of the protocol Π.

It is possible to relax this definition by replacing (2) by a requirement that View_T is statistically close, or computationally close to some ϵ-differentially private computation. The exact definition of differential privacy in the computational model requires some care; this definition will be given in the full version of the paper. The following informal lemma applies for such relaxations:

Lemma 1 (Informal). *Let $\hat{f}$ be ϵ-differentially private, and let Π be a t-secure protocol computing $\hat{f}$, then Π is (t, ϵ)-differentially private.*

Note 1. While a computational definition of differentially private protocols is probably the most appealing, we chose to present our work with Definition 2 – an *information theoretic* definition of differentially private protocols – because it allows us to prove bounds on protocols, demonstrating when constructing differentially private protocols is better than using the natural paradigm.

Note 2. We will only consider protocols computing a (randomized) function $\hat{f}(\cdot)$ resulting in all parties computing the *same* outcome of $\hat{f}(\mathbf{x})$. This can be achieved, e.g., by having one party compute the $\hat{f}(\mathbf{x})$ and send the outcome to all other parties.

2.3   The Local Model

The local model (previously discussed in [9, 14]) is a simplified distributed communication model where the parties communicate via a designated party – a *curator* – denoted C (with no local input). We will consider two types of differentially private local protocols. In *non-interactive* local protocols each party p_i applies an ϵ-differentially private algorithm $\mathcal{A}_i$ on its private input x_i and randomness r_i, and sends $\mathcal{A}_i(x_i; r_i)$ to C that then performs an arbitrary computation and publishes its result.

In *interactive* local protocols the input to each algorithm $\mathcal{A}_i$ includes x_i, r_i, and the history of messages received from the curator. The protocol proceeds in iterations where in each iteration C sends to party p_i a "query" message $q_{i,j}$ and party p_i responds with $\mathcal{A}_i(x_i; q_{i,1}, \ldots, q_{i,j}; r_i)$. It is required that the overall protocol preserves differential privacy, i.e., that the randomized function corresponding to the curator's view $\mathcal{A}_i(\cdot; q_{i,1}; r_i) \circ \mathcal{A}_i(\cdot; q_{i,1}, q_{i,2}; r_i) \circ \cdots \circ \mathcal{A}_i(\cdot; q_{i,1}, \ldots, q_{i,j}; r_i)$ preserves ϵ-differential privacy for every query messages $q_{i,1}, \ldots, q_{i,j}$ possible in the protocol. An immediate corollary is that $\mathcal{A}_i(\cdot; q_{i,1}, \ldots, q_{i,j}; r_i)$ should be ϵ-differentially private for all j.

2.4   Approximation

We will construct protocols whose outcome approximates a function $f : D^n \to \mathbb{R}$ by a probabilistic function. We say that a randomized function $\hat{f} : D^n \to \mathbb{R}$ is an *additive* (γ, τ)-approximation of f if $\Pr\left[|f(\mathbf{x}) - \hat{f}(\mathbf{x})| > \tau(n)\right] < \gamma(n)$ for all $\mathbf{x} \in D^n$. For example, Equation (1) yields an additive $(O(1), O(\mathrm{GS}_f/\epsilon))$-approximation to f.

3   Motivating Examples

We begin with two observations manifesting benefits of choosing an analysis together with a differentially private protocol for computing it. In the first example, this paradigm yields more efficient protocols than the natural paradigm; In the second example, it yields simpler protocols.

Binary Sum – $\sqrt{n}$ Additive Error. We begin with a simple observation regarding the binary sum function of Example 1: a very efficient $(n-1, \epsilon)$-differentially private protocol for approximating $\mathrm{SUM}(\mathbf{x}) = \sum_{i=1}^{n} x_i$ (where $x_i \in \{0, 1\}$) within $O(\sqrt{n}/\epsilon)$-additive approximation.

Let $\mathrm{flip}(x)$ be a randomized bit flipping operator returning x with probability $0.5 + \alpha$ and $1 - x$ otherwise (α will be determined later). Our protocol proceeds as follows: (i) Each party p_i with private input $x_i \in \{0, 1\}$ sends $z_i = \mathrm{flip}(x_i)$ to party p_1; (ii) Party p_1 sends $k = \sum_{i=1}^{n} z_i$ to all parties; (iii) Each party p_i locally outputs $\hat{f} = (k + (0.5 - \alpha)n)/2\alpha$. In this protocol, a total of $O(n)$ messages and $O(n \log n)$ bits of communication are exchanged.

To satisfy Definition 2, set $\alpha = \frac{\epsilon}{4+2\epsilon}$, yielding $\Pr[z_i = x_i]/\Pr[z_i = 1 - x_i] = (0.5 + \alpha)/(0.5 - \alpha) = 1 + \epsilon \le e^{\epsilon}$.

Note that $\mathbb{E}[k] = (0.5 + \alpha)\,\mathrm{SUM}(\mathbf{x}) + (0.5 - \alpha)(n - \mathrm{SUM}(\mathbf{x})) = 2\alpha\,\mathrm{SUM}(\mathbf{x}) + n(0.5 - \alpha)$, and hence, $\mathbb{E}[\hat{f}] = \mathrm{SUM}(\mathbf{x})$. By an application of the Chernoff bound, we get that $\hat{f}$ is an additive $(O(1), O(\sqrt{n}/\epsilon))$-approximation to $\mathrm{SUM}(\cdot)$.

It is natural to choose a *symmetric* approximation to $\mathrm{SUM}(\cdot)$ that only depends on $\mathrm{SUM}(\cdot)$. While the construction above yields an efficient protocol for such a function, we prove (using ideas from [4]) that no efficient SFE protocols for such functions exist. We leave the details for the full version.

Lemma 2. *Let* $\hat{f}$ *be a symmetric additive* $(O(1), n/10)$-*approximation to* $\mathrm{SUM}(\cdot)$. *Then any oblivious t-secure protocol computing $\hat{f}$ uses $\Omega(nt)$ messages*[5].

Distance from a Long Subsequence of 0's. Our second function measures how many bits in a sequence $\mathbf{x}$ of n bits should be set to zero to get an all-zero subsequence of length n^α. In other words, the minimum weight over all substrings of $\mathbf{x}$ of length n^α bits: $\mathrm{DIST}_\alpha(\mathbf{x}) = \min_i(\sum_{j=i}^{i+n^\alpha-1} x_j)$. For $t \leq n/2$, we present a (t, ϵ, δ)-differentially private protocol[6] approximating $\mathrm{DIST}_\alpha(\mathbf{x})$ with additive error $\tilde{O}(n^{\alpha/3}/\epsilon)$.

In our protocol, we treat the n-bit string $\mathbf{x}$ (where x_i is held by party p_i) as a sequence of $n^{1-\alpha/3}$ disjoint intervals, each $n^{\alpha/3}$ bit long. Let $i_1, \ldots, i_{n^{1-\alpha/3}}$ be the indices of the first bit in each interval, and observe that $\min_{i_k}(\sum_{j=i_k}^{i_k+n^\alpha-1} x_j)$ is an $n^{\alpha/3}$ additive approximation of DIST_α. The protocol for computing an approximation $\hat{f}$ to DIST_α is sketched below.

1. Every party p_i generates a random variable Y_i distributed according to the normal distribution $N(\mu = 0, \sigma^2 = 2R/n)$ where $R = \frac{2\log(\frac{2}{\delta})}{\epsilon^2}$, and shares $x_i + Y_i$ between parties $p_1, \ldots, p_{t+1}$ using an additive $(t+1)$-out-of-$(t+1)$ secret sharing scheme.
2. Every party p_i, where $1 \leq i \leq t+1$, sums, for every interval of length $n^{\alpha/3}$, the shares it got from the parties in the interval and sends this sum to p_1.
3. For every interval of length $n^{\alpha/3}$, party p_1 computes the sum of the $t+1$ sums it got for the interval. By the additivity of the secret sharing scheme, this sum is equal to $S_k = \sum_{j=i_k}^{i_k+n^{\alpha/3}-1}(x_j + Y_j) = \sum_{j=i_k}^{i_k+n^{\alpha/3}-1} x_j + Z_k$ where $Z_k = \sum_{j=i_k}^{i_k+n^{\alpha/3}-1} Y_j$ (notice that $Z_k \sim N(\mu = 0, \sigma^2 = 2R)$).
4. p_1 computes $\min_k \sum_{j=k}^{k+n^{2\alpha/3}} S_k$ and sends this output to all parties.

Using the analysis of [8], this protocol is a (t, ϵ, δ)-differentially private protocol when $2t < n$. Furthermore, it can be shown that with high probability the additive error is $\tilde{O}(n^{\alpha/3}/\epsilon)$. To conclude, we showed a simple 3 round protocol for DIST_α.

This protocol demonstrates two advantages of the paradigm of choosing what and how together. First, we choose an approximation of DIST_α (i.e., we compute the minimum of subsequences starting at a beginning of an interval). This approximation reduces the communication in the protocol. Second, we leak information beyond the output of the protocol, as p_1 learns the sums S_k's.[7]

---

[5] We note that the lemma does not hold for non-symmetric functions. For example, we can modify the bit flip protocol above to an SFE protocol for a non-symmetric function, retaining the number of messages sent (but not their length): in step (iii) let p_1 send $\mathbf{z} = (z_1, \ldots, z_n)$, and in step (iv) let p_i locally output $\hat{f} + \mathbf{z}2^{-n}$, treating $\mathbf{z}$ as an n-bit binary number.

[6] (ϵ, δ)-differential privacy is a generalization, defined in [8], of ϵ-differential privacy where it is only required that $\Pr[\hat{f}(\mathbf{x}) \in \mathcal{V}] \leq e^\epsilon \cdot \Pr[\hat{f}(\mathbf{x}') \in \mathcal{V}] + \delta$.

[7] One can use the techniques of [5] to avoid leaking these sums while maintaining a constant number of rounds, however the resulting protocol is less efficient.

4   Binary Sum – Below $\sqrt{n}$ Additive Error

We prove that in any ℓ-round, fixed-communication, (t, ϵ)-differentially private protocol computing the binary sum with additive error less than $\sqrt{n}/\tilde{O}(\ell)$, the number of messages sent in the protocol is $\Omega(nt)$.

Theorem 1. *In any ℓ-round, fixed-communication, (t, ϵ)-differentially private protocol for approximating SUM_n that sends at most $n(t + 1)/4$ messages the error is $\Omega(\sqrt{n}/(\epsilon\ell\sqrt{\log \ell}))$ with constant probability.*

We prove this lowerbound in steps. We first define a gap version of the threshold function, denoted GAP-TR, and observe that any differentially private protocol for SUM with error τ implies a differentially-private protocol for GAP-TR with gap $\tau/2$. Therefore, we prove impossibility of differentially-private computation of GAP-TR with small gap. In Section 4.1, we prove that if there is a protocol computing the GAP-TR function with at most $nt/4$ messages, then there is a protocol in the local model (i.e., with a curator) computing the GAP-TR function with the same gap. Thereafter, we prove that such a protocol in the local model has can only compute GAP-TR with gap $\Omega(\sqrt{n})$. In Section 4.2, we analyze properties of non-interactive protocols in the local model that compute GAP-TR and in Section 4.3 we generalize this analysis to interactive protocols in the local model that compute GAP-TR. In Section 4.4, we complete the proof of the lowerbound on the gap in the local model. Theorem 1 follows from the combination the transformation of the distributed protocol to the protocol in local model proved in Lemma 4 and the lowerbound for protocols in the local model proved in Theorem 2.

Theorem 1 suggests that whenever we require that the error of a differentially-private protocol for approximating $\mathrm{SUM}(\cdot)$ to be of magnitude smaller than $\sqrt{n}/\epsilon$, there is no reason to relinquish the simplicity of the natural paradigm for constructing protocols. In this case, it is possible to construct relatively simple efficient SFE protocols, which use $O(nt)$ messages, and compute an additive $(O(1/\epsilon), O(1))$-approximation of $\mathrm{SUM}(\cdot)$.

We next define the gap version of the threshold function:

Definition 3 (Gap Threshold Function). *We define the gap threshold function as follows: If $\mathrm{SUM}_n(x_1, \ldots, x_n) \leq \kappa$ then $\mathrm{GAP\text{-}TR}_{\kappa,\tau}(x_1, \ldots, x_n) = 0$ and if $\mathrm{SUM}_n(x_1, \ldots, x_n) \geq \kappa + \tau$ then $\mathrm{GAP\text{-}TR}_{\kappa,\tau}(x_1, \ldots, x_n) = 1$.*

In the above definition we consider a gap version of the threshold function and there are no requirements on the output of $\mathrm{GAP\text{-}TR}_{\kappa,\tau}$ when $\kappa < \mathrm{SUM}_n(x_1, \ldots, x_n) < \kappa + \tau$.

Claim. If there exists an ℓ-round, fixed-communication, (t, ϵ)-differentially private protocol that (γ, τ)-approximates SUM_n sending at most $n(t + 1)/4$ messages, then for every κ there exists an ℓ-round, (t, ϵ)-differentially private protocol that correctly computes $\mathrm{GAP\text{-}TR}_{\kappa,\tau/2}$ with probability at least γ sending at most $n(t + 1)/4$ messages.

Similarly, using "padding" arguments

Claim. If for some $0 \leq \kappa \leq n - \tau$ there exists an ℓ-round, fixed-communication, (t, ϵ)-differentially private n party protocol that correctly computes GAP-TR$_{\kappa, \tau}$ with probability at least γ sending at most $n(t+1)/4$ messages, then there exists an ℓ-round, (t, ϵ)-differentially private $n/2$-party protocol that correctly computes GAP-TR$_{0, \tau}$ with probability at least γ sending at most $n(t+1)/4$ messages.

4.1   Moving to the Local Model

We start with the transformation of a distributed protocol to a protocol in the local model. To analyze this transformation we will need the following simple lemma:

Lemma 3. *Fix a 3-party randomized protocol, assume that each p_i holds an inputs x_i, and fix some communication transcript c. Define α_i as the overall probability that in each round p_i with input x_i sends messages according to c provided that in previous rounds it gets messages according to c. Then, the probability that c is exchanged is $\alpha_1 \cdot \alpha_2 \cdot \alpha_3$.*

Lemma 4. *If there exists an ℓ-round, (t, ϵ)-differentially private protocol that correctly computes GAP-TR$_{\kappa, \tau}$ with probability at least γ sending at most $n(t+1)/4$ messages, then there exists a 2ℓ-round, ϵ-differentially private protocol in the local model that correctly computes GAP-TR$_{\kappa, \tau}$ with probability at least γ.*

Proof. Assume that there is a distributed protocol Π satisfying the conditions in the lemma. Recall that a party in Π is lonely if it has at most t neighbors and it is popular otherwise. As the protocol sends at most $n(t+1)/4$ messages, the protocol uses at most $n(t+1)/4$ channels. Since each channel connects two parties, there are at least $n/2$ lonely parties. We will construct a protocol in the local model which computes GAP-TR$_{\kappa, \tau}$ for $n/2$ parties in two stages: (i) We first construct a protocol $\mathcal{P}$ in the local model which computes GAP-TR$_{\kappa, \tau}$ for n parties and only protects the privacy of the lonely parties. (ii) We next fix the inputs of the popular parties and obtain a protocol $\mathcal{P}'$ for $n/2$ parties that protects the privacy of all parties.

First Stage. We convert the distributed protocol Π to a protocol $\mathcal{P}$ in the local model as follows: We have two rounds in $\mathcal{P}$ for every round of Π. For every message m that Party p_j sends to Party p_k in round i in Protocol Π, Party p_j sends m to the curator in round $2i - 1$ and the curator sends m to Party p_k in round $2i$. Finally, at the end of the protocol Party p_1 sends the output to the curator.

We next prove that $\mathcal{P}$ protects the privacy of lonely parties. Without loss of generality, let p_1 be a lonely party, T be the set of size at most t containing the neighbors of p_1, and $R = \{p_1, \ldots, p_n\} \setminus (T \cup \{p_1\})$. Fix any neighboring vectors of inputs $\mathbf{x}$ and $\mathbf{x}'$ which differ on x_1. The view of the curator in $\mathcal{P}$ contains all messages sent in the protocol. It suffices to prove that for every view v,

$$\Pr[\text{View}_{\mathcal{C}}(\mathbf{x}) = v] \leq e^{\epsilon} \cdot \Pr[\text{View}_{\mathcal{C}}(\mathbf{x}') = v]$$

(by simple summation it will follow for every set of views $\mathcal{V}$).

Fix a view v of the curator. For a set A, define α_A and α'_A as the probabilities in Π that in each round the set A with inputs from $\mathbf{x}$ and $\mathbf{x}'$ respectively sends messages according to v if it gets messages according to v in previous rounds (these probabilities are taken over the random inputs of the parties in A). Observe that if $p_1 \notin A$, then $\alpha_A = \alpha'_A$. By simulating p_1, T, R by three parties and applying Lemma 3, and by the construction of $\mathcal{P}$ from Π

$$\Pr\left[\text{View}_{\mathcal{C}}^{\mathcal{P}}(\mathbf{x}) = v\right] = \alpha_{\{p_1\}} \cdot \alpha_T \cdot \alpha_R, \quad \text{and}$$
$$\Pr\left[\text{View}_{\mathcal{C}}^{\mathcal{P}}(\mathbf{x}') = v\right] = \alpha'_{\{p_1\}} \cdot \alpha'_T \cdot \alpha'_R = \alpha'_{\{p_1\}} \cdot \alpha_T \cdot \alpha_R.$$

Thus, we need to prove that

$$\alpha_{\{p_1\}} \le e^\epsilon \alpha'_{\{p_1\}}. \tag{3}$$

We use the (t, ϵ) privacy of protocol Π to prove (3). Let v_T be the messages sent and received by the parties in T in v. As T separates p_1 from R, the messages in v_T are all messages in v except for the messages exchanged between parties in R. The view of T includes the inputs of T in $\mathbf{x}$, the messages v_T, and the random inputs $\mathbf{r_T} = \{r_i : p_i \in T\}$. For a set A, define β_A and β'_A as the probabilities that in Π in each round the set A with inputs from $\mathbf{x}$ and $\mathbf{x}'$ respectively sends messages according to v_T if it gets messages according to v_T in previous rounds. Note that $\beta_{\{p_1\}} = \alpha_{\{p_1\}}$ and $\beta'_{\{p_1\}} = \alpha'_{\{p_1\}}$ by the definition of $\mathcal{P}$. By simulating p_1, T, R by three parties, where the random inputs of T are fixed to $\mathbf{r_T}$, and by Lemma 3,

$$\Pr[\text{View}_T^{\Pi}(\mathbf{x}) = (\mathbf{x_T}, \mathbf{r_T}, v_T)] = \alpha_{\{p_1\}} \cdot \beta_R, \quad \text{and}$$
$$\Pr[\text{View}_T^{\Pi}(\mathbf{x}') = (\mathbf{x_T}, \mathbf{r_T}, v_T)] = \beta'_{\{p_1\}} \cdot \beta'_R = \alpha'_{\{p_1\}} \cdot \beta_R.$$

(recalling that $\mathbf{x_T} = \mathbf{x'_T}$). The above probabilities are taken over the random strings of R and p_1 when the random strings of T are fixed to $\mathbf{r_T}$. Therefore, the (t, ϵ) differential privacy of Π implies (3), and, thus, that $\mathcal{P}$ is ϵ-differentially private with respect to inputs of lonely parties.

Second Stage. There are at least $n/2$ lonely parties in Π, thus, w.l.o.g., parties $p_1, \ldots, p_{n/2}$ are lonely. We construct a protocol $\mathcal{P}'$ for computing GAP-TR$_{\kappa,\tau}$ for $n/2$ parties by executing Protocol $\mathcal{P}$ where (i) Party p_i, where $1 \le i \le n/2$, with input x_i sends messages in $\mathcal{P}'$ as Party p_i with input x_i sends them in $\mathcal{P}$; and (ii) Party p_1 in $\mathcal{P}'$ simulates all other $n/2$ parties in $\mathcal{P}$, that is, for every $n/2 < i \le n$, it chooses a random input r_i for p_i and in every round it sends to the curator the same messages as p_i would send with $x_i = 0$ and r_i. Since the curator sees the same view in $\mathcal{P}$ and $\mathcal{P}'$ and the privacy of lonely parties is protected in $\mathcal{P}$, the privacy of each of the $n/2$ parties in $\mathcal{P}'$ is protected. Protocol $\mathcal{P}'$ correctly computes GAP-TR$_{\kappa,\tau}$ with probability at least γ since we fixed $x_i = 0$ for $i < n/2 \le n$ and $\mathcal{P}'$ returns the same output distribution of Π, which correctly computes GAP-TR$_{\kappa,\tau}$ with probability at least γ. □

By Lemma 4 it suffices to prove lowerbounds on the gap τ for protocols in the local model.

4.2 GAP-TR in the Non-interactive Local Model

We consider the non-interactive local model where each party holds an input $x_i \in \{0,1\}$ and independently applies an algorithm A_i (also called a sanitizer) before sending the sanitized result c_i to the curator. We consider a differentially-private protocol computing GAP-TR$_{0,\tau}$ in this model and we wish to prove lowerbounds on τ. Notice that we take $\kappa = 0$, namely, we want to prove that the curator cannot distinguish between the all-zero input and inputs of weight at least τ (for small values of τ). More formally, we want to prove that if each A_i is ϵ-differentially private, then the curator errs with constant probability when computing GAP-TR$_{0,\tau}$ for $\tau = O(\sqrt{n})$. Towards this goal, we show that there are two inputs for which the curator sees similar distributions on the messages, thus, has to return similar answers. However, one input contains $\Omega(\sqrt{n})$ ones and the other is the all-zero input, and the algorithm errs on at least one of the inputs. We will prove the existence of such input with $\Omega(\sqrt{n})$ ones, by considering a distribution $\mathcal{A}$ on inputs and later proving that such input taken from the distribution $\mathcal{A}$ exists.

We note that in the local model randomness for the curator can be supplied by the parties and hence we assume, w.l.o.g., that the curator is deterministic. Thus, the curator, having received the sanitized input $\mathbf{c} = S(\mathbf{x}) \overset{\Delta}{=} (A_1(x_1), \ldots, A_n(x_n))$, applies a deterministic algorithm G to $\mathbf{c}$, where $G(\mathbf{c})$ is supposed to answer GAP-TR$_{\kappa,\tau}(x_1, \ldots, x_n)$ correctly.

Let $\alpha \overset{\Delta}{=} \frac{1}{\epsilon}\sqrt{\frac{d}{n}}$ for d to be determined later. We consider two distributions over which the input is chosen.

- Distribution $\mathcal{A}$: $x_i = 1$ with probability α, $x_i = 0$ with probability $(1 - \alpha)$ (the inputs of the different parties are chosen independently).
- Distribution $\mathcal{B}$: $x_i = 0$ with probability 1 (that is, $\mathcal{B}$ always chooses the all-zero input vector).

From here on, we use X to identify the random variable representing the input and X_i for its ith coordinate. When considering the random variable over $\mathcal{A}$ (respectively, $\mathcal{B}$), we use the notation $\Pr_{\mathcal{A}}[\cdot]$ (respectively, $\Pr_{\mathcal{B}}[\cdot]$). For a set D, we use the notation $\Pr_{\mathcal{A}}[D]$ (respectively, $\Pr_{\mathcal{B}}[D]$) to denote the probability of the event that $A_i(X_i) \in D$ when X_i is generated according to the probability distribution $\mathcal{A}$ (respectively, $\mathcal{B}$).

We denote for every possible output $\mathbf{c} = (c_1, \ldots, c_n)$ of S,

$$r(\mathbf{c}) \overset{\Delta}{=} \frac{\Pr_{\mathcal{A}}\left[S(X) = \mathbf{c}\right]}{\Pr_{\mathcal{B}}\left[S(X) = \mathbf{c}\right]} \quad \text{and} \quad r_i(c_i) \overset{\Delta}{=} \frac{\Pr_{\mathcal{A}}\left[A_i(X_i) = c_i\right]}{\Pr_{\mathcal{B}}\left[A_i(X_i) = c_i\right]}. \tag{4}$$

Define a random variable $\mathbf{C} = (C_1, \ldots, C_n)$ where $C_i = A_i(X_i)$ and X_i is chosen according to the distribution $\mathcal{A}$. We next bound $\Pr_{\mathcal{A}}[r(\mathbf{C}) > \delta]$.

Lemma 5. $\Pr_{\mathcal{A}}[r(\mathbf{C}) > \exp(\nu d)] < \exp\left(-(\nu - 8)^2 d/8\right)$ for every $\nu > 8$.

We prove Lemma 5 using the Hoeffding bound. Define the random variables $V_i \triangleq \ln r_i(C_i)$. For every $\eta > 0$, we have that

$$\Pr_{\mathcal{A}}[r(\mathbf{C}) > \eta] = \Pr_A\left[\prod_{i=1}^n r_i(C_i) > \eta\right] = \Pr_A\left[\sum_{i=1}^n V_i > \ln \eta\right], \qquad (5)$$

where the first equality holds since the X_is are chosen independently. To apply the Hoeffding bound, we need to compute bounds on each variable V_i, and to compute the expectation of V_i. Both tasks are achieved using the ϵ-differential privacy of the sanitizers, that is,

$$e^{-\epsilon} \leq \frac{\Pr[A_i(1) = c_i]}{\Pr[A_i(0) = c_i]} \leq e^{\epsilon}. \qquad (6)$$

Lemma 6. $-2\alpha\epsilon \leq V_i \leq 2\alpha\epsilon$ *for every* i.

Proof. For every i and every value c_i,

$$r_i(c_i) = \frac{\alpha\Pr[A_i(1) = c_i] + (1-\alpha)\Pr[A_i(0) = c_i]}{\Pr[A_i(0) = c_i]}$$
$$= 1 + \alpha\frac{\Pr[A_i(1) = c_i] - \Pr[A_i(0) = c_i]}{\Pr[A_i(0) = c_i]}.$$

Using $\Pr[A_i(1) = c_i] \leq e^{\epsilon}\Pr[A_i(0) = c_i]$ we get on one hand that

$$r_i(c_i) \leq 1 + \alpha\frac{\Pr[A_i(0) = c_i]e^{\epsilon} - \Pr[A_i(0) = c_i]}{\Pr[A_i(0) = c_i]} = 1 + \alpha(e^{\epsilon} - 1) \leq 1 + 2\alpha\epsilon$$

(since $e^{\epsilon} < 1 + 2\epsilon$ for every $0 < \epsilon \leq 1$). Thus, $V_i = \ln r_i(C_i) \leq \ln(1 + 2\alpha\epsilon) \leq 2\alpha\epsilon$, since $\ln(1+x) \leq x$ for every $0 \leq x \leq 1$. Using $e^{-\epsilon}\Pr[A_i(0) = c_i] \leq \Pr[A_i(1) = c_i]$ we get on the other hand that

$$r_i(c_i) \geq 1 + \alpha\frac{\Pr[A_i(0) = c_i]e^{-\epsilon} - \Pr[A_i(0) = c_i]}{\Pr[A_i(0) = c_i]} = 1 + \alpha(e^{-\epsilon} - 1) \geq 1 - \alpha\epsilon.$$

Thus, $V_i = \ln r_i(C_i) \geq \ln(1 - \alpha\epsilon) \geq -2\alpha\epsilon$, since $\ln(1 - x) \geq -2x$ for every $0 \leq x \leq 0.5$. $\qquad\square$

Lemma 7. $\mathbb{E}[V_i] \leq 8\alpha^2\epsilon^2$.

Proof. In this proof we assume that the output of A_i is a countable set. Denote $B_b \triangleq \{c_i : r_i(c_i) = 1 + b\}$ for every $-\alpha\epsilon \leq b \leq 2\alpha\epsilon$ (by Lemma 6, these are the only values possible for b). Note that by the definition of r_i, for every $c_i \in B_b$ $\Pr_{\mathcal{A}}[A_i(X_i) = c_i]/\Pr_{\mathcal{B}}[A_i(X_i) = c_i] = 1 + b$, thus, $\Pr_{\mathcal{B}}[B_b] = \frac{\Pr_{\mathcal{A}}[B_b]}{1+b} \leq (1 - b + 2b^2)\Pr_{\mathcal{A}}[B_b]$. Let $\beta = \alpha\epsilon$. We next bound $\mathbb{E}[V_i]$.

$$\mathbb{E}[V_i] = \mathbb{E}_{\mathcal{A}}[\ln r(C_i)] = \sum_{-\beta \leq b \leq 2\beta} \Pr_{\mathcal{A}}[B_b] \ln(1+b) \leq \sum_{-\beta \leq b \leq 2\beta} \Pr_{\mathcal{A}}[B_b] b$$

$$= \sum_{-\beta \leq b \leq 2\beta} \Pr_{\mathcal{A}}[B_b] - \sum_{-\beta \leq b \leq 2\beta} \Pr_{\mathcal{A}}[B_b](1 - b + 2b^2) + \sum_{-\beta \leq b \leq 2\beta} \Pr_{\mathcal{A}}[B_b](2b^2)$$

$$\leq \sum_{-\beta \leq b \leq 2\beta} \Pr_{\mathcal{A}}[B_b] - \sum_{-\beta \leq b \leq 2\beta} \Pr_{\mathcal{B}}[B_b] + \sum_{-\beta \leq b \leq 2\beta} \Pr_{\mathcal{A}}[B_b](2b^2)$$

$$\leq 1 - 1 + 8\beta^2 \sum_{-\beta \leq b \leq 2\beta} \Pr_{\mathcal{A}}[B_b] = 8\beta^2 = 8\alpha^2 \epsilon^2. \qquad \square$$

From Lemma 7, $\mathbb{E}(\sum_{i=1}^{n} V_i) \leq 8\alpha^2 \epsilon^2 n = 8d$. We next prove Lemma 5 which shows that $\sum_{i=1}^{n} V_i$ is concentrated around this value.

Proof (of Lemma 5). We apply the Hoeffding bound: Let $V_1, \ldots, V_n$ be independent random variables such that $V_i \in [a, b]$. Then, $\Pr[\sum_{i=1}^{n} V_i - \mu \geq t] \leq \exp\left(-\frac{2t^2}{n(b-a)^2}\right)$ for every $t > 0$ (where $\mu = \sum_{i=1}^{n} \mathbb{E}[X_i]$).

By (5), Lemma 6, and Lemma 7:

$$\Pr_{\mathcal{A}}[r(\mathbf{C}) > \exp(\nu d)] = \Pr_{\mathcal{A}}\left[\sum_{i=1}^{n} V_i > \nu d\right] = \Pr_{\mathcal{A}}\left[\sum_{i=1}^{n} V_i - \sum_{i=1}^{n} \mathbb{E}V_i > \nu d - \sum_{i=1}^{n} \mathbb{E}V_i\right]$$

$$\leq \Pr_{\mathcal{A}}\left[\sum_{i=1}^{n} V_i - \sum_{i=1}^{n} \mathbb{E}V_i > \nu d - n \cdot 8\alpha^2 \epsilon^2\right]$$

$$\leq \exp\left(-\frac{2(\nu d - n \cdot 8\alpha^2 \epsilon^2)^2}{16 n \alpha^2 \epsilon^2}\right) < \exp\left(-(\nu - 8)^2 d/8\right) \qquad \square$$

The following corollary is a rephrasing of Lemma 5 that follows from the definition of r in (4) and the fact that distribution $\mathcal{B}$ picks the all-zero input with probability 1.

Corollary 1. *Assume we sample $\mathbf{X}$ according to distribution $\mathcal{A}$ and compute $\mathbf{c} = S(\mathbf{X})$. Then, for every $\nu > 8$ with probability at least $1 - \exp\left(-(\nu - 8)^2 d/8\right)$*

$$\Pr_{\mathcal{A}}[S(\mathbf{Z}) = \mathbf{c}] \leq \exp(-\nu d) \Pr[S(\mathbf{0^n}) = \mathbf{c}],$$

where in the left hand side the probability is taken over the choice of Z according to the distribution $\mathcal{A}$ and the randomness of the sanitizers and in the right hand side the probability is taken over the randomness of the sanitizers.

4.3   GAP-TR$_{\kappa,\tau}$ in the Interactive Local Model

In this section we generalize Corollary 1 to interactive local protocols where each party holds an input $x_i \in \{0, 1\}$ and communicates with the curator in rounds. To achieve this goal, we decompose a 2ℓ-round $\epsilon/2$-differentially private protocol into ℓ protocols, and prove that each protocol is ϵ-differentially private. Thus, we can apply Corollary 1 to each protocol, and then apply a composition lemma.

Lemma 8. *Suppose we execute a ℓ-round, local, $\epsilon/2$-differentially private proto-col and we sample a vector $\mathbf{X}$ according to distribution $\mathcal{A}$ and compute $\mathbf{c} = S(\mathbf{X})$ (where $S(\mathbf{X})$ is the communication in the 2ℓ rounds). Then, for every $\nu > 8$ with probability at least $1 - \ell \exp\left(-(\nu - 8)^2 d/8\right)$*

$$\Pr_{\mathcal{A}}[S(\mathbf{Z}) = \mathbf{c}] \leq \exp\left(-\ell\nu d\right) \Pr[S(\mathbf{0^n}) = \mathbf{c}],$$

where in the left side the probability is taken over the choice of Z according to the distribution $\mathcal{A}$ and the randomness of the sanitizers and in the right side the probability is taken over the randomness of the sanitizers.

Proof (Sketch). Fix a 2ℓ-round, $\frac{\epsilon}{2}$-differentially private, local protocol $\mathcal{P}$. In the interactive local model, a protocol is composed of ℓ-interactions where in each interaction the curator sends a query to each party and the party sends an answer.

Our first goal is to make the parties stateless. Consider a party p_i. First, we assume that in interaction j the curator sends all queries and answers $q_{i,1}, a_{i,1}, \dots, a_{i,j-1}, q_{i,j}$ it sent and received from p_i in previous rounds. Second, we assume that party p_i chooses a fresh random string in each round, that is, in round j, party p_i chooses with uniform distribution a random string that is consistent with the queries and answers it got in the previous rounds, (since we assume that the parties are unbounded, such choice is possible). Party p_i uses this random string to answer the jth query. In other words, we can consider p_i as applying an algorithm $A_{i,j}$ to compute the jth answer; this algorithm depends on the previous queries and answers and uses an independent random string.

We next claim that $A_{i,j}$ is ϵ-differentially private. That is, we claim that the probability that $q_{i,j}$ is generated given the previous queries and answers is roughly the same when p_i holds the bit 0 and when p_i holds the bit 1. This follows from the following tow facts: (1) the probability of $q_{i,1}, a_{i,1}, \dots, a_{i,j-1}, q_{i,j}$ is roughly the same when p_i holds the bit 0 and when p_i holds the bit 1. (2) the probability of $q_{i,1}, a_{i,1}, \dots, a_{i,j-1}$ is roughly the same when p_i holds the bit 0 and when p_i holds the bit 1. The exact details are omitted. Thus, the answers of the n parties in interaction j are ϵ-private, and we can apply Corollary 1 to the concatenation of the n answers.

We now use the above protocol to construct a protocol $\mathcal{P}_1$ between a single party, holding a one bit input x and a curator. Throughout the execution of the protocol the party simulates all n parties as specified by the original protocol (i.e., sends messages to the curator with the same distribution as the n parties send them). If the bit of the party in $\mathcal{P}_1$ is 1 it chooses the n input bits of the n parties in $\mathcal{P}$ according to distribution $\mathcal{A}$. If the bit of the party in $\mathcal{P}_1$ is 0 it chooses the n input bits of the n parties in $\mathcal{P}$ to be the all-zero vector. By Corollary 1 we can apply the composition lemma – Lemma 10 – to the composition of the ℓ non-interactive protocols and the lemma follows. $\square$

Corollary 2. *For every $\nu > 8$ and for every set D of views in a 2ℓ-round protocol,*

$$\Pr_{\mathcal{B}}[D] \geq \frac{\Pr_{\mathcal{A}}[D] - \ell \exp\left(-(\nu - 8)^2 d/8\right)}{\exp\left(\ell\nu d\right)}.$$

Proof. Let

$$D_1 = \left\{ \mathbf{c} \in D : \Pr_{\mathcal{A}}[S(X) = \mathbf{c}] \leq \exp\left(\ell\nu d\right) \Pr_{\mathcal{B}}[S(X) = \mathbf{c}] \right\}$$

and

$$D_2 = \left\{ \mathbf{c} \in D : \Pr_{\mathcal{A}}[S(X) = \mathbf{c}] > \exp\left(\ell\nu d\right) \Pr_{\mathcal{B}}[S(X) = \mathbf{c}] \right\}.$$

By Lemma 8, $\Pr_{\mathcal{A}}[D_2] \leq \ell \exp\left(-(\nu - 8)^2 d/8\right)$. Furthermore, $\Pr_{\mathcal{B}}[D_1] \geq \frac{\Pr_{\mathcal{A}}[D_1]}{\exp(\ell\nu d)}$. Thus,

$$\Pr_{\mathcal{B}}[D] \geq \Pr_{\mathcal{B}}[D_1] \geq \frac{\Pr_{\mathcal{A}}[D_1]}{e^{\ell\nu d}} = \frac{\Pr_{\mathcal{A}}[D] - \Pr_{\mathcal{A}}[D_2]}{e^{\ell\nu d}} \geq \frac{\Pr_{\mathcal{A}}[D] - \ell e^{-(\nu-8)^2 d/8}}{e^{\ell\nu d}}. \qquad \square$$

4.4   Completing the Lowerbound for GAP-TR$_{0,\tau}$ in the Local Model

In this section we complete the proof that in any ℓ-round, local, ϵ-differentially private protocols for the gap-threshold function, namely, GAP-TR$_{0,\tau}$, the curator errs with constant probability when $\ell \ll \sqrt{n}$ and τ is small. For proving this result, we defined a distribution $\mathcal{A}$ which chooses each bit in the input independently at random where it is one with probability α and zero with probability $1 - \alpha$. Lemma 9, which follows from a standard Chernoff bound argument, states that when generating a vector $(X_1, \ldots, X_n)$ according to $\mathcal{A}$, the sum $\sum_{i=1}^n X_i$ is concentrated around its expected value, which is αn.

Lemma 9. $\Pr_{\mathcal{A}}\left[\sum_{i=1}^n X_i \leq (1 - \gamma)\alpha n\right] < \exp\left(-\sqrt{dn}\gamma^2/(2\epsilon)\right)$ *for every* $0 \leq \gamma < 1$.

By Corollary 2 the distributions on the outputs when the inputs are taken from $\mathcal{A}$ or $\mathcal{B}$ are not far apart. By Lemma 9, with high probability the number of ones in the inputs distributed according to $\mathcal{A}$ is fairly big, while in $\mathcal{B}$ the number of ones is zero. These facts are used in Theorem 2 to prove the lowerbound.

Theorem 2. *In any ℓ-round, local, ϵ-differentially private protocol for computing GAP-TR$_{0,\tau}$ for $\tau = O(\sqrt{n}/(\epsilon\ell\sqrt{\log \ell}))$ the curator errs with constant probability,*

Proof. Fix any ℓ-round, local, ϵ-differentially private protocol, and let G be the algorithm of the curator that given the communication computes the output of the protocol. Let $\tau = 0.5\alpha n = \sqrt{dn}/\epsilon$. We denote $D \triangleq \{\mathbf{c} : G(\mathbf{c}) = 1\}$, that is, D contains all vectors of communication for which the curator answers 1. There are two cases. If the probability of D under the distribution $\mathcal{A}$ is small, then the curator has a big error when the inputs are distributed according to $\mathcal{A}$. Otherwise, by Corollary 2, the probability of D under the distribution $\mathcal{B}$ is big, and the curator has a big error when the inputs are distributed according to $\mathcal{B}$. Formally, there are two cases:

Case 1: $\Pr_{\mathcal{A}}[D] < 0.99$. We consider the event that the sum of the inputs is at least $\tau = 0.5\alpha n$ and the curator returns an answer 0, that is, the curator errs. We next prove that when the inputs are distributed according to $\mathcal{A}$ the probability of the complementary of this event is bounded away from 1. By the union bound the probability of the complementary event is at most $\Pr_{\mathcal{A}}\left[\sum_{i=1}^{n} X_i < 0.5\alpha n\right] + \Pr_{\mathcal{A}}[D]$. By Lemma 9,

$$\Pr_{\mathcal{A}}[D] + \Pr_{\mathcal{A}}\left[\sum_{i=1}^{n} X_i < 0.5\alpha n\right] \leq 0.99 + \exp\left(-0.25\sqrt{dn}/(2\epsilon)\right) \approx 0.99.$$

Thus, in this case, with probability ≈ 0.01 the curator errs.

Case 2: $\Pr_{\mathcal{A}}[D] \geq 0.99$. In this case, we consider the event that the input is the all-zero string and the curator answers 1, that is, the curator errs. We next prove using Corollary 2 that when the inputs are distributed according to $\mathcal{B}$ (that is, they are the all-zero string), the probability of this event is bounded away from 0, that is, taking $\nu = \theta(\ell \log \ell)$ and $d = 1/(\ell \nu) = \theta(1/(\ell^2 \log \ell))$,

$$\Pr_{\mathcal{B}}[D] \geq \frac{\Pr_{\mathcal{A}}[D] - \ell \exp\left(-(\nu - 8)^2 d/8\right)}{\exp\left(\ell \nu d\right)} > \frac{0.99 - 0.5}{\exp\left(1\right)} > 0.01.$$

Thus, in this case, with probability at least 0.01, the curator errs. As $d = \theta(1/(\ell^2 \log \ell))$, we get that $\tau = \sqrt{dn}/\epsilon = \theta(\sqrt{n}/(\epsilon \ell \sqrt{\log \ell}))$. $\square$

Acknowledgments. We thank Adam Smith for conversations related to the topic of this paper. This research is partially supported by the Frankel Center for Computer Science, and by the Israel Science Foundation (grant No. 860/06).

References

1. Ben-Or, M., Goldwasser, S., Wigderson, A.: Completeness theorems for noncryptographic fault-tolerant distributed computations. In: The 20th STOC, pp. 1–10 (1988)
2. Blum, A., Dwork, C., McSherry, F., Nissim, K.: Practical privacy: the SuLQ framework. In: The 24th PODS, pp. 128–138 (2005)
3. Chaum, D., Crépeau, C., Damgård, I.: Multiparty unconditionally secure protocols. In: The 20th STOC, pp. 11–19 (1988)
4. Chor, B., Kushilevitz, E.: A communication-privacy tradeoff for modular addition. Inform. Process. Lett. 45(4), 205–210 (1993)
5. Damgård, I., Fitzi, M., Kiltz, E., Nielsen, J.B., Toft, T.: Unconditionally secure constant-rounds multi-party computation for equality, comparison, bits and exponentiation. In: Halevi, S., Rabin, T. (eds.) TCC 2006. LNCS, vol. 3876, pp. 285–304. Springer, Heidelberg (2006)
6. Dinur, I., Nissim, K.: Revealing information while preserving privacy. In: The 22nd PODS, pp. 202–210 (2003)
7. Dwork, C.: Differential privacy. In: Bugliesi, M., Preneel, B., Sassone, V., Wegener, I. (eds.) ICALP 2006. LNCS, vol. 4052, pp. 1–12. Springer, Heidelberg (2006)
8. Dwork, C., Kenthapadi, K., McSherry, F., Mironov, I., Naor, M.: Our data, ourselves: Privacy via distributed noise generation. In: Vaudenay, S. (ed.) EUROCRYPT 2006. LNCS, vol. 4004, pp. 486–503. Springer, Heidelberg (2006)

9. Dwork, C., McSherry, F., Nissim, K., Smith, A.: Calibrating noise to sensitivity in private data analysis. In: Halevi, S., Rabin, T. (eds.) TCC 2006. LNCS, vol. 3876, pp. 265–284. Springer, Heidelberg (2006)
10. Dwork, C., McSherry, F., Talwar, K.: The price of privacy and the limits of LP decoding. In: 39th STOC, pp. 85–94 (2007)
11. Dwork, C., Nissim, K.: Privacy-preserving datamining on vertically partitioned databases. In: Franklin, M. (ed.) CRYPTO 2004. LNCS, vol. 3152, pp. 528–544. Springer, Heidelberg (2004)
12. Evfimievski, A., Gehrke, J., Srikant, R.: Limiting privacy breaches in privacy preserving data mining. In: The 22nd PODS, pp. 211–222 (2003)
13. Goldreich, O., Micali, S., Wigderson, A.: How to play any mental game. In: 19th STOC, pp. 218–229 (1987)
14. Kasiviswanathan, S., Lee, H.K., Nissim, K., Raskhodnikova, S., Smith, A.: What can we learn privately? (manuscript, 2007)
15. McSherry, F., Talwar, K.: Mechanism design via differential privacy. In: The 48th FOCS, pp. 94–103 (2007)
16. Nissim, K., Raskhodnikova, S., Smith, A.: Smooth sensitivity and sampling in private data analysis. In: The 39th STOC, pp. 75–84 (2007)
17. Warner, S.L.: Randomized response: A survey technique for eliminating evasive answer bias. Journal of the American Statistical Association 60(309), 63–69 (1965)
18. Yao, A.C.: Protocols for secure computations. In: The 23th FOCS, pp. 160–164 (1982)

A   A Composition Lemma

Assume an interactive protocol where a (deterministic) curator C makes adaptive queries to a party p holding a private input $x \in \{0, 1\}$. I.e., for $0 \leq i \leq \ell$, in round $2i$ the curator sends p a message $A_i = C(i, \mathcal{V}_1, \ldots, \mathcal{V}_{i-1})$ computed over the transcript of messages $\mathcal{V}_1, \ldots, \mathcal{V}_{i-1}$ already received from p, and specifying a randomized algorithm A_i; in round $2i+1$ party p computes $\mathcal{V}_i = A_i(x)$ (using fresh random coins for each A_i) and sends $\mathcal{V}_i$ to C.

Definition 4. *A possible outcome $\mathcal{V}$ is ϵ-good for algorithm A if $\Pr[A(1) = \mathcal{V}] \leq e^{\epsilon} \Pr[A(0) = \mathcal{V}]$, where the probabilities are taken over the randomness of algorithm A. An algorithm A is (ϵ, δ)-good if $\Pr[A(1)$ is ϵ-good for $A] \geq 1 - \delta$, where the probability is taken over the randomness of A.*

Assume that the range of C only includes (ϵ, δ)-good algorithms. Define a randomized algorithm $\hat{A}$ that simulates the interaction between p and C, i.e., given input $x \in \{0, 1\}$ it outputs a transcript $(A_1, \mathcal{V}_1, A_2, \mathcal{V}_2, \ldots, A_\ell, \mathcal{V}_\ell)$ sampled according to the protocol above.

Lemma 10. *$\hat{A}$ is $(\ell\epsilon, 1 - (1 - \delta)^{\ell})$-good.*

Proof. Note first, that with probability at least $(1 - \delta)^{\ell}$, the result of $\hat{A}(1)$ is a transcript $\hat{\mathcal{V}} = (A_1, \mathcal{V}_1, A_2, \mathcal{V}_2, \ldots, A_\ell, \mathcal{V}_\ell)$ such that $\mathcal{V}_i$ is ϵ-good for A_i for all $i \leq \ell$. It suffices, hence, to prove that when that happens the transcript $\hat{\mathcal{V}}$ is $\ell\epsilon$-good for $\hat{A}$, and indeed: $\Pr[\hat{A}(1) = (A_1, \mathcal{V}_1, A_2, \mathcal{V}_2, \ldots, A_\ell, \mathcal{V}_\ell)] = \prod_{i=1}^{\ell} \Pr[A_i(1) = \mathcal{V}_i] \leq \prod_{i=1}^{\ell} e^{\epsilon} \cdot \Pr[A_i(0) = \mathcal{V}_i] = e^{\ell\epsilon} \cdot \Pr[\hat{A}(0) = (A_1, \mathcal{V}_1, A_2, \mathcal{V}_2, \ldots, A_\ell, \mathcal{V}_\ell)]$.   □

New Efficient Attacks on Statistical Disclosure Control Mechanisms

Cynthia Dwork and Sergey Yekhanin

Microsoft Research
{dwork,yekhanin}@microsoft.com

Abstract. The goal of a statistical database is to provide statistics about a population while simultaneously protecting the privacy of the individual records in the database. The tension between privacy and usability of statistical databases has attracted much attention in statistics, theoretical computer science, security, and database communities in recent years. A line of research initiated by Dinur and Nissim investigates for a particular type of queries, lower bounds on the distortion needed in order to prevent gross violations of privacy. The first result in the current paper simplifies and sharpens the Dinur and Nissim result.

The Dinur-Nissim style results are strong because they demonstrate insecurity of all low-distortion privacy mechanisms. The attacks have an all-or-nothing flavor: letting n denote the size of the database, $\Omega(n)$ queries are made before anything is learned, at which point $\Theta(n)$ secret bits are revealed. Restricting attention to a wide and realistic subset of possible low-distortion mechanisms, our second result is a more acute attack, requiring only a fixed number of queries for each bit revealed.

1 Introduction

The goal of a statistical database is to provide statistics about a population while simultaneously protecting the privacy of the individual records in the database. A natural example that highlights the tension between usability and preserving privacy is a hospital database containing medical records of the patients. On one hand, the hospital would like allow medical research that is based on the information in the database. On the other hand, the hospital is legally obliged to protect the privacy of its patients, i.e., leak no information regarding the medical condition of any specific patient that can be "traced back" to the individual.

The tension between privacy and usability of statistical databases has attracted considerable attention in the statistics, theoretical computer science, cryptography, security, and database communities since late 1970s. There is a a vast body of work on this subject (for references, see [1, 6, 10, 24, 25, 26]). However, the formal treatment of privacy has generally been unsatisfactory, either due to lack of specificity or because the notion of privacy compromise was not sufficiently general to capture many forms of leakage that ordinary people would still find unacceptable, or the schemes ensure security only against certain specific class of attacks.

D. Wagner (Ed.): CRYPTO 2008, LNCS 5157, pp. 469–480, 2008.
© International Association for Cryptologic Research 2008

In a seminal paper [12] Dinur and Nissim initiated a rigorous study of the trade-off between privacy and usability. They focused on a class of techniques that Adam and Wortmann, in their encyclopedic 1989 survey [1] of statistical disclosure control methods, call *output perturbation*. Roughly speaking, a query is a function that maps the database to a (real) number, and an output perturbation statistical disclosure control mechanism (curator) simply adds noise to the answers. Thus, the true answer of (say) 1910, may be reported as 1914 or 1907. The degree of distortion, that is, the magnitude of the noise, is an important measure of the utility of the mechanism. Dinur and Nissim formulated and investigated the question of how large the noise magnitude needs to be in order to ensure privacy in the following setting: each of the n rows in the database is a single bit, a query is specified by naming a subset $S \subseteq [n]$ of the rows, and the true answer to the query is the number of 1's in the specified set of rows: $\sum_{i \in S} d_i$, where d_i is the bit in the ith row, $1 \leq i \leq n$. They demonstrated a powerful attack on database curators, and concluded that *every* database curator that gives too accurate answers to too many queries inherently leaks private information [12].

The negative results of [12] have been highly influential. On one hand, they were a catalyst for a fruitful direction of obtaining provably secure statistical disclosure control mechanisms [3, 4, 9, 12, 16, 18, 20, 21, 23]. Provably secure mechanisms are now available [3, 9, 16, 18] for many standard data-mining tasks such as singular value decomposition, k-means clustering, principal component analysis, the perceptron algorithm, and contingency table release, several tasks in learning theory [4, 9, 20], and distributed implementations of these [15]. All these mechanisms have the property, shown to be inherent by [12], the magnitude of the noise increases with the number of questions asked. On the other hand, the results of Dinur and Nissim are important, unexpected, and rather disappointing for many research statisticians who often assumed, or at least hoped, that privacy can be achieved via a hypothetical clever "one-shot" procedure, that would turn the database into a sanitized object, permitting significantly accurate answers to be derived for queries that are not specified on the outset, without a risk of privacy violation.

In this paper we continue the line of research initiated by Dinur and Nissim. Our first result in the current paper simplifies and sharpens the Dinur and Nissim result. Our second result shows a limitation for a type of privacy mechanisms that includes tabular data release and synthetic data sets. We show a class of queries for which even adding *arbitrary* noise to a $(1/2-\epsilon)$ fraction of the answers fails to protect privacy against an adversary running in time independent of the database size. Thus, no mechanism of the specified type can safely provide very accurate answers to even a $(1/2 + \epsilon)$ fraction of these queries.

Before stating our contributions formally we discuss the results of Dinur and Nissim [12] and Dwork et al. [17] in more detail.

1.1   Results and Earlier Work

In an interactive privacy mechanism the database curator is a trusted entity that sits between the possibly adversarial user of the database and the actual data. Given a query, the curator computes the correct answer and adds some noise to the response. When the database is a vector of bits a mechanism is *blatantly non-private* if, after interacting with a database curator, an adversary can produce a candidate database c that agrees with the real database on all but $o(n)$ entries, i.e., $d_i = c_i$ for all but $o(n)$ values of $1 \le i \le n$. This model, while at first blush simplistic, is in fact sufficiently rich to capture many natural questions. A detailed discussion of the model can be found in [12, 14, 17].

Dinur and Nissim [12] showed that a mechanism in which curator that adds $o(\sqrt{n})$ noise to every response is blatantly non-private against a polynomial-time bounded adversary asking $O(n \log^2 n)$ questions[1].

At a high level, the attack of [12] proceeds in two steps. In the first step the adversary poses $O(n \log^2 n)$ random subset-sum queries, chosen by including each database record uniformly and independently with probability $1/2$. In the second step the adversary solves a linear system of equations with n variables and $O(n \log^2 n)$ constraints in order to find a candidate database that fits all available data. The second step of the attack carries most of the computational burden. The most efficient linear programming algorithm to date is due to Pravin M. Vaidya [27]. It requires $O(((m+n)n^2 + (m+n)^{1.5}n)L)$ arithmetic operations where m is the number of constraints, n is the number of variables, and L is bounded by the number of bits in the input. In the setting of [12] this yields an $O(n^5 \log^4 n)$ worst case running time.

Our first result sharpens the Dinur-Nissim attack. The new attack requires only n deterministically chosen queries, requires significantly less computation. Also of value, our analysis is much simpler, relying only on basic properties of the Fourier transform over the group $\mathbb{Z}_2^k$.

The key message of the Dinur-Nissim work is that any database curator that gives reasonably accurate answers to too many queries leaks private information. This however leaves open a possibility that some curator giving wildly inaccurate answers to a (small) fraction of the queries, and reasonably accurate answers to the rest may preserve privacy. Existence of such curators was studied by Dwork et al. [17], who have showed that if the query is now a vector of n standard normals, and the true answer is the inner product of the database with the vector, then any database mechanism adding noise bounded by $o(\sqrt{n})$ to at least 0.761 fraction of its responses is blatantly non-private[2]. Inspired by the LP decoding methods from the literature on compressed sensing of signals, e.g. [5, 7, 8], this attack also requires solving a random linear program with n variables and $O(n)$ constraints, and so has a worst case running time of $O(n^5)$.

Although the actual constant (≈ 0.761) is shown to be sharp threshold for LP decoding [17], other attacks may be possible. Indeed, it is plausible that every statistical disclosure control mechanism that adds low noise to $(1/2 + \epsilon)$

---

[1] Dinur and Nissim also showed that if the adversary can ask 2^n queries then the mechanism is blatantly non-private as long as the noise is magnitude is $o(n)$; however, here we restrict our attention to efficient adversaries.

[2] We think of this as a natural generalization of the Dinur-Nissim attack, in which the query vector is restricted to having binary entries.

fraction of its responses (and allows for sufficiently many queries) leaks private information, for any $\epsilon > 0$. Dwork et al. [17] have made a step towards proving this claim. Namely, they came up with an inefficient (i.e., $\exp(n)$-time) adversary that asks $O(n)$ questions from the curator and achieves blatant nonprivacy, in case the curator gives reasonably accurate responses to $(1/2 + \epsilon)$ fraction of queries.[3]

In our second result we address the question of whether a mechanism that adds unbounded noise to a $(1/2 - \epsilon)$ fraction of its responses can ensure privacy, and prove the contrary for a certain range of parameters. We obtain an attack running in $\text{poly}(e/\epsilon)$ time that can tolerate the optimal $(1/2 - \epsilon)$ fraction of unbounded noise provided the noise on the rest of the queries is at most e. As in the case of the previous attacks, the query is a vector of length n, and the true answer is the inner product of the (binary) database and the query vector. Note that the running time is independent of n; we are not counting the time needed to formulate the query (not hard, but depends on n) and to compute the response.

Note that one needs to be careful when specifying a fraction of queries to which a certain curator adds unbounded (or low) noise since curators can be of very different nature. In particular some curators may allow only for a certain (small) number of queries, and some may give different answers to the same query asked two times in a row.

Our attack applies to database curators that for certain values of p, given a (randomized) p-sized collection of queries coming from a 2-independent family add low noise to $(1/2 + \epsilon)p$ of their responses with a probability bounded away from $1/2$.

The class of curators that fall into the above category is quite broad. Specifically (as we later prove) it includes curators that may study the database and release an "object" that the adversary/analyst can study as desired. This captures, for example, completely non-interactive solutions such as: summary tables, statistics, synthetic data sets, and all other known forms of statistical data release in use today, but it also includes (hypothetical) programs for handling certain classes of queries with obfuscated data hard-wired in. Our model also (obviously) captures interactive curators (i.e., curators that keep a query log and adjust their responses to incoming queries based on such a log) that allow for p queries, and add unbounded noise to at most $(1/2 - \epsilon)p$ responses.

Our attack has important differences from the earlier attacks in the literature:

One Bit at a Time. Conceptually, our adversary attacks one bit at a time. That is, the adversary chooses a bit to attack and runs a simple program in which it forms queries and interacts with a database curator in order to obtain a

---

[3] Note that there are database curators that reveal no information about specific database records and give correct answers to exactly $1/2$ of the queries. For instance, consider a database curator that answers one half of the queries according to a database x and the other half of the queries according to a complement database $\bar{x}$. Clearly, an interaction with such a curator will not help an adversary distinguish an all-zeros database from an all-ones database.

(potentially wildly) noisy version of the true answer. The adversary can increase its success probability by running the attack multiple times. The adversary can attack the entire database by running the attack n times.

Small Noise is Very Small. The magnitude of the noise on the $(1/2+\epsilon)$ fraction of "good" responses will be bounded by something *smaller than* the maximum allowable coefficient in the query vector. This is the weakest aspect of our result. However, prior to this work no efficient attack was known even when a $(1/2+\epsilon)$ fraction of the responses have *zero* noise.

Viewed differently, the result says that if the "good" responses must, for reasons of utility, have noise bounded by some number p, then the system cannot safely permit $O(p)$ subset sum queries with coefficients even as large as $2p+1$.

Our attack is based on a new interplay between the basic properties of polynomials over reals and ideas coming from the theory of error-correcting codes [22].

2   Preliminaries

We start with the basic definitions. A database for us is simply an n-bit string $d = (d_1, \ldots, d_n) \in \{0,1\}^n$.

Definition 1. *Let d be an n-bit database. A query is a vector $q \in \mathbb{R}^n$. The true answer to a query q is the inner product $q \cdot d$, i.e., the weighted sum of database bits $a_q = \sum_{i \in q} q_i d_i$. A disclosure control mechanism $\mathcal{C}$ takes as input a query q and database d and provides a possibly noisy response in $\mathbb{R}$, for which $\mathcal{C}$ may employ randomness. We say that a response $\mathcal{C}(x, q)$ carries noise of at most σ if $|\mathcal{C}(x, q) - a_q| \leq \sigma$.*

The following formalization of *non-privacy*, due to Dinur and Nissim [12], has come to be called *blatant non-privacy*.

Definition 2. *Let $\mathcal{C}$ be a privacy mechanism. We say that $\mathcal{C}$ is blatantly non-private against a probabilistic algorithm $\mathcal{A}$ (an adversary) if after an interaction with $\mathcal{C}$, $\mathcal{A}$ recovers most of the database d with very high probability. Formally, for all $d \in \{0,1\}^n$,*

$$Pr[\mathcal{A}^{\mathcal{C}} \text{ outputs } y \in \{0,1\}^n \text{ such that } d_H(d, y) \in o(n)] \geq 1 - neg(n),$$

where the probability is taken over the randomness of $\mathcal{A}$, $neg(n)$ denotes a function that is asymptotically smaller than any inverse polynomial in n, and $d_H(x, y)$ stands for the Hamming distance.

Definition 3. *We also say that $\mathcal{C}$ is $(1 - \delta)$-non-private against an adversary $\mathcal{A}$ if for an arbitrary $i \in [n]$, $\mathcal{A}$ can recover the value of the bit d_i after an interaction with $\mathcal{C}$ with probability $1 - \delta$. Formally, $\forall d \in \{0,1\}^n$, $\forall 1 \leq i \leq n$,*

$$Pr[\mathcal{A}^{\mathcal{C}}(i) \text{ generates } z \in \{0,1\} \text{ such that } z = d_i] \geq 1 - \delta,$$

where the probability is taken over the random coin tosses of $\mathcal{C}$ and $\mathcal{A}$.

Clearly, the the definition above is useful only if $\delta < 1/2$. Note that the definition is very strong. It says that the curator $\mathcal{C}$ fails to protect *every* database record.

We measure the complexity of an attack on a statistical disclosure control mechanism with respect to: 1. the number of queries asked by an adversary; 2. the running time of an adversary. Our attacks (and all other attacks in the literature) are non-adaptive. They proceed in two steps. First an adversary asks all its questions from a curator; next the adversary processes the curator's responses in order to reveal some private information. We define the time complexity of an attack to be the time complexity of the second step.

3   Fourier Attack: $o(\sqrt{n})$ Noise, n Queries, $O(n \log n)$ Running Time

The goal of this section is to establish the following theorem.

Theorem 1. *There exists an adversary $\mathcal{A}$ that runs in $O(n \log n)$ time and achieves a blatant privacy violation against any database curator $\mathcal{C}$ that allows for n queries with integer $\{0, 1\}$ coefficients and adds $o(\sqrt{n})$ noise to every response.*

Our proof of theorem 1 relies on some standard properties of the Fourier transform over the finite group $\mathbb{Z}_2^k$. In the following subsection we briefly review the properties that are important for us.

3.1   Fourier Preliminaries

Characters of $\mathbb{Z}_2^k$ are homomorphisms from $\mathbb{Z}_2^k$ into the multiplicative group $\{\pm 1\}$. There exist 2^k characters. We denote characters by χ_a, where $a = (a_1, \ldots, a_k)$ ranges in $\mathbb{Z}_2^k$, and set $\chi_a(x) = (-1)^{\sum_{i=1}^n a_i x_i}$ for every $x = (x_1, \ldots, x_k) \in \mathbb{Z}_2^k$. Let $f(x)$ be a function from $\mathbb{Z}_2^k$ into reals. For an arbitrary $a \in \mathbb{Z}_2^k$ the Fourier coefficient $\hat{f}(\chi_a)$ is defined by $\hat{f}(\chi_a) = \sum \chi_a(x) f(x)$, where the sum is over all $x \in \mathbb{Z}_2^k$. For every $a \in \mathbb{Z}_2^k$ consider a set

$$S_a = \left\{ x \in \mathbb{Z}_2^k \ \Big| \ \sum_{i=1}^k a_i x_i = 0 \mod (2) \right\}. \tag{1}$$

It is easy to see that the size of S_a is equal to 2^k if $a = 0^k$, and is equal to 2^{k-1} otherwise. For every $a \in \mathbb{Z}_2^k$ consider the sum

$$\sigma_a(f) = \sum_{x \in S_a} f(x). \tag{2}$$

The Fourier coefficients of a function f can be easily expressed in terms of sums $\sigma_a(f)$:

$$\hat{f}(\chi_a) = \begin{cases} \sigma_0(f), & \text{if } a = 0; \\ 2\sigma_a(f) - \sigma_0(f), & \text{otherwise.} \end{cases} \tag{3}$$

Let $\hat{f} = (\hat{f}(\chi_a))_{a \in \mathbb{Z}_2^k}$ be a vector of Fourier coefficients of f. Consider a matrix $H \in \{\pm 1\}^{2^k \times 2^k}$. Rows and columns of H are labelled by elements of $\mathbb{Z}_2^k$ (taken in the same order). $H_{a,b} = \chi_a(b)$. H is a (Sylvester type) Hadamard matrix. It is not hard to verify that $HH = 2^k I$, where I is the identity matrix. Note that $\hat{f} = Hf$. Therefore

$$f = \frac{1}{2^k} H \hat{f}, \tag{4}$$

i.e., an inverse of a Fourier transform is simply another Fourier transform up to a scalar multiplication. The following classical (Parseval's) identity relates the absolute values of f to the absolute values of the Fourier coefficients of f :

$$\sum_{x \in \mathbb{Z}_2^k} |f(x)|^2 = \frac{1}{2^k} \sum_{a \in \mathbb{Z}_2^k} |\hat{f}(\chi_a)|^2. \tag{5}$$

3.2   The Attack

Proof of theorem 1. Let $d = (d_1, \ldots, d_n)$ be the database. Without a loss of generality assume that n is a power of two, $n = 2^k$. Consider an arbitrary bijection $g : \mathbb{Z}_2^k \to [n]$ between the group $\mathbb{Z}_2^k$ and the set $[n]$. Now database d defines a map f from $\mathbb{Z}_2^k$ to $\{0, 1\}$, where we set $f(x) = d_{g(x)}$, for every $x \in \mathbb{Z}_2^k$. Our attack proceeds in three steps. Firstly, the adversary $\mathcal{A}$ asks n queries from the curator $\mathcal{C}$ to obtain the noisy version of sums $\sigma_a(f)$. Secondly, $\mathcal{A}$ performs a simple computation to derive noisy Fourier coefficients of f from the curator's responses. Finally, $\mathcal{A}$ performs an inverse Fourier transform to (approximately) recover the function f from its noisy Fourier spectrum. Below is a more formal description.

-   For every $a \in \mathbb{Z}_2^k$, $\mathcal{A}$ asks for the sum of database bits in the set S_a, where S_a is defined by formula (1). $\mathcal{A}$ obtains the noisy values $\tilde{\sigma}_a(f)$ of sums $\sigma_a(f)$. Note that for every $a \in \mathbb{Z}_2^k$ we have $\tilde{\sigma}_a(f) = \sigma_a(f) + o(\sqrt{n})$.
-   $\mathcal{A}$ uses formula (3) to obtain a vector $\tilde{f}$ of noisy Fourier coefficients of f. Note that $\tilde{f} = \hat{f} + e$, where the absolute value of every coordinate of e is bounded by $o(\sqrt{n})$.
-   $\mathcal{A}$ applies formula (4) to obtain a noisy version of f from $\tilde{f}$. Specifically, $\mathcal{A}$ computes $h = \frac{1}{n} H \tilde{f}$, and for every coordinate $i \in [n]$, sets $y_i = 0$ if $h_i < 1/2$ and $y_i = 1$ otherwise.

Note that there are $O(n \log n)$ time algorithms to compute Fourier transform [11]. Therefore the overall running time of the attack is $O(n \log n)$. We now argue that the attacker always recovers a database y such that $d_H(d, y) = o(n)$. The linearity of the Fourier transform implies that it would suffice for us to show that the vector $\frac{1}{n} He$, can not have $\Omega(n)$ coordinates with absolute values above $1/2$. This follows immediately from the Parseval's identity (5) that asserts that the L_2 norm of e is n times larger than the L_2 norm of $\frac{1}{n} He$. $\square$

3.3   Summary of First Result

We presented a novel attack on statistical disclosure control mechanism that applies in the model considered earlier by Dinur and Nissim [12]. We believe that the most important feature of our attack is its conceptual simplicity; in addition, it is sharper than that of Dinur and Nissim [12] in the following respects:

- Our adversary makes fewer queries (n versus $O(n \log^2 n)$).
- Both algorithms first pose queries and then analyze the results. Our analysis is computationally more efficient ($O(n \log n)$ vs $\Omega(n^5 \log^4 n)$ worst case running time).
- Our adversary always achieves blatant non-privacy; previous attacks have a negligible probability of failure.

4   Interpolation Attack: $(1/2 - \epsilon)$ Fraction of Unbounded Noise, Poly(e/ϵ) Running Time

In this section the query vectors will have integer coefficients chosen from some range $[0, \ldots, p-1]$. Our goal is to establish the following theorem.

Theorem 2. *Let p be a prime. Suppose $0 < \epsilon \leq 1/2$, $e \geq 0$, and $\delta < 1/2$ are such that*

$$2\sqrt{(p-1)/\delta} + 8e + 3 \leq 2\epsilon(p-1) \quad and \quad e < (p-1)/8; \tag{6}$$

then any curator $\mathcal{C}$ that given a (randomized) $(p-1)$-sized collection of queries coming from a 2-independent family adds noise less than or equal to e to at least $(1/2 + \epsilon)(p-1) - \sqrt{(p-1)/\delta}$ of its responses with probability $(1 - \delta)$ is $(1 - \delta)$-non-private against an adversary that asks $p - 1$ queries and runs in $O(p^4)$ time.

We defer the discussion of the type of curators that are vulnerable to the attack above till later in this section, and we defer the proof of theorem 2 to the following subsection. Below we state some of the immediate corollaries of the theorem. The next corollary captures the asymptotic parameters of the attack from theorem 2. To obtain it, one simply needs to set $\delta = 1/4$ and use crude estimates for p to satisfy (6).

Corollary 1. *Let $0 < \epsilon \leq 1/2$ and $e \geq 0$ be arbitrary. Let p be a prime such that $p \geq 20/\epsilon^2$ and $p \geq 15e/\epsilon$. Suppose a database curator $\mathcal{C}$ allows queries with integer weights from $[0, \ldots, p-1]$. Also, assume that given a (randomized) $(p-1)$-sized collection of queries coming from a 2-independent family $\mathcal{C}$ adds noise less than or equal to e to at least $(1/2 + \epsilon)(p-1) - \sqrt{(p-1)/\delta}$ of its responses with probability $3/4$. Then $\mathcal{C}$ is $3/4$-non-private against an adversary that issues $O(p)$ queries and runs in $O(p^4)$ time.*

The corollary above may be somewhat unsatisfying since the adversary has a substantial $(1/4)$ probability of failing to correctly reveal private information. Note however, that (assuming the curator allows for more queries) the adversary can run its attack multiple times, to obtain independent estimates $y_i^{(1)}, \ldots, y_i^{(t)}$ for a ceratin specific bit d_i of the database d. Next the adversary can report the majority of $\{y_i^{(j)}\}_{j \in [t]}$ as a new estimate for d_i. A standard argument based on the Chernoff bound [2] shows that the new estimate has a vastly lower probability of an error.

Corollary 2. *Let $0 < \epsilon \leq 1/2$ and $e \geq 0$ be arbitrary. Let p be a prime such that $p \geq 20/\epsilon^2$ and $p \geq 15e/\epsilon$. Suppose a database curator $\mathcal{C}$ allows queries with integer weights from $[0, \ldots, p-1]$. Also, assume that given a (randomized) $(p-1)$-sized collection of queries coming from a 2-independent family $\mathcal{C}$ adds noise less than or equal to e to at least $(1/2 + \epsilon)(p-1) - \sqrt{(p-1)/\delta}$ of its responses with probability $3/4$. Then for every integer $t \geq 1$, $\mathcal{C}$ is $(1 - 2^{-t/12})$-non-private against an adversary that issues $O(tp)$ queries and runs in $O(tp^4)$ time.*

We now argue that the condition of theorem 2 (and corollaries 1 and 2) holds for non-interactive database curators whose responses to more than $(1/2+\epsilon)$ fraction of all possible queries carry low noise. Our argument relies on the following lemma that gives a well-known property of pairwise independent samples. The lemma follows from the fact that for pairwise independent random variables, the variance is the sum of the variances, and the Chebychev's inequality [19, lemma 2].

Lemma 1. *If S is a pairwise independent sample of elements from some domain D and I maps elements of D to the range $\{0, 1\}$; then for any $\delta > 0$,*

$$\Pr\left[\left|\frac{\sum_{x \in S} I(x)}{|S|} - E[I(x)]\right| \geq 1/\sqrt{\delta|S|}\right] \leq \delta.$$

Let $\mathcal{C}$ be a database curator such that $\mathcal{C}$'s responses to more than $(1/2 + \epsilon)$ fraction of all possible queries carry low noise. Let D be the domain of all possible queries and $I(w) : \{0, 1\}^n \to \{0, 1\}$ to be the incidence function of the set of queries that carry unbounded noise according to $\mathcal{C}$. Clearly, $E[I(x)] \leq 1/2 - \epsilon$. Therefore lemma 1 implies that with probability at least $1 - \delta$ the total number of points that carry unbounded noise in a random sample S of size $p - 1$ is at most $(1/2 - \epsilon)(p-1) + \sqrt{(p-1)/\delta}$ and theorem 2 applies.

We note that theorem 2 is weak in that the small noise is very small – considerably less than the maximum allowable coefficient in a query. In fact, this noise model even rules out a privacy mechanism that protects a single bit, say, d_i, by "flipping" it – replacing d_i with its complement $1 - d_i$, and then answering all queries accurately thereafter. On the other hand, to our knowledge, this is the first efficient adversary that successfully attacks any mechanism that can add arbitrary noise in a $(1/2 - \epsilon)$ fraction of the responses.

4.1   The Attack

The main idea behind the proof of Theorem 2 is to achieve error-correction via polynomial interpolation. This idea has been previously extensively used (in a related, yet distinct setting) of local decoding of Reed-Muller codes [19, 22]. Below is a high-level overview of our attack.

The attacker $\mathcal{A}$ thinks of its queries as points $q = (q_1, \ldots, q_n) \in \mathbb{F}_p^n$ in an n-dimensional linear space over a prime field $\mathbb{F}_p$. $\mathcal{A}$ reduces all responses to its queries modulo p, and treats the database $d = (d_1, \ldots, d_n) \in \{0, 1\}^n$ as an unknown n-variate linear form $f(q_1, \ldots, q_n) = \sum_{i=1}^{n} d_i q_i$ over $\mathbb{F}_p$. It is easy to see that in order to recover (say) the first bit of d, it would suffice for $\mathcal{A}$ to determine the value of $f(q)$ for $q = ((p-1)/2, 0, \ldots, 0)$ with an error of less than $(p-1)/4$.

The attacker does not directly ask the curator for the value of $f(q)$, since the response to the query q may carry unbounded noise (and therefore be misleading), but rather issues a randomized collection of $(p-1)$ queries $q^{(1)}, \ldots, q^{(p-1)}$ such that the value of $f(q)$ can (with high probability) be deduced from curator's responses to $q^{(1)}, \ldots, q^{(p-1)}$ up to a small error.

Below is the formal description and analysis of our attack. The attacker's goal is to obtain an approximately correct answer to $q = ((p-1)/2, 0, \ldots, 0)$ and thus recover the first bit of the database.

Proof of theorem 2

- $\mathcal{A}$ picks $u, v \in \mathbb{F}_p^n$ uniformly at random, and considers the parametric degree two curve $\chi = \{q + tu + t^2 v \mid t \in [1, \ldots, p-1]\}$ in the space $\mathbb{F}_p^n$ through the point q. Note that the points of χ form a pairwise independent sample of $\mathbb{F}_p^n$. A proof of this standard fact can be found for instance in [19, claim 1]. The condition of the theorem implies that with probability at least $1 - \delta$ the total number of points that carry unbounded noise on χ is at most $(1/2 - \epsilon)(p-1) + c\sqrt{p-1}$.

- $\mathcal{A}$ issues $p-1$ queries $\{q^{(t)} = q + tu + t^2 v\}_{t \in [1, \ldots, p-1]}$ corresponding to points of χ. Let $R = (r_1, \ldots, r_{p-1})$ be a sequence of curator's responses to those queries reduced modulo p. In what follows we assume the attacker $\mathcal{A}$ is lucky and at most $(1/2-\epsilon)(p-1)+c\sqrt{p-1}$ responses $\{r_t\}_{t \in [p-1]}$ carry unbounded noise.

  We say that $\alpha \in \mathbb{F}_p$ is e-small if either $\alpha \in [-e, \ldots, 0, \ldots, e] \mod (p)$.
  We say that a polynomial $h(t) \in \mathbb{F}_p[t]$ fits the sequence R if $(h(t) - r_t)$ is e-small for $(1/2 + \epsilon)(p-1) - c\sqrt{p-1}$ values of t. Note that the degree two polynomial $g(t) = f(q + tu + t^2 v) \in \mathbb{F}_p[t]$, that is a restriction of the linear function f to a degree two curve χ fits R. Also note that $g(0) = f(q)$ is 0 if the first bit of the database x is zero, and $(p-1)/2$ otherwise.
  We now argue that $g(t)$ is the only polynomial that fits R up to a $2e$-small additive factor. To see this suppose that some other polynomial $g_1(t)$ also fits R; then the polynomial $g(t) - g_1(t)$ has to take $2e$-small values at

$$(p-1) - 2((1/2 - \epsilon)(p-1) + c\sqrt{p-1}) = 2\epsilon(p-1) - 2c\sqrt{p-1} \geq 8e + 3$$

points in $\mathbb{F}_p^*$, where the inequality above follows from (6). Since a (non-constant) quadratic polynomial can take the same value in at most two points and $2(4e + 1) < 8e + 3$ we conclude that $g(t) - g_1(t)$ is a $2e$-small constant in $\mathbb{F}_p$.

- $\mathcal{A}$ applies a brute-force search over all quadratic polynomials $g_1(t) \in \mathbb{F}_p[t]$ to find a polynomial that fits the sequence R. $\mathcal{A}$ computes the value $g_1(0)$ and outputs 0 if $g_1(0)$ is $2e$-small, and 1 otherwise. According to (6), $2e < (p-1)/4$ and therefore $\mathcal{A}$ correctly recovers the first bit of the database.

Observe that the running time of the attack is $O(p^4)$. The attack involves a brute-force search over all $O(p^3)$ quadratic polynomials in $\mathbb{F}_p[t]$ and it takes $O(p)$ time to verify if a polynomial fits the sequence R. $\square$

4.2   Summary of Our Second Result

We presented a novel, efficient, attack on statistical disclosure control mechanisms with several unique features. The most novel feature of the attack is the use of polynomial interpolation in this context. The most interesting consequence of the attack is that it succeeds against privacy mechanisms that add unbounded noise to up to $(1/2 - \epsilon)$ fraction of their responses, provided the noise on other responses is sufficiently low; indeed, it even tolerates a small amount of noise in the remaining $(1/2 + \epsilon)$ responses. No efficient attacks with such property have been known earlier.

Acknowledgement

The authors thank Madhu Sudan for helpful discussions regarding the polynomial interpolation attack, and for providing the reference [19].

References

1. Adam, N., Wortmann, J.: Security-control methods for statistical databases: a comparative study. ACM Computing Surveys 21(4), 515–556 (1989)
2. Alon, N., Spencer, J.: The probabilistic method, 2nd edn. Wiley-Interscience [John Wiley and sons], New York (2000)
3. Barak, B., Chaudhuri, K., Dwork, C., Kale, S., McSherry, F., Talwar, K.: Privacy, accuracy, and consistency too: a holistic solution to contingency table release. In: Proc. of the 26th Symposium on Principles of Database Systems (PODS), pp. 273–282 (2007)
4. Blum, A., Ligett, K., Roth, A.: A learning theory approach to non-interactive database privacy. In: Proc. of the Symp. on the Theory of Computation (STOC) (2008)
5. Candes, E., Tao, T.: Near-optimal signal recovery from random projections: universal encoding strategies. IEEE Trans. Inform. Theory 52, 5406–5425 (2004)
6. Chawla, S., Dwork, C., McSherry, F., Smith, A., Wee, H.: Toward privacy in public databases. In: Kilian, J. (ed.) TCC 2005. LNCS, vol. 3378, pp. 363–385. Springer, Heidelberg (2005)

7. Chen, S., Donoho, D., Saunders, M.: Atomic decomposition via basis pursuit. SIAM Journal on Scientific Computing 48(1), 33–61 (1999)
8. Donoho, D., Johnstone, I.: Minimax estimation via wavelet shrinkage. Annals of Statistics 26(3), 879–921 (1998)
9. Blum, A., Dwork, C., McSherry, F., Nissim, K.: Practical privacy: the SuLQ framework. In: Proc. of the 24th Symposium on Principles of Database Systems (PODS), pp. 128–138 (2005)
10. Clifton, C., Kantarcioglu, M., Vaidya, J., Lin, X., Zhu, M.: Tool for privacy preserving data minining. SIGKDD Explorations 4(2), 28–34 (2002)
11. Cormen, T., Leiserson, C., Rivest, R., Stein, C.: Introduction to algorithms. MIT Press, Cambridge (2001)
12. Dinur, I., Nissim, K.: Revealing information while preserving privacy. In: Proc. of the 22nd Symposium on Principles of Database Systems (PODS), pp. 202–210 (2003)
13. Dwork, C.: Differential privacy. In: Bugliesi, M., Preneel, B., Sassone, V., Wegener, I. (eds.) ICALP 2006. LNCS, vol. 4052, pp. 1–12. Springer, Heidelberg (2006)
14. Dwork, C.: Ask a better question, get a better answer: a new approach to private data analysis. In: Schwentick, T., Suciu, D. (eds.) ICDT 2007. LNCS, vol. 4353, pp. 18–27. Springer, Heidelberg (2006)
15. Dwork, C., Kenthapadi, K., McSherry, F., Mironov, I., Naor, M.: Our data, ourselves: privacy via distributed noise generation. In: Vaudenay, S. (ed.) EUROCRYPT 2006. LNCS, vol. 4004, pp. 486–503. Springer, Heidelberg (2006)
16. Dwork, C., McSherry, F., Nissim, K., Smith, A.: Callibrating noise to sensitivity in private data analysis. In: Halevi, S., Rabin, T. (eds.) TCC 2006. LNCS, vol. 3876, pp. 265–284. Springer, Heidelberg (2006)
17. Dwork, C., McSherry, F., Talwar, K.: The price of privacy and the limits of LP decoding. In: Proc. of the 39th Symposium on the Theory of Computation (STOC), pp. 85–94 (2007)
18. Dwork, C., Nissim, K.: Privacy preserving data-mining on vertically partitioned databases. In: Franklin, M. (ed.) CRYPTO 2004. LNCS, vol. 3152, pp. 528–544. Springer, Heidelberg (2004)
19. Gemmell, P., Sudan, M.: Highly resilient correctors for polynomials. Information Processing Letters 43(4), 169–174 (1992)
20. Kasiviswanathan, S., Lee, H., Nissim, K., Raskhodnikova, S., Smith, A.: What Can We Learn Privately? (manuscript, 2007)
21. McSherry, F., Talwar, K.: Mechanism Design via Differential Privacy. In: Proc. of the 48th Symposium on the Foundations of Computer Science (FOCS) (2007)
22. MacWilliams, F., Sloane, N.: The theory of error-correcting codes. North-Holland, Amsterdam (1977)
23. Nissim, K., Raskhodnikova, S., Smith, A.: Smooth sensitivity and sampling in private data analysis. In: Proc. of the 39th Symposium on the Theory of Computation (STOC), pp. 75–84 (2007)
24. Slavkovic, A.: Statistical disclosure limitation beyond the margins: characterization of joint distributions for contingency tables. Ph.D. thesis, Department of statistics, Carnegie Mellon University (2004)
25. Shoshani, A.: Statistical databases: Characteristics, problems and some solutiuons. In: Proc. of the 8th International Conference on Very Large Databases (VLDB), pp. 208–222 (1982)
26. Sweeney, L.: Privacy-enchanced linking. SIGKDD Explorations 7(2), 72–75 (2005)
27. Vaidya, P.: An algorithm for linear programming which requires $O(((m + n)n^2 + (m + n)^{1.5}n)L)$ arithmetic opertaions. Mathematical Programming 47, 175–201 (1990)

Efficient Secure Linear Algebra in the Presence of Covert or Computationally Unbounded Adversaries

Payman Mohassel[*] and Enav Weinreb[**]

Abstract. In this work we study the design of secure protocols for linear algebra problems. All current solutions to the problem are either inefficient in terms of communication complexity or assume that the adversary is honest but curious. We design protocols for two different adversarial settings: First, we achieve security in the presence of a covert adversary, a notion recently introduced by [Aumann and Lindell, TCC 2007]. Roughly speaking, this guarantees that if the adversary deviates from the protocol in a way that allows him to cheat, then he will be caught with good probability. Second, we achieve security against arbitrary malicious behaviour in the presence of a computationally unbounded adversary that controls less than a third of the parties. Our main result is a new upper bound of $O(n^{2+1/t})$ communication for testing singularity of a shared $n \times n$ matrix in constant round, for any constant t in both of these adversarial environments. We use this construction to design secure protocols for computing the rank of a shared matrix and solving a shared linear system of equations with similar efficiency.

We use different techniques from computer algebra, together with recent ideas from [Cramer, Kiltz, and Padró, CRYPTO 2007], to reduce the problem of securely deciding singularity to the problem of securely computing matrix product. We then design new and efficient protocols for secure matrix product in both adversarial settings. In the two-party setting, we combine cut-and-choose techniques on random additive decomposition of the input, with a careful use of the random strings of a homomorphic encryption scheme to achieve simulation-based security. Thus, our protocol avoids general zero-knowledge proofs and only makes a black-box use of a homomorphic encryption scheme.

1 Introduction

Solving a set of linear equations is one of the most basic algorithmic tasks, with numerous applications to various fields. In a distributed system, linear constraints may reflect sensitive information and thus parties who wish to solve a joint set of equations are interested in revealing as little information as possible on their input. Research in secure multiparty computation (MPC) has been impressively successful in allowing secure computation of every function with

---

[*] Department of Computer Science, UC Davis. pmohassel@ucdavis.edu.
[**] CWI, Amsterdam, The Netherlands. e.n.weinreb@cwi.nl.

D. Wagner (Ed.): CRYPTO 2008, LNCS 5157, pp. 481–496, 2008.
© International Association for Cryptologic Research 2008

efficiency that is proportional to its circuit complexity [BGW88, CCD88, Yao86, GMW87]. However, for various linear algebraic tasks, these general constructions fall well short of giving optimal protocols in terms of communication and round complexity. Starting with the work of Cramer and Damgård [CD01], the task of designing secure protocols for linear algebraic problems has been the focus of several recent works in secure computation [NW06, KMWF07, CKP07].

We focus on the problem of deciding the singularity of a shared matrix $M \in F^{n \times n}$, where $\mathbb{F}$ is a finite field. Many linear algebraic tasks, e.g. solving a joint set of linear equations and computing the rank of a shared matrix, are efficiently reducible to this task. When no honest majority is assumed, as in the classic two-party setting, our protocols are secure in the presence of a *covert* adversary [AL07], assuming the existence of public key homomorphic encryption schemes. Previous communication efficient secure protocols for linear algebra were known only in the honest but curious setting. In case there is a guarantee that the adversary controls less than one third (one half) of the parties, our protocols are secure against malicious (honest but curious) behaviour relying on no computational assumptions. Our protocols are constant round and achieve communication complexity of $O(n^{2+1/t})$ for every constant t. This is the first constant round construction for secure linear algebra with nearly linear communication complexity in the input size.

Applying General Results Fails. Unfortunately, general results in MPC do not yield efficient protocols for linear algebra. This is due to the fact that the communication complexity linearly depends on the Boolean circuit complexity of the function to be computed (or on the number of multiplication gates in case of information theoretic security). Alas, the circuit complexity of matrix singularity, as well as that of many other linear algebraic problem, is tightly related to that of matrix product [BCS97]. The best known upper bound for circuits for matrix product is $O(n^\omega)$ [CW87] with $\omega \cong 2.38$, which is significantly larger than the input size. Moreover, in the information theoretic setting the round complexity of the protocols is related to the multiplicative depth of the corresponding circuit, preventing these general construction from yielding constant round protocols. This leads to the following approach: Design efficient protocols for matrix product and then use them to achieve protocols for other linear algebraic tasks.

Matrix Product in Realistic Adversary Models. Therefore, our first step is to design constant round secure protocols for matrix product in the covert setting and in the information theoretic malicious setting with communication complexity $O(n^2)$. In the honest but curious setting, simple solutions for this problems are achieved using homomorphic encryption in the computational setting and linear secret sharing schemes in the information theoretic setting. However, when arbitrary adversarial behavior is considered, this task becomes more involved. Since very few efficient secure protocols for interesting problems are known in the malicious setting when no honest majority is assumed, Aumann and Lindell [AL07] have defined an interesting compromise: protocols that are secure in the presence of a covert adversary. Roughly speaking, the security guarantee is that if the adversary deviates from the protocol in a way that allows him to "cheat", then the honest parties are guaranteed to detect this cheating with good probability (e.g. 1/2). The ability to tolerate cheating with some fixed probability turns out useful for secure computation of matrix product and, consequently, for other secure linear algebraic tasks.

Techniques for Matrix Product. The general idea behind our secure protocol for matrix product in the covert setting, is to compute a decomposition of the input matrices into additive shares and then use homomorphic encryption to perform the computation on these shares. We use the random strings of the encryption scheme to prove that the additive shares indeed sum up to the input matrix, avoiding expensive zero knowledge proofs. Towards this goal, we need the homomorphic encryption scheme to have a property that all the known schemes enjoy: when computing $C = E(m_1, r_1) + E(m_2, r_2)$, one can compute a string r such that $C = E(m_1 + m_2, r)$. We note that although not every homomorphic encryption scheme has this property, some well known encryption schemes, such as the one by Paillier [Pal99], are suitable for our purposes. After the computations take place, the parties reveal parts of their additive sharing of the input, catching cheating adversaries[1] with probability 1/2. Revealing parts of the decomposition of the input enables easy input extraction, which makes the simulation go through.

Constant Round Reduction From Singularity to Matrix Product. Unfortunately, in spite of the tight connection of matrix product and linear algebraic problems such as matrix singularity, efficient protocols for the former does not immediately imply efficient protocols for the latter. The reason is that the known reductions do not translate to protocols with constant round complexity. Therefore, we need to design a special purpose protocol for matrix singularity, equipped with our secure protocol for matrix product. We use ideas from [Wie86], [KP91], and [KS91] to reduce the problem of deciding the singularity of a general matrix M into deciding the singularity of a related *Toeplitz* matrix T. We then use a lemma by Leverrier [JáJ92] which reduces the problem into computing the traces of powers of T. Finally, we define the Toeplitz matrix of polynomials $(I - \lambda T)$ and use the Gohberg-Semencul formula for the inverse of a Toeplitz matrix, to compute the above traces efficiently. We rely on techniques for iterated matrix product [CKP07] (which, in turn, is based on techniques from [BIB89]), combined with some simple linear algebraic manipulations, to translate the above algorithmic ideas into a constant round secure protocol for matrix singularity with the above mentioned communication complexity.

---

[1] The cheating probability can be reduced to $1/k$ paying a factor of k in the communication complexity.

1.1 Related Work

Information theoretic setting. Cramer and Damgård initiated the study of se-cure protocols for solving various linear algebra problems [CD01]. Their work was done in the information theoretic multi-party setting, with the main fo-cus on achieving constant round complexity. The communication complexity[2] of their protocols is $\Omega(n^4)$ while the size of the inputs is just $O(n^2)$. Cramer et al. [CKP07] designed a constant round protocol for solving m equations in n vari-ables with communication complexity $O(m^4 + n^2 m)$ which improves on [CD01] for small values of m. The only protocol applicable to the information theoret-ical setting with communication complexity of roughly $O(n^2)$ is that of Nissim and Weinreb [NW06]. However, this protocol has polynomial round complexity ($\Omega(n^{0.27})$) and is proved secure only in the honest but curious model. (The pro-tocols of [CD01] and [CKP07] are secure in the malicious model, assuming the adversary controls less that a third of the parties.)

Computational Setting. As previously noted, using the general well-known gar-bled circuit method of Yao [Yao82], one can get a constant round protocol for various linear algebraic problems with communication complexity that is pro-portional to the Boolean circuit complexity of matrix multiplication, for which the best upper bound known is $O(n^\omega)$ [CW87] for $\omega \cong 2.38$. As discussed above, the protocol of [NW06] was the first to improve the communication complex-ity to roughly $O(n^2)$, in the price of large round complexity. Later, Kiltz et al. [KMWF07] improved on the round complexity to get a protocol with $O(\log n)$ rounds and communication complexity of roughly $O(n^2)$. However, this protocol is secure only in the honest but curious setting.

Organization In section 2 we introduce the necessary definitions and primitives. In sections 3 and 4, we design secure and efficient protocols for matrix product in the covert setting and information theoretic setting respectively. In section 5, we reduce the problem of securely testing singularity of a matrix to a secure matrix product, put everything together, give our main theorem and explain how to extend our results to other linear algebra problems.

2 Preliminaries

Notation. Our protocols work with elements over a finite field, which we denote by $\mathbb{F}$. We guarantee the security with probability $1 - O(n^2)/|\mathbb{F}|$, where $|\mathbb{F}|$ is size of the field and n is the size of matrices we deal with in our linear algebra problems. Letting $|\mathbb{F}|$ be superpolynomial in n, we can achieve protocols with negligible error probability.[3] We count the communication complexity of our protocols in terms of number of field elements communicated. By default, a vector $v \in \mathbb{F}^n$ is regarded as a column vector. By $\mathbb{F}[x]$ we refer to the ring

---

[2] The complexity of their protocols can be reduced to $O(n^3)$ using the matrix product protocol from this paper.

[3] For smaller fields, one can pay a polylogarithmic factor in the communication com-plexity and work in an extension field.

of polynomials over $\mathbb{F}$, and by $\mathbb{F}[[x]]$ to the field of rational functions over $\mathbb{F}$. Given a collection of matrices $C_1, \ldots, C_p$ over $\mathbb{F}$ where all the C_i's have the same number of rows, we denote by $|C_1|C_2|\ldots|C_p|$ the block matrix resulted by concatenating the $C_1, \ldots, C_p$.

2.1   Definitions for Security against Covert Adversaries

Aumann and Lindell, [AL07], give a formal definition of security against covert adversaries in the *ideal/real simulation paradigm*. This notion of adversary lies somewhere between those of semi-honest and malicious adversaries. Loosely speaking, the definition provides the following guarantee: Let $0 \leq \epsilon \leq 1$ be a value (called the deterrence factor). Then any attempts to cheat by an adversary is detected by the honest parties with probability at least ϵ. Thus provided that ϵ is sufficiently large, an adversary that wishes not to get caught cheating will refrain from attempting to cheat, lest it be caught doing so. Furthermore, in the strongest version of security against covert adversaries introduced in [AL07], the adversary will not learn any information about the honest parties' inputs if he gets caught. Please see [AL07] for the detailed definitions.

Homomorphic Encryption. We use a semantically-secure public-key encryption scheme that allows for simple computations on encrypted data. In particular, we use encryption schemes where given two encryptions $E(m_1)$ and $E(m_2)$, we can efficiently compute a *random* encryption of $m_1 + m_2$. We denote this by $E(m_1 + m_2) = E(m_1) +_h E(m_2)$. Note that this implies that given an encryption $E(m)$ and $c \in \mathbb{F}$, we can efficiently compute a random encryption $E(cm)$; we denote this by $E(cm) = c \times_h E(m)$. For a matrix M We denote by $E(M)$ an entry-wise encryption of the matrix. Note that the above implies that two encrypted matrices can be added, and that we can compute a multiplication of an encrypted matrix by a matrix in the clear.

As an example for a homomorphic encryption scheme, we can use Pallier's [Pal99] cryptosystem. One minor issue is that the domain of Pallier's cryptosystem is the ring Z_n, where n is the product of two large and secret primes. Though Z_n is in fact not a field, any operation which separates it from a field leads to a factoring of n, and thus is unlikely to occur during a computation.

2.2   Definitions for the Information Theoretic Model

Linear Secret Sharing Schemes. To design our secure multiparty protocols, we use a linear secret sharing scheme (LSSS) to share values over a finite field $\mathbb{F}$. Each party receives a share that contains one or more field elements from the dealer. Each share is computed as a fixed linear function of the secret and some random field elements chosen by the dealer. The *size* of an LSSS is the total number of field elements distributed by the dealer. We denote by $[a]$ a secret sharing of $a \in \mathbb{F}$. For a vector $v \in \mathbb{F}^n$, we denote by $[v]$ a coordinate-wise sharing of the vector. Similarly, for a matrix $M \in \mathbb{F}^{n \times n}$ we denote by $[M]$ an entry-wise sharing of the matrix, and for a polynomial $p(x) \in \mathbb{F}[x]$ we denote by $[p]$ a coefficient-wise secret sharing of the polynomial.

Due to the linearity of the secret sharing, given secret shares of $[a]$ and $[b]$ and a third field element $c \in \mathbb{F}$, parties can compute secret shares of $[a + b]$ and $[ca]$ non-interactively. Furthermore, we require the LSSS to be *multiplicative*. Roughly speaking, this means that party P_i can use his shares of $[a]$ and $[b]$ to non-interactively compute a value c_i. The product ab can then be computed from c_i's using a fixed reconstruction vector $(r_1, \ldots, r_k)$, where k is the number of parties[4]. In [CDM00], it is shown how to construct a multiplicative LSSS scheme from any LSSS scheme without sacrificing efficiency.

3   Matrix Product Secure against Covert Adversaries

Given shares of two $n \times n$ matrices, we design a protocol for securely computing shares of the product matrix in the presence of a covert adversary in a small constant number of rounds and communication complexity $O(n^2)$. It is possible to compile a $O(n^2)$ communication matrix product protocol that is secure against semi-honest adversaries into a protocol that is secure against malicious adversaries using generic (or specially designed) zero knowledge proofs. However, we do not know how to do so without adding a significant overhead to the communication complexity of the original protocol.

The main idea is to break the input matrices into multiple additive shares and use a homomorphic encryption along with cut-and-choose techniques to perform the required computation on the shares. We further use the random strings of the encryption scheme to prove that the additive shares in fact add up to the original inputs. For this purpose, we need the homomorphic encryption to have an extra property: If one knows the random strings that correspond to the encryptions $E(m_1)$, and $E(m_2)$, one can efficiently compute the random string that corresponds to the encryption $E(m_1 + m_2) = E(m_1) +_h E(m_2)$.

We proceed to discuss the way in which a shared matrix is held in our protocols. In a sharing of a matrix $M \in \mathbb{F}^{n \times n}$, Alice holds $(A, E_b(B, r_b), r_a)$ and Bob holds $(B, E_a(A, r_a), r_b)$, where A and B are random matrices subject to the condition that $A + B = M$, the strings r_a and r_b are uniformly random strings, and $E_a(\cdot, \cdot)$ $(E_b(\cdot, \cdot))$ denotes encryption under Alice's (Bob's) public key. The remainder of this section is organized as follows: We start by formally defining the matrix product functionality in terms of the above secret sharing representation of the matrices. Then, we present our protocol for efficiently implementing this functionality.

Definition 1 (Matrix Multiplication Functionality). *The matrix multiplication functionality (MatMul) is defined as follows:*

Input to Alice: $A_1, A_2, r_{a_1}, r_{a_2}, E_b(B_1, r_{b_1}), E_b(B_2, r_{b_2}), E_b(C, r_c), r_{a'}$
Input to Bob: $B_1, B_2, C, r_{b_1}, r_{b_2}, r_c, E_a(A_1, r_{a_1}), E_a(A_2, r_{a_2})$

---

[4] For ease of composition, we assume that each party holds a single field element as his share but note that our techniques automatically generalize to the case where the share sizes are larger.

Output of Alice: $(A_1 + B_1)(A_2 + B_2) + C$
Output of Bob: $E_a((A_1 + B_1)(A_2 + B_2) + C, r_{a'})$

A few remarks on this definition are in place. The inputs to the players contain two shared matrices $(A_1 + B_1)$ and $(A_2 + B_2)$, together with the matrix C (which is also Bob's share of the output) and the string $r_{a'}$, which is the random string according to which Alice encrypts her output. That is, we choose not to introduce randomness into the definition and leave the choices of C and $r_{a'}$ to the protocols that use matrix product as a sub-protocol. This design choice was made to simplify the presentation of the protocol. But, it is not hard to construct on top of our protocol for this functionality, a protocol for a functionality with random C and $r_{a'}$ as outputs. Hence we assume that we have access to a secure coin-tossing protocol such as the one given in [Lin01].

To simplify the composition, we divide the protocol into several parts, where the parts will be performed sequentially one after the other.

Alice's Computation

1. **Alice writes her inputs as sums.** For each $i \in \{1,2\}$, Alice chooses two random matrices $A_{i,1}$ and $A_{i,2}$ such that $A_{i,1} + A_{i,2} = A_i$, and generates two random strings $r_{i,1}$ and $r_{i,2}$. For each $j \in \{1,2\}$, Alice sends $D_{i,j} \overset{\text{def}}{=} E_a(A_{i,j}, r_{i,j})$ to Bob.

2. **Alice proves her sums.** For each $i \in \{1,2\}$, Alice computes a string $r_{0,i}$, such that

$$E_a(0, r_{0,i}) = E_a(A_i, r_{a_i}) -_h D_{i,1} -_h D_{i,2}.$$

   Alice sends $r_{0,1}$ and $r_{0,2}$ to Bob.

3. **Alice sends output parts that depend only on her input.** For every $i_1, i_2 \in \{1,2\}^2$, Alice generates a random element s_{i_1,i_2} and sends Bob $H_{i_1,i_2} \overset{\text{def}}{=} A_{1,i_1} \times_h D_{2,i_2} +_h E(0, s_{i_1,i_2})$.

4. **Bob's challenge.** Bob chooses a random number $c \in \{1,2\}$ and sends it to Alice.

5. **Alice proves knowledge of encrypted data.** For every $i \in \{1,2\}$ Alice sends Bob $A_{i,c}, r_{i,c}$. Moreover, Alice sends Bob $s_{c,1}$ and $s_{c,2}$ chosen at step 3.

6. **Bob verifies Alice's data**
   (a) **Alice's encryptions indeed sum to** $E_a(A_i, ra_i)$. For each $i \in \{1,2\}$, Bob verifies that indeed $E(0, r_0) = E(A_i, r_{a_c}) -_h D_{i,1} -_h D_{i,2}$.
   (b) **Alice knows her encrypted data.** For each $i \in \{1,2\}$, Bob verifies that $D_{i,c} = E_a(A_{i,c}, r_{i,c})$.
   (c) **The computations that depend only on Alice were performed correctly.** Bob verifies that indeed $H_{c,j} = A_{1,c} \times_h D_{2,j} +_h E(0, s_{c,j})$ for $j \in \{1,2\}$.

Bob's Computation

1. **Bob writes his inputs as sums.** For each $i \in \{1,2\}$, Bob randomly chooses two random matrices $B_{i,1}$ and $B_{i,2}$ such that $B_{i,1}+B_{i,2} = B_i$, and generates two random strings $q_{i,1}$ and $q_{i,2}$. For each $j \in \{1,2\}$, Bob sends $F_{i,j} \overset{\text{def}}{=} E_b(B_{i,j}, q_{i,j})$ to Alice. Similarly, Bob Chooses 12 random matrices C_{i_1,i_2}, C'_{i_1,i_2} and C''_{i_1,i_2} for every $i_1,i_2 \in \{1,2\}^2$, such that $\sum_{i_1,i_2} C_{i_1,i_2} + \sum_{i_1,i_2} C'_{i_1,i_2} = C$, and 12 random strings t_{i_1,i_2}, t'_{i_1,i_2}, and t''_{i_1,i_2}, and sends Alice $G_{i_1,i_2} = E_b(C_{i_1,i_2}, t_{i_1,i_2})$, $G'_{i_1,i_2} = E_b(C'_{i_1,i_2}, t'_{i_1,i_2})$, and $G''_{i_1,i_2} = E_b(C''_{i_1,i_2}, t''_{i_1,i_2})$, for every $i_1,i_2 \in \{1,2\}^2$.

2. **Bob proves his sums.** For each $i \in \{1,2\}$, Bob computes a string q_0, such that

$$E_b(0,q_0) = E(B_i, r_{b_i}) -_h F_{i,1} -_h F_{i,2}.$$

Bob sends q_0 to Alice. Similarly, Bob computes a string t_0 such that

$$E_b(0,t_0) = E(C, r_c) -_h \sum_{i_1,i_2} G_{i_1,i_2} -_h \sum_{i_1,i_2} G'_{i_1,i_2} -_h \sum_{i_1,i_2} G''_{i_1,i_2}.$$

Bob sends t_0 to Alice.

3. **Bob sends information that depends only on his input.** For every $i_1,i_2 \in \{1,2\}^2$, Bob sends Alice $L_{i_1,i_2} = B_{1,i_1} B_{2,i_2} + C''_{i_1,i_2}$.

4. **Bob performs computations on Alice's inputs.** For every $i_1,i_2 \in \{1,2\}$, computes $K_{i_1,i_2} = D_{1,i_1} B_{2,i_2} + C_{i_1,i_2}$ and $K'_{i_1,i_2} = D_{2,i_1} B_{1,i_2} + C'_{i_1,i_2}$. Bob sends K_{i_1,i_2} and K'_{i_1,i_2} to Alice for every $i_1,i_2 \in \{1,2\}^2$.

5. **Alice's challenge.** Alice chooses a two random numbers $d_1,d_2 \in \{1,2\}^2$ and sends them to Bob.

6. **Bob proves knowledge of encrypted data.** For every $i \in \{1,2\}$ Bob sends Alice B_{i,d_i} and q_{i,d_i}. Moreover for every $j \in \{1,2\}$ Bob sends Alice the matrices C_{j,d_j}, C'_{j,d_j}, and C''_{j,d_j} and the strings t_{j,d_j}, t'_{j,d_j}, and t''_{j,d_j}.

7. **Alice verifies Bob's data**

   (a) **Bob's encryptions indeed sum to $E_a(B_i, rb_i)$.** For every $i \in \{1,2\}$, Alice verifies that indeed $E_b(0,q_0) = E_b(B_i, r_{b_{d_i}}) - F_{i,1} - F_{i,2}$.

   (b) **Bob's encryptions indeed sum to $E_a(C, r_c)$.** Alice verifies that indeed $E_b(0,t_0) = E_b(C, r_c) - \sum_{i_1,i_2} G_{i_1,i_2} - \sum_{i_1,i_2} G'_{i_1,i_2}$.

   (c) **Bob knows his encrypted data.** For each $i \in \{1,2\}$, Alice verifies that $F_{i,d_i} = E_b(B_{i,d_i}, q_{i,d_i})$. Moreover, for every $j \in \{1,2\}$ Alice verifies that $G_{j,d_j} = E_b(C_{j,d_j}, t_{j,d_j})$ and that $G'_{j,d_j} = E_b(C'_{j,d_j}, t'_{j,d_j})$.

   (d) **The computations that depend only on Bob were computed correctly** Alice verifies that $L_{d_1,d_2} = B_{1,d_1} B_{2,d_2} + C''_{d_1,d_2}$

   (e) **Bob's homomorphic computations were correct** For every $j \in \{1,2\}$, Alice verifies that indeed $K_{j,d} = E_a(A_{1,j})B_{2,d} + C_{j,d}$ and $K'_{j,d} = E_a(A_{2,j})B_{1,d} + C'_{j,d}$.

Output Computation

1. **Output computation.** Alice decrypts all the values, sums everything up to get $(A_1+B_1)(A_2+B_2)+C$ and computes $O_A = E_a((A_1+B_1)(A_2+B_2)+ C, r_{a'})$. Bob on his side, computes $O_B = E_a((A_1+B_1)(A_2+B_2)+C, r_{a''})$ using homomorphic operations on the shares that he holds. Bob performs these computation in a deterministic way, such that Alice knows $r_{a''}$.

2. **Output Delivery.** Alice chooses two random matrix X_1 and X_2 such that $X_1 + X_2 = (A_1 + B_1)(A_2 + B_2) + C$. Alice chooses a random string x_1, and set the string x_2 to satisfy $E_a(X_1, x_1) +_h E_a(X_2, x_2) = O_A$. Alice sends $E_a(X_1, x_1)$ and $E_a(X_2, x_2)$ to Bob, together with a random string x_0 satisfying $O_B -_h O_A = E_a(0, x_0)$.

3. **Output Challenge.** Bob chooses a random number $y \in \{0, 1\}$ and sends it to Alice.

4. **Output Response.** Alice sends Bob X_y and x_y.

5. **Bob's output.** Bob verifies the encryption $E_a(X_y, x_y)$ and that $O_B -_h E_a(X_1, x_1) -_h E_a(X_2, x_2) = E_a(0, x_0)$. If this is the case, he outputs $E_a(X_1, x_1) +_h E_a(X_2, x_2)$.

6. **Alice's output.** Alice outputs $(A_1 + B_1)(A_2 + B_2) + C$.

The following theorem summarizes the results of this section.

Theorem 1. *Let n be a positive integer and consider matrices of dimension $n \times n$. Given a secure coin-tossing functionality, the above protocol securely realizes the matrix product functionality, in presence of a covert adversary with deterrence probability $\epsilon = 1/4$. The protocol is constant round and requires $O(n^2)$ communication.*

Due to lack of space, details of the simulators and proof of security appear in the full version. The above protocol in its current form is not secure under parallel composition due to the rewinding nature of the simulator in the proof. However, we can securely run many instances of the matrix product in parallel if (i) the same public key and encryption scheme is used for all instances and (2) the same challenges are used in all instances of the matrix product protocol. In fact, the description of the simulator in the proof for security of the parallel matrix product protocol will be almost identical to the simulator for the original protocol.

4 Secure Matrix Product in the Information Theoretic Model

Using standard techniques for multiplying two shared values leads to a simple protocol for matrix product in the information theoretic setting that requires $O(n^3)$ communication.

However, one can think of our element-wise secret sharing of the matrices as a secret sharing scheme over the matrix ring. In light of this observation, we can efficiently generalize the committed multiplication protocol of [CDM00] to a constant round committed matrix multiplication protocol that is secure against active adversaries, with $O(n^2)$ communication. A more detailed description of the protocols is deferred to the full version. The following theorem summarizes the result:

Theorem 2. *Given two shared matrices $[A]$ and $[B]$ where $A, B \in \mathbb{F}^{n \times n}$, there exists a multiparty protocol, secure against a malicious adversary that corrupts less than a third of the parties, for computing a secret sharing of the product $[C] = [AB]$ in a constant number of rounds and with $O(n^2)$ communication.*

As we will see in future sections, secure computation of other linear algebra problems is reduced to a secure matrix product protocol. The guaranteed security is in part due to the existing general composition theorems in the information-theoretic setting [KLR06].[5]

5   From Matrix Singularity to Matrix Product

In this section we design a secure protocol for deciding singularity of a shared matrix given an efficient implementation of a secure protocol for matrix product. Our techniques apply to both two-party protocols secure against covert adversaries and multiparty protocols secure against computationally unbounded malicious adversaries.

In Section 5.1, we design a constant round protocol for computing shares of the *linearly recurrent sequence* $v, Mv, \ldots, M^{2n-1}v$, given shares of a square matrix M and a vector v. Later, in Section 5.2, we apply this protocol to M and a random shared vector v, to reduce the problem of deciding the singularity of M into the problem of deciding the singularity of a related *Toeplitz* matrix T. In Section 5.3 we design a protocol for deciding singularity of a Toeplitz matrix, by applying the protocol for the linearly recurrent sequence on a different set of inputs. Finally, in Section 5.4, we connect all the above to get our main result: a secure constant round protocol for deciding shared matrix singularity with communication complexity $O(n^{2+1/t})$ for every positive integer t.

5.1   Secure Computation of the Sequence $\{M^i v\}$

Given secret shares of a matrix $M \in \mathbb{F}^{n \times n}$ and a vector $v \in \mathbb{F}^n$, our goal is to design a secure protocol for computing shares of the vector sequence $\{M^i v\}_{1 \le i \le 2n}$. This construction is an important building block for the design of our matrix singularity protocol. Using the methods of [CKP07] for computing powers of a matrix, one can securely compute the sequence $\{M^i v\}_{1 \le i \le 2n}$ in constant round and with $O(n^4)$ communication. In this section, we design a constant round protocol for the same task with communication complexity of $O(n^{2+1/t})$ for any arbitrary constant t.

---

[5] The general composition theorem we use from [KLR06], requires the composed protocols to perform an initial synchronization step. This synchronization step can easily be added to all of our protocols without any asymptotic overhead.

In [CKP07], the problem of computing sharings of $I, M, M^2, \ldots, M^d$ is reduced into (i) generating a sharing of $O(d)$ random matrices and their inverses, and (ii) executing $O(d)$ parallel[6] matrix multiplications. Using standard techniques that are available in both settings, both steps can be computed by performing $O(d)$ matrix multiplications of $n \times n$ matrices. Thus, using a constant round secure protocol for matrix product we get a secure protocol for computing these with $O(dn^2)$ communication which we refer to as $\mathrm{POWERS}_d(M)$. The following lemma summarizes this improvement.

Lemma 1 ($\mathrm{POWERS}_d(M)$). *Given shares of a matrix $M \in \mathbb{F}^{n \times n}$, there exist a protocol for securely computing shares of $\{I, M, M^2, \ldots, M^d\}$ in constant round and with $O(dn^2)$ communication.*

We are now ready to describe our protocol for computing shares of the sequence $\{M^i v\}_{1 \leq i \leq 2n}$. We introduce some notation. Denote $\ell \stackrel{\text{def}}{=} \lceil (2n)^{1/s} \rceil$, and for $0 \leq i \leq s - 1$, denote by POW_M^i, the following set of $\ell + 1$ powers of M: $\mathrm{POW}_M^i \stackrel{\text{def}}{=} \{I, M^{\ell^i}, M^{2\ell^i}, \ldots, M^{\ell^{i+1}}\}$. The following observation is easy to verify.

Observation 3. *For every $1 \leq t \leq 2n$, the matrix M^t can be represented as a product of s matrices $M^t = \prod_{0 \leq j \leq s-1} M_{t,j}$, where $M_{t,j} \in \mathrm{POW}_M^j$ for every $0 \leq j \leq s - 1$.*

Protocol $\mathrm{LIN\text{-}SEQ}_n(M, v)$

Input: Shares of a matrix $[M] \in \mathbb{F}^{n \times n}$, and vector $[v] \in \mathbb{F}^n$
Output: Shares of $2n$ vectors $[Mv], [M^2 v], \ldots, [M^{2n} v]$

1. Parties agree on a positive integer s, and let $\ell = \lceil (2n)^{1/s} \rceil$
2. Parties securely compute shares of the sets $\mathrm{POW}_M^1, \mathrm{POW}_M^2, \ldots, \mathrm{POW}_M^s$ by sequentially running $\mathrm{POWERS}_\ell(M), \mathrm{POWERS}_\ell(M^\ell), \ldots, \mathrm{POWERS}_\ell(M^{\ell^{s-1}})$ (See Lemma 1.)
3. Let $B_0 = v$
4. For $i = 0$ to $s - 1$:
   (a) For $j = 1, \ldots, \ell - 1$, parties compute $[C_j] = [M^{j\ell^i}][B_i]$ by performing the secure matrix product protocol of section 2.
   (b) As a result, parties hold secret shares of the $n \times \ell^{i+1}$ matrix $B_{i+1} = |C_{\ell-1}|C_{\ell-2}| \cdots |C_1|B_i|$.
5. Parties hold shares of the $n \times \ell^s$ matrix B_s which contains the sequence $M^i v$ as its columns, for $1 \leq i \leq l^s$.

Lemma 2. *Given a shared matrix $[M] \in \mathbb{F}^{n \times n}$ and a shared vector $[v] \in \mathbb{F}^n$, for any positive integer s, there exist a multiparty protocol for securely computing shares of the sequence $\{M^i v\}_{0 \leq i \leq 2n}$ in $O(s)$ rounds and with $O(sn^{2+1/s})$ communication.*

---

[6] Both in the covert setting and in the information theoretic setting, a secure protocol for computing d matrix products in parallel in constant round and communication complexity $O(dn^2)$ is a straightforward generalization of our matrix product protocols.

Proof. In view of Claim 3, it is easy to verify that after s iterations of the for loop, the matrix B_s contains the $2n$ vectors in the sequence $\{M^i v\}_{0 \leq i < 2n}$ as its columns. Based on Lemma 1, step 2 can be performed in $O(s)$ rounds and with $O(sn^{1/s}n^2)$ communication. In each iteration, step 4a requires ℓ multiplication of an $(n \times n)$ by an $(n \times \ell^i)$ matrix. Using an efficient matrix product protocol this requires only $O(n^2)$ communication, and leads to a total of $O(s\ell n^2) = O(sn^{1/s}n^2)$ communication for the s iterations of the loop.

5.2   From General Matrix Singularity to Toeplitz Matrix Singularity

In this section we reduce the problem of securely deciding the singularity of a general shared matrix M into the problem of securely deciding the singularity of a related shared matrix T.

Theorem 4. *For every integer s, there is an $O(s)$ rounds protocol that securely transforms a sharing of a matrix $M \in \mathbb{F}^{n \times n}$ into a sharing of a Toeplitz matrix $T \in \mathbb{F}^{n \times n}$, such that, with probability $1 - O(n^2)/|\mathbb{F}|$, matrix M is singular iff the matrix T is singular. The communication complexity of the protocol is $O(sn^{2+1/s})$.*

The proof of Theorem 4 relies on the following Lemmata. Due to lack of space, the full proof is defered to the final version of the paper.

Lemma 3 ([KS91]). *Consider the matrix $M \in \mathbb{F}^{n \times n}$ of rank r (unknown), and random non-singular matrices $V, W \in \mathbb{F}^{n \times n}$, let $M' = VMW$. Then, with probability greater than $1 - n(n+1)/|\mathbb{F}|$, the upper-left $i \times i$ submatrices of M' are invertible for $1 \leq i \leq r$.*

Lemma 4 ([KS91]). *Let $B \in \mathbb{F}^{n \times n}$ be a matrix with invertible upper left $i \times i$ submatrices for $1 \leq i \leq r$, where $r < n$ is the rank of B. Let D be a randomly chosen diagonal matrix in $\mathbb{F}^{n \times n}$. Then, $r = \deg(m_{DB}) - 1$ with probability greater than $1 - n^2/|\mathbb{F}|$.*

Lemma 5 ([Wie86]). *Let $A \in \mathbb{F}^{n \times n}$ and let m_A be the minimal polynomial of A. For $\mathbf{u}, \mathbf{v} \in \mathbb{F}^n$ chosen uniformly at random, consider the linearly recurrent sequence $\{a_i\} = \{u^t A^i v\}$. We have that the minimal polynomial of the sequence $\{a_i\}$ is equal to m_A with probability at least $1 - 2\deg(m_A)/|\mathbb{F}|$.*

For a linearly recurrent sequence $\{a_i\}_{i=0}^{\infty}$, and $\alpha > 0$, denote by T_α the following Toeplitz matrix:

$$
T_\alpha = \begin{pmatrix}
a_{\alpha-1} & a_{\alpha-2} & \cdots & a_1 & a_0 \\
a_\alpha & a_{\alpha-1} & \cdots & a_2 & a_1 \\
\vdots & a_\alpha & \ddots & \vdots & a_2 \\
& \vdots & & & \vdots \\
a_{2\alpha-3} & & & a_{\alpha-1} & \\
a_{2\alpha-2} & a_{2\alpha-3} & \cdots & a_\alpha & a_{\alpha-1}
\end{pmatrix}
$$

Lemma 6 ([KP91]). *Let* $\{a_i\}$ *be a linearly recurrent sequence, and let d be the degree of its minimum polynomial. For any* $\alpha \geq 0$, *let* T_α *be the Toeplitz matrix constructed as above. Then,* $Det(T_d) \neq 0$ *and for all* $\alpha > d$, $Det(T_\alpha) = 0$.

5.3 Deciding the Singularity of a Toeplitz Matrix

We test the singularity of a Toeplitz matrix by first computing its characteristic polynomial and then determining whether its constant coefficient is equal to zero or not. The well known lemma of Leverrier, connects the characteristic polynomial of a matrix to solution of a linear system that depends on the trace of powers of the matrix (see Appendix A). Particularly, if we compute the traces of matrices $T^1, \ldots, T^n$, computing the characteristic polynomial reduces to inverting an invertible matrix for which there exist simple and efficient protocols in both settings. Hence, our main challenge remains to compute traces of powers of T in a round and communication efficient manner.

Denote by $X \in \mathbb{F}[\lambda]^{n \times n}$ the matrix $X = I + \lambda T + \lambda^2 T^2 + \ldots + \lambda^n T^n$. Note that entries of X are polynomials of degree at most n in $\mathbb{F}[\lambda]$. Therefore, the trace of the matrix X is also a polynomial of degree at most n in λ. It is easy to see that coefficients of this trace polynomial are in fact the traces of powers of T. Hence, our goal is to compute the trace of X. However, as the naïve representation of the matrix X consists of n^3 field elements, we need to compute the trace of X in a more clever way.

Consider the matrix $(I - \lambda T) \in \mathbb{F}[\lambda]^{n \times n}$. Since T is a Toeplitz matrix, the matrix $(I - \lambda T)$ is Toeplitz as well. One may compute the power series extension

$$(I - \lambda T)^{-1} = I + \lambda T + \lambda^2 T^2 + \ldots \in \mathbb{F}[[\lambda]]^{n \times n}$$

Note that $X \equiv (I - \lambda T)^{-1} \pmod{\lambda^{n+1}}$. Hence, the matrix X is equivalent modulo λ^{n+1} to the inverse of a Toeplitz matrix from $\mathbb{F}[[\lambda]]^{n \times n}$.

The following Lemma, based on the Gohberg-Semencul formula (e.g. see [GS72, FMKL79, BGY80]), shows that X can be represented using only its first and last columns:

Lemma 7. *For any Toeplitz matrix* $T \in \mathbb{F}^{n \times n}$, *it holds that*

$$X \stackrel{def}{=} (I - \lambda T)^{-1} \ mod \ \lambda^{n+1} = I + \lambda T + \lambda^2 T^2 + \ldots + \lambda^n T^n$$

$$
= \frac{1}{u_1} \left(
\begin{pmatrix}
u_1 & & & \\
u_2 & u_1 & & \\
u_3 & u_2 & u_1 & \\
\vdots & & \ddots & \\
u_n & u_{n-1} & \ldots & u_2 & u_1
\end{pmatrix}
\begin{pmatrix}
v_1 & v_2 & \cdots & v_n \\
& u_1 & v_2 & & v_{n-1} \\
& & \ddots & \ddots & \vdots \\
& & & v_1 & v_2 \\
& & & & v_1
\end{pmatrix}
\right.
$$

$$
\left.
- \begin{pmatrix}
0 & & & \\
v_n & 0 & & \\
v_{n-1} & v_n & 0 & \\
\vdots & & \ddots & \ddots \\
v_2 & v_3 & \ldots & v_n & 0
\end{pmatrix}
\begin{pmatrix}
0 & u_n & u_{n-1} & \ldots & u_2 \\
& 0 & u_n & & u_3 \\
& & \ddots & \ddots & \vdots \\
& & & 0 & v_n \\
& & & & 0
\end{pmatrix}
\right) \ mod \ \lambda^n
$$

$$\tag{1}$$

where $u = (u_1, \ldots, u_n)^t$ and $v = (v_1, \ldots, v_n)^t$ are the first and last columns of $(I - \lambda T)^{-1}$ respectively.

This brief representation of X allows for an efficient computation of the trace of X. Particularly, we get the following equation from (1):

$$\text{Trace}(X) \equiv \text{Trace}((I - \lambda T)^{-1}) \equiv \frac{1}{u_1}(nu_1 v_1 + (n-2)u_2 v_2 + \ldots + (-n+2)u_n v_n) \tag{2}$$

To compute $\text{Trace}(X)$, it is sufficient to perform the following two steps **(i)** Compute the first and last columns of X: We apply Lemma 2 to compute the two sequences $\{T^i e_1\}_{1 \leq i \leq n}$ and $\{T^i e_n\}1 \leq i \leq n$ where e_1 and e_n are column vectors of $\langle 1, 0, \ldots, 0 \rangle$ and $\langle 0, \ldots, 0, 1 \rangle$, respectively. This automatically gives us the first and last columns of X, and requires $O(s)$ rounds and $O(sn^{2+1/s})$ communication. **(ii)** Compute the trace of X using these two columns: This can be done based on Equation 2. Since every polynomial among $u_1 \ldots, u_n, v_1, \ldots, v_n$ is of degree at most n, multiplying two of these polynomials can be done in constant round and communication $O(n)$ in both settings (see [MF06]). Finally, dividing by u_1, which is invertible modulo λ^{n+1} (since its constant coefficient is 1), can be done in constant round and with communication complexity $O(n)$ (see [MF06]). Hence, we conclude that computation of traces of $T, T^2, \ldots, T^n$ can be done in $O(s)$ rounds and communication complexity $O(sn^{2+1/s})$.

5.4   The Protocol for Matrix Singularity

The following protocol for testing singularity of a matrix combines all the techniques discussed in this section:

Input: Shares of a matrix $[M] \in \mathbb{F}^{n \times n}$
Output: Shares of a bit $[b]$, where $b = 1$ if M is singular and 0 otherwise.

1. Compute shares of the Toeplitz matrix $[T]$, as described in Section 5.2.
2. Execute $\text{LIN-SEQ}_n(T, (1, 0, \ldots, 0)^t)$ and $\text{LIN-SEQ}_n(T, (0, 0, \ldots, 0, 1)^t)$. The output of these two executions yield shares of the first and last columns ($[u]$ and $[v]$ respectively) of the matrix X.
3. Given shares of $[u]$ and $[v]$, compute shares of $TRACE(X)$, as described in Section 5.3.
4. Construct the matrix S from equation 3 without any interaction and solve the linear system of equation 3 using the protocol for inverting an invertible matrix (see e.g. [BIB89])[7]. As a result, they hold shares of coefficients of characteristic polynomial of T.
5. Securely test whether the constant coefficient of the characteristic polynomial of T is 0 using an equality testing protocol (e.g. [DFKNT06, Yao86]).

---

[7] Entries of S are traces of T^is which are computed in step 3.

Theorem 5. *Let n and $s < n$ be a positive integers and consider matrices of dimension $n \times n$. There exist secure protocols that realizes the Matrix Singularity functionality in the covert adversarial model, and the information-theoretic model in $O(s)$ rounds and with $O(sn^{2+1/s})$ communication.*

Other Linear algebra problems. Using existing techniques, one can efficiently and in constant round, reduce the task of secure computation of rank of a matrix and solving a linear system of equation to testing singularity of a matrix. It is easy to verify that the reductions will apply to both settings we considered in this paper. See [KMWF07] or the full version of this paper for more detail.

Acknowledgment. We would like to thank the anonymous reviewers for their helpful comments and suggestions.

References

[AL07]   Aumann, Y., Lindell, Y.: Security against covert adversaries: Efficient protocols for realistic adversaries. In: Vadhan, S.P. (ed.) TCC 2007. LNCS, vol. 4392, pp. 137–156. Springer, Heidelberg (2007)

[BCS97]   Bürgisser, P., Clausen, M., Shokrollahi, M.A.: Algebraic complexity theory. Springer, Berlin (1997)

[BGW88]   Ben-Or, M., Goldwasser, S., Wigderson, A.: Completeness theorems for noncryptographic fault-tolerant distributed computations. In: STOC, pp. 1–10 (1988)

[BGY80]   Brent, R.P., Gustavson, F.G., Yun, D.Y.Y.: Fast solution of Toeplitz systems of equations and computation of pade approximants. J. Algorithms, 259–295 (1980)

[BIB89]   Bar-Ilan, J., Beaver, D.: Non-cryptographic fault-tolerant computing in constant number of rounds of interaction. In: PODC, pp. 201–209. ACM Press, New York (1989)

[CCD88]   Chaum, D., Crépeau, C., Damgård, I.: Multiparty unconditionally secure protocols. In: STOC, pp. 11–19 (1988)

[CD01]   Cramer, R., Damgaard, I.: Secure distributed linear algebra in a constant number of rounds. In: Kilian, J. (ed.) CRYPTO 2001. LNCS, vol. 2139, pp. 119–136. Springer, Heidelberg (2001)

[CDM00]   Cramer, R., Damgård, I., Maurer, U.: General secure multi-party computation from any linear secret-sharing scheme. In: Preneel, B. (ed.) EUROCRYPT 2000. LNCS, vol. 1807, pp. 316–334. Springer, Heidelberg (2000)

[CKP07]   Cramer, R., Kiltz, E., Padró, C.: A note on secure computation of the Moore-Penrose pseudo-inverse and its application to secure linear algebra. In: Menezes, A. (ed.) CRYPTO 2007. LNCS, vol. 4622, p. 613. Springer, Heidelberg (2007)

[CW87]   Coppersmith, D., Winograd, S.: Matrix multiplication via arithmetic progressions. In: STOC, pp. 1–6. ACM Press, New York (1987)

[FMKL79]   Friedlander, B., Morf, M., Kailath, T., Ljung, L.: New inversion formulas for matrices classified in terms of their distance from Toeplitz matrices. Linear Algebra and Appl. 27, 31–60 (1979)

[GMW87]   Goldreich, O., Micali, S., Wigderson, A.: How to play any mental game. In: STOC, pp. 218–229 (1987)

[GS72]   Gohberg, I., Semencul, A.: On the inversion of finite Toeplitz matrices and their continuous analogs. Math. Issl., 201–233 (1972)

[JáJ92]   JáJá, J.: An introduction to parallel algorithms. Addison Wesley Longman Publishing Co., Inc., Redwood City (1992)

[KLR06]   Kushilevitz, E., Lindell, Y., Rabin, T.: Information-theoretically secure protocols and security under composition. In: STOC, pp. 109–118 (2006)

[KMWF07]   Kiltz, E., Mohassel, P., Weinreb, E., Franklin, M.: Secure linear algebra using linearly recurrent sequences. In: Vadhan, S.P. (ed.) TCC 2007. LNCS, vol. 4392, pp. 291–310. Springer, Heidelberg (2007)

[KP91]   Kaltofen, E., Pan, V.: Processor efficient parallel solution of linear systems over an abstract field. In: SPAA, pp. 180–191. ACM Press, New York (1991)

[KS91]   Kaltofen, E., Saunders, D.: On Wiedemann's method of solving sparse linear systems. In: AAECC-9, London, UK, pp. 29–38 (1991)

[DFKNT06] Damgaard, I., Fitzi, M., Kiltz, E., Nielsen, J.B., Toft, T.: Unconditionally Secure Constant-Rounds Multi-Party Computation for Equality, Comparison, Bits and Exponentiation. In: Halevi, S., Rabin, T. (eds.) TCC 2006. LNCS, vol. 3876, pp. 285–304. Springer, Heidelberg (2006)

[Lin01]   Lindell, Y.: Parallel Coin-Tossing and Constant-Round Secure Two-Party Computation. In: Kilian, J. (ed.) CRYPTO 2001. LNCS, vol. 2139, pp. 171–189. Springer, Heidelberg (2001)

[MF06]   Mohassel, P., Franklin, M.: Efficient Polynomial Operations in the Shared-Coefficient Setting. In: Yung, M., Dodis, Y., Kiayias, A., Malkin, T. (eds.) PKC 2006. LNCS, vol. 3958, pp. 44–57. Springer, Heidelberg (2006)

[NW06]   Nissim, K., Weinreb, E.: Communication efficient secure linear algebra. In: Halevi, S., Rabin, T. (eds.) TCC 2006. LNCS, vol. 3876, pp. 522–541. Springer, Heidelberg (2006)

[Pal99]   Pallier, P.: Public-key cryptosystems based on composite degree residuosity classes. In: Stern, J. (ed.) EUROCRYPT 1999. LNCS, vol. 1592, pp. 223–238. Springer, Heidelberg (1999)

[Wie86]   Wiedemann, D.H.: Solving sparse linear equations over finite fields. IEEE Trans. Inf. Theor. 32(1), 54–62 (1986)

[Yao82]   Yao, A.C.: Protocols for secure computations. In: FOCS, pp. 160–164 (1982)

[Yao86]   Yao, A.C.: How to generate and exchange secrets. In: FOCS, pp. 162–167 (1986)

A   Leverrier's Lemma

Lemma 8 (Leverrier's Lemma). *The coefficients $c_1, \ldots, c_n$ of the characteristic polynomial of an $n \times n$ matrix T satisfies the following system of equations:*

$$
S. \begin{pmatrix} c_1 \\ c_2 \\ c_3 \\ \vdots \\ c_n \end{pmatrix} = \begin{pmatrix} s_1 \\ s_2 \\ s_3 \\ \vdots \\ s_n \end{pmatrix} \; where, S = \begin{pmatrix} 1 & 0 & 0 & \ldots & 0 \\ s_1 & 2 & 0 & \ldots & 0 \\ s_2 & s_3 & 3 & \ldots & 0 \\ \vdots & \vdots & \vdots & \vdots & \vdots \\ s_{n-1} & s_{n-2} & s_{n-3} & \ldots & n \end{pmatrix} \tag{3}
$$

and $s_i = tr(T^i)$ for $1 \leq i \leq n$.

Collusion-Free Protocols in the Mediated Model

Joël Alwen[1], Abhi Shelat[2], and Ivan Visconti[3]

[1] New York University
251 Mercer St. New York, NY, 10012, USA
jalwen@cs.nyu.edu
[2] University of Virginia
Charlottesville 22904, USA
shelat@virginia.edu
[3] Dipartimento di Informatica ed Appl., Università di Salerno
84084 Fisciano (SA), Italy
visconti@dia.unisa.it

Abstract. Prior approaches [14, 15] to building collusion-free protocols require exotic channels. By taking a conceptually new approach, we are able to use a more *digitally*-friendly communication channel to construct protocols that achieve a stronger collusion-free property.

We consider a communication channel which can filter and rerandomize message traffic. We then provide a new security definition that captures collusion-freeness in this new setting; our new setting even allows for the mediator to be corrupted in which case the security gracefully fails to providing standard privacy and correctness. This stronger notion makes the property useful in more settings.

To illustrate feasibility, we construct a commitment scheme and a zero-knowledge proof of knowledge that meet our definition in its two variations.

Keywords: Secure Collusion Free, Mediated Communication, ZKPoK, Commitments.

1 Introduction

The Federal Communication Commission in the United States just finished auctioning several bands of communication spectrum. This was the seventeenth such auction run by the FCC and the commission has gained quite a bit of experience in running them. In particular, based on behavior in prior auctions, the commission employed several ad-hoc rules in order to maximize the revenue generated by this auction. One major concern for them is the problem of *collusion* between bidders. As documented in [9], in a prior auction, although many rules prohibited explicit collusion between the bidders, bidders have nonetheless devised clever signaling strategies *during* the auction in order to cheaply divide the auctioned rights. Thus, it is safe to say that an issue at the forefront of FCC's auction design team is to prevent bidders from engaging in such collaborative bidding strategies.

D. Wagner (Ed.): CRYPTO 2008, LNCS 5157, pp. 497–514, 2008.
© International Association for Cryptologic Research 2008

1.1 Does Cryptography Help the FCC?

It has long been held that secure multi-party computation protocols provide the most robust and secure way for implementing any collaborative task, including the running of an auction. Indeed, on face, secure multi-party protocols would mimic the role of a trusted-third party who ran the auction—without actually needing to trust a third party. While in practice, the FCC can function as such a third-party, doing so is not optimal. The FCC is a huge organization and the participants in high-value auctions might naturally question whether FCC "insiders" are helping the competition, by say leaking bid information during the auction. Thus, from both a practical and theoretical perspective, we ask how cryptography can improve the security of an auction.

There is unfortunately a serious problem which complicates the use of cryptography in such auctions. As pointed out by Lepinski, Micali, and shelat [15], the use of a cryptographic protocol can undo all of the carefully planned measures designed by the auctioneer to prevent collaborative bidding. In particular, since secure cryptographic protocols are randomized, a protocol may unintentionally produce a steganographic channel which bidders may use to communicate illegally about their bidding strategies. Thus, a careless use of cryptographic protocols would undo all of the mechanism rules designed to avoid collusion. This is one of the first examples of how cryptographic protocols may in fact be *worse* than trusted-third parties from a conceptual (as opposed to efficiency) point of view.

Fortunately, the authors of [15] do suggest a solution. They define and construct *collusion-free protocols* in a model in which players can exchange *physical envelopes*. Their security notion guarantees that no new method for players to collude are introduced by the protocol itself. Thus, their solution represents the best of both worlds: all of the auction rules which prevent collusive bidding remain effective, and yet all of the privacy and correctness properties implied by good cryptography are enjoyed as well.

To achieve the collusion-free property, the authors of [15] design a *verifiably deterministic* protocol in which at each point, a participant can only send a single message that the other players will accept: while each protocol message is unpredictable by the others, it is nonetheless, verifiable as the only appropriate next message. This property has also been called *forced action* to capture the idea that a player has only one (forced) action which it can take to continue the protocol. All other messages are interpreted by the other players as a signal to abort. Later, Izmalkov, Lepinski, and Micali [14] also use the idea of forced action and, by employing a new assumption concerning a physical randomizing ballot box, are able to construct protocols which are both information theoretically secure and collusion-free.

Notice, however, that both of these solutions employ exotic physical communication channels: the envelope and the ballot box. A principle criticism with these assumptions is that they have no analogues in the digital world. A physical ballot box relies on the forces of nature to provide randomness. But no party can ever be trusted to implement this in a digital (remote) setting because the moment some entity is trusted with generating the ballot box randomness, that entity essentially becomes entrusted with all of the security properties. Thus it might as well act as the trusted party computing the ideal functionality. Another (potentially less problematic) argument is that envelopes are both perfectly binding and perfectly hiding which again can not be implemented in a digital setting. Thus we are left with no choice but to implement these protocols in a physical setting.

Yet the engineering requirements of implementing such a *physical* communication model without introducing a subliminal channel seem daunting. Every last detail would have to be precisely controlled, from how exactly envelopes are transported from one player to another player, the timing of all events, marks on envelopes must be avoided and even their ambient temperature must be regulated. In the past, attacks such as reconstructing computer screens from reflection off of walls, extracting secret keys from smart cards meters away via the electric fields from card readers, or listening to private conversations by measuring the vibrations of window panes with lasers have shown that physical security and in particular physical isolation can be a *very* difficult to achieve.

Despite the difficulties with implementing these physical channels, some extra communication channel are provably necessary [15] in order to achieve collusion-freeness. Nonetheless, in a digital world, the (im)practicality of such physical channels makes it worthwhile to explore other solutions to the problem of collusion in cryptographic protocols, and our approach does precisely this.

1.2   Our New Approach

This paper addresses the problem of building collusion-free protocols without using physical channels.

Our insight comes from studying the *opposite* of verifiable determinism. Recall, the motivation behind forced-action was to remove the inherent entropy of cryptographic messages. Removing the entropy was necessary to remove the possibility of using steganography to collude during a protocol execution. Instead of taking this approach, we consider *adding* more randomness to each message so that any hidden message is smothered.

A protocol step in our approach may allow many acceptable messages and may require a lot of randomness—even randomness chosen by the prover and verifier. To avoid the previous problems of steganographic channels, our protocols only allow communication via a *mediator* communication channel. This channel has a very simple task: it removes any steganographic information from the messages sent by players by "rerandomizing" the messages. The mediator is similar to the "warden" model introduced by Simmons [20]. A first step in this direction was done in [1].

At first, it may seem that a mediator requires a substantial trust investment on behalf of the players. Let us first note that the amount of trust placed in the channel is no more than that placed in other such secure protocols. In particular, the private channel model used by Ben-Or, Goldwasser and Widgerson [4], the ballot box [14] and the communication model in [16] also assume that the adversary does not control or modify the messages sent by players. In fact, many protocols make implicit assumptions about their communication channels, be they private channels, common reference strings, synchronous channels, synchronous broadcast, envelopes [16], or ballot boxes [14]. Often the assumptions made on the communication channel are neither made explicit, nor understood explicitly.

While our channel is incomparable to these other ones, it remains plausible since any modern day router can implement the protocol instructions we require of the mediator. In particular, many modern Internet routers implementing IPsec already do many similar cryptographic operations on the data that they relay. In this sense, our mediator channel seems to be a *digital* rather than physical way to achieve the collusion-free property. As such we eliminate the need for stringent engineering requirements as players may be physically separated by great distances and any contact is on a purely informational basis.

Most importantly, unlike other protocols that employ exotic communication channels, we model the mediating communication channel as a party that can be corrupted just like other participants. (Indeed, in real life, the mediator would probably be an entity like the FCC and thus it is natural to consider what happens when it is corrupted.) As it turns out, our protocols become secure with respect to traditional multi-party protocol security notions when the mediator is corrupted. Thus, we only rely on the mediator to guarantee the collusion-free property—this is yet another reason our model is a more palatable one. If the FCC were the mediator, it is their natural interest to act honestly so as to prevent collusion. Even so, a cheating FCC would be unable to affect the privacy or correctness of the auction thereby mitigating insider threats.

Why Universal Composability Does not address the Issue. A natural question is to consider why the strong notion of universally composable security [6] is not sufficient for our purposes. Here the problem is one of modelling. Like prior security notions, UC (and its various extensions) continues to model the adversary as a monolithic entity. Thus, the security model already assumes that the corrupted parties can collude at-will among one-another during the protocol execution.

Our Contributions. To formalize our ideas, we present a new security definition for collusion-free security which models the communication channel as a corruptible participant. As discussed, our definition captures "graceful failures" with respect to the channel. Since this is an important part of our contribution, we discuss the communication model and formally define and discuss the security notions in Section 2. To illustrate the feasibility of our notions, we present a collusion-free commitment scheme in Section 3. The proof for this protocol illustrates technical issues of collusion-free security and serves as a warm-up for the secure collusion-free zero-knowledge proof of knowledge protocol we present in Section 4.

2   Model and Security Definition

We use the ideal versus real experiment indistinguishability paradigm introduced in [11] to formalize our new ideas about secure computation. Real world players are denoted with a san serif font like P, ideal players and functionalities are denoted with a calligraphic font like $\mathcal{P}$, and malicious (i.e., adversarial) players have a tilde like $\widetilde{P}$ and $\widetilde{\mathcal{P}}$. We introduce a new player M (and $\mathcal{M}$) who controls all communication between the players.

The standard security notion formalizes the idea that regardless of their input to f, for any set of corrupted real world players there exists a *single* ideal world simulator which can reconstruct everything the set of real players might learn during the execution of the secure protocol. This means *no* information beyond that exchanged in the ideal world can be exchanged between colluding real world parties.

We use a different adversarial model than the traditional monolithic approach. In particular we consider malicious parties $\widetilde{P}_1 \ldots \widetilde{P}_t$ who are playing a real world protocol with each other via the mediator M. In the ideal model, we place restrictions on the way in which $\widetilde{\mathcal{P}}_1 \ldots \widetilde{\mathcal{P}}_t$ can communicate with each other. In particular they can only communicate via a call to $\mathcal{F}$ or by using $\mathcal{M}$ as in intermediary. (This idea comes from [15], but they do not model $\mathcal{M}$.)

We wish to capture the intuition that the *distinct* adversaries $\widetilde{P}_1 \ldots \widetilde{P}_t$ should not be able to use an execution of the secure real world protocol as a means of computing any joint functionality beyond the intended ideal functionality implemented by the protocol. Clearly this goal will require a *steganography-free* real world protocol. Further we require that the real world protocol emulates $\mathcal{F}$ even in the face of a *set* of *coordinated malicious players*. Traditional security guarantees for multi-party computation are incomparable in this sense as they only constrain a monolithic adversary and how it can affect other players. For example although commitments are hiding if the sender is honest, a UC commitment can be used to compute any other functionality if sender and receiver are both corrupt and coordinated.

If mediator $\mathcal{M}$ is honest, it never sends messages to a player, and thus $\mathcal{F}$ is the only means of inter-player communication. If, on the other hand, the mediator $\mathcal{M}$ is corrupt, then all other corrupt parties in both the ideal and the real world models are able to perfectly coordinate their actions. Specifically, they (in the worst case) can use $\widetilde{M}$ in the real world and $\widetilde{\mathcal{M}}$ in the ideal world as perfectly secret and authenticated channels. Thus, in this case our security notion essentially collapses to the traditional monolithic model.

Per this discussion, our definition of a collusion free protocol comes in two parts. The first captures the case when the mediator is honest, and the second part adds the property of "secureness" by requiring that when the mediator is corrupt then any real world attack can also can be simulated in the ideal world by a monolithic adversary. In other words the mediator should only be trusted for avoiding collusions between players. If it is dishonest then the security of the protocol is (almost[1]) the same as traditional SFE security.

---

[1] Even in the ideal world the mediator is allowed to choose a forced abort set. We have chosen to make this explicit but we point out that in any model where the adversary is given full control of the network this power is implicit.

Authenticated Channels. Without a setup stage, it would be impossible for any player to establish the authenticity of information which supposedly originated from another player. This is particularly true in our model since all players communicate through M, and so a corrupted M could mount a classic man in the middle attack, hijack an identity mid protocol or route messages to the wrong destinations.

We note that [15] also requires a pre-processing phase to setup keys. In particular, they require a preprocessing round which must take place before inputs are distributed. Arguably this is an even greater problem as in certain settings (such as some auctions) it is unrealistic to expect players to meet before inputs are ever distributed. The protocol in [13, 14] implicitly uses authenticated channels since all of the participants are in physical proximity.

There are several approaches to handling this problem. The first is to assume a Public Key Infrastructure (PKI) and assume that players have registered keys and proven knowledge of the secret key. This is an approach implicitly taken by many protocols; however it is unsuitable in our work for two reasons. First, it implicitly requires that all parties perform some global action before receiving their private inputs. And secondly, it creates a technical problem for simulation. To handle a corrupt mediator, a simulator must feed messages on behalf of the honest players. If these honest players have registered public keys, then we must assume the mediator (who represents the communication channels) must also know the public keys of the honest users, and will therefore expect signatures under those keys for all messages. (Note, much like the GUC-model, we assume that the mediator can probe the *real* PKI and retrieve the real public keys of the players.) Thus, the simulator will need to know the secret keys of the honest players which seems unreasonable.

A second method is to add a setup phase to a protocol during which keys are shared. Since players know their inputs, to remain collusion-free, all communication must be passed through the mediator. In this case, however, a corrupt mediator can "fork" the honest parties into separate groups by creating several different public keys for some players. There is no broadcast channel when the mediator is corrupt, so an honest player must take on faith that the key it has received for some other player j is the one actually created and sent by player j. The authors of [3] study this forking model and suggest protocols and definitions for the situation. The protocol is complicated because it must handle non-malleability issues. We are currently working on using these tools for this task.

However, because our goal is to focus on the collusion-free property, we resolve this issue in this preliminary version by only considering broadcast-honest mediators. These are mediators who cheat arbitrarily, but perform one broadcast operation during the protocol honestly. Another way to consider such a mediator is to assume that there is a public bulletin board that all players can see, and that the mediator posts everyone's public key on this board. Such a notion is already considered in the electronic voting literature [5, 7, 8, 18]. Overall, restricting attention to these simpler cheating mediators clarifies the collusion-free issues; investigating ways to remove this restriction is a task for future work.

Aborts. An overlooked problem when considering collusion free protocols is signaling via aborts. The easiest solution is to simply assume no party will ever abort. For game theoretic applications (with punishment strategies) this remains an interesting security notion. Another approach might be to say that if one party aborts then all parties are forced to abort and no subsequent protocol is played. However in the interest of more general applications and a more robust security notion we take a different approach. We explicitly model aborts in the ideal world with greater detail.[2]

To pinpoint the issue, in prior ideal models for SFE, a party can simply abort; but in the real model instead a party can abort at a given round. Thus, it seems necessary that the ideal model allows a "round number" to be communicated during an abort in order to properly capture the full power of a real-world adversary. To specifically accommodate this in the ideal world, every abort message is accompanied by an integer. Let us emphasize that this phenomena only becomes an issue when considering the problem of steganographic communication and colluding parties since it might allow a second adversary to benefit. So we point out that this is not actually a weakening of our ideal model with respect to previous ones but rather a more detailed specification of the real security achieved. We also mention that abort messages in the ideal world are handled instantly. That is normally $\mathcal{F}$ will wait for all inputs before computing f and returning output. However if it receives an abort message it immediately sends this to all players and terminates. This corresponds to players knowing immediately (at least at the end of the current communication round) when a player has aborted in the real world rather then only finding out once the entire computation is complete. In the interest of clarity we will describe the effects of this design choice on our proofs (in the full version) in general terms. So as not to get lost in unnecessary details we make almost no further mention of aborts in the proofs and refer only to this section.

2.1   Notation

Let f be an n-input and n-output function $f : D \times D_1 \times \ldots \times D_n \to R \times R_1 \times \ldots \times R_n$. Here D_i is the domain of player i's private input and D is the domain of the public input. For example for zero knowledge D is the set of statements and D_{P} is the set of witnesses. Similarly R is the range of the public output while R_i is the range of player i's private input. (The public input and public output are just a special component of all players inputs and outputs which is the same.)

We denote by $\Pi = \langle \mathsf{P}_1(x_1, \mathsf{aux}_1), \ldots, \mathsf{P}_n(x_n, \mathsf{aux}_n) \rangle$ a protocol where player i has input x_i to f. (Note that honest players will faithfully use these as their inputs, but malicious parties may not.) The strings aux_1 through aux_n model prior information that players have at the beginning of an execution. Let $[n]$ to denote the set $\{1, 2, \ldots, n\}$. If $A \subseteq [n]$ then we write e_A to refer to the arbitrary output of all players in A. We write $\bar{A}$ to refer to the compliment of A in $[n]$.

---

[2] In particular in [15] aborts are modeled via a simple "abort flag" both in the ideal and real games.

A subscript of $\mathbb{I}$ denotes an ideal world protocol while a subscript of $\mathbb{R}$ denotes a real world protocol. For ideal world protocols a super script of $\mathcal{F}$ implies that all parties have access to the ideal functionality $\mathcal{F}$. Thus $\langle S, R, M \rangle_{\mathbb{R}}$ is a real world protocol between S and R with mediator M while $\langle \mathcal{S}, \mathcal{R}, \mathcal{M} \rangle_{\mathbb{I}}^{\mathcal{F}}$ is an ideal world protocol between $\mathcal{S}$ and $\mathcal{R}$ who have access to ideal functionality $\mathcal{F}$ (and ideal mediator $\mathcal{M}$). We will use $\mathtt{VU}(\mathsf{P}(x))$ to denote the entire view of P for an execution with input x (including random tape, inputs and messages recieved) and we use $\mathtt{IO}(\mathsf{P}(x))$ to denote only the input and output pair of an execution.

Ideal execution. In the ideal execution, all players, including the ideal mediator $\mathcal{M}$, have private channels to the ideal functionality $\mathcal{F}$. In addition, $\mathcal{M}$ has bidirectional channels to all ideal players.

A round of an ideal protocol begins with all players $\mathcal{P}_1, \ldots, \mathcal{P}_n$ and $\mathcal{M}$ sending their input to $\mathcal{F}$. The mediator's input to $\mathcal{F}$ is a tuple $(S_1, \ldots, S_\ell, I_i, \ldots, I_\ell)$ which allows the corrupt mediator to "fork" the different players into separate groups. Namely, $\mathcal{M}$ inputs a list of non-empty disjoint subsets $S_1, \ldots, S_\ell$ that partition the set of players $\{1, \ldots, n\}$. In addition, for each S_i, $\mathcal{M}$ specifies the inputs I_i to $\mathcal{F}$ for the other players $\bar{S}_i$.

For each S_i, the ideal functionality evaluates $\mathcal{F}$ on the inputs provided by the players in S_i and the inputs for the players in $\bar{S}_i$ specified by I_i. It then begins to deliver the outputs to the players in S_i in some canonical order. When it is player j's turn to receive output, the ideal functionality queries $\mathcal{M}$ about whether to deliver the output or whether to broadcast an integer $a \in [\mathtt{poly}(k)]$ and halt. If $\mathcal{M}$ agrees to delivery, then the ideal functionality sends the output to that player and continues to the next player in order. The output values for the players in $\bar{S}_i$ are sent to $\mathcal{M}$. If instead $\mathcal{M}$ chooses to broadcast an integer a, then the ideal functionality sends $(\bot, a)$ to all players and halts.

$\mathcal{F}$ can maintain internal state between rounds allowing for such functionalities as commitment schemes. In the ideal model:

- an honest player never sends a message to $\mathcal{M}$;
- an honest $\mathcal{M}$ never sends a message to a player, always inputs $S_0 = [n]$ and $I_0 = \emptyset$ to $\mathcal{F}$ and always agrees to output delivery;
- a corrupt ideal mediator $\mathcal{M}$ does not see the private outputs of $\mathcal{F}$ (at least for honest players). The only values related to the computation of honest parties it learns is the public input and public output and the output values for each of the corrupted players.

Real execution. In the real world, there is no $\mathcal{F}$ and instead players have authenticated private channels connecting them to a special party M called the (real world) mediator. Thus M sits at the center of a star network connected to P_1 through P_n "mediating" all communication. As a result, M controls many aspects of the computation which necessitates the complicated provisions given to the ideal mediator described above.

Real world executions begin with each player and the mediator selecting fresh independent uniform random strings and jointly executing an interactive protocol. At the end of both a real and an ideal execution all players privately output an arbitrary string which captures information they have learned. The honest mediator always outputs the special message $\perp$. Honest parties output the results of computing f, but corrupted players can output whatever they wish.

2.2   Collusion Free Protocols

As usual with ideal/real paradigm definitions, a protocol is considered secure if the probability ensemble representing a real execution is indistinguishable from an ensemble representing an ideal execution. The definition below, however, changes the order of some quantifiers to capture a specific concern.

Definition 1. *Let* $\Pi = \langle P_1, \ldots, P_n, M \rangle_{\mathbb{R}}$ *be an n player protocol with security parameter k and let $\mathcal{F}$ be an ideal functionality computing the n-input and n-output function f. A protocol Π is a* collusion free *protocol for $\mathcal{F}$ if for any efficient real player $\widetilde{P}$ (including those playing the honest protocol) there exists an efficient ideal simulator $\widetilde{\mathcal{P}}$ with access to a common random tape R such that for any vector of players $(\widetilde{P}_1, \ldots, \widetilde{P}_n)$ for all inputs $x_i \in \{0,1\}^*$ and for all* $\mathsf{aux}_i \in \{0,1\}^*$, *the following two ensembles are indistinguishable:*

$$\left\{ \left\langle \left\{ \mathtt{VU}(\widetilde{P}_i(x_i, \mathsf{aux}_i)) \right\}_{i \in [n]}, \mathtt{IO}(M(\mathsf{aux_M})) \right\rangle_{\mathbb{R}} \right\}_k$$

and

$$\left\{ \left\langle \left\{ \mathtt{VU}(\widetilde{\mathcal{P}}_i(x_i, \mathsf{aux}_i, R)) \right\}_{i \in [n]}, \mathtt{IO}(\mathcal{M}(\mathsf{aux_M})) \right\rangle_{\mathbb{I}}^{\mathcal{F}} \right\}_{R,k}$$

Further, we call Π a secure collusion free *protocol if in addition, every mediator $\widetilde{M}$ also has a simulator $\widetilde{\mathcal{M}}$ for which the indistinguishability holds.*

In the above definition, the ensembles are taken over the values of k, R and the random choices of the players and mediator. By "efficient" adversaries we mean a probabilistic polynomial time one in the size of the input and security parameter.

Comments. Since the experiments' outputs include the complete vector of arbitrary output strings, the components must have a similar *joint* distributions in both experiments.

For the case of collusion freeness, M is honest, so the ideal adversaries must be able to produce the same joint output distribution as in the real world, but only by communicating via at most one call to $\mathcal{F}$. Therefore these sets of *distinct* adversaries can not jointly compute anything more then what is revealed by f. In particular they can not use the real world protocol to exchange information steganographically or jointly compute anything beyond what one call to f reveals.

Secure collusion-free protocols are more general. If mediator is corrupt then since honest parties output their results from f, even a monolithic adversary (i.e., distinct adversaries perfectly coordinating through the corrupt mediator) can not alter the output of the joint functionality nor learn anything about it beyond what their own output reveals. (The worst they can do is cause aborts as $\widetilde{\mathcal{M}}$ gets to choose the forced abort set.) This guarantees the traditional security properties such as privacy and correctness for general SFE in much the same way the UC framework captures these notions. (Though with weaker composability properties since a distinguisher for multi-party collusion freeness only sees inputs and outputs but does not have all the powers of the environment in UC definitions.)

Another subtlety is a consequence of the order of the quantifiers and the intuitive attack scenario we model. For every corrupt player, there exists a simulator with access to a common random tape R shared amongst all such simulators. This means that the ideal simulators do not know which other parties are corrupt. The only joint information they are given is the random tape R. One might argue that it is meaningful for simulators to know the complete set of corrupt parties and their code. After all colluding parties may well be perfectly coordinated going into the protocol. However such a definition would not preclude real world player from using an execution of Π to exchange information unrelated to the current execution. So for example it might be that a pair of corrupt parties which have had no previous contact could use an execution to exchange phone numbers. Although they might have done this anyway in advance we still would like to *force* them to have exchanged this information before hand. This allows for applications such as the online poker house or security in a large intelligence agencies network where individual players may never had contact before.

By requiring that all simulators only share the (read only) random tape R, we model the fact that in the real world protocols, players observe common randomness although they can not influence what this randomness is (via secure secret sharing for example). However beyond this publicly visible randomness no information is exchanged. We note that this is a strengthening with respect to Definition 1 in [15]. Further, the fact that players can agree on common randomness after a real execution is also a property of the ballot-box protocol in [14] and so technically breaks the central theorem of perfectly emulating the ideal game. In the context of game theoretic applications this may in fact present a real problem as it is known that correlated randomness can be used to achieve equilibria with better expected payoffs then otherwise possible [2]. Thus at a minimum perfect composability is lost. The problem in [14] has been addressed in follow up work [13]; we also resolve it for this protocol in the full version.

Another improvement over the collusion free definition in [15] is that we no longer require a "reconciliation function" to compare the real and ideal world outputs. This simplifies both the definition of collusion-free protocols and simplifies the security proof strategy.

2.3   Authenticated Channels

Typically, key registration is handled by a trusted-third party. As mentioned above, since we do not want to assume such a PKI and also want to model security when external trust assumptions fail, we describe a pre-processing protocol using the mediator. Naturally, our process provides a somewhat weaker form of security then might be achieved with a trusted key registration server. In particular there is no guarantee that an honest player will successfully register his key when the mediator has been corrupted. But on the other hand even if identities were established perfectly once computation starts $\tilde{\mathsf{M}}$ can always cut off a player i and no other player j can tell whether this is because of a corrupt $\tilde{\mathsf{M}}$ or corrupt player i. In some sense, this is unavoidable.

The other weakness of our preprocessing is that $\tilde{\mathsf{M}}$ can fake the presence of a player by including a public key for them. However this too is unavoidable since even with perfect authentication $\tilde{\mathsf{M}}$ could simply setup a dummy player with an identity and take part in the authentication and joint computation stages through the dummy. As such, this is a problem which is outside the scope of this work.

Initialization of the set of parties. A first approach to key-setup might be to have parties generate a key pair, register their public key with M and run a ZKPoK to prove they know the corresponding secret keys. However, as noted by [15], if public keys are selected by the parties, the keys may not be chosen honestly. For example, they can be chosen from intentionally "weak" or special distributions which allow other parties to break the key. This could allow parties to exchange information during the protocol. To solve this problem, a more elaborate registration procedure is executed. First each player runs a coin-flipping protocol with M (without the final decommitment) such that only the player knows the result of the coin flipping. The player then uses this randomness as input to the algorithm that generates the key pair. Then the public key is sent to M along with a ZKPoK that it used the result of the coin flipping to generate the pair. Once all public keys have been registered with M it broadcasts the set and the joint computation phase can begin. We call the resulting initialization stage Δ_n^{M}.

There is one technicality to handle. We require that the ZKPoK enjoys the concurrent zero knowledge property [10, 19]. Indeed, an adversarial M could run as verifier of the ZKPoKs by coordinating the executions in an adaptive and malicious way with all honest parties. In order to avoid such attacks we require a concurrent ZKPoK protocol [17, 19].

Collusion Free Authentication. Once keys have been shared, there are various ways in which they can be used to emulate authenticated channels. Since it is not our principle focus, we describe the following correct but not necessarily most efficient method for doing so. Let us briefly note the subtle requirements which prevent standard protocols from being adopted. On the one hand, we cannot allow one player to send a signature to another (since that would enable steganography), but we must provide enough evidence to the second player to accept authenticated messages.

At the beginning of each real world collusion free-protocol execution e each player P_i sends a random k-bit string r_e to M along with a signature σ of r_e (where k is the security parameter). The string r_e will act as P_i's session identifier for e. M sends $\mathsf{Com}(r_e)$ and $\mathsf{Com}(\sigma)$ to all other players P_j and proves in ZK that $\mathsf{Com}(\sigma)$ decommits to a signature of the decommitment of $\mathsf{Com}(r_e)$

which verifies under svk_i. A subsequent round t, message m_e to player j of the execution e is accompanied by a signature of $\sigma' = \mathsf{Sig}_i(r_e, t, \mathsf{svk}_j, m_e)$. M proves in zero knowledge to player j that there is a signature σ' of a message with the following prefix. σ (i.e., the decommitment of the first message $\mathsf{Com}(\sigma)$) is a valid signature for the first part of the prefix, and the second part of the prefix t, is the current round number, and the third part of the prefix equals player j's verification key. Thus the player j knows that m_e came from P_i playing a session with the current random identifier, that it is the t-th round message in such a session and that it is the intended recipient. With overwhelming probability (in k) there will only be a single session with this session identifier so the message must be from P_i during this session.

Calling Subroutines. One issue remains with this approach. The collusion free protocol e may use another collusion free protocols e' as subroutines. To make sure that M can not run replay attacks emulating P_i for the entire l-th call of the subroutine (by playing an old $r_{e'}$ and σ) the subroutine is essentially treated as a message in the calling protocol. In particular all signatures (including the first σ of the subroutine) are of messages prefixed by l, the number of the call and r_e, the session identifier of the calling protocol. For each message in e' M additionally proves in ZK that this extra prefix corresponds to the session identifier of e and the current call number. If M can substitute any previous sub-protocol for P_i then M can be used for existential forgery attacks against the signature scheme since either r_e or l will be different (with overwhelming probability in k).

  For most of the rest of this paper we will not explicitly discuss the authentication mechanism as this will needlessly complicate exposition. The only exception to this is when we discuss how ideal world simulators internally emulate honest players when running their experiments as this will require them to be able to create signed messages.

3   Collusion Free Commitments

The first primitive we consider is a commitment scheme. That is we consider the functionality $\mathcal{F}_{\mathsf{Com}}$ which works in two rounds. In the first round it takes a message m as input from the (ideal world) sender $\mathcal{S}$ and produces output *committed* for the (ideal world) receiver $\mathcal{R}$. In the second round it takes input *decommit* from $\mathcal{S}$ and gives the output m to $\mathcal{R}$. All other inputs and outputs are the special message $\perp$.

The Difficulty. Before continuing, let us discuss why this is a non-trivial task. Players can not be allowed to directly exchange messages as this would enable steganography. So the mediator is used to rerandomize communication. However there is a more subtle attack. Consider the last round of the real world protocol to be played by a particular player. Suppose they can tell the value of even just two (non-constant) bits in another parties view (or even a function there of).

Then by choosing whether or not to abort depending on the value of those bits in the current execution the player can signal information via the real world protocol (beyond the mere abort message modeled in the ideal world). Therefore having the mediator ensure messages are randomized correctly is not enough. In fact there can not be any (computable) correlation—be it chosen in advance or random but fixed—between players' views. I.e. it is not enough to ensure one player can not fix a function of some bits in another's view, but it must even be impossible for one player to subsequently guess any function of some bits of another's view after they have been fixed.

Yet at the same time we do not want to require the mediator to be honest for the security of the protocol (i.e., hiding and binding for a commitment scheme). In the following protocol we resolve these conflicting requirements.

The Protocol. In figure 1 we give a protocol which makes use of a statistically binding commitment scheme (Com, Dec) and a zero knowledge proof of knowledge ZKPoK. We then prove it to be a collusion free protocol for the $\mathcal{F}_{\mathsf{Com}}$ functionality. In particular we prove that corrupt $\widetilde{\mathsf{R}}$ and $\widetilde{\mathsf{S}}$ can be simulated by a pair $\widetilde{\mathcal{R}}$ and $\widetilde{\mathcal{S}}$ in the ideal world. Further we prove the protocol to be hiding if S is honest and statistically binding if R is honest[3]. We note that one-way permutations suffice to construct all three primitives.

Theorem 1. *Assuming the existence of one way-functions the protocol in figure 1 is a statistically binding commitment scheme. Further it is a collusion free protocol for the functionality $\mathcal{F}_{\mathsf{Com}}$.*

Common Input: Signature verification key $\mathsf{svk_S}$.
Committer Input: Signature key $\mathsf{ssk_S}$ and message m

Commitment Phase

C.1 S selects random coins r, and sends $c = \mathsf{Com}(m, r)$ to M.
C.2 S runs ZKPoK with M proving knowledge of m and r such that $c = \mathsf{Com}(m, r)$.
C.3 M selects random coins r' and sends $c' = \mathsf{Com}(c, r')$ to R.

Decommitment Phase

D.1 S sends $(m, r) = \mathsf{Dec}(c, r)$ to M.
D.2 M sends m to R.
D.3 M runs ZKPoK with R proving that there exist coins r and r' such that $c' = \mathsf{Com}(\mathsf{Com}(m, r), r')$.

Fig. 1. Collusion-Free Perfectly-Binding Commitments

---

[3] We note that if a statistically hiding commitment scheme $(\mathsf{Com'}, \mathsf{Dec'})$ is used instead of $(\mathsf{Com}, \mathsf{Dec})$ and the zero-knowledge argument ZK is statistical zero knowledge then the resulting protocol is statistically hiding. The proof remains largely the same.

Proof. (High-level proof idea) The idea behind the following proof can be summarized as follows. The hiding property of the protocol is reduced to the hiding of (Com, Dec) and the zero knowledge property of ZKPoK in steps C.1 and C.2. The binding property of the protocol in figure 1 is reduced to the binding property of (Com, Dec) and the soundness of the ZKPoK in step D.3. Both of these proofs appear in the full version.

To prove collusion freeness we construct two simulators, one for a corrupt sender and one for a corrupt receiver. The simulator for the Sender extracts the values m, r and forwards appropriate messages to the ideal functionality. The simulator for the Receiver sends a commitment to 0 during the commit protocol, sends the committed message (received from the ideal functionality) during the decommit protocol and uses the simulator for the ZKPoK to equivocate the commitment. The hiding property of the commitment scheme is necessary to prove that these simulators meet the definition.

Lemma 1. *Assuming the existence of one-to-one one-way functions, the protocol in figure 1 is a collusion-free protocol for $\mathcal{F}_{\mathsf{Com}}$.*

Proof. Omitted for space.

Achieving Secure Collusion-Free Commitments. For lack of space we only mention here how to achieve secure collusion-free commitments. First of all, the setup phase requires that each player knows the public key of each other player, where the public key consists of a pair of public keys, one for signature, and one for rerandomizable CPA encryption (e.g. El Gamal).

In the protocol, M receives a commitment of m from S, sends to R the commitment of the commitment and proves to R in zero knowledge that the committed message has been sent by S. This ends the commitment phase.

The opening phase is instead more complex and is played as follows. M and S play a simulatable coin-tossing protocol such that only M obtains the output. M receives from S an encryption of m under R's public-key. The ciphertext is rerandomized by M and sent to R, using as randomness the output of the coin-tossing protocol. M also proves to R that the ciphertext is a rerandomization of a ciphertext sent by S, using as randomness the one established by the coin-tossing protocol. So far, m has been read by R and not by M, however a consistency check with the commitment phase has still to be played. The opening phase continues with S that proves in zero-knowledge to M, using a specific implementation of Blum's protocol that the encrypted message corresponds to the committed one. The specific implementation goes as follows. Consider a triple of messages (a, e, z) of Blum's protocol. After receiving a, M computes a commitment c of a and sends it to R. Then M plays again with R a simulatable coin-tossing protocol, so that only it obtains the output e, and this is played in the second round of Blum's protocol. Then M receives z from S and proves to R in zero-knowledge that it knows an accepting third message z for Blum's protocol, where the first message is committed in c and the second round is the output of the coin-tossing protocol.

4   Secure Collusion-Free ZKPoK

The second primitive we construct is a *secure* collusion free zero knowledge proof of knowledge for any language in NP. We consider the protocol $\langle \mathcal{P}, \mathcal{V}, \mathcal{M} \rangle_{\mathbb{I}}^{\mathcal{F}_{\mathsf{ZKPoK}}}$ where the functionality $\mathcal{F}_{\mathsf{ZKPoK}}$ works in one round. As public input it receives a graph G and from $\mathcal{P}$ it takes private input a 3-coloring w. If w is a valid 3-coloring of G it produces public output *true* and otherwise the public output is $(false, \lambda)$. All other inputs and outputs not specified above are the special message $\perp$. Intuitively λ models the specific round at which the proof failed. As with aborts this is important since we need to capture every bit of communication in the real world between prover and verifier and failing a proof in the i-th round can be used to signal $\log(i)$ bits.

The Protocol. In figure 2 we give a protocol which is based on the zero knowledge proof of knowledge for graph 3-colorability (G3C) in [12]. To do this we make use of a statistically binding commitment scheme (Com, Dec) and a zero knowledge proof of knowledge ZKPoK. We then prove this protocol be a secure collusion free protocol for $\mathcal{F}_{\mathsf{ZKPoK}}$. Note that apart from collusion freeness this also implies all the usual properties of a zero knowledge proof of knowledge. In particular the protocol is complete; if V is honest then it is a proof of knowledge with respect to any $(\widetilde{\mathsf{P}}, \widetilde{\mathsf{M}})$; and if P is honest then it is zero knowledge with respect to any $(\widetilde{\mathsf{M}}, \widetilde{\mathsf{V}})$. We also note that one way permutations are sufficient for constructing all primitives used by this protocol.

We prove secure collusion freeness by showing that for any non-empty subset in $\{\widetilde{\mathsf{P}}, \widetilde{\mathsf{M}}, \widetilde{\mathsf{V}}\}$ of corrupt parties we can construct a set of simulators such that their joint output together with that of the honest ideal parties is (jointly) indistinguishable from that of all real world players (corrupt and honest).

Theorem 2. *Assuming the existence of one-way permutations the protocol in figure 2 is a secure collusion free protocol for the functionality $\mathcal{F}_{\mathsf{ZKPoK}}$.*

If all parties are corrupt then there is no security proof to give. Thus now concentrate on the remaining 6 cases.

The rest of the proof is structured as follows (with some intuition behind each step in the bullets):

1. We construct 3 simulators, one for each corrupt party. The simulator for a corrupt verifier is based on the zero knowledge simulator for underlying ZKPoK for G3C. The simulator for a corrupt prover is based on the knowledge extractor for the underlying ZKPoK. The simulator for the mediator is a combination of the previous two.

2. We describe how to handle aborts and failed proofs. Essentially these are coordinated via messages to and from $\mathcal{F}_{\mathsf{ZKPoK}}$ and the value λ which indicates at which round aborts and failure have occurred.

3. We prove a lemma stating that assuming the existence of one-way permutations if the mediator and either verifier or prover is corrupt then the pair

---

Common Input: Verification keys svk_P and svk_V (from file F), graph $G = (V, E)$.
Prover P Input: Signature key ssk_P, 3-coloring w of G.
Verifier V Input: Signature key ssk_V.

ZK.1 M sends $\phi = \text{Com}(\pi, r)$ to V where π is a random permutation over $[|E|]$.
ZK.2 P selects a random permutations of $\{red, green, blue\}$ and applies it to w resulting in the 3-coloring w'. It colors G according to w' resulting in set $E' = w'(E)$ of the colors for the end points of each edge in G. P computes $\nu = \{\nu_e = \text{Com}((c_1, c_2)_e, r_e)\}_{e \in E'}$ where $c_1, c_2 \in \{red, green, blue\}$. P sends the set of commitments of the 3-coloring ν to M.
ZK.3 M sends $\nu' = \{\nu'_e = \text{Com}(\nu_e, r'_e)\}_{e \in E'}$ to V.
ZK.4 V sends $\eta \in E$, a challenge edge of G to M.
ZK.5 M sends $\eta' = \pi(\eta)$ to P.
ZK.6 P sends $((c_1, c_2)_{\eta'}, r_{\eta'}) = \text{Dec}(v_{\eta'})$, decommitment to the coloring of edge η' to M.
ZK.7 If $c_1 = c_2$ or either one is not in $\{red, green, blue\}$ then M sends $failed$ to V. Then it uses witness $(r_{\eta'}, r'_{\eta'}, \pi)$ to run ZKPoK with V proving that $\nu'_{\pi(\eta)}$ is a commitment of a commitment of an edge with different (valid) colored ends and moreover the edge is the one selected by V according to the permutation π committed in ϕ. If V is not convinced by the proof it rejects.
ZK.8 Repeat steps ZK.1-ZK.7 $|E|^2$ times with new random tapes. V accepts G is 3-colorable if and only if it accepts all proofs in steps ZK.7.

Fig. 2. Collusion-Free Zero Knowledge

of simulators and honest party have the desired output. The proof resembles the proofs of correctness of the underlying simulator and knowledge extractors for the ZKPoK of G3C.

4. We prove a lemma stating that assuming the existence of one-way permutations, if one party is corrupt then the simulator and honest parties have the correct output distribution. These cases are special cases of the lemma proven in the previous step.

5. We prove a lemma stating that assuming the existence of one-way permutations the protocol is collusion free. This lemma covers the final case when verifier and prover are corrupt playing through an honest mediator. The proof works because an honest mediator "re-randomizes" messages which causes the view of any one corrupt party to be relatively unchanged by modifications to any other (non-aborting) parties algorithm. Thus, the simulator for the sender can use the honest receiver algorithm in its internal emulation.

6. We draw on all three lemmas to conclude the proof of theorem 2. We point out that all cases of possible sets of corrupt parties have been dealt with and so the theorem holds.

The detailed proof can be found in the full version.

Acknowledgments

We wish to thank Giuseppe Persiano for his invaluable contributions which helped kick start this paper. In particular the idea of a mediated communication model is due to him. Also we would like to thank Yevgeniy Dodis and Daniel Wichs for there many thoughtful discussions on the topic. Finally we are grateful for Jesper Nielsen's comments concerning the intricacies of applying MPC to game theoretic situations.

We thank the anonymous reviewers for their suggestions. The work of the authors has been supported in part by the European Commission through the IST program under Contract IST-2002-507932 ECRYPT and the one of the last author through the FP6 program under contract FP6-1596 AEOLUS.

References

[1] Alparone, M., Persiano, G.: Staganography-free implementation of yao's protocol. Technical report (unpublished, 2006)

[2] Aumann, R.J.: Subjectivity and Correlation in Randomized Strategies. Journal of Mathematical Economics 1(1), 67–96 (1974)

[3] Barak, B., Canetti, R., Lindell, Y., Pass, R., Rabin, T.: Secure computation without authentication. In: CRYPTO 2006. LNCS, vol. 4622, pp. 361–377. Springer, Heidelberg (2005)

[4] Ben-Or, M., Goldwasser, S., Wigderson, A.: Completeness Theorems for Non-Cryptographic Fault-Tolerant Distributed Computation. In: STOC 1988: Proceedings of the twentieth annual ACM symposium on Theory of computing, pp. 1–10. ACM, New York (1988)

[5] Boneh, D., Golle, P.: Almost Entirely Correct Mixing with Applications to Voting. In: CCS 2002, pp. 68–77 (2002)

[6] Canetti, R.: Universally composable security: A new paradigm for cryptographic protocols. In: Proc. 42nd FOCS, pp. 136–145 (2001)

[7] Chaum, D., Ryan, P., Schneider, S.: A Practical Voter-Verifiable Election Scheme. In: de Capitani di Vimercati, S., Syverson, P.F., Gollmann, D. (eds.) ESORICS 2005. LNCS, vol. 3679, pp. 118–139. Springer, Heidelberg (2005)

[8] Cramer, R., Gennaro, R., Schoenmakers, B.: A Secure and Optimally Efficient Multi-Authority Election Scheme. In: EUROCRYPT 2006. LNCS, pp. 103–118. Springer, Heidelberg (2006)

[9] Cramton, P., Schwartz, J.: Collusive bidding in the fcc spectrum auctions. Technical Report 02collude, University of Maryland, Department of Economics (December 2002), http://ideas.repec.org/p/pcc/pccumd/02collude.html

[10] Dwork, C., Naor, M., Sahai, A.: Concurrent Zero-Knowledge. In: 30th ACM Symposium on Theory of Computing (STOC 1998), pp. 409–418. ACM, New York (1998)

[11] Goldreich, O., Micali, S., Wigderson, A.: How to Play Any Mental Game. In: STOC 1987: Proceedings of the nineteenth annual ACM conference on Theory of computing, pp. 218–229. ACM Press, New York (1987)

[12] Goldreich, O., Micali, S., Wigderson, A.: Proofs that Yield Nothing but their Validity or all Languages in NP have Zero-Knowledge Proof Systems. J. ACM 38(3), 690–728 (1991)

[13] Izmalkov, S., Lepinski, M., Micali, S.: Verifiably secure devices. In: Canetti, R. (ed.) TCC 2008. LNCS, vol. 4948, pp. 273–301. Springer, Heidelberg (2008)

[14] Izmalkov, S., Micali, S., Lepinski, M.: Rational Secure Computation and Ideal Mechanism Design. In: FOCS 2005: Proceedings of the 46th Annual IEEE Symposium on Foundations of Computer Science, Washington, DC, USA, pp. 585–595. IEEE Computer Society, Los Alamitos (2005)

[15] Lepinksi, M., Micali, S., Shelat, A.: Collusion-Free Protocols. In: STOC 2005: Proceedings of the thirty-seventh annual ACM symposium on Theory of computing, pp. 543–552. ACM, New York (2005)

[16] Lepinski, M., Micali, S., Peikert, C., Shelat, A.: Completely Fair SFE and Coalition-Safe Cheap Talk. In: PODC 2004: Proceedings of the twenty-third annual ACM symposium on Principles of distributed computing, pp. 1–10. ACM Press, New York (2004)

[17] Micciancio, D., Petrank, E.: Simulatable Commitments and Efficient Concurrent Zero-Knowledge. In: Biham, E. (ed.) EUROCRYPT 2003. LNCS, vol. 2656, pp. 140–159. Springer, Heidelberg (2003)

[18] Neff, C.: A Verifiable Secret Shuffle and its Application to e-Voting. In: CCS 2001, pp. 116–125 (2001)

[19] Prabhakaran, M., Rosen, A., Sahai, A.: Concurrent Zero-Knowledge with Logarithmic Round Complexity. In: 43th IEEE Symposium on Foundations of Computer Science (FOCS 2002), pp. 366–375 (2002)

[20] Simmons, G.J.: The prisoners' problem and the subliminal channel. In: CRYPTO 1983, pp. 51–67 (1983)

Efficient Constructions of Composable Commitments and Zero-Knowledge Proofs

Yevgeniy Dodis[1], Victor Shoup[1], and Shabsi Walfish[2]

[1] New York University
{dodis,shoup}@cs.nyu.edu
[2] Google
walfish@cs.nyu.edu

Abstract. Canetti et al. [7] recently proposed a new framework — termed *Generalized Universal Composability* (GUC) — for properly analyzing concurrent execution of cryptographic protocols in the presence of a global setup, and constructed the first known GUC-secure implementations of commitment (GUCC) and zero-knowledge (GUC ZK), which suffice to implement any two-party or multi-party functionality under several natural and relatively mild setup assumptions. Unfortunately, the feasibility results of [7] used rather inefficient constructions.

In this paper, we dramatically improve the efficiency of (adaptively-secure) GUCC and GUC ZK assuming data erasures are allowed. Namely, using the same minimal setup assumptions as those used by [7], we build
- a direct and efficient constant-round GUC ZK for R from any "dense" Ω-protocol [21] for R. As a corollary, we get a semi-efficient construction from any Σ-protocol for R (*without doing the Cook-Levin reduction*), and a very efficient GUC ZK for proving knowledge of a discrete log representation.
- the first *constant-rate* (and constant-round) GUCC scheme.

Additionally, we show how to properly model a random oracle in the GUC framework without losing *deniability*, which is one of the attractive features of the GUC framework. In particular, by adding the random oracle to the setup assumptions used by [7], we build the first two-round (which we show is optimal), deniable, straight-line extractable and simulatable ZK proof for any NP relation R.

1 Introduction

UC FRAMEWORK. The *Universal Composability* (UC) framework introduced by Canetti [6] is an increasingly popular framework for analyzing cryptographic protocols that are expected to be concurrently executed with other, possibly malicious protocols. The UC framework has many attractive properties, one of which is a powerful composition theorem enabling the design of complex protocols to be split into simpler sub-protocols. In particular, Canetti, Lindell, Ostrovsky and Sahai [13] showed that, under well established cryptographic assumptions, UC-secure commitments and zero-knowledge (ZK) proofs are sufficient to implement any other functionality, confirming our long-standing

D. Wagner (Ed.): CRYPTO 2008, LNCS 5157, pp. 515–535, 2008.
© International Association for Cryptologic Research 2008

intuition that commitments and ZK proofs are fundamental cryptographic primitives.[1]

Unfortunately, a series of sweeping impossibility results [6, 9, 12] showed that most useful cryptographic functionalities, including commitment and ZK, are impossible to realize in the "plain UC" framework. This means that some form of a "trusted setup", such as a common reference string (CRS) or a public-key infrastructure (PKI), is necessary to build UC-secure protocols (unless one is willing to relax some important properties of UC-security). To address this issue, the original UC framework was augmented to allow trusted setup. However, until the recent work of [7], this extension only allowed one to model such setup as a *local setup*. This means that the setup cannot be seen by the environment or other protocols, and, as a consequence, it only exists meaningfully in the real model. In particular, the simulator had complete control over the setup in the ideal model. For example, in the CRS model the simulator had the freedom to choose its own CRS and embed some trapdoor information into it. As was argued in a series of papers [3, 7, 9, 14], this modeling creates several serious problems not present in the "plain UC" framework. Two of the most significant such problems are *lack of deniability* and *restrictive composition*. For example, an ideal ZK proof is "deniable", since the verifier only learns that the statement is true, but cannot reliably prove it to a third party. Unfortunately, it was argued in [7] that any UC-secure realization of ZK in the CRS model is *never deniable*. The composition problem is a bit more subtle to explain. In essence, one can only compose several instances of *specially-designed protocols*. In particular, it is not safe to use protocols which can depend on the setup information (*e.g.*, the CRS), even if these protocols are perfectly secure in the ideal model. We refer the reader to [7, 18, 27] for more discussion of this issue.

GUC FRAMEWORK. Motivated by solving the problems caused by modeling the setup as a local subroutine, Canetti et al. [7] introduced a new extension of the UC framework — termed *Generalized Universal Composability* (GUC) — for properly analyzing concurrent execution of cryptographic protocols in the presence of a *global setup*. We stress that GUC is a general *framework* strictly more powerful than UC. Namely, one can still model local setup as before. However, the GUC framework also allows one to model *global setup* which is directly accessible to the environment. More precisely, the GUC framework allows one to design protocols that share state via *shared functionalities* (such as a *global CRS* or *global* PKI). Since the same shared functionality will exist in multiple sessions, the environment effectively has direct access to the functionality, meaning that the simulator cannot "tamper" with the setup in the ideal model. In fact, the same setup exists both in the real *and in the ideal models*. As the result, modeling the global setup in this manner regains the attractive properties of the "plain UC", including deniability and general composition. This was formally

---

[1] Although [13] presented their results in the common reference string (CRS) model using the JUC theorem [14], one can extract a general implication which is independent of the CRS and does not use JUC. See page 131 of Walfish's thesis [27] for details.

shown by [7] for the case of composition, and informally argued for deniability (since the simulator no longer has any "unfair" advantage over the real-model attacker, so the real-model attacker can run the simulator "in its head" to make up transcripts of conversation which never happened in real life). To put this (convincing but) informal argument on firmer ground, in the full version [18] we give a very strong definition of deniable zero-knowledge (much stronger than previous notions appearing in the literature), and show that GUC-security implies this notion, as long as the setup is modeled as a shared functionality.

Of course, having introduced GUC, a natural question is whether one can actually build GUC-secure protocols under *natural* setup assumptions. On the positive side, one can always artificially model local setup as "global setup" by ensuring that a fresh instance of a setup is run for every protocol instance, and, more importantly, that only the participants of a given protocol have reliable access to this setup information. For example, the CRS setup of UC could be equivalently modeled in GUC as a kind of one-time "*session* reference string" (SRS) functionality: the SRS will pick a fresh reference string for each protocol instance, and will make this string available precisely to the parties running this session (and the adversary), but nobody else. It is easily shown that *UC+CRS is equivalent to GUC+SRS*, so the feasibility result of [13] would apply to the "global SRS" setup. Of course, such a "session reference string" model is very unrealistic and difficult to implement, and one may wonder if a *truly global* CRS setup would suffice as well. Unfortunately, [7] showed that the (global) CRS model (as well as other global setup which only provides *public* information, such as the random oracle model [22]) is *not* enough to sidestep the impossibility results of [6, 9, 12]. (In particular, the protocols of [13, 22] are insecure in the GUC framework with the global CRS/random oracle.) This means that any setup sufficient for GUC feasibility must provide some unpublished information, as was the case with the SRS model (where the SRS was hidden from the environment and other protocols).

ACRS MODEL. Luckily, Canetti et al. [7] introduced a new setup assumption, called *Augmented CRS* (ACRS), and showed that it can be used to GUC-realize commitment and ZK (and, thus, any other functionality), even in the presence of adaptive adversaries.[2] The ACRS model is very close to the (global) CRS model, but is (necessarily) augmented so as to circumvent the impossibility result for plain CRS. As in the CRS setup, all parties have access to a short reference string that is taken from a pre-determined distribution. In addition, the ACRS setup allows corrupted parties to obtain "personalized" secret keys that are derived from the reference string, their public identities, and some "global secret" that is related to the public string and remains unknown. It is stressed that *only corrupted parties* may obtain their secret keys. This may sound strange at first, but is actually a huge advantage of the ACRS model over the more traditional

---

[2] [7] also showed similar results in a variant of a PKI-like "key registration with knowledge (KRK)" setup from [3]. However, since the ACRS model is more minimal and all our results easily extend to the KRK model, we only concentrate on the ACRS model.

"identity-based" setup, where even honest parties *need* to obtain (and, therefore, safeguard) their keys. Namely, the ACRS setup implies that the protocol may not include instructions that require knowledge of the secret keys, and, thus, honest parties do not need their secret keys. In fact, they can only lose *their own* security by obtaining these keys and using them carelessly. This is consistent with any secret-key cryptosystem, where a party will lose its security by publishing its secret key. Luckily, though, the ACRS model permits the luxury of never worrying about losing one's secret key, since one should not get it in the first place. In contrast, malicious parties provably cannot gain anything by obtaining their keys (*i.e.*, they cannot break the security of honest parties). Hence, as a practical matter, one expects that ACRS model is very similar to the CRS model, where parties cannot access any secret information. However, the *mere ability* to get such information is what gives us security, even though we expect that a "rational" party, *either honest or malicious*, will not utilize this ability: honest parties do not need it, and malicious parties do not gain from it.

Of course, one may justifiably criticize the ACRS model because of the need for a trusted party who is always available, as opposed to the (global) CRS model, where no party is needed after the CRS is generated. Indeed, it is a non-trivial setup to realize (although *much* more natural than the SRS model, and seemingly minimal in light of the impossibility result mentioned above). However, as pointed out by [8], the ACRS model has the following "win-win" guarantee. Assume that one proves some protocol secure in the GUC+ACRS model, but in reality the trusted party will only generate a CRS, and will be unavailable afterwards. Then, from a syntactic point of view, *we are back in the (global) CRS model*. In particular, the protocol is still secure in the "old UC+CRS" setting! On an intuitive level, however, it seems to be *more secure* than a protocol proven secure in the "old UC+CRS" setting. This is because the simulator does not need to know a global trapdoor (which is deadly for the security of *honest* parties in the *real* model), but only the secret keys of the *corrupted* parties, which are guaranteed to never hurt the security of honest parties in the real model. For example, the CRS can be safely reused by other protocols, completely solving the "restricted composition" problem of UC that we mentioned earlier. Essentially, properties associated with deniability/non-transferability appear to be the only security properties lost by "downgrading" ACRS into CRS.

EFFICIENCY IN THE GUC FRAMEWORK. Thus, from the security and functionality perspectives, the GUC+ACRS model appears to be strictly superior to the UC+CRS model. The question, however, is what is the price in terms of efficiency? Unfortunately, the GUC-feasibility results of [7] are quite inefficient: the commitment scheme committed to the message in a bit-by-bit manner, while the zero-knowledge proof for a relation R was implemented using the generic Cook-Levin reduction to a canonical NP-complete problem. Thus, now that the GUC-feasibility of secure computation has been established, it is natural to ask if one can build *efficient*, GUC-secure commitment and ZK proofs in the ACRS (resp. KRK; see Footnote 2) model. In this paper, we provide such efficient GUC-secure commitment and ZK proofs which are secure against adaptive corruptions, therefore making the ARCS model an attractive alternative to the CRS model on nearly (see below) all fronts.

The only drawback of our solution is that we rely on *data erasures*, which is not the case for most efficient UC protocols, such as that of Damgård and Nielsen [17] (or the inefficient GUC feasibility results of [7]). However, unlike sacrificing adaptive security, which is a *critical* concern (addressed in our work) given the highly dynamic nature of protocols concurrently running on the Internet,[3] we believe that the assumption of data erasures is realistic. Furthermore, this assumption is widely used in practice (for example, for analyzing most key exchange protocols, such as Diffie-Hellman), and was already used in several works on UC security as well (*e.g.*, [10, 11, 21, 24], although it was often hidden deep within the paper). Coupled with the fact that erasures allow us to obtain dramatically more efficient (in fact, *practical*) protocols, we believe that use of this assumption here is justified. Of course, we hope that future research will remove/weaken this restriction, and comment on this more in the last paragraph of the introduction, where we discuss the random oracle model.

OUR RESULTS ON GUC ZK. We present an efficient compiler giving a direct, efficient, constant-round and GUC-secure ZK proof (GUC ZK) for any NP relation R from any "dense Ω-protocol" [21] for R. The notion of Ω-protocols was introduced by Garay, MacKenzie and Yang [21]. Briefly, Ω-protocols are Σ-protocols (*i.e.*, they satisfy special soundness and ZK properties of Σ-protocols), with an extra property that one can generate the public parameter ρ of the system together with a trapdoor information τ, such that knowledge of τ allows one to extract the witness from any valid conversation between the prover and the verifier (as opposed to the usual special soundness, where one needs two different transcripts with the same first flow). [21, 24] used Ω-protocols for the similar task of building UC-secure ZK proofs in the CRS model, which are inherently not GUC-secure. As a result, our compiler is *considerably* more involved than the compiler of [21, 24] (which also used erasures). For example, in the GUC setting the simulator is not allowed to know τ, so we have to sample the public ρ in the ACRS model using a special coin-flipping protocol introduced by [7]. As a result, our compiler requires Ω-protocols whose reference parameters are "dense" (*i.e.*, indistinguishable from random), and none of the previous Ω-protocols of [21, 24] are suitable for our purpose.

Thus, of independent interest, we show several novel *dense* Ω-protocols. First, we show how to build a direct, but only semi-efficient dense Ω-protocol for any NP relation R from any Σ-protocol for R. Although this Ω-protocol uses the cut-and-choose technique (somewhat similar to the technique of Pass [26], but in a very different setting), it is quite general and gives a much more efficient Ω-protocol than the technique of [7, 13] which requires a generic Cook-Levin reduction. Second, we show a *very efficient* number-theoretically based dense Ω-protocol for proving knowledge of a discrete log representation. Once again, this Ω-protocol had to use some interesting additional tools on top on the prior

---

[3] We remark that adaptive security with erasures trivially implies static security, and is usually much harder to achieve than the latter.

"non-dense" Ω-protocol of [21], such as a special "projective Paillier encryption" of Cramer and Shoup [15]. As a result, we get a semi-efficient GUC ZK for any R having an efficient Σ-protocol, and a very efficient GUC ZK for proving the knowledge of discrete log representation.

OUR RESULTS ON GUC COMMITMENTS. Using the techniques developed for ZK, we proceed to build the first *constant-rate* (and constant-round) GUC-secure commitments (GUCC) in the ACRS model. In spirit our result is similar to the result of Damgård and Nielsen [17], who constructed the first constant-rate UC-secure commitments in the "old" CRS framework. However, our techniques are very different, and it seems hopeless to adapt the protocol of [17] to the GUC framework. Instead, we essentially notice that the required GUCC would easily follow from our techniques for GUC ZK, provided we can build an efficient Ω-protocol for a special relation on R on *identity-based trapdoor commitments* (IBTCs) — a notion introduced by [7] to implement the ACRS setup. Intuitively, a prover needs to show that he knows the message being committed by a value c (w.r.t. a particular identity). In particular, if one can build an IBTC scheme where the required relation R would involve the proof of knowledge of some discrete log representation, our previous GUC ZK protocol would complete the job. Unfortunately, the IBTCs constructed by [7] had a much more complicated form. Therefore, of independent interest, we build a new IBTC scheme which is based on Water's signature [28]. The resulting IBTC not only has the needed form for its relation R, but is also much simpler and more efficient than prior IBTCs built in the standard model. Combining these results, we finally build the required GUCC.

RESULTS ON MODELING RANDOM ORACLE IN GUC. Finally, we briefly comment on using the random oracle (RO) model in conjunction with the GUC framework. The RO is simply modeled as a shared functionality available both in the real and in the ideal model. As such, the simulator cannot "reprogram" the RO. Even more counter-intuitively, it cannot even "prematurely extract" the values used by the real-model attacker! This is because we can assume that all such queries are made by the environment (which the simulator cannot control), and the inputs are only given to the attacker on a "need-to-know" basis. Correspondingly, the RO model is much more restricted in the GUC framework (in particular, by itself it is provably insufficient to GUC-realize most functionalities [7, 8]). However, we still show that one *can* meaningfully use it in the conjunction with the ACRS model, because we *are allowed* to extract and reprogram the RO in the proof of security. In particular, by applying the Fiat-Shamir heuristic to our GUC ZK protocols, we obtain an *efficient*, two-round (which we show is optimal; see Theorem 4), straight-line extractable and simulatable (in fact, GUC-secure!) ZK proof for any relation R having an efficient dense Ω-protocol. Moreover, in this protocol one only needs to erase some short data during a *local computation* (*i.e.*, no sensitive data needs to be stored while waiting for some network traffic), making the need for data erasures extremely minimal. Of course, we can get a less efficient 2-round GUC ZK protocol with these properties that does *not* rely on data erasures at all, by applying the Fiat-Shamir heuristics to the inefficient protocol of [7]. This means that we get a general feasibility of round-optimal GUC ZK for NP in the ACRS+RO model which does not rely on data erasures.

We briefly compare the resulting deniable ZK protocol to previous related work on deniable ZK (*e.g.*, [23, 26]) in Section 6.

2   Definitions and Tools

2.1   GUC Security

At a high level, the UC security framework formalizes the following emulation requirement:

> *A protocol π that emulates protocol ϕ does not affect the security of anything else in the environment differently than ϕ would have – even when π is composed with arbitrary other protocols that may run concurrently with π.*

Unfortunately, the UC security framework requires that parties running in a session of π do not share state with any other protocol sessions at all, limiting the legitimate applicability of that framework. In particular, *global setups* such as a Common Reference String (CRS) or Public Key Infrastructure (PKI) are not accurately modeled. The GUC security framework, introduced in [7], formalizes the same intuitive emulation requirement as the UC framework. However, the GUC framework does so even for protocols π that make use of shared state information that is common to multiple sessions of π, as well as other protocols in the environment (running concurrently with π).

More formally, the security framework of [6] defines a notion called "UC-emulation". A protocol π is said to UC-emulate another protocol ϕ if, for every *adversary* $\mathcal{A}$ attacking ϕ, there exists a *simulator* $\mathcal{S}$ attacking π such that no *environment* $\mathcal{Z}$ can distinguish between $\mathcal{A}$ attacking ϕ, and $\mathcal{S}$ attacking π. In the distinguishing experiment, the environment is *constrained* to interact only with parties participating in a single session of a challenge protocol (either π or ϕ), along with its corresponding attacker (either $\mathcal{A}$ or $\mathcal{S}$, respectively) in a "black-box" manner. This limited interaction prevents the model from capturing protocols that share state with other protocols running in the environment, since the distinguishing experiment does not allow the environment to access any state information used by the parties it is interacting with.

The Generalized Universal Composability (GUC) security framework of [7] extends the original UC security framework of [6] to incorporate the modeling of protocols that share state in an arbitrary fashion. In particular, the GUC framework provides mechanisms to support direct modeling of global setups such as a CRS or PKI. This is done by first defining the notion of *shared functionalities* that can maintain state and are accessible to any party, in any protocol session. The distinguishing experiment of GUC allows the environment to access

---

Functionality $\mathcal{G}_{\mathrm{acrs}}$

Initialization Phase: At the first activation, run an algorithm Setup to generate a public key/master secret key pair (PK, MSK).

Providing the public value: When activated by any party requesting the CRS, return PK to the requesting party and to the adversary.

Dormant Phase: Upon receipt of a message $(\mathtt{retrieve}, sid, ID)$ from a *corrupt* party P whose identity is ID, return the value $SK_{ID} \leftarrow \mathsf{Extract}(PK, ID, MSK)$ to P. (Receipt of this message from honest parties is ignored.)

Fig. 1. The Identity-Based Augmented CRS Functionality

any shared functionalities. GUC also removes the constraint on the protocols invoked by the environment, allowing it to interact with any (polynomial) number of parties running arbitrary protocols, including multiple sessions of the protocol being attacked. That is, GUC allows the environment to directly invoke and observe arbitrary protocols that run alongside the distinguishing "challenge protocol" – and those arbitrary protocols may even share state information with the challenge protocol and with the environment itself (via shared functionalities). If a protocol π (that may share state in this fashion) "UC-emulates" a protocol ϕ with respect to such *unconstrained environments*, we say that π GUC-emulates ϕ. We say that a protocol π is a GUC-secure *realization* of a particular functionality $\mathcal{F}$ if π GUC-emulates the ideal protocol for $\mathcal{F}$. Further details of the formal modeling for UC and GUC security can be found in [6] and [7, 27]. In this work, we will focus on the construction of efficient GUC-secure realizations of commitments and zero knowledge, with security even against adversaries capable of adaptive corruptions. As is common throughout the UC literature, we will assume the availability of secure (*i.e.*, private and authenticated) channels. The realization of such secured channels over insecure networks (such as the Internet) is a non-trivial problem studied in [27], but is beyond the scope of this work.

2.2   The ACRS Model

Unfortunately, it is impossible to GUC-realize most useful two-party functionalities in the plain model, or even in the CRS model (see [7]). To avoid this impossibility, we make use of a special *Augmented Common Reference String* (ACRS) trusted setup (which we denote by the functionality $\mathcal{G}_{\mathrm{acrs}}$), as was first proposed in [7]. Another possible alternative would be to use a PKI model supporting "Key Registration with Knowledge" [3, 7] (which we denote by the functionality $\mathcal{G}_{\mathrm{krk}}$) – indeed, our efficient protocols can easily be transformed to use the $\mathcal{G}_{\mathrm{krk}}$ setup – but the more minimal ACRS model suffices and is clearly less costly to implement than a PKI. Thus, we will focus on the ACRS setting. The shared functionality $\mathcal{G}_{\mathrm{acrs}}$ describing ACRS setup, which is parameterized by the algorithms Setup and Extract, is given in Figure 1.

Intuitively, the ACRS setup provides a simple CRS to all parties, and also agrees to supply an identity-based trapdoor for identity P to any "corrupt" party P that asks for one. The provision that only corrupt parties can get their trapdoors is used to model the restriction that protocols run by honest parties should not use the trapdoor – *i.e.* honest parties should never *have* to obtain their trapdoors in order to run protocols. In reality, a trusted party will perform the ACRS initialization phase, and then supply the trapdoor for P to any party P that asks for its trapdoor. Of course, in practice, most parties will never bother to request their trapdoors since the trapdoors are not useful for running protocols. (Ultimately, these trapdoors will be used to enable corrupt parties to simulate attacks by using $\mathcal{S}$, a task that no honest party should need to perform.)

In the following sections, we show how to construct efficient GUC-secure realizations of commitments and zero knowledge using this instantiation of the $\mathcal{G}_{\mathrm{acrs}}$ shared functionality. (As explained in Section 4 of [8], this is enough to GUC-realize any other well-formed functionality.) We then show how to optimize the round complexity of these protocols by using $\mathcal{G}_{\mathrm{acrs}}$ in conjunction with the RO model.

2.3   Omega Protocols

The notion of an Ω-protocol was introduced in [21], and we recall the basic idea here. While our notion of an Ω-protocol is the same in spirit as that in [21], we also introduce some new properties, and there are a few points where the technical details of our definition differ. Details can be found in the full version [18].

Let *ParamGen* be an efficient probabilistic algorithm that takes as input 1^λ, where λ is a security parameter, and outputs a *system parameter* Λ. The system parameter Λ determines finite sets X, $L \subset X$, W, and a relation $R \subset L \times W$, where for all $x \in L$, we have $(x, w) \in R$ for some $w \in W$. The sets X and W, and the relation R should be efficiently recognizable (given Λ). An element $x \in X$ is called an *instance*, and for $(x, w) \in R$, w is called a *witness* for x.

There is also an efficient probabilistic algorithm *RefGen* that takes as input a system parameter Λ and outputs a pair (ρ, τ), where ρ is called a *reference parameter*, and τ is called a *trapdoor*.

An Ω-protocol Π is played between a *prover* P and a *verifier* V. Both P and V take as common input a system parameter Λ, a reference parameter ρ, and an instance $x \in X$. An honest prover P is only run for $x \in L$, and always takes a witness w for x as an additional, private input. Execution runs in three steps: in the first step, P sends a message a to V; in the second, V sends a random challenge c to P; in the third, P sends a response z to V, whereupon V either *accepts* or *rejects* the *conversation* (a, c, z).

Of course, there is a basic *completeness* requirement, which says that if both prover and verifier follow the protocol then the verifier always accepts.

We say that Π is *trapdoor sound* if there exists an efficient *trapdoor extractor algorithm* $\mathcal{E}_{\mathrm{td}}$ such that the following holds: for every efficient cheating prover $\tilde{P}$, it should be infeasible for $\tilde{P}$ (given input (Λ, ρ)) to make V (given input (Λ, ρ, x)) accept a conversation (a, c, z) for an instance x such that execution of $\mathcal{E}_{\mathrm{td}}$ on input $(\Lambda, \tau, x, a, c, z)$ fails to produce witness w for x. Here, (Λ, ρ, τ) are generated by the algorithms *ParamGen* and *RefGen*; c is generated by V; and x, a, and z are generated adversarially.

We shall also make use of the following variant of trapdoor soundness. Very roughly, we say that Π is *partial trapdoor sound for a function f*, if it is a proof of knowledge (in the traditional, rewinding sense) of a witness w of the instance x, such that the value calculated by the trapdoor extractor $\mathcal{E}_{\mathrm{td}}$ (on the same inputs as above) is equal to $f(w)$. As we will see, partial trapdoor soundness is sufficient for some applications, and can be realized using a somewhat more efficient protocol.

We say that Π is *honest verifier zero-knowledge (HVZK)* if there is a *simulator algorithm* ZKSim that on input (Λ, ρ, x, c) can produce a simulation of the conversation (a, c, z) that would arise from an interaction between an honest prover P with input (Λ, ρ, x, w), and a cheating verifier $\tilde{V}$, subject to the constraint that $\tilde{V}$'s challenge c must be generated before it sees a. Here, (Λ, ρ) are generated by the algorithms *ParamGen* and *RefGen*; and x, w, and c are generated by $\tilde{V}$. The requirement is that $\tilde{V}$ should not be able to distinguish the output of the simulator from the output of the real prover.

We note that the notion of an Ω-protocol extends that of a Σ-protocol ([16, 17]). The distinguishing feature is the reference parameter, and the trapdoor soundness property that says that a witness may be extracted using a trapdoor in the reference parameter, rather than by rewinding. The notion of trapdoor soundness is closely related to that of *verifiable encryption* [1, 5]. Indeed, all known constructions of Ω-protocols boil down to using a public key for a semantically secure encryption scheme as reference parameter, where the trapdoor is the secret key; the prover encrypts a witness, and then proves that it did so using a Σ-protocol.

For our application to GUC ZK and GUC commitments, we introduce an additional property that we require of an Ω-protocol. A given system parameter Λ determines a set $\widehat{\Phi}$ of possible reference parameters. Suppose there is some set Φ that contains $\widehat{\Phi}$, with the following properties: (i) the uniform distribution on Φ is efficiently samplable; (ii) membership in Φ is efficiently determined; (iii) Φ is an abelian group (which we write multiplicatively), such that the group and group inverse operations are efficiently computable; (iv) it is hard to distinguish a random element of Φ (generated uniformly), from a random element of $\widehat{\Phi}$ (as generated by *RefGen*). If all of these conditions obtain, we say Π has *dense reference parameters*, and we call Φ the set of *extended reference parameters*.

2.4   Identity-Based Trapdoor Commitments

The notion of an identity-based trapdoor commitment scheme (IBTC) was introduced in [2] (as ID-based Chameleon Hash functions), with some additional refinements appearing in [7]. We recall the basic idea here, leaving the formal definition to the full version [18].

An IBTC scheme has a Setup algorithm that takes as input 1^λ, where λ is the security parameter, and outputs a *public key PK* and a *master secret key MSK*. The public key *PK* determines a set $\mathcal{D}$ of *decommitment values*. To generate a commitment to a message m, a user computes $d \xleftarrow{\$} \mathcal{D}$ and $\kappa \leftarrow$ $\mathsf{Com}_{ID}(d, m)$. Here, Com_{ID} is a deterministic algorithm (which implicitly takes

---

Functionality $\mathcal{F}_{\mathrm{zk}}^{R}$

$\mathcal{F}_{\mathrm{zk}}$, parameterized by a binary relation R and running with a prover P, a verifier V, and an adversary $\mathcal{S}$, proceeds as follows upon receipt of a message (ZK-prover, sid, P, V, x, w) from the prover P:

If $(x, w) \in R$, then send (ZK-proof, sid, P, V, x) to V and $\mathcal{S}$ and halt. Otherwise halt.

Fig. 2. The Zero-Knowledge Functionality for Relation R

PK as a parameter, but we shall in general omit this). The value κ is called a *commitment* to m, while the pair (d, m) is called an *opening* of κ.

Like any commitment, a IBTC should be *binding*: it should be hard to open a commitment under some ID to two different messages; that is, it should be hard to find ID, d, m, d', m' such that $m \neq m'$ and $\mathsf{Com}_{ID}(d, m) = \mathsf{Com}_{ID}(d', m')$. In addition, there should be an *identity-based trapdoor*, which allows for *identity-based equivocation* of commitments. More precisely, there are three algorithms Extract, ECom, and Eqv, which work as follows. Given (PK, ID, MSK) as input, Extract computes a trapdoor SK_{ID} for the identity ID. Using this trapdoor, algorithm ECom may be invoked with input (PK, ID, SK_{ID}) to produce a pair (κ, α), where κ is a "fake" commitment, and α is a trapdoor specifically tuned to κ. Finally, running algorithm Eqv on input $(PK, ID, SK_{ID}, \kappa, \alpha, m)$ for any message m produces a decommitment d, such that (d, m) is an opening of κ. The security property for equivocation is that is should be hard to distinguish a value d produced in this way from a random decommitment. Moreover, this equivocation property should not interfere with the binding property *for identities whose trapdoors have not been extracted*.

3   GUC Zero-Knowledge in the ACRS Model

The ideal Zero-Knowledge functionality for relation R, $\mathcal{F}_{\mathrm{zk}}$, is described in Figure 2.[4]

Here we give a general transformation from any Ω-protocol Π for a relation R to a GUC-secure zero-knowledge proof for the relation R in the augmented CRS ($\mathcal{G}_{\mathrm{acrs}}$) model. We need to assume that the Ω-protocol satisfies the correctness, trapdoor soundness, honest verifier zero knowledge (HVZK), and dense reference parameters properties. We denote by Φ the space of extended reference parameters for Π. We also need an identity-based trapdoor commitment (IBTC) scheme. Commitments in this scheme are written $\mathsf{Com}_{ID}(d, m)$.

The augmented CRS is instantiated using the public key (and trapdoor extractor) of the IBTC. In addition, any system parameters Λ for the Ω-protocol

---

[4] Technically, the relation R may be determined by system parameters, which form part of a CRS. Here, we note that the same CRS must be used in both the "ideal" and "real" settings.

are placed in the public value of the augmented CRS. Note that there is no trapdoor associated with the system parameter for the Ω-protocol, so this system parameter is essentially a "standard" CRS. A critical difference between our approach and that of Garay et al. [21] is that the reference parameter for the Ω-protocol are not placed in the CRS; rather, a fresh reference parameter ρ is generated with every run of the protocol using a three-move "coin toss" protocol (which, in turn, makes use of the IBTC).

Here is how the GUC ZK protocol between a prover P and verifier V works. The common input is an instance x (in addition to PK and the identities of the players). Of course, P also has a witness w for x as a private input.

1. V computes $\rho_1 \xleftarrow{\$} \Phi$, forms commitment $\kappa_1 = \mathsf{Com}_P(d_1, \rho_1)$, and sends κ_1 to P.
2. P computes $\rho_2 \xleftarrow{\$} \Phi$ and sends ρ_2 to V.
3. V first verifies that $\rho_2 \in \Phi$, and then sends the opening (d_1, ρ_1) to P.
4. P verifies that (d_1, ρ_1) is a valid opening of κ_1, and that $\rho_1 \in \Phi$.
   Both P and V locally compute $\rho \leftarrow \rho_1 \cdot \rho_2$.
5. P initiates the Ω-protocol Π, in the role of prover, using its witness w for x. P computes the first message a of that protocol, forms the commitment $\kappa' = \mathsf{Com}_V(d', a)$, and sends κ' to V.
6. V sends P a challenge c for protocol Π.
7. P computes a response z to V's challenge c, and sends (d', a, z) to V.
   P then **erases** the random coins used by Π.
8. V verifies that (d', a) is a valid opening of κ' and that (a, c, z) is an accepting conversation for Π.

Theorem 1. *The protocol described above GUC-emulates the $\mathcal{F}_{\mathrm{zk}}^R$ functionality in the secure-channels model, with security against adaptive corruptions (with erasures).*

Proof (sketch). We first observe that the protocol above only makes use of a single shared functionality, $\mathcal{G}_{\mathrm{acrs}}$. Therefore, we are free to make use of the equivalence theorem and EUC model of [7]. This allows us to prove the GUC security of the protocol using the familiar techniques of the UC framework, with only a single (but crucial) modification – we will allow the environment access to the shared functionality.

Let $\mathcal{A}$ be any PPT adversary attacking the above protocol. We describe an ideal adversary $\mathcal{S}$ attacking the ideal protocol for $\mathcal{F}_{\mathrm{zk}}^R$ that is indistinguishable from $\mathcal{A}$ to any distinguishing environment $\mathcal{Z}$, in the presence of a shared setup $\mathcal{G}_{\mathrm{acrs}}$. In standard fashion, $\mathcal{S}$ will run a copy of $\mathcal{A}$ internally. We now formally describe how $\mathcal{S}$ interacts with its internal copy of $\mathcal{A}$. We focus here on the non-trivial aspects of the simulator.

Simulating a proof between an honest P and corrupt V. The following simulation strategy is employed whenever P is honest and V is corrupted at any point prior to, or during, the execution of the protocol. $\mathcal{S}$, upon notification from $\mathcal{F}_{\mathrm{zk}}^R$ of a successful proof from P of statement x, proceeds as follows. First, acting on behalf of the corrupt party V, $\mathcal{S}$ obtains the trapdoor SK_V from $\mathcal{G}_{\mathrm{acrs}}$.

Next, $\mathcal{S}$ runs the coin-tossing phase of the protocol with the corrupt party V (being controlled by $\mathcal{S}$'s internal copy of $\mathcal{A}$) normally. Upon completion of the coin-tossing phase at Step 5, rather than sending a commitment to the first message sent by Π (which would require the witness w as an input) as per the protocol specification, $\mathcal{S}$ obeys the following procedure for the next 3 steps of the protocol:

5. $\mathcal{S}$ computes $(\hat{\kappa}', \alpha) \leftarrow \mathsf{ECom}(V, SK_V)$. $\mathcal{S}$ then sends the equivocable commitment $\hat{\kappa}'$ to the corrupt verifier V (which is part of $\mathcal{S}$'s internal simulation of $\mathcal{A}$).
6. $\mathcal{S}$ receives a challenge c from the corrupt verifier V.
7. $\mathcal{S}$ runs the HVZK simulator ZKSim for protocol Π on input (Λ, ρ, x, c), obtaining messages a and z. $\mathcal{S}$ then equivocates $\hat{\kappa}'$, by computing $d' \leftarrow \mathsf{Eqv}(V, SK_V, \hat{\kappa}', \alpha, a)$, and sends d', a, z to the corrupt verifier V.

Observe that this simulation is done entirely in a straight-line fashion, and requires only the trapdoor SK_V belonging to corrupt party V.

If P is also corrupted at some point during this simulation, $\mathcal{S}$ must generate P's internal state information and provide it to $\mathcal{A}$. If P is corrupted prior to Step 5, then $\mathcal{S}$ can easily provide the random coins used by P in all previous steps of the protocol (since those are simply executed by $\mathcal{S}$ honestly). A corruption after Step 5 but before Step 7 is handled by creating an honest run of protocol Π using witness w (which was revealed to $\mathcal{S}$ immediately upon the corruption of P), and computing the internal value d' via $d' \leftarrow \mathsf{Eqv}(V, SK_V, \kappa', \alpha, a)$, where a is now the honestly generated first message of Π. Finally, if corruption of P occurs after Step 7 of the simulation, the internal state is easily generated to be consistent with observed protocol flows, since they already contain all relevant random coins, given the erasure that occurs at the end of Step 7.

Intuitively, the faithfulness of this simulation follows from the equivocability and binding properties of commitments, and the HVZK and dense reference parameters properties of the Ω-protocol Π. We stress that while the *proof* of this requires a rewinding argument (see the full version [18]), the simulation itself is straight-line.

Simulating a proof between a corrupt P and honest V. The following simulation strategy is employed whenever V is honest, and P is corrupted at any point prior to or during the execution of the protocol. First, acting on behalf of the corrupt party P, $\mathcal{S}$ obtains the trapdoor SK_P from $\mathcal{G}_{\mathrm{acrs}}$. Then $\mathcal{S}$ generates a pair (ρ, τ) using the *RefGen* algorithm for Π, and "rigs" the coin-tossing phase of the protocol by playing the role of V (communicating with the internal simulation of the corrupt party P) and modifying the initial steps of the protocol as follows:

1. $\mathcal{S}$ computes $(\hat{\kappa}_1, \alpha) \leftarrow \mathsf{ECom}(P, SK_P)$, and sends $\hat{\kappa}_1$ to P.
2. P replies by sending some string ρ_2 to V.
3. $\mathcal{S}$ computes $\rho_1 \leftarrow \rho \cdot \rho_2^{-1}$, and $d_1 \leftarrow \mathsf{Eqv}(P, SK_P, \hat{\kappa}_1, \alpha, \rho_1)$.
   $\mathcal{S}$ first verifies that $\rho_2 \in \Phi$. Then $\mathcal{S}$ sends the opening (d_1, ρ_1) to P.

The remainder of the protocol is simulated honestly.

Functionality $\mathcal{F}_{\text{com}}$

Functionality $\mathcal{F}_{\text{com}}$ proceeds as follows, with committer P and recipient V. .

Commit Phase: Upon receiving a message (commit, sid, P, V, m) from party P, record the value m and send the message (receipt, sid, P, V) to V and the adversary. Ignore any future commit messages.

Reveal Phase: Upon receiving a message (reveal, sid) from P: If a value m was previously recorded, then send the message (reveal, sid, m) to V and the adversary and halt. Otherwise, ignore.

Fig. 3. The Commitment Functionality $\mathcal{F}_{\text{com}}$ (see [9])

Observe that the outcome of this coin-flipping phase will be the same ρ generated by $\mathcal{S}$ at the start of the protocol (along with its corresponding trapdoor information τ). If and when the verifier accepts, $\mathcal{S}$ runs the trapdoor extractor $\mathcal{E}_{\text{td}}$ for Π on input $(\Lambda, \tau, x, a, c, z)$ to obtain a witness w for x. $\mathcal{S}$ then sends the pair (x, w) to the ideal functionality $\mathcal{F}_{\text{zk}}^R$ on behalf of the corrupt prover P.

In the event that V is also corrupted at any point prior to completion of the protocol, $\mathcal{S}$ simply produces internal state for V consistent with the visible random coins in the transcript (none of the verifier's random coins are hidden by the honest protocol).

Intuitively, the faithfulness of this simulation follows from the equivocability and binding properties of commitments, and the trapdoor soundness and dense reference parameters properties of the Ω-protocol Π. Again, we stress that while the *proof* of this requires a rewinding argument (*e.g.*, the Reset Lemma of [4]), the simulation itself is straight-line.

Now that we have fully described the behavior of $\mathcal{S}$, it remains to prove that $\mathcal{S}$ interacting with $\mathcal{F}_{\text{zk}}^R$ (the ideal world interaction) is indistinguishable from $\mathcal{A}$ interacting with the protocol (the real-world interaction), from the standpoint of any environment $\mathcal{Z}$ with access to $\mathcal{G}_{\text{acrs}}$. We stress that even $\mathcal{Z}$ cannot obtain trapdoor information from $\mathcal{G}_{\text{acrs}}$ for any honest parties, since $\mathcal{G}_{\text{acrs}}$ will not respond to requests for such trapdoors. The proof of indistinguishability follows from a relatively straightforward argument, using the security properties of the IBTC and Ω-protocol. See the full version [18].

4   GUC Commitments in the ACRS Model

The ideal functionality for a commitment scheme is shown in Figure 3. Messages m may be restricted to some particular *message space*.

Our protocol makes use of an Ω-protocol for the IBTC opening relation; here, a witness for a commitment κ with respect to an identity ID is a valid opening (d, m) (*i.e.*, $\text{Com}_{ID}(d, m) = \kappa$). Instead of trapdoor soundness, we only require partial trapdoor soundness with respect to the function $f(d, m) := m$.

Our new GUC commitment protocol has two phases. The commit phase is the same as the ZK protocol in the previous section, except that Step 5 now runs as follows:

5.' P generates a commitment $\kappa = \mathsf{Com}_V(d, m)$, and then initiates the Ω-protocol Π, in the role of prover, using its witness (d, m).
P computes the first message a of that protocol, forms the commitment $\kappa' = \mathsf{Com}_V(d', a)$, and sends κ and κ' to V.

In the reveal phase, P simply sends the opening (d, m) to V, who verifies that (d, m) is a valid opening of κ.

Theorem 2. *The protocol described above GUC-emulates the $\mathcal{F}_{\mathrm{com}}$ functionality in the secure-channels model, with security against adaptive corruptions (with erasures).*

The proof is analogous to that of our zero knowledge protocol, but entails some minor changes that include the partial trapdoor soundness requirement for Π. See [18] for more detail.

5 Efficient Implementations

5.1 Constructing Ω Protocols from Σ Protocols

We now briefly sketch how to efficiently construct an Ω-protocol Π for a relation R, given any efficient Σ-protocol Ψ for relation R. Intuitively, we must ensure that the dense reference parameter and trapdoor extractability properties of Π will hold, in addition to carrying over Σ-protocol Ψ's existing properties.

Let the reference parameter for Π be the public key pk for a "dense" semantically secure encryption E (where the dense property of the encryption scheme simply satisfies the requirements of the Dense Reference Parameter property of Ω protocols). Standard El-Gamal encryption will suffice for this purpose (under the DDH assumption). Let $\psi = E_{pk}(s, m)$ denote an encryption of message m with random coins s.

Let a, z^c denote the first and last messages (respectively) of the prover in protocol Ψ when operating on input (x, w, r) and with challenge c, where $(x, w) \in R$ and r denotes the random coins of the prover. The three messages to be sent in protocol Π will be denoted as a', c', z'.

Intuitively, we will use a cut-and-choose technique to provide extractability, and then amplify the soundness by parallel repetition k times. The first message a' of Π is constructed as follows:

1. For $i = 1, \ldots, k$, choose random coins r_i and compute a_i, z_i^0, and z_i^1 using the prover input (x, w, r_i).
2. For $i = 1, \ldots, k$, compute ciphertexts $\psi_i^0 = E_{pk}(s_i^0, z_i^0)$ and $\psi_i^1 = E_{pk}(s_i^1, z_i^1)$.
3. Set $a' := (\psi_1^0, \psi_1^1, \ldots, \psi_k^0, \psi_k^1)$.

The challenge c' sent to the prover in Π is a k-bit string $c' = c'_1 c'_2 \ldots c'_k$. The last message z' of protocol Π is then constructed as follows.

1. For $i = 1, \ldots, k$, set $z'_i := (s_i^{c'_i}, z_i^{c'_i})$.
2. Set $z' := (z'_1, \ldots, z'_k)$.

The verifier's algorithm for Π is simply constructed accordingly, verifying that all the ciphertexts were correctly constructed, and that the corresponding conversations for Ψ are valid. The proof of the following theorem is standard and is therefore omitted.

Theorem 3. *Π constructed as above is an Ω-protocol for relation R, provided that Ψ is a Σ-protocol for relation R and E is a dense one-time semantically secure public key encryption scheme.*

5.2   An Efficient Identity-Based Trapdoor Commitment with Ω-Protocol

While the protocol in §5.1 is certainly much more efficient than that in [7] (at least for languages with efficient Σ-protocols), we would like to get an even more efficient protocol that avoids the cut-and-choose paradigm altogether. In this section, we briefly show how we can obtain such a protocol for GUC commitments. Unlike the GUC commitment scheme in [7], which could commit bits, our GUC commitment scheme can be used to commit to values in a much larger set. Moreover, because of the special algebraic structure of the scheme, our GUC commitment protocol can be combined with other, well-known protocols for proving properties on committed values (*e.g.*, the that product of two committed integers is equal to a third committed integer).

To achieve this goal, we need an IBTC scheme that supports an efficient Ω-protocol, so that we can use this scheme as in §4. As observed in [7], based on a variation of an idea in [19], to build an IBTC scheme, one can use a secure signature scheme, along with a Σ-protocol for proof of knowledge of a signature on a given message. Here, the message to be signed is an identity *ID*. Assuming the Σ-protocol is HVZK, we can turn it into a commitment scheme, as follows. For a conversation (a, c, z), the commitment is a, the value committed to is c, and the decommitment is z. To commit to a value c, one runs the HVZK simulator. The trapdoor for a given *ID* is a signature on *ID*, and using this signature, one can generate equivocable commitments just by running the actual Σ-protocol.

For our purposes, we suggest using the Waters signature scheme [28]. Let $\mathbb{G}$ and $\mathbb{H}$ be a groups of prime order q, let $e : \mathbb{G} \to \mathbb{H}$ be an efficiently computable, non-degenerate bilinear map, and let $\mathbb{G}^* := \mathbb{G} \setminus \{1\}$. A public reference parameter consists of random group elements $\mathbf{g}_1, \mathbf{g}_2, \mathbf{u}_0, \mathbf{u}_1, \ldots, \mathbf{u}_k \in \mathbb{G}$, a description of a collision-resistant hash function $H : \{0,1\}^* \to \{0,1\}^k$, and a group element $\mathbf{h}_1$. A signature on a message m is a pair $(\mathbf{s}_1, \mathbf{s}_2) \in \mathbb{G} \times \mathbb{G}$, such that $e(\mathbf{s}_1, \tilde{\mathbf{u}}_m^{-1}) \cdot e(\mathbf{s}_2, \mathbf{g}_1) = e(\mathbf{h}_1, \mathbf{g}_2)$, where $\tilde{\mathbf{u}}_m := \mathbf{u}_0 \prod_{b_i=1} \mathbf{u}_i$ and $H(m) = b_1 \cdots b_k \in \{0,1\}^k$. Waters' signature is secure assuming the CDH for the group $\mathbb{G}$. With overwhelming probability, the signing algorithm will produce a signature $(\mathbf{s}_1, \mathbf{s}_2)$ where neither $\mathbf{s}_1$ nor $\mathbf{s}_2$ are 1, so we can effectively assume this is always the case.

To prove knowledge of a Waters signature $(\mathbf{s}_1, \mathbf{s}_2) \in \mathbb{G} \times \mathbb{G}$ on a message $m \in \{0,1\}^*$, we may use the following protocol. The prover chooses $w_1, w_2 \in \mathbb{Z}_q^*$ at random, and computes $\bar{\mathbf{s}}_1 \leftarrow \mathbf{s}_1^{1/w_1}$ and $\bar{\mathbf{s}}_2 \leftarrow \mathbf{s}_2^{1/w_2}$. The prover then sends

$\bar{s}_1$ and $\bar{s}_2$ to the verifier, and uses a standard Σ-protocol to prove knowledge of exponents $w_1, w_2 \in \mathbb{Z}_q$ such that $\gamma_1^{w_1} \gamma_2^{w_2} = \gamma$ where $\gamma_1 := e(\bar{s}_1, \tilde{\mathbf{u}}_m^{-1})$, $\gamma_2 := e(\bar{s}_2, \mathbf{g}_1)$, and $\gamma := e(\mathbf{h}_1, \mathbf{g}_2)$.

The identity-based commitment scheme derived from the above Σ-protocol works as follows. Let $ID \in \{0,1\}^*$ be the identity, and let $m \in \mathbb{Z}_q$ be the message to be committed. The commitment is computed as follows: $\bar{s}_1, \bar{s}_2 \xleftarrow{\$} \mathbb{G}^*$, $d_1, d_2 \xleftarrow{\$} \mathbb{Z}_q$, $\gamma_1 \leftarrow e(\bar{s}_1, \tilde{\mathbf{u}}_{ID}^{-1})$, $\gamma_2 \leftarrow e(\bar{s}_2, \mathbf{g}_1)$, $\gamma \leftarrow e(\mathbf{h}_1, \mathbf{g}_2)$, $\bar{\gamma} \leftarrow \gamma_1^{d_1} \gamma_2^{d_2} \gamma^m$. The commitment is $(\bar{s}_1, \bar{s}_2, \bar{\gamma})$.

A commitment $(\bar{s}_1, \bar{s}_2, \bar{\gamma}) \in \mathbb{G}^* \times \mathbb{G}^* \times \mathbb{H}$ is opened by revealing d_1, d_2, m that satisfies the equation $\gamma_1^{d_1} \gamma_2^{d_2} \gamma^m = \bar{\gamma}$, where $\gamma_1, \gamma_2, \gamma$ are computed as in the commitment algorithm, using the given values $\bar{s}_1, \bar{s}_2$.

The trapdoor for such a commitment is a Waters signature on the identity ID. Using such a signature, one can just run the Σ-protocol, and open the commitment to any value. The commitment will look the same as an ordinary commitment, unless either component of the signature is the identity element, which happens with negligible probability.

As the opening of a commitment is essentially just a representation of a group element relative to three bases, there is a standard Σ-protocol for proving knowledge of an opening of a given commitment. Moreover, using techniques from Camenisch and Shoup [5], we can actually build an Ω-protocol for such a proof of knowledge, which avoids the cut-and-choose paradigm.

Garay et al. [21] give an Ω-protocol for a very similar task, which could easily be adapted for our purposes, except that the protocol in [21] does not satisfy the dense reference parameters property, which is crucial for our construction of a GUC commitment. To appreciate the technical difficulty, the MacKenzie et al. protocol is based on Paillier encryption, using an RSA modulus N. The secret key for this encryption scheme is the factorization of N, and this is used as "global" trapdoor to a CRS in their proof of security in the UC/CRS model. However, in the GUC framework, we cannot have such a global trapdoor, which is why we make use of Camenisch and Shoup's approach.[5]

The Camenisch and Shoup approach is based on a variant of Paillier encryption, introduced in Cramer and Shoup [15], which we call here *projective Paillier encryption*. While the goal in [5] and [15] was to build a chosen ciphertext secure encryption scheme, and we only require semantic security, it turns out that their schemes do not require the factorization of the RSA modulus N to be a part of the secret key. Indeed, the modulus N can be generated by a trusted party, who then erases the factorization and goes away, leaving N to be used as a shared system parameter. We can easily "strip down" the scheme in [5], so that it only provides semantic security. The resulting Ω-protocol will satisfy all the properties we need to build a GUC commitment, under standard assumptions (the Quadratic Residuosity, Decision Composite Residuosity, and Strong RSA).

---

[5] It should be noted that the "mixed commitments" of Damgård and Nielsen [17] also have a very similar global extraction trapdoor, which is why we also cannot use them to build GUC commitments.

Due to lack of space, all the remaining details for the IBTC scheme and the Ω-protocol for proof of knowledge of a representation are relegated to the full version [18].

6   Achieving Optimal Round Complexity with Random Oracles

While our constructions for GUC zero knowledge and commitments are efficient in both computational and communication complexity, and the constant round complexity of 6 messages is reasonable, it would be nice improve the round complexity, and possibly weaken the data erasure assumption. In this section we address the question if such improvements are possible in the random oracle (RO) model. We first remark that even the RO model, without any additional setup, does not suffice for realizing GUC commitments or zero knowledge (see [7, 8]). However, we may still obtain some additional efficiency benefits by combining the ACRS and RO models. Ideally, we would like to achieve non-interactive zero knowledge (NIZK), and, similarly, a non-interactive commitment. Unfortunately, this is not possible if we insist upon adaptive security, even if we combine the ACRS or KRK setup models with a random oracle.

Theorem 4. *There do not exist* adaptively secure *and* non-interactive *protocols for GUC-realizing* $\mathcal{F}_{\mathrm{com}}$ *and* $\mathcal{F}_{\mathrm{zk}}^R$ *(for most natural and non-trivial NP relations R) in the ACRS or KRK setup models. This impossibility holds even if we combine the setup with the random oracle model, and even if we allow erasures.*

We give a more formal statement and proof of this result in the full version [18]. Intuitively, there are two conflicting simulation requirements for GUC-secure commitments/ZK proofs that pose a difficulty here: a) given knowledge of the sender/prover's secret key, they must be "extractable" to the simulator, yet b) given knowledge of the recipient/verifier's secret key, they must be "simulatable" by the simulator. It is impossible for a single fixed message to simultaneously satisfy *both* of these conflicting requirements, so an adversary who can later obtain both of the relevant secret keys via an adaptive corruption will be able to test them and see which of these requirements was satisfied. This reveals a distinction between simulated interactions and real interactions, so we must resort to an interactive protocol if we wish to prevent the adversary from being able to detect this distinction. Accordingly, we will now show that it is possible to achieve *optimal* 2-round ZK and commitment protocols in the GUC setting using both the ACRS and RO setups.

ROUND-OPTIMAL ZK USING RANDOM ORACLES. We achieve our goal by simply applying the Fiat-Shamir heuristic [20] to our efficient zero knowledge and commitment protocols, replacing the first three and last three messages of each protocol with a single message. We defer a more formal discussion and analysis of GUC security in the combined ACRS and RO model with the Fiat-Shamir heuristic to the full version [18] (additional details can be also be found in [27]), but briefly comment on three important points. First, note that the only erasure required by our protocols now occurs entirely during a *single local computation*, without delay – namely, during the computation of the second message, where an entire run of three-round protocol is computed and the local randomness used to generate that run is then immediately erased. Thus, the need for data erasures is really minimal for these protocols.

Second, the proof of security for the modified protocols is virtually unaltered by the use of the Fiat-Shamir heuristic. In particular, observe that the GUC simulator $\mathcal{S}$ uses identical simulation strategies, and *does not* need to have access to a transcript of oracle queries, nor does it require the ability to "program" oracle responses. Thus, only in the *proof of security* (namely, that the environment cannot tell the real and the ideal worlds) do we use the usual "extractability" and "programmability" tricks conventionally used in the RO model.

Third, we stress that since the GUC modeling of a random oracle (accurately) allows the oracle to be accessed directly by all entities – including the environment – the aforementioned feature that $\mathcal{S}$ does not require a transcript of all oracle queries, nor the ability to program oracle responses, is *crucial* for deniability. It was already observed by Pass [26] that deniable zero knowledge simulators must not program oracle queries. However, we observe that even using a "nonprogrammable random oracle" for the simulator is still not sufficient to ensure truly deniable zero knowledge. In particular, if the modeling allows the simulator to observe interactions with the random oracle (even without altering any responses to oracle queries), this can lead to attacks on deniability. In fact, there is a very practical attack (sketched in Appendix 6) stemming from precisely this issue that will break the deniability of the protocols proposed by Pass [26]. Our GUC security modeling precludes the possibility of any such attacks.

Of course, unlike the model of [26], we superimpose the ACRS model on the RO model, providing all parties with implicit secret keys. This bears a strong resemblance to the model of [23], which employs the following intuitive approach to provide deniability for the prover P: instead proving the statement, P will prove "either the statement is true, or I know the verifier's secret key". Indeed, our approach is quite similar in spirit. However, we achieve a much stronger notion of deniability than that of [23]. Our zero knowledge protocols are the first constant round protocols to simultaneously achieve straight-line extractability (required for concurrent composability) and deniability against an adversary who can perform adaptive corruptions. In contrast, the protocol of [23] is not straightline extractable, and is not deniable against adaptive corruptions (this is easy to see directly, but also follows from Theorem 4, by applying the Fiat-Shamir heuristics to the 3-round protocol of [23]).

Finally, if one does not care about efficiency, applying our techniques to the inefficient protocols of [7], we get a general, round-optimal feasibility result for all of NP:

Theorem 5. *Under standard cryptographic assumptions, there exists a (deniable) 2-round GUC ZK protocol for any language in NP in the ACRS+RO model, which does not rely on data erasures.*

References

1. Asokan, N., Shoup, V., Waidner, M.: Optimistic Fair Exchange of Digital Signatures. In: Proc. of Eurocrypyt (1998)
2. Ateniese, G., de Medeiros, B.: Identity-based Chameleon Hash and Applications. In: Juels, A. (ed.) FC 2004. LNCS, vol. 3110. Springer, Heidelberg (2004)
3. Barak, B., Canetti, R., Nielsen, J., Pass, R.: Universally Composable Protocols with Relaxed Set-up Assumptions. In: Proc. of FOCS (2004)
4. Bellare, M., Palacio, A.: GQ and Schnorr Identification Schemes: Proofs of Security against Impersonation under Active and Concurrent Attacks. In: Yung, M. (ed.) CRYPTO 2002. LNCS, vol. 2442. Springer, Heidelberg (2002)
5. Camenisch, J., Shoup, V.: Practical Verifiable Encryption and Decryption of Discrete Logarithms. In: Boneh, D. (ed.) CRYPTO 2003. LNCS, vol. 2729. Springer, Heidelberg (2003)
6. Canetti, R.: Universally Composable Security: A New Paradigm for Cryptographic Protocols. In: Proc. of FOCS (2001)
7. Canetti, R., Dodis, Y., Pass, R., Walfish, S.: Universal Composability with Global Setup. In: Proc. of TCC (2007)
8. Canetti, R., Dodis, Y., Pass, R., Walfish, S.: Universal Composability with Global Setup (full version), http://eprint.iacr.org/2006/432
9. Canetti, R., Fischlin, M.: Universally Composable Commitments. In: Kilian, J. (ed.) CRYPTO 2001. LNCS, vol. 2139. Springer, Heidelberg (2001)
10. Canetti, R., Halevi, S., Katz, J., Lindell, Y., MacKenzie, P.: Universally Composable Password-Based Key Exchange. In: Cramer, R. (ed.) EUROCRYPT 2005. LNCS, vol. 3494. Springer, Heidelberg (2005)
11. Canetti, R., Krawczyk, H.: Universally Composable Notions of Key Exchange and Secure Channels. In: Knudsen, L.R. (ed.) EUROCRYPT 2002. LNCS, vol. 2332. Springer, Heidelberg (2002)
12. Canetti, R., Kushilevitz, E., Lindell, Y.: On the Limitations of Universally Composable Two-Party Computation Without Set-Up Assumptions. In: Biham, E. (ed.) EUROCRYPT 2003. LNCS, vol. 2656. Springer, Heidelberg (2003)
13. Canetti, R., Lindell, Y., Ostrovsky, R., Sahai, A.: Universally Composable Two-Party and Multi-Party Secure Computation. In: Proc. of STOC (2002)
14. Canetti, R., Rabin, T.: Universal Composition with Joint State. In: Boneh, D. (ed.) CRYPTO 2003. LNCS, vol. 2729. Springer, Heidelberg (2003)
15. Cramer, R., Shoup, V.: Universal Hash Proofs and a Paradigm for Adaptive Chosen Ciphertext Secure Public Key Encryption. In: Knudsen, L.R. (ed.) EUROCRYPT 2002. LNCS, vol. 2332. Springer, Heidelberg (2002)
16. Damgård, I.: Efficient Concurrent Zero-Knowledge in the Auxiliary String Model. In: Preneel, B. (ed.) EUROCRYPT 2000. LNCS, vol. 1807. Springer, Heidelberg (2000)
17. Damgård, I., Nielsen, J.: Perfect Hiding and Perfect Binding Universally Composable Commitment Schemes with Constant Expansion Factor. In: Yung, M. (ed.) CRYPTO 2002. LNCS, vol. 2442. Springer, Heidelberg (2002)
18. Dodis, Y., Shoup, V., Walfish, S.: Efficient Constructions of Composable Commitments and Zero-Knowledge Proofs (full version), http://www.shoup.net/papers/gucc.pdf
19. Feige, U.: Alternative Models for Zero Knowledge Interactive Proofs. Ph.D. thesis, Weizmann Institute of Science, Rehovot, Israel (1990)

20. Fiat, A., Shamir, A.: How to Prove Yourself: Practical Solutions to Identification and Signature Problems. In: Proc. of Crypto (1987)
21. Garay, J., MacKenzie, P., Yang, K.: Strengthening Zero-Knowledge Protocols Using Signatures. In: Biham, E. (ed.) EUROCRYPT 2003. LNCS, vol. 2656. Springer, Heidelberg (2003)
22. Hofheinz, D., Muller-Quade, J.: Universally Composable Commitments Using Random Oracles. In: Naor, M. (ed.) TCC 2004. LNCS, vol. 2951. Springer, Heidelberg (2004)
23. Jakobsson, M., Sako, K., Impagliazzo, R.: Designated Verifier Proofs and their Applications. In: Maurer, U.M. (ed.) EUROCRYPT 1996. LNCS, vol. 1070. Springer, Heidelberg (1996)
24. MacKenzie, P., Yang, K.: On Simulation-Sound Trapdoor Commitments. In: Cachin, C., Camenisch, J.L. (eds.) EUROCRYPT 2004. LNCS, vol. 3027. Springer, Heidelberg (2004)
25. Paillier, P.: Public-key Cryptosystems Based on Composite Degree Residuosity Classes. In: Stern, J. (ed.) EUROCRYPT 1999. LNCS, vol. 1592. Springer, Heidelberg (1999)
26. Pass, R.: On Deniabililty in the Common Reference String and Random Oracle Model. In: Boneh, D. (ed.) CRYPTO 2003. LNCS, vol. 2729. Springer, Heidelberg (2003)
27. Walfish, S.: Enhanced Security Models for Network Protocols. Ph.D. thesis. New York University (2007),
    http://www.cs.nyu.edu/web/Research/Theses/walfish_shabsi.pdf
28. Waters, B.: Efficient Identity-Based Encryption Without Random Oracles. In: Cramer, R. (ed.) EUROCRYPT 2005. LNCS, vol. 3494. Springer, Heidelberg (2005)

A   An Attack on "Deniable" ZK Protocol of [26]

Consider a prover P, a verifier V, and a third party Z who wishes to obtain evidence that P has interacted with V in the 2-round "deniable" ZK protocol of [26]. The third party Z uses RO to prepare a valid verifier's first message α for the protocol, asks V to forward α to P, and then relay back P's response β. In this case, its clear that V cannot know the transcript of RO queries issued by Z during the creation of α, and therefore V cannot run the zero knowledge simulator of [26]. In fact, the soundness of the protocol ensures that V cannot efficiently construct an accepting reply β to α without P's help. Therefore, if V is later able to obtain a valid response β, Z is correctly convinced that P has interacted with V, and P cannot deny that the interaction took place. Thus, the protocol is not deniable in "real life", despite meeting the definition of [26].

Noninteractive Statistical Zero-Knowledge Proofs for Lattice Problems

Chris Peikert[1,*] and Vinod Vaikuntanathan[2,**]

[1] SRI International
cpeikert@alum.mit.edu
[2] MIT
vinodv@mit.edu

Abstract. We construct *noninteractive statistical zero-knowledge* (NISZK) proof systems for a variety of standard approximation problems on lattices, such as the shortest independent vectors problem and the complement of the shortest vector problem. Prior proof systems for lattice problems were either interactive or leaked knowledge (or both).

Our systems are the first known NISZK proofs for any cryptographically useful problems that are not related to integer factorization. In addition, they are proofs of knowledge, have reasonable complexity, and generally admit efficient prover algorithms (given appropriate auxiliary input). In some cases, they even imply the first known *interactive* statistical zero-knowledge proofs for certain cryptographically important lattice problems.

We also construct an NISZK proof for a special kind of disjunction (i.e., OR gate) related to the shortest vector problem. This may serve as a useful tool in potential constructions of noninteractive (computational) zero knowledge proofs for NP based on lattice assumptions.

1 Introduction

A central idea in computer science is an *interactive proof system*, which allows a (possibly unbounded) prover to convince a computationally-limited verifier that a given statement is true [7, 29, 30]. The beautiful notion of *zero knowledge*, introduced by Goldwasser, Micali, and Rackoff [29], even allows the prover to convince the verifier while revealing *nothing more than* the truth of the statement.

Many of the well-known results about zero knowledge, e.g., that NP (and even all of IP) has zero-knowledge proofs [10, 24], refer to *computational* zero

* This material is based upon work supported by the National Science Foundation under Grants CNS-0716786 and CNS-0749931. Any opinions, findings, and conclusions or recommedations expressed in this material are those of the author(s) and do not necessarily reflect the views of the National Science Foundation.
** Work performed while at SRI International. Supported in part by NSF Grant CNS-0430450.

© International Association for Cryptologic Research 2008

knowledge, where security holds only against a bounded cheating verifier (typically under some complexity assumption). Yet there has also been a rich line of research concerning proof[1] systems in which the zero-knowledge property is *statistical.* The advantages of such systems include security against even *unbounded* cheating verifiers, usually without any need for unproved assumptions. Much is now known about the class SZK of problems possessing statistical zero-knowledge proofs; for example, it does not contain NP unless the polynomial-time hierarchy collapses [2, 20], it is closed under complement and union [38], it has natural complete (promise) problems [28, 42], and it is insensitive to whether the zero-knowledge condition is defined for arbitrary *malicious* verifiers, or only for *honest* ones [26].

Removing interaction. Zero-knowledge proofs inherently derive their power from interaction [25]. In spite of this, Blum, Feldman, and Micali [14] showed how to construct meaningful *noninteractive* zero-knowledge proofs (consisting of a single message from the prover to the verifier) if the parties simply share access to a uniformly random string. Furthermore, noninteractive *computational* zero-knowledge proofs exist for all of NP under plausible cryptographic assumptions [13, 14, 19, 31].

Just as with interactive proofs (and for the same reasons), it is also interesting to consider noninteractive proofs where the zero-knowledge condition is statistical. Compared with SZK, much less is known about the class NISZK of problems admitting such proofs. Clearly, NISZK is a (possibly proper) subset of SZK. It is also known to have complete (promise) problems [17, 27], but unlike SZK, it is not known whether NISZK is closed under complement or disjunction (OR).[2] Some conditional results are also known, e.g., NISZK = SZK if and only if NISZK is closed under complement [27] (though it seems far from clear whether this condition is true or not).

Applying NISZK proofs. In cryptographic schemes, the benefits of NISZK proofs are manifold: they involve a minimal number of messages, they remain secure under parallel and concurrent composition, and they provide a very strong level of security against unbounded cheating provers and verifiers alike, typically without relying on any complexity assumptions. However, the only *concrete* problems of cryptographic utility known to be in NISZK are all related in some way to integer factorization, i.e., variants of quadratic residuosity [14, 15, 16] and the language of "quasi-safe" prime products [21].[3]

Another important consideration in applying proof systems (both interactive and noninterative) is the complexity of the prover. Generally speaking, it is *not*

---

[1] In this work, we will be concerned exclusively with *proof* systems (as opposed to *argument* systems, in which a cheating prover is computationally bounded).

[2] An earlier version of [17] claimed that NISZK was closed under complement and disjunction, but the claims have since been retracted.

[3] The language of graphs having trivial automorphism group is in NISZK, as are the (NISZK-complete) "image density" [17] and "entropy approximation" [27] problems, but these problems do not seem to have any immediate applications to cryptographic schemes.

enough simply to have a proof system; one also needs to be able to implement the prover *efficiently* given a suitable witness or auxiliary input. For interactive SZK, several proof systems for specific problems (e.g., those of [29, 36]) admit efficient provers, and it was recently shown that *every* language in SZK ∩ NP has an efficient prover [37]. For *noninteractive* statistical zero knowledge, prover efficiency is not understood so well: while the systems relating to quadratic residuosity [14, 15, 16] have efficient provers, the language of quasi-safe prime products [21] is known to have an efficient prover only if interaction is allowed in one component of the proof.

1.1 Lattices and Proof Systems

Ever since the foundational work of Ajtai [4] on constructing hard-on-average cryptographic functions from *worst-case* assumptions relating to *lattices*, there has been significant interest in characterizing the complexity of lattice problems. Proof systems have provided an excellent means of making progress in this endeavor. We review some recent results below, after introducing the basic notions.

An n-dimensional lattice in $\mathbb{R}^n$ is a periodic "grid" of points consisting of all integer linear combinations of some set of linearly independent vectors $\mathbf{B} = \{\mathbf{b}_1, \dots, \mathbf{b}_n\} \subset \mathbb{R}^n$, called a *basis* of the lattice. Two of the central computational problems on lattices are the *shortest vector* problem SVP and the *closest vector* problem CVP. The goal of SVP is to find a (nonzero) lattice point whose length is minimal, given an arbitrary basis of the lattice. The goal of CVP, given an arbitrary basis and some target point $\mathbf{t} \in \mathbb{R}^n$, is to find a lattice point closest to $\mathbf{t}$. Another problem, whose importance to cryptography was first highlighted in Ajtai's work [4], is the *shortest independent vectors* problem SIVP. Here the goal (given a basis) is to find n linearly independent lattice vectors, the longest of which is as short as possible. All of these problems are known to be NP-complete in the worst case (in the case of SVP, under randomized reductions) [3, 12, 44], so we do not expect to obtain NISZK (or even SZK) proof systems for them.

In this work, we are primarily concerned with the natural *approximation* versions of lattice problems, phrased as promise (or "gap") problems with some approximation factor $\gamma \geq 1$. For example, the goal of GapSVP_γ is to accept any basis for which the shortest nonzero lattice vector has length at most 1, and to reject those for which it has length at least γ. One typically views the approximation factor as a function $\gamma(n)$ of the dimension of the lattice; problems become easier (or at least no harder) for increasing values of γ. Known polynomial-time algorithms for lattice problems obtain approximation factors $\gamma(n)$ that are only slightly subexponential in n [5, 6, 33, 43]. Moreover, obtaining a $\gamma(n) = \mathrm{poly}(n)$ approximation requires exponential time and space using known algorithms [5, 6, 11]. Therefore, lattice problems appear quite difficult to approximate to within even moderately-large factors.

Proof systems. We now review several proof systems for the above-described lattice problems and their complements. Every known system falls into one of two categories: *interactive* proofs that generally exhibit some form of statistical zero knowledge, or *noninteractive* proofs that are *not zero knowledge* (unless, of course, the associated lattice problems are trivial).

First of all, it is apparent that GapSVP_γ, GapCVP_γ, and $\mathsf{GapSIVP}_\gamma$ have trivial NP proof systems for any $\gamma \geq 1$. (E.g., for GapSVP_γ one can simply give a nonzero lattice vector of length at most 1.) Of course, the proofs clearly leak knowledge.

Goldreich and Goldwasser [23] initiated the study of interactive proof systems for lattice problems, showing that the complement problems $\mathsf{coGapSVP}_\gamma$ and $\mathsf{coGapCVP}_\gamma$ have AM proof systems for $\gamma(n) = O(\sqrt{n/\log n})$ factors. In other words, there are interactive proofs that *all* nonzero vectors in a given lattice are long, and that a given point in $\mathbb{R}^n$ is *far* from a given lattice.[4] Moreover, the protocols are perfect zero knowledge for *honest* verifiers, but they are not known to have efficient provers. Aharonov and Regev [1] showed that for slightly looser $\gamma(n) = O(\sqrt{n})$ factors, the same two problems are even in NP. In other words, for such γ the interactive proofs of [23] can be replaced by a *noninteractive* witness, albeit one that leaks knowledge. Building upon [1, 23], Guruswami, Micciancio, and Regev [32] showed analogous AM and NP proof systems for $\mathsf{coGapSIVP}_\gamma$.

Micciancio and Vadhan [36] gave (malicious verifier) SZK proofs with *efficient provers* for GapSVP_γ and GapCVP_γ, where $\gamma(n) = O(\sqrt{n/\log n})$. To our knowledge, there is no known zero-knowledge proof system for the cryptographically important $\mathsf{GapSIVP}_\gamma$ problem (even an interactive one), except by a reduction to $\mathsf{coGapSVP}$ using so-called "transference theorems" for lattices [8]. This reduction introduces an extra n factor in the approximation, resulting in fairly loose $\gamma(n) = O(n^{1.5}/\sqrt{\log n})$ factors. The same applies for the *covering radius* problem GapCRP [32], where the goal is to estimate the maximum distance from the lattice over all points in $\mathbb{R}^n$, and for the $\mathsf{GapGSMP}$ problem of approximating the *Gram-Schmidt minimum* of a lattice.

1.2   Our Results

We construct (without any assumption) *noninteractive statistical zero-knowledge* proof systems for a variety of lattice problems, for reasonably small approximation factors $\gamma(n)$. These are the first known NISZK proofs for lattice problems, and more generally, for any cryptographically useful problem not related to integer factorization. In addition, they are proofs of knowledge, have reasonable communication and verifier complexity, and admit efficient provers. They also imply the first known *interactive* statistical zero-knowledge proofs for certain lattice problems. Specifically, we construct the following:

- NISZK proofs (with efficient provers) for the $\mathsf{GapSIVP}_\gamma$, GapCRP_γ, and $\mathsf{GapGSMP}_\gamma$ problems, for any factor $\gamma(n) = \omega(\sqrt{n \log n})$.[5]

---

[4] Because GapSVP_γ and GapCVP_γ are in NP∩coAM for $\gamma(n) = O(\sqrt{n/\log n})$, the main conclusion of [23] is that these problems are *not* NP-hard, unless the polynomial-time hierarchy collapses.

[5] Recall that a function $g(n) = \omega(f(n))$ if $g(n)$ grows faster than $c \cdot f(n)$ for every constant $c > 0$.

In particular, this implies the first known (even interactive) SZK proof systems for these problems with approximation factors tighter than $n^{1.5}/\sqrt{\log n}$.

– An NISZK proof for coGapSVP$_\gamma$ for any factor $\gamma(n) \geq 20\sqrt{n}$. This is essentially the best we could hope for (up to constant factors) given the state of the art, because coGapSVP$_\gamma$ is not even known to be in NP for any factor $\gamma(n) < \sqrt{n}$.

For this proof system, we are able to give an efficient prover for $\gamma(n) = \omega(n \cdot \sqrt{\log n})$ factors, and an efficient *quantum* prover for slightly tighter $\gamma(n) = O(n/\sqrt{\log n})$ factors. (The prover's advice and the proof itself are still entirely classical; only the algorithm for generating the proof is quantum.)

– An NISZK proof for a special *disjunction* problem of two or more coGapSVP$_\gamma$ instances. As we describe in more detail below, this system may serve as an important ingredient in an eventual construction of noninteractive (computational) zero knowledge proofs for all of NP under lattice-related assumptions.

Our systems are also *proofs of knowledge* of a full-rank set of relatively "short" vectors in the given lattice. This is an important property in some of the applications to lattice-based cryptography we envision, described next.

Applications

Public key infrastructure. It is widely recognized that in public-key infrastructures, a user who presents her public key to a certification authority should also prove knowledge of a corresponding secret key (lest she present an "invalid" key, or one that actually belongs to some other user). A recent work of Gentry, Peikert, and Vaikuntanathan [22] constructed a variety of cryptographic schemes (including "hash-and-sign" signatures and identity-based encryption) in which the secret key can be any full-rank set of suitably "short" vectors in a public lattice. Our NISZK proof systems provide a reasonably efficient and statistically-secure way to prove knowledge of such secret keys. Implementing this idea requires some care, however, due to the exact nature of the knowledge guarantee and the fact that we are dealing with proof systems for *promise* problems.

To be more specific, a user generates a public key containing some basis $\mathbf{B}$ of a lattice Λ, and acts as the prover in the GapSIVP$_\gamma$ system for (say) $\gamma \approx \sqrt{n}$. In order to satisfy the completeness hypothesis, an honest user needs to generate $\mathbf{B}$ along with a full-rank set of lattice vectors all having length at most ≈ 1. The statistical zero-knowledge condition ensures that nothing about the user's secret key is leaked to the authority. Now consider a potentially malicious user. By the soundness condition, we are guaranteed only that Λ contains a full-rank set of lattice vectors all *of length at most* γ (otherwise the user will not be able to give a convincing proof). Under this guarantee, our knowledge extractor is able to extract a full-rank set of lattice vectors of somewhat larger length $\approx \gamma \cdot \sqrt{n} \approx n$. Therefore, the extracted secret key vectors may be somewhat longer than the honestly-generated ones. Fortunately, the schemes of [22] are parameterized by a value L, so that they behave identically on any secret key consisting of vectors of length at most L. Letting L be a bound on the length of the *extracted* vectors ensures that the proof of knowledge is useful in the broader context, e.g., to a simulator that needs to generate valid signatures under the presented public key. We also remark that our NISZK proofs can be made more compact in size when applied to the hard-on-average *integer* lattices used in [22] and related works, by dealing only with integer vectors rather than high-precision real vectors.

NICZK *for all of* NP?. Our proof systems may also be useful in constructing noninteractive *computational* zero-knowledge proof systems for all of NP based on the hardness of lattice problems. We outline a direction that follows the general approach of Blum, De Santis, Micali, and Persiano [13], who constructed an NICZK for the NP-complete language 3SAT under the quadratic residuosity assumption.

In [13], the common input is a 3SAT formula, and the auxiliary input to the prover is a satisfying assignment. The prover first chooses N, a product of two distinct primes. He associates, in a certain way, each true literal with a quadratic nonresidue from $\mathbb{Z}_N^*$, and each false literal with a quadratic residue. He proves in zero knowledge that (a) for each variable, either it or its negation is associated with a quadratic residue (thus, a variable and its negation cannot both be assigned true), and (b) for each clause, at least one of its three literals is associated with a quadratic nonresidue (thus, each clause is true under the implicit truth assignment). Thus, the entire proof involves zero-knowledge proofs of a disjunction of quadratic residuosity instances (for case (a)) and a disjunction of quadratic nonresiduosity instances (for case (b)).

We can replicate much of the above structure using lattices. Briefly, the modulus N translates to a suitably-chosen lattice Λ having *large* minimum distance, a quadratic nonresidue translates to a superlattice Λ_i of Λ also having *large* minimum distance, and a quadratic residue translates to a superlattice having *small* minimum distance. It then suffices to show in zero knowledge that (a) for each variable, the lattice associated to either it or its negation (or both) has small minimum distance, and (b) for each clause, the lattice associated to one of the variables in the clause has large minimum distance. In Section 3.2, we show how to implement part (b) by constructing an NISZK proof for a special disjunction of coGapSVP instances. However, we do not know how to prove noninteractively that one or more lattices has *small* minimum distance, i.e., a disjunction of GapSVP instances (see Section 1.3 for discussion). This seems to be the main technical barrier for obtaining NICZK for all of NP under lattice assumptions.

Finally, our NISZK proofs immediately imply statistically-secure *zaps*, as defined by Dwork and Naor [18], for the same problems. Zaps have a number of applications in general, and we suspect that they may find equally important applications in lattice-based cryptography.

Techniques. The main conceptual tool for achieving zero knowledge in our proof systems is a lattice quantity called the *smoothing parameter*, introduced by Micciancio and Regev [35] (following related work of Regev [40]). The smoothing parameter was introduced for the purpose of obtaining worst-case to average-case reductions for lattice problems, but more generally, it provides a way to generate an (almost-)uniform random variable related to an arbitrary given lattice.

In more detail, let $\Lambda \subset \mathbb{R}^n$ be a lattice, and imagine "blurring" all the points of Λ according to a Gaussian distribution. With enough blur, the discrete structure of the lattice is entirely destroyed, and the resulting picture is (almost) uniformly-spread over $\mathbb{R}^n$. Technically, this intuitive description corresponds to choosing a noise vector $\mathbf{e}$ from a Gaussian distribution (centered at the origin) and reducing $\mathbf{e}$ modulo any basis $\mathbf{B}$ of the lattice. (The value $\mathbf{e} \bmod \mathbf{B}$ is the unique point $\mathbf{t} \in \mathcal{P}(\mathbf{B}) = \{\sum_i c_i \mathbf{b}_i \, : \, \forall\, i, c_i \in [0, 1)\}$ such that $\mathbf{t} - \mathbf{e} \in \Lambda$; it can be computed efficiently given $\mathbf{e}$ and $\mathbf{B}$.) Informally, the smoothing parameter of the lattice is the amount of noise needed to obtain a nearly uniform distribution over $\mathcal{P}(\mathbf{B})$ via this process.

Our NISZK proofs all share a common structure regardless of the specific lattice problem in question. It is actually most instructive to first consider the zero-knowledge *simulator*, and then build the prover and verifier around it. In fact, we have already described how the simulator works: given a basis $\mathbf{B}$, it simply chooses a Gaussian noise vector $\mathbf{e}'$ and computes $\mathbf{t}' = \mathbf{e}' \bmod \mathbf{B}$. The vector $\mathbf{t}' \in \mathcal{P}(\mathbf{B})$ is the simulated common random "string," and $\mathbf{e}'$ is the simulated proof.[6] In the real proof system, the random string is a uniformly random $\mathbf{t} \in \mathcal{P}(\mathbf{B})$, and the prover (suppose for now that it is unbounded) generates a proof $\mathbf{e}$ by sampling from the Gaussian distribution *conditioned on* the event $\mathbf{e} = \mathbf{t} \bmod \mathbf{B}$. The verifier simply checks that indeed $\mathbf{t} - \mathbf{e} \in \Lambda$ and that $\mathbf{e}$ is "short enough."

For statistical zero knowledge, suppose that YES instances of the lattice problem have small smoothing parameter. Then the simulated random string $\mathbf{t}' = \mathbf{e}' \bmod \mathbf{B}$ is (nearly) uniform, just as $\mathbf{t}$ is in the real system; moreover, the distribution of the simulated proof $\mathbf{e}'$ conditioned on $\mathbf{t}'$ is the exactly the same as the distribution of the real proof $\mathbf{e}$. For completeness, we use the fact (proved in [35]) that a real proof $\mathbf{e}$ generated in the specified way is indeed relatively short. Finally, for soundness, we require that in NO instances, a significant fraction of random strings $\mathbf{t} \in \mathcal{P}(\mathbf{B})$ are simply too far away from the lattice to admit any short enough proof $\mathbf{e}$. (The soundness error can of course be attentuated by composing several independent proofs in parallel.)

The two competing requirements for YES and NO instances (for zero knowledge and soundness, respectively) determine the resulting approximation factor for the particular lattice problem. For the GapSIVP, GapCRP, and GapGSMP problems, the factor is $\approx \sqrt{n}$, but for technical reasons it turns out to be only $\approx n$ for the coGapSVP problem. To obtain tighter $O(\sqrt{n})$ factors, we design a system that can be seen as a zero-knowledge analogue of the NP proof system of Aharonov and Regev [1]. Our prover simply gives many independent proofs $\mathbf{e}_i$ (as above) in parallel, for uniform and independent $\mathbf{t}_i \in \mathcal{P}(\mathbf{B})$. The verifier, rather than simply checking the *lengths* of the individual $\mathbf{e}_i$s, instead performs an "eigenvalue test" on the entire collection. Although the eigenvalue test and its purpose (soundness) are exactly the same as in [1], we use it in a technically

---

[6] A random *binary string* can be used to represent a uniformly random $\mathbf{t}' \in \mathcal{P}(\mathbf{B}) \subset \mathbb{R}^n$ by its n coefficients $c_i \in [0, 1)$ relative to the given basis $\mathbf{B}$, to any desired level of precision.

different way: whereas in [1] it bounds a certain quantity computed by the verifier (which leaks knowledge, but *guarantees* rejection), here it bounds the volume of "bad" random strings that could potentially allow for false proofs.

We now turn to the issue of prover efficiency. Recall that the prover must choose a Gaussian noise vector $\mathbf{e}$ *conditioned on* the event that $\mathbf{e} = \mathbf{t}$ mod $\mathbf{B}$. Such conditional distributions, called *discrete Gaussians* over lattices, have played a key role in several recent results in complexity theory and cryptography, e.g., [1, 35, 39, 41]. The recent work of [22] demonstrated an algorithm that can use any suitably "short" basis of the lattice as advice for *efficiently sampling* from a discrete Gaussian. Applying this algorithm immediately yields efficient provers for the tightest $\gamma(n) = \omega(\sqrt{n \log n})$ factors for GapSIVP and related problems, and $\gamma(n) = \omega(n \cdot \sqrt{\log n})$ factors for coGapSVP. We also describe a *quantum* sampling algorithm (using different advice) that yields an efficient quantum prover for coGapSVP, for slightly tighter $\gamma(n) = O(n/\sqrt{\log n})$ factors.

Finally, we add that all of our proof systems easily generalize to arbitrary ℓ_p norms for $p \geq 2$, under essentially the same approximation factors $\gamma(n)$. The proof systems themselves actually remain exactly the same; their analysis in ℓ_p norms relies upon general facts about discrete Gaussians due to Peikert [39].

1.3   Open Questions

Recall that SZK is closed under complement and union [38] and that every langauge in SZK ∩ NP has a statistical zero-knowledge proof with an efficient prover [37]. Whether NISZK has analogous properties is a difficult open problem with many potential consequences. Our work raises versions of these questions for *specific* problems, which may help to shed some light on the general case.

We have shown that $coGapSVP_\gamma$ has NISZK proofs for certain $\gamma(n) = \mathrm{poly}(n)$ factors; does its complement $GapSVP_\gamma$ have such proofs as well? As described above, we suspect that a positive answer to this question, combined with our proofs for the special coGapSVP disjunction problem, could lead to noninteractive (computational) zero knowledge proofs for all of NP under worst-case lattice assumptions. In addition, because the *closest* vector problem GapCVP and its complement coGapCVP both admit SZK proofs, it is an interesting question whether they also admit NISZK proofs. The chief technical difficulty in addressing any of these questions seems to be that a short (or close) lattice vector guarantees nothing useful about the smoothing parameter of the lattice (or its dual). Therefore it is unclear how the simulator could generate a uniformly random string together with a meaningful proof.

The factors $\gamma(n)$ for which we can demonstrate *efficient* provers are in some cases looser than those for which we know of *inefficient* provers. The gap between these factors is solely a consequence of our limited ability to sample from discrete Gaussians. Is there some succinct (possibly quantum) advice that permits efficient sampling from a discrete Gaussian with a parameter close to the smoothing parameter of the lattice (or close to the tightest known bound on the smoothing parameter)? More generally, does every problem in NISZK ∩ NP have an NISZK proof with an efficient prover?

Finally, although we construct an NISZK proof for a problem that is structurally similar to the disjunction (OR) of many coGapSVP instances, there are additional technical constraints on the problem. It would be interesting to see if these constraints could be relaxed or lifted entirely.

2   Preliminaries

For any positive integer n, $[n]$ denotes the set $\{1, \ldots, n\}$. The function log always denotes the natural logarithm. We extend any function $f(\cdot)$ to a countable set A in the following way: $f(A) = \sum_{x \in A} f(x)$. A positive function $\epsilon(\cdot)$ is *negligible* in its parameter if it decreases faster than the inverse of any polynomial, i.e., if $\epsilon(n) = n^{-\omega(1)}$. The *statistical distance* between two distributions X and Y over a countable set A is $\Delta(X, Y) = \frac{1}{2} \sum_{a \in A} |\Pr[X = a] - \Pr[Y = a]|$.

Vectors are written using bold lower-case letters, e.g., $\mathbf{x}$. Matrices are written using bold capital letters, e.g., $\mathbf{X}$. The ith column vector of $\mathbf{X}$ is denoted $\mathbf{x}_i$. We often use matrix notation to denote a set of vectors, i.e., $\mathbf{S}$ also represents the set of its column vectors. We write $\mathrm{span}(\mathbf{v}_1, \mathbf{v}_2, \ldots)$ to denote the linear space spanned by its arguments. For a set $S \subseteq \mathbb{R}^n$, $\mathbf{v} \in \mathbb{R}^n$, and $c \in \mathbb{R}$, we let $S + \mathbf{x} = \{\mathbf{y} + \mathbf{x} : \mathbf{y} \in S\}$ and $cS = \{c\mathbf{y} : \mathbf{y} \in S\}$.

The symbol $\|\cdot\|$ denotes the Euclidean norm on $\mathbb{R}^n$. We say that the norm of a set of vectors is the norm of its longest element: $\|\mathbf{X}\| = \max_i \|\mathbf{x}_i\|$. For any $\mathbf{t} \in \mathbb{R}^n$ and set $V \subseteq \mathbb{R}^n$, the distance from $\mathbf{t}$ to V is $\mathrm{dist}(\mathbf{t}, V) = \inf_{\mathbf{v} \in V} \mathrm{dist}(\mathbf{t}, \mathbf{v})$.

2.1   Noninteractive Proof Systems

We consider proof systems for promise problems $\Pi = (\Pi^{\mathrm{YES}}, \Pi^{\mathrm{NO}})$ where each instance of the problem is associated with some value of the security parameter n, and we partition the instances into sets Π_n^{YES} and Π_n^{NO} in the natural way. In general, the value of n might be different from the length of the instance; for example, the natural security parameter for lattice problems is the dimension n of the lattice, but the input basis might be represented using many more bits. In this work, we assume for simplicity that instances of lattice problems have lengths bounded by some fixed polynomial in the dimension n, and we treat n as the natural security parameter.

Definition 1 (Noninteractive Proof System). *A pair (P, V) is a noninteractive proof system for a promise problem $\Pi = (\Pi^{YES}, \Pi^{NO})$ if P is a (possibly unbounded) probabilistic algorithm, V is a deterministic polynomial-time algorithm, and the following conditions hold for some functions $c(n), s(n) : \mathbb{N} \to [0, 1]$ and for all $n \in \mathbb{N}$:*

  - *Completeness: For every $x \in \Pi_n^{YES}$, $\Pr[V(x, r, P(x, r))$ accepts$] \geq 1 - c(n)$.*
  - *Soundness: For every $x \in \Pi_n^{NO}$, $\Pr[\exists \pi : V(x, r, \pi)$ accepts$] \leq s(n)$.*

The probabilities are taken over the choice of the random input r and the random choices of P. The function $c(n)$ is called the completeness error, and the function $s(n)$ is called the soundness error. For nontriviality, we require $c(n) + s(n) \leq 1 - 1/poly(n)$.

The random input r is generally chosen uniformly at random from $\{0,1\}^{p(n)}$ for some fixed polynomial $p(\cdot)$. For notational simplicity, we adopt a model in which the random input r is chosen from an efficiently-sampleable set R_x that may depend on the instance x. This is without loss of generality, because given a random string $r' \in \{0,1\}^{p(n)}$, both prover and verifier can generate $r \in R_x$ simply by running the sampling algorithm with randomness r'.

By standard techniques, completeness and soundness errors can be reduced via parallel repetition. Note that our definition of soundness is *non-adaptive*, that is, the NO instance is fixed in advance of the random input r. Certain applications may require *adaptive* soundness, in which there do not exist *any* instance $x \in \Pi_n^{NO}$ and valid proof π, except with negligible probability over the choice of r. For proof systems, a simple argument shows that non-adaptive soundness implies adaptive soundness error $2^{-p(n)}$ for any desired $p(n) = \text{poly}(n)$: let $B(n) = \text{poly}(n)$ be a bound on the length of any instance in Π_n^{NO}, and compose the proof system in parallel some $\text{poly}(n)$ times to achieve (non-adaptive) soundness $2^{-p(n)-B(n)}$. Then by a union bound over all $x \in \Pi_n^{NO}$, the resulting proof system has adaptive soundness $2^{-p(n)}$.

Definition 2 (NISZK). *A noninteractive proof system (P,V) for a promise problem $\Pi = (\Pi^{YES}, \Pi^{NO})$ is statistical zero knowledge if there exists a probabilistic polynomial-time algorithm S (called a simulator) such that for all $x \in \Pi^{YES}$, the statistical distance between $S(x)$ and $(r, P(x,r))$ is negligible in n:*

$$\Delta(\, S(x)\, , \, (r, P(x,r))\,) \leq negl(n).$$

The class of promise problems having noninteractive statistical zero knowledge proof systems is denoted NISZK.

For defining proofs of knowledge, we adapt the general approach advocated by Bellare and Goldreich [9] to our noninteractive setting. In particular, the definition is entirely distinct from that of a proof system, and it refers to *relations* (not promise problems). Let $R \subseteq \{0,1\}^* \times \{0,1\}^*$ be a binary relation where the first entry x of each $(x,y) \in R$ is associated with some value of the security parameter n, and partition the relation into sub-relations R_n in the natural way. Let $R_x = \{y : (x,y) \in R\}$ and $\Pi_n^R = \{x : \exists\, y \text{ such that } (x,y) \in R_n\}$.

Definition 3 (Noninteractive proof of knowledge). *Let R be a binary relation, let V be a determinstic polynomial time machine, and let $\kappa(n), c(n) : \mathbb{N} \to [0,1]$ be functions. We say that V is a* knowledge verifier *for the relation R with* nontriviality error c *and* knowledge error κ *if the following two conditions hold:*

1. Nontriviality (with error c): *there exists a probabilistic function P such that for all $x \in \Pi_n^R$, $\Pr[V(x,r,P(x,r)) \text{ accepts}] \geq 1 - c(n)$.*
2. Validity (with error κ): *there exists a probabilistic oracle machine E such that for for every probabilistic function P^* and every $x \in \Pi_n^R$ where*

$$p_x = \Pr[V(x,r,P^*(x,r)) \text{ accepts}] > \kappa(n),$$

$E^{P^*}(x)$ *outputs a string from R_x in expected time $\text{poly}(n)/(p_x - \kappa(n))$.*

2.2   Lattices

For a matrix $\mathbf{B} \in \mathbb{R}^{n \times n}$ whose columns $\mathbf{b}_1, \ldots, \mathbf{b}_n$ are linearly independent, the n-dimensional *lattice*[7] Λ generated by the *basis* $\mathbf{B}$ is

$$\Lambda = \mathcal{L}(\mathbf{B}) = \{\mathbf{Bc} = \sum\nolimits_{i \in [n]} c_i \cdot \mathbf{b}_i \ : \ \mathbf{c} \in \mathbb{Z}^n\}.$$

The *fundamental parallelepiped* of $\mathbf{B}$ is the half-open set

$$\mathcal{P}(\mathbf{B}) = \{\sum_i c_i \mathbf{b}_i \ : \ 0 \le c_i < 1, i \in [n]\}.$$

For any lattice basis $\mathbf{B}$ and point $\mathbf{x} \in \mathbb{R}^n$, there is a unique vector $\mathbf{y} \in \mathcal{P}(\mathbf{B})$ such that $\mathbf{y} - \mathbf{x} \in \mathcal{L}(\mathbf{B})$. This vector is denoted $\mathbf{y} = \mathbf{x} \bmod \mathbf{B}$, and it can be computed in polynomial time given $\mathbf{B}$ and $\mathbf{x}$.

For any (ordered) set $\mathbf{S} = \{\mathbf{s}_1, \ldots, \mathbf{s}_n\} \subset \mathbb{R}^n$ of linearly independent vectors, let $\tilde{\mathbf{S}} = \{\tilde{\mathbf{s}}_1, \ldots, \tilde{\mathbf{s}}_n\}$ denote its Gram-Schmidt orthogonalization, defined iteratively in the following way: $\tilde{\mathbf{s}}_1 = \mathbf{s}_1$, and for each $i = 2, \ldots, n$, $\tilde{\mathbf{s}}_i$ is the component of $\mathbf{s}_i$ orthogonal to $\mathrm{span}(\mathbf{s}_1, \ldots, \mathbf{s}_{i-1})$. Clearly, $\|\tilde{\mathbf{s}}_i\| \le \|\mathbf{s}_i\|$.

Let $\mathcal{C}_n = \{\mathbf{x} \in \mathbb{R}^n \ : \ \|\mathbf{x}\| \le 1\}$ be the closed unit ball. The *minimum distance* of a lattice Λ, denoted $\lambda_1(\Lambda)$, is the length of its shortest nonzero element: $\lambda_1(\Lambda) = \min_{0 \ne \mathbf{x} \in \Lambda} \|\mathbf{x}\|$. More generally, the *ith successive minimum* $\lambda_i(\Lambda)$ is the smallest radius r such that the closed ball $r\mathcal{C}_n$ contains i linearly independent vectors in Λ: $\lambda_i(\Lambda) = \min\{r \in \mathbb{R} : \dim \mathrm{span}(\Lambda \cap r\mathcal{C}_n) \ge i\}$. The *Gram-Schmidt minimum* $\tilde{bl}(\Lambda)$ is $\tilde{bl}(\Lambda) = \min_{\mathbf{B}} \|\tilde{\mathbf{B}}\| = \min_{\mathbf{B}} \max_{i \in [n]} \|\tilde{\mathbf{b}}_i\|$, where the minimum is taken over all (ordered) bases $\mathbf{B}$ of Λ. The definition is restricted to bases without loss of generality, because for any (ordered) full-rank set $\mathbf{S} \subset \Lambda$, there is an (ordered) basis $\mathbf{B}$ of Λ such that $\|\tilde{\mathbf{B}}\| \le \|\tilde{\mathbf{S}}\|$ (see [34, Lemma 7.1]). The *covering radius* $\mu(\Lambda)$ is the smallest radius r such that closed balls $r\mathcal{C}_n$ centered at every point of Λ cover all of $\mathbb{R}^n$: $\mu(\Lambda) = \max_{\mathbf{x} \in \mathbb{R}^n} \mathrm{dist}(\mathbf{x}, \Lambda)$.

The *dual lattice* $\Lambda*$ of Λ, is the set $\Lambda^* = \{\mathbf{x} \in \mathbb{R}^n \ : \ \forall \, \mathbf{v} \in \Lambda, \langle \mathbf{x}, \mathbf{v} \rangle \in \mathbb{Z}\}$ of all vectors having integer inner product with *all* the vectors in Λ. It is routine to verify that this set is indeed a lattice, and if $\mathbf{B}$ is a basis for Λ, then $\mathbf{B}^* = (\mathbf{B}^{-1})^T$ is a basis for Λ^*. It also follows from the symmetry of the definition that $(\Lambda^*)^* = \Lambda$.

Lemma 4 ([8]). *For any n-dimensional lattice Λ, $1 \le 2 \cdot \lambda_1(\Lambda) \cdot \mu(\Lambda^*) \le n$.*

Lemma 5 ([34, Theorem 7.9]). *For any n-dimensional lattice Λ,*

$$\tilde{bl}(\Lambda) \le \lambda_n(\Lambda) \le 2\mu(\Lambda).$$

A random point in $\mathcal{P}(\mathbf{B})$ is unlikely to be "close" to the lattice, where the notion of closeness is relative to the covering radius.

---

[7] Technically, this is the definition of a *full-rank* lattice, which is all we will be concerned with in this work.

Lemma 6 ([32, Lemma 4.1]). *For any lattice* $\Lambda = \mathcal{L}(\mathbf{B})$,

$$\Pr_{\mathbf{t} \in \mathcal{P}(\mathbf{B})} \left[\mathrm{dist}(\mathbf{t}, \Lambda) < \frac{\mu(\Lambda)}{2} \right] \leq \frac{1}{2},$$

where the probability is taken over $\mathbf{t} \in \mathcal{P}(\mathbf{B})$ *chosen uniformly at ranodm.*

We now define some standard approximation problems on lattices, all of which ask to estimate (to within some factor γ) the value of some geometric lattice quantity. We define promise (or "gap") problems $\Pi = (\Pi^{\mathrm{YES}}, \Pi^{\mathrm{NO}})$, where the goal is to decide whether the instance belongs to the set Π^{YES} or the set Π^{NO} (these two sets are disjoint, but not necessarily exhaustive; when the input belongs to neither set, any output is acceptable). In the complement of a promise problem, Π^{YES} and Π^{NO} are simply swapped.

Definition 7 (Lattice Problems). *Let* $\gamma = \gamma(n)$ *be an approximation factor in the dimension n. For any function ϕ from lattices to the positive reals, we define an approximation problem where the input is a basis $\mathbf{B}$ of an n-dimensional lattice. It is a YES instance if $\phi(\mathcal{L}(\mathbf{B})) \leq 1$, and is a NO instance if $\phi(\mathcal{L}(\mathbf{B})) > \gamma(n)$.*

In particular, we define the following concrete problems by instantiating ϕ:

- *The* Shortest Vector Problem GapSVP_γ, *for* $\phi = \lambda_1$.
- *The* Shortest Independent Vectors Problem $\mathsf{GapSIVP}_\gamma$, *for* $\phi = \lambda_n$.
- *The* Gram-Schmidt Minimum Problem $\mathsf{GapGSMP}_\gamma$, *for* $\phi = \tilde{bl}$.
- *The* Covering Radius Problem GapCRP_γ, *for* $\phi = \mu$.

Note that the choice of the quantities 1 and γ above is arbitrary; by scaling the input instance, they can be replaced by β and $\beta \cdot \gamma$ (respectively) for any $\beta > 0$ without changing the problem.

Gaussians on Lattices. Our review of Gaussian measures over lattices follows the development by prior works [1, 35, 40]. For any $s > 0$ define the Gaussian function centered at $\mathbf{c}$ with parameter s as:

$$\forall \mathbf{x} \in \mathbb{R}^n, \ \rho_{s,\mathbf{c}}(\mathbf{x}) = e^{-\pi \|\mathbf{x}-\mathbf{c}\|^2 / s^2}.$$

The subscripts s and $\mathbf{c}$ are taken to be 1 and $\mathbf{0}$ (respectively) when omitted. The total measure associated to $\rho_{s,\mathbf{c}}$ is $\int_{\mathbf{x} \in \mathbb{R}^n} \rho_{s,\mathbf{c}}(\mathbf{x}) \, d\mathbf{x} = s^n$, so we can define a continuous Gaussian distribution centered at $\mathbf{c}$ with parameter s by its probability density function $\forall \mathbf{x} \in \mathbb{R}^n, \ D_{s,\mathbf{c}}(\mathbf{x}) = \rho_{s,\mathbf{c}}(\mathbf{x})/s^n$.

It is possible to sample from $D_{s,\mathbf{c}}$ efficiently to within any desired level of precision. For simplicity, we use real numbers in this work and assume that we can sample from $D_{s,\mathbf{c}}$ exactly; all the arguments can be made rigorous by using a suitable degree of precision.

For any $\mathbf{c} \in \mathbb{R}^n$, real $s > 0$, and lattice Λ, define the *discrete Gaussian distribution over Λ* as:

$$\forall \mathbf{x} \in \Lambda, \ D_{\Lambda,s,\mathbf{c}}(\mathbf{x}) = \frac{\rho_{s,\mathbf{c}}(\mathbf{x})}{\rho_{s,\mathbf{c}}(\Lambda)}.$$

(As above, we may omit the parameters s or $\mathbf{c}$.) Intuitively, $D_{\Lambda,s,\mathbf{c}}$ can be viewed as a "conditional" distribution, resulting from sampling $\mathbf{x} \in \mathbb{R}^n$ from a Gaussian centered at $\mathbf{c}$ with parameter s, and conditioning on the event $\mathbf{x} \in \Lambda$.

Definition 8 ([35]). *For an n-dimensional lattice Λ and positive real $\epsilon > 0$, the smoothing parameter $\eta_\epsilon(\Lambda)$ is defined to be the smallest s such that $\rho_{1/s}(\Lambda^*\backslash\{\mathbf{0}\}) \leq \epsilon$.*

The name "smoothing parameter" is due to the following (informally stated) fact: if a lattice Λ is "blurred" by adding Gaussian noise with parameter $s \geq \eta_\epsilon(\Lambda)$ for some $\epsilon > 0$, the resulting distribution is $\epsilon/2$-close to uniform over the entire space. This is made formal in the following lemma.

Lemma 9 ([35, Lemma 4.1]). *For any lattice $\mathcal{L}(\mathbf{B})$, $\epsilon > 0$, $s \geq \eta_\epsilon(\mathcal{L}(\mathbf{B}))$, and $\mathbf{c} \in \mathbb{R}^n$, the statistical distance between $(D_{s,\mathbf{c}} \bmod \mathbf{B})$ and the uniform distribution over $\mathcal{P}(\mathbf{B})$ is at most $\epsilon/2$.*

The smoothing parameter is related to other important lattice quantities.

Lemma 10 ([35, Lemma 3.2]). *Let Λ be any n-dimensional lattice, and let $\epsilon(n) = 2^{-n}$. Then $\eta_\epsilon(\Lambda) \leq \sqrt{n}/\lambda_1(\Lambda^*)$.*

Lemma 11 ([22, Lemma 3.1]). *For any n-dimensional lattice Λ and $\epsilon > 0$, we have*

$$\eta_\epsilon(\Lambda) \quad \leq \quad \tilde{bl}(\Lambda) \cdot \sqrt{\log(2n(1+1/\epsilon))/\pi}.$$

In particular, for any $\omega(\sqrt{\log n})$ function, there is a negligible function $\epsilon(n)$ for which $\eta_\epsilon(\Lambda) \leq \tilde{bl}(\Lambda) \cdot \omega(\sqrt{\log n})$.

Note that because $\tilde{bl}(\Lambda) \leq \lambda_n(\Lambda)$, we also have $\eta_\epsilon(\Lambda) \leq \lambda_n(\Lambda) \cdot \omega(\sqrt{\log n})$; this is Lemma 3.3 in [35].

The smoothing parameter also influences the behavior of *discrete* Gaussian distributions over the lattice. When $s \geq \eta_\epsilon(\Lambda)$, the distribution $D_{\Lambda,s,.\mathbf{c}}$ has a number of nice properties: it is highly concentrated within a radius $s\sqrt{n}$ around its center $\mathbf{c}$, it is not concentrated too heavily in any single direction, and it is not concentrated too heavily on any fixed hyperplane. We refer to [35, Lemmas 4.2 and 4.4] and [41, Lemma 3.13] for precise statements of these facts.

3 Noninteractive Statistical Zero Knowledge

Here we demonstrate NISZK proofs for several natural lattice problems. Due to lack of space, we give intuitive proof sketches here and defer complete proofs to the full version.

We first introduce an intermediate lattice problem (actually, a family of problems parameterized by a function $\epsilon(n)$) called SOS, which stands for "smooth-or-separated." The SOS problem exactly captures the two properties we need for our first basic NISZK proof system: in YES instances, the lattice can be completely smoothed by a Gaussian with parameter 1, and in NO instances, a random point is at least $\sqrt{n}$ away from the lattice with good probability. Moreover, the SOS problem is at least as expressive as several standard lattice problems of interest, by which we mean that there are simple (deterministic) reductions to SOS from $\mathsf{GapSIVP}_\gamma$, GapCRP_γ, $\mathsf{GapGSMP}_\gamma$, and $\mathsf{coGapSVP}_\gamma$ (for appropriate approximation factors γ).

Definition 12 (Smooth-Or-Separated Problem). *For any positive function $\epsilon = \epsilon(n)$, an input to $\epsilon\text{-}\mathsf{SOS}_\gamma$ is a basis $\mathbf{B}$ of an n-dimensional lattice. It is a YES instance if $\eta_\epsilon(\mathcal{L}(\mathbf{B})) \leq 1$, and is a NO instance if $\mu(\mathcal{L}(\mathbf{B})) > \gamma(n)$.*[8]

The NISZK proof system for SOS is described precisely in Figure 1. For the moment, we ignore issues of efficiency and assume that the prover is unbounded. To summarize, the random input is a uniformly random point $\mathbf{t} \in \mathcal{P}(\mathbf{B})$, where $\mathbf{B}$ is the input basis. The prover samples a vector $\mathbf{e}$ from a Gaussian (centered at the origin), *conditioned* on the event that $\mathbf{e}$ is congruent to $\mathbf{t}$ modulo the lattice, i.e., $\mathbf{e} - \mathbf{t} \in \mathcal{L}(\mathbf{B})$. In other words, the prover samples from a *discrete* Gaussian distribution. The verifier accepts if $\mathbf{e}$ and $\mathbf{t}$ are indeed congruent modulo $\mathcal{L}(\mathbf{B})$, and if $\|\mathbf{e}\| \leq \sqrt{n}$.

In the YES case, the smoothing parameter is at most 1. This lets us prove that the sampled proof $\mathbf{e}$ is indeed shorter than $\sqrt{n}$ (with overwhelming probability), ensuring completeness. More interestingly, it means that the simulator can first choose $\mathbf{e}$ from a *continuous* Gaussian, and then set the random input $\mathbf{t} = \mathbf{e} \bmod \mathbf{B}$. By Lemma 9, this $\mathbf{t}$ is almost-uniform in $\mathcal{P}(\mathbf{B})$, ensuring zero knowledge. In the NO case, the covering radius of the lattice is large. By Lemma 6, with good probability the random vector $\mathbf{t} \in \mathcal{P}(\mathbf{B})$ is simply too far away from the lattice to admit any short enough $\mathbf{e}$, hence no proof can convince the verifier. (A complete proof of Theorem 13 below is given in the full version.)

NISZK proof system for SOS

Common Input: A basis $\mathbf{B}$ of an n-dimensional lattice $\Lambda = \mathcal{L}(\mathbf{B})$.
Random Input: A vector $\mathbf{t} \in \mathbb{R}^n$ chosen uniformly at random from $\mathcal{P}(\mathbf{B})$.
Prover P: Sample $\mathbf{v} \sim D_{\Lambda, -\mathbf{t}}$, and output $\mathbf{e} = \mathbf{t} + \mathbf{v} \in \mathbb{R}^n$ as the proof.
Verifier V: Accept if $\mathbf{e} - \mathbf{t} \in \Lambda$ and if $\|\mathbf{e}\| \leq \sqrt{n}$, otherwise reject.

Fig. 1. The noninteractive zero-knowledge proof system for the SOS problem

Theorem 13. *For any $\gamma(n) \geq 2\sqrt{n}$ and any negligible function $\epsilon(n)$, the problem $\epsilon\text{-}\mathsf{SOS}_\gamma \in \mathsf{NISZK}$ via the proof system described in Figure 1. The completeness error of the system is $c(n) = 2^{-n+1}$ and the soundness error is $s(n) = 1/2$.*

By deterministic reductions to the $\epsilon\text{-}\mathsf{SOS}_\gamma$ problem, several standard lattice problems are also in NISZK. The proof of the following corollary is a straightforward application of Lemmas 4, 5, 10, and 11, and is deferred to the full version.

---

[8] Using techniques from [35], it can be verified that the YES and NO sets are disjoint whenever $\gamma \geq \sqrt{n}$ and $\epsilon(n) \leq 1/2$.

Corollary 14. *For every $\gamma(n) \geq 1$ and any fixed $\omega(\sqrt{\log n})$ function, there is a deterministic polynomial-time reduction from each of the following problems to ϵ-SOS$_\gamma$ (for some negligible function $\epsilon(n)$):*

- *GapSIVP$_{\gamma'}$, GapCRP$_{\gamma'}$, and GapGSMP$_{\gamma'}$ for any $\gamma'(n) \geq 2\omega(\sqrt{\log n}) \cdot \gamma(n)$,*
- *coGapSVP$_{\gamma'}$ for any $\gamma'(n) \geq 2\sqrt{n} \cdot \gamma(n)$.*

In particular, the problems GapSIVP$_{\gamma'}$, GapCRP$_{\gamma'}$, and GapGSMP$_{\gamma'}$ for $\gamma'(n) = \omega(\sqrt{n \log n})$ and coGapSVP$_{4n}$ are in NISZK.

We now turn to the knowledge guarantee for the protocol. For a function $\epsilon = \epsilon(n)$, we define a relation R_ϵ where an instance (for security parameter n) is a basis $\mathbf{B} \subset \mathbb{R}^{n \times n}$ of a lattice having smoothing parameter η_ϵ bounded by 1 (without loss of generality), and a witness for $\mathbf{B}$ is a full-rank set $\mathbf{S} \subset \mathcal{L}(\mathbf{B})$ of lattice vectors having length at most $2\sqrt{n}$.

Theorem 15. *For any positive $\epsilon(n) \leq 1/3$, the verifier described in Figure 1 is a knowledge verifier for relation R_ϵ with nontriviality error $c(n) = 2^{-n+1}$ and knowledge error $\kappa(n) = \epsilon(n)/2$.*

Now consider the complexity of the prover in the protocol from Figure 1. Note that the prover has to sample from the discrete Gaussian distribution $D_{\Lambda, -\mathbf{t}}$ (with parameter 1). For this purpose, we use a recent result of Gentry, Peikert and Vaikuntanathan [22, Theorem 4.1], which shows how to sample (within negligible statistical distance) from $D_{\mathcal{L}(\mathbf{B}), s, \mathbf{t}}$ for any $s \geq \|\tilde{\mathbf{B}}\| \cdot \omega(\sqrt{\log n})$. The next corollary immediately follows (proof in the full version).

Corollary 16. *The following problems admit NISZK proof systems with efficient provers: GapSIVP$_{\omega(\sqrt{n \log n})}$, GapCRP$_{\omega(\sqrt{n \log n})}$, and coGapSVP$_{\omega(n^{1.5}\sqrt{\log n})}$.*

3.1   Tighter Factors for coGapSVP

For coGapSVP, Corollaries 14 and 16 give NISZK proof systems only for $\gamma(n) \geq 4n$; with an efficient prover, the factor $\gamma(n) = \omega(n^{1.5} \log n)$ is looser still.

Here we give a more sophisticated NISZK proof specifically for coGapSVP. The proof of the next theorem is given in the full version.

Theorem 17. *For any $\gamma(n) \geq 20\sqrt{n}$, the problem coGapSVP$_\gamma$ is in NISZK, via the proof system described in Figure 2.*

Furthermore, for any $\gamma(n) \geq \omega(n\sqrt{\log n})$, the prover can be implemented efficiently with an appropriate succinct witness. For any $\gamma(n) \geq n/\sqrt{\log n}$, the prover can be implemented efficiently as a quantum *algorithm with a succinct* classical *witness.*

3.2   NISZK for a Special Disjunction Language

Here we consider a special language that is structurally similar to the disjunction of many coGapSVP$_\gamma$ instances. For simplicity, we abuse notation and identify lattices with their arbitrary bases in problem instances.

NISZK proof system for coGapSVP

Common Input: A basis $\mathbf{B}$ of an n-dimensional lattice $\Lambda = \mathcal{L}(\mathbf{B})$. Let $N = 10n^3 \log n$.

Random Input: Vectors $\mathbf{t}_1, \ldots, \mathbf{t}_N \in \mathcal{P}(\mathbf{B}^*)$ chosen independently and uniformly at random from $\mathcal{P}(\mathbf{B}^*)$, defining the matrix $\mathbf{T} \in (\mathcal{P}(\mathbf{B}^*))^N \subset \mathbb{R}^{n \times N}$.

Prover P: For each $i \in [N]$, choose $\mathbf{v}_i \sim D_{\Lambda^*, -\mathbf{t}_i}$, and let $\mathbf{e}_i = \mathbf{t}_i + \mathbf{v}_i$. The proof is the matrix $\mathbf{E} \in \mathbb{R}^{n \times N}$.

Verifier V: Accept if both of the following conditions hold, otherwise reject:
1. $\mathbf{e}_i - \mathbf{t}_i \in \Lambda^*$ for all $i \in [N]$, and
2. All the eigenvalues of the $n \times n$ positive semidefinite matrix $\mathbf{E}\mathbf{E}^T$ are at most $3N$.

Fig. 2. The noninteractive zero-knowledge proof system for coGapSVP

Definition 18. *For a prime q, an input to* $\mathsf{OR\text{-}coGapSVP}^k_{q,\gamma}$ *is an n-dimensional lattice Λ such that $\lambda_1(\Lambda) > \gamma(n)$, and k superlattices $\Lambda_j \supset \Lambda$ for $j \in [k]$ such that the quotient groups Λ^*/Λ_j^* are all isomorphic to the additive group $G = \mathbb{Z}_q$.*

It is a YES instance if $\lambda_1(\Lambda_i) > \gamma(n)$ for some $i \in [k]$, and is a NO instance if $\lambda_1(\Lambda_i) \leq 1$ for every $i \in [k]$.

Theorem 19 below relates to the $\mathsf{OR\text{-}coGapSVP}^2_{q,\gamma}$ problem; it generalizes to any $k > 2$ with moderate changes (mainly, the $\sqrt{q}$ factors in the statement of Theorem 19 become $q^{(k-1)/k}$ factors).

Theorem 19. *For prime $q \geq 100$ and $\gamma(n) \geq 40\sqrt{qn}$,* $\mathsf{OR\text{-}coGapSVP}^2_{q,\gamma}$ *is in NISZK.*

Furthermore, if $\gamma(n) \geq 40\sqrt{q} \cdot \omega(n\sqrt{\log n})$, then the prover can be implemented efficiently with appropriate succinct witnesses.

References

1. Aharonov, D., Regev, O.: Lattice problems in NP ∩ coNP. J. ACM 52(5), 749–765 (2005); Preliminary version in FOCS 2004
2. Aiello, W., Håstad, J.: Statistical zero-knowledge languages can be recognized in two rounds. J. Comput. Syst. Sci. 42(3), 327–345 (1991); Preliminary version in FOCS 1987
3. Ajtai, M.: The shortest vector problem in L_2 is NP-hard for randomized reductions (extended abstract). In: STOC, pp. 10–19 (1998)
4. Ajtai, M.: Generating hard instances of lattice problems. Quaderni di Matematica 13, 1–32 (2004); Preliminary version in STOC 1996
5. Ajtai, M., Kumar, R., Sivakumar, D.: A sieve algorithm for the shortest lattice vector problem. In: STOC, pp. 601–610 (2001)
6. Ajtai, M., Kumar, R., Sivakumar, D.: Sampling short lattice vectors and the closest lattice vector problem. In: IEEE Conference on Computational Complexity, pp. 53–57 (2002)
7. Babai, L.: Trading group theory for randomness. In: STOC, pp. 421–429 (1985)

8. Banaszczyk, W.: New bounds in some transference theorems in the geometry of numbers. Mathematische Annalen 296(4), 625–635 (1993)

9. Bellare, M., Goldreich, O.: On defining proofs of knowledge. In: Brickell, E.F. (ed.) CRYPTO 1992. LNCS, vol. 740, pp. 390–420. Springer, Heidelberg (1993)

10. Ben-Or, M., Goldreich, O., Goldwasser, S., Håstad, J., Kilian, J., Micali, S., Rogaway, P.: Everything provable is provable in zero-knowledge. In: Goldwasser, S. (ed.) CRYPTO 1988. LNCS, vol. 403, pp. 37–56. Springer, Heidelberg (1988)

11. Blömer, J., Naewe, S.: Sampling methods for shortest vectors, closest vectors and successive minima. In: Arge, L., Cachin, C., Jurdziński, T., Tarlecki, A. (eds.) ICALP 2007. LNCS, vol. 4596, pp. 65–77. Springer, Heidelberg (2007)

12. Blömer, J., Seifert, J.-P.: On the complexity of computing short linearly independent vectors and short bases in a lattice. In: STOC, pp. 711–720 (1999)

13. Blum, M., De Santis, A., Micali, S., Persiano, G.: Noninteractive zero-knowledge. SIAM J. Comput. 20(6), 1084–1118 (1991); Preliminary version in STOC 1998

14. Blum, M., Feldman, P., Micali, S.: Non-interactive zero-knowledge and its applications (extended abstract). In: STOC, pp. 103–112 (1988)

15. De Santis, A., Di Crescenzo, G., Persiano, G.: The knowledge complexity of quadratic residuosity languages. Theor. Comput. Sci. 132(2), 291–317 (1994)

16. De Santis, A., Di Crescenzo, G., Persiano, G.: Randomness-efficient non-interactive zero-knowledge (extended abstract). In: Degano, P., Gorrieri, R., Marchetti-Spaccamela, A. (eds.) ICALP 1997. LNCS, vol. 1256, pp. 716–726. Springer, Heidelberg (1997)

17. De Santis, A., Di Crescenzo, G., Persiano, G., Yung, M.: Image density is complete for non-interactive-SZK (extended abstract). In: Larsen, K.G., Skyum, S., Winskel, G. (eds.) ICALP 1998. LNCS, vol. 1443, pp. 784–795. Springer, Heidelberg (1998)

18. Dwork, C., Naor, M.: Zaps and their applications. In: FOCS, pp. 283–293 (2000)

19. Feige, U., Lapidot, D., Shamir, A.: Multiple noninteractive zero knowledge proofs under general assumptions. SIAM J. Comput. 29(1), 1–28 (1999); Preliminary version in FOCS 1990

20. Fortnow, L.: The complexity of perfect zero-knowledge (extended abstract). In: STOC, pp. 204–209 (1987)

21. Gennaro, R., Micciancio, D., Rabin, T.: An efficient non-interactive statistical zero-knowledge proof system for quasi-safe prime products. In: ACM Conference on Computer and Communications Security, pp. 67–72 (1998)

22. Gentry, C., Peikert, C., Vaikuntanathan, V.: Trapdoors for hard lattices and new cryptographic constructions. In: STOC, pp. 197–206 (2008)

23. Goldreich, O., Goldwasser, S.: On the limits of nonapproximability of lattice problems. J. Comput. Syst. Sci. 60(3), 540–563 (2000); Preliminary version in STOC 1998

24. Goldreich, O., Micali, S., Wigderson, A.: Proofs that yield nothing but their validity or all languages in NP have zero-knowledge proof systems. J. ACM 38(3), 691–729 (1991); Preliminary version in FOCS 1986

25. Goldreich, O., Oren, Y.: Definitions and properties of zero-knowledge proof systems. J. Cryptology 7(1), 1–32 (1994)

26. Goldreich, O., Sahai, A., Vadhan, S.P.: Honest-verifier statistical zero-knowledge equals general statistical zero-knowledge. In: STOC, pp. 399–408 (1998)

27. Goldreich, O., Sahai, A., Vadhan, S.P.: Can statistical zero knowledge be made non-interactive? or on the relationship of SZK and NISZK. In: Wiener, M. (ed.) CRYPTO 1999. LNCS, vol. 1666, pp. 467–484. Springer, Heidelberg (1999)

28. Goldreich, O., Vadhan, S.P.: Comparing entropies in statistical zero knowledge with applications to the structure of SZK. In: IEEE Conference on Computational Complexity, pp. 54–73 (1999)
29. Goldwasser, S., Micali, S., Rackoff, C.: The knowledge complexity of interactive proof systems. SIAM J. Comput. 18(1), 186–208 (1989); Preliminary version in STOC 1985
30. Goldwasser, S., Sipser, M.: Private coins versus public coins in interactive proof systems. In: STOC, pp. 59–68 (1986)
31. Groth, J., Ostrovsky, R., Sahai, A.: Perfect non-interactive zero knowledge for NP. In: Vaudenay, S. (ed.) EUROCRYPT 2006. LNCS, vol. 4004, pp. 339–358. Springer, Heidelberg (2006)
32. Guruswami, V., Micciancio, D., Regev, O.: The complexity of the covering radius problem. Computational Complexity 14, 90–121 (2004); Preliminary version in CCC 2004
33. Lenstra, A.K., Lenstra Jr., H.W., Lovász, L.: Factoring polynomials with rational coefficients. Mathematische Annalen 261(4), 515–534 (1982)
34. Micciancio, D., Goldwasser, S.: Complexity of Lattice Problems: a cryptographic perspective. The Kluwer International Series in Engineering and Computer Science, vol. 671. Kluwer Academic Publishers, Boston (2002)
35. Micciancio, D., Regev, O.: Worst-case to average-case reductions based on Gaussian measures. SIAM J. Comput. 37(1), 267–302 (2007); Preliminary version in FOCS 2004
36. Micciancio, D., Vadhan, S.P.: Statistical zero-knowledge proofs with efficient provers: Lattice problems and more. In: Boneh, D. (ed.) CRYPTO 2003. LNCS, vol. 2729, pp. 282–298. Springer, Heidelberg (2003)
37. Nguyen, M.-H., Vadhan, S.P.: Zero knowledge with efficient provers. In: STOC, pp. 287–295 (2006)
38. Okamoto, T.: On relationships between statistical zero-knowledge proofs. J. Comput. Syst. Sci. 60(1), 47–108 (1996); Preliminary version in STOC 1996
39. Peikert, C.: Limits on the hardness of lattice problems in ℓ_p norms. Computational Complexity 17(2), 300–351 (2008); Preliminary version in CCC 2007
40. Regev, O.: New lattice-based cryptographic constructions. J. ACM 51(6), 899–942 (2004); Preliminary version in STOC 2003
41. Regev, O.: On lattices, learning with errors, random linear codes, and cryptography. In: STOC, pp. 84–93 (2005)
42. Sahai, A., Vadhan, S.P.: A complete problem for statistical zero knowledge. J. ACM 50(2), 196–249 (2003); Preliminary version in FOCS 1997
43. Schnorr, C.-P.: A hierarchy of polynomial time lattice basis reduction algorithms. Theor. Comput. Sci. 53, 201–224 (1987)
44. van Emde Boas, P.: Another NP-complete problem and the complexity of computing short vectors in a lattice. Technical Report 81-04, University of Amsterdam (1981)

A Framework for Efficient and Composable Oblivious Transfer

Chris Peikert[1,*], Vinod Vaikuntanathan[2,**], and Brent Waters[1,***]

[1] SRI International
cpeikert@alum.mit.edu, bwaters@csl.sri.com
[2] MIT
vinodv@mit.edu

Abstract. We propose a simple and general framework for constructing oblivious transfer (OT) protocols that are *efficient, universally composable,* and *generally realizable* under any one of a variety of standard number-theoretic assumptions, including the decisional Diffie-Hellman assumption, the quadratic residuosity and decisional composite residuosity assumptions, and *worst-case* lattice assumptions.

Our OT protocols are round-optimal (one message each way), quite efficient in computation and communication, and can use a single common string for an unbounded number of executions between the same sender and receiver. Furthermore, the protocols can provide *statistical* security to either the sender or the receiver, simply by changing the distribution of the common string. For certain instantiations of the protocol, even a common *uniformly random* string suffices.

Our key technical contribution is a simple abstraction that we call a *dual-mode* cryptosystem. We implement dual-mode cryptosystems by taking a unified view of several cryptosystems that have what we call "messy" public keys, whose defining property is that a ciphertext encrypted under such a key carries *no information* (statistically) about the encrypted message.

As a contribution of independent interest, we also provide a multi-bit *amortized* version of Regev's lattice-based cryptosystem (STOC 2005) whose time and space complexity are improved by a linear factor in the security parameter n. The resulting amortized encryption and decryption times are only $\tilde{O}(n)$ bit operations per message bit, and the ciphertext expansion can be made as small as a constant; the public key size and underlying lattice assumption remain essentially the same.

* This material is based upon work supported by the National Science Foundation under Grants CNS-0716786 and CNS-0749931. Any opinions, findings, and conclusions or recommendations expressed in this material are those of the author(s) and do not necessarily reflect the views of the National Science Foundation.

** Work performed while at SRI International.

*** Supported by NSF CNS-0749931, CNS-0524252, CNS-0716199; the US Army Research Office under the CyberTA Grant No. W911NF-06-1-0316; and the U.S. Department of Homeland Security under Grant Award Number 2006-CS-001-000001. The views and conclusions contained in this document are those of the authors and should not be interpreted as necessarily representing the official policies, either expressed or implied, of the U.S. Department of Homeland Security.

D. Wagner (Ed.): CRYPTO 2008, LNCS 5157, pp. 554–571, 2008.
© International Association for Cryptologic Research 2008

1   Introduction

Oblivious transfer (OT), first proposed by Rabin [30], is a fundamental primitive in cryptography, especially for secure two-party and multiparty computation [16, 34]. Oblivious transfer allows one party, called the receiver, to obtain exactly one of two (or more) values from another other party, called the sender. The receiver remains oblivious to the other value(s), and the sender is oblivious to which value was received. Since its introduction, OT has received an enormous amount of attention in the cryptographic literature (see [11, 13, 23], among countless others).

OT protocols that are secure against semi-honest adversaries can be constructed from (enhanced) trapdoor permutations and made robust to malicious adversaries using zero-knowledge proofs for NP [16]; however, this general approach is quite inefficient and is generally not suitable for real-world applications.

For practical use, it is desirable to have efficient OT protocols based on concrete assumptions. Naor and Pinkas [28] and independently, Aiello, Ishai and Reingold [1] constructed efficient two-message protocols based on the decisional Diffie-Hellman (DDH) assumption. Abstracting their approach via the the projective hash framework of Cramer and Shoup [10], Tauman Kalai [22] presented analogous protocols based on the quadratic residuosity and decisional composite residuosity assumptions. The primary drawback of these constructions is that their security is proved only according to a "half-simulation" definition, where an ideal world simulator is shown only for a cheating receiver. Therefore, they are not necessarily secure when integrated into a larger protocol, such as a multiparty computation. Indeed, as pointed out in [28], such protocols may fall to selective-failure attacks, where the sender causes a failure that depends upon the receiver's selection.

Very recently (and independently of our work), Lindell [25] used a cut-and-choose technique (in lieu of general zero knowledge) to construct fully-simulatable OT protocols under the same set of assumptions as in [1, 22, 28]. For this level of security, Lindell's result is the most efficient protocol that has appeared (that does not rely on random oracles), yet it still adds a few communication rounds to prior protocols and amplifies their computational cost by a statistical security parameter (i.e., around 40 or so). There are also several recent works on constructing fully-simulatable protocols for other variants of OT, and on obtaining efficiency through the use of random oracles; see Section 1.3 for details.

We point out that all of the above fully-simulatable protocols are proved secure in the plain *stand-alone* model, which allows for secure sequential composition, but not necessarily *parallel* or *concurrent* composition. For multiparty computation or in complex environments like the Internet, composability can offer significant security and efficiency benefits (e.g., by saving rounds of communication through parallel execution).

In addition, while there is now a significant body of literature on constructing cryptographic primitives from *worst-case lattice assumptions* (e.g., [2, 3, 15, 26, 29, 31, 32]), little is known about obtaining oblivious transfer. Protocols for semi-honest adversaries can be easily inferred from the proof techniques of [3, 31, 32], and made robust to malicious adversaries by standard coin-flipping and zero-knowledge techniques. However, for efficiency it appears technically difficult to instantiate the projective hash framework for OT [10, 22] under lattice assumptions, so a different approach may be needed.

1.1   Our Approach

Our goal is to construct oblivious transfer protocols enjoying all of the following desirable properties:

1. **Secure and composable:** we seek OT protocols that are secure according to a full simulation-based definition, and that compose securely (e.g., in parallel) with each other and with other protocols.
2. **Efficient:** we desire protocols that are efficient in computation, communication, and usage of any external resources.
3. **Generally realizable:** we endeavor to design an *abstract framework* that is realizable under a variety of concrete assumptions, including those related to lattices. Such a framework would demonstrate the conceptual generality of the approach, and could protect against future advances in cryptanalysis, such as improved algorithms for specific problems or the development of practical quantum computers (which are known to defeat factoring- and discrete log-based cryptosystems [33], but not those based on lattices).

We present a simple and novel framework for reaching all three of the above goals, working in the universal composability (UC) model of Canetti [5] with *static* corruptions of parties.

Our protocol is based on a new abstraction that we call a *dual-mode* cryptosystem. The system is initialized in a setup phase that produces a common string CRS, which is made available to all parties (we discuss this setup assumption in more detail below). Depending on the instantiation, the common string may be uniformly random, or created according to some prescribed distribution.

Given a dual-mode cryptosystem, the concrete OT protocol is very simple: the receiver uses its selection bit (and the CRS) to generate a "base" public key and secret key, and delivers the public key to the sender. The sender computes two "derived" public keys (using the CRS), encrypts each of its values under the corresponding derived key, and sends the ciphertexts to the receiver. Finally, the receiver uses its secret key to decrypt the appropriate value. Note that the protocol is message-optimal (one in each direction), and that it is essentially as efficient as the underlying cryptosystem. The security of the protocol comes directly from the dual-mode properties, which we describe in more detail next.

Dual-mode cryptosystems. The setup phase for a dual-mode cryptosystem creates the CRS according to one of two chosen modes, which we call the *messy* mode and the *decryption* mode.

The system has three main security properties. In messy mode, for *every* (possibly malformed) base public key, at least one of the derived public keys hides its encrypted messages *statistically*. Moreover, the CRS is generated along with a "trapdoor" that makes it easy to determine (given the base public key) which of the derived keys does so. These properties imply *statistical* security against even an unbounded cheating receiver.

In decryption mode, the honest receiver's selection bit is likewise hidden *statistically* by its choice of base key. In addition, there is a (different) trapdoor for the CRS that makes it easy to generate a base public key together with *two* properly-distributed secret keys corresponding to each potential selection bit. This makes it possible to decrypt both of the sender's ciphertexts, and implies statistical security against even an unbounded cheating sender.

Finally, a dual-mode system has the property that messy mode and decryption mode are *computationally* indistinguishable; this is the only computational property in the definition. The OT protocol can therefore provide statistical security for *either* the sender or the receiver, depending on the chosen mode (we are not aware of any other OT protocol having this property). This also makes the security proof for the protocol quite modular and symmetric: computational security for a party (e.g., against a bounded cheating receiver in decryption mode) follows directly from statistical security in the other mode, and by the indistinguishability of modes.[1]

The dual-mode abstraction has a number of nice properties. First, the definition is quite simple: for any candidate construction, we need only demonstrate three simple properties (two of which are statistical). Second, the same CRS can be used for an *unbounded* number of OT executions between the same sender and receiver (we are not aware of any other OT protocol having this property). Third, we can efficiently realize the abstraction under any one of several standard assumptions, including the DDH assumption, the quadratic residuosity and decisional composite residuosity assumptions, and *worst-case* lattice assumptions (under a slight relaxation of the dual-mode definition).

Of course, the security of our protocol depends on a trusted setup of the CRS. We believe that in context, this assumption is reasonable (or even quite mild). First, it is known that OT in the plain UC model *requires* some type of trusted setup [6]. Second, as we have already mentioned, a single CRS suffices for any number of OT executions between the same sender and receiver. Third, several of our instantiations require only a common *uniformly random* string, which may be obtainable without relying on a trusted party via natural processes (e.g., sunspots).

1.2 Concrete Constructions

To construct dual-mode systems from various assumptions, we build upon several public-key cryptosystems that admit what we call *messy* public keys (short for "message-lossy;" these are also called "meaningless" keys in a recent work of Kol and Naor [24]). The defining property of a messy key is that a ciphertext produced under it carries *no information* (statistically) about the encrypted

---

[1] Groth, Ostrovsky, and Sahai [19] used a similar "parameter-switching" argument in the context of non-interactive zero knowledge.

message. More precisely, the encryptions of any two messages m_0, m_1 under a messy key are statistically close, over the randomness of the encryption algorithm. Prior cryptosytems based on lattices [3, 31, 32], Cocks' identity-based cryptosystem [9], and the original OT protocols of [1, 28] all rely on messy keys as a key conceptual tool in their security proofs.

Messy keys play a similarly important role in our dual-mode constructions. As in prior OT protocols [1, 22, 28], our constructions guarantee that for any base public key (the receiver's message), at least one of the derived keys is messy. A novel part of our constructions is in the use of a trapdoor for efficiently *identifying* messy keys (this is where the use of a CRS seems essential).

For our DDH- and DCR-based constructions, we obtain a dual-mode cryptosystem via relatively straightforward abstraction and modification of prior protocols (see Section 5). For quadratic residuosity, our techniques are quite different from those of [10, 22]; specifically, we build on a modification of Cocks' identity-based cryptosystem [9] (see Section 6). In both of these constructions, we have a precise characterization of messy keys and a trapdoor algorithm for identifying them.

Our lattice-based constructions are more subtle and technically involved. Our starting point is a cryptosytem of Regev [32], along with some recent "master trapdoor" techniques for lattices due to Gentry, Peikert, and Vaikuntanathan [15]. We do not have an *exact* characterization of messy keys for this cryptosystem, nor an error-free algorithm for identifying them. However, [15] presents a trapdoor algorithm that correctly identifies *almost every* public key as messy. By careful counting and a quantifier-switching argument, we show that for almost all choices of the CRS, *every* (potentially malformed) base public key has at least one derived key that is both messy and is correctly identified as such by the trapdoor algorithm.

As an additional contribution of independent interest, we give a multi-bit version of Regev's lattice-based cryptosystem [32] whose time and space efficiency are smaller by a linear factor in the security parameter n. The resulting system is very efficient (at least asymptotically): the amortized runtime per message bit is only $\tilde{O}(n)$ bit operations, and the ciphertext expansion is as small as a constant. The public key size and underlying lattice assumption are essentially the same as in [32]. Due to space constraints, we defer the exposition of our lattice-based constructions to the full version.

Our DDH and DCR constructions transfer (multi-bit) strings, while the QR and lattice constructions allow for essentially single-bit transfers. It is an interesting open question whether string OT can be achieved under the latter assumptions. Simple generalizations of our framework and constructions can also yield 1-out-of-k OT protocols.

1.3   Related Work

Under assumptions related to bilinear pairings, Camenisch, Neven, and shelat [4] and Green and Hohenberger [17] recently constructed fully-simulatable protocols in the plain stand-alone model (and the UC model, in [18]) for k-out-of-n *adaptive selection* OT, as introduced by Naor and Pinkas [27]. Adaptive selection can be useful for applications such as oblivious database retrieval.

Jarecki and Shmatikov [21] constructed UC-secure *committed string OT* under the decisional composite residuosity assumption, using a common *reference string*; their protocol is four rounds (or two in the random oracle model). Garay, MacKenzie, and Yang [14] constructed (enhanced) committed OT using a constant number of rounds assuming both the DDH and strong RSA assumptions. Damgard and Nielsen [12] constructed UC-secure OT against *adaptive corruptions*, under a public key infrastructure setup and the assumption that threshold homomorphic encryption exists.

There are other techniques for achieving efficiency in oblivious transfer protocols that are complementary to ours; for example, Ishai *et al.* [20] showed how to amortize k oblivious transfers (for some security parameter k) into many more, without much additional computation in the random oracle model.

2   Preliminaries

We let $\mathbb{N}$ denote the natural numbers. For $n \in \mathbb{N}$, $[n]$ denotes the set $\{1, \ldots, n\}$. We let $\mathsf{poly}(n)$ denote an unspecified function $f(n) = O(n^c)$ for some constant c. The security parameter will be denoted by n throughout the paper. We let $\mathsf{negl}(n)$ denote some unspecified function $f(n)$ such that $f = o(n^{-c})$ for *every* fixed constant c, saying that such a function is *negligible* (in n). We say that a probability is *overwhelming* if it is $1 - \mathsf{negl}(n)$.

2.1   The Universal Composability Framework (UC)

We work in the standard universal composability framework of Canetti [5] with *static* corruptions of parties. We use the definition of computational indistinguishability, denoted by $\overset{c}{\approx}$, from that work. The UC framework defines a probabilistic poly-time (PPT) *environment* machine $\mathcal{Z}$ that oversees the execution of a protocol in one of two worlds. The "ideal world" execution involves "dummy parties" (some of whom may be corrupted by an *ideal adversary* $\mathcal{S}$) interacting with a *functionality* $\mathcal{F}$. The "real world" execution involves PPT parties (some of whom may be corrupted by a PPT *real world adversary* $\mathcal{A}$) interacting only with each other in some protocol π. We refer to [5] for a detailed description of the executions, and a definition of the real world ensemble $\mathsf{EXEC}_{\pi,\mathcal{A},\mathcal{Z}}$ and the ideal world ensemble $\mathsf{IDEAL}_{\mathcal{F},\mathcal{S},\mathcal{Z}}$. The notion of a protocol π *securely emulating* a functionality $\mathcal{F}$ is as follows:

Definition 1. *Let $\mathcal{F}$ be a functionality. A protocol π UC-realizes $\mathcal{F}$ if for any adversary $\mathcal{A}$, there exists a simulator $\mathcal{S}$ such that for all environments $\mathcal{Z}$,*

$$\mathsf{IDEAL}_{\mathcal{F},\mathcal{S},\mathcal{Z}} \overset{c}{\approx} \mathsf{EXEC}_{\pi,\mathcal{A},\mathcal{Z}}.$$

The common reference string functionality $\mathcal{F}_{\mathrm{CRS}}^{\mathcal{D}}$ produces a string with a fixed distribution that can be sampled by a PPT algorithm $\mathcal{D}$.

Functionality $\mathcal{F}_{\mathrm{OT}}$

$\mathcal{F}_{\mathrm{OT}}$ interacts with a sender **S** and a receiver **R**.

- Upon receiving a message $(\mathsf{sid}, \mathsf{sender}, x_0, x_1)$ from **S**, where each $x_i \in \{0,1\}^{\ell}$, store (x_0, x_1) (The lengths of the strings ℓ is fixed and known to all parties).
- Upon receiving a message $(\mathsf{sid}, \mathsf{receiver}, \sigma)$ from **R**, check if a $(\mathsf{sid}, \mathsf{sender}, \ldots)$ message was previously sent. If yes, send (sid, x_σ) to **R** and (sid) to the adversary $\mathcal{S}$ and halt. If not, send nothing to **R** (but continue running).

Fig. 1. The oblivious transfer functionality $\mathcal{F}_{\mathrm{OT}}$ [7]

Oblivious Transfer (OT) is a two-party functionality, involving a sender **S** with input x_0, x_1 and a receiver **R** with an input $\sigma \in \{0,1\}$. The receiver **R** learns x_σ (and nothing else), and the sender **S** learns nothing at all. These requirements are captured by the specification of the OT functionality $\mathcal{F}_{\mathrm{OT}}$ from [7], given in Figure 1.

Our OT protocols operate in the common reference string model, or, in the terminology of [5], the $\mathcal{F}_{\mathrm{CRS}}$-hybrid model. For efficiency, we would like to reuse the same common reference string for distinct invocations of oblivious transfer whenever possible. As described in [8], this can be achieved by designing a protocol for the *multi-session extension* $\hat{\mathcal{F}}_{\mathrm{OT}}$ of the OT functionality $\mathcal{F}_{\mathrm{OT}}$. Intuitively, $\hat{\mathcal{F}}_{\mathrm{OT}}$ acts as a "wrapper" around any number of independent executions of $\mathcal{F}_{\mathrm{OT}}$, and coordinates their interactions with the parties via subsessions (specified by a parameter ssid) of a single session (specified by sid).

The *UC theorem with joint state* (JUC theorem) [8] says that any protocol π operating in the $\mathcal{F}_{\mathrm{OT}}$-hybrid model can be securely emulated in the real world by appropriately composing π with a *single* execution of a protocol ρ implementing $\hat{\mathcal{F}}_{\mathrm{OT}}$. This single instance of ρ might use fewer resources (such as common reference strings) than several independent invocations of some other protocol that only realizes $\mathcal{F}_{\mathrm{OT}}$; in fact, the protocols ρ that we specify will do exactly this.

3   Dual-Mode Encryption

Here we describe our new abstraction, called a *dual-mode* cryptosystem. It is initialized in a trusted setup phase, which produces a common string crs known to all parties along with some auxiliary "trapdoor" information t (which is only used in the security proof). The string crs may be either uniformly random or selected from a prescribed distribution, depending on the concrete instantiation. The cryptosystem can be set up in one of two modes, called *messy* mode and *decryption* mode. The first crucial security property of a dual-mode cryptosystem is that no (efficient) adversary can distinguish, given the crs, between the modes.

Once the system has been set up, it operates much like a standard public-key cryptosystem, but with an added notion that we call *encryption branches*.

The key generation algorithm takes as a parameter a chosen *decryptable* branch $\sigma \in \{0, 1\}$, and the resulting secret key sk corresponds to branch σ of the public key pk. When encrypting a message under pk, the encrypter similarly specifies a branch $b \in \{0, 1\}$ on which to encrypt the message. Essentially, messages encrypted on branch $b = \sigma$ can be decrypted using sk, while those on the other branch cannot. Precisely what this latter condition means depends on the mode of the system.

When the system is in messy mode, branch $b \neq \sigma$ is what we call *messy*. That is, encrypting on branch b *loses all information* about the encrypted message — not only in the sense of semantic security, but even *statistically*. Moreover, the trapdoor for crs makes it easy to find a messy branch of *any* given public key, even a malformed one that could never have been produced by the key generator.

In decryption mode, the trapdoor circumvents the condition that only one branch is decryptable. Specifically, it allows the generation of a public key pk and corresponding secret keys sk_0, sk_1 that enable decryption on branches 0 and 1 (respectively). More precisely, for both values of $\sigma \in \{0, 1\}$, the distribution of the key pair (pk, sk_σ) is statistically close to that of an honestly-generated key pair with decryption branch σ.

We now proceed more formally. A dual-mode cryptosystem with message space $\{0, 1\}^\ell$ consists of a tuple of probabilistic algorithms having the following interfaces:

- Setup($1^n, \mu$), given security parameter n and mode $\mu \in \{0, 1\}$, outputs (crs, t). The crs is a common string for the remaining algorithms, and t is a trapdoor value that enables either the FindMessy or TrapKeyGen algorithm, depending on the selected mode.
  For notational convenience, we define a separate messy mode setup algorithm SetupMessy($\cdot$) := Setup($\cdot, 0$) and a decryption mode setup algorithm SetupDec($\cdot$) := Setup($\cdot, 1$).
  All the remaining algorithms take crs as their first input, but for notational clarity, we usually omit it from their lists of arguments.
- KeyGen(σ), given a branch value $\sigma \in \{0, 1\}$, outputs (pk, sk) where pk is a public encryption key and sk is a corresponding secret decryption key for messages encrypted on branch σ.
- Enc(pk, b, m), given a public key pk, a branch value $b \in \{0, 1\}$, and a message $m \in \{0, 1\}^\ell$, outputs a ciphertext c encrypted on branch b.
- Dec(sk, c), given a secret key sk and a ciphertext c, outputs a message $m \in \{0, 1\}^\ell$.
- FindMessy(t, pk), given a trapdoor t and some (possibly even malformed) public key pk, outputs a branch value $b \in \{0, 1\}$ corresponding to a messy branch of pk.
- TrapKeyGen(t), given a trapdoor t, outputs (pk, sk_0, sk_1), where pk is a public encryption key and sk_0, sk_1 are corresponding secret decryption keys for branches 0 and 1, respectively.

Definition 2 (Dual-Mode Encryption). *A dual-mode cryptosystem is a tuple of algorithms described above that satisfy the following properties:*

1. **Completeness for decryptable branch:** *For every* $\mu \in \{0, 1\}$, $(\text{crs}, t) \leftarrow$ Setup$(1^n, \mu)$, $\sigma \in \{0, 1\}$, $(pk, sk) \leftarrow$ KeyGen(σ), *and* $m \in \{0, 1\}^{\ell}$, *decryption is correct on branch* σ: Dec$(sk, \text{Enc}(pk, \sigma, m)) = m$.

2. **Indistinguishability of modes:** *the first outputs of* SetupMessy *and* SetupDec *are computationally indistinguishable, i.e.,* SetupMessy$_1(1^n) \overset{c}{\approx}$ SetupDec$_1(1^n)$.

3. **(Messy mode) Trapdoor identification of a messy branch:** *For every* $(\text{crs}, t) \leftarrow$ SetupMessy(1^n) *and every (possibly malformed)* pk, FindMessy (t, pk) *outputs a branch value* $b \in \{0, 1\}$ *such that* Enc$(pk, b, \cdot)$ *is messy. Namely, for every* $m_0, m_1 \in \{0, 1\}^{\ell}$, Enc$(pk, b, m_0) \overset{s}{\approx}$ Enc(pk, b, m_1).

4. **(Decryption mode) Trapdoor generation of keys decryptable on both branches:** *For every* $(\text{crs}, t) \leftarrow$ SetupDec(1^n), TrapKeyGen(t) *outputs* (pk, sk_0, sk_1) *such that for every* $\sigma \in \{0, 1\}$, $(pk, sk_\sigma) \overset{s}{\approx}$ KeyGen(σ).

It is straightforward to generalize these definitions to larger sets $\{0, 1\}^k$ of branches, for $k > 1$ (in this generalization, FindMessy would return $2^k - 1$ different branches that are all messy). Such a dual-mode cryptosystem would yield a 1-out-of-2^k oblivious transfer in an analogous way. All of our constructions can be suitably modified to satisfy the generalized definition; for simplicity, we will concentrate on the branch set $\{0, 1\}$ throughout the paper.

4 Oblivious Transfer Protocol

Here we construct a protocol dm that emulates the multi-session functionality $\hat{\mathcal{F}}_{\text{OT}}$ functionality in the $\mathcal{F}_{\text{CRS}}$-hybrid model. The dm protocol, which uses any dual-mode cryptosystem, is given in Figure 2.

The protocol can actually operate in either mode of the dual-mode cryptosystem, which affects only the distribution of the CRS that is used. In messy mode, the receiver's security is computational and the sender's security is *statistical*, i.e., security is guaranteed even against an *unbounded* cheating receiver. In decryption mode, the security properties are reversed.

To implement the two modes, we define two different instantiations of $\mathcal{F}_{\text{CRS}}^{\mathcal{D}}$ that produce common strings according to the appropriate setup algorithm: $\mathcal{F}_{\text{CRS}}^{\text{ext}}$ uses $\mathcal{D} = $ SetupMessy$_1$, and $\mathcal{F}_{\text{CRS}}^{\text{dec}}$ uses $\mathcal{D} = $ SetupDec$_1$.

Theorem 3. *Let* mode $\in \{\text{ext}, \text{dec}\}$. *Protocol* dm^{mode} *securely realizes the functionality* $\hat{\mathcal{F}}_{\text{OT}}$ *in the* $\mathcal{F}_{\text{CRS}}^{\text{mode}}$-*hybrid model, under static corruptions.*

For mode $= \text{ext}$, *the sender's security is statistical and the receiver's security is computational; for* mode $= \text{dec}$, *the security properties are reversed.*

Proof. Given all the properties of a dual-mode cryptosystem, the proof is conceptually quite straightforward. There is a direct correspondence between completeness and the case that neither party is corrupted, between messy mode and statistical security for the sender, and between decryption mode and statistical security for the receiver. The indinstinguishability of modes establishes computational security for the appropriate party in the protocol.

Protocol dm^{mode} for Oblivious Transfer

The dm^{mode} protocol is parameterized by mode $\in \{\mathsf{ext}, \mathsf{dec}\}$ indicating the type of crs to be used.

SENDER INPUT: $(\mathsf{sid}, \mathsf{ssid}, x_0, x_1)$, where $x_0, x_1 \in \{0,1\}^{\ell}$.
RECEIVER INPUT: $(\mathsf{sid}, \mathsf{ssid}, \sigma)$, where $\sigma \in \{0,1\}$.
When activated with their inputs, the sender $\mathbf{S}$ queries $\mathcal{F}_{\mathrm{CRS}}^{\mathsf{mode}}$ with $(\mathsf{sid}, \mathbf{S}, \mathbf{R})$ and gets back $(\mathsf{sid}, \mathsf{crs})$. The receiver $\mathbf{R}$ then queries $\mathcal{F}_{\mathrm{CRS}}^{\mathsf{mode}}$ with $(\mathsf{sid}, \mathbf{S}, \mathbf{R})$ and gets $(\mathsf{sid}, \mathsf{crs})$.
$\mathbf{R}$ computes $(pk, sk) \leftarrow \mathsf{KeyGen}(\mathsf{crs}, \sigma)$, sends $(\mathsf{sid}, \mathsf{ssid}, pk)$ to $\mathbf{S}$, and stores $(\mathsf{sid}, \mathsf{ssid}, sk)$.
$\mathbf{S}$ gets $(\mathsf{sid}, \mathsf{ssid}, pk)$ from $\mathbf{R}$, computes $y_b \leftarrow \mathsf{Enc}(pk, b, x_b)$ for each $b \in \{0,1\}$, and sends $(\mathsf{sid}, \mathsf{ssid}, y_0, y_1)$ to $\mathbf{R}$.
$\mathbf{R}$ gets $(\mathsf{sid}, \mathsf{ssid}, y_0, y_1)$ from $\mathbf{S}$ and outputs $(\mathsf{sid}, \mathsf{ssid}, \mathsf{Dec}(sk, y_\sigma))$, where $(\mathsf{sid}, \mathsf{ssid}, sk)$ was stored above.

Fig. 2. The protocol for realizing $\hat{\mathcal{F}}_{\mathrm{OT}}$

Let $\mathcal{A}$ be a static adversary that interacts with the parties $\mathbf{S}$ and $\mathbf{R}$ running the dm^{mode} protocol. We construct an ideal world adversary (simulator) $\mathcal{S}$ interacting with the ideal functionality $\hat{\mathcal{F}}_{\mathrm{OT}}$, such that no environment $\mathcal{Z}$ can distinguish an interaction with $\mathcal{A}$ in the above protocol from an interaction with $\mathcal{S}$ in the ideal world. Recall that $\mathcal{S}$ interacts with both the ideal functionality $\hat{\mathcal{F}}_{\mathrm{OT}}$ and the environment $\mathcal{Z}$.

$\mathcal{S}$ starts by invoking a copy of $\mathcal{A}$ and running a simulated interaction of $\mathcal{A}$ with $\mathcal{Z}$ and the players $\mathbf{S}$ and $\mathbf{R}$. More specifically, $\mathcal{S}$ works as follows:

Simulating the communication with $\mathcal{Z}$: Every input value that $\mathcal{S}$ receives from $\mathcal{Z}$ is written into the adversary $\mathcal{A}$'s input tape (as if coming from $\mathcal{A}$'s environment). Every output value written by $\mathcal{A}$ on its output tape is copied to $\mathcal{S}$'s own output tape (to be read by the environment $\mathcal{Z}$).

Simulating the case when only the receiver $\mathbf{R}$ is corrupted: Regardless of the mode of the protocol, $\mathcal{S}$ does the following. Run the messy mode setup algorithm, letting $(\mathsf{crs}, t) \leftarrow \mathsf{SetupMessy}(1^n)$. When the parties query the ideal functionality $\mathcal{F}_{\mathrm{CRS}}^{\mathsf{mode}}$, return $(\mathsf{sid}, \mathsf{crs})$ to them. (Note that when mode $= \mathsf{ext}$, the crs thus returned is identically distributed to the one returned by $\mathcal{F}_{\mathrm{CRS}}^{\mathsf{mode}}$, whereas when mode $= \mathsf{dec}$, the simulated crs has a different distribution from the one returned by $\mathcal{F}_{\mathrm{CRS}}^{\mathsf{mode}}$ in the protocol).

When $\mathcal{A}$ produces a protocol message $(\mathsf{sid}, \mathsf{ssid}, pk)$, $\mathcal{S}$ lets $b \leftarrow \mathsf{FindMessy}(\mathsf{crs}, t, pk)$. $\mathcal{S}$ then sends $(\mathsf{sid}, \mathsf{ssid}, \mathsf{receiver}, 1 - b)$ to the ideal functionality $\hat{\mathcal{F}}_{\mathrm{OT}}$, receives the output $(\mathsf{sid}, \mathsf{ssid}, x_{1-b})$, and stores it along with the value b.

When the dummy $\mathbf{S}$ is activated for subsession $(\mathsf{sid}, \mathsf{ssid})$, $\mathcal{S}$ looks up the corresponding b and x_{1-b}, computes $y_{1-b} \leftarrow \mathsf{Enc}(pk, 1 - b, x_{1-b})$ and $y_b \leftarrow \mathsf{Enc}(pk, b, 0^{\ell})$ and sends the adversary $\mathcal{A}$ the message $(\mathsf{sid}, \mathsf{ssid}, y_0, y_1)$ as if it were from $\mathbf{S}$.

Simulating the case when only the sender **S** *is corrupted:* Regardless of the mode of the protocol, $\mathcal{S}$ does the following. Run the decryption mode setup algorithm, letting $(\mathsf{crs}, t) \leftarrow \mathsf{SetupDec}(1^n)$. When the parties query the ideal functionality $\mathcal{F}_{\mathsf{CRS}}^{\mathsf{mode}}$, return $(\mathsf{sid}, \mathsf{crs})$ to them.

When the dummy **R** is activated on $(\mathsf{sid}, \mathsf{ssid})$, $\mathcal{S}$ computes $(pk, sk_0, sk_1) \leftarrow \mathsf{TrapKeyGen}(\mathsf{crs}, t)$, sends $(\mathsf{sid}, \mathsf{ssid}, pk)$ to $\mathcal{A}$ as if from **R**, and stores $(\mathsf{sid}, \mathsf{ssid}, pk, sk_0, sk_1)$. When $\mathcal{A}$ replies with a message $(\mathsf{sid}, \mathsf{ssid}, y_0, y_1)$, $\mathcal{S}$ looks up the corresponding (pk, sk_0, sk_1), computes $x_b \leftarrow \mathsf{Dec}(sk_b, y_b)$ for each $b \in \{0, 1\}$ and sends to $\hat{\mathcal{F}}_{\mathsf{OT}}$ the message $(\mathsf{sid}, \mathsf{ssid}, \mathsf{sender}, x_0, x_1)$.

Simulating the remaining cases: When both parties are corrupted, the simulator $\mathcal{S}$ just runs $\mathcal{A}$ internally (who itself generates the messages from both **S** and **R**).

When neither party is corrupted, $\mathcal{S}$ internally runs the honest **R** on input $(\mathsf{sid}, \mathsf{ssid}, \sigma = 0)$ and honest **S** on input $(\mathsf{sid}, \mathsf{ssid}, x_0 = 0^\ell, x_1 = 0^\ell)$, activating the appropriate algorithm when the corresponding dummy party is activated in the ideal execution, and delivering all messages between its internal **R** and **S** to $\mathcal{A}$.

We complete the proof using the following claims, which are straightforward to prove using the dual-mode properties (so we defer the proofs to the full version):

1. *(Statistical security for* **S** *in messy mode.)* When $\mathcal{A}$ corrupts the receiver **R**,

$$\mathsf{IDEAL}_{\hat{\mathcal{F}}_{\mathsf{OT}}, \mathcal{S}, \mathcal{Z}} \overset{s}{\approx} \mathsf{EXEC}_{\mathsf{dm}^{\mathsf{ext}}, \mathcal{A}, \mathcal{Z}}.$$

2. *(Statistical security for* **R** *in decryption mode.)* When $\mathcal{A}$ corrupts the sender **S**,

$$\mathsf{IDEAL}_{\hat{\mathcal{F}}_{\mathsf{OT}}, \mathcal{S}, \mathcal{Z}} \overset{s}{\approx} \mathsf{EXEC}_{\mathsf{dm}^{\mathsf{dec}}, \mathcal{A}, \mathcal{Z}}.$$

3. *(Parameter switching.)* For any protocol π^{mode} in the $\mathcal{F}_{\mathsf{CRS}}^{\mathsf{mode}}$-hybrid model, any adversary $\mathcal{A}$ and any environment $\mathcal{Z}$,

$$\mathsf{EXEC}_{\pi^{\mathsf{ext}}, \mathcal{A}, \mathcal{Z}} \overset{c}{\approx} \mathsf{EXEC}_{\pi^{\mathsf{dec}}, \mathcal{A}, \mathcal{Z}}.$$

We now complete the proof as follows. Consider the protocol $\mathsf{dm}^{\mathsf{ext}}$. When $\mathcal{A}$ corrupts **R**, by item 1 above we have statistical security for **S** (whether or not **S** is corrupted). When $\mathcal{A}$ corrupts **S**, by items 2 and 3 above we have

$$\mathsf{IDEAL}_{\hat{\mathcal{F}}_{\mathsf{OT}}, \mathcal{S}, \mathcal{Z}} \overset{s}{\approx} \mathsf{EXEC}_{\mathsf{dm}^{\mathsf{dec}}, \mathcal{A}, \mathcal{Z}} \overset{c}{\approx} \mathsf{EXEC}_{\mathsf{dm}^{\mathsf{ext}}, \mathcal{A}, \mathcal{Z}},$$

which implies computational security for **R**.

It remains to show computational security when neither the sender nor the receiver is corrupted. Let $\mathsf{EXEC}_{\mathsf{dm}^{\mathsf{ext}}, \mathcal{A}, \mathcal{Z}}(x_0, x_1, b)$ (resp, $\mathsf{EXEC}_{\mathsf{dm}^{\mathsf{dec}}, \mathcal{A}, \mathcal{Z}}(x_0, x_1, b)$) denote the output of an environment in the protocol $\mathsf{dm}^{\mathsf{ext}}$ (resp, $\mathsf{dm}^{\mathsf{dec}}$) that sets the inputs of the sender **S** to be (x_0, x_1) and the input of the receiver **R** to be the bit b. The following sequence of hybrids establishes what we want.

$$\mathsf{EXEC}_{\mathsf{dm}^{\mathsf{ext}}, \mathcal{A}, \mathcal{Z}}(x_0, x_1, 1) \overset{s}{\approx} \mathsf{EXEC}_{\mathsf{dm}^{\mathsf{ext}}, \mathcal{A}, \mathcal{Z}}(0^\ell, x_1, 1) \overset{c}{\approx}$$
$$\mathsf{EXEC}_{\mathsf{dm}^{\mathsf{ext}}, \mathcal{A}, \mathcal{Z}}(0^\ell, x_1, 0) \overset{s}{\approx} \mathsf{EXEC}_{\mathsf{dm}^{\mathsf{ext}}, \mathcal{A}, \mathcal{Z}}(0^\ell, 0^\ell, 0)$$

The first two and the last two experiments are statistically indistinguishable because of the messy property of encryption, and the second and third experiments are computationally indistinguishable because of the computational hiding of the receiver's selection bit. The first experiment corresponds to the real world execution, whereas the last experiment is what the simulator runs. Furthermore, by the completeness of the dual-mode cryptosystem, the first experiment is statistically indistinguishable from the ideal world exection with inputs (x_0, x_1, b).

The proof of security for protocol $\mathsf{dm}^{\mathsf{dec}}$ follows symmetrically, and we are done. $\qquad\square$

5 Realization from DDH

5.1 Background

Let $\mathcal{G}$ be a an algorithm that takes as input a security parameter 1^n and outputs a group description $\mathbb{G} = (G, p, g)$, where G is a cyclic group of prime order p and g is a generator of G.

Our construction will make use of groups for which the DDH problem is believed to be hard. The version of the DDH assumption we use is the following: for random *generators* $g, h \in G$ and for *distinct* but otherwise random $a, b \in \mathbb{Z}_p$, the tuples (g, h, g^a, h^a) and (g, h, g^a, h^b) are computationally indistinguishable.[2] This version of the DDH assumption is equivalent to another common form, namely, that $(g, g^a, g^b, g^{ab}) \overset{c}{\approx} (g, g^a, g^b, g^c)$ for independent $a, b, c \leftarrow \mathbb{Z}_p$, because g^a is a generator and $c \neq ab$ with overwhelming probability.

5.2 Cryptosystem Based on DDH

We start by presenting a cryptosystem based on the hardness of the Decisional Diffie-Hellman problem, which slightly differs from the usual ElGamal cryptosystem in a few ways. The cryptosystem depends on a randomization procedure that we describe below. We note that the algorithm Randomize we describe below is implicit in the OT protocol of Naor and Pinkas [28].

Lemma 4 (Randomization). *Let G be an arbitrary multiplicative group of prime order p. For each $x \in \mathbb{Z}_p$, define $\mathrm{DLOG}_G(x) = \{(g, g^x) : g \in G\}$. There is a probabilistic algorithm* Randomize *that takes generators $g, h \in G$ and elements $g', h' \in G$, and outputs a pair $(u, v) \in G^2$ such that:*

- *If $(g, g'), (h, h') \in \mathrm{DLOG}_G(x)$ for some x, then (u, v) is uniformly random in $\mathrm{DLOG}_G(x)$.*
- *If $(g, g') \in \mathrm{DLOG}_G(x)$ and $(h, h') \in \mathrm{DLOG}_G(y)$ for some $x \neq y$, then (u, v) is uniformly random in G^2.*

---

[2] To be completely formal, the respective ensembles of the two distributions, indexed by the security parameter n, are indistinguishable.

Proof. Define $\mathsf{Randomize}(g, h, g', h')$ to do the following: Choose $s, t \leftarrow \mathbb{Z}_p$ independently and let $u = g^s h^t$ and $v = (g')^s (h')^t$. Output (u, v).

Since g and h are generators of G, we can write $h = g^r$ for some nonzero $r \in \mathbb{Z}_p$. First suppose (g, g') and (h, h') belong to $\mathrm{DLOG}_G(x)$ for some x. Now, $u = g^s h^t = g^{s+rt}$ is uniformly random in G, since g is a generator of G and s is random in $\mathbb{Z}_p$. Furthermore, $v = (g')^s (h')^t = (g^s h^t)^x = u^x$ and thus, $(u, v) \in \mathrm{DLOG}_G(x)$.

Now suppose $(g, g') \in \mathrm{DLOG}_G(x)$ and $(h, h') \in \mathrm{DLOG}_G(y)$ for some $x \neq y$. Then $u = g^s h^t = g^{s+rt}$ and $v = g^{xs+ryt}$. Because $r(x - y) \neq 0 \in \mathbb{Z}_p$, the expressions $s + rt$ and $xs + ryt$ are linearly independent combinations of s and t. Therefore, over the choice of $s, t \in \mathbb{Z}_p$, u and v are uniform and independent in G.

We now describe the basic cryptosystem.

- $\mathsf{DDHKeyGen}(1^n)$: Choose $\mathbb{G} = (G, p, g) \leftarrow \mathcal{G}(1^n)$. The message space of the system is G.
  Choose another generator $h \leftarrow G$ and exponent $x \leftarrow \mathbb{Z}_p$. Let $pk = (g, h, g^x, h^x)$ and $sk = x$. Output (pk, sk).
- $\mathsf{DDHEnc}(pk, m)$: Parse pk as (g, h, g', h'). Let $(u, v) \leftarrow \mathsf{Randomize}(g, h, g', h')$. Output the ciphertext $(u, v \cdot m)$.
- $\mathsf{DDHDec}(sk, c)$: Parse c as (c_0, c_1). Output c_1 / c_0^{sk}.

Now consider a public key pk of the form (g, h, g^x, h^y) for distinct $x, y \in \mathbb{Z}_p$ (and where g, h are generators of G). It follows directly from Lemma 4 that $\mathsf{DDHEnc}(pk, \cdot)$ is messy. Namely, for every two messages $m_0, m_1 \in \mathbb{Z}_p$, $\mathsf{DDHEnc}(pk, m_0) \stackrel{s}{\approx} \mathsf{DDHEnc}(pk, m_1)$.

5.3   Dual-Mode Cryptosystem

We now construct a dual-mode encryption scheme based on the hardness of DDH.

- Both $\mathsf{SetupMessy}$ and $\mathsf{SetupDec}$ start by choosing $\mathbb{G} = (G, p, g) \leftarrow \mathcal{G}(1^n)$.
  $\mathsf{SetupMessy}(1^n)$: Choose random generators $g_0, g_1 \in G$. Choose *distinct nonzero* exponents $x_0, x_1 \leftarrow \mathbb{Z}_p$. Let $h_b = g_b^{x_b}$ for $b \in \{0, 1\}$. Let $\mathsf{crs} = (g_0, h_0, g_1, h_1)$ and $t = (x_0, x_1)$. Output (crs, t).
  $\mathsf{SetupDec}(1^n)$: Choose a random generator $g_0 \in G$, a random nonzero $y \in \mathbb{Z}_p$, and let $g_1 = g_0^y$. Choose a random nonzero exponent $x \in \mathbb{Z}_p$. Let $h_b = g_b^x$ for $b \in \{0, 1\}$, let $\mathsf{crs} = (g_0, h_0, g_1, h_1)$ and $t = y$. Output (crs, t).
  In the following, all algorithms are implicitly provided the crs and parse it as (g_0, h_0, g_1, h_1).
- $\mathsf{KeyGen}(\sigma)$: Choose $r \leftarrow \mathbb{Z}_p$. Let $g = g_\sigma^r$, $h = h_\sigma^r$ and $pk = (g, h)$. Let $sk = r$. Output (pk, sk).
- $\mathsf{Enc}(pk, b, m)$: Parse pk as (g, h). Let $pk_b = (g_b, h_b, g, h)$. Output $\mathsf{DDHEnc}(pk_b, m)$ as the encryption of m on branch b.
- $\mathsf{Dec}(sk, c)$: Output $\mathsf{DDHDec}(sk, c)$.

- FindMessy(t, pk): Parse the messy mode trapdoor t as (x_0, x_1) where $x_0 \neq x_1$. Parse the public key pk as (g, h). If $h \neq g^{x_0}$, then output $b = 0$ as a (candidate) messy branch. Otherwise, we have $h = g^{x_0} \neq g^{x_1}$ because $x_0 \neq x_1$, so output $b = 1$ as a (candidate) messy branch.
- TrapKeyGen(t): Parse the decryption mode trapdoor t as a nonzero $y \in \mathbb{Z}_p$. Pick a random $r \leftarrow \mathbb{Z}_p$ and compute $pk = (g_0^r, h_0^r)$ and output $(pk, r, r/y)$.

We remark that SetupMessy actually produces a crs that is statistically close to a common *random* (not reference) string, because it consists of four generators that do not comprise a DDH tuple.

Theorem 5. *The above scheme is a dual-mode cryptosystem, assuming that DDH is hard for $\mathcal{G}$.*

Proof. Completeness follows by inspection from the correctness of the basic DDH cryptosystem.

We now show indistinguishability of the two modes. In messy mode, crs $= (g_0, h_0 = g_0^{x_0}, g_1, h_1 = g_1^{x_1})$, where g_0, g_1 are random generators of G and x_0, x_1 are distinct and nonzero in $\mathbb{Z}_p$. Let $a = \log_{g_0} g_1$, which is nonzero but otherwise uniform in $\mathbb{Z}_p$. Then $b = \log_{h_0}(h_1) = a \cdot x_1/x_0$ is nonzero and distinct from a, but otherwise uniform. Therefore crs is statistically close to a random DDH non-tuple (g_0, h_0, g_0^a, h_0^b), where $a, b \leftarrow \mathbb{Z}_p$ are distinct but otherwise uniform. Now in decryption mode, crs $= (g_0, h_0 = g_0^x, g_1, h_1 = g_1^x)$, where x is nonzero and random in $\mathbb{Z}_p$. Since $\log_{h_0}(h_1) = \log_{g_0}(g_1) = y$ is nonzero and random in $\mathbb{Z}_p$, crs is statistically close to a random DDH tuple. Under the DDH assumption, the two modes are indistinguishable.

We now demonstrate identification of a messy branch. By inspection, FindMessy(t, pk) computes a branch b for which (g_b, h_b, g, h) (the key used when encrypting under pk on branch b) is not a DDH tuple. By Lemma 4, this b is therefore a messy branch.

We conclude with trapdoor key generation. Let $(\text{crs}, y) \leftarrow \text{SetupDec}(1^n)$. Note that crs is a DDH tuple of the form $(g_0, h_0 = g_0^x, g_1 = g_0^y, h_1 = g_1^x)$, where x and y are nonzero. TrapKeyGen(crs, y) outputs $(pk, sk_0, sk_1) = ((g_0^r, h_0^r), r, r/y)$. The output of KeyGen$(\sigma)$, on the other hand, is $((g_\sigma^r, h_\sigma^r), r)$. We now show that (pk, sk_σ) and KeyGen(σ) are identically distributed.

Indeed, $(pk, sk_0) = (g_0^r, h_0^r, r)$ is identically distributed to KeyGen(0), by definition of KeyGen. By a renaming of variables letting $r = r'y$, we have that $(pk, sk_1) = (g_0^r, h_0^r, r/y)$ is identical to $(g_0^{r'y}, h_0^{r'y}, r') = (g_1^{r'}, h_1^{r'}, r')$, which is distributed identically to KeyGen(1), since $r' = r/y \in \mathbb{Z}_p$ is uniformly distributed.

Larger branch sets. We briefly outline how the dual-mode cryptosystem is modified for larger branch sets $\{0, 1\}^k$. Essentially, the scheme involves k parallel and independent copies of the one above, but all using the same group $\mathbb{G}$. The encryption algorithm Enc computes a k-wise secret sharing of the message, and encrypts each share under the corresponding copy of the scheme. This ensures that decryption succeeds only for the one specific branch selected to be decryptable. The FindMessy algorithm includes a branch $b \in \{0, 1\}^k$ in its output list of messy branches if *any* branch b_i is messy for its corresponding scheme.

6   Realization from QR

6.1   Cryptosystem Based on QR

We start by describing a (non-identity-based) variant of Cocks' cryptosystem [9], which is based on the conjectured hardness of the quadratic residuosity problem.

For $N \in \mathbb{N}$, let $\mathbb{J}_N$ denote the set of all $x \in \mathbb{Z}_N^*$ with Jacobi symbol 1. Let $\mathbb{QR}_N \subset \mathbb{J}_N$ denote the set of all quadratic residues (squares) in $\mathbb{Z}_N^*$. The message space is $\{\pm 1\}$. Let $\left(\frac{t}{N}\right)$ denote the Jacobi symbol of t in $\mathbb{Z}_N^*$.

- CKeyGen(1^n): Choose two random n-bit safe primes[3] p and q and let $N = pq$. Choose $r \leftarrow \mathbb{Z}_N^*$ and let $y \leftarrow r^2$. Let $pk = (N, y)$, and $sk = r$. Output (pk, sk).
- CEnc(pk, m): Parse pk as (N, y). Choose $s \leftarrow \mathbb{Z}_N^*$ at random such that $\left(\frac{s}{N}\right) = m$, and output $c = s + y/s$.
- CDec(sk, c): Output the Jacobi symbol of $c + 2 \cdot sk$.

The following lemma is implicit in the security proof from [9].

Lemma 6 (Messy Characterization). *Let N be a product of two random n-bit safe primes p and q, let $y \in \mathbb{Z}_N^*$ and let $pk = (N, y)$. If $y \notin \mathbb{QR}_N$, then* CEnc($pk, \cdot$) *is messy. Namely,* CEnc($pk, +1$) $\overset{s}{\approx}$ CEnc($pk, -1$).

Proof. Consider the equation $c = s + y/s \bmod N$ in terms of s (for fixed c and y), and say $s = s_0$ is one of the solutions. Then we have $c = s_0 + y/s_0 \bmod p$ and $c = s_0 + y/s_0 \bmod q$. The other solutions are s_1, s_2, and s_3, where

$$s_1 = s_0 \bmod p \qquad\qquad s_2 = y/s_0 \bmod p \qquad\qquad s_3 = y/s_0 \bmod p$$
$$s_1 = y/s_0 \bmod q \qquad\qquad s_2 = s_0 \bmod q \qquad\qquad s_3 = y/s_0 \bmod q.$$

If $y \notin \mathbb{QR}_N$, then at least one of $\alpha_1 = \left(\frac{y}{p}\right)$ or $\alpha_2 = \left(\frac{y}{q}\right)$ is -1. Then $\left(\frac{s_1}{N}\right) = \alpha_2\left(\frac{s_0}{N}\right)$, $\left(\frac{s_2}{N}\right) = \alpha_1\left(\frac{s_0}{N}\right)$ and $\left(\frac{s_3}{N}\right) = \alpha_1\alpha_2\left(\frac{s_0}{N}\right)$. Thus, two of $\left(\frac{s_i}{N}\right)$ are $+1$ and the other two are -1. It follows that c hides $\left(\frac{s}{N}\right)$ perfectly.

6.2   Dual-Mode Cryptosystem

We now describe a dual-mode cryptosystem that is based on the above cryptosystem.

- SetupMessy(1^n): Choose two random n-bit safe primes p and q and let $N = pq$. Choose $y \leftarrow \mathbb{J}_N \setminus \mathbb{QR}_N$. Let crs $= (N, y)$, and $t = (p, q)$. Output (crs, t).
  SetupDec(1^n): Let $N = pq$ for random n-bit safe primes as above. Choose $s \leftarrow \mathbb{Z}_N^*$, and let $y = s^2 \bmod N$. Let crs $= (N, y)$, amd $t = s$. Output (crs, t). In the following, all algorithms are implicitly provided the crs and parse it as (N, y), and all operations are performed in $\mathbb{Z}_N^*$.
- KeyGen(σ): Choose $r \leftarrow \mathbb{Z}_N^*$, and let $pk = r^2/y^\sigma$. Let $sk = r$. Output (pk, sk).

---

[3] Safe primes are primes p such that $\frac{p-1}{2}$ is also prime.

- $\mathsf{Enc}(pk, b, m)$: Let $pk_b = (N, pk \cdot y^b)$. Output $\mathsf{CEnc}(pk_b, m)$.
- $\mathsf{Dec}(sk, c)$: Output $\mathsf{CDec}(sk, c)$.
- $\mathsf{FindMessy}(t, pk)$: Parse the trapdoor t as (p, q) where $N = pq$. If $pk \in \mathbb{QR}_N$ (this can be checked efficiently using p and q), then output $b = 1$ as the (candidate) messy branch; otherwise, output $b = 0$.
- $\mathsf{TrapKeyGen}(t)$: Choose a random $r \leftarrow \mathbb{Z}_N^*$ and let $pk = r^2$ and $sk_b = r \cdot t^b$ for each $b \in \{0, 1\}$. Output (pk, sk_0, sk_1).

Theorem 7. *The above scheme is a dual-mode cryptosystem, assuming the hardness of the quadratic residuosity problem.*

Proof. We first show completeness. Say $(pk, sk) \leftarrow \mathsf{KeyGen}(\sigma)$. Thus, $pk = r^2 y^{-\sigma}$ for some r. $\mathsf{Enc}(pk, \sigma, m)$ runs $\mathsf{CEnc}(pk \cdot y^\sigma, m) = \mathsf{CEnc}(r^2, m)$. Thus, the public key used in the Cocks encryption algorithm is a quadratic residue. By the completeness of the Cocks cryptosystem, the decryption algorithm recovers m.

We now show indistinguishability of the two modes. In messy mode, $\mathsf{crs} = (N, y)$, where y is a uniform element in $\mathbb{J}_N \setminus \mathbb{QR}_N$. In decryption mode, $\mathsf{crs} = (N, y)$, where y is a uniform element in $\mathbb{QR}_N$. By the QR assumption, these are indistinguishable.

We now demonstrate identification of a messy branch. Let pk be the (possibly malformed) public key. Since $y \notin \mathbb{QR}_N$, either pk or $pk \cdot y$ is not a quadratic residue. Lemma 6 implies that one of the branches of pk is messy; it can be found using the factorization $t = (p, q)$ of N.

We conclude with trapdoor key generation. Let $y = t^2$. $\mathsf{TrapKeyGen}(\mathsf{crs}, t)$ outputs $(r^2, r, r \cdot t)$. The output of $\mathsf{KeyGen}(\sigma)$, on the other hand, is $(r^2 y^{-\sigma}, r)$. Now, $(pk, sk_0) = (r^2, r)$ is distributed identically to $\mathsf{KeyGen}(0)$, by definition of KeyGen. By a renaming of variables letting $r = r'/t$, we have $(pk, sk_1) = ((r')^2/t^2, r') = ((r')^2/y, r')$, which is distributed identically to $\mathsf{KeyGen}(1)$, since $r' = r/t \in \mathbb{Z}_N^*$ is uniformly distributed.

For larger branch sets $\{0, 1\}^k$, the scheme is modified in a manner similar to the one from Section 5, where all k parallel copies of the scheme use the same modulus N.

Acknowledgments

We thank Susan Hohenberger, Yuval Ishai, Oded Regev, and the anonymous reviewers for helpful comments on earlier drafts of this paper.

References

1. Aiello, W., Ishai, Y., Reingold, O.: Priced oblivious transfer: How to sell digital goods. In: Pfitzmann, B. (ed.) EUROCRYPT 2001. LNCS, vol. 2045, pp. 119–135. Springer, Heidelberg (2001)
2. Ajtai, M.: Generating hard instances of lattice problems. Quaderni di Matematica 13, 1–32 (2004); Preliminary version in STOC 1996

3. Ajtai, M., Dwork, C.: A public-key cryptosystem with worst-case/average-case equivalence. In: STOC, pp. 284–293 (1997)
4. Camenisch, J., Neven, G., Shelat, A.: Simulatable adaptive oblivious transfer. In: Naor, M. (ed.) EUROCRYPT 2007. LNCS, vol. 4515, pp. 573–590. Springer, Heidelberg (2007)
5. Canetti, R.: Universally composable security: A new paradigm for cryptographic protocols. In: FOCS, pp. 136–145 (2001)
6. Canetti, R., Fischlin, M.: Universally composable commitments. In: Kilian, J. (ed.) CRYPTO 2001. LNCS, vol. 2139, pp. 19–40. Springer, Heidelberg (2001)
7. Canetti, R., Lindell, Y., Ostrovsky, R., Sahai, A.: Universally composable two-party and multi-party secure computation. In: STOC, pp. 494–503 (2002)
8. Canetti, R., Rabin, T.: Universal composition with joint state. In: Boneh, D. (ed.) CRYPTO 2003. LNCS, vol. 2729, pp. 265–281. Springer, Heidelberg (2003)
9. Cocks, C.: An identity based encryption scheme based on quadratic residues. In: IMA Int. Conf, pp. 360–363 (2001)
10. Cramer, R., Shoup, V.: Universal hash proofs and a paradigm for adaptive chosen ciphertext secure public-key encryption. In: Knudsen, L.R. (ed.) EUROCRYPT 2002. LNCS, vol. 2332, pp. 45–64. Springer, Heidelberg (2002)
11. Crépeau, C.: Equivalence between two flavours of oblivious transfers. In: Pomerance, C. (ed.) CRYPTO 1987. LNCS, vol. 293, pp. 350–354. Springer, Heidelberg (1988)
12. Damgård, I., Nielsen, J.B.: Universally composable efficient multiparty computation from threshold homomorphic encryption. In: Boneh, D. (ed.) CRYPTO 2003. LNCS, vol. 2729, pp. 247–264. Springer, Heidelberg (2003)
13. Even, S., Goldreich, O., Lempel, A.: A randomized protocol for signing contracts. Commun. ACM 28(6), 637–647 (1985)
14. Garay, J.A., MacKenzie, P.D., Yang, K.: Efficient and universally composable committed oblivious transfer and applications. In: Naor, M. (ed.) TCC 2004. LNCS, vol. 2951, pp. 297–316. Springer, Heidelberg (2004)
15. Gentry, C., Peikert, C., Vaikuntanathan, V.: Trapdoors for hard lattices and new cryptographic constructions. In: STOC (to appear, 2008), http://eprint.iacr.org/2007/432
16. Goldreich, O., Micali, S., Wigderson, A.: How to play any mental game or a completeness theorem for protocols with honest majority. In: STOC, pp. 218–229 (1987)
17. Green, M., Hohenberger, S.: Blind identity-based encryption and simulatable oblivious transfer. In: Kurosawa, K. (ed.) ASIACRYPT 2007. LNCS, vol. 4833, pp. 265–282. Springer, Heidelberg (2007)
18. Green, M., Hohenberger, S.: Universally composable adaptive oblivious transfer. Cryptology ePrint Archive, Report 2008/163 (2008), http://eprint.iacr.org/
19. Groth, J., Ostrovsky, R., Sahai, A.: Perfect non-interactive zero knowledge for NP. In: Vaudenay, S. (ed.) EUROCRYPT 2006. LNCS, vol. 4004, pp. 339–358. Springer, Heidelberg (2006)
20. Ishai, Y., Kilian, J., Nissim, K., Petrank, E.: Extending oblivious transfers efficiently. In: Boneh, D. (ed.) CRYPTO 2003. LNCS, vol. 2729, pp. 145–161. Springer, Heidelberg (2003)
21. Jarecki, S., Shmatikov, V.: Efficient two-party secure computation on committed inputs. In: Naor, M. (ed.) EUROCRYPT 2007. LNCS, vol. 4515, pp. 97–114. Springer, Heidelberg (2007)

22. Kalai, Y.T.: Smooth projective hashing and two-message oblivious transfer. In: Cramer, R. (ed.) EUROCRYPT 2005. LNCS, vol. 3494, pp. 78–95. Springer, Heidelberg (2005)
23. Kilian, J.: Founding cryptography on oblivious transfer. In: STOC, pp. 20–31 (1988)
24. Kol, G., Naor, M.: Cryptography and game theory: Designing protocols for exchanging information. In: Canetti, R. (ed.) TCC 2008. LNCS, vol. 4948, pp. 320–339. Springer, Heidelberg (2008)
25. Lindell, Y.: Efficient fully simulatable oblivious transfer. In: Malkin, T. (ed.) CT-RSA 2008. LNCS, vol. 4964. Springer, Heidelberg (2008)
26. Micciancio, D., Regev, O.: Worst-case to average-case reductions based on Gaussian measures. SIAM J. Comput. 37(1), 267–302 (2007); Preliminary version in FOCS 2004
27. Naor, M., Pinkas, B.: Oblivious transfer with adaptive queries. In: CRYPTO. LNCS, pp. 573–590. Springer, Heidelberg (1999)
28. Naor, M., Pinkas, B.: Efficient oblivious transfer protocols. In: SODA, pp. 448–457 (2001)
29. Peikert, C., Waters, B.: Lossy trapdoor functions and their applications. In: STOC (to appear, 2008), http://eprint.iacr.org/2007/279
30. Rabin, M.O.: How to exchange secrets by oblivious transfer. Technical report, Harvard University (1981)
31. Regev, O.: New lattice-based cryptographic constructions. J. ACM 51(6), 899–942 (2004)
32. Regev, O.: On lattices, learning with errors, random linear codes, and cryptography. In: STOC, pp. 84–93 (2005)
33. Shor, P.W.: Polynomial-time algorithms for prime factorization and discrete logarithms on a quantum computer. SIAM J. Comput. 26(5), 1484–1509 (1997)
34. Yao, A.C.-C.: How to generate and exchange secrets (extended abstract). In: FOCS, pp. 162–167 (1986)

Founding Cryptography on Oblivious Transfer – Efficiently

Yuval Ishai[1,*], Manoj Prabhakaran[2,**], and Amit Sahai[3,***]

[1] Technion, Israel and University of California, Los Angeles
yuvali@cs.technion.il
[2] University of Illinois, Urbana-Champaign
mmp@cs.uiuc.edu
[3] University of California, Los Angeles
sahai@cs.ucla.edu

Abstract. We present a simple and efficient compiler for transforming secure multi-party computation (MPC) protocols that enjoy security only with an honest majority into MPC protocols that guarantee security with no honest majority, in the oblivious-transfer (OT) hybrid model. Our technique works by combining a secure protocol in the honest majority setting with a protocol achieving only security against *semi-honest* parties in the setting of no honest majority.

Applying our compiler to variants of protocols from the literature, we get several applications for secure two-party computation and for MPC with no honest majority. These include:

– **Constant-rate two-party computation in the OT-hybrid model.** We obtain a statistically UC-secure two-party protocol in the OT-hybrid model that can evaluate a general circuit C of size s and depth d with a total communication complexity of $O(s) + poly(k, d, \log s)$ and $O(d)$ rounds. The above result generalizes to a constant number of parties.

– **Extending OTs in the malicious model.** We obtain a computationally efficient protocol for generating many string OTs from few string OTs with only a *constant amortized communication overhead* compared to the total length of the string OTs.

– **Black-box constructions for constant-round MPC with no honest majority.** We obtain general computationally UC-secure MPC protocols in the OT-hybrid model that use only a constant number of rounds, and only make a *black-box* access to a pseudorandom generator. This gives the first constant-round protocols for three or more parties that only make a black-box use of cryptographic primitives (and avoid expensive zero-knowledge proofs).

* Supported in part by ISF grant 1310/06, BSF grant 2004361, and NSF grants 0205594, 0430254, 0456717, 0627781, 0716389.

** Supported in part by NSF grants CNS 07-16626 and CNS 07-47027.

*** Research supported in part from NSF grants 0627781, 0716389, 0456717, and 0205594, a subgrant from SRI as part of the Army Cyber-TA program, an equipment grant from Intel, an Alfred P. Sloan Foundation Fellowship, and an Okawa Foundation Research Grant.

D. Wagner (Ed.): CRYPTO 2008, LNCS 5157, pp. 572–591, 2008.
© International Association for Cryptologic Research 2008

1   Introduction

Secure multiparty computation (MPC) [4, 11, 22, 40] allows several mutually distrustful parties to perform a joint computation without compromising, to the greatest extent possible, the privacy of their inputs or the correctness of the outputs. MPC protocols can be roughly classified into two types: (1) ones that only guarantee security in the presence of an honest majority, and (2) ones that guarantee security[1] against an arbitrary number of corrupted parties.

A qualitatively important advantage of protocols of the second type is that they allow each party to trust nobody but itself. In particular, this is the only type of security that applies to the case of secure two-party computation. Unfortunately, despite the appeal of such protocols, their efficiency significantly lags behind known protocols for the case of an honest majority. (For the potential efficiency of the latter, see the recent practical application of MPC in Denmark [5].) This is the case even when allowing parties to use idealized cryptographic primitives such as bit commitment and oblivious transfer.

In this work we revisit the problem of founding secure two-party computation and MPC with *no honest majority* on oblivious transfer. Oblivious transfer (OT) [19, 38] is a two-party protocol that allows a receiver to obtain one out of two strings held by a sender, without revealing to the sender the identity of its selection. More precisely, OT is a secure implementation of the functionality which takes inputs s_0, s_1 from the sender and a choice bit b from the receiver, and outputs s_b to the receiver. Kilian [33] showed how to base general secure two-party computation on OT. Specifically, Kilian's result shows that given the ability to call an ideal oracle that computes OT, two parties can securely compute an arbitrary function of their inputs with unconditional security. We refer to secure computation in the presence of an ideal OT oracle as secure computation in the *OT-hybrid model*. Kilian's result was later generalized to the multi-party setting (see [15] and the references therein). Unfortunately, these constructions are quite inefficient and should mainly be viewed as feasibility results.

When revisiting the problem of basing cryptography on OT, we take a very different perspective from the one taken in the original works. Rather than being driven primarily by the goal of obtaining *unconditional security*, we are mainly motivated by the goal of achieving better *efficiency* for MPC in "the real world", when unconditional security is typically impossible or too expensive to achieve.[2]

Advantages of OT-based cryptography. There are several important advantages to basing cryptographic protocols on oblivious transfer, as opposed to concrete number-theoretic or algebraic assumptions.

---

[1] Concretely, in this type of protocols it is generally impossible to guarantee output delivery or even fairness, and one has to settle for allowing the adversary to abort the protocol after learning the output.

[2] Our results still imply efficient unconditionally secure protocols under physical assumptions, such as off-line communication with a trusted dealer, secure hardware, or noisy channels.

- PREPROCESSING. OTs can be pre-computed in an off-line stage, before the actual inputs to the computation or even the function to be computed are known, and later very cheaply converted into actual OTs [1].
- AMORTIZATION. The cost of pre-computing OTs can be accelerated by using efficient methods for *extending OTs* [2, 27, 29]. In fact, the results of the current paper imply additional improvement to the asymptotic cost of extending OTs, and thus further strengthen this motivation.
- SECURITY. OTs can be realized under a variety of computational assumptions, or even with unconditional security under physical assumptions. (See [37] for efficient realizations of *UC-secure* OT in the CRS model under various standard assumptions.) Furthermore, since the methods for extending OTs discussed above only require protocols to use a relatively small number of OTs, one could potentially afford to diversify assumptions by combining several candidate OT implementations [28].

1.1   Our Results

Motivated by the efficiency gap between the two types of MPC discussed above, we present a simple and efficient general compiler that transforms MPC protocols with security in the presence of an honest majority into secure two-party protocols in the OT-hybrid model. More generally and precisely, our compiler uses the following two ingredients:

- An "outer" MPC protocol Π with security against a constant fraction of *malicious* parties. This protocol may use secure point-to-point and broadcast channels. It realizes a functionality f whose inputs are received from and whose outputs are given to two distinguished parties.
- An "inner" two-party protocol ρ for a (typically simple) functionality g^Π defined by the outer protocol, where the security of ρ only needs to hold against *semi-honest* parties. The protocol ρ can be in the *OT-hybrid model*.

The compiler yields a two-party protocol $\Phi_{\Pi,\rho}$ which realizes the functionality f of the outer protocol with security against malicious parties in the OT-hybrid model. If the outer protocol Π is UC-secure [9] (as is the case for most natural outer protocols) then so is $\Phi_{\Pi,\rho}$. It is important to note that $\Phi_{\Pi,\rho}$ only makes a *black-box* use of the outer protocol Π and the inner protocol ρ,[3] hence the term "compiler" is used here in a somewhat unusual way. This black-box flavor of our compiler should be contrasted with the traditional GMW compiler [21, 22] for transforming a protocol with security in the semi-honest model into a protocol with security in the malicious model. Indeed, the GMW compiler needs to apply (typically expensive) zero-knowledge proofs that depend on the code of the protocol to which it applies. Our compiler naturally generalizes to yield MPC protocols with more than two parties which are secure (in the OT-hybrid model) in the presence of an arbitrary number of malicious parties.

---

[3] Furthermore, the functionality g^Π realized by ρ is also defined in a black-box way using the next-message function of Π. This rules out the option of allowing the compiler access to the code of f by, say, incorporating it in the output of g^Π.

Combining our general compiler with variants of protocols from the literature, we get several applications for secure two-party computation and MPC with no honest majority.

Revisiting the classics. As a historically interesting example, one can obtain a conceptually simple derivation of Kilian's result [33] by using the BGW protocol [4] (or the CCD protocol [11]) as the outer protocol, and the simple version of the GMW protocol in the *semi-honest OT-hybrid model* [21, 22, 23] as the inner protocol. In fact, since the outer protocol is not required to provide optimal resilience, the BGW protocol can be significantly simplified. The resulting protocol has the additional benefits of providing full simulation-based (statistical) UC-security and an easy generalization to the case of more than two parties.

Constant-rate two-party computation in the OT-hybrid model. Using a variant of an efficient MPC protocol of Damgård and Ishai [17] combined with secret sharing based on algebraic geometric codes due to Chen and Cramer [12] as the outer protocol, we obtain a statistically UC-secure two-party protocol *in the OT-hybrid model* that can evaluate a general circuit C of size s with a total communication complexity of $O(s)$. (For simplicity, we ignore from here on *additive* terms that depend polynomially on the security parameter k, the circuit depth, and $\log s$. These terms become dominated by the leading term in most typical cases of large circuits.) This improves over the $O(k^3 s)$ complexity of the best previous protocol of Crépeau et al. [15], and matches the best asymptotic complexity in the semi-honest model.

By using preprocessing to pre-compute OTs on random inputs, the protocol in the OT-hybrid model gives rise to a (computationally secure) protocol of comparable efficiency in the plain model. Following off-line interaction that results in each party storing a string of length $O(s)$, the parties can evaluate an arbitrary circuit of size s on their inputs using $O(s)$ bits of communication and *no cryptographic computations*. Note that the preprocessing stage can be carried out offline, before the actual inputs are available or even the circuit C is known. Furthermore, the cost of efficiently implementing the off-line stage can be significantly reduced by using techniques for amortizing the cost of OTs on which we improve. The above results extend to the case of more than two parties, with a multiplicative overhead that grows polynomially with the number of parties.

Unlike two-party protocols that are based on Yao's garbled circuit method [40], the above protocols cannot be implemented in a constant number of rounds and require $O(d)$ rounds for a circuit of depth d. It seems that in most typical scenarios of large-scale secure computation, the overall efficiency benefits of our approach can significantly outweigh its higher round-complexity.

Extending OTs in the malicious model. Somewhat unexpectedly, our techniques for obtaining efficient cryptographic protocols which *rely* on OT also yield better protocols for *realizing* the OTs consumed by the former protocols. This is done by using an outer protocol that efficiently realizes a functionality which implements many instances of OT. More concretely, we obtain a protocol for generating many OTs from few OTs whose amortized cost in communication and cryptographic computation is a constant multiple of the efficient protocol for the semi-honest model given by Ishai, Kilian, Nissim, and Petrank [29]. Using the protocol from [29] inside the inner protocol, we can upgrade the security of this OT extension protocol to the malicious model with only a constant communication and cryptographic overhead. This improves over a recent result from [27] that obtains similar efficiency in terms of the number of hash functions being invoked, but worse asymptotic communication complexity. Our OT extension protocol can be used for efficiently implementing the off-line precomputation of all the OTs required by our protocols in the OT-hybrid model.

Black-box constructions for constant-round MPC with no honest majority. We combine our general compiler with a variant of a constant-round MPC protocol of Damgård and Ishai [16] to obtain general *computationally* UC-secure MPC protocols in the OT-hybrid model that use only a constant number of rounds, and only make a *black-box* access to a pseudorandom generator. This provides a very different alternative to a similar result for the two party case that was recently obtained by Lindell and Pinkas [35], and gives the first constant-round protocols for three or more parties that only make a black-box use of cryptographic primitives (and avoid expensive zero-knowledge proofs).

Additional results. In Section 5 we describe two additional applications: a constant-rate black-box construction of OT for malicious parties from OT for semi-honest parties (building on a recent black-box feasibility result of [26, 31]), and a construction of asymptotically optimal OT combiners [28] (improving over [27]). In the full version we present a two-party protocol in the OT-hybrid model that uses only a *single* round of OTs and no additional interaction. (This applies to functionalities in which only one party receives an output.) The protocol only makes $n + o(n)$ OT calls, where n is the size of the input of the party which receives the output.

1.2   Techniques

Our main compiler was inspired by the "MPC in the head" paradigm introduced by Ishai, Kushilevitz, Ostrovsky, and Sahai [32] and further developed by Harnik, Ishai, Kushilevitz, and Nielsen [27]. These works introduced the idea of having parties "imagine" the roles of other parties taking part in an MPC (which should have honest majority), and using different types of cross-checking to ensure that an honest majority really is present in the imagined protocol. Our approach is similar to the construction of OT combiners from [27] in that it uses an outer MPC protocol to add privacy and robustness to an inner two-party protocol which may potentially fail.[4] A major difference, however, is that our approach provides security in the malicious model while only requiring the inner protocol to be secure in the *semi-honest* model.

---

[4] This idea is also reminiscent of the player virtualization technique of Bracha [6] and the notion of concatenated codes from coding theory.

The central novelty in our approach is a surprisingly simple and robust enforcement mechanism that we call the "watchlist" method (or more appropriately, the *oblivious* watchlist method). In describing our approach, we will refer for simplicity to the case of two-party computation involving two "clients" A and B. In our compiler, an outer MPC protocol requiring an honest majority of servers is combined with an inner two-party computation protocol with security against only *semi-honest* adversaries. This is done by having the outer MPC protocol jointly "imagined" by the two clients. Each server's computation is jointly simulated by the two clients, using the inner semi-honest two-party protocol to compute the next-message-functions for the servers. The only method we use to prevent cheating is that both clients maintain a watchlist of some fraction of the servers, such that client A will have full knowledge of the internal state of all servers in A's watchlist, while client B has no idea which servers are on A's watchlist. Then client A simply checks that the watchlisted servers behave as they should in the imagined outer MPC protocol. If a dishonest client tries to cheat for too many servers, then he will be caught because of the watchlist with overwhelming probability. On the other hand, since the outer MPC protocol is robust against many bad servers, a dishonest client *must* attempt to cheat in the computation of many servers in order to be able to gain any unfair advantage in the execution of the protocol. Our watchlist-based method for enforcing honest behavior should be contrasted with the non-black-box approach of the GMW compiler [22] that relies on zero-knowledge proofs.

It is instructive to contrast our approach with "cut-and-choose" methods from the literature. In standard cut-and-choose protocols, one party typically prepares many instances of some object, and then the other party asks for "explanations" of several of these objects. A central difficulty in such an approach is to prevent the compromised instances from leaking information about secrets, while combining the un-compromised instances in a useful way (see e.g. [35]). In contrast, our approach achieves these goals seamlessly via the privacy and robustness of the outer MPC protocol. To see how our approach leads to efficiency improvements as well, we will make an analogy to error-correcting codes. In traditional cut-and-choose, one has to prepare many copies of an object that will only be used once, analogous to a repetition-based error-correcting code. Underlying our approach are the more sophisticated error-correcting codes that can be used in MPC protocols in the honest majority setting. While we have to sacrifice some working components (our servers) due to the watchlists, the others perform useful work that is not wasted, and this allows us to get more "bang for the buck", especially in settings where amortization is appropriate.

2   Preliminaries

Model. We use the Universal Composition (UC) framework [9], although our protocols can also be instantiated in the stand-alone setting using the composability framework of [8, 21]. The parties in the protocols have access to (private, point-to-point) communication channels, as well as possibly one or more ideal functionalities such as OT or broadcast.

Oblivious Transfer. The basic oblivious transfer primitive we rely on is a $\binom{2}{1}$ string-OT, referred to as OT. Below we will also employ $\binom{q}{1}$ string-OT. There are efficient and unconditionally UC-secure reductions with constant communication overhead of these primitives to $\binom{2}{1}$ bit-OT (implicit in [7, 13, 14, 18]). Hence, one could also assume bit-OT as our basic primitive. When settling for computational security, OT on long strings can be efficiently reduced to a single instance of OT on short strings via the use of a pseudorandom generator.

Our watchlist initialization protocol will use Rabin string-OT, which delivers an input string from the sender to the receiver with a fixed probability δ. We point out how a Rabin-string-OT with rational erasure probability p/q (for positive integers $p < q$) can be securely realized using $\binom{q}{1}$ string-OT with constant communication overhead. The sender inputs q strings to the $\binom{q}{1}$ string-OT, of which a random subset of p are the message being transferred and the rest are arbitrary (say the zero string); the receiver picks up one of the q strings uniformly at random; then the sender reveals to the receiver which p-sized subset had the string being transferred; if the receiver picked a string not belonging to this set, it outputs erasure, and else outputs the string it received.[5]

Our model of OT is asynchronous: multiple OT's executed in the same round can be executed in an arbitrary, adversarially controlled order. (We note, however, that synchronous OT can be easily and efficiently reduced to asynchronous OT via simultaneous message exchange.)

3   Protocol Compiler

In this section we describe how to build a protocol $\Phi^{OT}_{\Pi,\rho}$ that securely realizes a functionality $\mathcal{F}$ against active corruptions, using two component protocols Π and ρ^{OT} of weaker security. Π is a protocol for $\mathcal{F}$ itself, but uses several servers and depends on all but a constant fraction of them being honest. ρ^{OT} is a protocol for a functionality $\mathcal{G}$ (which depends on Π), but is secure only against passive corruptions. Below we describe the requirements on Π, ρ^{OT} and the construction of $\Phi^{OT}_{\Pi,\rho}$.

3.1   The Outer Protocol Π

Π is a protocol among $n + m$ parties (we will use $n = \Theta(m^2 k)$, k being the security parameter for Π), with m parties $\overline{C}_i$ $(i = 1, \ldots, m)$ designated as the *clients*, and the other parties $\overline{P}_j$ $(i = 1, \ldots, n)$ designated as the *servers*.

– *Functionality:* Π is a protocol for some functionality $\mathcal{F}$ (which could be deterministic or randomized, and possibly reactive) among the m clients. The servers do not have any inputs or produce any outputs.

– *Security:* Π UC-securely realizes the functionality $\mathcal{F}$, against *adaptive* corruption of up to t servers, and either static or adaptive corruption of any number

---

[5] Note that the sender can "cheat" by using arbitrary inputs to the $\binom{p}{q}$ string-OT and declaring an arbitrary set as the p-sized subset containing the message. But this simply corresponds to picking one of the messages in the declared p-sized subset (considered as a multi-set) uniformly at random, and using it as the input to the p/q-Rabin-string-OT.

of clients (see Remark 2). We assume static client corruption by default. We will require $t = \Omega(n)$. The corruptions are active (i.e., the corrupt parties can behave arbitrarily) and the security could be statistical or computational.

- *Protocol Structure:* The protocol Π proceeds in rounds where in each round each party sends messages to the other parties (over secure point-to-point channels) and updates its state by computing on its current state, and then also incorporates the messages it receives into its state. Each server $\overline{P}_j$ maintains a state $\overline{\Sigma}_j$. For the sake of an optimization in our applications, we will write $\overline{\Sigma}_j$ as $(\overline{\sigma}_j, \overline{\mu}_{1 \leftrightarrow j}, \ldots, \overline{\mu}_{m \leftrightarrow j})$, where $\overline{\mu}_{i \leftrightarrow j}$ is just the collection of messages between $\overline{C}_i$ and $\overline{P}_j$. We will refer to $\overline{\mu}_{i \leftrightarrow j}$ as the "local" parts of the state and $\overline{\sigma}_j$ as the "non-local" part of the state. Note that client $\overline{C}_i$ is allowed to know the local state $\overline{\mu}_{i \leftrightarrow j}$ of each server $\overline{P}_j$.

The servers' program in Π is specified by a (possibly randomized) function π which takes as input a server's current state and incoming messages from clients and servers, and outputs an updated state as well as outgoing messages for the clients and other servers. That is,[6]

$$\pi(\overline{\sigma}_j; \overline{\boldsymbol{\mu}}_j; \overline{\mathbf{w}}._{\rightarrow j}; \overline{\mathbf{u}}._{\rightarrow j}) \rightarrow (\overline{\sigma}'_j, \overline{\mathbf{m}}'_{j \rightarrow .}, \overline{\mathbf{u}}'_{j \rightarrow .}).$$

where $\overline{\boldsymbol{\mu}}_j = (\overline{\mu}_{1 \rightarrow j}, \ldots, \overline{\mu}_{m \rightarrow j})$ is the vector of local states, $\overline{\mathbf{w}}._{\rightarrow j} = (\overline{w}_{1 \rightarrow j}, \ldots, \overline{w}_{m \rightarrow j})$ is messages received in this round by server $\overline{P}_j$ from the clients, and similarly $\overline{\mathbf{u}}._{\rightarrow j} = (\overline{u}_{1 \rightarrow j}, \ldots, \overline{u}_{n \rightarrow j})$ is the set of messages $\overline{P}_j$ received from the other servers. The outputs $\overline{\mathbf{m}}'_{j \rightarrow .} = (\overline{m}'_{j \rightarrow 1}, \ldots, \overline{m}'_{j \rightarrow m})$ and $\overline{\mathbf{u}}'_{j \rightarrow .} = (\overline{u}'_{j \rightarrow 1}, \ldots, \overline{u}'_{j \rightarrow n})$ stand for messages to be sent by $\overline{P}_j$ to the clients and to the servers respectively. The output $\overline{\sigma}'_j$ is the updated (non-local) state of the server $\overline{P}_j$. The local states are updated (by definition) as $\overline{\mu}'_{i \leftrightarrow j} := \overline{\mu}_{i \leftrightarrow j} \circ (\overline{w}_{i \rightarrow j}, \overline{m}'_{j \rightarrow i})$.

Finally, if Π is in the broadcast-hybrid model, one can efficiently implement each broadcast by having the broadcasting party send the message to all clients. While this isn't equivalent to broadcast in the MPC model, our compiler will provide robustness against inconsistent messages.

3.2 The Inner Functionality $\mathcal{G}$ and the Inner Protocol ρ^{OT}

We define a (possibly randomized) m-party functionality $\mathcal{G}_j$ which will be used to "implement" server $\overline{P}_j$ by the clients $\overline{C}_i$ $(i = 1, \ldots, m)$. $\mathcal{G}_j$ works as follows:

- From each client $\overline{C}_i$ get input $(S_i, M_i, \overline{\mu}_{i \rightarrow j}, \overline{w}_{i \rightarrow j})$, where S_i will be considered an additive share of the non-local state $\overline{\sigma}_j$ of the server $\overline{P}_j$, and M_i an additive share of $\overline{\mathbf{u}}._{\rightarrow j}$, all the messages received by $\overline{P}_j$ from the other servers in the previous round.[7]

---

[6] For the sake of brevity we have omitted the round number, server number, and number of servers as explicit inputs to π. We shall implicitly use the convention that these are part of each component in the input.

[7] By default, this additive sharing uses bitwise XOR. However, it is sometimes beneficial to use a different finite abelian group for this purpose. This allows to implement group additions performed by the outer protocol non-interactively, by having clients directly add their shares.

– Compute $S_1 + \ldots + S_m$, and $M_1 + \ldots + M_m$ to reconstruct $\overline{\sigma}_j$ and $\overline{\mathbf{u}}_{.\to j}$. Evaluate $\overline{\pi}$ (as given in the above displayed equation) to obtain $(\overline{\sigma}'_j, \overline{\mathbf{m}}'_{j\to.}, \overline{\mathbf{u}}'_{j\to.})$.

– To $\overline{C}_i$ give output $(S'_i, \overline{m}'_{j\to i}, M'_i)$ where $(S'_1, \ldots, S'_m)$ form a random additive sharing of the updated state $\overline{\sigma}'_j$ and $(M'_1, \ldots, M'_m)$ form a random additive sharing of the messages to the servers $\overline{\mathbf{u}}'_{j\to.}$.

We will need a protocol ρ^{OT} (in the OT-hybrid model) to carry out this computation. But the security requirement on this protocol is quite mild: ρ^{OT} *securely realizes $\mathcal{G}_j$ against passive corruption (i.e., honest-but-curious adversaries).* The security could be statistical or computational. Also, the security could be against adaptive corruption or static corruption.

In all our applications, we shall exploit an important optimization in an inner protocol to implement $\mathcal{G}_j$. Suppose an invocation of $\overline{\pi}$ (i.e., for some server $\overline{P}_j$ and some round number) depends only on the local state $\overline{\mu}_{i\leftrightarrow j}$ and possibly $\overline{w}_{i\to j}$, does not change the state $\overline{\sigma}_j$, and is deterministic. We call such a computation a *type I computation* (all other computations are called type II computations). A simple secure implementation of $\mathcal{G}_j$ for type I computations involves the client $\overline{C}_i$ computing $(\overline{\mathbf{m}}'_{j\to.}, \overline{\mathbf{u}}'_{j\to.})$ itself, and sending each client $C_{i'}$ as output $(X_{i'}, \overline{m}'_{j\to i'}, M'_{i'})$ for each party, where $X_{i'}$ is a random sharing of 0 and $M'_{i'}$ is a random sharing of $\overline{\mathbf{u}}'_{j\to.}$. The client $C_{i'}$ sets $S'_{i'} := S_{i'} + X_{i'}$. (This last step of adding a share of 0 is in fact redundant in our compiled protocol; we include it only for the sake of modular exposition.)

Thus what the compiler needs to be given as the inner protocol is an implementation of $\mathcal{G}_j$ only for type II computations. Then it is the computational complexity of type II computations that will be reflected in the communication complexity of the compiled protocol.

3.3   The Compiled Protocol

At a high-level, the compiled protocol $\Phi^{\mathsf{OT}}_{\Pi,\rho}$ has the following structure.

1. *Watchlists initialization:* Using OT, the following infrastructure is set up first: each honest client randomly chooses a set of k servers to put on its watchlist (which only that client knows). For each client i and server $\overline{P}_j$ there is a "watchlist channel" W_{ij} such that any of the clients can send a message in W_{ij}, and client $\overline{C}_i$ will receive this message if and only if server $\overline{P}_j$ is on its watchlist. As we shall see, the implementation of this will allow a corrupt client to gain access (albeit partial) to the watchlist channels of more than k servers. Nevertheless, we note that the total number of servers for which the adversary will have access to the watchlist channel will be $O(km^2) < t/2$.

We shall also require another variant of watchlist channel (that can be set up on top of the above watchlist channel infrastructure): for each server $\overline{P}_j$ there is a "watchlist broadcast channel" $\mathbf{W}_j$ such that any client can send a message on $\mathbf{W}_j$ and *all the clients* who have server $\overline{P}_j$ on their watchlists will receive this message. (Note that when there are only two clients, this variant is no different from the previous one.)

If the adversary has access to the watchlist channel for server $\overline{P}_j$, then we allow the adversary to learn which other clients have access to their watchlist channels for server $\overline{P}_j$. Jumping ahead, we remark that in this case we will consider server $\overline{P}_j$ as corrupted. By the choice of parameters this will corrupt at most $t/2$ servers (except with negligible probability).

2. *Simulating the execution of Π:* Each client $\overline{C}_i$ plays the role of $\overline{C}_i$ in Π. In addition, the clients will themselves implement the servers in Π as follows. At the beginning of each round of Π, the clients will hold a secret sharing of the state of each server. Then they will use the inner protocol to execute the server's next-message and state-evolution functions and update the shared state.

The purpose of the watchlists is two-fold: firstly it is used to force (to some extent) that the clients do not change their inputs to the inner protocol *between* invocations; secondly it is used to force honest behavior *within* the inner protocol executions. The actual use of watchlists is quite simple:

(a) To enforce consistency between invocations of the inner protocol, each client $\overline{C}_i$ is required to report over the watchlist broadcast channel $\mathbf{W}_j$ every message that it provides as input to or receives as output from every invocation of the inner protocol for $\mathcal{G}_j$.

(b) To enforce honest behavior within the protocol execution, each client is required to report over watchlist channels W_{ij} (for all i) every message that it receives within the invocation of the inner protocol for $\mathcal{G}_j$. Further, for each invocation of the inner protocol j, the watchlist broadcast channel $\mathbf{W}_j$ is used to carry out a "coin-tossing into the well" to generate the coins for each client to be used in that protocol. (This coin-tossing step is not necessary when certain natural protocols with a slightly stronger security guarantee — like the basic "passive-secure" GMW protocol in the OT-hybrid model — are used. See Remark 1 below.)

Any honest client who has server $\overline{P}_j$ in its watchlist must check that the reported values from all clients are according to the protocol and are consistent with the other messages received in the protocol. Note that at the beginning of the execution of the inner protocol, all clients are already committed to their inputs and randomness during the protocol. Further, all honest clients honestly report the messages received from the other protocols. As such a client watching server $\overline{P}_j$ has an almost complete view of the protocol execution, and it knows ahead of time exactly what messages should be reported over the watchlist channels in an honest execution. This is sufficient to catch any deviation in the execution, if the protocol uses only communication channels. *However*, if the protocol involves the use of OT channels (or more generally, other ideal functionalities) then it creates room for an adversary to actively cheat and possibly gain an advantage over passive corruption. Then the adversary can change its inputs to the OT functionality without being detected (or arrange the probability of being detected to depend on the inputs of honest clients). To prevent this kind of cheating, we shall force that if the adversary changes its input to the OT functionality, then with at least a constant probability this will produce a different output for an honest client (if the adversary is the sender in the OT), or (if the adversary is the receiver in the OT) the adversary will end up reporting a different output over the watchlist. This is easily enforced by using a simple standard reduction of OT to OT with random inputs from both parties.

Remark 1 (On tossing coins.). A protocol which is secure against passive corruptions is not necessarily secure when the adversary can maliciously choose the random tape for the corrupt players. This is the reason our compiler needs to use a coin-tossing in the well step to generate the coins for the inner protocols. However, most natural protocols remain secure even if the adversary can choose the coins. This is the case for perfectly secure protocols like the basic "passive-secure" GMW protocol (in the OT-hybrid model). When using such an inner protocol, the compiler can simply omit the coin-tossing into the well step.

Setting up the Watchlist Channels and Broadcast Channels. First we describe how the watchlist channels described above are set up using OTs, and then how to obtain watchlist broadcast channels using them. The basic idea is for the clients to pick up sufficiently long one-time pads from each other using OT, and later send messages masked with a fresh part of these one-time pads.

For this we shall be using Rabin-string-OT (i.e., erasure channel with a fixed erasure probability, and adequately long binary strings being the alphabet). See Section 2 for implementation details.

The construction of the watchlist channels is as follows: First each client randomly chooses a set of k servers to put on its watchlist. Next, each pair of clients (i', i) engages in n instances of δ-Rabin-string-OTs where client $C_{i'}$ sends a random string r_j (of length ℓ) to $\overline{C}_i$. By choice of $\delta = \Omega(k/n)$, we ensure that except with negligible probability $\overline{C}_i$ obtains the string in more than k of the n instances. (By the union bound, this will hold true simultaneously for all pairs (i', i), except with negligible probability.) Now, client $\overline{C}_i$ specifies to client $C_{i'}$ a random permutation σ on $[n]$ conditioned on the following: if j is in the watchlist of $\overline{C}_i$ and $\sigma(j) = j'$, then $r_{j'}$ was received by $\overline{C}_i$. Now, to send a message on the watchlist channel W_{ij}, the client $C_{i'}$ will use (a fresh part of) $r_{\sigma(j)}$ to mask the message and send it to $\overline{C}_i$. Note that if j is in the watchlist of client $\overline{C}_i$, then this construction ensures that $\overline{C}_i$ can read all messages sent on W_{ij} by any client. If the strings r_j are ℓ bits long then at most ℓ bits can be sent to the watchlist channel constructed this way.

Finally, we consider obtaining watchlist broadcast channel $\mathbf{W}_j$ from watchlist channels W_{ij} set up as above. This is similar to how broadcast is obtained from point-to-point channels in [24]. To send a message on $\mathbf{W}_j$ first a client sends the message on W_{ij} for every i. Then each client $\overline{C}_i$ on receiving a message on a watchlist channel W_{ij} sends it out on $W_{i'j}$ for every $i' \neq i$. (If $\overline{C}_i$ does not have access to W_{ij}, it sends a special message (of the same length) to indicate this.) Then it checks if all the messages it receives in this step over W_{ij} are the same as the message it received in the previous step, and if not aborts.

It can be verified that the above constructions indeed meet the specification of the watchlist infrastructure spelled out in the beginning of this section.

Theorem 1. *Let $\mathcal{F}$ be a (possibly reactive) m-party functionality. Suppose Π is an outer MPC protocol realizing $\mathcal{F}$, as specified in Section 3.1, with $n = \Theta(m^2 k)$ and $t = \Theta(k)$, for a statistical security parameter k. Let $\mathcal{G}$ be the functionality defined in Section 3.2 and ρ^{OT} a protocol that securely realizes $\mathcal{G}$ in the OT-hybrid model against passive (static) corruptions. Then the compiled protocol $\Phi_{\Pi,\rho}^{\mathsf{OT}}$ described above securely realizes $\mathcal{F}$ in the OT-hybrid model against active (static) corruptions. If both Π and ρ^{OT} are statistically/computationally secure, then the compiled protocol inherits the same kind of security.*

$\Phi_{\Pi,\rho}^{\mathsf{OT}}$ has communication complexity $\mathrm{poly}(m) \cdot (C_\Pi + nr_\Pi C_\rho)$, round complexity $O(r_\Pi r_\rho)$, and invokes OT $\mathrm{poly}(m) \cdot nr_\Pi q_\rho$ times, where C_Π is the communication complexity of Π, r_Π is the number of rounds of Π, C_ρ is the communication plus randomness complexity of ρ^{OT}, r_ρ is the round complexity of ρ^{OT}, and q_ρ is the number of invocations of OT in ρ^{OT}.

Here by communication complexity of a protocol in the OT-hybrid model we include the communication with the OT functionality. By randomness complexity of a protocol we mean the total number of random bits used by (honest) parties executing the protocol. We remark that the complexity bounds given above can typically be tightened when analyzing specific inner and outer protocols.

Remark 2 (On adaptive security.). Above we assumed that the inner protocol ρ^{OT} is secure against static corruptions, and Π is secure against static client corruptions (and up to t adaptive server corruptions). Then the compiled protocol $\Phi_{\Pi,\rho}^{\mathsf{OT}}$ is secure against static corruptions. However, if ρ^{OT} is secure against adaptive corruptions, depending on the security of Π we can get $\Phi_{\Pi,\rho}^{\mathsf{OT}}$ to be secure against adaptive corruptions. If Π is secure against an adversary who can adaptively corrupt up to $m-1$ clients and up to t servers, then $\Phi_{\Pi,\rho}^{\mathsf{OT}}$ is secure against adaptive corruption up to $m-1$ clients. All known constant-round protocols are restricted to this type of adaptive security, unless honest parties are allowed to erase data. If Π is secure against an adversary which could in addition, after the protocol execution ends, corrupt all the remaining honest clients and servers together, then $\Phi_{\Pi,\rho}^{\mathsf{OT}}$ is secure against adaptive corruption of up to all m clients. This is the typical adaptive security feature of outer protocols whose round complexity depends on the circuit depth, and even of constant-round protocols if data erasure is allowed.

Proof sketch: The proof of security for our compiler follows from a conceptually very simple simulator. Full details will be given in the full version of this paper; here we sketch a high-level overview of how our simulator works. At a very high level, the simulator's job is very simple: Since it simulates the OT channels that the adversary uses in the protocol, the simulator will have full knowledge of everything that is sent over the watchlists, as well as in every invocation of OT used within the inner protocol. Thus, the simulator will know immediately if the adversary causes any of the imagined servers to behave dishonestly. It is easy to argue that if the adversary cheats with respect to any server that is on an honest party's watchlist, then it will be caught with constant probability (this is enforced in part by the reduction of OT to OT with random inputs). Since each honest party's watchlist is large, this shows that if the adversary causes too many servers to behave dishonestly, it will be caught by an honest party with overwhelming probability.

To make this formal, the simulator will invoke Sim_{outer}, the simulator for the outer MPC protocol. The simulator will be allowed to corrupt up to t servers when interacting with Sim_{outer}. When the simulator observes that the adversary is trying to cause dishonest behavior by some server, then it corrupts that server (thereby learning the state and history of that server, allowing the simulator to finish the interaction with the adversary and provide appropriate output to it). As argued above, if the adversary causes dishonest behavior in too many servers, it will get caught with overwhelming probability, and therefore our simulator will not need to exceed t corruptions. The only caveat here is if the adversary simultaneously tries to cause cheating in too many servers (e.g. all the servers at once). To deal with this situation, we ensure that the adversary is caught *before* it receives any output, and so we can simulate the interaction with the adversary before we have to corrupt the corresponding server in the outer protocol. This follows in a straightforward way from the way that the watchlists are used and the fact that OT's are only used with random inputs.   □

4   Instantiating the Building Blocks

For concrete applications of our compiler, we need to choose outer and inner protocols to which the compiler can be applied. The requirements on these components can be considered much easier to meet than security against active corruption in the case of no honest majority. As such the literature provides a wide array of choices that we can readily exploit.

Instances of the Outer Protocol. For the purpose of feasibility results, the classical BGW protocol [4, 9] can be used as the outer protocol. But in our applications, we shall resort to two efficient variants obtained from more recent literature [16, 17].[8]

Using a combination of [12, 17] (as described below) a boolean circuit C of size s and depth d (with bounded fan-in) can be evaluated with a total communication complexity of $O(s) + poly(n, k, d, \log s)$ bits, where k is a statistical security parameter, for n servers and any constant number of clients.[9] The protocol requires $O(d)$ rounds. For this protocol the only type II functions in the servers' program (see Section 3.1) consist of evaluating multiplications in a finite field $\mathbb{F}$ whose size is independent of the number of servers. (Here we do not consider linear functions over $\mathbb{F}$, which can be handled "for free" by the inner

---

[8] Efficiency aside, by using UC-secure outer protocols, our compiled protocols are also UC-secure.

[9] While we do not attempt here to optimize the additive term, we note that a careful implementation of the protocol seems to make this term small enough for practical purposes. In particular, the dependence of this term on d can be eliminated for most natural instances of large circuits.

protocol provided that the servers' states are additively shared over $\mathbb{F}$ among the clients.) The total number of multiplications computed by all servers throughout the protocol execution is $O(s) + poly(n, d)$ (for any constant number of clients).

An MPC protocol as above can be obtained by combining a version of an MPC protocol from [17] with algebraic geometric secret sharing over fields of constant size [12].[10] This combination directly yields a protocol with the above properties for $\mathbf{NC}^0$ circuits, which was recently used in [32] to obtain constant-rate zero-knowledge proofs and in [27] to obtain constant-rate OT combiners. In the full version we present the (natural) extension of this protocol that can be applied to arbitrary depth-d circuits, at the cost of requiring $O(d)$ rounds.

Another useful instance of an outer protocol is obtained from the *constant-round* protocol from [16], as described in Section 5.2. Unlike the previous constant-round MPC protocol from [3], this protocol only makes a black-box use of a pseudorandom generator.

Instances of the Inner Protocol. The main choice of the inner protocol, which suffices for most of our applications, is the simple version of the GMW protocol [21, 22] that provides perfect security against a *passive* adversary in the OT-*hybrid* model. The communication complexity is $O(m^2 s)$ where m is the number of clients and s is the size of the boolean circuit being evaluated (excluding XOR gates). The round complexity is proportional to the circuit depth (where here again, XOR gates are given for free). When evaluating functions in $\mathbf{NC}^1$ (which will always be the case in our applications) the inner protocol can be implemented using a single round of OTs in the two-party case, or a constant number of rounds in the general case, without compromising unconditional security. This is done by using a suitable randomized encoding of the function being computed, e.g., one based on an unconditionally secure variant of Yao's garbled circuit technique [30, 40]. In the two-party case, the protocol needs to use only as many OTs as the length of the *shorter* input. This will be useful for some applications.

5   Applications

In this section we describe the main applications of our general compiler. These are mostly obtained by applying the compiler to variants of efficient MPC protocols and two-party protocols from the literature.

5.1   Constant-Rate Secure Computation in the OT-Hybrid Model

Our first application is obtained by instantiating the general compiler with the following ingredients. The outer protocol is the constant-rate MPC protocol described in Section 4. The inner protocol can be taken to be the "passive-secure GMW" protocol in the OT-hybrid model.

---

[10] Using Franklin and Yung's variant of Shamir's secret sharing scheme [20, 39], as originally done in [17], would result in logarithmic overhead to the communication complexity of the protocol, and a polylogarithmic overhead in the complexity of the applications.

Theorem 2. *Let C be a boolean circuit of size s, depth d and constant fan-in representing an m-party deterministic functionality f for some constant $m \geq 2$. Then there is a statistically UC-secure m-party protocol realizing f in the OT-hybrid model whose total communication complexity (including communication with the OT oracle) is $O(s) + \text{poly}(k, d, \log s)$, where k is a statistical security parameter, and whose round complexity is $O(d)$. Security holds against an adaptive adversary corrupting an arbitrary number of parties.*

The OTs required by the above protocol can be generated during a preprocessing stage at no additional cost. The above theorem extends to the case of a non-constant number of parties m, in which case the communication complexity grows by a multiplicative factor of $poly(m)$. The theorem applies also to *reactive* functionalities, by naturally extending the outer protocol to this case, and to *randomized* functionalities, provided that they are adaptively well-formed [10] or alternatively if honest parties are trusted to erase data.

Finally, it can be extended to the case of *arithmetic* circuits (at the cost of settling for computational security) by using an inner protocol based on homomorphic encryption. We defer further details to the full version.

5.2   Black-Box Constructions for Constant-Round MPC with No Honest Majority

Traditional MPC protocols for the case of no honest majority followed the so-called GMW paradigm [21, 22], converting protocols for the semi-honest model into protocols for the malicious model using zero-knowledge proofs. Since such proofs are typically expensive and in particular make a non-black-box use of the underlying cryptographic primitives, it is desirable to obtain alternative constructions that avoid the general GMW paradigm and only make a black-box use of standard cryptographic primitives.

The protocols of [15, 33] (as well as the more efficient constructions from Section 5.1) achieve this goal, but at the cost of round complexity that depends on the depth of the circuit. The question of obtaining constant-round protocols with the same features remained open.

In the case of MPC with honest majority, this problem was solved by Damgård and Ishai [16], providing a black-box alternative to a previous protocol of Beaver, Micali, and Rogaway [3] that made a non-black-box use of a pseudorandom generator. The case of two-party computation was recently resolved by Lindell and Pinkas [35] (see also [34, 36]), who presented a constant-round two-party protocol that makes a black-box use of (parallel) OT as well as a statistically hiding commitment. The question of extending this result to three or more parties remained open, as the technique of [35] does not seem to easily extend to more than two parties. Partial progress in this direction was recently made in [25].

By applying our compiler to a variant of the MPC protocol from [16], we obtain the following theorem:

Theorem 3. *For any $m \geq 2$ there exists an m-party constant-round MPC protocol in the OT-hybrid model which makes a black-box use of a pseudorandom* *generator and achieves computational UC-security against an active adversary which may adaptively corrupt at most $m - 1$ parties.*

Note that unlike the protocol of [35] our protocol is UC-secure and does not rely on statistically hiding commitments. On the down side, it requires a larger number of OTs which is comparable to the circuit size rather than the input size, though the latter cost may be amortized using efficient methods for extending OTs (see Section 5.3) and moved to a preprocessing phase. We defer further optimizations of the protocol to the full version.

Proof sketch: The protocol from [16] is a general constant-round protocol involving n servers and m clients. It is adaptively, computationally UC-secure against an adversary that may corrupt an arbitrary strict subset of the clients and a constant fraction of the servers. Furthermore, players in this protocol only make a black-box use of a PRG, or alternatively a one-time symmetric encryption scheme. If all the invocations of the encryption scheme were done by clients, the claimed result would follow by directly applying our compiler with this protocol as the outer protocol (since local computations performed by clients remain unmodified by the compiler). While the protocol from [16] inherently requires servers to perform encryptions, it can be easily modified to meet the form required by our compiler. This is done by making the servers only perform encryptions where both the key and the message to be encrypted are known to *one* of the clients. Using the watchlist approach, the protocol produced by the compiler will make the corresponding client perform the encryption instead of the server.

For simplicity, we describe this modification for the case of two clients, Alice and Bob. This easily generalizes to any number of clients m. In any case where a server in the protocol of [16] needs to broadcast an encryption of the form $E_k(m)$, it will instead do the following. The server parses the key k as a pair of keys $k = (k_A, k_B)$ and additively secret-shares the message m as $m = m_A + m_B$. Now it sends k_A, m_A to Alice and k_B, m_B to Bob (this is a dummy operation that is only used to argue security). Finally, the server broadcasts $E_{k_A}(m_A)$ and $E_{k_B}(m_B)$. Note that each of these two computations is of Type I, namely it is done on values already known to one of the clients. Moreover, it is easy to see that the above distributed encryption scheme is still semantically secure from the point of view of an adversary that corrupts just one of the clients. Thus, the simulation argument from [16] (that only relies on the semantic security of E) applies as is. □

5.3   OT Extension in the Malicious Model

Beaver [2] suggested a technique for extending OTs using a one-way function. Specifically, by invoking k instances of OT one can implement a much larger number n of OTs by making use of an arbitrary one-way function. A disadvantage of Beaver's approach is that it makes a non-black-box use of the one-way function, which typically makes his protocol inefficient. A black-box approach for extending OTs was suggested by Ishai, Kilian, Nissim, and Petrank [29]. In the semi-honest model their protocol has the following features. Following an initial seed of k string OTs (where k is a computational security parameter), each additional string OT only requires to make a couple of invocations of a cryptographic hash function (that satisfies a certain property of "correlation robustness"[11] as well as a PRG. The amortized communication complexity of this protocol is optimal up to a constant factor, assuming that each of the sender's strings is (at least) of the size of the input to the hash function. To obtain a similar result for the malicious model, [29] employed a cut-and-choose approach which multiplies the complexity by a statistical security parameter. A partial improvement was recently given in [27], where the overhead in terms of the use of the hash function was reduced to a constant, but the overhead to the communication remained the same. This result was obtained via the use of efficient OT combiners [28]. We improve the (amortized) communication overhead to be constant as well. While our result could be obtained via an improvement to the construction of OT combiners in [27] (see Section 5.4), we sketch here a simple derivation of the result by applying our compiler to the protocol for the semi-honest model in [29]. In the full version we will show an alternative, and self-contained, approach for obtaining a similar result by applying our general secure two-party protocol to an appropriate $\mathbf{NC}^0$ functionality.

The efficient OT extension protocol is obtained as follows. The outer protocol will be the MPC protocol from Section 4 with two clients, called a sender and a receiver, and k servers. The protocol will be applied to the following multi-OT functionality. The sender's input is an n-tuple of pairs of k-bit strings, and the receiver's input is an n-tuple of choice bits. The receiver's output is the n-tuple of chosen k-bit strings. This outer protocol can be implemented so that each of the k servers performs just a single Type II computation, consisting of an $\mathbf{NC}^0$ function with one input of length $O(n)$ originating from the sender and another input of length $O(n/k)$ originating from the receiver. Using a suitable randomized encoding (see Section 4), each of these inner computations can be securely implemented (in the semi-honest model) using $O(n/k)$ OTs on k-bit strings. However, instead of directly invoking the OT oracle for producing the required OTs, we use the OT extension protocol for the *semi-honest* model from [29]. The two-party protocol obtained in this way realizes the multi-OT functionality with computational UC-security, and only makes a black-box use of a correlation-robust hash function as well as a seed of $O(k^2)$ OTs (which also includes the OTs for initializing the watchlists). Its constant communication overhead (for $n \gg k$) is inherited from the outer and inner components. We defer further optimizations to the full version.

Black-Box Constructions of OT. Note that the above construction (before plugging in the protocol from [29]) has the feature that the inner protocol can make a *black-box* use of any OT protocol for the *semi-honest* model. This implies the following black-box approach for converting "semi-honest OTs" into "malicious OTs". First, make $O(k)$ black-box invocations of an arbitrary malicious OT

---

[11] The correlation robustness property defined in [29] is satisfied by a random function. Arguably, it is sufficiently natural to render practical hash functions insecure if they are demonstrated not to have this property.

to generate the watchlists. (Here and in the following, we allow a free black-box use of a PRG to extend a single OT on short strings, or few bit OTs, into OT on a long strings.) Then, make $O(n)$ *black-box* calls to any OT protocol for the semi-honest model to generate n instances of OT in the malicious model. The above black-box approach applies both to the UC and to the standalone model. Together with the black-box constructions of OT of Ishai, Kushilevitz, Lindell, and Petrank [31] and Haitner [26], we get a black-box construction of malicious OT in the standalone model from semi-honest OT with a *constant* amortized OT production rate. The constant rate applies both to the cases of bit-OT and string-OT.

5.4   OT Combiners

An OT combiner [28] allows one to obtain a secure implementation of OT from n OT candidates, up to t of which may be faulty. The efficiency of OT combiners was recently studied by Harnik, Ishai, Kushilevitz, and Nielsen [27], who obtained a construction for the semi-honest model that tolerates $t = \Omega(n)$ bad candidates and has a constant production rate, namely produces m good instances of OT using a total of $O(m)$ calls to the candidates. They also present a similar variant for the malicious model, but this variant has two weaknesses. First, the OTs being produced are only computationally secure (even if the good OT candidates have unconditional security, say by using semi-trusted parties or physical assumptions). Second, the communication complexity of the combiner protocol has a multiplicative overhead that grows polynomially with a cryptographic security parameter. Our approach can be used to eliminate both of these weaknesses, obtaining unconditionally secure OT combiners in the malicious model that tolerate $t = \Omega(n)$ bad candidates and have a constant production rate and a constant communication overhead.

We achieve the above by applying the protocol of Theorem 2 such that each OT which is associated with server i (both during the actual protocol and during the watchlist initialization) is implemented by invoking the i-th OT candidate. Unlike Theorem 2, here we need to rely on the robustness of the outer protocol (rather than settle for the weaker notion of "security with abort"). Another modification to the protocol of Theorem 2 is that the protocol is not aborted as soon as the first inconsistency is detected, but rather only aborts when there are inconsistencies involving at least, say, $t/10$ servers. This is necessary to tolerate incorrect outputs provided by faulty OT candidates. Since the faulty candidates can be emulated by an adversary corrupting the corresponding servers, we can afford to tolerate a constant fraction faulty candidates.

References

1. Beaver, D.: Precomputing oblivious transfer. In: Coppersmith, D. (ed.) CRYPTO 1995. LNCS, vol. 963, pp. 97–109. Springer, Heidelberg (1995)
2. Beaver, D.: Correlated pseudorandomness and the complexity of private computations. In: Proc. 28th STOC, pp. 479–488. ACM, New York (1996)

3. Beaver, D., Micali, S., Rogaway, P.: The round complexity of secure protocols (extended abstract). In: STOC, pp. 503–513. ACM, New York (1990)
4. Ben-Or, M., Goldwasser, S., Wigderson, A.: Completeness theorems for non-cryptographic fault-tolerant distributed computation. In: Proc. 20th STOC, pp. 1–10. ACM, New York (1988)
5. Bogetoft, P., Christensen, D.L., Damgård, I., Geisler, M., Jakobsen, T., Krøigaard, M., Nielsen, J.D., Nielsen, J.B., Nielsen, K., Pagter, J., Schwartzbach, M., Toft, T.: Multiparty computation goes live. Cryptology ePrint Archive, Report 2008/068 (2008), http://eprint.iacr.org/
6. Bracha, G.: An o(log n) expected rounds randomized byzantine generals protocol. J. ACM 34(4), 910–920 (1987)
7. Brassard, G., Crépeau, C., Santha, M.: Oblivious transfers and intersecting codes. IEEE Transactions on Information Theory 42(6), 1769–1780 (1996)
8. Canetti, R.: Security and composition of multiparty cryptographic protocols. Journal of Cryptology: the journal of the International Association for Cryptologic Research 13(1), 143–202 (2000)
9. Canetti, R.: Universally composable security: A new paradigm for cryptographic protocols. Electronic Colloquium on Computational Complexity (ECCC) TR01-016, 2001. Previous version A unified framework for analyzing security of protocols availabe at the ECCC archive TR01-016. Extended abstract in FOCS 2001 (2001)
10. Canetti, R., Lindell, Y., Ostrovsky, R., Sahai, A.: Universally composable two-party computation. In: Proc. 34th STOC, pp. 494–503. ACM, New York (2002)
11. Chaum, D., Crépeau, C., Damgård, I.: Multiparty unconditionally secure protocols. In: Proc. 20th STOC, pp. 11–19. ACM, New York (1988)
12. Chen, H., Cramer, R.: Algebraic geometric secret sharing schemes and secure multi-party computations over small fields. In: Dwork, C. (ed.) CRYPTO 2006. LNCS, vol. 4117, pp. 521–536. Springer, Heidelberg (2006)
13. Crépeau, C.: Equivalence between two flavours of oblivious transfers. In: Pomerance, C. (ed.) CRYPTO 1987. LNCS, vol. 293, pp. 350–354. Springer, Heidelberg (1988)
14. Crépeau, C., Savvides, G.: Optimal reductions between oblivious transfers using interactive hashing. In: Vaudenay, S. (ed.) EUROCRYPT 2006. LNCS, vol. 4004, pp. 201–221. Springer, Heidelberg (2006)
15. Crépeau, C., van de Graaf, J., Tapp, A.: Committed oblivious transfer and private multi-party computation. In: Coppersmith, D. (ed.) CRYPTO 1995. LNCS, vol. 963, pp. 110–123. Springer, Heidelberg (1995)
16. Damgård, I., Ishai, Y.: Constant-round multiparty computation using a black-box pseudorandom generator. In: Shoup, V. (ed.) CRYPTO 2005. LNCS, vol. 3621, pp. 378–394. Springer, Heidelberg (2005)
17. Damgård, I., Ishai, Y.: Scalable secure multiparty computation. In: Dwork, C. (ed.) CRYPTO 2006. LNCS, vol. 4117, pp. 501–520. Springer, Heidelberg (2006)
18. Dodis, Y., Micali, S.: Parallel reducibility for information-theoretically secure computation. In: Bellare, M. (ed.) CRYPTO 2000. LNCS, vol. 1880, pp. 74–92. Springer, Heidelberg (2000)
19. Even, S., Goldreich, O., Lempel, A.: A randomized protocol for signing contracts. Commun. ACM 28(6), 637–647 (1985)
20. Franklin, M.K., Yung, M.: Communication complexity of secure computation (extended abstract). In: STOC, pp. 699–710. ACM, New York (1992)
21. Goldreich, O.: Foundations of Cryptography: Basic Applications. Cambridge University Press, Cambridge (2004)

22. Goldreich, O., Micali, S., Wigderson, A.: How to play ANY mental game. In: ACM (ed.) Proc. 19th STOC, pp. 218–229. ACM, New York (1987); See [21, Chap. 7] for more details

23. Goldreich, O., Vainish, R.: How to solve any protocol problem - an efficiency improvement. In: Pomerance, C. (ed.) CRYPTO 1987. LNCS, vol. 293, pp. 73–86. Springer, Heidelberg (1988)

24. Goldwasser, S., Lindell, Y.: Secure computation without agreement. In: Malkhi, D. (ed.) DISC 2002. LNCS, vol. 2508, pp. 17–32. Springer, Heidelberg (2002)

25. Goyal, V., Mohassel, P., Smith, A.: Efficient two party and multi party computation against covert adversaries. In: Smart, N. (ed.) EUROCRYPT 2008. LNCS, vol. 4965, pp. 289–306. Springer, Heidelberg (2008)

26. Haitner, I.: Semi-honest to malicious oblivious transfer - the black-box way. In: Canetti, R. (ed.) TCC 2008. LNCS, vol. 4948, pp. 412–426. Springer, Heidelberg (2008)

27. Harnik, D., Ishai, Y., Kushilevitz, E., Nielsen, J.B.: OT-combiners via secure computation. In: Canetti, R. (ed.) TCC 2008. LNCS, vol. 4948, pp. 393–411. Springer, Heidelberg (2008)

28. Harnik, D., Kilian, J., Naor, M., Reingold, O., Rosen, A.: On robust combiners for oblivious transfer and other primitives. In: Cramer, R. (ed.) EUROCRYPT 2005. LNCS, vol. 3494, pp. 96–113. Springer, Heidelberg (2005)

29. Ishai, Y., Kilian, J., Nissim, K., Petrank, E.: Extending oblivious transfers efficiently. In: Boneh, D. (ed.) CRYPTO 2003. LNCS, vol. 2729, pp. 145–161. Springer, Heidelberg (2003)

30. Ishai, Y., Kushilevitz, E.: Perfect constant-round secure computation via perfect randomizing polynomials. In: Widmayer, P., Triguero, F., Morales, R., Hennessy, M., Eidenbenz, S., Conejo, R. (eds.) ICALP 2002. LNCS, vol. 2380, pp. 244–256. Springer, Heidelberg (2002)

31. Ishai, Y., Kushilevitz, E., Lindell, Y., Petrank, E.: Black-box constructions for secure computation. In: STOC, pp. 99–108. ACM, New York (2006)

32. Ishai, Y., Kushilevitz, E., Ostrovsky, R., Sahai, A.: Zero-knowledge from secure multiparty computation. In: STOC, pp. 21–30. ACM, New York (2007)

33. Kilian, J.: Founding cryptography on oblivious transfer. In: STOC, pp. 20–31. ACM, New York (1988)

34. Kiraz, M., Schoenmakers, B.: A protocol issue for the malicious case of Yao's garbled circuit construction. In: Yung, M., Dodis, Y., Kiayias, A., Malkin, T. (eds.) PKC 2006. LNCS, vol. 3958, pp. 283–290. Springer, Heidelberg (2006)

35. Lindell, Y., Pinkas, B.: An efficient protocol for secure two-party computation in the presence of malicious adversaries. In: Naor, M. (ed.) EUROCRYPT 2007. LNCS, vol. 4515, pp. 52–78. Springer, Heidelberg (2007)

36. Mohassel, P., Franklin, M.K.: Efficiency tradeoffs for malicious two-party computation. In: Yung, M., Dodis, Y., Kiayias, A., Malkin, T. (eds.) PKC 2006. LNCS, vol. 3958, pp. 458–473. Springer, Heidelberg (2006)

37. Peikert, C., Vaikuntanathan, V., Waters, B.: A framework for efficient and composable oblivious transfer. In: These proceedings available from Cryptology ePrint Archive, Report 2007/348 (2008), http://eprint.iacr.org/

38. Rabin, M.: How to exchange secrets by oblivious transfer. Technical Report TR-81, Harvard Aiken Computation Laboratory (1981)

39. Shamir, A.: How to share a secret. Communications of the ACM 11 (November 1979)

40. Yao, A.C.: How to generate and exchange secrets. In: Proc. 27th FOCS, pp. 162–167. IEEE, Los Alamitos (1986)

Author Index

Printing: Mercedes-Druck, Berlin
Binding: Stein+Lehmann, Berlin

Lecture Notes in Computer Science

Sublibrary 4: Security and Cryptology

For information about Vols. 1– 3995
please contact your bookseller or Springer

Vol. 4677: A. Aldini, R. Gorrieri (Eds.), Foundations of Security Analysis and Design IV. VII, 325 pages. 2007.

Vol. 4657: C. Lambrinoudakis, G. Pernul, A.M. Tjoa (Eds.), Trust, Privacy and Security in Digital Business. XIII, 291 pages. 2007.

Vol. 4637: C. Kruegel, R. Lippmann, A. Clark (Eds.), Recent Advances in Intrusion Detection. XII, 337 pages. 2007.

Vol. 4631: B. Christianson, B. Crispo, J.A. Malcolm, M. Roe (Eds.), Security Protocols. IX, 347 pages. 2007.

Vol. 4622: A. Menezes (Ed.), Advances in Cryptology - CRYPTO 2007. XIV, 631 pages. 2007.

Vol. 4593: A. Biryukov (Ed.), Fast Software Encryption. XI, 467 pages. 2007.

Vol. 4586: J. Pieprzyk, H. Ghodosi, E. Dawson (Eds.), Information Security and Privacy. XIV, 476 pages. 2007.

Vol. 4582: J. López, P. Samarati, J.L. Ferrer (Eds.), Public Key Infrastructure. XI, 375 pages. 2007.

Vol. 4579: B.M. Hämmerli, R. Sommer (Eds.), Detection of Intrusions and Malware, and Vulnerability Assessment. X, 251 pages. 2007.

Vol. 4575: T. Takagi, T. Okamoto, E. Okamoto, T. Okamoto (Eds.), Pairing-Based Cryptography – Pairing 2007. XI, 408 pages. 2007.

Vol. 4567: T. Furon, F. Cayre, G. Doërr, P. Bas (Eds.), Information Hiding. XI, 393 pages. 2008.

Vol. 4521: J. Katz, M. Yung (Eds.), Applied Cryptography and Network Security. XIII, 498 pages. 2007.

Vol. 4515: M. Naor (Ed.), Advances in Cryptology - EUROCRYPT 2007. XIII, 591 pages. 2007.

Vol. 4499: Y.Q. Shi (Ed.), Transactions on Data Hiding and Multimedia Security II. IX, 117 pages. 2007.

Vol. 4464: E. Dawson, D.S. Wong (Eds.), Information Security Practice and Experience. XIII, 361 pages. 2007.

Vol. 4462: D. Sauveron, K. Markantonakis, A. Bilas, J.-J. Quisquater (Eds.), Information Security Theory and Practices. XII, 255 pages. 2007.

Vol. 4450: T. Okamoto, X. Wang (Eds.), Public Key Cryptography – PKC 2007. XIII, 491 pages. 2007.

Vol. 4437: J.L. Camenisch, C.S. Collberg, N.F. Johnson, P. Sallee (Eds.), Information Hiding. VIII, 389 pages. 2007.

Vol. 4392: S.P. Vadhan (Ed.), Theory of Cryptography. XI, 595 pages. 2007.

Vol. 4377: M. Abe (Ed.), Topics in Cryptology – CT-RSA 2007. XI, 403 pages. 2006.

Vol. 4356: E. Biham, A.M. Youssef (Eds.), Selected Areas in Cryptography. XI, 395 pages. 2007.

Vol. 4341: P.Q. Nguyên (Ed.), Progress in Cryptology - VIETCRYPT 2006. XI, 385 pages. 2006.

Vol. 4332: A. Bagchi, V. Atluri (Eds.), Information Systems Security. XV, 382 pages. 2006.

Vol. 4329: R. Barua, T. Lange (Eds.), Progress in Cryptology - INDOCRYPT 2006. X, 454 pages. 2006.

Vol. 4318: H. Lipmaa, M. Yung, D. Lin (Eds.), Information Security and Cryptology. XI, 305 pages. 2006.

Vol. 4307: P. Ning, S. Qing, N. Li (Eds.), Information and Communications Security. XIV, 558 pages. 2006.

Vol. 4301: D. Pointcheval, Y. Mu, K. Chen (Eds.), Cryptology and Network Security. XIII, 381 pages. 2006.

Vol. 4300: Y.Q. Shi (Ed.), Transactions on Data Hiding and Multimedia Security I. IX, 139 pages. 2006.

Vol. 4298: J.K. Lee, O. Yi, M. Yung (Eds.), Information Security Applications. XIV, 406 pages. 2007.

Vol. 4296: M.S. Rhee, B. Lee (Eds.), Information Security and Cryptology – ICISC 2006. XIII, 358 pages. 2006.

Vol. 4284: X. Lai, K. Chen (Eds.), Advances in Cryptology – ASIACRYPT 2006. XIV, 468 pages. 2006.

Vol. 4283: Y.Q. Shi, B. Jeon (Eds.), Digital Watermarking. XII, 474 pages. 2006.

Vol. 4266: H. Yoshiura, K. Sakurai, K. Rannenberg, Y. Murayama, S.-i. Kawamura (Eds.), Advances in Information and Computer Security. XIII, 438 pages. 2006.

Vol. 4258: G. Danezis, P. Golle (Eds.), Privacy Enhancing Technologies. VIII, 431 pages. 2006.

Vol. 4249: L. Goubin, M. Matsui (Eds.), Cryptographic Hardware and Embedded Systems - CHES 2006. XII, 462 pages. 2006.

Vol. 4237: H. Leitold, E.P. Markatos (Eds.), Communications and Multimedia Security. XII, 253 pages. 2006.

Vol. 4236: L. Breveglieri, I. Koren, D. Naccache, J.-P. Seifert (Eds.), Fault Diagnosis and Tolerance in Cryptography. XIII, 253 pages. 2006.

Vol. 4219: D. Zamboni, C. Krügel (Eds.), Recent Advances in Intrusion Detection. XII, 331 pages. 2006.

Vol. 4189: D. Gollmann, J. Meier, A. Sabelfeld (Eds.), Computer Security – ESORICS 2006. XI, 548 pages. 2006.

Vol. 4176: S.K. Katsikas, J. López, M. Backes, S. Gritzalis, B. Preneel (Eds.), Information Security. XIV, 548 pages. 2006.

Vol. 4117: C. Dwork (Ed.), Advances in Cryptology - CRYPTO 2006. XIII, 621 pages. 2006.

Vol. 4116: R. De Prisco, M. Yung (Eds.), Security and Cryptography for Networks. XI, 366 pages. 2006.

Vol. 4107: G. Di Crescenzo, A. Rubin (Eds.), Financial Cryptography and Data Security. XI, 327 pages. 2006.

Vol. 4083: S. Fischer-Hübner, S. Furnell, C. Lambrinoudakis (Eds.), Trust and Privacy in Digital Business. XIII, 243 pages. 2006.

Vol. 4064: R. Büschkes, P. Laskov (Eds.), Detection of Intrusions and Malware & Vulnerability Assessment. X, 195 pages. 2006.

Vol. 4058: L.M. Batten, R. Safavi-Naini (Eds.), Information Security and Privacy. XII, 446 pages. 2006.

Vol. 4047: M. Robshaw (Ed.), Fast Software Encryption. XI, 434 pages. 2006.

Vol. 4043: A.S. Atzeni, A. Lioy (Eds.), Public Key Infrastructure. XI, 261 pages. 2006.

Vol. 4004: S. Vaudenay (Ed.), Advances in Cryptology - EUROCRYPT 2006. XIV, 613 pages. 2006.